877 GABRIEL ST.
ORLÉANS

introductory physical chemistry

introductory physical chemistry

DONALD H. ANDREWS

PROFESSOR EMERITUS
Department of Chemistry
The Johns Hopkins University

DISTINGUISHED PROFESSOR OF BIOPHYSICS
Adjunct Faculty
Department of Biological Sciences
Florida Atlantic University

McGRAW-HILL BOOK COMPANY
New York   St. Louis   San Francisco   Düsseldorf
London   Mexico   Panama   Sydney   Toronto

This book was set in News Gothic by Graphic Services, Inc., and
printed on permanent paper and bound by Von Hoffman Press, Inc.
The designers were B. Kann and Marsha Cohen;
the drawings were done by B. Handelman Associates, Inc.
The editors were James L. Smith and Eva Marie Strock.   William P. Weiss supervised
the production.

To my colleagues

Manley L. Boss
Richard G. Domey
Harrison A. Hoffman
Vincent R. Saurino
Peter L. Sguros

preface

This book has been written as a text for an introductory physical chemistry course extending through two semesters or three quarters of a normal college or university year.

The most desirable level of sophistication for such a course depends to a great extent on the previous level of training of the students. Some may have had advanced courses at the high school level followed by a general chemistry course which covered a large part of basic physical chemistry. Others may have had a less comprehensive training and may need a more elementary approach in their first confrontation with thermodynamics and atomic and molecular theory. Some already may have acquired ability to apply mathematical rigor to chemical thinking. Others may find this kind of applied mathematics almost impossible to understand. This is frequently true of those who are taking physical chemistry not as part of a curriculum for chemistry majors but as a preparation for careers in the life sciences, geology, or even engineering.

Both for chemistry majors without special advance preparation and for majors in other sciences the best physical chemistry course is often one where basic principles are stressed, even at the expense of a reduced coverage of far-flung applications. For, even though at many institutions physical chemistry gets the largest number of votes as the most difficult course, the basic principles of physical chemistry really are simple. The mathematics required to state these principles logically and precisely is also simple. The problem is to get the student to have a clear mental picture of the basic pattern of relationships and not to try to get answers by the mere manipulation of formulas. To achieve such a goal is the objective of this text.

As one can see from the Contents, an effort has been made to present physical chemistry in terms of a strictly limited number of fundamental concepts. The first few chapters picture the various states of matter. For chemistry itself is based on the idea that matter is composed of particles; electrons, protons, and neutrons are organized into atoms; these atoms are organized into molecules; and the atoms and molecules are found in various states of organization among themselves, such as the rather disordered gases, the semiordered liquids, and the highly ordered crystals and macromolecules. A clear understanding of the nature of these states is essential in order that the states may in turn be used to illustrate and make concrete the more intangible concepts of thermodynamics.

After he has acquired a clear picture of the states of matter, the student is shown that chemistry is more than a descriptive science, a depicting of static patterns. Chemical reactions represent dynamic modes of change; and the quantitative description of change rests on the concept of energy. It is only in the coupling of the flow of matter with the flow of energy that an understanding of the fundamental nature of chemical change can be achieved. To this end, in Chaps. 6 to 9 the student is introduced to the concept of the conservation of energy and the role that probability plays in determining the pattern of energy flow, summed up in the concept of entropy. The broad significance of ideas of energy and entropy is driven home in Chaps. 10 to 14, with the discussions of phase equilibria, solutions, chemical equilibria, and electrochemistry. Finally, after a survey of the nonequilibrium aspects of energy flow, as presented in the basic theories of chemical kinetics in Chap. 15, the student is ready to turn his attention back to structure with new insight into its meaning.

The journey so far makes up roughly the first half of the book; in the second half the student explores structure as a way of gaining acquaintance with much of the current frontier research while deepening his understanding of the roles of energy and entropy not only in chemistry but in many other areas such as biology, geology, metallurgy, even oceanography and meteorology. Particularly in writing this second half of the text, the author has been influenced by the thinking of the Advisory Council on College Chemistry as set forth with special clarity in its newsletter of May, 1967, where the recommendation is made that in courses of general, organic, analytical, and physical chemistry the relations of these disciplines to biochemistry should be made more explicit by specific examples. Certainly the interdisciplinary field shared by chemistry, physics, and biology can assert a valid claim to being one of the most, if not the most, exciting areas of research today.

Every science major ought to be familiar with the outstanding developments in this field which are employing the basic concepts of physical chemistry along with the mathematics of quantum theory and the insights of information theory and cybernetics to give us our first tantalizing hints of the fundamental nature of life itself. Therefore after an elementary survey of the quantum aspects of matter and light in Chaps. 16 to 21, the last six chapters present extensive illustrations of the principles of physical chemistry at work in a biochemical milieu. This includes our new knowledge of the role of energy in biochemistry involving ATP and ADP and our new insight into the role of entropy and negentropy (information) in governing biogrowth through DNA and RNA. Our ventures into this new micro universe are matched only in excitement by our first ventures into extraterrestrial space in the macro universe.

After such an experience with physical chemistry and its applications, some students, it is hoped, will emerge with a real facility in mathematical thinking in these areas. Others may have to be content with a broader perspective but a still limited perspicacity. To help as much as possible in the acquisition of skills for quantitative thinking, examples have been inserted at the end of each section with the solutions explained in detail. Problems for the student to solve on his own are found at the end of each chapter, including varying degrees of challenge.

The author takes this opportunity to thank his many friends and associates who have been so helpful during various stages of the preparation. Special thanks are offered to his colleagues at Florida Atlantic University: to Professor Theodore I. Bieber, of the Department of Chemistry; to Professors John Blakemore and James McGuire, of the Department of Physics; to Professor Peter L. Sguros of the Department of Biological Sciences; and most of all to Professor Manley L. Boss, Chairman of the Department of Biological Sciences. The author thanks Lucie Geckler, science librarian of the Johns Hopkins University, and her staff for invaluable assistance and advice. The author warmly thanks James L. Smith, Eva Marie Strock, and William P. Weiss, of the McGraw-Hill Book Company, as well as Frances M. Alley and Elizabeth P. Richardson for essential help with the manuscript; above all thanks are due the author's wife for continuing encouragement and invaluable counsel from inception to conclusion.

Donald H. Andrews

contents

The gaseous state is the simplest state of matter. For this reason gases have a special importance in physical chemistry, and a knowledge of the behavior pattern of gases is essential for understanding the physicochemical nature of many other more complex forms of matter, both animate and inanimate.

For example, nearly all kinds of living matter depend directly or indirectly on a flow of gas to maintain the life process. During a year an average human being must draw into his lungs from 2 to 5 million liters of air in order to obtain the 20 percent of life-sustaining oxygen gas which it contains. These oxygen molecules must impinge on the microthin walls of the lung capillaries with a sufficient number of collisions to ensure the diffusion of $\frac{1}{4}$ liter of oxygen gas into the bloodstream per minute. Marine forms of life must obtain a comparable supply of oxygen from the air dissolved in water. And even though the molecular environment of an oxygen molecule in aqueous solution is very different from the environment when it is in the gaseous state, the laws governing the behavior of gases still apply with an astonishing degree of accuracy to the behavior of molecules in dilute solution. Since almost all the life processes—metabolism, growth, and even replication—involve the transfer of molecules into and out of solution, a knowledge of the thermodynamic and kinetic nature of gases is an essential foundation for the understanding of these fundamental vital processes.

Chapter One
ideal gases

1-1 THE GASEOUS STATE

Gases, liquids, and *solids* constitute the three principal *states of matter.* From earliest childhood everyone becomes familiar with these three states. Everyone has handled *solid* objects, like rods of steel, which retain their shape unless violently twisted or deformed; everyone has poured *liquids,* like water, which do not retain any one fixed form but adjust their shapes to those of the containers in which they are placed; and nearly everyone has seen a toy balloon inflated, a process which leads to the conclusion that *gases* tend to fill completely any container which encloses them.

There are, of course, intermediate states of matter,

e.g., liquid crystals, which lie somewhere between the liquid and solid state (the cytoplasm inside a biological cell resembles a liquid crystal in many ways), and there are other forms of matter, e.g., aerosols, which lie between the liquid and the gaseous state. But in spite of these anomalies, the division according to the pattern of physical behavior into solids, liquids, and gases is a highly useful way of classifying matter.

Descriptions of matter in these three principal states are based on patterns of *physical properties,* such as mass, density, pressure, temperature, and the like. Both physics and physical chemistry are concerned with the measurement of these physical properties. In physics, answers are sought to questions like: If 7 g of air at a temperature of 38°C is confined in a bulb at a pressure of 1 atm, what is the effective bulb volume? Such questions involve the relationship of *physical* properties to each other. Physical chemistry pushes the study of physical properties farther and investigates their relationship to chemical and other physical properties. Thus, in physical chemistry one may ask: If 100 g of hydrogen and 100 g of oxygen, both at a pressure of 1 atm and a temperature of 2000°C, react chemically with each other, how much water will be formed under equilibrium conditions as a product of the reaction?

In seeking answers to such questions, the inquiry almost always focuses on the role of energy in the process under investigation. Precise quantitative measurements are made of properties in order to determine the exact amount and the specific location of energy in the molecules of the reactants and of the products and the energy gained or lost as the reaction takes place. Thus, in addition to the interrelationships of directly measurable physical properties like pressure, volume, and temperature, the relations between these properties and energy are important in physical chemistry.

A century ago, chemistry was far more qualitative than quantitative. Chemicals were likely to be described in such qualitative terms as having a pink color, an acrid odor, or a corrosive action. Today, chemistry is becoming highly quantitative; and especially in physical chemistry the concern is with precise quantitative measurements of mass, density, temperature, pressure, and the amount of heat given out or absorbed

when a physical or chemical change takes place. As a first step toward achieving an understanding in quantitative terms of the behavior of gases, it is therefore helpful to survey first their physical properties and how they are measured.

1-2   SYSTEMS

The properties of a substance are those characteristic aspects which serve to identify its condition or state as well as its particular nature. Properties that can be measured quantitatively with precision are basic in physical chemistry. As an example, consider a simple laboratory device by which the properties of a gas can be studied. Figure 1-1 shows a cylinder of glass closed at both ends containing 32 g of gaseous oxygen, confined under a gastight piston that can slide up and down. The upward thrust due to the pressure of the gas is balanced by the downward force produced by the weight at the top of the piston rod. There is a scale on the right, so that the height of the piston indicates the volume in which the gas is confined. The manometer on the left measures the pressure of the gas, and the thermometer shows its temperature. Under the conditions described above, the 32 g of oxygen inside the cylinder constitute a *physicochemical system.*

A physicochemical system is a portion of matter so isolated from its surroundings that meaningful observations of the relationships between its properties can be made.

In a *closed* system none of the material under observation enters or leaves the system during the period of observation. In an *open* system, some of the material under observation does enter or leave the system during observation. Thus, the system shown in Fig. 1-1 is a closed system. In general, a system must be isolated from its surroundings in such a way that if there is any flow of matter or energy into or out of the system, it takes place *only* through channels where it can both be controlled and quantitatively measured. The gas cannot diffuse through the walls or escape past the piston in undetermined amounts if measurements are to be

Fig. 1-1   A physicochemical system.

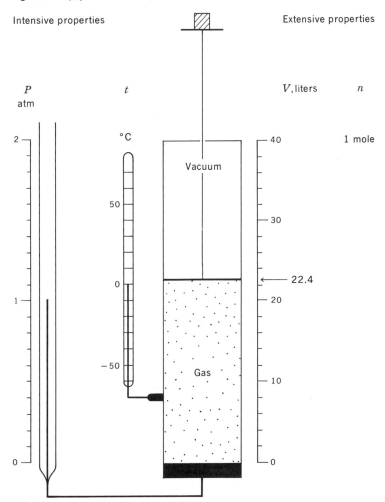

Intensive properties

Extensive properties

meaningful; heat must not flow into or out of the system at random.

Of course, no actual experimental arrangement for studying a system is completely ideal in this respect. Therefore in evaluating the degree of meaning in measurements, it is necessary to take into account the fact that there may be a small uncontrolled escape both of matter and energy. For example, gaseous helium diffuses through glass at a rate sufficiently high for the amount of helium in a glass vessel to be reduced by as much as 0.1 percent within a matter of weeks. The

care with which a device to contain a system must be constructed thus depends on the ultimate desired accuracy of the measurements.

1-3   EXTENSIVE AND INTENSIVE PROPERTIES

In general, properties fall into two distinct classes. There are the properties whose value depends on *how much* of the particular substance is studied. These are called *extensive properties*. Given two different

Fig. 1-2   A mercury thermometer.

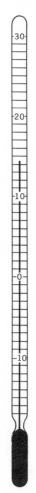

temperature, the temperature of the second sample will not be higher just because it is a larger sample. Temperature does not depend on the size of the sample; by way of contrast, the volume occupied by the second sample is twice that of the first, all other factors being maintained the same.   Intensive properties include *temperature, pressure, density, concentration, surface tension,* and *refractive index.*

1-4   HEAT AND TEMPERATURE

A form of energy that plays an important role in physical chemistry is *heat.*   Everyone is familiar with the sensation expressed by the words hot and cold.   Hold in the left hand a cube of metal that feels hot; at the same time, hold in the right hand a cube of metal that feels cold.   The cube in the left hand feels hot because it contains more heat (thermal energy) than the one in the right hand.   If these same two cubes are then placed in contact, in less than a minute the feel of the two becomes the same, because energy in the form of heat has flowed from the hot cube to the cold one.

The qualitative definition of *temperature* is based on the flow of heat.   Temperature is always defined so that heat flows from a material body like a cube of metal at a *higher temperature* to another body at a *lower temperature* when the two bodies are brought into contact.   In early scientific investigations it was also established that when a body changes its state from a *lower* temperature to a *higher* temperature, the body normally expands.   The thermal expansion of fluid in an instrument called a thermometer is the basis for the *quantitative* definition of temperature.

The most widely used device of this sort is the mercury thermometer (Fig. 1-2).   It consists of a capillary tube, generally of glass, with a bulb sealed on one end and filled with mercury.   As the mercury is heated, it expands into the capillary.   The height of the mercury is then taken as a measure of the temperature.   Thus, if two similar thermometers are placed in blocks of metal at different temperatures and the blocks are brought into contact, heat flows from the block where the temperature reading is higher to the block where the temperature reading is lower.

systems of the type shown in Fig. 1-1, if the second has twice as many molecules of oxygen as the first, it will have twice the mass also.   If the oxygen of the second system and the oxygen of the first have the same temperature, pressure, and so on, the volume of the second system will be twice that of the first.   Extensive properties include mass, volume, and heat content.

The properties that do *not* depend on the amount of matter in the system are called *intensive* properties. Suppose that there are 1,000 liters of oxygen in a tank; if one removes 10 g of oxygen and measures its temperature and then takes out 20 g and also measures its

In order to set up a *scale* of temperature, it is necessary to select a *zero point* of temperature defined in an experimentally reproducible way and another point of temperature that will determine the size of the *unit* of temperature, which is the *degree*. In the centigrade, or Celsius, scale of temperature, the zero point is selected as the temperature of a mixture of liquid water and ice crystals when the water is saturated with the gases of the atmosphere at normal atmospheric pressure. The size of the degree of temperature is defined by assigning the value of 100° to the temperature of an equilibrium mixture of liquid water and steam under 1-atm pressure. Since an understanding of the precise way in which these temperature points are determined by experimentation requires a rather advanced knowledge of physical chemistry, detailed presentation of this topic is postponed to a later chapter.

1-5   STATE

The term *state* is used in two different ways in physical chemistry. In the broader usage, chemists refer to the three different *states* of matter, gaseous, liquid, and crystalline, thus distinguishing between the strikingly different patterns of arrangement and movement of molecules in these three different states. In the narrower usage, the term *state* refers to the precisely specified pattern of arrangement and movement of molecules that is defined when precise numerical values for each of the physical properties are specifically given for a physicochemical system. For example, Fig. 1-1 shows a system consisting of 32 g (1 mole) of oxygen gas. The state of this gas is defined by the values given for its properties, pressure P as 1 atm, volume V as 22.4 liters, and temperature t as 0°C.

For these figures to be meaningful, it is necessary to be sure that the values of the properties are constant, that they are not changing with time, or at least are changing slowly. If one observes the pressure first, then the volume, and finally the temperature, it is necessary that any rate of change be small enough so that one can say that the values of the observed properties all refer to one and the same state, within the degree of precision of the experiment. It is also necessary that the same value of a property be found

uniformly throughout the whole volume of the system under observation. Of course, many systems are studied in physical chemistry where the conditions of constancy and uniformity do not obtain, but these require a special kind of logical analysis which does not apply to the considerations of this chapter. Unless otherwise specified, the assumption is always made that the discussion refers to a steady and uniform state.

Many important experiments in physical chemistry are carried out by changing a system from one state to another. As an illustration, suppose that the piston in the apparatus in Fig. 1-3 is pushed down so that the volume of the gas is reduced from 22.4 to 11.2 liters while the cylinder is immersed in a mixture of ice and water so that the temperature is maintained constant at 0°C. For purposes of analysis, the experiment may be regarded as consisting of three parts: (1) the observation of the properties of the gas to determine the state *before* the piston is pushed down; (2) the change brought about by pushing the piston down; and (3) the observation of the properties to determine the new state *after* the piston has been pushed down to its new position.

The state of the system before the change takes place is usually called the *initial state*. The properties of this state are denoted by symbols with the numeral 1 as a subscript, P_1, V_1, t_1. The state of the system after the change takes place is usually called the *final state*. The properties of this state are denoted by symbols with the numeral 2 as a subscript, P_2, V_2, t_2.

1-6   BOYLE'S LAW

Experiments of the sort described in the preceding section were first carried out systematically by the British scientist Robert Boyle 300 years ago. Instead of using a pump equipped with a piston, he used a mercury column both to compress the gas and to measure the pressure, as shown in Fig. 1-4a. The system shown in Fig. 1-4b is the equivalent of Boyle's experimental device and has the advantage of displaying the different factors involved somewhat more clearly. Here it may be assumed that the surface of the mercury in the cylinder at the right does not vary appreciably in height

Fig. 1-3 **The change of state from P_1, V_1, t_1, to P_2, V_2, t_2.**

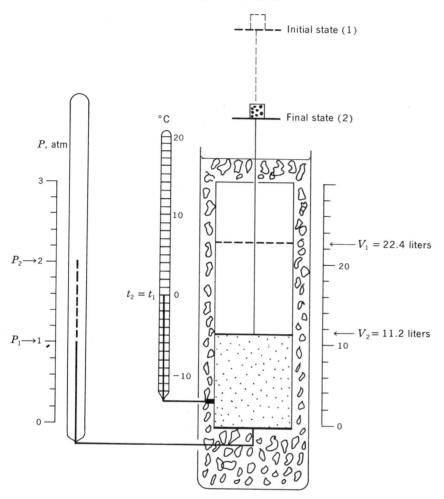

because of the larger diameter. The pressure is read directly off the scale measuring the height of the mercury in the small bore of the capillary tube. The volume is read directly off the scale that shows the height of the piston.

Boyle demonstrated that when the temperature is held constant, the product of pressure times volume is a constant. This is expressed

$$PV = \text{const} \qquad (1\text{-}1)$$

For example, suppose that the initial pressure P_1 is 1 atm and that the initial volume V_1 is 22.4 liters. If the pressure is increased to 2 atm (P_2), the value of the

volume decreases to 11.2 liters (V_2). Thus, an alternative way of expressing the relationship is

$$P_1 V_1 = P_2 V_2 \qquad (1\text{-}2)$$

This relationship is known as *Boyle's law*. Many gases exhibit behavior closely in accord with this law. However, if the pressure is very high, the volume relatively small, or the temperature extremely low, wide deviations from this law are likely. The nature of these deviations will be discussed in Chap. 3.

If observations like those of Boyle were made with the system shown in Fig. 1-1, a series of related values for P and V would be found (Table 1-1). A plot of the

Boyles law
$P_1 U_1 = P_2 U_2$

values of P against corresponding values of V is shown in Fig. 1-5; the graph is a hyperbola, where the value of one variable changes in inverse proportion to the value of the other variable.

Example 1-1

A bicycle pump has a volume of 1,500 cm³ when the piston is at the highest possible position. The air under the piston is at 1 atm. If the piston is pushed down, forcing the air into a volume of 500 cm³, what will the pressure be after the temperature comes back to its initial value?

Boyle's law can be used in the form

$$P_1 V_1 = P_2 V_2$$

Substituting the given numerical values

$$1 \text{ atm} \times 1{,}500 \text{ cm}^3 = P_2 \times 500 \text{ cm}^3$$

gives

$$P_2 = \frac{1 \text{ atm} \times 1{,}500 \text{ cm}^3}{500 \text{ cm}^3} = 3 \text{ atm} \qquad Ans.$$

1-7   THE LAW OF GAY-LUSSAC

In 1787 J. A. C. Charles, a French physicist, observed that equal volumes of hydrogen, air, carbon dioxide, and oxygen all expand by an equal amount when heated from 0 to 80°C. After making more extensive studies on a larger number of gases, J. L. Gay-Lussac, also French, reported in 1802 that the expansion of gases is related to the temperature by the equation

$$V = a(1 + bt) \tag{1-3}$$

where a is the volume of the gas at 0°C and b has a value approximately equal to $\frac{1}{273}$. This relationship is generally called the law of Gay-Lussac, although it is sometimes referred to as Charles' law in recognition of the earlier observations.

Just as deviations are found from Boyle's law when gases are studied under conditions of high pressure and low temperature, similar deviations are also found from the law of Gay-Lussac. When conditions are such that both the laws of Boyle and of Gay-Lussac are

obeyed, the gas is said to exhibit ideal behavior and is referred to as an *ideal gas*.

The law of Gay-Lussac can be demonstrated experimentally by using an apparatus like that in Fig. 1-6, where pressure is maintained at a constant value by keeping a fixed weight on top of the piston; then, as heat is transferred into or out of the gas through the copper rod at the bottom, the volume of the gas increases or decreases. A typical series of observations

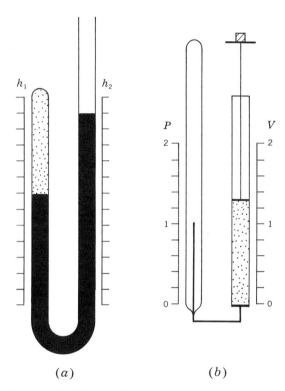

(a)   (b)

Fig. 1-4   Boyle's experiment: (a) actual apparatus and (b) equivalent apparatus.

TABLE 1-1

V, liters	P, atm
44.8	0.500
22.4	1.000
11.2	2.000
5.6	4.000

Fig. 1-5   The graph of P against V for 32 g of O_2 at 0°C.

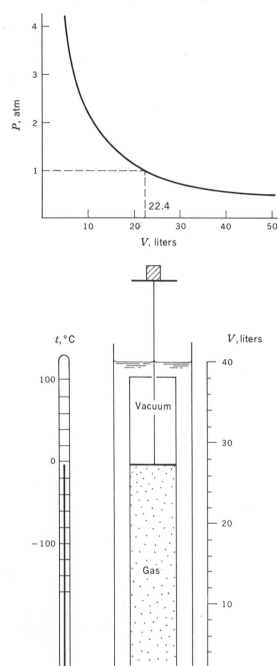

Fig. 1-6   The Gay-Lussac experiment: volume as a function of temperature.

for such a system is given in Table 1-2, and the plot of these values is shown in Fig. 1-7.

In this apparatus a fluid (the gas) expands with temperature, just as in the mercury thermometer a fluid (the mercury) expands with temperature; thus such a device is a *gas thermometer*. If the position of the piston at the temperature of a mixture of ice and liquid water is taken as the zero point of the temperature scale and the position of the piston at the temperature of a mixture of water and water vapor at 1 atm is taken as 100°, the readings of the gas thermometer coincide closely with the readings of a mercury thermometer when the two devices are brought into contact with a series of bodies at different temperatures. In other words, the expansion of a gas obeying the laws of Boyle and Gay-Lussac closely resembles the expansion of mercury with increasing temperature. Moreover, a gas thermometer has a wider usefulness than a mercury thermometer since it can be used at temperatures below the freezing point and above the boiling point of mercury, i.e., in ranges where the mercury thermometer cannot function.

By studying the behavior of gases at increasingly low pressures, one can make observations that permit accurate calculation of the relation between volume and temperature for an *ideal* gas. A temperature scale established in this way has a far broader and more significant meaning than one based on the expansion of a liquid like mercury.

1-8   ABSOLUTE TEMPERATURE

If the plot of observations of V against t for an ideal gas is extrapolated to the point where $V = 0$, as shown in Fig. 1-8, it is found that the temperature of $-273.15°$ corresponds to zero volume for the ideal gas. The British physicist Lord Kelvin proposed that a new scale of temperature be established in which the zero point corresponds to the value of temperature at which the volume of an *ideal* gas becomes equal to zero. He suggested the name *absolute temperature* and the symbol T to denote temperature values on this scale. It is now called the Kelvin scale in his honor, and the abbreviation K is used to distinguish numerical values in degrees; thus, the value of 0°C on the Kelvin scale is 273.15°K, as shown in Figs. 1-9a and b, which depict an idealized gas thermometer.

The relationship between the older centigrade, or Celsius, scale t and the absolute, or Kelvin, temperature scale T is given by

$$T = t + 273.15 \qquad (1\text{-}4)$$

where the numerical constant is the latest precise figure adopted by the International Commission on temperature scales.

Using the Kelvin scale, the law of Gay-Lussac can be written

$$\frac{V}{T} = \text{const} \qquad (1\text{-}5)$$

or in the alternative form

$$\frac{V_1}{T_1} = \frac{V_2}{T_2} \qquad (1\text{-}6)$$

1-9 AVOGADRO'S PRINCIPLE

When samples of different ideal gases each containing the same number of moles are brought to the same pressure and temperature, all the gases occupy the same volume. This relationship is called *Avogadro's principle* in honor of the Italian scientist who first called attention to it. One mole is the amount of the gas that contains 6.02×10^{23} molecules; this constant is called *Avogadro's number* N_A. Since volume is an *extensive* property,

$$V = n\text{v} \qquad (1\text{-}7)$$

where v is the volume of 1 *mole* of gas under the same conditions of T and P at which V is measured and n is the number of moles of gas contained in volume V. At 0°C and 1 atm, 1 mole of an ideal gas occupies the volume of 22.41 liters, as determined by experimental observations of real gases that closely obey the laws of Boyle and Gay-Lussac.

1-10 THE IDEAL-GAS LAW

In order to see how Boyle's law, Charles' law, and Avogadro's principle can be combined into a single relationship called the ideal-gas law, consider the change of 1 mole of an ideal gas from an initial state where the values of pressure, temperature, and volume are de-

TABLE 1-2

t, °C	V, liters
0	22.4
273	44.8
546	67.2
819	89.6

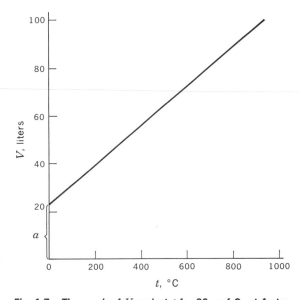

Fig. 1-7 The graph of V against t for 32 g of O_2 at 1 atm.

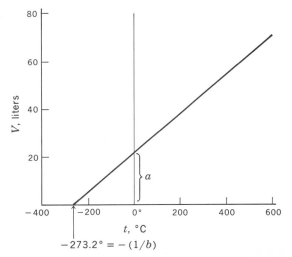

Fig. 1-8 The graph of V against t extrapolated to zero volume. $V = a(1 + bt)$.

Fig. 1-9 (*a*) Volume as a function of absolute temperature T; (*b*) a constant-pressure gas thermometer.

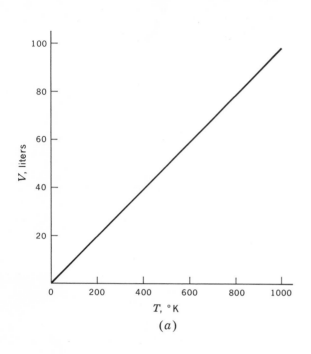

(*a*)

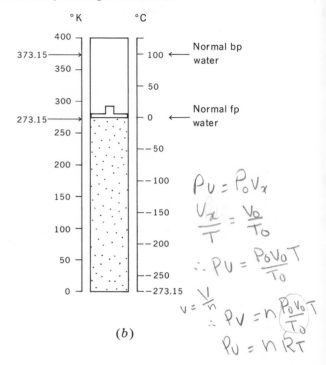

(*b*)

noted by the symbols P, v, and T to a final state where they are denoted by P_0, v_0, and T_0 and have the numerical values $P_0 = 1$ atm, $v_0 = 22.41$ liters, and $T_0 = 273.15°K$.

This change can be carried out in two steps, symbolically shown by

$$\overset{\text{step I}}{PvT \longrightarrow} \overset{\text{step II}}{P_0v_xT \longrightarrow} P_0v_0T_0 \qquad (1\text{-}8)$$

In step I, the temperature is kept constant at T while P and v change to P_0 and v_x; in step II, the pressure is kept constant at P_0 while v_x and T change to v_0 and T_0. According to Boyle's law,

$$Pv = P_0v_x \qquad (1\text{-}9)$$

According to Charles' law,

$$\frac{v_x}{T} = \frac{v_0}{T_0} \qquad (1\text{-}10)$$

Elimination of v_x from the two equations gives

$$Pv = \frac{P_0v_0}{T_0}T \qquad (1\text{-}11)$$

From Avogadro's principle the relation is obtained that

$$v = \frac{V}{n} \qquad (1\text{-}12)$$

and insertion of this expression for v in Eq. (1-11) gives

$$PV = n\frac{P_0v_0}{T_0}T \qquad (1\text{-}13)$$

Putting in the numerical values, it is found that

$$\frac{P_0v_0}{T_0} = \frac{1 \text{ atm} \times 22.41 \text{ liters mole}^{-1}}{273.15°}$$

$$= 0.08205 \text{ liter atm deg}^{-1} \text{ mole}^{-1} \qquad (1\text{-}14)$$

This constant is denoted by R, called the ideal-gas constant. With it Eq. (1-13) takes the form

$$PV = nRT \qquad (1\text{-}15)$$

This expression is known as the *ideal-gas law*. An equation expressing a relationship between P, V, and T is called an *equation of state*.

The numerical value of the constant R depends on the units employed for P, V, n, and T. Some of the most useful values are given in Table 1-3. Since the product of pressure and volume has the dimension of work or energy, it is frequently helpful to express R so that it includes units of energy.

Example 1-2

A gas is at a pressure of 1 atm, its volume is 44.8 liters, and its temperature is 273°K. Calculate the number of moles.

To make the equation dimensionally correct the value of R is set equal to 0.08205 liter atm deg^{-1} mole^{-1}. With these numerical values the equation is solved for n:

$$n = \frac{PV}{RT}$$

$$= \frac{1 \text{ atm} \times 44.8 \text{ liters}}{0.08205 \text{ liter atm deg}^{-1} \text{mole}^{-1} \times 273.2°}$$

$$= 2 \text{ moles}$$

Example 1-3

A system contains 3,000 cm^3 of gas under a pressure of 400 mm Hg at a temperature of 300°C. Calculate the number of moles of gas in the system.

Here the value for R of 82.055 cm^3 atm deg^{-1} mole^{-1} is used, and the equation is again solved for n:

$$n = \frac{PV}{RT}$$

$$= \frac{400 \text{ mm Hg}/760 \text{ mm Hg atm}^{-1} \times 3,000 \text{ cm}^3}{82.055 \text{ cm}^3 \text{ atm deg}^{-1} \text{mole}^{-1} \times 573°}$$

$$= 3.36 \times 10^{-2} \text{ mole}$$

Example 1-4

A system contains 56 g of nitrogen gas at a volume of 50 liters and a temperature of 350°C. Calculate the pressure.

First, the weight of the gas must be divided by the molecular weight

$$n = \frac{56 \text{ g}}{28 \text{ g mole}^{-1}} = 2 \text{ moles}$$

When this value of n together with the values of the other variables is inserted into the ideal-gas law, the pressure is

$$P = \frac{nRT}{V}$$

$$= \frac{2 \text{ moles} \times 0.082 \text{ liter atm deg}^{-1} \text{mole}^{-1} \times 623°}{50 \text{ liters}}$$

$$= 2.0 \text{ atm}$$

1-11   THE CALCULATION OF MOLECULAR WEIGHT

These examples suggest the possibility of determining the molecular weight of an unknown gas by making observations of the weight of a gas, together with its pressure, volume, and temperature. As an illustration of this method, consider a bulb of glass equipped with a stopcock (Fig. 1-10) with a volume of 0.500 liter. The bulb is first evacuated and weighed. A sample of unknown gas is then admitted into the bulb at a pressure of 2 atm and a temperature of 27°C. The bulb is then weighed again and found to have increased in weight by 1.32 g. To calculate the molecular weight of this gas, insert the values of the variables into the expression for the ideal-gas law:

$$n = \frac{PV}{RT} = \frac{2 \text{ atm} \times 0.5 \text{ liter}}{0.082 \text{ liter atm deg}^{-1} \text{mole}^{-1} \times 300°}$$

$$= 0.041 \text{ mole} \qquad (1\text{-}16)$$

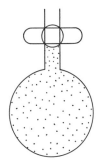

Fig. 1-10   A bulb for determining the molecular weight of a gas.

TABLE 1-3   Values of the ideal-gas constant R

82.055 cm^3 atm deg^{-1} mole^{-1}
0.082054 liter atm deg^{-1} mole^{-1}
8.3143 joules deg^{-1} mole^{-1}
1.9872 defined cal deg^{-1} mole^{-1}

The molecular weight M will then be related to the weight W of the gas and the value of n by

$$M = \frac{W}{n} = \frac{1.15 \text{ g}}{0.041 \text{ mole}} = 28 \text{ g mole}^{-1} \quad (1\text{-}17)$$

Thus the gas might be N_2, CO, or DCN.

1-12   DALTON'S LAW OF PARTIAL PRESSURES

The relationships between P, V, T, and n in a *mixture* of ideal gases was first studied systematically by an English chemist, John Dalton, in the first decade of the nineteenth century. By placing known amounts of different gases one at a time in a container, he measured the pressure of each gas at a temperature which was maintained constant throughout the series of measurements.   He then mixed all the samples together in the same container and at the same temperature as before and found the pressure to be equal to the sum of the values of the pressures measured on the separate samples.   Like the ideal-gas law, this relationship is valid when the pressure is not too high and the temperature is not too low.   The relationship is called Dalton's law of partial pressures:

The total pressure exerted by a mixture of gases is equal to the sum of the pressures which each component would exert if placed separately in the same container provided that no chemical reaction takes place and that all components exhibit ideal behavior.

This relationship can be expressed symbolically as:

$$P_{\text{tot}} = P_1 + P_2 + P_3 + \cdots \quad (1\text{-}18)$$

where P_{tot} is the total pressure, P_1 is the pressure of the gas designated as component 1 when measured separately, P_2 the pressure of a gas designated as component 2, and so on.   The values of P_1, P_2, P_3, and other similar terms are called the *partial pressures* of the respective gases.   Stated in another way, this relationship indicates that if gases behave ideally, the same relationship exists between the partial pressure of the gas, its temperature, the volume in which it is contained, and the number of moles of that gas present whether the gas is in a container all by itself or in a container with other ideal gases.   The presence of the other gases in the container does not affect the be-

havior of any one of the gases under consideration.   Thus, the following equations all hold simultaneously:

$$P_1 V = n_1 RT$$
$$P_2 V = n_2 RT \quad (1\text{-}19)$$
$$P_3 V = n_3 RT$$

If the values of the partial pressures as expressed in Eqs. (1-19) are substituted in Eq. (1-18), the relation is

$$P_{\text{tot}} = \frac{(n_1 + n_2 + n_3 + \cdots)RT}{V} \quad (1\text{-}20)$$

If n_{tot} designates the sum of all the moles of gas present,

$$n_{\text{tot}} = n_1 + n_2 + n_3 + \cdots \quad (1\text{-}21)$$

then the ideal-gas law for the mixture may be expressed

$$P_{\text{tot}} V = n_{\text{tot}} RT \quad (1\text{-}22)$$

Making the assumption that in an ideal gas no molecule has any influence on neighboring molecules, one can conclude that the behavior of the molecules of any one kind of gas will not be influenced by the presence of molecules of any other kind of gas as long as all the gases behave ideally.

Example 1-5

Consider the behavior of a system of gases enclosed in the glass cylinder with piston (Fig. 1-1). Suppose it contains 1 mole of oxygen, 2 moles of nitrogen, and 3 moles of hydrogen with the volume V 22.4 liters and temperature equal to 273°K.   What total pressure will be observed on a manometer attached to this system?
   The calculation is

$$P = \frac{(n_1 + n_2 + n_3)RT}{V}$$
$$= \frac{(1+2+3 \text{ moles}) \times 0.082 \text{ liter atm deg}^{-1} \text{mole}^{-1} \times 273°}{22.4 \text{ liters}}$$
$$= 6 \text{ atm}$$

1-13   APPLICATIONS IN THE BIOLOGICAL SCIENCES

As pointed out at the beginning of this chapter, gases play a crucial role in maintaining life.   In the higher forms of life, oxygen gas is brought into the body by breathing. An average pair of human lungs holds about 3 liters of air at 1-atm pressure and 37°C (normal body temperature) when at the point of normal deflation. During inhalation an additional $\frac{1}{2}$ liter of air is drawn into the lungs in a normal breath. A man at rest breathes about 7 liters of air per minute.

Figure 1-11 is a cross section of a typical pair of human lungs. In its overall action, the lung behaves very much like a rubber balloon. It inflates and stretches as the air enters and collapses as the air leaves, but the lung is very different from a thin-walled elastic bag. To the naked eye, lung tissue may appear solid, but an average pair of lungs weighs only about 1 kg and holds about 3 liters of air at the end of expiration. Since the density of human tissue is about 1 g cm^{-3}, these figures indicate that the lungs are actually three-fourths air space. When muscular action causes the volume of the lung to expand, according to Boyle's law, the pressure in the lung is reduced; under this pressure gradient, air flows in through the nasal passages, the larynx, the trachea, and the bronchial tubes until the pressure of 1 atm is established throughout the interior of the lung. Then when the muscles contract and reduce the volume of the lung the pressure rises above 1 atm again in accordance with Boyle's law. This establishes a pressure gradient in the opposite direction between the interior of the lungs and the exterior atmosphere so that air now flows out until 1-atm pressure is reestablished throughout the lung space.

This action can be expressed in terms of the ideal-gas law as follows:

	atm		liters		moles		liters atm deg^{-1} mole^{-1}		°K
Complete exhalation:	P_1 1	$\times$	V_1 3.0	$=$	n_1 0.118	$\times$	R 0.0821	$\times$	T 310
Muscular expansion:	P_2 0.86	$\times$	V_2 3.5	$=$	n_1 0.118	$\times$	R 0.0821	$\times$	T 310
Complete inhalation:	P_1 1	$\times$	V_2 3.5	$=$	n_2 0.137	$\times$	R 0.0821	$\times$	T 310
Muscular contraction:	P_1 1.17	$\times$	V_1 3.0	$=$	n_2 0.137	$\times$	R 0.0821	$\times$	T 310
Complete exhalation:	P_1 1	$\times$	V_1 3.0	$=$	n_1 0.118	$\times$	R 0.0821	$\times$	T 310

$$(1\text{-}23)$$

In this analysis, it is assumed that the volume of the lung changes almost instantaneously from that at complete exhalation to that at complete inhalation, but actually the volume changes slowly, and these equations are only an approximation to show the relation between the ideal-gas law and this important biological process.

The body regulates the rate of breathing through a complex chemical cybernetic mechanism. Some of the physicochemical aspects of this mechanism will be discussed in later chapters of this book. The principal factor in regulating the rate of breathing is the need to eliminate carbon dioxide from the bloodstream. Through metabolism, the oxygen that has been introduced into the body through the act of breathing combines with carbon and forms the carbon dioxide that emerges as a gas in the interior of the lungs. During exercise, the volume of air that circulates through breathing is increased in proportion to the amount of carbon dioxide that must be eliminated from the blood. The nerve centers in the brain control the rate of breathing so that about 20 liters of fresh air reach the gas-exchanging surface of the lung for every liter of carbon dioxide which is set loose there from the bloodstream. It is only when oxygen is in short supply that the control mechanism changes to regulate breathing in terms of oxygen flow rather than carbon dioxide elimination. Thus it is the *partial* pressure of the CO_2 and the *partial* pressure of the oxygen which, acting under Dalton's law of partial pressures, control the rate of breathing.

Problems

Note: In all these problems assume that the gases are obeying the ideal-gas law.

1. If 1 mole of helium gas at 25°C is confined in a 10-liter container, calculate the pressure.
2. Nitrogen gas is stored in a steel cylinder with a volume of 100 liters at a pressure of 150 atm and a temperature of 27°C. Calculate the number of moles of gas in the cylinder.
3. Gas from the cylinder described in Prob. 2 is used to test the bursting point of a weather balloon. Gas flows from the cylinder through a reducing valve into the balloon, which is expanded until it bursts. Just before gas was withdrawn from the storage cylinder, the pressure read 150.0 atm and the temperature was

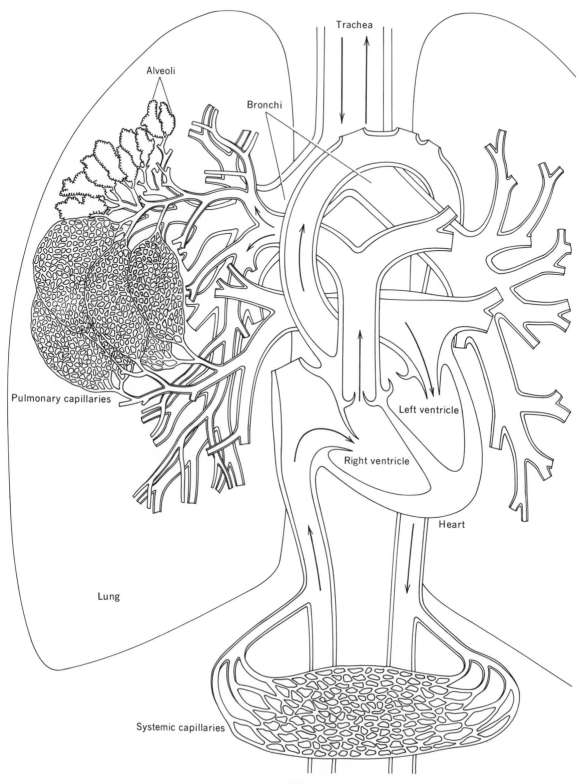

Fig. 1-11   The lung-heart oxygen-transport system.   (From "The Lung" by J. H. Comroe, Jr.   Copyright © 1966 by Scientific American, Inc.   All rights reserved.   February, 1966 issue, p. 56.)

27°C. At the instant when the balloon burst the pressure in the storage tank was 145.2 atm and the temperature had not changed. When the balloon burst, the pressure of the gas inside the balloon was 5.02 atm and the temperature was 23°C. Calculate the volume of the balloon when it burst.

4. A U-shaped tube is constructed of glass with the left arm sealed off at the top. Mercury is poured into the right arm of the tube, trapping an amount of gas above the surface of the mercury in the left arm. The tube is filled with mercury to the point where the level of the mercury in the right arm is 20.0 cm above the level of the mercury in the left arm. The volume of the space in which the gas is trapped is 26.0 cm^3. The temperature of the gas is 26.8°C. Calculate the number of moles of gas entrapped. Assume that the pressure of the atmosphere on the right-hand leg of the tube is exactly 1 atm.

5. A small boy blows up a toy balloon with air expelled from his lungs. One deep breath of air blown into the balloon gives it a volume of 50.3 cm^3 at a pressure of 1.20 atm and the temperature of 38.1°C. Calculate the amount of air in cubic centimeters that the boy inhales in taking a comparable deep breath when the pressure remains constant at 1.00 atm.

6. A common remedy for acid indigestion is sodium bicarbonate, $NaHCO_3$. Assuming that this reacts with the acid in the stomach according to the equation $NaHCO_3 + HCl = NaCl + H_2O + CO_2$, calculate the volume of gaseous CO_2 produced from 1 g of sodium bicarbonate when the gas is at a pressure of 1.20 atm and a temperature of 37°C.

7. In underwater construction projects men frequently work in caisson chambers containing air at a sufficient pressure to keep the water from flowing in. Calculate the percentage increase in the number of moles of oxygen drawn into the lungs at each breath in such a chamber 100 ft under the surface of water as compared with a similar breath in air at 1-atm pressure.

8. A constant-volume gas thermometer can be constructed by sealing a bulb on the end of a glass capillary tube where the cross-sectional area of the capillary is so small that its volume can be neglected with respect to that of the bulb. If such a bulb has a volume of 8.765 cm^3 and the pressure is 786.5 mm Hg when the bulb is at a temperature of 0°C, calculate the number of moles of gas in the bulb.

9. When placed in liquid nitrogen boiling at 1-atm pressure, the pressure in the thermometer bulb (in Prob. 8) is 223.0 mm. Calculate the normal boiling-point temperature of nitrogen both in degrees Kelvin and in degrees Celsius.

10. For the gas thermometer described in Prob. 8, make a plot of temperature as a function of pressure.

11. If the same amount of gas used in the constant-volume gas thermometer described above was placed in a constant-pressure gas thermometer operating at 1.000 atm, the volume would vary as a function of the temperature. Make a plot of this relationship.

12. If the gas used in the above problems is placed in a cylinder equipped with a sliding piston and is then maintained at constant temperature, the pressure will be a function of the volume. Plot the pressure against the volume when the temperature is maintained at 300°K.

13. Make a plot of the logarithm of the pressure as a function of the logarithm of the volume when temperature is maintained constant at 300°K and the quantity of gas is the same as in the above problems.

14. The atmosphere that we breathe at sea level has about the composition shown in the table. A boy

Component	Mole %
N_2	78.09
O_2	20.95
Ar	0.93
CO_2	0.03

breathes in 50 cm^3 of air and holds it in his lungs for a few moments. When he breathes the air out, the composition of CO_2 has risen to 0.1 mole percent. If the air is at a pressure of 1.00 atm and at a temperature of 30.0°C when the volume was measured before inhalation, calculate the number of moles of carbon that have been removed from the body by the exhalation.

15. In order to make air more suitable for breathing at the high pressures encountered in undersea caissons, it is customary to add helium to the air that is pumped to the caisson. If 1 mole of helium is added to every mole of air fed to the pump, calculate the partial pressure of nitrogen, oxygen, argon, carbon dioxide, and helium in the air that the workers are breathing if the total pressure is 4 atm. Assume that the air initially had the composition given in Prob. 14.

16. Calculate the number of moles of helium that must be added to 1 mole of air in order for a worker in breathing to draw into his lungs the same number of molecules of oxygen that are available in each breath at the surface. Take surface pressure as 1.00 atm and the pressure in the caisson as 4.00 atm.

17. Calculate the number of molecules that enter the lungs when one breathes in 50 cm³ of air at a pressure of 1.00 atm and a temperature of 25.0°C.

18. A glass bulb equipped with stopcock when evacuated weighs 50.107 g and has a volume of 949.2 cm³. When filled with a sample of a certain gas to a pressure of 1 atm at a temperature of 27.3°C, it has a weight of 51.339 g. Calculate the molecular weight of the gas contained in the bulb. If the sample is a pure substance, what common gas is it?

19. When 2 liters of hydrogen gas is mixed with 1 liter of oxygen gas, each gas being at a temperature 400°C and at a pressure of 0.100 atm, the two gases react completely; the remaining gas is brought to the same temperature as the original gases while being kept in the original container. Calculate the final pressure.

20. When 0.1 g of carbon is introduced into a combustion bomb containing oxygen at a pressure of 50.0 atm and a temperature of 25.8°C in a volume of 0.503 liter, the carbon is ignited and burns completely to CO_2. After the gases are returned to the original temperature, calculate the total pressure in the bomb and the partial pressure of each kind of gas present.

21. If 10 g of sucrose, $C_{12}H_{22}O_{11}$, is burned in the bomb described in the previous problem and all water is removed from the resultant gases by drying, calculate the total pressure and the partial pressures of each of the gases in the product at 25°C.

REFERENCES

Glasstone, S.: "Textbook of Physical Chemistry," sec. IV, The Gaseous State, D. Van Nostrand Company, Inc., New York, 1947.

Hunt, R. M.: "Robert Boyle," The University of Pittsburgh Press, Pittsburgh, 1955.

Hall, R.: "From Galileo to Newton, 1630–1720," Harper & Row, Publishers, Incorporated, New York, 1963.

Hildebrand, J. H.: "An Introduction to Kinetic Theory," Reinhold Publishing Corporation, New York, 1963.

Rowlinson, J. S.: "The Perfect Gas," The Macmillan Company, New York, 1963.

Comroe, J. H., Jr.: The Lung, *Sci. Am.,* February, 1966, p. 56.

Hall, M. B.: Robert Boyle, *Sci. Am.,* August, 1967, p. 96.

2-1 THE KINETIC NATURE OF GASES

At the time of the earliest experimental observation of gases by Boyle, Charles, Gay-Lussac, Dalton, and the other pioneers in this field, many scientists must have asked the question: What would a gas look like if we could really see it? As evidence accumulated showing that all matter is made up of small particles, the atoms, and combinations of these particles, the molecules, investigators of the nature of the gaseous state saw more and more clearly that many of the properties of gases could be accounted for by the assumption that a gas consists of molecular particles moving back and forth at

Chapter Two

the kinetic theory of gases

very high speeds and colliding both with each other and with the walls of the vessel containing the gas. Thus, there emerged a kinetic model of an ideal gas with the following features:

1. The gas consists of a large number of very small particles.
2. In their motions, these particles obey Newton's laws of mechanics.
3. As a result, these particles have kinetic energy of motion.
4. There is no force of interaction between the particles.
5. The collisions of these particles are elastic; i.e., the total kinetic energy of the colliding bodies is the same after collision as before.

On the basis of these assumptions it is possible to picture a model ideal gas with the properties closely resembling those of many actual gases.

2-2 THE KINETIC NATURE OF PRESSURE

To see how a moving particle can exert pressure, think first of a single gas molecule confined in a cubical box, as shown in Fig. 2-1. In order to make

Fig. 2-1   Gas molecule in a box.

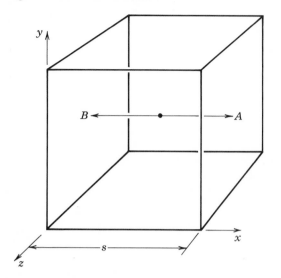

Before the collision, the velocity is $+u$; after the collision, the velocity is $-u$, as the molecule moves back in the opposite *direction* but at the same speed u. The *change* in velocity is equal to the velocity after the collision (final state) *minus* the velocity before the collision (initial state)

$$\Delta u = -u - +u = -2u \qquad (2\text{-}2)$$

$\qquad$ final $\qquad$ initial $\qquad$ change in
$\qquad$ velocity $\quad$ velocity $\quad$ velocity

(*Note:* According to conventional usage the term Δu is read "the change in u"; it is *not* the product of Δ and u.) If the change in velocity at *each* collision is $-2u$ and there are r collisions per second, the total change in velocity *per second* at the right-hand wall is $-2ru$; this quantity is the time rate of change of velocity

$$-2ru = \frac{du}{dt} \qquad (2\text{-}3)$$

Collisions per second

The number of collisions per second at the right-hand wall can be calculated from the number of round trips per second which the molecule makes, since the molecule will collide with the right-hand wall once during each round trip. Denote the distance between the walls by s cm; a round trip over and back covers $2s$ cm; the length of path traveled in 1 sec (or the number of centimeters per second traveled) is the velocity u cm $sec^{-1} \times 1$ sec. Thus

$$u \text{ cm sec}^{-1} = r \text{ trips sec}^{-1} \times 2s \text{ cm trip}^{-1} \qquad (2\text{-}3a)$$

and

$$r = \frac{u}{2s} \qquad (2\text{-}3b)$$

so that

$$F = m\frac{du}{dt} = m(-2ru) = -\frac{2mu^2}{2s} = -\frac{mu^2}{s} \qquad (2\text{-}4)$$

The force exerted on the wall by the particle F_w is equal in magnitude and opposite in sign to F

$$F_w = -F = \frac{mu^2}{s} \qquad (2\text{-}5)$$

The pressure on the wall is equal to the force per unit area and is therefore

this model of a gas as simple as possible, imagine that the walls of the box are ideal plane surfaces and that the molecule is moving in such a way that it strikes the right-hand wall at the central point marked A; it then rebounds elastically, moving back along its original path to strike the left-hand wall at the central point marked B. Thus, the molecule goes back and forth along the line joining the center points of these opposite walls. Because the collisions with the walls are elastic, the molecule neither gains nor loses any energy and its speed remains constant.

When the molecule collides with the right-hand wall and its direction of motion is reversed, it exerts a force on the wall and the wall exerts an equal and opposite force on the molecule. Numerically this force F acting on the molecule is equal to the time rate of change of momentum of the molecule. The momentum of the molecule is equal to its mass m multiplied by its velocity u. Therefore, the time rate of change of momentum is equal to the mass multiplied by the time rate of change of velocity

$$F = \frac{d(mu)}{dt} = m\frac{du}{dt} \qquad (2\text{-}1)$$

The time rate of change of velocity can be calculated by considering the change in velocity which takes place at the instant when the molecule collides with the wall.

$$P_1 = \frac{F_w}{A} = \frac{F_w}{s^2} \qquad (2\text{-}6)$$

where P_1 is the pressure exerted by this single molecule and A is the area of the right-hand wall, which is equal to s^2. If the value of F_w given by Eq. (2-5) is substituted in Eq. (2-6), the latter is changed to

$$P_1 = \frac{mu^2}{s^3} = \frac{mu^2}{V} \qquad (2\text{-}7)$$

since s^3 is equal to the volume V of the cubical box.

Pressure per mole

Now imagine that the single molecule in the box shown in Fig. 2-1 is replaced by 1 g mole of gas. Where originally there was only one molecule, there are now 6.02×10^{23} molecules (Avogadro's number N_A). Next imagine the artificial situation where a third of these molecules move back and forth between the left- and right-hand faces on paths similar to that shown in Fig. 2-1; another third move between the front and back faces along similar paths; and the last third move between the top and bottom faces; none of the molecules collide with each other. Under these circumstances the total pressure P on the right-hand face of the box will be due to collisions involving $\frac{1}{3}N_A$ molecules and will be equal to the sum of the pressures of the individual molecules:

$$P = P_1 + P_2 + P_3 + \cdots P_{(1/3)N_A} \qquad (2\text{-}8)$$

where P_1 denotes the pressure due to molecule 1, P_2 that due to molecule 2, and so on. Denoting the velocity of molecule 1 by u_1, molecule 2 by u_2, etc., Eq. (2-7) can be written in the form

$$P = \frac{m}{v}(u_1^2 + u_2^2 + u_3^2 + \cdots u^2_{(1/3)N_A}) \qquad (2\text{-}9)$$

where V is replaced by v since this is the volume containing 1 mole of gas. Since the assumption is made that each wall of the box has the same number of molecules striking it per second as any other wall and that all molecules strike the wall at a right angle to the surface, the pressure on any one wall is equal to the pressure on any other wall, and this pressure is the pressure P of the total gas as normally measured.

Root-mean-square velocity

A special average molecular velocity may be defined, called the root-mean-square velocity u_{rms}:

$$u^2_{rms} = \frac{u_1^2 + u_2^2 + u_3^2 + \cdots + u^2_{(1/3)N_A}}{\frac{1}{3}N_A} \qquad (2\text{-}10)$$

Thus

$$\tfrac{1}{3}N_A u^2_{rms} = u_1^2 + u_2^2 + u_3^2 + \cdots + u^2_{(1/3)N_A} \qquad (2\text{-}11)$$

Substituting this in Eq. (2-9) gives

$$P = \frac{mN_A}{3v} u^2_{rms} \qquad (2\text{-}12)$$

The weight m of an individual molecule multiplied by Avogadro's number N_A is just the molecular weight M of the gas

$$mN_A = M \qquad (2\text{-}13)$$

and so Eq. (2-12) may be written

$$Pv = \tfrac{1}{3}Mu^2_{rms} \qquad (2\text{-}14)$$

Kinetic energy

The kinetic energy E_k of the molecules will be the sum of the kinetic energy of the individual molecules

$$E_k = \tfrac{1}{2}m(u_1^2 + u_2^2 + u_3^2 + \cdots u_{N_A}{}^2)$$
$$= \tfrac{1}{2}mN_A u^2_{rms} = \tfrac{1}{2}Mu^2_{rms} \qquad (2\text{-}15)$$

so that

$$Mu^2 = 2E_k \qquad (2\text{-}16)$$

Substituting this expression in Eq. (2-14) gives

$$Pv = \tfrac{2}{3}E_k \qquad (2\text{-}17)$$

When experimental measurements are made on real gases like helium and hydrogen, which approximate ideal gases in their behavior, it is found that the relationship expressed in Eq. (2-17) holds very precisely. This confirms the overall correctness of the derivation of Eq. (2-17) from the kinetic model of the ideal gas. Even though the assumption was made that the gas particles moved in an artificially restricted way, the motions in a real gas produce an overall effect similar to the hypothesized motions in the model. The underlying reason for this agreement is associated with the principle of equipartition of energy, which will be discussed in Chap. 6.

In this kinetic model the assumptions are also made that there are no forces of interaction between the gas molecules and that they do not collide; i.e., there is no appreciable volume occupied by the molecules themselves. In other words, the gas obeys the ideal-gas law

$$Pv = RT \tag{2-18}$$

Combining Eqs. (2-17) and (2-18) shows that

$$E_k = \tfrac{3}{2}RT = \tfrac{1}{2}Mu^2_{\text{rms}} \tag{2-19}$$

2-3   MOLECULAR VELOCITIES

Equation (2-19) can be put in a form which expresses the rms velocity u_{rms} as a function of the molecular weight and the temperature

$$u_{\text{rms}} = \left(\frac{3RT}{M}\right)^{1/2} \tag{2-20a}$$

In using this equation it is necessary to express the value of R in units consistent with the dimensions of the other terms in the equation.

The simple average molecular velocity (u_{avg} or $\bar{u}$) is given by the expression

$$\bar{u} = \frac{1}{N_A} \sum_{n=1}^{N_A} u_n = \left(\frac{8RT}{\pi M}\right)^{1/2} = 0.9213\, u_{\text{rms}} \tag{2-20b}$$

Example 2-1

Calculate the rms velocity of oxygen molecules in the lungs at normal body temperature, 37.0°C.

Body temperature on the absolute scale is $37.0 + 273.2° = 310.2°K$. The molecular weight of oxygen is 32 g mole^{-1}. Substituting these values in Eqs. (2-20),

$$u_{\text{rms}} = \Big(3 \times 8.314$$

$$\times 10^7 \text{ g cm}^2 \text{ sec}^{-2} \text{ deg}^{-1} \text{ mole}^{-1} \frac{310.2 \text{ deg}}{32 \text{ g mole}^{-1}}\Big)^{1/2}$$

$$= 492 \text{ m sec}^{-1} \qquad Ans.$$

An oxygen molecule in a capillary tube in the lungs behaves much like the molecule pictured in the box in

Fig. 2-1. A knowledge of the average velocity of oxygen molecules makes it possible to calculate the frequency with which such a molecule strikes one side of the capillary if it bounces back and forth between opposite walls like the molecule in the box.

Example 2-2

Calculate the number of times per second that a molecule of oxygen strikes one side of a capillary tube in the lung when the molecule is at body temperature; assume that the capillary diameter s is 0.1 mm and that the molecule moves on a path similar to that in Fig. 2-1.

Using the value of $\bar{u}$ listed in Table 2-1 and Eq. (2-3a), the number of collisions per second r can be calculated

$$r \text{ trips sec}^{-1} = \frac{\bar{u} \text{ cm sec}^{-1}}{2s \text{ cm trip}^{-1}}$$

$$r = \frac{4.53 \times 10^4 \text{ cm sec}^{-1}}{2 \times 10^{-2} \text{ cm trip}^{-1}}$$

$$= 2.27 \times 10^6 \text{ trip sec}^{-1}$$

$$= 2.27 \times 10^6 \text{ collisions sec}^{-1}$$

$$Ans.$$

Values of the avg velocities of a few different kinds of molecules are given in Table 2-1. As implied by Eqs. (2-20), the lower the molecular weight, the higher the molecular velocity. It is interesting to compare the figures in this table with values observed for the velocity of sound in various gases as given in Table 2-2. Sound is transmitted by a kind of chain reaction of molecules, A hitting B, B moving a short distance and hitting C, C moving and hitting D. While the pattern of movement obviously differs from the movement of a single molecule, one might expect the velocity of propagation of sound to be of the same order of magnitude as the molecular velocity, and this is found to be the case. Moreover, the faster moving molecules also transmit sound more rapidly, so that a knowledge of the velocity of sound propagation provides at least qualitative information about molecular velocity.

Effusion

In the early nineteenth century, the Scottish chemist Thomas Graham carried out an extensive series of studies on the rates at which different gases will pass out of, or effuse, from small openings, as shown in Fig. 2-2. He concluded from his experiments that the rate of effusion is in inverse proportion to the square root of the molecular weight

$$\frac{r_1}{r_2} = \sqrt{\frac{M_2}{M_1}} \qquad (2\text{-}21)$$

where r_1 and r_2 are the rates of effusion of gases 1 and 2, which may be expressed in terms of moles per second, or other similar units, and M_1 and M_2 are the molecular weights of the two gases expressed in grams per mole.

Theoretically, the speed with which gases effuse through a small opening should be proportional to their molecular velocities and therefore inversely proportional to the square root of their molecular weights. This may be seen by expressing the velocity for each gas in terms of temperature and molecular weight by an equation similar to Eq. (2-20b):

$$(u_{\text{avg}})_1 = \sqrt{\frac{8RT}{\pi M_1}} \qquad (2\text{-}22a)$$

$$(u_{\text{avg}})_2 = \sqrt{\frac{8RT}{\pi M_2}} \qquad (2\text{-}22b)$$

Dividing Eq. (2-22a) by (2-22b) gives

$$\frac{(u_{\text{avg}})_1}{(u_{\text{avg}})_2} = \sqrt{\frac{M_2}{M_1}} \qquad (2\text{-}23)$$

Thus, it is seen that the average velocities of the gases are in inverse ratio to the molecular weights, in accord with Graham's observations. Values obtained from experimental measurements are given in Table 2-3.

The transfer of gases into and out of the lungs in man and in animals involves many processes similar to effusion where gases must pass through small openings. Not only in normal breathing but in the administration of anesthetics through the lungs, the molecular weight of the gas involved significantly affects the rate of gaseous transfer.

Example 2-3

Calculate the ratio of the rate of effusion of nitrous oxide, N_2O, to that of diethyl ether, $C_4H_{10}O$, using Graham's law [Eq. (2-21)].

The molecular weight M_1 of N_2O is 44; the molecular weight M_2 of $C_4H_{10}O$ is 74. Putting these values in Eq. (2-21) gives the numerical ratio of the rates of diffusion, r_1/r_2

$$\frac{r_1}{r_2} = \sqrt{\frac{M_2}{M_1}} = \sqrt{\frac{74}{44}} = \sqrt{1.68} = 1.30$$

$$Ans.$$

The gases listed in Tables 2-1 and 2-2 include several of special physiological interest. Inhaled air consists mostly of nitrogen and oxygen with some water vapor and some carbon dioxide. When divers or caisson workers breathe air under a pressure several times greater than atmospheric, the concentration of oxygen and nitrogen in the bloodstream goes far above normal. If the pressure is reduced too quickly, bubbles of gas form in the bloodstream, resulting in an illness known as the *bends*. This pathological reaction can be avoided by "decompressing" the workers more slowly.

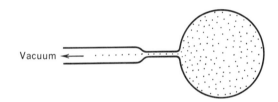

Fig. 2-2   Effusion of a gas through a pinhole into a vacuum.

TABLE 2-1   Average velocities ($\bar{u}$) of molecules in meters per second

Molecule	At 0°C	At 25°C	At 37°C
He	1,204	1,258	1,283
A	381	398	406
Hg	170	178	181
H_2	1,692	1,768	1,803
D_2	1,196	1,250	1,276
N_2	454	475	484
CO	455	475	484
O_2	425	444	453
H_2O	567	592	604
CO_2	363	379	386
NH_3	583	609	621
CH_4	601	628	640
C_6H_6	272	284	290

TABLE 2-2   Velocity of sound at 0°C

Medium	Velocity, m sec^{-1}
Dry air	331
Ammonia	415
Carbon monoxide	337
Carbon dioxide	258
Hydrogen	1,270
Oxygen	317
Ethyl ether	179
Water vapor	401

TABLE 2-3   Ratios of rates of effusion obtained experimentally[a] and calculated theoretically

	Experiment	Theory
$\dfrac{\text{Oxygen}}{\text{Hydrogen}}$	3.82	3.98
$\dfrac{\text{Carbon dioxide}}{\text{Hydrogen}}$	4.36	4.67

[a] T. Graham, *Phil. Trans. Royal Soc. (London),* **136:**573 (1846).

The safe time needed for decompression can be reduced by substituting helium for the nitrogen in the air, since helium, having a much lower molecular weight, both comes out of the blood faster and effuses out of the lung capillaries faster.

2-4   MOLECULAR COLLISIONS

In a gas at body temperature (37°C) and 1-atm pressure, a molecule can travel only a very short distance before colliding with another molecule. The number of collisions in a centimeter path of flight can be calculated with a fair degree of accuracy by considering the number of molecules per cubic centimeter in the gas under these conditions and the effective diameter of the molecule itself. Assume that the molecule is something like a tiny Ping-Pong ball that collides elastically with its neighbors and with the wall. If the molecule is moving with the velocity u_{avg} and there are no

collisions, it travels along a straight line and in 1 sec covers a distance of u_{avg} cm sec$^{-1} \times 1$ sec $= s$ cm. This path is pictured in Fig. 2-3; the molecule under consideration is shown as the black dot just above the arrow at the top of the central tube. The molecule has the radius r and the diameter $2r$. As the molecule moves down along the path of s cm, it sweeps out a volume of space shown as the tube of length s and diameter $2r$ which contains the molecule at the top. The volume of this tube V_1 is

$$V_1 = s\pi r^2 \tag{2-24}$$

where s is the length of the tube, which is equal numerically to u_{avg}, and πr^2 is the cross-sectional area of the tube.

If the other molecules are uniformly distributed in space, the number of them which at any moment penetrate in whole or in part into the volume V_1 will be the number with centers lying in the volume of a tube with a length of s cm and a radius of $2r$ cm. The volume V_2 of such a tube is given by

$$V_2 = s\pi(2r)^2 = u_{\text{avg}}4\pi r^2 \tag{2-25}$$

The number of moles per cubic centimeter n^* in the gas can be determined from the ideal-gas law

$$n^* = \frac{n}{V} = \frac{P}{RT} \tag{2-26}$$

The number of molecules per cubic centimeter N^* is obtained by multiplying n^* by N_A, Avogadro's number:

$$N^* = n^* N_A \tag{2-27}$$

The number of molecules N_2 in volume V_2 is given by

$$N_2 = N^* V_2 = N^* u_{\text{avg}}4\pi r^2 \tag{2-28}$$

N_2 is thus the number of light molecules shown in Fig. 2-3. Since the dark molecule collides with each of these light molecules in its passage down the volume V_1, which takes 1 sec, the number of collisions it makes per second z_1^1 according to this model will be equal to N_2.

$$z_1^1 = N_2 = N^* u_{\text{avg}}4\pi r^2 \tag{2-29}$$

The model from which the above formula is derived is somewhat oversimplified. The molecule for which path length is calculated actually is deflected more or less by each collision, and the molecules with which it collides are not stationary but moving. More detailed calculations show that these considerations increase the number of collisions by a factor of $\sqrt{2}$, so that the *actual* number of collisions z_1 made by one molecule per second is

$$z_1 = \sqrt{2}(4\pi r^2)u_{avg}N^* \tag{2-30}$$

and so the number of collisions per second per cm^3 z which all the molecules make with each other is

$$z = \sqrt{2}(4\pi r^2)u_{avg}N^*(\tfrac{1}{2}N^*) = 2\sqrt{2}\pi r^2 u_{avg}(N^*)^2 \tag{2-31}$$

The factor of $\tfrac{1}{2}$ is introduced into the expression because since each collision involves a *pair* of molecules, each molecule is counted twice when z_1 is multiplied by N^*.

Example 2-4

Calculate the number of collisions per second in 1 cm^3 of O_2 in the lungs at body temperature and 1-atm pressure.

The answer is obtained from Eq. (2-31):

$$z = 2\sqrt{2}\pi r^2 u_{avg}(N^*)^2$$

The value of N^* is calculated by first obtaining the value of n^* from Eq. (2-26):

$$n^* = \frac{P}{RT}$$

$$= \frac{1\ \text{atm}}{82.06\ cm^3\ \text{atm deg}^{-1} \times \text{mole}^{-1}\ 310.2°}$$

$$= 3.9285 \times 10^{-5}\ \text{mole cm}^{-3}$$

$N^* = n^*N_A = 3.929 \times 10^{-5} \times 6.02 \times 10^{23}$

$\quad = 2.365 \times 10^{19}$ molecules cm^{-3}

The value of u_{avg} is taken from Table 2-1, and the value of r is taken from Table 2-4, giving

$z = 2\sqrt{2}(3.1416)(1.81 \times 10^{-8}\ cm)^2\ 4.53$

$\qquad \times 10^4\ cm\ sec^{-1}\ (2.365 \times 10^{19})^2$

$\quad = 8.89 \times 3.28 \times 10^{-16}$

$\qquad\qquad \times 4.53 \times 10^4 \times 5.59 \times 10^{38}$

$\quad = 7.38 \times 10^{28}$ collisions sec^{-1} *Ans.*

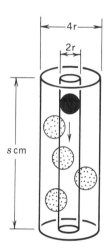

Fig. 2-3 Molecular collisions. In moving along the path length of s cm in 1 sec, the black molecule, which lies with its center in the tube with diameter $= 2r$, collides with every other molecule (shaded).

TABLE 2-4   Molecular diameters of respiratory gases[a]

Compound	Molecular weight	Molecular radius r, Å	Molecular form	Rotational degrees of freedom[b]	
CO_2	44	2.31	O=C=O	Linear	2
H_2O	18	2.29	H–O–H	Nonlinear	3
N_2	28	1.89	N≡N	Linear	2
O_2	32	1.81	O=O	Linear	2
He	4	1.10	He	Point	0

[a] 1 Å $= 10^{-8}$ cm.
[b] See Sec. 2.7.

Mean free path

The vertical length s of the cylinder in Fig. 2-3 is the distance which the molecule travels in 1 sec. If this distance is divided by the number of collisions z_1 the molecule has along this path, the result is the average length of the path between collisions. This quantity is designated L and is called the *mean free path*:

$$L \text{ cm collision}^{-1} = \frac{u_{\text{avg}} \text{ cm sec}^{-1}}{z_1 \text{ collisions sec}^{-1}} \qquad (2\text{-}32)$$

Combining this equation with the equation for z_1 gives

$$L = \frac{1}{4\pi r^2 N^*} \qquad (2\text{-}33)$$

A more sophisticated calculation shows that a more accurate relation is

$$L = \frac{1}{4\sqrt{2}\pi r^2 N^*} \qquad (2\text{-}34)$$

2-5   COLLISION DIAMETERS

In the preceding calculations the assumption is made that a molecule is a spherical particle much like a Ping-Pong ball, with a hard, elastic surface and a definite radius. Actually real molecules rebound elastically when they collide, not because of a contact between hard surfaces but because of interaction between extended fields of force, as shown in Fig. 2-4. In Chap. 19, dealing with molecular structure, the nature of the forces brought into play when molecules collide will be discussed in greater detail. These forces involve electrical repulsion and the interaction of the waves associated with the fundamental particles like electrons and

protons of which the molecules are composed; so that instead of picturing a sharply defined molecular diameter like that of a Ping-Pong ball, it is necessary to think of a diameter which the molecules *appear* to have as their fields of force gradually interact. At this point there is no need to be concerned about the variations in this effective diameter, which depend on how such a diameter is calculated. Actually the values obtained differ considerably, depending on the kinds of data used for the calculation. Another uncertainty in the calculation of molecular diameter arises from the fact that so many molecules are not spherical but oblong or V-shaped or formed in a variety of other geometrical patterns. Table 2-4 lists the effective molecular radius for a number of molecules that play a significant part in respiration under different conditions; the sketches at the right approximate the shape of the molecule.

2-6   GAS VISCOSITY

As gases flow through tubes, as in the passage from the atmosphere into the lungs, there is a tendency for the molecules close to the walls of the tube to move more slowly and for the molecules near the center of the tube to move more rapidly. As the air flows through such a tube, therefore, the velocity of molecules with respect to the wall varies from a maximum at the center of the tube to zero at the wall, where there is an adsorbed layer of gas which has zero velocity with respect to the wall. The value of the maximum relative velocity at the center of the tube depends upon the *viscosity* of the gas.

Viscosity is a measure of the resistance to flow. For liquids, differences in resistance to flow are obvious. Water flows through a funnel into a bottle rapidly compared with molasses, which has a high viscosity and flows slowly. Since flow of gases into or out of the lungs is of considerable importance both from the point of view of normal breathing and of anesthesia, it is of interest to see the relation between viscosity and some of the other properties which were studied in the discussion of the kinetic model of a gas.

Figure 2-5 shows a cross section of a tube like one of the bronchial tubes of the lung through which air is flowing upward with a velocity v. This overall velocity v of the gross flow of the gas is to be distinguished from

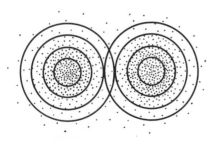

Fig. 2-4   Collision by interpenetration of force fields between statistical electron clouds.

the speeds at which the individual molecules of the gas dash around in typical random fashion, with some molecules moving very rapidly and some very slowly, so that a whole range of speed is represented. And the average of the individual velocities of the molecules is u_{avg}, which for oxygen at body temperature is 453 m sec^{-1}. In the situation described here, it is assumed that the average upward component of the molecular velocity is slightly greater than the downward component, so that the air as a whole is moving upward and outward at a gross velocity v, which is 10 cm sec^{-1} at the center of the tube.

Consider now the molecules that lie on the surfaces of two imaginary concentric cylinders shown in Fig. 2-5 as light lines marked a and b. These surfaces lie at a distance apart L equal to the *mean free path* of the gas. On the average, the upward motion of the molecules on layer a will be slower than the upward motion of the molecules on layer b, since the former lie nearer to the wall of the tube, where the average velocity upward is zero. When a molecule from layer a jumps over to layer b, the average upward velocity of layer b is decreased. A molecule from b jumping over to layer a has the opposite effect. Thus, there is a continual flow of energy toward the wall; as molecules finally collide with the stationary wall, their kinetic energy flows into the wall as heat and is lost from the motion. It is this phenomenon which produces the resistance to flow called *viscosity*.

The difference between the velocity v_b of the flow in layer b and v_a of layer a can be expressed by

$$v_b - v_a = L \frac{dv}{dx} \tag{2-35}$$

where dv/dx is the velocity gradient, or increase in velocity per centimeter out from the wall.

The momentum carried over when a molecule moves from layer a to layer b is therefore $mL(dv/dx)$, where m is the mass of the molecule.

As in the model showing the kinetic nature of gas pressure, it may be assumed here, with respect to the *molecular kinetic* motion, that one-third of all the molecules in the tube are moving along lines parallel with the x direction, one-third along the y direction, and one-third along the z direction; it also may be assumed that one-half of the molecules moving along the x direction

Fig. 2-5   Flow of viscous gas.

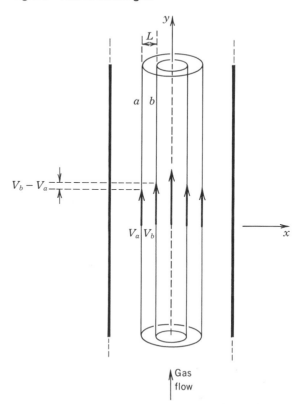

are moving to the right and one-half are moving to the left. Thus the number leaving a square centimeter of surface a and passing to the corresponding square centimeter of surface b per second will be one-sixth of the number of molecules per cubic centimeter N^* multiplied by the avg molecular velocity, $\frac{1}{6} N^* u_{avg}$; the corresponding number passing from surface b to surface a will be the same, so that the total number of molecules per second exchanging momentum between layer a and b will be $\frac{1}{3} N^* u_{avg}$.

According to Newton's law, the force f acting between layer a and layer b will be equal to the amount of momentum change per second, i.e., the time rate of change of momentum. Thus,

$$f = \frac{1}{3} N^* u_{avg} mL \frac{dv}{dx} \tag{2-36}$$

Viscosity is defined as force per unit velocity gradient and is usually denoted by the small Greek letter eta η.

Thus,

$$\eta = \frac{f}{dv/dx} \tag{2-37}$$

Combining the above expressions, the relation is found for the above model

$$\eta = \tfrac{1}{3}N^*u_{avg}mL(dv/dx)/(dv/dx) \tag{2-38}$$

Canceling reciprocal terms and substituting the symbol $\bar{u}$ for $\bar{u}_{avg}$, we get

$$\eta = \tfrac{1}{3}N^*\bar{u}mL \tag{2-39}$$

where $\bar{u}$ is the average velocity, which is $1.085u_{avg}$.

Using Eq. (2-34), L can be replaced by a term containing the effective collision radius

$$\eta = \frac{m\bar{u}}{8\sqrt{2}\pi r^2} \tag{2-40}$$

The units of viscosity can be found by filling in the units of the terms in this equation

$$\eta = \frac{\text{g cm sec}^{-1}}{\text{cm}^2} = \frac{\text{g}}{\text{cm sec}} = \frac{\text{dynes}}{\text{cm}^2} \tag{2-41}$$

The unit viscosity corresponding to 1 g cm^{-1} sec^{-1} is called a *poise* to honor Poiseuille, the French scientist who did so much of the original basic research in this field. The smaller units are also frequently used: centipoise $= 0.01$ poise, millipoise $= 0.001$ poise, and micropoise $= 10^{-6}$ poise. Observed values of gas viscosities are given in Table 2-5.

TABLE 2-5   Viscosity at 37°C for gases important in respiration and anesthesia

Gas	η, micropoises
Air	190
CO_2	156
Ethyl ether	79
He	204
H_2	91
O_2	208
N_2	183

Example 2-5

Calculate the ratio of the viscosities of helium and oxygen at 37°C from the molecular weights, molecular velocities, and molecular radii.

Using Eq. (2-40), the ratio is given by

$$\frac{\eta_{He}}{\eta_{O_2}} = \frac{m_{He}\bar{u}_{He}r^2_{O_2}}{m_{O_2}\bar{u}_{O_2}r^2_{He}}$$

The ratio of the weights of the two kinds of molecules is the same as the molecular weights. The ratio of the average velocities is found from the ratio of the values of u_{avg} given in Table 2-1. The values of r are given in Table 2-4.

$$\frac{\eta_{He}}{\eta_{O_2}} = \frac{4}{32}\frac{1,283}{453}\left(\frac{1.81}{1.10}\right)^2 = 0.96 \qquad Ans.$$

The ratio of the observed viscosities from Table 2-5 is 0.98.

2-7   THE EQUIPARTITION OF ENERGY

For a monatomic gas similar to the kinetic model discussed in Sec. 2-2, the kinetic energy of 1 mole E_k is equal to $\tfrac{3}{2}RT$. For a single molecule the average kinetic energy e_k is equal to $E_k/N_A = \tfrac{3}{2}kT$, where k is the Boltzmann constant, equal to R/N_A.

The principle of *equipartition of energy* asserts that the energy of motion of the gas is divided equally among the components of motion in each of the three dimensions of space. If every external force like that due to gravity is negligible, then as the gas molecules swarm around inside a container, there is no physical relationship between their motions and the directions x, y, and z. Since normally the specification of the actual direction of the coordinate system is completely arbitrary, it is impossible to imagine that nature would favor any direction with a larger proportion of energy than is given to any other direction, and so it is clear why energy is partitioned equally between the components of motion in each of the three directions of space. Since the total energy is $\tfrac{3}{2}kT$, and since there are three dimensions of space, this means that the kinetic energy associated with the components of motion in any single dimension must be one-third of the total, or $\tfrac{1}{2}kT$.

Degrees of freedom

In order to specify in terms of cartesian coordinates exactly where a given molecule is at a given instant in space, it is necessary to specify the value of each of these three coordinates relative to some chosen origin. As the molecule moves freely in space, the value of each coordinate varies. For this reason, a molecule moving in a three-dimensional space has 3 *degrees of freedom*.

As will be discussed later in Chap. 6, the rotation of a *monatomic* molecule does not enter into the calculation of energy at normal temperatures, and there is no need to specify angular coordinates. On the other hand, in order to denote completely the position of a rigid *diatomic* molecule in space, one must specify the values not only of the x, y, and z coordinates for the center of mass of the molecule but also of two angles such as θ and ϕ for the position of the line joining the nuclei of the two atoms. Because of considerations similar to those which eliminate rotation of a monatomic molecule from energy calculations, it is not necessary to specify the third angle for rotation around the axis of the molecule. Because there are now five variables to be specified (x, y, z, θ, and ϕ), the diatomic molecule has 5 degrees of freedom. Under the principle of equipartition of energy there is $\frac{1}{2}kT$ of energy associated with each of these degrees of freedom, making the total average kinetic energy E_k equal to $\frac{5}{2}kT$. This is also true for all *linear* molecules like CO_2 and HCN when they are regarded as rigid bodies.

In order to specify completely the position of a *triatomic* molecule like H_2O, it is necessary to specify the values of the three space coordinates x, y, and z and also the values of three angular coordinates such as θ, ϕ, and ψ, shown in Fig. 2-6. There are now 6 degrees of freedom, and the value of E_k is $\frac{6}{2}kT$.

When a molecule is regarded not as a rigid body but as flexible, there are still other degrees of freedom. Actually, in real molecules there is generally quite a bit of flexibility. A hundred years ago there was no awareness of its importance. When models of molecules were constructed for classroom demonstrations, the atoms frequently were shown as small wooden balls and the chemical bonds joining these atoms together were represented by rigid wooden sticks. Evidence from the study of the spectra of molecules (the way

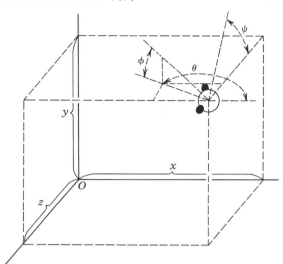

Fig. 2-6 The six coordinates necessary to locate a nonlinear molecule (H_2O) in space. Translational coordinates: x, y, z; rotational coordinates: θ, ϕ, ψ.

in which light is absorbed or emitted by molecules) has shown incontrovertibly that molecules are not like these rigid ball-and-stick models but resemble much more models where the bonds are represented by spiral springs. Thus, the atoms can vibrate toward and away from each other; and a molecule has not only energy of translational motion and energy of rotational motion but also an energy of internal vibration.

In mechanics it is shown that each point mass (the nucleus of an atom) has 3 degrees of freedom. In a molecule like water, H_2O, there are 3 degrees of freedom for each of the three atoms of which the molecule is composed; therefore, the total number of degrees of freedom for the water molecule is 9. Thus, the general equation relating the number of atoms in a molecule n to the total number of degrees of freedom d_{tot} is given by

$$3n = d_{tot} \tag{2-42}$$

For a nonlinear molecule like water this expression becomes:

$$d_{tot} = d_{trans} + d_{rot} + d_{vib} \tag{2-43a}$$

$$9 = 3 + 3 + 3 \tag{2-43b}$$

For a linear molecule like CO_2, which has the structural formula O:C:O, there are $3n = 3 \times 3 = 9$ total degrees of freedom, and, as shown above, three of these are translational (x, y, z) and two are rotational (θ, ϕ); thus, there are 4 degrees of vibrational freedom, and the relation is

$$d_{tot} = d_{trans} + d_{rot} + d_{vib} \qquad (2\text{-}44a)$$

$$9 = 3 + 2 + 4 \qquad (2\text{-}44b)$$

The general formula for a nonlinear molecule is

$$d_{tot} = 3n = d_{trans} + d_{rot} + d_{vib} \qquad (2\text{-}45a)$$

$$3n = 3 + 3 + 3n - 6 \qquad (2\text{-}45b)$$

Table 2-6 lists the degrees of freedom for several molecules, and Fig. 2-7 shows the corresponding modes of motion.

Example 2-6

For the linear molecule acetylene, H—C≡C—H, calculate the number of each kind of degree of freedom.

Using Eq. (2-44a),

$$12 = 3 \times 4 = 3 + 2 + (3 \times 4) - 5$$

$$= 3 + 2 + 7$$

2-8   THE HEAT CAPACITY OF GASES AT CONSTANT VOLUME

On the basis of the kinetic model of an ideal gas, extended with the help of the ideas discussed in the last section, it is possible to predict the values of the heat capacity when various gases behaving like rigid bodies are heated in a container at constant volume. The *molar heat capacity at constant volume* c_v is defined as the amount of heat absorbed by 1 mole of gas when heated through a temperature increase of 1 degree at constant volume. This is the rate at which the energy of 1 mole of the gas increases with temperature. In the case of a monatomic gas, Eq. (2-19) shows that the energy is equal to $\frac{3}{2}RT$. Using the notation of the calculus for partial derivatives;

$$c_v = \left(\frac{\partial E}{\partial T} \right)_v = \left[\frac{\partial (\frac{3}{2}RT)}{\partial T} \right]_v = \frac{3}{2}R \qquad (2\text{-}46)$$

Since the heat capacity is thus $\frac{1}{2}R$ per degree of freedom, and since each degree of freedom involves the same amount of energy, the contributions to the heat capacity from the different degrees of freedom can be summarized as shown in Table 2-7.

Example 2-7

Calculate the heat capacity at constant volume for acetylene, H—C≡C—H, regarding this as a linear rigid molecule.

The degrees of freedom are

$$d_{tot} = d_{trans} + d_{rot} + d_{vib}$$

$$3n = 3 + 2 + 3n - 5$$

$$12 = 3 + 2 + 7$$

The heat capacity at constant volume is $\frac{1}{2}R$ for each translational and rotational degree of freedom; and so $c_v = \frac{5}{2}R = 5$ cal deg^{-1} mole^{-1}.

TABLE 2-6   Degrees of freedom in gaseous molecules

		d_{trans}	d_{rot}	d_{vib}	d_{tot}
Monatomic	He	3	0	0	3
Diatomic	N≡N	3	2	1	6
Triatomic:					
Linear	O=C=O	3	2	4	9
Nonlinear	H₂O (bent)	3	3	3	9

Fig. 2-7   Rotation and translation showing total number of degrees of freedom (d).

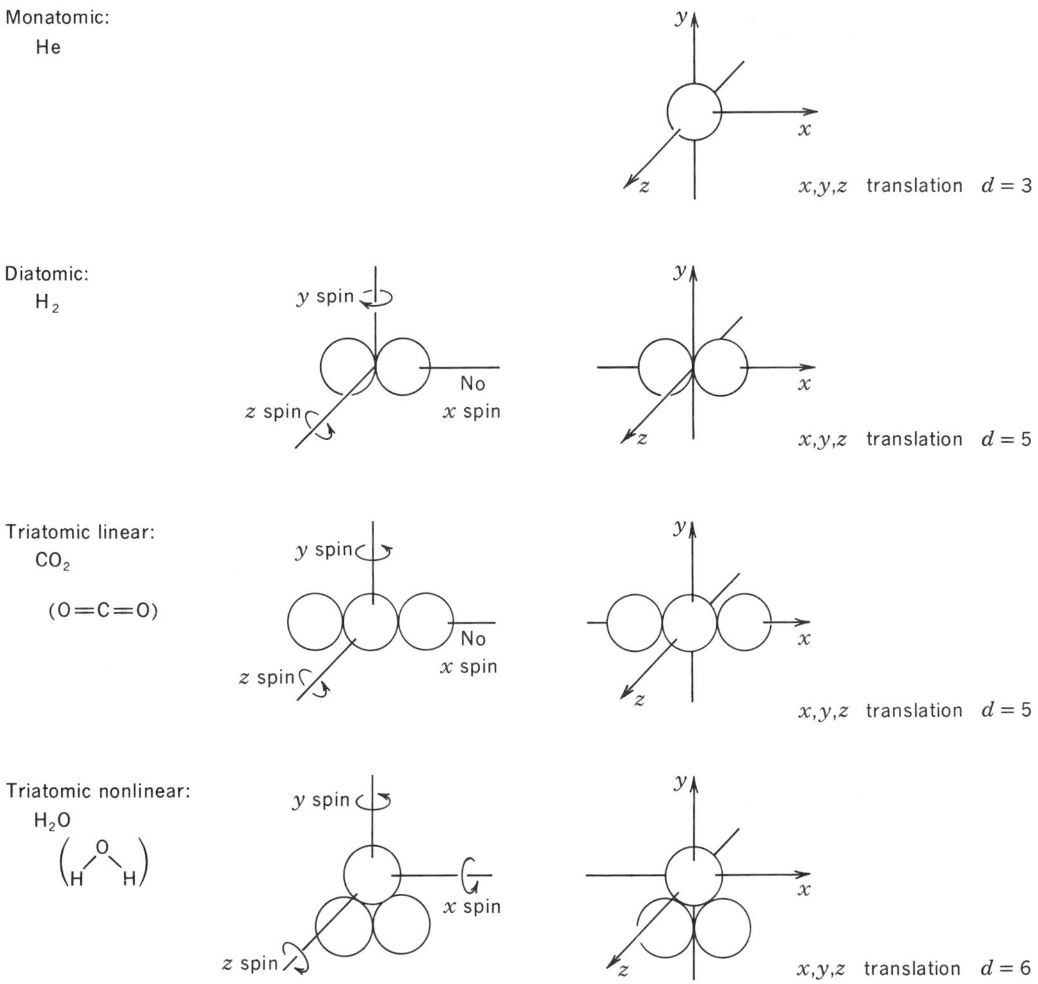

Monatomic:
He

x,y,z translation $d = 3$

Diatomic:
H_2

y spin

z spin   No
          x spin

x,y,z translation $d = 5$

Triatomic linear:
CO_2

$(O{=}C{=}O)$

y spin

z spin   No
          x spin

x,y,z translation $d = 5$

Triatomic nonlinear:
H_2O

$\left(\begin{smallmatrix} & O & \\ H & & H \end{smallmatrix}\right)$

y spin

          x spin

z spin

x,y,z translation $d = 6$

2-9   THE HEAT CAPACITY OF GASES AT CONSTANT PRESSURE

When a gas is heated not in a closed fixed volume but in something like a cylinder equipped with a sliding piston which exerts a constant pressure on the gas, it is clear that the gas will expand.   From the ideal-gas law

$$P \times V = n \times R \times T \tag{2-47}$$

const variable   const   variable

so that if the temperature increases, the volume must increase.   Experimentally it is found that for an ideal gas R cal mole^{-1} per degree of freedom is required to effect this expansion.   Therefore R must be added to the heat capacity at constant volume in order to give the amount of heat necessary to raise 1 mole of the gas by $1°$ at constant pressure, the heat capacity at constant pressure c_p.   The theoretical reason for this will be discussed in Chap. 7.   Since R has the value of 1.9872 cal deg^{-1} mole^{-1}, the values of c_p can be calculated easily for the different gases discussed in the previous section.   They are shown in Table 2-8, together with some of the observed values of the heat capacity at $15°C$.

TABLE 2-7   Heat capacity at constant volume of rigid gas molecules

Type	Molecule	c_{trans}	c_{rot}	c_{vib}	$c_{tot} = c_v$
Monatomic	He	$\frac{3}{2}R$	0	0	$\frac{3}{2}R$
Diatomic	N≡N	$\frac{3}{2}R$	$\frac{2}{2}R$	0	$\frac{5}{2}R$
Triatomic:					
Linear	O=C=O	$\frac{3}{2}R$	$\frac{2}{2}R$	0	$\frac{5}{2}R$
Nonlinear	H₂O (bent)	$\frac{3}{2}R$	$\frac{3}{2}R$	0	$\frac{6}{2}R$

TABLE 2-8   Heat capacity at constant pressure at 15°C

Type	Gas		c_p, cal deg⁻¹ mole⁻¹		Molecular shape
			Theory	Observed	
Monatomic	Ar	$\frac{5}{2}R$	4.97	5.01	Ar
Diatomic	N₂	$\frac{7}{2}R$	6.94	6.95	N≡N
	O₂	$\frac{7}{2}R$	6.94	6.97	O=O
	CO	$\frac{7}{2}R$	6.94	6.95	C=O
	HCl	$\frac{7}{2}R$	6.94	7.05	H—Cl
	H₂	$\frac{7}{2}R$	6.94	6.78	H—H
	Cl₂	$\frac{7}{2}R$	6.94	8.15	Cl—Cl
Triatomic:					
Linear	CO₂	$\frac{7}{2}R$	6.94	8.76	OCO
Nonlinear	H₂S	$\frac{8}{2}R$	7.93	8.64	S with H H (bent)
	CH₄	$\frac{8}{2}R$	7.93	8.45	HCH with H H

For the monatomic gas argon the agreement between the value of the heat capacity calculated on the basis of theory and that actually observed is good. For the three diatomic gases nitrogen, oxygen, and carbon monoxide, the agreement between theory and experiment is also remarkably close. Recall that in making the theoretical calculation it was assumed that the molecule is rigid and that no heat is being absorbed to set elastic vibrations in motion within the molecule. The data for hydrogen chloride show that the observed value lies somewhat higher than the experimental value. This may be due to internal vibration. In the first three diatomic compounds, N_2, O_2, and CO, multiple bonds connect the atoms. This causes more rigidity than in hydrogen chloride, where there is only a single bond connecting the hydrogen and the chlorine. For chlorine gas there is definite evidence that internal vibration absorbs some energy when the gas is heated at 15°C. The value reported of 8.15 cal deg⁻¹ lies so far above the theoretical value of 6.94 cal deg⁻¹ that the difference cannot be ascribed to experimental error. Because the two chlorine atoms are connected by a single bond, and because they are relatively heavy atoms compared with the others in the table, the more detailed theory of heat capacity predicts some absorption of energy by internal vibration at this temperature.

Among the diatomic molecules listed in Table 2-8, hydrogen is the only one for which the observed heat capacity lies below the predicted value. In the more detailed theory of heat capacity, it may be shown that this is due to the fact that at 15°C the rotational motion of hydrogen is somewhat restricted by quantization in a way that reduces its absorption of heat. This will be discussed further in Chap. 6.

The linear triatomic molecule CO_2 should have the same heat capacity as a diatomic molecule *if it is rigid.* Note that the observed heat capacity is considerably higher than any values observed for diatomic molecules. Again, in the more detailed theory of heat capacity, this can be shown to be due to the absorption of energy by internal vibration; i.e., the CO_2 molecule is not rigid. Study of chemical bonds makes it clear that it is much easier to bend a chemical bond than to stretch it. The elastic bending of the CO_2 molecule accounts for the fact that the observed heat capacity lies somewhat above the theoretical heat capacity.

The same explanation holds for the value of the heat capacity for H_2S and for CH_4. These are both nonlinear molecules, and if they were completely rigid, c_p should have a value of about 8 cal deg^{-1}. c_p is slightly above this value because of the absorption of energy by internal vibration, and this can be accounted for quantitatively by the more detailed theory of heat capacity, which can be developed on the basis of the quantum theory.

In conclusion, the agreement between the observed heat capacities and the values calculated on the basis of the kinetic model of gases provides convincing evidence that this model comes close to portraying the actual behavior of real gas molecules.

of heat. Such a measurement is normally carried out in an instrument called a *calorimeter,* the name being derived from the root words for heat and for measurement. In the design and operation of a calorimeter the principal problem is to measure accurately the amount of heat that goes into increasing the temperature of the gas. The weight of the sample can be determined accurately; temperature can be measured easily with a precision of 0.01°C and sometimes to a precision of 0.0001°C without too much difficulty. But the thermal energy frequently escapes through various types of leaks so that one cannot be sure that all the energy put into the calorimeter actually goes into heating the sample.

The difficulty of controlling thermal leakage increases with the size of the surface of the vessel in which the sample is confined. Since gases are relatively low in density, a large volume is required to get a reasonable mass of gas; of course, one can put the gas under high pressure, but then one risks the danger of the calorimeters exploding.

One practical way of measuring the heat capacity of a gas is to let the gas flow through a tube which has a thermometer at the entrance and at the exit (Fig. 2-8). The tube is thermally insulated by a high vacuum, and an electric heater is placed in the part of the tube lying between the two thermometers. When a measured constant rate of gas flow is established, the heat input produces a temperature difference between the two ends of the tube. Then a knowledge of the rate of heat

Example 2-8

Calculate the heat capacity c_p for the compound cyclopropane, C_3H_6, used as an anesthetic; treat it as a rigid nonlinear molecule.

$$c_p = \tfrac{1}{2}R(d_{\text{trans}} + d_{\text{rot}}) + R$$
$$= \tfrac{1}{2}R(3 + 3) + R$$
$$= 4R = 8 \text{ cal deg}^{-1} \text{ mole}^{-1}$$

2-10   THE CALORIMETRIC MEASUREMENT OF GAS HEAT CAPACITIES

To measure heat capacity, one must increase by a known amount the temperature of a known mass of gas through the introduction of a measured quantity

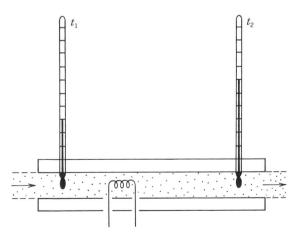

Fig. 2-8   Gas-flow calorimeter.

input, the rate of gas flow, and the temperature difference makes it possible to calculate the heat capacity of the gas.

Example 2-9

In a gas calorimeter the rate of gas flow r is 0.01 mole sec^{-1} of nitrogen gas; the rate of heat input j is 0.0695 cal sec^{-1}; the thermometer at the entrance to the tube reads $14.500°C$ (t_1); and the thermometer at the exit of the tube reads $15.500°C$ (t_2). Calculate the value of c_p.

$(c_p)_{mean}$

$$= \frac{heat\ input}{no.\ of\ moles \times temperature\ change}$$

$$= \frac{j}{r(t_2 - t_1)}$$

$$= \frac{0.0695\ cal\ sec^{-1}}{0.0100\ mole\ sec^{-1}(15.500 - 14.500°)}$$

$$= 6.95\ cal\ deg^{-1}\ mole^{-1}$$

From this experiment the *mean* heat capacity of 1 mole of nitrogen gas between 14.5 and 15.5° is found to be 6.95 cal deg^{-1} $mole^{-1}$. Since the heat capacity changes so slightly with temperature, this may be taken as the molal heat capacity c_p at $15.000°C$.

Problems

1. A molecule of oxygen is confined in a cubical box 1 m^3 in volume. The molecule bounces back and forth on a straight line between opposite faces of the cube at a rate of 5×10^4 cm sec^{-1}. How many collisions with the right-hand wall does the molecule make per second?

2. If a molecule of hydrogen gas is confined under the same conditions as outlined in Prob. 1 and has the kinetic energy of motion of the oxygen molecule, calculate the speed at which it is traveling and the number of collisions per second which it makes with the right-hand wall of the box.

3. Calculate the average kinetic energy of a gas at a temperature of $298°K$ and express the result in calories.

4. Calculate the rms velocity of an atom of helium at a temperature of $298°K$.

5. On a mole basis the composition of the normal air is approximately 21 percent O_2 and 79 percent N_2. When helium is substituted for nitrogen in the air breathed in deep-sea diving, calculate the average molecular weight of this artificial air.

6. If the rate of absorption or desorption of a gas from the bloodstream is proportional to its rms velocity, calculate how much more rapidly helium will be desorbed than nitrogen when pressure is reduced.

7. Calculate the number of collisions per square centimeter in a caisson where the temperature is $320°K$, the pressure is 4 atm, and the composition of the "air" is 21 percent oxygen and 79 percent helium on a mole basis.

8. Calculate the avg velocity of molecules of carbon tetrachloride, CCl_4, at $298°K$. Calculate the ratio of the velocity of diffusion of carbon tetrachloride as compared with air at the same temperature.

9. A certain gas is found to diffuse out through a pinhole at a rate that is 0.811 times that of air. Calculate the molecular weight of the gas.

10. A gas is found to diffuse through a hole at a rate only 0.00652 that of air at room temperature, $298°K$. Calculate the molecular weight of this gas, its avg velocity, and the number of collisions per second which molecules of this gas would make on 1 cm^2 of skin at room temperature ($298°K$) and a pressure of 1 atm.

11. Assuming that the distance from a lung capillary to the external air is 27 cm for a small child, calculate the length of time necessary for a carbon dioxide molecule to traverse this distance if the molecule moves in a straight line.

12. Calculate the number of impacts of oxygen molecules in air at a pressure of 1 atm (molal composition 20 percent O_2) when the air is in contact with 1 mm^2 of lung tissue.

13. Calculate the number of molecular impacts on the same tissue by carbon dioxide molecules from air at 1 atm if the molal concentration of carbon dioxide is 0.1 percent.

14. Calculate the number of impacts per second made by a carbon dioxide molecule on the walls of a capillary 0.01 cm in diameter if the molecule bounces directly back and forth between opposite walls and the capillary is at body temperature.

15. Calculate the number of impacts on the capillary wall per second of a molecule of ethyl ether, $C_4H_{10}O$, under the same conditions as in Prob. 14.

16. The gas cyclopropane, C_3H_6, has occasionally been used as an anesthetic. Calculate the relative rate of the effusion of this gas as compared with ethyl ether.

17. A tunnel construction worker comes out of a caisson where the pressure of air has been maintained at 4 atm. The pressure is reduced so quickly that a bubble of gas 1 mm in diameter forms in one of his veins. Assume that the temperature of the gas in the bubble is 37°C and that the pressure is 1.20 atm. Calculate the number of impacts made on the wall of the bubble by a molecule of nitrogen if it bounces directly back and forth across the bubble during each second. Calculate the number of impacts per second under the same conditions for a molecule of helium.

18. Calculate the total number of molecules in the gas bubble described in Prob. 7 assuming that it is spherical.

19. Calculate the average distance that a molecule of nitrogen travels between collisions with other molecules in the bubble described in Prob. 7 assuming that all the gas in the bubble is nitrogen.

20. Calculate the average length of time between collisions for a single molecule of nitrogen moving under the conditions described in Prob. 19.

21. When a man inhales 1 liter of air into his lungs and the air is at 1 atm and 0°C, by how many calories is the heat content of the lung decreased when the air is expelled from the lungs at a temperature of 37°C? Assume that the mass of lung tissue with which a breath of air comes directly into contact is 100 g and that the heat capacity is 1.00 cal deg^{-1} g^{-1}. Calculate the final equilibrium temperature when 0.500 liter of air at 0°C and this lung tissue come into thermal equilibrium.

22. If a man is trapped in a room in a burning building and the temperature of the air around him is momentarily raised to 500°C, what will be the increase in temperature of the lung tissue when equilibrium is established between a breath of 0.500 liter of air and the tissue? See Prob. 21 for data.

REFERENCES

Jeans, J. H.: "The Dynamical Theory of Gases," Cambridge University Press, London, 1921.

Kennard, E. H.: "Kinetic Theory of Gases," McGraw-Hill Book Company, New York, 1938.

Jeans, J. H.: "An Introduction to the Kinetic Theory of Gases," The Macmillan Company, New York, 1940.

Present, R. D.: "The Kinetic Theory of Gases," McGraw-Hill Book Company, New York, 1958.

Guggenheim, E. A.: "Elements of the Kinetic Theory of Gases," Pergamon Press, New York, 1960.

Loeb, L. B.: "Kinetic Theory of Gases," Dover Publications, Inc., New York, 1961.

Hildebrand, J. H.: "An Introduction to Kinetic Theory," Reinhold Publishing Corporation, New York, 1963.

Chapter Three

real gases

3-1 DEVIATIONS FROM THE IDEAL-GAS LAW

As later investigators extended the studies of the gaseous state made by Boyle, Charles, Gay-Lussac, and Dalton and used more refined methods of measurement, they soon realized that real gases exhibit many significant deviations from the ideal-gas law. If the familiar form of this law

$$Pv = nRT \tag{3-1a}$$

is written with all the terms on the left side

$$\frac{Pv}{nRT} = 1 \tag{3-1b}$$

it is clear that the quotient Pv/nRT should be equal to unity when the law is precisely obeyed. But careful experimental observations of P, v, and T indicate that this quotient often exhibits both positive and negative deviations from unity. The nature of these deviations is most clearly seen by plotting the quotient against pressure, as shown in Fig. 3-1; as an example, the graphs are drawn for experimental data obtained with methane, CH_4. In this figure the graph for an *ideal* gas is the horizontal dashed line, for which the value of Pv/nRT is unity at all pressures. Note that the measurements for methane have been extended to a rather high pressure (1,000 atm), so that one must focus attention on the extreme left side of the diagram to see the magnitude of the deviations at ordinary pressures. Thus, at about room temperature (20°C) and up to pressures of about 10 atm, the quotient Pv/nRT deviates from unity by less than 1 percent, indicating that in this range the ideal-gas law is quite accurately obeyed. At 200°C, the deviation is less than 1 percent between 0 and 20 atm. On the other hand, at −70°C there is a 10 percent negative deviation at 20 atm; and at a really high pressure (1,000 atm), the deviation is over 90 percent positive even at room temperatures.

A careful consideration of the theory of the constant-pressure gas thermometer provides a clue to the causes of these deviations. A schematic drawing of such a thermometer is shown in Fig. 3-2. In this device the gas is confined under a movable piston; the pressure is maintained by the weight on top of the piston; as the gas is heated or cooled, it ex-

Fig. 3-1 Deviations from the ideal-gas law. [After H. M. Kvalnes and V. L. Gaddy, *J. Am. Chem. Soc.*, 53:394 (1931).]

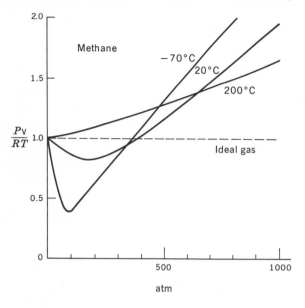

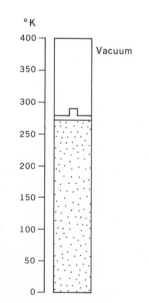

Fig. 3-2 Constant-pressure gas thermometer.

pands or contracts, and the piston indicates the temperature on a scale that is placed at the side of the cylinder. The mathematical relation between temper-

ature and volume is obtained by transposing terms in the ideal-gas law [Eq. (3-1*a*)]

$$T = \frac{P}{nR} \; \mathrm{v} \tag{3-2}$$

Since the term P/nR is a constant, T is proportional to v. By noting the value of v at the normal freezing point and boiling point of water, the numerical scale is determined, as explained in Chap. 1.

Molecular volume

For an ideal gas obeying this relation, obviously v must be equal to zero when $T = 0°$K, because of the proportionality between v and T. Therefore at $0°$K the piston comes right down to the bottom of the cylinder, the gas occupies no volume at all, and consequently the molecules of which the ideal gas is composed must have no volume. But as contrasted with this picture of an *ideal* gas, everyday experience with matter indicates that *real* molecules do have volume. Familiar matter like wood, metal, or water obviously occupies space. The implication is that these relatively incompressible forms of matter the molecules themselves do have considerable volume; and this conclusion is supported by the evidence from the more detailed investigations of the solid and liquid states to be considered in subsequent chapters. It is logical therefore that as temperature drops toward $0°$K, the value of v for a real gas does not approach zero but instead reaches a positive value which represents the volume of the molecules themselves.

Intermolecular attraction

Another aspect of the behavior of gases as temperature is lowered is the phenomenon of condensation. All gases without exception condense into the liquid or solid state if the temperature is lowered sufficiently. The most logical explanation for this change from the gaseous state, where the molecules are relatively far apart, to the liquid or solid state, where they are relatively close together, is that molecules *attract* each other. But if a gas is to obey the ideal-gas law at all temperatures, including the region close to $0°$K, condensation must not take place; in other words, there must be no significant force of attraction between the molecules of an ideal gas.

Thus the condition of *no intermolecular force* together with the condition of *no molecular volume* constitutes the physical basis for the concept of an ideal gas; and the deviations of real gases from the ideal-gas law can be understood in terms of these two factors: molecular volume and molecular interaction.

Turning back to Fig. 3-1, it may be seen that the value of Pv/nRT is *greater* than unity at very high pressures, where the molecules are forced very close together and the volume of the molecules themselves becomes a significantly large fraction of the total volume occupied by the gas. By way of contrast, as the temperature is lowered and the thermal energy in the gas is less, the force of attraction between the molecules plays a more significant part in reducing the pressure below the value predicted by the ideal-gas law; consequently, Pv/nRT takes on values *less* than unity.

Figure 3-3 shows graphs of the values of Pv/nRT plotted against values of P for the three gases, Ne, O_2, and CO_2, all observed at 273°K. The horizontal dotted line represents the plot for the ideal gas. Note that the deviations for neon are all positive. In this case, as indicated by the low boiling point (27°K), the force of attraction between the molecules is extremely small, so that the deviation is due essentially to the volume occupied by the molecules. In the case of oxygen (bp = 90°K) the force of attraction is larger, and the consequent reduction of the pressure produces a negative deviation. For carbon dioxide, which condenses to the solid state at 195°K and has a far higher force of attraction between the molecules than oxygen, the negative deviation is still greater.

Real isotherms

Another informative way to show the nature of the deviations from the ideal-gas law is to make a plot of the graphs of *pressure* against *volume* at a number of different temperatures; such graphs are called *isotherms* since the temperature is constant along each line; for ideal gases, these lines constitute a family of hyperbolas. In Fig. 3-4 the two upper curves (dashed lines) are the isotherms of 1 mole of *ideal* gas at 0 and 50°C; the lower curves (solid lines) are the isotherms based on experimental observations of the variation of pressure with volume for 1 mole of carbon dioxide at six different temperatures between 0 and 50°C. At 50° the *observed* isotherm has the hyperbolic shape characteristic of *ideal-gas* behavior; at the right of the diagram, where the volume is 0.8 liter, the ideal and the observed isotherms actually coincide, but at the left of the diagram the observed isotherm lies below the ideal-gas isotherm because the forces of attraction between the molecules reduce the observed pressure, making it less than the pressure of an ideal gas. At 40°C an abnormal bend appears in the isotherm, and below 31.04°C part of the gas liquefies when the volume is sufficiently reduced.

Liquefaction

In order to see how the process of liquefaction alters the shape of the isotherm, consider the various changes in the state of 1 mole of the gas at 0°C as its volume is reduced from 0.7 liter (point 1 on Fig. 3-4) to 0.049 liter (point 5) along the 0°C isotherm. These changes of state are shown schematically in Fig. 3-5; each drawing indicates the state of the gas at the corresponding point on the isotherm in Fig. 3-4 to which the drawing number refers. The cylinder marked 1 at

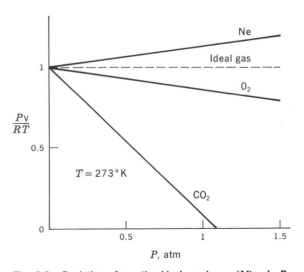

Fig. 3-3 Deviations from the ideal-gas law. (After L. P. Hammett, "Introduction to the Study of Physical Chemistry," McGraw-Hill Book Company, New York, 1952.)

Fig. 3-4   Isotherms of CO_2 from ideal gas equation (dashed line) and experimental values (solid line).

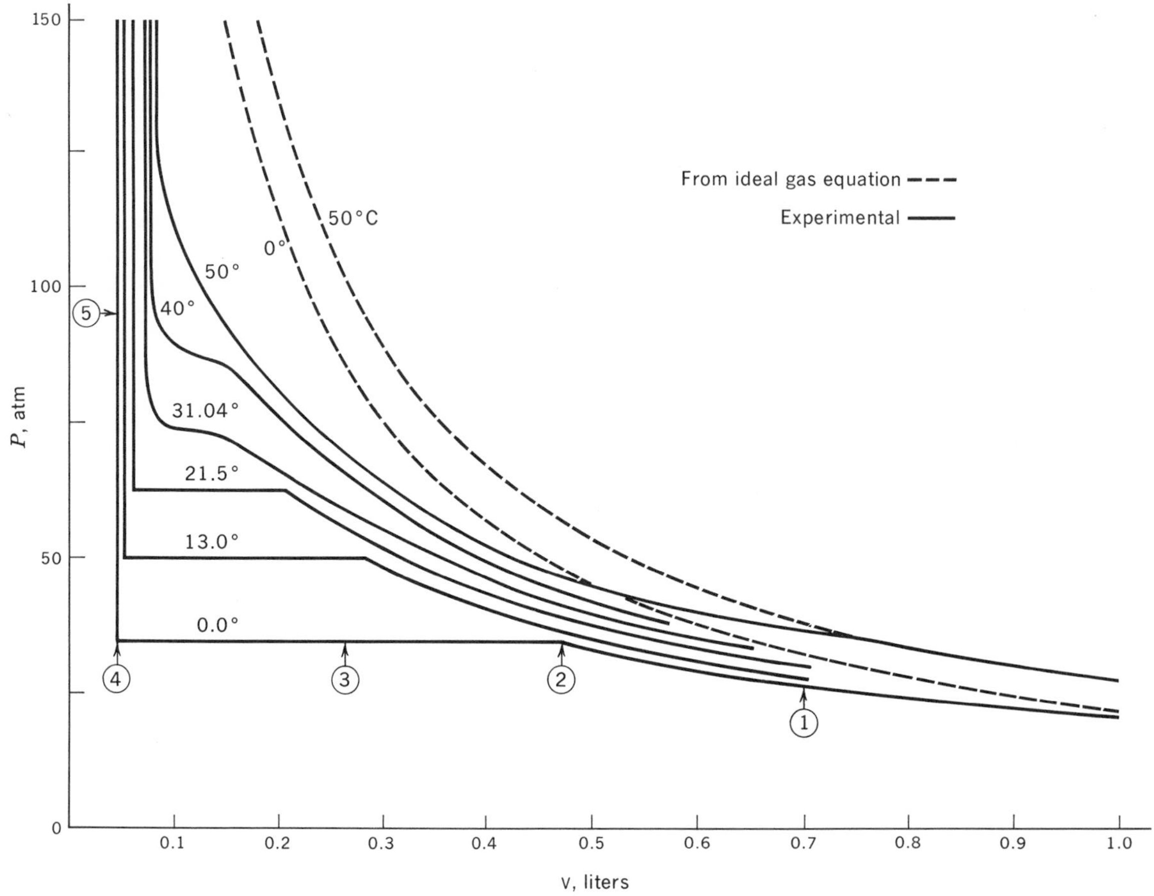

the extreme left of Fig. 3-5 shows the system as completely gaseous when in the state marked 1 on the isotherm for 0°C. When volume is reduced to point 2, the first trace of liquid appears; a further reduction in volume does not produce any increase in pressure; instead, it merely causes more gas to condense into liquid while pressure remains constant. Consequently, there is a *discontinuity* in the slope of the isotherm at the point where liquefaction begins, and the isotherm becomes a horizontal line along which no increase in pressure takes place as volume is reduced. At point 3 one-half of the gas has condensed into the liquid state. At point 4, where v = 0.05 liter, the gas is completely liquefied. Consequently any further reduction in volume now produces a large *increase* in pressure

since the liquid is highly incompressible; thus, at point 4, where the last trace of gas disappears, there is another sharp discontinuity in the slope of the isotherm, and the isotherm rises abruptly for lower values of the volume.

Similar discontinuities in slope at each end of the horizontal portion may be seen in the isotherms for 13.0 and 21.5°C.

3-2   THE VAN DER WAALS EQUATION

Shortly after isotherms like those in Fig. 3-4 were obtained by Thomas Andrews in 1869 from experimental observations of carbon dioxide, the Dutch physicist van der Waals in 1873 proposed a modification of the ideal-gas law to provide an equation which displayed relations between P, v, and T analogous to those experimentally observed. In order to explain van der Waals' approach to this problem, let us suppose that the cylinder with a volume of 1 liter shown in cross section in Fig. 3-6 is filled with 1 mole of carbon dioxide at 40°C. Maintaining the temperature constant, the piston is lowered so that the volume is reduced to 0.2 liter. As the volume is reduced, the pressure rises because the molecules are now trapped in a smaller space; and in moving around under the influence of their kinetic energy, they collide more frequently with the walls of the vessel and with each other. Thinking along these lines, van der Waals pointed out that the total volume v of the cylinder containing 1 mole of gas is divided significantly into two parts. There is the unshaded portion v_{free}, which is free space and which is therefore available for molecules to enter and move about, and there is the other portion $v_{molecules}$, which is the "solid" volume of the molecules *themselves*, shown in cross

section by the black dots in Fig. 3-6, where each dot represents a single molecule. When the total volume is relatively large, as in Fig. 3-6a, the portion of the volume $v_{molecules}$ occupied by the molecules themselves shown as black dots may be almost negligible compared with the total volume v of the cylinder. On the other hand, when the total volume is greatly reduced by lowering the piston as shown in Fig. 3-6b, the part of the total volume v occupied by the molecules themselves $v_{molecules}$ can become even greater than the remaining part of the total volume that is free space v_{free} and available for molecular motion.

Free volume

Now functionally the volume v in the ideal-gas law, $Pv = RT$, is the volume v_{free} in which the molecules are effectively free to move about. For an ideal gas, where the molecules themselves occupy *no* volume ($v_{molecules} = 0$), v_{free} is equal to the total volume v of the container; but when the molecules do occupy an

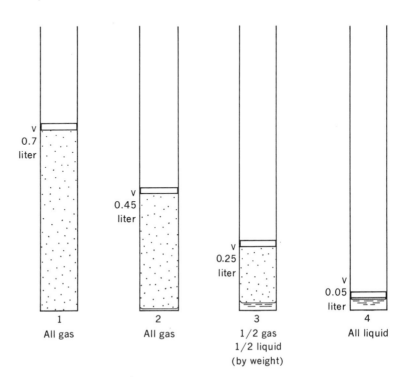

Fig. 3-5   Compression of 1 mole of CO_2 at 0°C.

Fig. 3-6   Compression of CO$_2$ at 40°C.

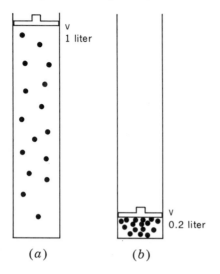

(a)   (b)

appreciable part of the total volume, v_{free} should be set equal to the difference between v and $v_{molecules}$:

$$v_{free} = v - v_{molecules} \qquad (3\text{-}3)$$

Actually, the volume $v_{molecules}$ occupied by the molecules themselves is very difficult to define and even harder to determine by direct measurement, since the "surface" of a molecule is actually a zone of varying density of electric charge. When two molecules collide, the phenomenon does not resemble the collision of two billiard balls with hard surfaces and relatively unchangeable volumes but is more like the collision of two spherical sponges which exert an appreciable force of repulsion on each other only when the "surface" of each sponge has been considerably indented. Van der Waals therefore proposed that the molecular volume, effectively causing deviations from the ideal-gas law, be designated by b and that v in the ideal-gas law be changed to $v - b$, to improve the agreement between the theoretical and experimentally determined values. Values of b are experimentally determined by measuring the deviation from the ideal-gas law at high pressures and relatively high temperatures, where the effect due to the molecular volume predominates.

Kinetic pressure

In order to take into account the effect of intermolecular attraction, van der Waals also suggested that a cor-

rection be applied to the pressure P. In the ideal-gas law the quantity P is the ideal *kinetic* pressure exerted by the molecules; it is the value of the pressure which multiplied by the value of the volume in which the molecules move gives the temperature term for the gas RT. As explained above, if there is a force of attraction between the molecules, part of the ideal kinetic pressure is "used up" in overcoming the intermolecular attraction, and the part remaining (the actual pressure observed on a manometer) will be correspondingly less than the ideal kinetic pressure:

$$P_{ob} = P_k - P_{\text{against intermolecular attraction}} \qquad (3\text{-}4)$$

Thus, the true kinetic pressure is equal to the sum of the observed pressure and the pressure used up against the attraction of the molecules

$$P_k = P_{obs} + P_{\text{against intermolecular attraction}} \qquad (3\text{-}5)$$

Van der Waals reasoned that the part of the pressure used up against intermolecular attraction should decrease as the volume increases and suggested the expression

$$P_{\text{against intermolecular attraction}} = \frac{a}{v^2} \qquad (3\text{-}6)$$

where a is a constant characteristic of each nonideal gas. Thus, the effective kinetic pressure is given by

$$P_k = P_{obs} + \frac{a}{v^2} \qquad (3\text{-}7)$$

Values of a are calculated by observing the deviations from the ideal-gas law at relatively low temperatures, where the effect of intermolecular attraction predominates.

Van der Waals equation for n moles

Van der Waals concluded that in any gas law the product RT thus should be equated to the product of the *kinetic* pressure and the actual *free* volume in which the molecules can move. Therefore, he proposed that the ideal-gas law should be modified for nonideal gases in the following way:

$$\textit{Ideal gas:} \quad \underset{\substack{\text{kinetic pressure}}}{\overset{\substack{\text{observed} \\ \text{pressure} \\ \downarrow}}{P}} \times \underset{\substack{\text{free volume}}}{\overset{\substack{\text{observed} \\ \text{volume} \\ \downarrow}}{v}} = RT \qquad (3\text{-}8)$$

Real gas:
$$\underbrace{P + \frac{a}{v^2}}_{\substack{\text{kinetic} \\ \text{pressure for} \\ \text{real gas}}} \times \underbrace{v - b}_{\substack{\text{free volume} \\ \text{for real gas}}} = RT \quad (3\text{-}9)$$

observed pressure ↓   observed volume ↓

As written above [Eq. (3-9)], the van der Waals equation gives the relation between P, v, and T when the system consists of 1 mole of gas. If the system contains n moles, the term n must be introduced at every point where the increase in the number of moles affects any of the terms in the equation. Such a change gives

$$\left(P + \frac{an^2}{V^2}\right)(V - nb) = nRT \quad (3\text{-}10)$$

Van der Waals constants

These van der Waals constants a and b are essentially *empirical* constants; values for them are chosen which result in the best agreement between the points calculated from the van der Waals equation and the points experimentally observed. Values for a and b are shown for a number of gases in Table 3-1. As one might expect, the larger molecules have the larger values of b. Of course, these larger molecules also have larger surfaces, and therefore it is reasonable that, other factors being the same, there will be larger intermolecular attraction, and consequently a will have larger values also. If a molecule, instead of being roughly spherical like CH_4, has an elongated or sausage shape like CO_2, there also will be a larger surface and a will be larger. Finally, if there are chemical forces of interaction like hydrogen bonding, the value of a will be larger; this is shown by comparing the value for CO_2 with the value for H_2O.

Example 3-1

For 1 mole of CO_2 gas confined in a volume of 0.500 liter at 50°C, use van der Waals equation to calculate the pressure.

Solve van der Waals equation (3-9) for the pressure and insert the appropriate values

$$P = \frac{RT}{v - b} - \frac{a}{v^2}$$

$$= \frac{0.0821 \text{ liter atm deg}^{-1} \times 323°}{0.500 - 0.043 \text{ liter}}$$

$$- \frac{3.60 \text{ atm liter}^{-2}}{(0.500 \text{ liter})^2}$$

$$= 58.0 - 14.4 \text{ atm} = 43.6 \text{ atm} \qquad Ans.$$

Note: In Eq. (3-9) for 1 mole of gas, the dimension of mole cancels out.

Example 3-2

At what temperature will 10.00 moles of H_2O in a volume of 20.00 liters exert a pressure of 50.0 atm according to van der Waals equation? At what temperature will the same amount of an ideal gas exert the same pressure in the same volume?

$$T = \frac{\left(P + \frac{n^2a}{v^2}\right)(v - nb)}{nR}$$

$$= \frac{\left[50.0 \text{ atm} + \frac{(10.00 \text{ moles})^2(5.46 \text{ atm mole}^{-1} \text{ liter}^{-2})}{(20.0 \text{ liters})^2}\right]}{10.00 \text{ moles} \times 0.0821 \text{ liter atm deg}^{-1} \text{ mole}^{-1}}$$
$$\times (20.0 \text{ liters} - 10.00 \text{ moles} \times 0.0305 \text{ liter mole}^{-1})$$

$$= \frac{51.4 \text{ atm} \times 19.70 \text{ liters}}{0.821 \text{ liter atm deg}^{-1}} = 1233°\text{K} \qquad Ans.$$

For the ideal gas:

$$T = \frac{Pv}{nR}$$

$$= \frac{50.0 \text{ atm} \times 20.0 \text{ liters}}{10.0 \text{ mole} \times 0.0821 \text{ liter atm deg}^{-1} \text{ mole}^{-1}}$$

$$= 1218°\text{K} \qquad Ans.$$

TABLE 3-1   Van der Waals constants

Gas	a, liters² atm mole⁻²	b, liters mole⁻¹
He	0.0341	0.0237
H_2	0.244	0.0266
N_2	1.39	0.0391
CO	1.49	0.0399
Ar	1.35	0.0330
O_2	1.36	0.0318
CH_4	2.25	0.0343
CO_2	3.60	0.0427
NH_3	4.17	0.0371
$n\text{-}C_5H_{12}$	19.01	0.1460
CH_3OH	9.52	0.0670
CCl_4	20.4	0.1383
C_6H_6	18.0	0.1154
H_2O	5.46	0.0305

3-3   VAN DER WAALS ISOTHERMS

In Fig. 3-7 the dotted line is the graph of the theoretical isotherm for CO_2 at 50°C calculated from the van der Waals equation using the values of a and b given in Table 3-1.   The solid line in this figure is the isotherm at 50°C plotted from experimental measurements. The dashed line is the theoretical isotherm at 50°C calculated from the ideal-gas law.   It is clear that at relatively high pressures the van der Waals equation agrees with the experimentally observed values far better than the ideal-gas law.

In Fig. 3-8 the same comparison of isotherms is made at 0°C.   At this temperature the experimental isotherm exhibits discontinuities in slope at either end of the horizontal portion corresponding to the zone in which liquefaction takes place.   The theoretical isotherm calculated from the van der Waals equation

shows a reversal of sign of the slope in this same region; this is due to the algebraic nature of the equation.

Van der Waals cubic equation

Although it may not be apparent at first glance, the van der Waals equation is the equivalent of a cubic algebraic equation in terms of v.   This may be shown in the following way.   First, multiply the two terms on the left of the equation together, giving

$$P v + \frac{a}{v} - Pb - \frac{ab}{v^2} - RT = 0 \qquad (3\text{-}11)$$

This expression is then multiplied by v^2 and divided by P, giving

$$v^3 - \left(b + \frac{RT}{P}\right)v^2 + \frac{av}{P} - \frac{ab}{P} = 0 \qquad (3\text{-}12)$$

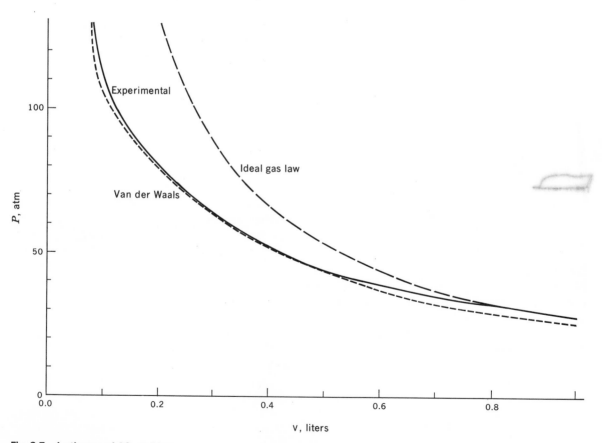

Fig. 3-7   Isotherms of CO_2 at 50°C.

Fig. 3-8   Isotherms of CO₂ at 0°C.

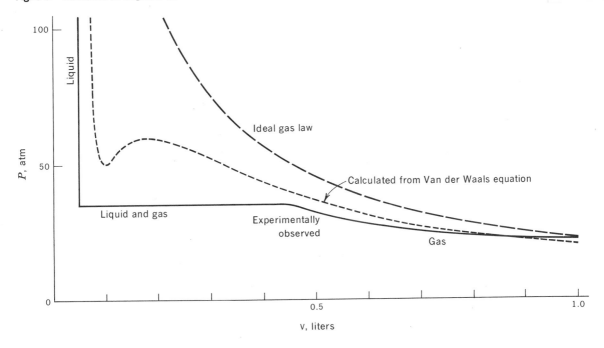

The S shape is typical of the graph of a cubic equation in the range where the solution has three real roots. The experimental isotherm instead of showing this S shape displays the horizontal portion where pressure stays constant during liquefaction.

It is interesting to note that with increasing temperature there is a decrease in the length of the horizontal portion of the isotherm that represents the volume zone in which liquefaction takes place as the gas is compressed at constant temperature.

Figure 3-9a shows the experimental isotherms for CO_2, with a dashed line drawn to bring out this point more clearly. Thus the right ends of the horizontal portions of the isotherms have been joined by the dashed line marked v_g^*. On each isotherm this line passes through the point representing the volume at which gas first begins to turn partially to liquid as volume is decreased at constant temperature. The left ends of the horizontal portions of the isotherms have been joined by a dashed line marked v_l^*. These points represent the volume at which the system becomes completely liquid.

3-4   THE CRITICAL POINT

In Fig. 3-10 there is shown the graph of the length of the volume range $v_g^* - v_l^*$ in which liquefaction takes place plotted against temperature, using the data from Fig. 3-9. This graph is a straight line which intersects the temperature axis at 31°C. It is at this temperature that the isotherm has zero slope only at a single point; above this temperature the isotherm never becomes horizontal. This temperature is called the *critical temperature* T_c; the value of the volume at the point where the slope is zero is called the *critical volume* v_c; and the corresponding value of the pressure is called the *critical pressure* P_c. The point itself is called the *critical point*.

The physical meaning of the critical point may be seen by considering the appearance of a gas confined in a capillary tube and maintained at a pressure and volume such that the state of the gas is represented by a point lying on the dashed line shown in Fig. 3-9a and marked M. Below the critical temperature, in a system represented by a point on this line both liquid and

Fig. 3-9 (*a*) Isotherms of CO_2 showing P_c, V_c, and t_c; (*b*) isotherms in the immediate neighborhood of the critical point.   True critical temperature T_c; temperature above which phase boundary disappears T_m; volume per molecule in saturated vapor v_g^*; volume per molecule in the condensed phase v_l^*.   (From J. E. Mayer and M. G. Mayer, "Statistical Mechanics," John Wiley & Sons, Inc., New York, 1940.)

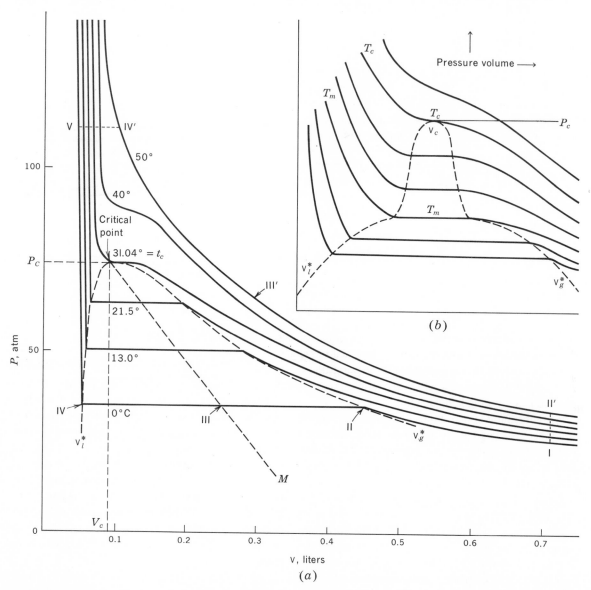

(*a*)

gas will always be present.   At the physical boundary between these two forms of the substance, called *phases*, there will be a *meniscus*, which is the visible surface of the liquid.   When the temperature of the system is increased to the critical temperature, the meniscus disappears, because the distinction between the liquid phase and the gaseous phase vanishes.   This is shown by graphs of the densities of the two phases as a function of the temperature as plotted in Fig. 3-11; at the critical temperature the density of the gas becomes exactly equal to the density of the liquid.   In other words, at this temperature the dispersing influence of the kinetic motion of the gas molecules becomes so great in contrast to the condensing influence of the intermolecular attraction that the molecules can no longer cling together to maintain the form of a liquid. Above this temperature one can no longer speak of a *liquid* phase as distinct from a *gas* phase; one can only speak of a *fluid* phase.

J. E. Mayer and M. G. Mayer (see References) have suggested that slightly below the critical temperature T_c there is a short range of temperature over which the distinction between the liquid state and the vapor state is still not sharp enough to make formation of a meniscus possible; the latter appears at $T_m < T_c$.   Experimental observations confirm this theory.   The theoretical shapes of the isotherms just below the critical point are shown in Fig. 3-9*b*.

Another informative aspect of the relationship between the liquid and the gaseous state appears when one considers what happens as a system is carried by a series of changes through a closed path like that shown in Fig. 3-9*a*.   Suppose that the system is initially in the completely gaseous state indicated by point I on the figure, corresponding to $v = 0.7$ liter, $t = 0°C$, and $P = 26$ atm.   Maintaining v constant, pressure and temperature are increased to bring the system to point II', where $t = 50°C$.   Then the system is changed by decreasing the volume at constant temperature, through point III' to point IV', where $P = 110$ atm. Then volume is still further reduced by lowering the temperature to 0°C while pressure is kept constant at 110 atm, a state corresponding to point V.   Finally, with temperature constant the volume is increased so that the pressure drops to 35 atm at 0°C and the system is now at point IV.

Recall that the system started as completely gaseous at point I.   During this whole series of changes, I → II' → IV' → V → IV, there has been no change of state, no appearance of a meniscus to indicate liquid formation.   From this, one might conclude that the system is still a *gas*.   But suppose that the change from I to IV were made along the different path consisting of steps I → II → III → IV.   At II the meniscus appears; at III the gas is half condensed into the liquid state; and at IV the system is completely liquid.   Yet the system

brought to point IV by the *upper* path, where it apparently stayed as a gas, must be in precisely the *same* state as the system brought to point IV by the *lower* path, where it changes from gas to liquid; i.e., the system must be a *liquid* at IV.   Thus it is clear that, by passing through the *fluid* state above the critical temperature, a gas can be changed into a liquid without passing through a two-phase condensation process. Therefore the distinction between a gas and a liquid is significant only at temperatures where the condensing influence of the intermolecular force can predominate over the dispersing influence of the thermal motion.   The fluid state above the critical temperature can be designated specifically as the *hyperfluid state*.

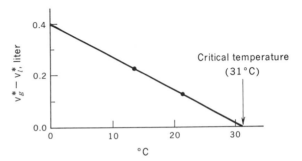

Fig. 3-10   The determination of the critical temperature of CO_2 by extrapolating $v_g^* - v_l^*$ to zero.

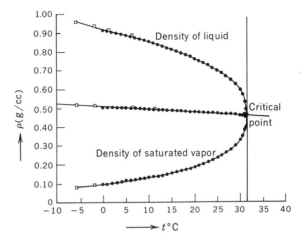

Fig. 3-11   Variation of density ρ with temperature in the vicinity of the critical point of CO_2.   [After Cailletet and Mathias, *Compt. rend.*, 102:1202 (1886).]

Fig. 3-12   Isothermals of isopentane.   (After S. Young, "Stoi-chiometry," p. 116, Longmans, Green & Co., Ltd., London, 1918.)

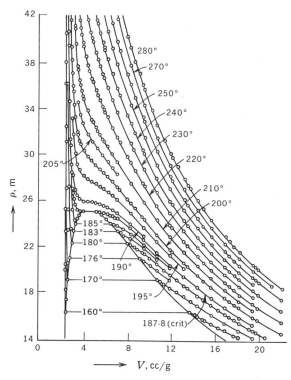

The isothermals for isopentane in Fig. 3-12 show the areas for the liquid, gas, and hyperfluid states.   Table 3-2 lists the values of the critical temperature, the critical pressure, and the critical volume for a number of gases which show a wide variation in the amount of attraction between the molecules.

Van der Waals roots

An interesting relation exists between these critical constants and the van der Waals constants a and b.   At the critical point the three roots v_1, v_2, v_3 of the cubic form of the van der Waals equation (3-12) are all equal to v_c.   Thus the root form of Eq. (3-12),

$$(v - v_1)(v - v_2)(v - v_3) = 0 \qquad (3\text{-}13)$$

can be written

$$(v - v_c)^3 = 0 \qquad (3\text{-}14)$$

The expansion of this equation is

$$v^3 - 3v_c v^2 - 3v_c^2 v - v_c^3 = 0 \qquad (3\text{-}15)$$

Comparing the coefficients in Eq. (3-15) with those in Eq. (3-12), it is clear that

$$3v_c = b + \frac{RT_c}{P_c} \qquad 3v_c^2 = \frac{a}{P_c} \qquad v_c^3 = \frac{ab}{P_c}$$

$$(3\text{-}16)$$

TABLE 3-2   Critical constants

Gas	T_c, °K	P_c, atm	v_c, cm^3 mole^{-1}
He	5.3	2.26	57.6
H$_2$	33.3	12.8	65.0
N$_2$	126.1	33.5	90.0
CO	134.0	35.0	90.0
Ar	150.7	48.0	77.1
O$_2$	153.4	49.7	74.4
CH$_4$	190.2	45.6	98.8
CO$_2$	304.3	73.0	95.7
H$_2$S	373.5	88.9	97.5
NH$_3$	405.6	111.5	72.4
n-C$_5$H$_{12}$	470.3	33.0	310.2
H$_2$Se	411.0	88.0	
CH$_3$OH	513.1	78.5	117.1
CCl$_4$	556.2	45.0	275.8
C$_6$H$_6$	561.6	47.9	256.4
H$_2$O	647.2	217.7	45.0

When these equations are solved for a and b, it is found that

$$a = 3P_c v_c^2 \qquad b = \frac{v_c}{3} \tag{3-17}$$

and

$$\frac{8P_c v_c}{3T_c} = R \tag{3-18}$$

These equations express the relationships between the van der Waals constants and the critical constants of a gas which obeys *precisely* the van der Waals equation having these particular constants. Since the van der Waals equation only approximates the actual behavior of the gas, the relation between the van der Waals constants and the *experimentally observed* critical constants is also only approximate. Nevertheless, the relation is of great interest in studying the nature of the factors which cause deviations from the ideal-gas law.

Example 3-3

Using the values of the critical constants from Table 3-2, calculate the value of the van der Waals constants a and b for CH_4 and compare with the values of a and b in Table 3-1.

$$b = \frac{v_c}{3} = \frac{0.0988 \text{ liter mole}^{-1}}{3}$$

$$= 0.0329 \text{ liter mole}^{-1}$$

Table 3-1: $\quad b = 0.0343 \text{ liter mole}^{-1}$

$$a = 3P_c v_c^2 = 3 \times 45.6 \text{ atm}$$
$$\times (0.0988 \text{ liter mole}^{-1})^2$$
$$= 1.34 \text{ liter}^2 \text{ atm mole}^{-2}$$

Table 3-1: $\quad a = 2.25 \text{ liter}^2 \text{ atm mole}^{-2}$

Water

There is a special significance in the abnormally high critical temperature for water, and a knowledge of its causes is of great importance in interpreting action in aqueous solution. Of all the molecules found in nature, water occupies a unique and significant place because it is the fluid of life. Water is not only the basic substance of the blood, it is the essential component of the cytoplasm, the fluid medium inside the biological cell in which the processes of cell metabolism and rep-

lication are carried out. Water vapor also plays an important role in cooling the body through the process of perspiration. It is therefore interesting to compare the values of P_c, T_c, and v_c for water with those for the other substances listed in Table 3-2.

If the normal rule were followed relating properties of molecules with similar structure so that all the critical constants increased with increasing molecular weight, in the series H_2O, H_2S, and H_2Se one would expect H_2Se to have the highest values. Instead, the value of the critical temperature for H_2S is 373.5°K and H_2Se is 411°K while H_2O has the much higher value of 647.2°K. The critical pressure for H_2O is also abnormally high. This anomaly is undoubtedly due to the great intermolecular attraction between water molecules that results from *hydrogen bonding*.

This type of intermolecular attraction is of great importance in many areas of biochemistry. In body fluids like blood, lymph, spinal fluid, and cytoplasm, large molecules like sugar, protein, RNA, and DNA are in solution. Atoms are added to these solute molecules or removed from them, and the large molecules themselves may leave the fluid and become part of the solid tissue with which the fluid is in contact, as in the process of growth. Intermolecular forces play a large part in determining the course of all these biochemical reactions. The van der Waals constants and the critical constants provide important evidence bearing on the nature of these intermolecular forces.

In a liquid like water, where the forces are acting both between the water molecules and between the water and the larger molecules in solution, a remarkable parallelism emerges between the kinetic behavior of the large molecules and the behavior of gaseous molecules. The water provides a medium for the large molecules in which they behave in many ways *as if* they were gas molecules moving in free space! For this reason the physicochemical behavior of nonideal gases can be a valuable model in interpreting the behavior in body fluids of many molecules of biochemical interest.

3-5   THE LAW OF CORRESPONDING STATES

An interesting modification of van der Waals equation can be obtained through the concept of *corresponding states*. When the van der Waals constants are replaced by their value in terms of the critical constants [Eq. (3-17)], the van der Waals equation

$$P = \frac{RT}{v - b} - \frac{a}{v^2} \tag{3-19}$$

becomes

$$P = \frac{8P_c v_c T}{3T_c(v - v_c/3)} - \frac{3P_c v_c^2}{v^2} \tag{3-20}$$

which can be rearranged to

$$\frac{P}{P_c} = \frac{8(T/T_c)}{3(v/v_c) - 1} - \frac{3}{(v/v_c)^2} \tag{3-21}$$

Note that in Eq. (3-21) each of the variables P, T, and v is divided by the corresponding critical constant, giving three dimensionless ratios, called the *reduced pressure* π, *reduced volume* φ, and *reduced temperature* τ and are defined as

$$\frac{P}{P_c} = \pi \qquad \frac{T}{T_c} = \tau \qquad \frac{v}{v_c} = \phi \tag{3-22}$$

When for two gases A and B, $\pi_A = \pi_B$, $\tau_A = \tau_B$, and $\phi_A = \phi_B$, the two gases are said to be in *corresponding states*.

Equation (3-21) can thus be written as

$$\pi = \frac{8\tau}{3\phi - 1} - \frac{3}{\phi^2} \tag{3-23}$$

This equation is called the *law of corresponding states* for the van der Waals equation. Within the accuracy of the van der Waals equation, this relationship holds for *all* gases. It shows that if the state variables are taken as P/P_c, T/T_c, and v/v_c, these *new state variables* are related in the same way for all gases in corresponding states. At constant volume ($\phi = $ const) there is a linear relationship between π and τ

$$\pi = \frac{8}{3\phi - 1}\tau - \frac{3}{\phi^2} \tag{3-24}$$

The use of these reduced properties is illustrated by the plot of Pv/RT against the reduced pressure π for various values of the reduced temperature τ shown in Fig. 3-13 for a number of gases. The values for widely different gases lie remarkably close to a single curve. The logarithm of Pv/RT plotted against the logarithm of the reduced pressure π is shown for various values of the reduced temperature τ in Fig. 3-14.

Example 3-4

Calculate the value of the pressure for 1 mole of CO_2 at 40°C when confined in a volume of 0.107 liter, using (*a*) the law of corresponding states and (*b*) the ideal-gas law.

(*a*) $\pi = \dfrac{8}{(3v/v_c) - 1}\dfrac{T}{T_c} - \dfrac{3}{(v/v_c)^2}$

$= \dfrac{8}{[3(0.107)/(0.0957 \text{ liter})] - 1}$

$\times \dfrac{313.2°}{304.2°} - \dfrac{3}{(0.107/0.0957 \text{ liter})^2}$

$= \dfrac{8}{3 \times 1.118 - 1} 1.03 - \dfrac{3}{1.1^2}$

$= \dfrac{8}{2.35} 1.03 - \dfrac{3}{1.21}$

$= 3.51 - 2.40 = 1.11$

$P = \pi P_c = 73.0 \times 1.11 = 81 \text{ atm} \quad Ans.$

(*b*) $P = \dfrac{RT}{v}$

$= \dfrac{0.0821 \text{ liter atm deg}^{-1} \text{mole}^{-1} \times 313.2}{0.107 \text{ liter mole}^{-1}}$

$= 240 \text{ atm}$

Note: The experimentally observed pressure is 88 atm.

3-6   OTHER EQUATIONS OF STATE

Another useful equation of state is the general *virial equation* obtained by putting the ratio Pv/RT equal to unity plus a series of correction terms:

$$\frac{Pv}{RT} = 1 + \frac{B}{v} + \frac{C}{v^2} + \frac{D}{v^3} + \cdots \tag{3-25}$$

These correction terms correspond reasonably well to the deviations from unity shown in graphs like those in Fig. 3-1 when the coefficients B, C, D, etc., are made appropriate functions of temperature.

J. A. Beattie and P. W. Bridgeman proposed the virial equation

$$pv = RT + \frac{\beta}{v} + \frac{\gamma}{v^2} + \frac{\delta}{v^3} \tag{3-26}$$

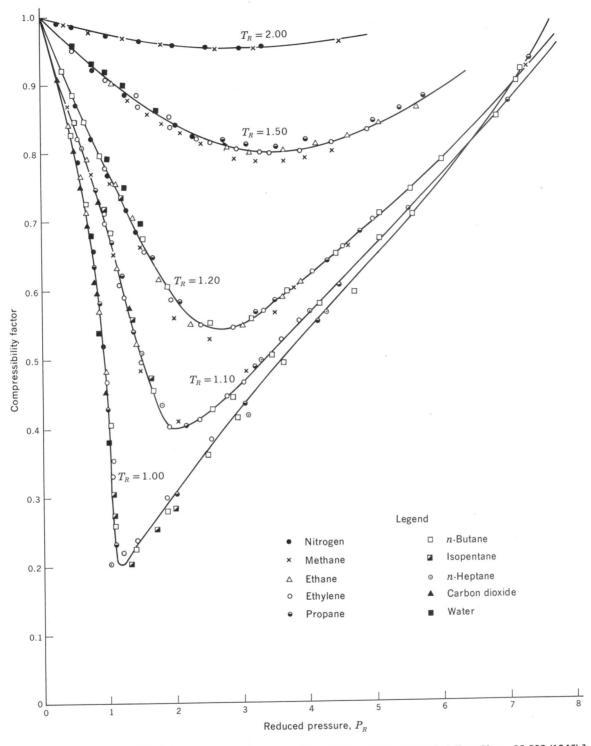

Fig. 3-13 Compressibility factor $P v/RT$ as a function of reduced variables. [After Jen Su, *Ind. Eng. Chem.*, **38**:803 (1946).]

Fig. 3-14 Compressibility factor as a function of reduced pressure π on a log-log plot. (After O. A. Hougen, K. M. Watson, and R. A. Ragatz, "Chemical Process Principles," pt. 2, John Wiley & Sons, Inc., New York, 1959.)

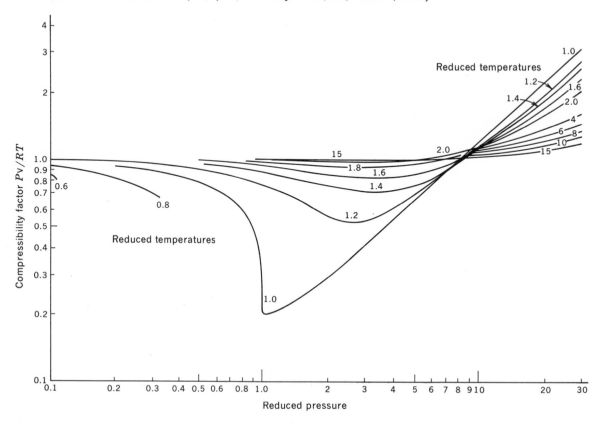

where

$$\beta = RTB_0 - A_0 - RcT^{-2} \tag{3-27a}$$

$$\gamma = -RTB_0 b + A_0 a - RB_0 cT^{-2} \tag{3-27b}$$

$$\delta = \frac{RB_0 bc}{T^2} \tag{3-27c}$$

Several other equations of state have been devised which are modifications of the van der Waals equation. P. A. D. Berthelot suggested the relation

$$P = \frac{RT}{v - b} - \frac{a}{Tv^2} \tag{3-28}$$

where b has the same significance as in the van der Waals equation and a here contains the dimension of temperature.

For this equation the law of corresponding states is

$$\pi = \frac{8\tau}{3\phi - 1} - 3\tau\phi^2 \tag{3-29}$$

Another modification of the van der Waals equation has been proposed by C. Dieterici:

$$P = RT \frac{\exp(-a/vRT)}{v - b} \tag{3-30}$$

where the correction for intermolecular attraction assumes an exponential form and the correction for molecular volume is the same as in the van der Waals equation.

For this equation the law of corresponding states is

$$\pi = \frac{\tau}{2\phi - 1} e^{2 - 2/\tau\phi} \tag{3-31}$$

The experimental values are compared with values calculated from the equations of van der Waals,

Berthelot, and Dieterici for nitrogen at 50°C in Fig. 3-15 and for ethylene at 40°C in Fig. 3-16.

Example 3-5

At 50°C, 1 mole of CO_2 exerts a pressure of 52.2 atm when confined in a volume of 0.400 liter. Using the van der Waals constant b for this gas in the Berthelot equation (3-28), calculate the value of a for this equation.

Solving Eq. (3-28) for a results in

$$a = Tv^2\left(\frac{RT}{v - b} - P\right)$$

Putting in numerical values, this becomes

$a = 323.2°(0.400 \text{ liter})^2$

$\quad \times \left(\dfrac{0.0821 \text{ liter atm deg}^{-1} \text{ mole}^{-1} \times 323.2°}{0.400 - 0.0427 \text{ liter}} - 52.2 \text{ atm}\right)$

$\quad = 51.7\left(\dfrac{26.5}{0.357} - 52.2\right) = 51.7(74.3 - 52.2)$

$\quad = 51.7 \times 22.1$

$\quad = 1.143 \times 10^3 \text{ liter}^2 \text{ atm deg mole}^{-2}$

Example 3-6

Calculate the value of a in the Dieterici equation from the experimental data given in Example 3-5 and the value of b in Table 3-1.

Solve the Dieterici equation for a

$$\frac{P(v - b)}{RT} = \exp^{(-a/vRT)}$$

$a = -vRT \times 2.303 \log \dfrac{P(v - b)}{RT}$

$a = -0.400 \text{ liter} \times 0.0821 \text{ liter atm deg}^{-1} \text{ mole}^{-1}$

$\times 323.2° \times 2.303 \log \dfrac{52.2 \text{ atm}(0.400 - 0.047 \text{ liter})}{0.0821 \text{ liter atm deg}^{-1} \text{ mole}^{-1} \times 323.2°}$

$\quad = -24.4 \log \dfrac{18.42}{26.5} = 24.4 \log 1.44$

$\qquad\qquad\qquad\qquad = 24.4 \times 0.1584$

$\quad = 3.86 \text{ liter}^2 \text{ atm mole}^{-1} \qquad Ans.$

Problems

1. Helium is used frequently to replace some of the nitrogen in the gas breathed by deep-sea divers. If the temperature is 10.0°C and the volume per mole is 2.00 liters for the helium in a caisson, calculate the pressure (a) if the ideal-gas law is obeyed and (b) if the van der Waals equation is obeyed; calculate the percentage deviation from the gas law for (b).

2. Oxygen is frequently stored in tanks under high pressure for use in enriching the air breathed by hospital patients. If the temperature is 30°C and the volume per mole is 0.300 liter, calculate the pressure of the oxygen from the van der Waals equation and from the ideal-gas law.

3. The volume of CO_2 at 21.5°C and at a pressure of 62 atm is 0.05 liter mole^{-1} in the liquid state. If 10 kg of CO_2 is stored in a fire extinguisher having a volume of 20 liters, what fraction of the CO_2 is in the liquid state?

4. Calculate from the van der Waals constants the

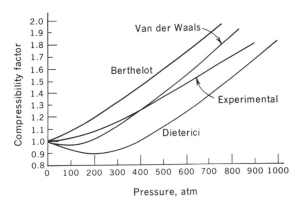

Fig. 3-15 A comparison of values of the compressibility factor Pv/RT of N_2 as a function of temperature calculated from various equations of state and experimentally observed. (From U.S. Natl. Bur. *Std. Circ.* 279, 1925.)

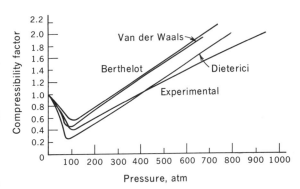

Fig. 3-16 Values of the compressibility factor Pv/RT of ethylene as a function of pressure, calculated from various equations of state and experimentally observed. (From U.S. Natl. Bur. *Std. Circ.* 279, 1925.)

critical volume and critical pressure for CO_2. Compare with the observed critical volume and temperature and calculate the percentage error in the calculated values.

5.   The critical temperature of a substance is related to the van der Waals constants by the equation $T_c = 8a/27Rb$, where R is the ideal-gas constant in appropriate units. Calculate the critical temperature for water and compare with the observed critical temperature.

6.   Using van der Waals equation, calculate the values of the pressure for 1 mole of water at 500°C when the molar volume is respectively 40, 50, and 60 cm³. Make a plot of v against P for the above data.

7.   Calculate the volume required to store 100 kg of water at a pressure of 200 atm if the vapor obeys the ideal-gas law. From the graph obtained in Prob. 6, estimate the value of the above volume at 500°C if the vapor obeys the van der Waals equation.

8.   A mixture of 2 g of hydrogen gas and 16 g of oxygen gas is pumped into a combustion bomb with a volume of 1 liter and brought to a temperature of 700°K. Combustion then takes place, effectively changing the mixture to water. Calculate the pressure before and after combustion, provided that temperature is maintained constant and the gases obey van der Waals equation.

9.   Calculate the pressure of 1 mole of hydrogen at the critical volume and critical temperature if the ideal-gas law is obeyed. Calculate the ratio of the ideal pressure to the observed critical pressure.

10.   Calculate the pressure of 1 mole of NH_3 at the critical volume and temperature if the ideal-gas law is obeyed. Calculate the ratio of this ideal pressure to the observed critical pressure.

11.   Using van der Waals equation, calculate the pressure of 1 mole of hydrogen gas at the critical temperature and volume and compare with the result obtained in Prob. 9 and with observed critical pressure.

12.   Using van der Waals equation, calculate the pressure of 1 mole of ammonia at the critical temperature and volume and compare with the result obtained in Prob. 10.

13.   Assume that at 0°K, 1 mole of molecules occupies a volume equal to the van der Waals constant b; assume also that each molecule accounts for a cubical element of this volume, where the edge of the cube equals the molecular diameter; on the basis of these approximate assumptions, calculate the molecular diameters of He, Ar, CH_4, and CCl_4 from the data in Table 3-1.

14.   From the graph in Fig. 3-1, estimate the value of Pv/RT at −70°C and 100 atm and calculate the value of v at this point; using this value and the value of b from Table 3-1, calculate the value of a in the Berthelot equation for methane.

15.   Using the value of a obtained in the above problem and the value of b from Table 3-1, calculate the value of Pv/RT at −70°C when v = 0.200 liter mole⁻¹.

16.   From the graph in Fig. 3-3, estimate the value of Pv/RT for O_2 at 1 atm. Using these data and the value of b from Table 3-1, calculate the value of a in the Dieterici equation.

17.   Using the values of a and b from Prob. 16, calculate the value of P for O_2 at 0°C when v = 0.0100 liter, using the Dieterici equation.

18.   Using Eq. (3-23), calculate the pressure exerted by 1 mole of benzene vapor when confined in a volume of 0.500 liter at a temperature of 300°C.

19.   Using Eq. (3-23), calculate the temperature at which 1 mole of carbon tetrachloride vapor will exert a pressure of 100 atm when confined in a volume of 0.300 liter.

20.   The values of the critical constants of acetic acid are $P_c = 57.11$ atm, $v_c = 0.1712$ liter, and $T_c = 594.7$°K. Calculate the values of the van der Waals constants a and b for acetic acid.

21.   Using the van der Waals equation, make a plot of the isotherm for acetic acid which passes through the critical temperature.

REFERENCES

Mayer, J. E., and M. G. Mayer: "Statistical Mechanics," chap. 12, John Wiley & Sons, Inc., New York, 1940.

Dodge, B. F.: "Chemical Engineering Thermodynamics," pp. 150–181, McGraw-Hill Book Company, New York, 1944.

Su, Jen: Corresponding States, *Ind. Eng. Chem.,* **38:** 803 (1946).

Taylor, H. S., and S. Glasstone: "A Treatise on Physical Chemistry," pp. 187–209, D. Van Nostrand Company, Inc., New York, 1951.

Hirschfelder, J. O., C. F. Curtiss, and R. B. Bird: "Molecular Theory of Gases and Liquids," pp. 1–8, John Wiley & Sons, Inc., New York, 1954.

Swinbourne, E. S.: The van der Waals Gas Equation, *J. Chem. Educ.,* **32:**366 (1955).

Pings, C. J., and B. H. Sage: Equations of State, *Ind. Eng. Chem.,* **49:** 1315 (1957).

Winter, S. S.: A Simple Model for van der Waals, *J. Chem. Educ.,* **33:** 459 (1959).

Hougen, O. A., K. M. Watson, and R. A. Ragatz: "Chemical Process Principles," 2d ed., chap. 14, John Wiley & Sons, Inc., New York, 1959.

Eyring, H., D. Henderson, B. J. Stover, and E. M. Eyring: "Statistical Mechanics and Dynamics," chap. 11, John Wiley & Sons, Inc., New York, 1964.

Rice, O. K.: "Statistical Mechanics, Thermodynamics and Kinetics," chap. 8, W. H. Freeman and Company, San Francisco, 1967.

If all molecules were like those postulated in the theory of ideal gases, the world around us would be far different from the one we actually know. For if there were no forces of attraction between molecules, many familiar solids and liquids could not exist. Water would not be found in the crystalline or liquid state, no matter how low the temperature; it would always be a gas. We ourselves could not exist in our present bodily form because our blood would be gaseous instead of liquid. Thousands of different solid and liquid components that make up the living body owe their unique properties to the interaction of molecules of which they are com-

Chapter Four

crystals and liquids

posed. It is in the energy changes resulting from this interaction that we find the key for understanding the basic differences between solids, liquids, and gases.

4-1 THE ROLE OF ENERGY IN CONDENSATION

As contrasted with gases, solids and liquids represent *condensed* states. In 1 cm³ of a liquid or a crystal at 1-atm pressure there are many times more molecules than in the same volume of a gas at the same pressure. To illustrate this, let us suppose that a glass tube (Fig. 4-1) is filled with nitrogen gas. The gas is cooled, and as the volume contracts, more gas is fed into the tube. At $-196°C$ part of the gas starts to liquefy, and this process is continued until the tube is about half filled with liquid. The tube is then cooled to $-209.88°C$, and the liquid begins to freeze. The freezing continues until the bottom quarter of the tube is filled with crystals of nitrogen. Now the pressure of the gas above the surface of the liquid is measured as the temperature is held constant at $-209.88°C$ and is found to be almost exactly 100 mm Hg. If measurements of density are also performed on these three states of nitrogen, the density of the gas is found to be 7.12×10^{-4} g cm⁻³; that of the liquid is 0.81 g cm⁻³; and that of the

Fig. 4-1   Gas, liquid, and crystal phases in equilibrium at the triple point of nitrogen.   (From D. H. Andrews and R. J. Kokes, "University Chemistry," vol. 1, John Wiley & Sons, Inc., New York, 1959.)

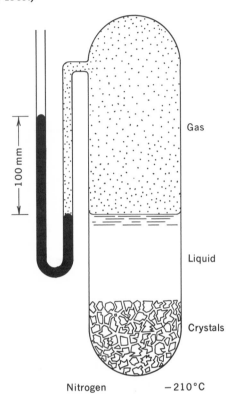

100 mm

Gas

Liquid

Crystals

Nitrogen   −210°C

solid is 1.03 g cm⁻³.   The liquid and solid states are roughly 1,000 times denser than the gaseous state because molecules in liquids and solids strongly attract each other and as a result are much more tightly packed.

At the temperature of −209.88°C, called the *triple-point temperature,* these three states can exist indefinitely in contact with each other.   The states are in *equilibrium* with each other.   To the naked eye, the system appears to be static; there is no observable net change in the fraction of the system in each state; no motion is visible.   But if we were able to watch individual molecules, the dynamism of this equilibrium would be evident.   We would see the molecules of the gas moving rapidly in space and striking the surface of the liquid, and we would observe that a considerable number of them penetrate the liquid and remain there. A certain number of molecules shoot off the surface of

the liquid and wander away into space as gaseous molecules.   If we could count the number of molecules from the gas hitting the surface of the liquid at each second and remaining as part of the liquid and the number of molecules shooting upward from the liquid surface and going out into space as gaseous molecules, we would find that the number leaving the gas and entering the liquid per second is closely the same as the number leaving the liquid and entering the gas per second. This is the essence of the equilibrium between two physical states.   In the dynamic exchange of molecules which takes place constantly between any two states in contact, as shown in Fig. 4-2, there must always be the same number going each way if equilibrium is to be maintained.   If more molecules leave the liquid per second than enter it, then ultimately the liquid completely evaporates and the whole system becomes gas. By the same token, if more molecules leave the gaseous state per second than enter it, ultimately the gas condenses completely and the whole system becomes liquid.   The same considerations apply to the equilibrium between the liquid state and the solid state.   A system like this at its *triple point* is an example of dynamic *physical equilibrium* between three different states of a substance.   Each state is called a *phase* of the system; at the top is the gaseous phase, at the center, the liquid phase, and at the bottom the crystalline, or solid, phase.

If the pressure on the gas is increased to 1 atm and the temperature is raised from −209.88 to −209.86°C, all the gaseous nitrogen condenses, and the system consists only of liquid and crystalline nitrogen in equilibrium.   This temperature is the *melting point,* or, alternatively, the *freezing point.*   When crystalline nitrogen is warmed slowly under a pressure of 1 atm, melting begins at this temperature.   When liquid nitrogen is cooled slowly under a pressure of 1 atm, conversion to the crystalline state, or freezing, begins at this temperature.

In a later discussion of the kinetics of changes between the liquid and crystalline states in Chap. 22, the necessity for *seeding* in order to induce crystallization will be explained, together with the factors which make the melting point slightly higher than the freezing point when the change of state takes place at a finite rate. In this chapter, the melting-point temperature and the freezing-point temperature may be taken as effectively the same. It is the temperature at which the liquid state and the crystalline state are in equilibrium with each other when under a pressure of 1 atm.

Heats of vaporization and fusion

As shown in Chap. 2, the number of molecules of gas colliding with the surface is proportional to the density of the gas and the average velocity of the gas molecules. For nitrogen gas at a pressure of 100 mm and at the triple-point temperature there are approximately 10^{23} molecules sec^{-1} colliding with each square centimeter of surface. A comparison shows that the liquid is about 1,000 times denser than the gas. Because of the greater density of molecules in the liquid, one might expect about 100 times more molecules to leave the liquid and go into the gas than come from the gas and enter the liquid. But since the number of molecules moving out of the liquid into the gas *equals* the number moving from gas to liquid, it is obvious that only *one out of many* molecules at the surface of the liquid will actually leave the liquid and go off into the gaseous state, although on the average, molecules in the gas moving toward the liquid and within a distance less than the mean free path of motion in the gas will strike the surface of the liquid.

The reason such a small percentage of the molecules at the surface of the liquid can move into the gas is that evaporation requires energy much greater than the average kinetic energy of the molecules in the liquid. Any molecule escaping from the surface of the liquid must overcome the force of attraction pulling it toward its neighbors. An energy of 1.11×10^{-21} cal is needed to lift one molecule of nitrogen from the surface of the liquid and to send it out into the space as a gas molecule. This is analogous to the energy required to lift a rocket off the face of the earth against the force of gravitational attraction between the rocket and the atoms that make up the substance of our planet and to send the rocket shooting off into space. When 1.11×10^{-21} cal molecule^{-1} is multiplied by Avogadro's number, 6.02×10^{23} molecules mole^{-1}, it gives 666 cal mole^{-1}, the amount of energy necessary to convert 1 mole of nitrogen from the liquid to the gaseous state. This quantity is called the *molal heat of vaporization* H_v.

Energy is also required to melt the nitrogen, to change the molecules of nitrogen from their compact solid state to the somewhat less dense liquid state; 85.3 cal is needed to change 1 mole of liquid nitrogen from the solid state to the liquid state. This quantity is called the *molal heat of fusion* H_f.

Table 4-1 lists values of molal heats of fusion and heats of vaporization for a number of compounds. In order to understand the nature of this change from solid to liquid, it is necessary first to understand the

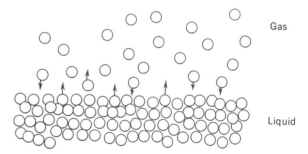

Fig. 4-2   Gas-liquid equilibrium.

TABLE 4-1	Heats of fusion H_f and vaporization H_v				
Compound	Formula	t_m, °C	H_f, cal mole^{-1}	t_b, °C	H_v, cal mole^{-1}
Ammonia	NH_3	−75	1840	−33.4	5607
Carbon dioxide	CO_2	−56.2	1993	−60	3837
Carbon monoxide	CO	−206	224	−192	1411
Oxygen	O_2	−219	106	−182.9	1629
Hydrogen	H_2	−259.1	28	−252.8	218
Water	H_2O	0	1435	100	9721

nature of the *arrangement* of the atoms in these two states.

4-2   MOLECULAR ORDER IN CONDENSED PHASES

During the last 50 years great strides have been made in understanding the details of the orderly way in which closely packed atoms are arranged in the liquid and solid states. The simplest way to study this order would be to use a microscope so powerful that one could actually see the atoms lined up side by side in solids and liquids, but because the wavelength of visible light is more than a 1,000 times greater than the diameter of atoms and molecules, it is impossible to get the kind of resolution needed to see or photograph atoms and molecules directly. However, indirect ways have been found to observe atoms and molecules with the help of x-rays, where the wavelength is about the same as atomic and molecular diameters. The details of these methods will be discussed in Chap. 22. At this point, only the basic features of liquid and crystal order are presented.

If one could put a grain of common table salt, sodium chloride, NaCl, under an imaginary microscope permitting the direct observation of atoms, the image would be something like that shown in Fig. 4-3. Note that the atoms are arranged in a regular order. Along any one horizontal line there are found, alternately, sodium ion, chloride ion, sodium ion, chloride ion, sodium ion, this pattern being repeated over and over again until the surface of the crystal is reached. Along this same line positive and negative charges alternate, positive on the sodium ion and negative on the chloride ion. The same alternating order is found along any vertical line. Moving a centimeter along such a path one would encounter millions of ions and would find always this same regularity of order persisting, except at flaw lines.

A solid with this type of regularity is a *crystal*. Atoms arranged in such a regular order are said to be in the form of a *crystal lattice*. The lattice found in a crystal of sodium chloride is one of the simpler types. This simplicity results from the spherical shape of each ion. The rigidity of the lattice is produced by the force of attraction each positive ion exerts on all the negative

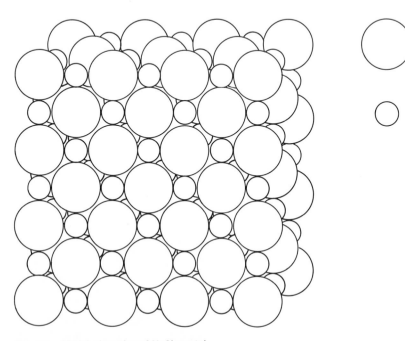

Chloride ion

Sodium ion

Fig. 4-3   Model of portion of NaCl crystal.

ions surrounding it and by the force of attraction each negative ion also exerts on all the positive ions surrounding it. These attractive forces act approximately like spiral springs, as shown in Fig. 4-4. A lattice made up not of spherical ions but of molecules with more complicated shapes is somewhat more complex; e.g., Fig. 4-5 is a diagram of a lattice made up of molecules of the organic compound strychnine sulfate pentahydrate. In biological structures we find many solids that are still more complex in character; e.g., keratin, a basic part of hair and toenails, may have a structure as complex as that shown in Fig. 4-6.

The rigidity of crystals as contrasted with the fluidity of liquids is due to the regular order which permits closer packing in crystals than in liquids. Because of the forces of attraction between the molecules the whole crystal structure resists any tendency to deform it, since such deformation would stretch these forces and require considerable energy. By contrast, the molecules in a liquid are much less ordered with respect to each other and in general are more loosely packed. The displacement of a molecule or an ion with respect to another requires far less energy. Consequently, a liquid can flow readily while a crystal cannot.

The degree of order in a liquid depends to a considerable extent on the shape of the molecule. If a liquid is composed primarily of spherical molecules, such as one finds in substances like helium, neon, and argon, there is only a moderate tendency for these molecules to maintain any orderly arrangement. On the other

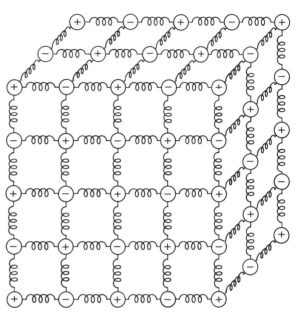

Fig. 4-4 Positive and negative ions in crystal lattice. The force of attraction between positive and negative ions balanced against the repulsion between electron clouds produces a net force resembling a spiral spring.

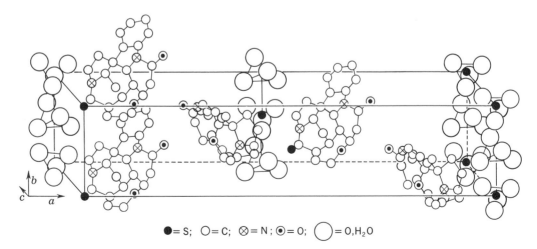

● = S; ○ = C; ⊗ = N; ⊙ = O; ◯ = O,H₂O

Fig. 4-5 Molecules of strychnine sulfate pentahydrate in a lattice, with a few atoms omitted to clarify drawing. [After C. Bokhaven, J. C. Schoone, and J. H. Bijvoet, *Proc. Koninkl. Ned. Akad. Wetenschap.,* **50:825 (1947); 51:990 (1948); 52:120 (1949).]**

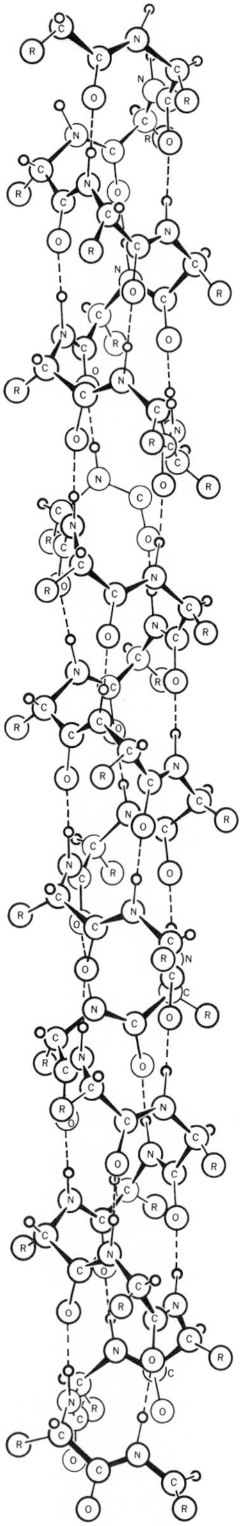

Fig. 4-6   Alpha-helix structure in keratin.   [After L. Pauling and R. Corey, *Proc. Natl. Acad. Sci. U.S.,* 37:729 (1951).]

hand, long-chain organic compounds like stearic acid or dodecane tend to line up lengthwise when packed together in either solid or liquid form (Fig. 4-7). Thus, one can find a series of substances forming an almost continuous transitional pattern from rigid crystals, like salt, through jellylike viscous matter, like the cytoplasm in biological cells, to simple liquids, like water and alcohol.

4-3   DYNAMIC EQUILIBRIUM BETWEEN PHASES

In thinking of these differences in molecular order in crystals, liquids, and gases, it is important to keep the dynamic aspect of the picture in mind. At temperatures above $0°K$ all substances have thermal energy. Particularly in the ranges of temperature of special interest in the applications of physical chemistry to biology and medicine, all the molecules have considerable kinetic energy of motion. A vivid portrayal can be found through statistical calculations of the paths along which molecules move in these three states of matter. The computed tracings of such paths are shown in Fig. 4-8*a*, *b*, and *c*, the computation being based on a system of 32 particles in a box making 3,000 collisions with each other. In Fig. 4-8*a* the trajectories are shown of particles bouncing against each other in the gaseous state. The paths are very irregular, and the particles wander almost at random throughout the space in which the gas is confined. In this calculation the gas is assumed to be in a relatively dense form. Figure 4-8*b* shows the motion of molecules in the liquid. Molecules wander every now and then from one place to another in the liquid, swapping places, and eventually moving quite a distance from their original location; but they spend *most* of the time vibrating back and forth around a single location. Figure 4-8*c* shows the paths of the molecules when the forces acting between them are those normally found in the crystalline state. Here each molecule spends practically all its time vibrating around a single point in the crystal lattice. An exchange of place with another molecule is far less frequent than in the liquid.

Potential-energy difference between phases

In this perspective turn back and consider again the nature of the interaction between the phases of nitrogen shown in Fig. 4-1. At the interface between the liquid and the gaseous states a constant interchange of molecules between the two phases takes place. Molecules of the gas are constantly hitting the surface of the liquid. Some rebound back into the gas, others penetrate into the liquid and remain there, giving up some of their kinetic energy. In the gas the molecules move with translational energy $\frac{1}{2}mu^2$, where m is the mass of the molecule and u is its velocity. They also have potential energy with respect to the interior of the liquid because of the forces of attraction exerted when the molecule gets near the liquid. This potential energy is similar to the potential energy of any object in space above the surface of the earth. If a cannonball were dropped from an airplane flying 1,000 ft over the ocean, the potential energy the cannonball has because of its altitude is gradually converted into kinetic energy as the ball falls swiftly toward the surface of the water. When the ball penetrates the water and its rapid motion is slowed down through the forces of friction, the potential energy and the kinetic energy it had when falling freely in space are largely transformed into thermal energy, the energy of vibration of the atoms of the cannonball itself and of the atoms of the water, which the ball has warmed by its impact.

At the same time, at the interface between the phases of nitrogen the opposite process is also occurring. From time to time, a few of the molecules of liquid at the surface acquire more energy of vibration as energy is transferred back and forth among molecules in the liquid in the complex series of collisions the molecules are constantly making. When a molecule at the surface happens to get enough kinetic energy in the direction away from the surface, it can rise into the gas; and if the energy is great enough for it to get so far out into the gas that the force of attraction by the liquid surface is not sufficient to make it return, it remains a gaseous molecule. When the rate of evaporation equals the rate of condensation, the two phases are in equilibrium.

Exactly the same considerations apply to the equilibrium between the liquid and the crystal, the principal difference here being that the molecules striking the

Fig. 4-7 Liquid structure with long-chain molecules.

crystal surface cannot penetrate inside the crystal as easily as molecules striking the liquid because of the greater rigidity of the crystal. Because the molecules in the crystal are far more ordered than those of a liquid, a molecule approaching the surface of the crystal must also attain the right *orientation* with respect to the crystal if it is to stay there as a part of the ordered lattice.

The question of molecular orientation is especially important in biological phase changes. Many aspects of growth in the life process consist of the transfer of molecules from the liquid phase to the solid phase. The more complex the shape of the molecule, the more important it is for a molecule to be oriented in exactly the right position if it is to remain as a permanent part of the solid surface. Thus, the motion of molecules in the liquid state, the extent to which they are constantly changing their orientation, and the extent to which the liquid itself may be ordered—all these factors are important in many life processes.

4-4 THE NATURE OF PHASE CHANGES

In the preceding sections, factors were examined that distinguish the states of matter from one another. The gaseous state differs from the liquid and solid states primarily by having a far lower density; its molecules also have a relatively high potential energy acquired in escaping from solid or liquid against the forces of attraction. The liquid state differs from the crystalline state by having its molecules arranged in a far less orderly manner and generally in a more loosely packed array.

In the temperature range below 63°K there is an imbalance of rates of exchange between the crystalline

Fig. 4-8 Trajectories of molecules produced by a computer from statistical relations. [T. W. Wainwright and B. J. Alder, *Nuovo Cimento*, **9 (Suppl. 1):116 (1958).** For a description of the computation, see also *References*: D. Dreisbach, p. 69 and J. A. Barker, pp. 20–22. Photographs courtesy of B. J. Alder.]

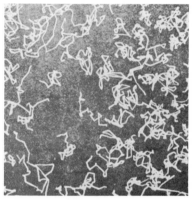

(*a*) Gaseous state

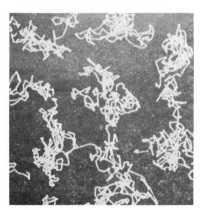

(*b*) Liquid state

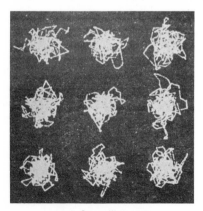

(*c*) Crystalline state

state on the one hand and the liquid and gaseous states on the other, an imbalance which favors the crystalline state. Suppose, for example, that at 50°K a few molecules on the surface of a crystal of nitrogen suddenly shake themselves loose and become disordered so that they constitute a minute portion of liquid. At this temperature, far more molecules pass from the liquid to the crystal state per second than pass from the crystal to the liquid. Consequently, in 1 msec or less the liquid disappears. As temperature rises, the rate at which molecules pass from crystal to liquid increases more rapidly with temperature than the rate at which molecules pass from liquid to crystal, until at 63°K the two rates are equal and the two states are in equilibrium. As heat continues to flow into the system, the material in the crystalline state melts, until finally the whole substance has liquefied.

Looking now at the difference between rate of change from liquid to vapor and vapor to liquid at a temperature above the melting point but below the boiling point, we find that the rate of change from liquid to gas is less than the rate of change from gas to liquid if the gas is at a pressure of 1 atm. For the crystal and the liquid the pressure makes relatively little difference in the rates at which the two phases interchange. For the gas the rate of change is almost directly proportional to the pressure. The pressure determines the density of the gas, the number of molecules per cubic centimeter. The number of molecules per cubic centimeter, in turn, determines the number of molecules striking each unit area of the liquid per second. If the liquid is confined in a vessel under a piston and the piston has a weight adjusted to maintain a pressure of 1 atm, there will be no gaseous phase above the liquid nitrogen below 77°K. If a minute bubble of gas is formed momentarily by a few molecules shaking themselves loose and pushing the piston up by a slight amount, the rate at which those molecules return to the liquid is considerably greater than the rate at which liquid molecules would move out into the temporary gaseous space. Consequently, the gaseous phase can exist at 1 atm only temporarily, infrequently, and over very small areas of the liquid surface. A portion of gas cannot be maintained in permanent equilibrium with the liquid. However, as the liquid is heated to still higher temperatures, at 77°K the point is reached where under 1-atm pressure the rate of passage of molecules from the gaseous state to the liquid states equals the rate of passage from the liquid to the gaseous state; the liquid is at its normal *boiling point*. Consequently, as heat flows into the liquid, molecules of the liquid pass into the gas phase and stay there, more and more liquid vaporizing until finally all the liquid has turned into gas. Only then will the temperature begin to rise again as more heat is fed into the substance, which is now completely gas.

4-5 VAPOR PRESSURE

If crystals of nitrogen are sealed into an evacuated glass tube at 20°K, part of the nitrogen evaporates and forms a gas phase. The number of molecules in the gaseous phase automatically adjusts to that density where the number of molecules striking and remaining on the crystal surface equals the number leaving the crystal and going into the gas phase per second. The pressure corresponding to the density of gas at this equilibrium point is called the *vapor pressure* of the crystal at this particular temperature.

In the same way a sample of liquid nitrogen at 70°K placed in an evacuated and sealed glass capsule vaporizes just enough to provide a density of molecules in the space above the liquid which makes the rate of change from the gas to the liquid equal to the rate of change from the liquid to the gas. The pressure corresponding to this equilibrium density is called the *vapor pressure* of the liquid.

The ratio of the vapor pressure P_2 of a liquid at temperature T_2 to the vapor pressure P_1 of the same liquid at temperature T_1 is given by

$$\frac{P_2}{P_1} = \exp\left[-\frac{H_v}{R}\left(\frac{1}{T_2} - \frac{1}{T_1}\right)\right] \qquad (4\text{-}1)$$

where H_v is the heat of vaporization per mole of the liquid expressed in calories and R is the gas constant expressed in calories per degree per mole.

This equation is derived from the laws of thermodynamics in Chap. 10; it is introduced here as a relationship helpful in understanding the properties of gas-liquid and gas-solid equilibria even though the basis for its derivation has not been established.

If the temperature T_1 is the normal boiling point of the liquid at 1 atm, it may be written as T_b; the pressure P_1 will, of course, be 1 atm. Under these conditions the numerical subscripts are not needed and the equation can be rewritten

$$\frac{P}{P_1} = \exp\left[\frac{H_v}{R}\left(\frac{1}{T} - \frac{1}{T_b}\right)\right] \qquad (4\text{-}2)$$

It is frequently useful to have this equation in logarithmic form

$$\ln \frac{P}{P_1} = -\frac{H_v}{R}\left(\frac{1}{T} - \frac{1}{T_b}\right) \qquad (4\text{-}3)$$

This equation can be put into the form where the logarithm has the base 10, a common logarithm, by substituting $\ln P = 2.303 \log P$:

$$\log \frac{P}{P_1} = -\frac{H_v}{2.303R}\left(\frac{1}{T} - \frac{1}{T_b}\right) \qquad (4\text{-}4)$$

From these equations it can be seen that the vapor pressure is an exponential function, so that as temperature rises, vapor pressure rises even faster; if temperature is doubled, the vapor pressure far more than doubles. The relationship of vapor pressure to temperature for water is shown in Fig. 4-9a.

Since $P_1 = 1$ atm, Eq. (4-4) may be written

$$\log P = -\frac{H_v}{2.303R}\frac{1}{T} + \frac{H_v}{2.303RT_b} \qquad (4\text{-}5)$$

This equation is in the algebraic form

$$y = ax + b \qquad (4\text{-}6)$$

where

$$y = \log P \qquad a = -\frac{H_v}{2.303R} \qquad x = \frac{1}{T}$$

$$b = \frac{H_v}{2.303RT_b}$$

Since Eq. (4-6) is the equation for a straight line, where a is the slope of the line and b is the intercept on the y axis when x is equal to 0, if instead of the coordinates P and T, $\log P$ and $1/T$ are used, the graph of Eq. (4-5) is a straight line, as shown in Fig. 4-9b.

Example 4-1

The heat of vaporization of ethyl alcohol is 6936 cal mole^{-1} at the normal boiling point,

Fig. 4-9   (*a*) **Vapor pressure of water as a function of** T.   (*b*) **Vapor pressure of water (log P and $1/T$ plot).**

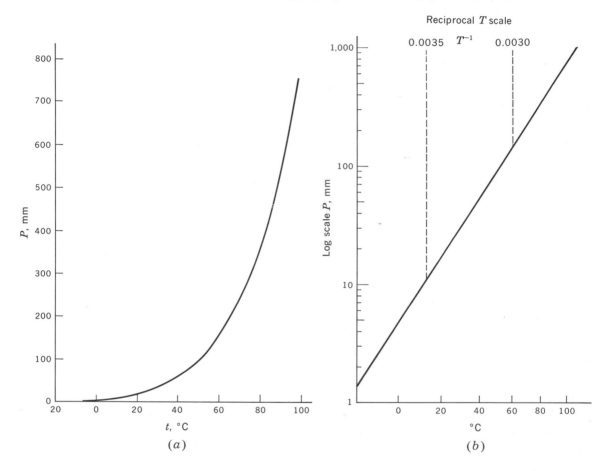

(a)   *(b)*

78.3°C.   Calculate the vapor pressure at body temperature, 37.0°C.

Transform temperature to degrees Kelvin and put the numerical values in Eq. (4-4).

$$\log \frac{P}{P_1} = -\frac{H_v}{2.303R}\left(\frac{1}{T} - \frac{1}{T_b}\right) \qquad P_1 = 1 \text{ atm}$$

$$= \frac{-6936 \text{ cal}}{2.303 \times 1.9872} \text{ cal deg}^{-1} \text{mole}^{-1}$$

$$\times \left(\frac{1}{310.2} - \frac{1}{351.5}\right)$$

$$= -1515.7 \times (0.0032237 - 0.0028449)$$

$$= -1515.7 \times 0.0003788 = -0.5741$$

$$P = 0.267 \text{ atm} = 203 \text{ mm Hg} \qquad Ans.$$

Exactly the same equations hold for the vapor pressure of a crystal P_c where the molal heat of sublimation H_s is substituted for the molal heat of vaporization H_v and the normal sublimation temperature T_s is substituted for the normal boiling temperature T_b:

$$\log P_c = -\frac{H_s}{2.303R}\left(\frac{1}{T} - \frac{1}{T_s}\right) \qquad (4\text{-}7)$$

If the ratio of the vapor pressure at one temperature to that at another is wanted, the equations may be put in the form

Liquid: $$\log \frac{P_2}{P_1} = -\frac{H_v}{2.303R}\left(\frac{1}{T_2} - \frac{1}{T_1}\right) \qquad (4\text{-}7a)$$

Crystal: $$\log \frac{P_{c2}}{P_{c1}} = -\frac{H_s}{2.303R}\left(\frac{1}{T_2} - \frac{1}{T_1}\right) \qquad (4\text{-}7b)$$

Example 4-2

The vapor pressure of crystalline benzene is 1.00 mm at $-36.7°C$ and 10.00 mm at $-11.5°C$. Calculate the molal heat of sublimation.

Put the numerical values in Eq. (4-7b):

$$\log \frac{10.00}{1.00} = \frac{H_s}{2.303R} \left(\frac{1}{261.7} - \frac{1}{236.5} \right)$$

$$1 = - \frac{H_s}{4.576} (0.0038211 - 0.004228)$$

$$1 = \frac{H_s}{4.576} 0.000407$$

$$H_s = \frac{4.576}{0.000407} = 11.2 \text{ kcal} \quad Ans.$$

Figure 4-10 shows the vapor pressure of liquid water and of ice in the vicinity of the melting point, $0°C$. Figure 4-11 shows graphically the effect of vapor pressure on the exchange of molecules between the two phases. The inverted U tube at the left (Fig. 4-11a) has crystals of ice at the bottom of one leg and *supercooled* liquid water at the bottom of the other; i.e., the temperature $(-2°C)$ is below the melting point. As can be seen from Fig. 4-10, the vapor pressure P_L of the supercooled liquid is higher than the vapor pressure of the crystal P_C; and so vapor passes from liquid to crystal, and ultimately all the molecules in the right leg flow over the hump and condense in the left leg. The U tube at the center of Fig. 4-11b is at $0°C$. At this temperature $P_C = P_L$; consequently, there is no net flow of molecules, and the two phases can remain in equilibrium with each other indefinitely. The tube in Fig. 4-11c shows the two phases at $+2°C$. If the dashed curve in Fig. 4-10 were extrapolated to this temperature, it would be found that P_C is higher than P_L; therefore the net flow is from crystal to liquid. Eventually all the water passes into the liquid state.

Actually, it is extremely difficult to heat ice above $0°C$. In theory it can be done momentarily by suddenly irradiating the ice with intense infrared rays, but because of the tendency of the molecules at the surface of the crystals to go spontaneously into the liquid state, there is no easily attained *superheated* state of the crystal analogous to the supercooled state of the liquid.

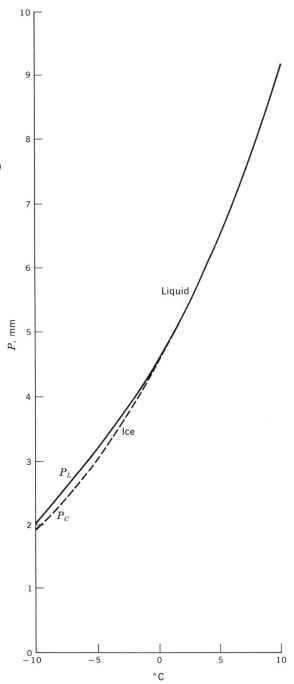

Fig. 4-10   Vapor pressure of liquid water and ice in the neighborhood of the freezing point.

Fig. 4-11 Exchange of water vapor between the crystalline and liquid states. The relative length of the arrows denotes the relative values of the vapor pressures.

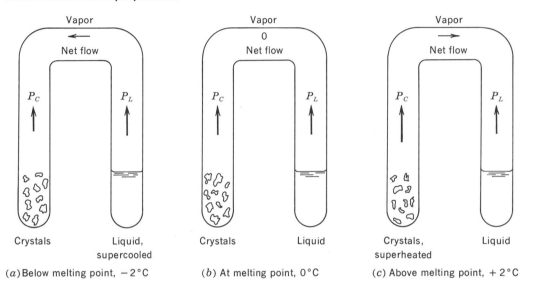

(a) Below melting point, $-2°C$ (b) At melting point, $0°C$ (c) Above melting point, $+2°C$

Many liquids are readily cooled below the melting point without crystallizing. In general, the more complex the shape of the molecule, the harder it is for the molecules to rearrange themselves into a crystal lattice when the temperature is reached (the freezing point) at which, in theory, the vapor pressure of the liquid equals the vapor pressure of the crystal. In such supercooled liquids the molecules remain randomly oriented in the liquid state even when the liquid is cooled so far below the theoretical freezing point that it becomes more and more viscous and finally assumes a rigidity as great as that of a true crystal. Such rigid supercooled liquids are called *glasses*. Window glass is an example of such a supercooled liquid.

Since the vapor pressure is a measure of the tendency of molecules to leave condensed states and go into the gaseous state, it is a very important property from which to obtain clues about the physical and even chemical behavior of molecules. Phase changes are involved in many life processes. There is a constant interchange of water molecules between the liquid state and the gaseous state in the lungs, and on the surface of the skin water molecules are constantly passing into the gaseous state; they evaporate, thereby absorbing heat and keeping the skin cool. But by far

the most important phase changes in the life processes are those thousands which involve the transfer of molecules between the liquid and the solid state. After becoming familiar with the principles of thermodynamics, we shall be able to understand in far greater detail the nature of these biophase changes and their relationships to vapor pressure.

4-6 LIQUID CRYSTALS

For nearly all molecules, the distinction between the highly ordered crystalline state and the more random molecular arrangement in the liquid state is sharp. The H_2O molecules of water in the condensed state are arranged either in an almost perfect lattice order, as in *ice*, or much more randomly, as in liquid water. When ice is heated and melts, the molecules move at once from essentially complete order (ice) to comparative disorder (liquid water).

By way of contrast, when crystalline cholesteryl benzoate crystals are warmed up to the melting point, they first melt, to form a milky liquid, and at a much higher temperature change sharply into a clear liquid. Both changes appear to be true phase changes. The same changes take place in the opposite direction and at the same temperatures when the clear liquid is cooled down through the range where the material ultimately becomes crystalline. Such phases found between crystal and liquid phases are called *mesomorphic* or *paracrystalline*.

This phenomenon was first observed by the Austrian botanist Friedrich Reinitzer in 1888. Shortly thereafter, the German physicist O. Lehmann showed that the opaque liquid had certain crystallike properties and proposed the name *liquid crystal* for it. Studies of the optical properties of this opaque liquid suggest an intermediate kind of order, where the fluid is quasi-crystalline. The different kinds of order of the molecules in the three phases (crystal, liquid crystal, and liquid) are roughly comparable to the kinds of order that can be attained by stacking sewing needles. If the needles are stacked vertically parallel and in layers so that all the points lie in a series of parallel planes, one above the other, the kind of order is analogous to *crystalline* order. If the needles are stacked on an uneven surface, so that they are all parallel vertically but the points are at random heights, it is roughly com-

parable to certain types of *liquid crystal*. If the needles are tossed in a random crisscrossed pile, it is comparable to the *liquid* state.

Cross-sectional drawings of these three states of order for an unsymmetrical molecule are shown in Fig. 4-12. Many substances exhibit liquid-crystal behavior far more complex than that suggested by this figure. Extensive studies have been made to determine the kind of molecular geometry and interaction which produce a tendency to form these mesomorphic phases. All this emphasizes the fact that in nature patterns of order range over a wide spectrum and that order in the liquid or quasi-liquid state may be a most complex factor influencing both physical and chemical behavior, especially in biofluids.

Cells

Among all the hundreds of kinds of liquids that occur in various parts of living organisms, the liquid in the interior of a biological cell, *cytoplasm (protoplasm)*, probably exhibits the most significant kind of complex fluid order observed anywhere in nature. The basic constituent of the cytoplasm is water; but mingled throughout

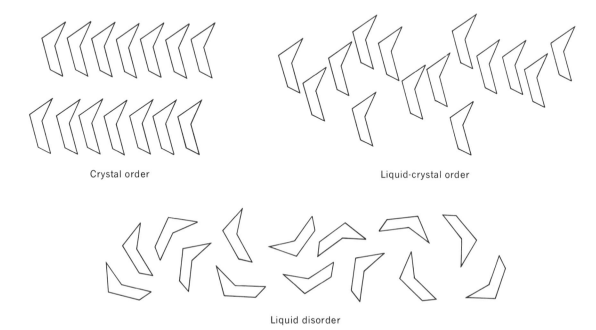

Crystal order

Liquid-crystal order

Liquid disorder

Fig. 4-12 Types of order.

the water are hundreds of kinds of molecules both large and small. The chemical composition of typical cytoplasm is shown in Table 4-2.

For many years the general opinion seemed to be that all these various kinds of molecules moved around in the water solvent almost as easily as molecules of sugar circulating in a dilute aqueous solution. Recently, there has been a growing awareness that cytoplasm is much more like a liquid crystal than a true liquid; in fact, it may be so highly organized that it bears a far greater resemblance to the crystalline than to the liquid state with regard to the degree of order, although it has fluidity of a sort and allows the passage of many kinds of molecules through it. There is increasing evidence, however, that this passage is far less random and far more directed than most scientists realized some years ago.

In their classic book on plant physiology, Meyer and Anderson summarized the nature of cytoplasm as follows:

Although cytoplasm appears to be a simple liquid, no simple liquid could possibly possess the remarkable powers of synthesis, assimilation, reproduction, growth, and sensitivity that characterize the cytoplasm of living plant cells. The properties and behavior of cytoplasm clearly show that it is not a substance and that is must be regarded as a complex system of substances. This system is dynamic; it is constantly undergoing changes yet at the same time the changes are so regulated and controlled that the system is not disrupted. A cell is alive only so long as the organization of this dynamic, cytoplasmic system is maintained.[1]

In the light of its function and from the evidence especially from the electron microscope, cytoplasm appears to have a far more complex structure than an ordinary aqueous solution. The elucidation of this structure is one of the most interesting challenges to physical chemistry today. Can this intermediate state between the crystal and the liquid be explained in terms of the present laws of physical chemistry? There is a great deal of evidence that for this purpose the concepts of thermodynamics must be expanded to deal with nonequilibrium systems and that the principles of quantum mechanics must be extended to deal with interactions throughout systems composed of many millions of molecules in order to understand the nature of the life processes that take place within the biological cell. The Hungarian biochemist Albert Szent-Györgyi has suggested that highly intricate and extended energy patterns are produced by the interactions in the complex biological ordering of molecules, patterns that are analogous to the energy patterns in the crystalline state when impurities alter the perfect crystalline order. This offers a promising perspective in which to understand more fully the interplay between energy and order, often called the *cybernetic* aspect of the life process, a word stemming from the Greek word meaning *steersman*. This aspect of biodynamics will be examined in some detail in Chap. 27.

4-7 COMPRESSIBILITY

One of the major differences between gases on the one hand and liquids and solids on the other lies in their

TABLE 4-2 Constituents of a plant cytoplasm including both soluble and insoluble matter[a]

	Approximate percentage
A. Water	90.00
B. Soluble substances:	
Monosaccharides	1.42
Proteins	0.22
Amino acids	2.43
C. Relatively insoluble substances:	
Nucleoproteins	3.23
Nucleic acids	0.25
Globulin	0.05
Lipoproteins	0.48
Neutral fats	0.68
Phytosterol	0.32
Phosphatides	0.13
Other organic matter	0.35
Mineral matter	0.44

[a] After W. W. Lepeschkin, *Ber. Deut. Botan. Ges.*, **41:** 179–187 (1923).

[1] B. S. Meyer and D. B. Anderson, "Plant Physiology," 2d ed., p. 58, D. Van Nostrand Company, Inc., New York, 1952. In the original, *protoplasm* is used instead of *cytoplasm*.

relative compressibility. When a typical gas like nitrogen is maintained at constant temperature, doubling the pressure halves the volume, a decrease in volume of 50 percent. When a typical liquid like water is kept at constant temperature, doubling the pressure decreases the volume by only 0.0005 percent. Thus gases are readily compressible while nearly all liquids and solids are highly incompressible under ordinary conditions.

These striking differences in behavior are due primarily to the differences in the degree of separation of the molecules composing these different states of matter. In air at room temperature and under 1-atm pressure the molecules are separated on the average by a distance in empty space that is over 10 times the diameter of the molecule itself. In water or in ice the molecules are in effect actually touching one another. It is true that the thermal energy of the water molecules causes vibration, so that there is a slightly greater degree of separation at room temperature than there would be if liquid water could be cooled to absolute zero, where there would be essentially no vibratory molecular motion; but in the vast majority of cases, the molecules of liquids and solids are so closely juxtaposed that any compression brings into play forces of repulsion that make the material strongly resist any substantial decrease in volume.

For liquids the relation at constant temperature between the volume V of a quantity of liquid and the pressure P imposed on it can be expressed quite accurately by the equation

$$V = V°[1 + \beta(P - 1)] \qquad (4\text{-}8)$$

where $V°$ is the volume of the quantity of liquid under 1-atm pressure and β is the coefficient of compressibility, defined by

$$\beta \equiv \frac{1}{V}\left(\frac{\partial V}{\partial P}\right)_T \qquad (4\text{-}9)$$

Table 4-3 lists values of these coefficients for seven liquids. The low value of β for water is an important factor in the mechanics of the circulation of the blood. If blood were as compressible as air, the circulatory system in the body would have to have a very different design.

TABLE 4-3   Isothermal coefficients of compressibility β

Compound	t, °C	β, atm^{-1} × 10^{-5}
Acetic acid	40	−10.51
Benzene	40	−11.50
Chloroform	40	−12.00
Ethyl alcohol	40	−12.91
Methyl alcohol	40	−14.03
n-Octane	40	−13.54
Water	0	−5.08
	35	−4.54
	45	−4.50

Example 4-3

Calculate the decrease in volume in cubic centimeters when 1 liter of blood is put under a pressure of 5 atm at body temperature. Assume that β for blood equals that for water.

Interpolating from the values of β in Table 4-3, β for water at 37°C has the value -4.53×10^{-5} atm^{-1}. The decrease in volume will be $V° - V$. Using the value for V from Eq. (4-8),

$$
\begin{aligned}
V° - V &= V° - V°[1 + \beta(P - 1)] \\
&= V°\beta(P - 1) \\
&= -1.0000 \text{ liter} \\
&\quad \times (-4.53 \times 10^{-5} \text{ atm}^{-1})(5 \text{ atm} - 1) \\
&= -1{,}000.0 \text{ cm}^3(-1.720 \times 10^{-4}) \\
&= 0.1812 \text{ cm}^3 \qquad Ans.
\end{aligned}
$$

Elasticity

As a rule, a liquid like water has no tendency to retain its shape; it assumes the shape of the vessel into which it is poured. By way of contrast, a bar of spring steel has a strong tendency to retain its shape: try to bend it, try to twist it, try to squeeze it. It may be possible to produce a slight and momentary deformation of the bar, but when one stops applying force, the bar springs back to its original shape; it exhibits *elasticity*. Elasticity is defined as the property by virtue of which a body resists and recovers from deformation produced by force.

There is, in general, a limit to the amount of deformation any given body can undergo and still return to its original shape when the deforming force is removed. The smallest value of the stress producing permanent alteration is known as the *elastic limit*.

For crystals the precise specification of compressibility or elasticity is complicated by the fact that the same force applied as a squeeze on one pair of opposite faces of a crystal may produce a distortion different from that when the same squeeze is applied to a different pair of opposite faces. In other words, the direction with respect to the crystal lattice in which the force is applied influences its effect. For nearly all liquids the direction of application of the force has no effect on the resultant change in volume; these liquids are *isotropic* with respect to compressibility. Crystals, where the direction of application of the force does affect the result, are *nonisotropic*.

Crystals like rock salt, NaCl, have little elasticity; they tend to break when deformed even slightly. On the other hand, many solids composed of long chain-like molecules can be deformed 100 times more than a crystal like rock salt and still return to their original shape. In many solids the basic molecules are of the chain type and are interconnected to form a three-dimensional molecular network. These solids are known collectively as *polymers*.

Rubber is an example of a type of polymer having high elasticity. It has a structure in which the molecules probably act like spiral springs. A section of rubber can be stretched to 10 times its normal length and yet spring back to its original shape when the stretching force is removed.

Nearly all body tissues have considerable elasticity. The human skin is quite elastic. Even the "skin," or sac, forming the outer layer of every biological cell is elastic, an important property when the cell grows. Since work must be done to stretch elastic fibers, energy is introduced into the material, and the relation between this energy and the change of state of the material is important in many biological processes. These relationships will be discussed in later chapters, after the fundamental laws between energy and elongation have been related to the laws of thermodynamics in Chaps. 7 to 9.

TABLE 4-4 Volume coefficients of thermal expansion for solids

	α, deg^{-1} × 10^{-5}	t, °C
Diamond	0.354	40
Emerald	0.168	40
Glass (approx.)	3	40
Ice	11.25	−10
Iron	3.6	50
Sulfur	22.3	30
Tin	6.9	50

TABLE 4-5 Volume coefficients of of thermal expansion for liquids at 20°C

	α, deg^{-1} × 10^{-3}
Acetic acid	1.07
Methanol	1.2
Chloroform	1.27
Ether	1.66
Mercury	0.182
Phenol	1.09
Water	0.207

4-8 THERMAL EXPANSION

When a liquid is heated under a constant pressure such as that of the atmosphere, we see that the volume of the liquid increases. This phenomenon, called *thermal expansion,* is the basis of familiar bulb-type thermometers. When the thermometer is moved from a colder to a warmer environment, the liquid in the bulb increases in volume. This expansion is much greater than that of the bulb itself, and consequently some of the liquid is forced into the capillary of the thermometer stem, thereby raising the end of the thread of liquid in the capillary further up the thermometer scale.

For almost all liquids the volume can be expressed quite accurately as a function of the temperature by a linear equation

$$V = V°(1 + \alpha t) \tag{4-10}$$

where t = temperature, °C

$V° =$ volume of liquid at 0°C

$\alpha =$ const characteristic of substance

α is defined by

$$\alpha \equiv \frac{1}{V}\left(\frac{\partial V}{\partial T}\right)_p \qquad (4\text{-}11)$$

In general, α increases with temperature to the extent of about 0.1 percent per degree. Values of this quantity for a few typical solids and liquids are given in Tables 4-4 and 4-5. The value of $V°$ for 1 g of substance can be calculated from the values of density in Table 4-5a.

From the point of view of atomic and molecular mechanisms, solids and liquids expand with increasing temperature because the increasing thermal energy results in a greater amplitude of vibration of the individual atoms. As a result of the nonlinear relation of force to distance between neighboring atoms, the atoms move farther apart as the temperature increases and the amplitude of the thermal vibration also increases.

Example 4-4

The volume of 1,000 g of H_2O at 10°C is 1,000.300 cm³, and at 30°C the volume is 1,004.373 cm³. Calculate the mean coefficient of thermal expansion at 20°C and compare with the value given in Table 4-5 for α at 20°C.

$$\alpha = \frac{1}{V}\left(\frac{\partial V}{\partial T}\right)_p$$

$$\alpha_{\text{mean}} = \frac{1}{V_{\text{av}}}\frac{V_{30°} - V_{10°}}{30 - 10°}$$

$$= \frac{1}{1{,}002 \text{ cm}^3}\frac{4.073 \text{ cm}^3}{20°}$$

$$= 0.203 \times 10^{-3} \text{ deg}^{-1} \qquad Ans.$$

Note: The value in Table 4-5 for α is 0.207×10^{-3} deg⁻¹. The difference between α and α_{mean} is due to the nonlinear increase of α with temperature over this range; this difference is not generally significant in physical chemistry.

4-9   HEAT CAPACITY

The heat capacity per mole at constant volume c_v is

TABLE 4-5a   Densities

Substance	Density, g cm⁻³	t, °C
Solids:		
Bone	1.8	25
Cork	0.24	25
Dolomite	2.84	25
Ice	0.92	0
Quartz	2.65	25
Sodium chloride	2.18	25
Glucose	1.59	25
Tallow	0.94	25
Iron	7.90	20
Aluminum	2.70	20
Graphite	2.25	25
Sodium	0.97	25
Liquids:		
Acetic acid	1.05	20
Acetone	0.79	20
Ethyl alcohol	0.79	20
Methyl alcohol	0.81	0
Benzene	0.90	0
Chloroform	1.49	20
Water	0.9998	0
	1.0000	4
	0.997	25
	0.993	37

defined as the temperature rate of absorption of heat by 1 mole as temperature is increased. In nearly all cases this is effectively equal to the amount of heat absorbed when 1 mole of the substance is heated through a temperature interval of 1°, the *mean* heat capacity for this interval. As noted in Chap. 2, for monatomic gases like helium and the other noble gases, which behave like ideal gases, c_v is effectively a constant with the value $\frac{3}{2}R$ cal deg⁻¹ mole⁻¹, except at very high or very low temperatures. Two French scientists, P. L. Dulong and A. T. Petit, pointed out in 1819 that for many solids with monatomic crystal lattices the heat capacity at room temperature has a value a little over twice that for monatomic gases, a relation usually referred to as the rule of Dulong and Petit. Their conclusion was based on measurements not of c_v but of c_p. However, the Austrian theoretical physicist Lud-

TABLE 4-6   Heat capacity of monatomic solids at 20°C

Element	c_p, cal deg^{-1} mole^{-1}	c_v, cal deg^{-1} mole^{-1}
Al	5.77	5.6
Bi	6.14	6.0
Cr	5.72	5.6
Co	5.90	5.8
Cu	5.85	5.7
Au	6.24	5.9
Fe	5.98	5.9
Pb	6.34	5.9
Mg	5.98	5.8
Ni	6.16	6.0
Pt	6.32	6.1
Ag	6.02	5.7
Na	6.78	6.3
Sn	6.43	6.1
Zn	6.05	5.8
$3R = 5.96$		Av 5.9

TABLE 4-7   Molal heat capacity at constant pressure of some liquids

Liquid	t, °C	Heat capacity c_p, cal deg^{-1} mole^{-1}
Acetic acid	0	28.10
Chloroform	20	27.93
Ethyl ether	0	39.21
	30	40.54
Ethanol	0	24.65
	25	26.77
Glycerol	0	49.73
Methanol	0	18.13
	20	19.22
Salol	44.1	83.76
Water	0	18.14
	15	18.02
	30	17.98
	37	17.98
	50	18.00

wig Boltzmann pointed out some years later (1871) an explanation for this regularity, and the American physical chemist G. N. Lewis employed an equation relating c_p and c_v to provide a quantitative basis for it. The equation is

$$c_p - c_v = \frac{-\alpha^2 vT}{\beta} \qquad (4\text{-}12)$$

where α = coefficient of thermal expansion defined in Eq. (4-8)

β = coefficient of compressibility defined in Eq. (4-9)

v = molal volume

T = temperature, °K

Table 4-6 gives the experimentally observed values of c_p at 20°C for a number of metals with monatomic crystal lattices and the values of c_v calculated from them with the help of Lewis's equation. The average of the values of c_v in this table is close to the value of $3R$, just twice the value of c_v for monatomic gases. Effectively, the absorption of heat to increase *potential* energy as well as *kinetic* energy is like adding 3 more degrees of freedom for each atom. The effective number of degrees of freedom is raised from 3 per atom to 6 per atom, and the heat capacity c_v is doubled.

The decrease of c_v at temperatures below room temperature was first explained by Albert Einstein and will be discussed in Sec. 6-8.

Table 4-7 lists values of the heat capacity at constant pressure c_p for a number of liquids. Note that these values are much higher than the values of c_p for metals in Table 4-6. The molecules of all the substances in Table 4-7 contain three or more atoms, so that in addition to the 3 degrees of freedom of translation for the molecule there are also degrees of freedom of rotational oscillation and of internal vibration. Energy is absorbed in all these additional degrees of freedom, making the value of c_p correspondingly higher.

In nearly all substances of biochemical interest, in the food we eat and the fluids, tissues, and bone synthesized by the body from this food, the basic thermal vibrators are not *individual* atoms but molecules with many degrees of freedom. The factors contributing to heat capacity are consequently far more complex than for metals.

As has been noted before, the life process can be understood much better as a pattern of energy exchange than as a pattern of atomic rearrangement. For this reason, a knowledge of the ways in which the atoms of the body's fluids and solids possess and exchange energy is vital in order to achieve an insight into the body's functioning in metabolism, replication, growth, and intrabody communication. The next five chapters discuss the basis for understanding the role of energy in influencing the behavior of complex molecules.

Example 4-5

Calculate the approximate amount of heat necessary to raise the temperature of 100 g of ethyl alcohol from 20 to 30°C at 1-atm pressure.

The value of c_p for ethanol in Table 4-7 is 26.77 cal deg^{-1} mole^{-1}, which may be taken as the average heat capacity over the range 20 to 30°C. Therefore, the amount of heat q to raise 1 mole of ethanol from 20 to 30° is

$$q = c_p(t_2 - t_1)$$
$$= 26.77 \text{ cal deg}^{-1} \text{ mole}^{-1}(30 - 20°)$$
$$= 267.7 \text{ cal mole}^{-1}$$

Since the molecular weight of ethanol is 46.07 g mole^{-1}, 100 g is thus equal to (100 g)/(46.07 g mole^{-1}), or 2.17 moles. The heat necessary to raise the temperature of 100 g of ethanol from 20 to 30°C is thus 2.17 moles $\times$ 267.7 cal mole^{-1} = 581 cal.   *Ans.*

Problems

1. The heat of vaporization of water averages about 10 kcal mole^{-1} between room temperature and 100°C. Taking this as a constant value over this range, calculate the vapor pressure of water at body temperature, 37.0°C.

2. According to experimental measurements, the vapor pressure of water is 40 mm at 34.1°C and 100 mm at 51.6°C. From these values calculate the heat of vaporization of water.

3. The vapor pressure of liquid silver is 400 mm at 2090°C and 760 mm at 2212°C. Calculate the aver-

age heat of vaporization of silver over this range of temperature.

4. The heat of vaporization of ethyl alcohol is 204 cal g^{-1} at the boiling point, 78.3°C. Calculate the vapor pressure at 37.0°C.

5. According to experimental measurements the vapor pressure of diethyl ether has the following values at the temperatures indicated:

P, mm	t, °C
1	−74.3
10	−48.1
40	−27.7
100	−11.5
400	+17.9
760	+34.6

Make a plot of the logarithm of the vapor pressure against the corresponding reciprocal absolute temperature.

6. From the plot in Prob. 5 estimate the vapor pressure of diethyl ether at 37.0°C and the heat of vaporization at that temperature.

7. From the data in Tables 4-5 and 4-5*a* calculate the volume at 40°C of 1 mole of acetic acid, of chloroform, and of water.

8. From the values in Prob. 7 and the data in Table 4-3, calculate the change in volume when pressure changes from 1 to 10 atm for 1 mole of acetic acid, of chloroform, and of water at 40°C.

9. Calculate the number of calories of heat required to warm an aluminum pan filled with 200 g of H_2O from 30 to 100°C if the pan weighs 200 g. Mean c_p of Al = 5.77 cal deg^{-1} mole^{-1}; mean c_p of H_2O = 18 cal deg^{-1} mole^{-1}.

10. A 1,000-g lead weight with a temperature of 50°C is dropped into an iron pot weighing 1,500 g and filled with 2 kg of water, both pot and water being at 25°C. Assuming that no heat is gained or lost to the surroundings, calculate the final temperature when equilibrium is attained. Use appropriate c_p values in Tables 4-6 and 4-7 as average values for these intervals.

11. A metal bar weighing exactly 1 kg and at 100.00°C is dropped into an iron vessel weighing 800 g and containing 2 kg of water, all at 20.00°. When thermal equilibrium is attained, the temperature is 27.44°. Identify the metal as one of those listed in Table 4-6.

12. When a 100-lb boy drinks three glasses of ice water, to what temperature is his body lowered? Assume that the heat capacity of the body is the same as water; normal body temperature is 37.0°C; a glass of ice water is the equivalent of 1 lb at 0°C.

13. Assuming the heat capacity of the human body to be roughly the same as the equivalent weight of water, calculate the amount of ice that must be melted to cool a 150-lb man from 37.0 to 20.0°C in preparation for low-temperature surgery.

14. For exploring the surface of the moon on foot, an astronaut must wear a spacesuit with thermal insulation. In such activity the body may generate roughly 1 kcal of heat per kilogram weight per hour. If no energy were lost to the surroundings, and if no energy were lost through perspiration or other similar processes, how much would body temperature increase per hour due to this rate of heat production?

15. When water contained in a tightly stoppered glass bottle freezes, the bottle frequently breaks. Calculate the increase in volume of 1 liter of water when it freezes.

16. Calculate the increase in volume of a spherical biological cell 10^{-3} cm in diameter when the liquid content freezes, using the density values for water and ice. What is the percentage change in volume? What is the increase in the radius?

17. Calculate the pressure in atmospheres in a freshwater lake at a depth of (a) 10 m; (b) 100 m; (c) 1,000 m if the density of water is not appreciably changed by the pressure and the temperature is 25°C.

18. Calculate the pressure at 1,000-m depth in a freshwater lake at 25°C if the density changes linearly with pressure, using the values in Table 4-3 to obtain an interpolated value for the coefficient of compressibility at 25°C.

19. Calculate the composition of a mixture of ethanol and water which will have the same density as tallow at 25°C, assuming a linear variation of density with composition. (Take the density of ethanol as 0.77 at 25°C.)

20. Calculate the fraction of an iceberg that remains above the surface of water when it is floating in the freshwater mouth of a river at 0°C.

REFERENCES

Crystals

Bunn, C.: "Chemical Crystallography: An Introduction to Optical and X-ray Methods," Oxford University Press, Fair Lawn, N.J., 1961.

Phillips, F. C.: "An Introduction to Crystallography," Longmans, Green & Co., Ltd., London, 1963.

Robertson, J. M.: "Organic Crystals and Molecules," Cornell University Press, Ithaca, N.Y., 1953.

Bunn, C.: "Crystals: Their Role in Nature and in Science," Academic Press, Inc., New York, 1964.

Mandelkern, L.: "Crystallization of Polymers," McGraw-Hill Book Company, New York, 1964.

Ubbelohde, A. R.: "Melting and Crystal Structure," Oxford University Press, London, 1965.

Liquid crystals

Meyer, B. S., and D. B. Anderson: "Plant Physiology," pp. 58–69, D. Van Nostrand Company, Inc., Princeton, N.J., 1952.

Gray, G. W.: "Molecular Structure and the Properties of Liquid Crystals," Academic Press, Inc., New York, 1962.

Fergason, J. L.: "Liquid Crystals," *Sci. Am.,* August, 1964, pp. 76–85.

Liquids

Hala, E., J. Pick, V. Fried, and O. Vilim: "Vapor-Liquid Equilibrium," Pergamon Press, New York, 1958.

Rowlinson, J. S.: "Liquids and Liquid Mixtures," Academic Press, Inc., New York, 1959.

Bernal, J. D.: "The Structure of Liquids," *Sci. Am.,* August, 1960, pp. 124–134.

Barker, J. A.: "Lattice Theories of the Liquid State," The Macmillan Company, New York, 1963.

Dreisbach, D.: "Liquids and Solutions," Houghton Mifflin Company, Boston, 1966.

Pryde, J. A.: "The Liquid State," Hutchinson University Library, London, 1966.

Heat capacity

Dulong, P. L., and A. T. Petit: *Ann. Chim. Phys.,* **10:**395 (1819).

Boltzmann, L.: *Sitz. Kgl. Akad. Wiss. Wien,* **63**(2):679 (1871).

Lewis, G. N.: *J. Am. Chem. Soc.,* **29:**1165, 1516 (1907).

Lewis, G. N., and M. Randall: ''Thermodynamics,'' 1st ed., pp. 71–75, McGraw-Hill Book Company, New York, 1923; 2d ed., rev. by K. S. Pitzer and L. Brewer, pp. 53–55, 1961.

5-1   HEATS OF REACTION

In Chap. 4 there was a discussion of the energy change associated with a *physical* change of state, such as ice melting to form water.   In this chapter the discussion is concerned with the energy change associated with a *chemical* change of state such as the oxidation of carbon to form carbon dioxide. The chemical equation for this reaction is

$$C + O_2 \rightarrow CO_2 \qquad (5\text{-}1a)$$

The energy change associated with this reaction depends on how much carbon reacts and under what conditions.   Normally, the equation as written

Chapter Five

thermochemistry

above signifies that 1 mole of carbon reacts with 1 mole of oxygen to form 1 mole of carbon dioxide. The conditions under which the reaction takes place can be specified in the following way:

At 25°C: $C(gr) + O_2(g,\ 1\ atm) \rightarrow CO_2(g,\ 1\ atm)$
$$(5\text{-}1b)$$

where (*gr*) stands for graphite.   Under these conditions, 94.1 kcal flows out of the system into the surroundings.   In some texts, this is regarded as the production of energy, and the amount of the energy change is written with a positive sign.   In this text, the *system* is regarded as the basis for energy calculations; the system has less energy after the change than it had before; therefore, the energy change is regarded as *negative:*

At 25°C: $C(gr) + O_2(g,\ 1\ atm) \rightarrow$
$$CO_2(g,\ 1\ atm) - \underset{\text{heat change}}{94.1\ \text{kcal}} \quad (5\text{-}1c)$$

In writing thermochemical equations the physical state is often abbreviated by *g* for gas, *s* for solid, and *l* for liquid.

If instead of oxidizing 1 mole of carbon in the form of graphite, 1 mole of carbon in the form of diamonds (*di*) is oxidized, the equation reads

At 25°*C*:   C(*di*) + O$_2$(*g*, 1 atm) →

$$CO_2(g, 1 \text{ atm}) - 94.5 \text{ kcal} \quad (5\text{-}2)$$

heat change

Because 1 mole of diamond has 0.4 kcal more energy than 1 mole of graphite, the reaction produces an additional 0.4 kcal of heat.

These changes in energy are shown diagrammatically in Fig. 5-1, where energy is plotted on the vertical scale. Note that the system goes from a higher to a lower energy level when the reaction takes place. Note also that the line in the diagram indicating the energy level for diamond lies at a level higher than the line for graphite.

The whole life process is intimately linked with the flow of energy. In living organisms, a network of closely knit chemical reactions is taking place, atoms are constantly moving from one molecular configuration to another, and the driving force behind this movement comes from energy flowing from a higher to a lower potential, like the flow of water from a higher to a lower level in driving a mill wheel. In more strictly physicochemical language, the origin of this life dynamism is in the change of the system from a higher to a lower *energy level*. A major factor producing this energy flow is the conversion of the linkages of carbon from patterns that contain relatively high amounts of

energy to linkages that contain relatively low amounts; but instead of taking place in a single transformation involving a large energy change like that shown in Fig. 5-1, the oxidation of carbon in the life process takes place through a series of reactions consisting of many small steps like a descending flight of stairs. As the system passes through each stage, energy is provided and used to keep this life process moving forward. In order to depict the energy aspects of these complicated networks of reactions accurately, it is necessary to use precisely defined concepts for specifying energy changes; these concepts make it possible to show more clearly the laws that relate the energy changes to each other.

5-2   ENTHALPY

One of the most useful thermochemical concepts is the defined property called *enthalpy*. The nature of enthalpy is illustrated by the energy change in Eqs. (5-1) and (5-2). The mole of graphite and the mole of oxygen contain energy of two different kinds. First, there is the *internal* energy of the molecules, denoted by E. The internal energy consists both of the kinetic energy of motion of the atoms and of the potential energy which can be increased or decreased when the electron-cloud stresses in the atoms or molecules are changed as the linkages are changed in a chemical reaction. In the second place, because the gas occupies a definite volume in space and is under 1 atm pressure, the system contains "volume" potential energy.

Suppose that the solid carbon and gaseous oxygen are placed in the space under the piston shown in Fig. 5-2. The volume of the carbon granules (5.34 cm^3) may be regarded as negligible compared with the volume of the oxygen gas (24,464 cm^3). In Fig. 5-2a the gas is shown supporting a piston 100 cm^2 in area on which a weight of 103.32 kg has been placed to balance exactly the upward thrust due to the 1-atm pressure of the gas. In Fig. 5-2b the upward thrust of the 1-atm pressure of the oxygen in the cylinder is balanced by the downward normal pressure of the earth's atmosphere. In Fig. 5-2c the weight alone is shown suspended at the same height as in a.

When the weight is at a height of 244.64 cm, it has a

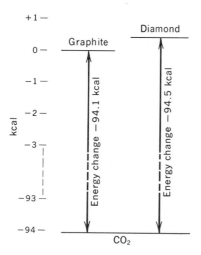

Fig. 5-1   The heat changes in the oxidation of graphite and diamond.

Fig. 5-2   The PV factor in enthalpy.

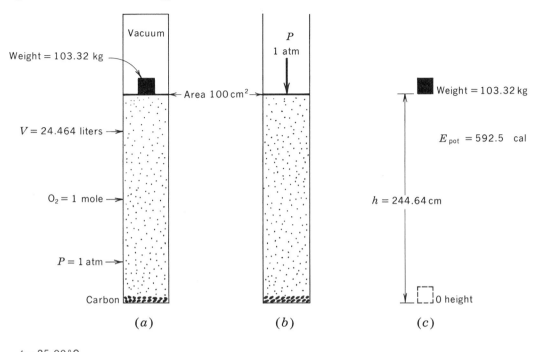

$t = 25.00°C$
$T = 298.15°K$
$PV = 24.464$ liter atm $= 592.5$ cal

potential energy of 592.5 cal, compared with the zero potential energy of the weight at zero height; for if the weight falls from the height of 244.64 cm to zero height, it can do work that is the equivalent of this much energy.

If the gas in Fig. 5-2a is cooled at constant pressure so that it contracts to zero volume, the weight on top of the piston will fall to zero height and will lose its potential energy of 592.5 cal.   This is exactly the value in calorie units of the energy equivalent of pressure times volume, the PV product for 1 mole of gas when it occupies the volume of 24.464 liters at 1-atm pressure and at a temperature of 25°C (298.15°K).   The system has 592.5 cal of volume potential energy either when it is holding up the piston against the force of the weight, as in Fig. 5-2a, or against the force of the atmosphere, as in Fig. 5-2b.

It is useful to combine the internal energy and the volume potential energy as a sum.   This is achieved by the use of the quantity called *enthalpy* denoted by the symbol H and defined by the equation

$$H = E + c \times P \times V \qquad (5\text{-}3)$$

enthalpy   internal   conversion   pressure   volume
        energy   factor

cal   cal   cal (liter atm)$^{-1}$ $\dfrac{atm}{volume}$ $\dfrac{liter}{potential\ energy}$

In using this equation the units must be consistent. If pressure is expressed in atmospheres and volume in liters, their product must be multiplied by a conversion factor that expresses the number of calories per liter atmosphere (24.2179); then the total product can be added to the internal energy expressed as calories to give the enthalpy expressed as calories.

To assign numerical values to potential energy we must define the height at which the potential energy is equal to zero.   For example, in Fig. 5-2c the weight is 244.64 cm above the surface of the earth; the potential energy of the weight at the surface of the earth is defined as zero; then at the height shown in the figure it has a potential energy of 592.5 cal, equivalent to the loss of potential energy when it falls from that height to zero height at the surface of the earth. The weight is said to have *positive* potential energy if it is at a height *above* the surface of the earth. If the weight is *below* the earth's surface at the bottom of a well, it has *negative* potential energy.

In making thermochemical calculations, it is customary to define the enthalpy of all chemical elements in their standard state at 25°C as equal to zero. This provides a systematic way of expressing energy changes when atoms combine with each other to form molecules containing different kinds of atoms. In the system in Fig. 5-2, the enthalpy $H°$ of 1 mole of oxygen under standard conditions (25°C and 1-atm pressure) is defined as zero, where the superscript circle indicates the standard state:

$$H° = E° + c × P° × v° \quad (5-4)$$

| 0 | −592.47 cal | 24.218 cal (liter atm)$^{-1}$ | 1 atm | 24.464 liters |

$$592.47 \text{ cal}$$

Thus the standard molal internal energy $E°$ is arbitrarily taken as −592.47 cal in order to balance out the potential energy of +592.47 cal and give $H°$ the value of zero.

It is permissible to assign the value of zero to the $H°$ of each element at 25°C because in ordinary chemical reactions no chemical element is ever converted into another chemical element. Of course, in considering nuclear reactions, e.g., one in which carbon changes into nitrogen, it is impossible to assign the value zero to $H°$ for all the chemical elements.

When pressure is constant and the system loses or gains work energy only by changing volume, the heat flowing in or out when a reaction takes place is numerically equal to the sum of the enthalpies of the products minus the sum of the enthalpies of the reactants. It is customary to use the symbol $\Delta H°$ to designate the molal *change* in enthalpy when the system changes from the standard state where atoms are combined together as reactants to the standard state where atoms are combined together as products. This relationship can be expressed in the form of an equation:

At 25°C:

$$C + O_2 \rightarrow CO_2$$
$$-(H_C° + H_{O_2}°) + H_{CO_2}° = \Delta H° \quad (5-5)$$

| −(0 kcal | 0 kcal) | −94.1 kcal | −94.1 kcal |

Since the enthalpy of carbon and of oxygen in their standard states is defined as zero, the molal enthalpy of carbon dioxide in its standard state must be −94.1 kcal.

When a reaction consists of the *formation of a compound* (in this case CO_2) from its elements (in this case C and O_2 in their standard states), it is customary to denote the molal enthalpy change as $\Delta H_f°$, called the enthalpy of formation under standard conditions. A number of standard molal enthalpies of formation will be given later in this chapter. The standard enthalpy of formation is equal to the value of the heat flowing into or out of the system when elements in their standard states combine at 25°C to form the compound in the standard state.

5-3 MEASUREMENT OF HEATS OF REACTION

There are many ways of measuring heats of formation and other heats of reaction. By far the commonest method uses a *calorimeter*. The most familiar type of calorimeter used for measuring heats of reaction involving combustion is a bomb calorimeter of the sort shown in Fig. 5-3. In the center of the calorimeter is a metal vessel made with relatively heavy walls so that high pressures generated by the reaction will not explode the whole apparatus. For example, if a measurement is to be made of the heat given out when carbon combines with oxygen, the carbon is placed in the bottom of the bomb, usually in the form of fine granules, and the upper part of the bomb is filled with excess oxygen so that the reaction will go to completion. A small coil of platinum wire is placed in contact with the carbon and connected to copper wires leading to an external source of electric power. When a relatively heavy current is passed through the coil it becomes red hot and ignites the carbon. The reaction then takes place, generating carbon dioxide gas and liberating a relatively large amount of heat.

Before the reaction the bomb is immersed in a vessel filled with water, and after thermal equilibrium has been established, the initial temperature is measured. Then the contents of the calorimeter are ignited, and the heat generated by the reaction is absorbed by the water, raising the temperature both of the water and its container. After a few minutes the heat spreads uniformly throughout the water and the other objects in contact with it so that the whole inner assembly comes to thermal equilibrium at a final temperature, usually about 5 to 10° above the initial temperature.

In determining, for example, the heat of combustion of graphite, one measures the amount of heat given out when a predetermined amount of graphite is oxidized completely to carbon dioxide. If the heat capacity of the bomb, the water surrounding it, and the vessel containing the water is known, the product of this heat capacity by the temperature increase is the amount of heat released by the reaction, provided that no heat has escaped to the surroundings or passed from the surroundings into the bomb and its holder during the course of the experiment. In order to ensure against heat exchange the bomb is surrounded by a jacket containing water initially at the same temperature as the bomb and its own container. As the temperature of the bomb and its immediate container rises following the reaction, heat is supplied electrically to the outer container at just the right rate to keep the temperatures of the outer and inner containers equal. Since there is little net flow of heat between two objects at nearly the same temperature, this ensures that the unwanted gain or loss of heat from the inner container is reduced to a minimum.

It is possible to determine the *heat capacity* of the bomb, the inner container, and the products of the reaction by passing a measured current of electricity through the platinum coil and then measuring the resultant increase of the temperature of the bomb and the inner container.

Example 5-1

Suppose that the bomb and the inner container are warmed by 5.1365°C when 105.60 kcal of heat is generated electrically. Then the heat capacity of the calorimeter is 20.559 kcal deg⁻¹.

Fig. 5-3   Bomb calorimeter (schematic).

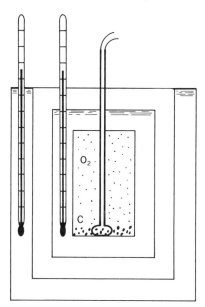

$$\frac{105.60 \text{ kcal}}{5.1365°} = 20.559 \text{ kcal deg}^{-1}$$

If 10.165 g of carbon in the form of graphite is burned in the calorimeter and a temperature increase of 3.8713° is produced, calculate the heat of formation of carbon dioxide.

$$\frac{3.8713° \times 20.559 \text{ kcal deg}^{-1}}{(10.165 \text{ g})/(12.0112 \text{ g mole}^{-1})}$$
$$= 94.05 \text{ kcal mole}^{-1} \qquad Ans.$$

For measurements of heats of reaction to be made with precision and within a convenient length of time, the reaction must take place with reasonable rapidity. It is also essential that the reactants be brought to a precisely defined state with respect to temperature, pressure, and crystal form before the combustion takes place and that it is possible for the products at the end to be similarly brought to a precisely defined state. It is also necessary that all the materials react according to the specified equation and that there are no side reactions. As a rule, combustion reactions are ideal in these respects, other types of reactions being frequently much less suitable for thermal study. Just as it is possible to add one chemical reaction to another algebraically and get a third reaction, one thermochemical equation can be added to another to get the equation for a third kind of reaction. By this indirect procedure, heats of combustion can be combined to obtain values of the heats of reaction for many other kinds of reactions where direct thermochemical measurements would be difficult or impossible.

Values of some heats of combustion are given in Table 5-1.

5-4   THERMOCHEMICAL EQUATIONS

When two chemical equations are added or subtracted to yield a third equation, the heats of reaction also can be added or subtracted with the usual algebraic conventions to yield the heat of reaction for the third equation. This relationship is known as Hess' law. For example, subtracting Eq. (5-7) from Eq. (5-6), as shown below, gives Eq. (5-8), which expresses the heat change when carbon in the form of diamond changes to carbon in the form of graphite.

$$C(di) + O_2(g) \rightarrow CO_2 \qquad \Delta H° = -94.5 \text{ kcal} \qquad (5\text{-}6)$$

$$-[C(gr) + O_2(g) \rightarrow CO_2(g) \quad \Delta H° = -94.1 \text{ kcal}] \qquad (5\text{-}7)$$

$$C(di) \rightarrow C(gr) \qquad \Delta H° = -0.4 \text{ kcal} \qquad (5\text{-}8)$$

In combining equations algebraically, it is customary in the result to transfer any terms representing atoms or molecules with negative sign to the opposite side of the equation. In the example, after making the subtraction, since graphite would appear on the left-hand side of the equation with a negative sign, it is transferred to the right-hand side, where it is positive. The terms for oxygen cancel out, and those for carbon dioxide also cancel out. The difference between the two heat terms is -0.4 kcal. Since heat is evolved in this reaction, the equation shows that diamond has more internal energy than graphite.

To find the heat of formation of propane, C_3H_8, under standard conditions one can combine the equation for the combustion of graphite multiplied by 3 and the equation for the combustion of hydrogen multiplied by 4 and subtract the equation for the combustion of

propane. This gives Eq. (5-12), which shows that the heat of formation of propane under standard conditions is -24.9 kcal.

$$3[C(gr) + O_2(g) \rightarrow CO_2(g)$$
$$\Delta H° = -94.1 \text{ kcal}] \qquad (5\text{-}9)$$

$$4[H_2(g) + \tfrac{1}{2}O_2(g) \rightarrow H_2O(l)$$
$$\Delta H° = -68.3 \text{ kcal}] \qquad (5\text{-}10)$$

$$-[C_3H_8(g) + 5O_2(g) \rightarrow 3CO_2(g) + 4H_2O(l)$$
$$\Delta H° = -530.6 \text{ kcal}] \qquad (5\text{-}11)$$

$$4H_2(g) + 3C(gr) \rightarrow C_3H_8(g)$$
$$\Delta H° = -24.9 \text{ kcal} \qquad (5\text{-}12)$$

It is frequently of interest to calculate the heat change when a reaction results in the production rather than the consumption of oxygen. For example, the overall equation for the formation of glucose and oxygen from carbon dioxide and water is of great importance in biochemistry. As shown in Table 5-1, the enthalpy change for the oxidation of glucose is -673.0 kcal. This reaction is

$$6O_2 + C_6H_{12}O_6 \rightarrow 6CO_2 + 6H_2O$$
$$\Delta H° = -673.0 \text{ kcal} \qquad (5\text{-}13)$$

The action of sunlight on plants results in a reaction which captures the energy in the light and stores it in the form of chemical energy in glucose, $C_6H_{12}O_6$. This reaction is the reverse of Eq. (5-13):

$$6CO_2 + 6H_2O \rightarrow C_6H_{12}O_6 + 6O_2$$
$$\Delta H° = +673.0 \text{ kcal} \qquad (5\text{-}14)$$

In general, when references are made to thermodynamic quantities like *heat of formation, the implication is that the quantity in question refers to 1 mole* and the term *molal* will not be written explicitly unless needed to avoid confusion. Small capital letters like H are used to denote these molal quantities.

Actually the synthesis of glucose takes place in a number of small steps (as shown later in Chap. 21 in the discussion of photochemistry); however, in the overall process 673.0 kcal must enter the reacting material in order to make the reaction take place and form a product far higher on the energy scale than the original reactants, as shown in Fig. 5-4. The value of the enthalpy change when a compound is formed from the elements under standard conditions is called the

TABLE 5-1 Enthalpy changes for combustion at 25°C

End products are $CO_2(g)$, $N_2(g)$, and $H_2O(l)$ at 1 atm

Compound	Formula	$\Delta H°$, kcal mole^{-1}
Hydrogen	$H_2(g)$	−68.3
Carbon (graphite)	$C(gr)$	−94.1
Carbon (diamond)	$C(di)$	−94.5
Carbon monoxide	$CO(g)$	−67.6
Methane	$CH_4(g)$	−212.8
Ethane	$C_2H_6(g)$	−372.8
Propane	$C_3H_8(g)$	−530.6
n-Butane	$C_4H_{10}(g)$	−688.0
n-Pentane	$C_5H_{12}(g)$	−838.3
Ethylene	$C_2H_4(g)$	−337.2
Acetylene	$C_2H_2(g)$	−310.6
Benzene	$C_6H_6(l)$	−789.1
Sucrose	$C_{12}H_{22}O_{11}(s)$	−1349.6
Methanol	$CH_3OH(l)$	−170.9
Ethanol	$C_2H_5OH(l)$	−327.6
Acetic acid	$CH_3COOH(l)$	−208.5
Lactose	$C_{12}H_{22}O_{11}(s)$	−1350.5
Fructose	$C_6H_{12}O_6(s)$	−675.4
Glucose	$d\text{-}C_6H_{12}O_6(s)$	−673.0
Urea	$(NH_2)_2CO(s)$	−151.6
Cystine	$(SCH_2CHNH_2COOH)_2(s)$	−993.6
Leucine	$(CH_3)_2CHCH_2CHNH_2COOH(s)$	−855.3
Alanine	$CH_3CH(NH_2)COOH(s)$	−387.4
Valine	$(CH_3)_2CHCH(NH_2)COOH(s)$	−700.6
Glycine	$CH_2(NH_2)COOH(s)$	−234.5
Ethyl acetate	$CH_3COOC_2H_5(l)$	−536.9
Cyclohexane	$(CH_2)_6(l)$	−937.8

standard enthalpy of formation or the *standard heat of formation*. A number of these values are listed in Table 5-2 and shown graphically in Fig. 5-5.

Example 5-2

Using the data in Table 5-2, calculate the enthalpy change in the reaction

$$3C_2H_2 \rightarrow C_6H_6$$
$$\text{acetylene} \qquad \text{benzene}$$

The reaction for the formation of acetylene under standard conditions at 25°C is

$$2C(gr) + H_2(g) = C_2H_2(g)$$
$$\Delta H_f° = +54.19 \text{ kcal mole}^{-1}$$

The reaction for the formation of benzene under standard conditions at 25°C is

$$6C(gr) + 3H_2(g) = C_6H_6(l)$$
$$\Delta H_f° = +19.82 \text{ kcal mole}^{-1}$$

Multiplying the first equation by 3 gives

$$6C(gr) + 3H_2(g) = 3C_2H_2(g)$$
$$\Delta H° = 162.57 \text{ kcal mole}^{-1}$$

Subtracting this from the equation directly above it gives

$$3C_2H_2(g) \rightarrow C_6H_6(l)$$
$$\Delta H° = -142.75 \text{ kcal mole}^{-1} \qquad Ans.$$

Example 5-3

Calculate the heat of formation of benzene, C_6H_6, from acetylene, C_2H_2, under standard conditions at 25°C from heats of combustion.

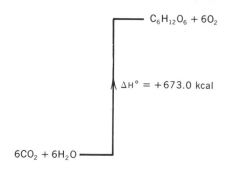

$C_6H_{12}O_6 + 6O_2$

$\Delta H° = +673.0$ kcal

$6CO_2 + 6H_2O$

Fig. 5-4   Energy change in photosynthesis.

The equation for the reaction is

$$3HC \equiv CH \rightarrow C_6H_6$$

Using the data in Table 5-1, the following thermochemical equations can be written:

$$3C_2H_2 + 7\tfrac{1}{2}O_2 \rightarrow 6CO_2 + 3H_2O$$
$$\Delta H° = -931.8 \text{ kcal mole}^{-1}$$

$$C_6H_6(l) + 7\tfrac{1}{2}O_2 \rightarrow 6CO_2 + 3H_2O$$
$$\Delta H° = -789.1 \text{ kcal mole}^{-1}$$

Subtracting the second equation from the first,

$$3C_2H_2 \rightarrow C_6H_6$$
$$\Delta H° = -142.7 \text{ kcal mole}^{-1} \qquad Ans.$$

TABLE 5-2   Standard heats of formation at 25°C[a]

Compound	Formula	$\Delta H_f°$, kcal mole^{-1}
Carbon monoxide	$CO(g)$	−26.416
Carbon dioxide	$CO_2(g)$	−94.052
Water (liquid)	$H_2O(l)$	−68.32
Water (gaseous)	$H_2O(g)$	−57.80
Ammonia	$NH_3(g)$	−11.04
Carbonyl sulfide	$COS(g)$	−32.8
Carbon disulfide	$CS_2(g)$	+27.55
Methane	$CH_4(g)$	−17.89
Chloroform	$CHCl_3(g)$	−24
Carbon tetrachloride	$CCl_4(g)$	−25.5
Methanol	$CH_3OH(g)$	−48.08
Methyl mercaptan	$CH_3SH(g)$	−3.7
Formaldehyde	$HCHO(g)$	−27.7
Formic acid	$HCOOH(g)$	−90.39
Ethanol	$C_2H_5OH(g)$	−56.625
Acetaldehyde	$CH_3CHO(g)$	−39.67
Acetic acid	$CH_3COOH(g)$	−103.8
Acetylene	$C_2H_2(g)$	+54.19
Ethylene	$C_2H_4(g)$	+12.50
Ethane	$C_2H_6(g)$	−20.24
Benzene	$C_6H_6(g)$	+19.82
Cyclohexene	$C_6H_{10}(g)$	−1.70
Cyclohexane	$C_6H_{12}(g)$	−29.43

[a] G. N. Lewis and M. Randall, "Thermodynamics," 2d ed., rev. by K. S. Pitzer and L. Brewer. Copyright 1961. McGraw-Hill Book Company. Used by permission.

5-5 CHANGE OF ENTHALPY WITH TEMPERATURE

It is frequently helpful to know the value of a heat of reaction at temperatures other than the standard temperature of 25°C. Since the heat, or enthalpy, of reaction is the difference between the enthalpies of the products and the enthalpies of the reactants, the change of the enthalpy of reaction with temperature can be calculated from the changes of the temperature of the reactants and products.

Consider a reaction like that for the formation of glucose

$$6CO_2 + 6H_2O \rightarrow C_6H_{12}O_6 + 6O_2$$
$$\Delta H^\circ = +673.0 \text{ kcal mole}^{-1} \quad (5\text{-}15)$$

As the system changes from carbon dioxide and water to glucose and oxygen, heat flows into the system and the enthalpy increases by 673.0 kcal for every mole of glucose formed. If heat flows into a system, not to change the chemical nature of the system but merely to raise the temperature, enthalpy also increases. For example, the heat capacity of gaseous CO_2 in the neighborhood of 25°C is 8.88 cal deg^{-1} mole^{-1}. This means that 8.88 cal will be absorbed by the mole of CO_2 when it is heated from 25 to 26°C. This, in turn, means that the enthalpy of the 1 mole of CO_2 has increased by 8.88 cal when it is heated through this 1° interval. The general form of this relationship may be expressed

$$H_{T_2} - H_{T_1} = c_p(T_2 - T_1) \quad (5\text{-}16)$$

when the interval through which the substance is heated is small enough for the heat capacity to remain effectively constant.

If the interval through which the substance is heated is so large that there is a variation in heat capacity, this relation must be in the form

$$H_{T_2} - H_{T_1} = \int_{T_1}^{T_2} c_p \, dT \quad (5\text{-}17)$$

To apply these equations to the calculation of the change with temperature of ΔH° for a reaction, consider the general reaction

$$\underset{\text{reactants}}{dD + eE} \rightarrow \underset{\text{products}}{fF + gG} \quad (5\text{-}18)$$

Fig. 5-5 Values of ΔH_f° at 25°C in the gaseous state. (Data from G. N. Lewis and N. Randall, "Thermodynamics," 2d ed., rev. by K. S. Pitzer and L. Brewer, McGraw-Hill Book Company, New York, 1961.)

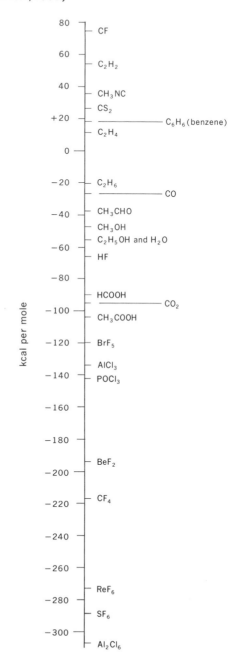

The heat capacity of the combined reactants will be

$$d(c_p^\circ)_D + e(c_p^\circ)_E = (c_p^\circ)_{reactants} \tag{5-19}$$

The heat capacity of the combined products will be

$$f(c_p^\circ)_F + g(c_p^\circ)_G = (c_p^\circ)_{products} \tag{5-20}$$

The symbol Δc_p° is used to denote the change in the heat capacity of the system under standard conditions brought about when the reaction proceeds to use up d moles of D and e moles of E and to form f moles of F and g moles of G

$$\Delta c_p^\circ = [f(C_p^\circ)_F + g(C_p^\circ)_G] - [d(C_p^\circ)_D + e(C_p^\circ)_E] \tag{5-21}$$

$$\underbrace{}_{\text{products}} \quad \underbrace{}_{\text{reactants}}$$

Similar symbolism is used to denote the corresponding change in the enthalpy of the system

$$\Delta H^\circ = (fH_F + gH_G) - (dH_D + eH_E) \tag{5-22}$$

$$\underbrace{}_{\text{products}} \quad \underbrace{}_{\text{reactants}}$$

and so Eq. (5-17) can be put into the form applicable to a reaction

$$\Delta H_{T_2}^\circ - \Delta H_{T_1}^\circ = \int_{T_1}^{T_2} \Delta c_p^\circ \, dT \tag{5-23}$$

For a number of substances c_p can be expressed as a function of temperature by an algebraic equation

$$c_p = a + bT + cT^{-2} \tag{5-24}$$

The range over which this equation is applicable varies from substance to substance and must be checked carefully with the original source of the data in order to determine the accuracy of the calculation. Within this limitation, the equation for Δc_p can be put in the form

$$\Delta c_p = [(a_F + a_G) - (a_D + a_E)]$$
$$+ [(b_F + b_G) - (b_D + b_E)]T$$
$$+ [(c_F + c_G) - (c_D + c_E)]T^{-2} \tag{5-25}$$

Then Eq. (5-23) may be put in the following form by inserting Eq. (5-25) and integrating:

$$\Delta H_{T_2}^\circ = \Delta H_{T_1}^\circ + \Delta a(T_2 - T_1)$$
$$+ \frac{1}{2}\Delta b(T_2^2 - T_2^2) - \Delta c\left(\frac{1}{T_2} - \frac{1}{T_1}\right) \tag{5-26}$$

Values of a, b, and c are listed in Table 5-3.

Example 5-4

Calculate the value of ΔH_f° for $H_2O(l)$ at 55°C.

Values of Δa, Δb, and Δc are calculated from the values in Table 5-3. Putting these values in Eq. (5-26) and letting $T_1 = 298.2°K$ and $T_2 = 328.2°K$ gives

$$\Delta H_{T_2} = -68{,}320 + 7.94(30°)$$
$$- \tfrac{1}{2}(0.00128)(328.2^2 - 298.2^2)$$
$$- 3 \times 10^4\left(\frac{1}{328.2} - \frac{1}{298.2}\right)$$
$$= -68{,}320 + 238 - (0.00064)$$
$$\times (107{,}715 - 88{,}923)$$
$$- 3 \times 10^4(0.003047 - 0.003353)$$
$$= -68{,}320 + 238 - 12 - 2$$
$$= -68{,}096 \text{ cal mole}^{-1} \qquad Ans.$$

5-6   BOND ENERGY

When a mixture of 1 mole of hydrogen and $\frac{1}{2}$ mole of oxygen at room temperature and 1-atm pressure is ignited, water is formed by a reaction which releases so much energy and takes place so rapidly that an explosion results. The thermochemical equation[1] is

$$At \ 25°C: \quad H_2 + \tfrac{1}{2}O_2 = H_2O(g) \quad \Delta H_f^\circ = -57.8 \text{ kcal} \tag{5-27}$$

This value of ΔH_f° is based on calorimetric measurements.

Nearly all this energy comes from the rearrangement of the chemical bonds. Schematically the reaction may be thought of as taking place in three steps:

$$Step \ 1: \quad H_2(g) \rightarrow 2H(g) \tag{5-28}$$
$$Step \ 2: \quad \tfrac{1}{2}O_2(g) \rightarrow O(g) \tag{5-29}$$
$$Step \ 3: \quad 2H(g) + O(g) \rightarrow H_2O(g) \tag{5-30}$$

In the first step 1 mole of hydrogen molecules is dissociated; the H—H bonds are broken. In the second step $\frac{1}{2}$ mole of oxygen molecules is dissociated; the O=O bonds are broken. In the third step the elementary atoms of H and O combine to form 1 mole of

---

[1] In order to simplify the discussion, the product is assumed to be water in the meta-stable gaseous state.

TABLE 5-3 Values of the heat-capacity-equation coefficients for elements and compounds in the standard state where c_p° is in calories per mole per degree[a]

$$c_p = a + bT + cT^{-2}$$

	a	$b, \times 10^{-3}$	$c, \times 10^5$
Gases:			
Monatomic	4.97	0	0
H_2	6.52	+0.78	+0.12
O_2	7.16	1.00	−0.40
N_2	6.83	0.90	−0.12
S_2	8.72	0.16	−0.90
CO	6.79	0.98	−0.11
F_2	8.26	0.60	−0.84
Cl_2	8.85	0.16	−0.68
Br_2	8.92	0.12	−0.30
I_2	8.94	0.14	−0.17
CO_2	10.57	2.10	−2.06
H_2O	7.30	2.46	0
NH_3	7.11	6.00	−0.37
CH_4	5.65	11.44	−0.46
Liquids:			
I_2	19.20	0	0
H_2O	18.04	0	0
NaCl	16.0	0	0
Solids:			
C (graphite)	4.03	1.14	−2.04
Al	4.94	2.96	
Cu	5.41	1.50	
Pb	5.29	2.80	+0.23
I_2	9.59	11.90	
NaCl	10.98	3.90	

[a] Values from G. N. Lewis and M. Randall, "Thermodynamics," 2d ed., rev. by K. S. Pitzer and L. Brewer. Copyright 1961. McGraw-Hill Book Company, New York.

water molecules; the O—H bonds are formed. The breaking of the H—H and the O=O bonds requires much less energy than the amount released when the O—H bonds are formed, so that the net result is a loss of energy from the system.

Generally speaking, in all chemical reactions the energy change is due primarily to bond rearrangement. For this reason, there is considerable interest in seeing whether energy values can be assigned to different types of bonds and how great the accuracy is with which such *bond energies* can be used to make calculations of heats of reaction.

As discussed in Sec. 20-7, observations of molecular spectra make possible the accurate calculation of the amount of energy necessary to break the bond in many diatomic molecules. For example, 103.22 kcal mole[−1] is needed to dissociate H_2. If this reaction were carried out at 0°K, this much energy would be required

At 0°K: $H_2 \rightarrow 2H$

$$\Delta E_0^\circ = \Delta H_0^\circ = 103.22 \text{ kcal mole}^{-1} \quad (5\text{-}31)$$

At 0°K the energy change is equal to the enthalpy change since $\Delta H_0^\circ = \Delta E_0^\circ + P\Delta v_0^\circ$ and at this temperature $\Delta v_0^\circ = 0$; the subscript 0 denotes at 0°K.

It is possible to calculate from thermal data the increase in ΔH° when the temperature is changed from 0 to 298.15°K (abbreviated in the subscript to 298) and it is found that

$$H_2 \rightarrow 2H \quad (\Delta H_{298}^\circ)_f = 104.2 \text{ kcal mole}^{-1} \quad (5\text{-}32)$$

This is the *heat of atomization* or more precisely the *enthalpy of atomization* of H_2 at 25°C. It is the value assigned to the "enthalpy" of the H—H bond.

In strict terminology, the energy of the H—H bond $E(H—H)$ is 103.7 kcal mole^{-1}, and the enthalpy of the H—H bond $H(H—H)$ is 104.2 kcal mole^{-1}, but in many texts the distinction is not made between bond energy and bond enthalpy since for many bonds the data are not sufficiently accurate to make the distinction significant. For 1 mole of an ideal gas at 1 atm and 25°C, the Pv term in the enthalpy $(E + Pv)$ has the value of only 0.5 kcal. For 1 mole of a liquid or a solid it is generally less than 10^{-3} kcal, so that in making rough calculations the difference between ΔH and ΔE is negligible for most reactions and the values obtained from the spectra of diatomic molecules for the dissociation energy can be used directly to calculate heats of reaction. The problem of obtaining values of the energy needed to break bonds in polyatomic molecules is somewhat more complicated. Sometimes spectroscopic and thermodynamic data can be combined to obtain bond energies. For example, the dissociation of water into hydrogen and oxygen atoms can be written in two steps:

At 25°C: $H_2O(g) \rightarrow H(g) + OH(g)$

$$\Delta H_{298}^\circ = 120 \text{ kcal} \quad (5\text{-}33)$$

At 25°C: $OH(g) \rightarrow O(g) + H(g)$

$$\Delta H_{298}^\circ = 101 \text{ kcal} \quad (5\text{-}34)$$

From this it appears that breaking an O—H bond in H_2O takes 19 percent more energy than dissociating O—H itself. Undoubtedly the dissociation energy of the O—H bond in other molecules will have still other

values. In view of this it is customary to assign the average of the two values of ΔH_{298}° given above as the value for the O—H bond in tables of bond energies.

Values of energy for bonds occurring in solids are calculated either from thermal data of a combination of thermal data, spectroscopic data, and electron-impact data from observations with a mass spectrometer. For example, values for bonds such as Si—Si can be obtained from the standard enthalpy of sublimation of silicon adjusted to 25°C.

One of the most difficult problems is the calculation of the enthalpy for the C—C bond. One of the most recent values[1] is

At 25°C: $C(gr) \rightarrow C(g)$ $\Delta H_{298}^\circ = 170.9 \text{ kcal mole}^{-1}$

$$(5\text{-}35)$$

The conclusion is that the sublimation of graphite is roughly the *equivalent* of breaking two C—C bonds and that the value of $\Delta H_{298}^\circ = 83$ kcal assigned for a single C—C bond is approximately one-half ΔH_{298}° for the sublimation of graphite.

A list of bond energies is given in Table 5-4. Where the accuracy of the calculation permits, these are true bond enthalpies.

Example 5-5

From the data in Table 5-4 calculate ΔH_{298}° for the reaction

This reaction involves breaking the H—H bond and the C=C bond and forming two C—H bonds and a C—C bond. The value of ΔH_{298}° is given by

$$\Delta H_{298}^\circ = [\underbrace{H(H—H) + H(C=C)}_{\text{bonds broken}}] - [\underbrace{2H(C—H) + H(C—C)}_{\text{bonds formed}}]$$

$$= \underbrace{(103 + 145)}_{\text{energy in}} - \underbrace{(2 \times 99 + 83)}_{\text{energy out}}$$

$$= -33 \text{ kcal mole}^{-1} \quad Ans.$$

[1] D. R. Stull (ed.), "(JANAF) Thermochemical Tables," Dow Chemical Co., Midland, Mich., 1964.

TABLE 5-4   Bond enthalpies ΔH_{298}°[a]

H—H	104.2	103.2[c]		F—F	36.6	
H—N	93.4	92.2[c]		Cl—Cl	58.0[d]	57.1[c]
H—O	110.6	109.4[c]		Br—Br	46.1	
H—F	134.6			I—I	36.1	
H—Cl	103.2	102.1[c]		Cl—F	60.6	
H—Br	87.5	86.7[c]		Cl—Br	52.3	52.2[b]
H—I	71.4			Cl—I	50.3	
H—C	98.8	98.2[c]		Br—I	42.5	42.4[b]
H—S	81.1			Ge—Ge	37.6	
H—Si	70.4			Ge—Cl	97.5	
H—P	76.4			Sn—Sn	34.2	
H—As	81.1			P—P	51.3	
H—Se	66.1			As—As	32.1	
H—Te	57.5			Sb—Sb	30.2	
C—C	83.1	80.5[c]		Bi—Bi	25	
C—O	84.0	79[c]		Se—Se	44.0	
C—N	69.7			Te—Te	33	
C—F	105.4			Si—Si	42.2	
C—Cl	78.5			Si—O	88.2	
C—Br	65.9			Si—S	54.2	
C—I	57.4			Si—F	129.3	
C—Si	69.3			Si—Cl	85.7	
C=C	145[c]	148.4[b]		Si—Br	69.1	
C=O	173[c]			Si—I	50.9	
C≡C	198[c]			P—Cl	79.1	
N—N	38.4	37[c]		P—Br	65.4	
N≡N	226[d]			P—I	51.4	
N—F	64.5			As—F	111.3	
N—Cl	47.7			As—Cl	68.9	
O—O	33.2	34[c]		As—Br	56.5	
O=O	118[d]			As—I	41.6	
O—F	44.2			S—Cl	59.7	
O—Cl	48.5			S—Br	50.7	
S—S	50.9					

[a] The values without footnotes are from Linus Pauling, "The Nature of the Chemical Bond," 3d ed., p. 85, Cornell University Press, Ithaca, N.Y., 1960.
[b] From R. Daudel, R. Lefebvre, and C. Moser, "Quantum Chemistry," p. 169, Interscience Publishers, Inc., New York, 1959.
[c] From Kenneth S. Pitzer, "Quantum Chemistry," p. 170. © 1953. By permission of Prentice-Hall, Inc., Englewood Cliffs, New Jersey.
[d] From L. E. Strong and W. J. Stratton, "Chemical Energy," p. 48, Reinhold Publishing Corporation, New York, 1965.

Example 5-6

From the data in Table 5-2 for enthalpies of formation calculate the value of ΔH_{298}° for the reaction in Example 5-5.

$$\Delta H_{298}^{\circ} = -\Delta H_{298}^{\circ}(C_2H_4) - \Delta H_{298}^{\circ}(H_2)$$
$$+ \Delta H_{298}^{\circ}(C_2H_6)$$
$$= -(+12.5) - 0 + (-17.9)$$
$$= -30.4 \text{ kcal mole}^{-1} \quad Ans.$$

Note the close agreement with the answer in Example 5-5.

Example 5-7

Calculate ΔH_f° for CH_3OH from bond enthalpies.

Enthalpy in:

$$
\begin{array}{llll}
2(C\!-\!C)\dagger & 2 \times 83 & = & 166 \\
2(H\!-\!H) & 2 \times 104 & = & 208 \\
\tfrac{1}{2}O_2 & \tfrac{1}{2} \times 118 & = & \underline{59} \\
& \text{Total in} & \overline{433} & \text{kcal mole}^{-1}
\end{array}
$$

$$
\begin{array}{lllll}
\textit{Enthalpy out:} & 3(C\!-\!H) & 3 \times 99 & = & 297 \\
& 1(C\!-\!O) & 1 \times 79 & = & 79 \\
& 1(O\!-\!H) & 1 \times 111 & = & \underline{111} \\
& & \text{Total out} & & 487 \text{ kcal mole}^{-1} \\
\end{array}
$$

$$
\Delta H_f^{\circ}\,(25°C) = \underset{\text{enthalpy in}}{433} - \underset{\text{enthalpy out}}{487} = -54 \text{ kcal mole}^{-1}
$$

Note: The thermodynamically calculated value of ΔH_f° (25°C) is -49 kcal mole^{-1}.

5-7   THERMOCHEMISTRY AND BIOCHEMISTRY

The most important structural unit in living matter is the biological cell. The simplest kind of living matter consists of unicellular organisms, in which each cell acts primarily as an independent unit and the interchange of energy between a cell and its surroundings is secondary to the flow of energy within each cell. Contrasted with this simpler type of life there are multicellular organisms, where the exchange of energy between the different cells is a dominant factor in the life process of the total organism.

At the molecular level, the biochemical reactions are principally of two kinds, *synthesis* and *degradation*. As a broad generalization, the processes of degradation supply energy to the processes of synthesis. The degradation processes also supply energy for both inter- and intracellular communication. In a complex organism like the human body, the maintenance of the total life process depends on a constant stream of messages exchanged between different cells and within cells.

† The equivalent of $2E(C\!-\!C)$ is needed to get one free carbon atom from graphite.

The transmission of these messages requires energy. Besides direct transmission by electrochemical means, as in the network of the nervous system, there is a constant flow of molecules of different kinds throughout the body, partly for the purpose of carrying chemical messages to direct chemical reactions and partly for the purpose of providing the reactants necessary to maintain life and growth. In addition to all these forms of activity that require energy, there are the somewhat grosser needs for energy to make possible the movement of muscles and to maintain the body at relatively constant temperature. Except in the warmest of tropical climates, the body is constantly losing energy to the cooler surroundings. Since nearly all kinds of chemical actions vary in rate with temperature, body temperature must be maintained within a relatively narrow range if the trillions of chemical reactions taking place in the body are to go forward in an orderly and integrated fashion.

From this brief survey the importance of the role of energy in biochemistry is clear; its study is based primarily on physical chemistry.

Just as calorimeters are used to measure the change associated with simple chemical reactions like the combustion process where carbon combines with oxygen to produce carbon dioxide, biocalorimeters have been designed to measure energy changes in far more complex biochemical reactions. Usually the measurement of the total energy output of the body as heat is coupled with a measurement of oxygen consumption, of the carbon dioxide produced, and of the nitrogen excretion.

The average human being uses daily about 50 kcal per kilogram weight up to the age of three years. The rate grows less as age increases, leveling off to 20 kcal day^{-1} kg^{-1} at the age of forty years for those in sedentary occupations. During a day of hard muscular work a man may consume energy at the rate of over 50 kcal kg^{-1}. Rats sometimes use energy 4 times as fast as human beings per kilogram of weight, and elephants need only about one-fifth as much as human beings per kilogram of weight.

Plants can absorb energy directly from sunlight and use it to bring about synthetic reactions which result in the storage of photoenergy as chemical energy in the molecules serving as food. In animals the life process is maintained largely by taking in these food substances with a relatively high content of energy; the energy is then distributed through a myriad of reactions in the life process and ultimately leaves the body, largely in the form of heat. As an example, the gross energy change in plants can be expressed

$$6CO_2 + 6H_2O \rightarrow C_6H_{12}O_6 + 6O_2 \qquad \Delta H = +673 \text{ kcal}$$
$$\text{glucose}$$
$$(5\text{-}36)$$

The gross energy change in animals can be expressed

$$C_6H_{12}O_6 + 6O_2 \rightarrow 6CO_2 + 6H_2O \qquad \Delta H = -673 \text{ kcal}$$
$$\text{glucose}$$
$$(5\text{-}37)$$

Thus, the photosynthesis of glucose per mole produces six oxygen molecules and stores within the plant 673 kcal of energy, while the oxidation of glucose per mole in the animal yields 673 kcal, which leaves the body ultimately in the form of heat. However, instead of this *direct* burning, or oxidation, of glucose to form carbon dioxide and water, the actual pattern of the flow of energy in animals is far more complex; the degradation processes take place in a series of many small steps providing the driving force that keeps going all the reactions necessary to maintain the functioning of the total life process.

Not only must the energy changes be considered jointly with respect to the changes in the linkages of atoms, but the *efficiency* of biochemical processes must also be taken into account. These relationships will be explored further in Chaps. 7 to 9 on thermodynamics.

Problems

1. Using the data in Table 4-8, calculate the value in calories of the PV factor in the enthalpy of water at $0°C$ and 1 atm when water is in the form of ice and of liquid; repeat for vapor, assuming ideal-gas behavior.

2. Calculate the change in enthalpy at $25°C$ when (*a*) 1 mole of propane combines with 1 mole of hydrogen to form 1 mole of methane and 1 mole of ethane all at 1 atm in the gaseous state; (*b*) when 1 mole of *n*-pentane combines with 1 mole of hydrogen to form 1 mole of *n*-butane and 1 mole of methane.

3. From the heats of combustion in Table 5-1, calculate the standard enthalpy of formation of 1 mole of benzene from its elements at $25°C$.

4. Calculate the standard heat of formation of 1 mole of $(CH_2)_6$ at $25°C$ from 3 moles of ethylene from the data in Table 5-1.

5. From the data in Tables 5-1 and 5-2, calculate the heat of formation of alanine from its elements under standard conditions at $25°C$.

6. From Table 5-1 calculate the enthalpy change under standard conditions at $25°C$ for the reaction:

$$C_2H_5OH + CH_3COOH = CH_3COOC_2H_5 + H_2O$$

All substances are in the liquid state; neglect heats of solution.

7. Calculate the standard heat of formation of liquid H_2O at $37°C$ using the data in Tables 5-2 and 5-3.

8. Calculate the standard heat of formation of NH_3 at $37°C$ using the data in Tables 5-2 and 5-3.

9. Calculate the standard heat of formation of CO_2 at $37°C$ using the data in Table 5-2.

10. Calculate the enthalpy change in the reaction

$$CO + \tfrac{1}{2}O_2 = CO_2$$

under standard conditions at $25°C$ using the data in Table 5-2.

11. From the data in Table 5-4 calculate the standard enthalpy of formation of ethane at $25°C$ and compare with the value given in Table 5-2.

12. From bond enthalpies calculate the amount of photoenergy necessary to synthesize glucose and compare with the value given in Eq. (5-36). The structural formula for glucose is:

```
    H—C=O
      |
    H—C—O—H
      |
H—O—C—H
      |
    H—C—O—H
      |
    H—C—O—H
      |
    H—C—O—H
      |
      H
```

13. From the data in Table 5-4 calculate the enthalpy change for the reaction

$$(CH_3)_2CHCH(NH_2)COOH + CH_2(NH_2)COOH =$$
$$(CH_3)_2CHCH(NH_2)(CO)NHCH_2COOH + H_2$$

14. If glucose were the sole source of energy for human life, calculate the grams of glucose necessary for a 220-lb man engaged in hard physical labor for a day.

15. Calculate the volume of CO_2 under standard conditions at 37°C produced per day by the man described in Prob. 14.

16. Calculate the glucose necessary to sustain a 2-ton elephant engaged in hard labor for a day.

17. How many grams of butane must be burned to heat 10 liters of water from 25 to 100°C? (Average c_p for H_2O is 18 cal deg^{-1} mole^{-1}.)

18. How many grams of butane must be burned to convert 10 liters of liquid water into steam at 1 atm at 100°C?

19. If a human body were powered by the combustion of butane, how many grams per day would be burned by a 220-lb man in performing hard muscular labor?

20. How many liters of ethyl alcohol at 20°C would be required to provide the energy necessary to sustain a 2-ton elephant during a working day at hard labor?

REFERENCES

Gaydon, A. G.: "Dissociation Energies," Dover Publications, Inc., New York, 1950.

Rossini, F. A., et al. (eds.): "Tables of Selected Chemical Thermodynamic Properties," *Natl. Bur. Std. U.S. Circ.* 500, 1952.

Cottrell, T. L.: "The Strengths of Chemical Bonds," Butterworth & Co. (Publishers), Ltd., London, 1958.

Pauling, L.: "The Nature of the Chemical Bond," 3d ed., pp. 79–88, Cornell University Press, Ithaca, N.Y., 1960.

Lewis, G. N., and M. Randall: "Thermodynamics," 2d ed., rev. by K. S. Pitzer and L. Brewer, pp. 53–74, McGraw-Hill Book Company, New York, 1961.

Moelwyn-Hughes, E. A.: "Physical Chemistry," 2d ed., pp. 1035–1044, Pergamon Press, New York, 1961.

Stull, D. R.: "Joint Army-Navy-Air Force (JANAF) Thermochemical Tables," Dow Chemical Co., Midland, Mich., 1964.

Kerr, J. A., and A. F. Trotman-Dickenson: Strengths of Chemical Bonds, "Handbook of Chemistry and Physics," 45th ed., p. F-94, Chemical Rubber Co., Cleveland, Ohio, 1964.

Strong, L. E., and W. J. Stratton: "Chemical Energy," Reinhold Publishing Corporation, New York, 1965.

Benson, S. W.: "Bond Energies," *J. Chem. Educ.,* **42:** 502 (1965).

6-1 THE STATISTICAL THEORY OF THERMAL ENERGY

In the study of the kinetic theory of gases in Chap. 3, attention was called to the wide variety of velocities with which gaseous molecules move. For example, in oxygen gas at room temperature there are a few molecules with velocities greater than 10,000 m sec^{-1}, there are a few with velocities less than 1 cm sec^{-1}, and many have velocities in between, with the majority in the vicinity of the average velocity, about 500 m sec^{-1}. Since the kinetic energy per molecule e_k is

$$e_k = \tfrac{1}{2}mu^2 \tag{6-1}$$

Chapter Six

thermal energy

where m is the mass of the molecule and u is the velocity with which it is moving, it is clear that the energy possessed by individual molecules covers a wide range of values, just like the velocities.

The average kinetic energy E_k of 1 mole of gas is given by

$$E_k = \tfrac{1}{2}M\overline{u^2} = \tfrac{3}{2}RT \tag{6-2}$$

where M = molecular weight of gas
$\overline{u^2}$ = average squared velocity
R = gas constant
T = temperature, °K

Thus the *average* energy in 1 mole of gas can be related to a number of the other properties of the gas such as temperature, pressure, and density. There are, however, many aspects of molecular behavior both in the gaseous state and in the solid and liquid states which depend not on the *average* velocity of many molecules but on the individual velocity of single molecules. It is helpful therefore to know exactly how many molecules have small amounts of energy, medium amounts of energy, and large amounts of energy; i.e., it is helpful to have a formula showing the statistical distribution of energy among molecules. Of course, keep in mind that a molecule may have a small amount of energy at one moment and then, a fraction of a second later, after it has collided with several other molecules, have a relatively large amount of energy; then, after a few more collisions, its individual value of energy may lie close to the average energy for all the molecules. Thus, not only does the energy vary from molecule to molecule; it also varies for a single molecule with the passage of time.

This constant fluctuation of energy distribution is particularly significant in the chemical processes of biological growth, in which molecules are constantly passing from the fluid state, where they move quite randomly much like gaseous molecules, into more ordered states, where they may be adsorbed on a surface like that of an enzyme or permanently grouped in a crystalline array like that found in bone. In the liquid state the molecule moves back and forth and turns and twists with varying amounts of kinetic energy; when the molecule attaches itself to a surface, the motion is far more restricted than in the liquid state and the energy is reduced.

6-2   THE HARMONIC OSCILLATOR

In order to have a simple example to show the effect of the statistical distribution of thermal energy in solids and liquids in such processes, consider first an idealized situation where a single atom such as hydrogen is momentarily bound by a chemical force to a surface like that of an enzyme. Such an atom vibrates approximately like a *harmonic oscillator*.

Figure 6-1 shows a hydrogen atom attached to an enzyme surface. Actually, as depicted in Fig. 6-1a, the forces holding this atom to the surface are of a complex nature. Part of the force is due to the overlapping of the electron orbital of the hydrogen atom and the electron orbitals of the atoms forming the surface, which will be discussed in Chap. 19. There may be an electrical attraction between the hydrogen atom and the surface if the atom has a net positive charge and the surface a net negative charge. The neighboring atoms in the surrounding fluid also exert forces on the hydrogen atom to a lesser degree. However, the net result of all these forces usually resembles the force produced by a spiral spring, as shown in Fig. 6-1b. Thus, for small displacements of the hydrogen atom toward and away from the surface, the force between the hydrogen atom and the surface varies in a manner expressed by Hooke's law

$$F = -k_h x \qquad (6\text{-}3)$$

where $F =$ force exerted on atom,
  $x =$ displacement of atom from position where $F = 0$
  $k_h =$ constant relating displacement to the force it produces

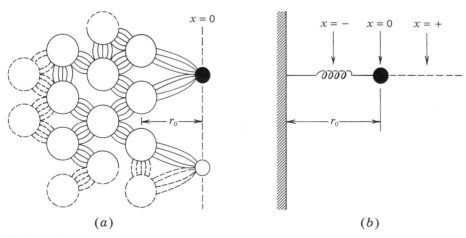

(a) (b)

Fig. 6-1   Hydrogen atom (solid circle) on the surface of an enzyme as an oscillator: (a) **attached by orbital force;** (b) **as idealized harmonic oscillator.**

F is regarded as positive if it pushes the atom away from the surface and negative if it pulls the atom toward the surface. In this discussion, the only forces considered are those produced when the atom moves along a line that is *normal,* or *perpendicular,* to the surface.

If the atom were held by a spiral spring, as shown in Fig. 6-1*b*, and if no other external force were acting on it, there would be an equilibrium position for the atom at a distance r_0 from the enzyme; i.e., when the spring is neither stretched nor compressed but has its normal, unstressed length, no force acts on the atom. When the atom is moved to the right, away from the enzyme, the spring is stretched and the force of the spring tends to pull the atom back to the equilibrium position. When the atom is pushed closer to the enzyme, the spring is compressed and the force of the spring tends to push the atom away from the enzyme.

Take the line through the atom and the center of the spring perpendicular to the enzyme surface as the x axis and designate as $x = 0$ the point where the atom is at rest when the spring is neither stretched nor compressed (Fig. 6-2). Since Eq. (6-3) is the equation of a straight line, the force varies linearly with the displacement; moving the atom further away from the enzyme gives a positive value of x and consequently a negative value of F, or a force pulling the atom back; moving the atom toward the enzyme gives a negative value of x and consequently a force of repulsion, i.e., a positive force, which tends to move the atom back away from the enzyme.

There is a variety of experimental evidence to show that the actual force in many kinds of chemical bonding closely resembles that of a spiral spring. The observations of spectra confirm this for the force of covalent chemical bonds (Chap. 20), and there is every reason to believe that the force acting on an adsorbed molecule has a similar character when the atom is moved out of its equilibrium position toward or away from the surface through a distance of 0.2 Å or less. However, if the atom is moved further and further away, displaced by an amount of 1 Å or more, the force does not continue to increase but goes through a maximum negative value and falls back to zero when the atom is sufficiently far from the enzyme surface. The displacement necessary to reduce the force to zero depends on

Fig. 6-2 Force as a function of displacement from equilibrium position r_0 where $X = 0$, for an oscillator where the force varies according to Hooke's law.

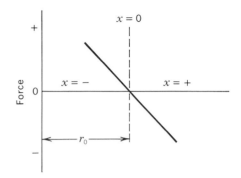

Fig. 6-3 Force as a function of displacement from equilibrium position r_0 where $X = 0$, for an oscillator where the force is caused by orbital binding. Shaded area is the energy of dissociation.

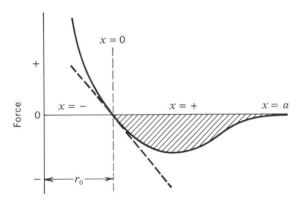

the nature of the bonding of the atom to the surface, i.e., chemical bonding, hydrogen bonding, coulombic attraction, van der Waals' attraction, or a combination of forces. Usually, when removed to 10 Å, the atom is effectively free from surface forces. The present discussion is concerned with only small displacements, where Hooke's law is obeyed.

For displacements sufficiently small, the experimental curve shown in Fig. 6-3 as a solid line may be treated theoretically as the equivalent of the dotted straight line tangent to the curve at $x = 0$. A mass particle vibrating under the influence of a restoring force that obeys Hooke's law is called a *harmonic oscillator.*

6-3 ENERGY LEVELS

When a hydrogen atom is held by a springlike force on the surface of an enzyme in a biological cell at body temperature, the atom vibrates toward and away from the surface. The energy associated with this motion is of two kinds. The *kinetic* energy e_k is given by

$$e_k = \tfrac{1}{2}mu^2 \qquad (6\text{-}4)$$

The *potential* energy, when the atom is at position x, is usually defined as the work done against the force in moving the atom to position x from a position away from the surface x_a, where $F = 0$:

$$e_p = \int_0^x F\,dx \qquad (6\text{-}5)$$

At this point in the discussion there is no need to consider the complicated mathematical equation required to express F as a function of x. Graphically, the integral in Eq. (6-5) is equal to the shaded area between the x axis and the curve representing F as a function of x, as shown in Fig. 6-3. When the atom moves from the equilibrium position x_0 to the position x_a, work has to be done on the system to pull the atom away from the surface against the force tending to hold it there. This means that the potential energy of the atom increases when it is pulled away from the surface. According to the same reasoning, when the atom moves from position x_a to position x_0, it is moving in the direction toward which the force is pulling it and consequently potential energy decreases. If the potential energy at position x_a is taken equal to zero by definition, the potential energy at position x_0 must therefore be less than the potential energy at x_a; in other words, the potential energy at the equilibrium position x_0 has a negative value. In Fig. 6-4a the graph represents the potential energy of the atom. It is zero when the atom is sufficiently removed from the surface; it has the lowest value when the atom is at the equilibrium position ($x = 0$); it again becomes higher as the atom moves in closer to the surface.

An atom of hydrogen adsorbed on an enzyme surface at body temperature has vibrational energy. As a result, the atom moves back and forth, compressing the chemical bond when it moves closer to the surface and stretching it when it moves farther away. As the atom oscillates between the maximum of compression and the maximum of extension of the bond, the energy

changes from potential to kinetic energy and vice versa. When the atom moves toward the surface and finally is at the point of maximum compression, its motion ceases momentarily when it stops moving to the left and then starts moving back to the right. When the motion ceases, the atom has no kinetic energy; all the energy is in the form of potential energy. As the atom moves more and more toward the right, it passes the equilibrium point where $x = 0$; at this point, there is a minimum of potential energy. Then, as the atom moves further to the right to the point at which the bond has the maximum stretching, the motion of the atom again momentarily ceases and all the energy is again in the form of potential energy.

From the point of view of mechanics, this motion is the equivalent of a ball rolling back and forth in a concave dish, sometimes called a *well*, as shown at the bottom of Fig. 6-4. As the ball rolls to the left, it climbs up the left side of the dish until finally the motion ceases. At this point the ball reaches a maximum altitude and its energy consists wholly of potential energy. The ball then rolls back down toward the bottom of the dish. When it reaches the bottom, its potential energy is at a minimum. Then as it rolls up the right side of the dish, it loses kinetic energy and gains potential energy until, at the point where it stops at maximum altitude, the energy is again wholly potential energy.

Whether the system consists of a ball attached by a spring to a wall or a ball rolling back and forth in a dish, the total amount of energy remains constant if there is no mechanism by which the system can gain or lose energy from the surroundings. In such a system the amount of energy is a function of the amplitude of vibration. In order to have the behavior of the system as simple as possible, we postulate that in its variation with distance the force obeys Hooke's law. This is approximately true for the bottom portion of the dish (Fig. 6-4a). A Hooke's law parabola is shown in Fig. 6-4b. From the point of view of classical mechanics, it is possible for the system to have energy ranging all the way from the minimum value, if there is no vibration at all and the ball remains stationary at the bottom of the well, to any of the continuous series of larger values ranging upward as high as the height of the walls permit. Contrasted with this situation, where there is a continuous range of possible energy values for the os-

Fig. 6-4 Quantized energy levels in (*a*) oscillator with orbital force; (*b*) oscillator with force varying according to Hooke's law; (*c*) potential well.

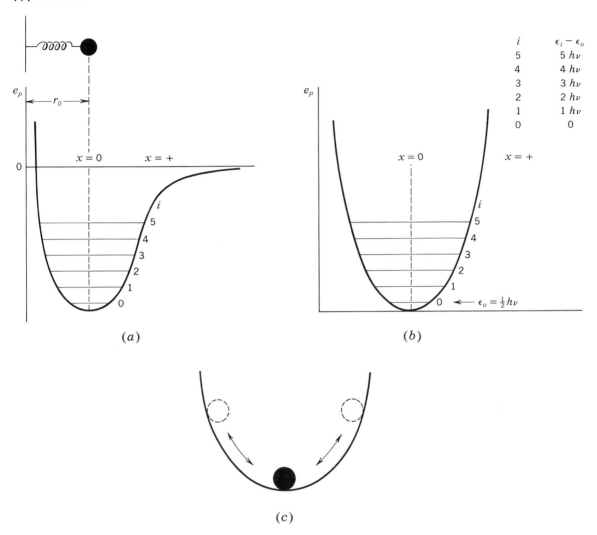

i	$\epsilon_i - \epsilon_o$
5	$5\,h\nu$
4	$4\,h\nu$
3	$3\,h\nu$
2	$2\,h\nu$
1	$1\,h\nu$
0	0

(*a*)

(*b*)

(*c*)

cillator, a hydrogen atom bound by a springlike force to an enzyme surface can take on only certain *restricted* values of the energy; it is a *quantized* oscillator.

Albert Einstein was the first to call attention to the evidence (from the observed heat capacities of substances) supporting the conclusion that only certain values of the energy are permitted for such an atomic oscillator and that in consequence the amplitude with which the oscillator vibrates is restricted to certain discrete values which bear a simple relation to the fre-

quency of vibration. The frequency of vibration is defined as the number of times per second the oscillator executes the total cycle of oscillation passing from the central equilibrium point to the maximum displacement at the left, then back again through the equilibrium point to the maximum displacement to the right and, finally, returning to the equilibrium point. Frequency has the dimension of *hertz* (Hz) (cycles per second), generally shortened to the form sec^{-1}. A small Greek nu (ν) is used to denote frequency.

For a harmonic oscillator the frequency remains constant and is independent of the amplitude of vibration. The detailed nature of quantization will be discussed in Chap. 19. At this point, only the rules for finding these *quantized* values of the energy will be given.

As will be shown later, the quantum theory leads to the conclusion that an atom behaving like a harmonic oscillator never can be in a state of zero energy. Even at the absolute zero of temperature, the atom has a *zero-point energy* equal to $\frac{1}{2}h\nu$. In this expression h is a universal constant called *Planck's constant* in honor of the German physicist Max Planck who first pointed out its significance in connection with his quantum theory of radiation; h has the value of 6.6256×10^{-27} erg sec. When this constant is multiplied by the value of the frequency ν with the unit sec^{-1}, the product has the dimensions of ergs, or energy. When the oscillator has the minimum amount of energy, it is said to be in the *zero energy level,* shown as the horizontal line marked 0 in Fig. 6-4. In this state it oscillates back and forth between the position on the left and the position on the right at the ends of the line marked zero. When the oscillator moves into states of greater energy by vibrating at greater amplitude, it can increase its energy only by multiples of $h\nu$. The amount of energy $h\nu$ or any integral multiple thereof is called a *quantum* of energy. When an oscillator in the zero energy level absorbs one such quantum of energy, it moves into the first energy level, which lies $h\nu$ ergs above the zero energy level. The second energy level is at a distance $h\nu$ ergs above the first energy level. Thus the formula relating the energy of the oscillator to its frequency and the number i of the energy level is

$$\epsilon_i = \tfrac{1}{2}h\nu + ih\nu = (i + \tfrac{1}{2})h\nu \qquad (6\text{-}6)$$

Following Planck's epoch-making hypothesis of the quantized nature of light emission and absorption, Einstein pointed out the evidence for believing that light consists of energy which does not flow as a continuous stream or wave but behaves in many respects as if it consisted of energy packed in tiny bundles called *light quanta* or *photons*. The energy in each photon ϵ_q is given by

$$\epsilon_q = h\nu_q \qquad (6\text{-}7)$$

where ν_q is the frequency of vibration of the light and the subscript q denotes *quantum*. Suppose that an oscillator absorbs a photon of light so that the energy is raised from the zero level to the first level; then the energy of the photon is related to the energy increase in the oscillator by the expression

$$h\nu_q = \epsilon_q = \epsilon_1 - \epsilon_0 = (1 + \tfrac{1}{2})h\nu_a - \tfrac{1}{2}h\nu_a = h\nu_a \qquad (6\text{-}8)$$

where ϵ_1 = energy of atom in first energy level
ϵ_0 = energy in zero energy level
ν_a = frequency of vibration of atom

From this it is clear that the frequency of the light absorbed ν_q is equal to the frequency of vibration of the atom ν_a when the vibrator moves from energy level i to energy level $i + 1$. The same relation holds for light emission.

The principles of classical mechanics lead to the conclusion that the frequency with which a true harmonic oscillator vibrates is a constant independent of the amount of energy in the oscillator. This frequency is related to the Hooke's law constant k_h and the mass of the oscillator by the expression

$$\nu = \frac{1}{2\pi} \sqrt{\frac{k_h}{m}} \qquad (6\text{-}9)$$

for an oscillator attached to a fixed object, like a ball attached by a spiral spring to an immovable wall. When a hydrogen atom is attached to an enzyme, there is a slight motion of the enzyme mass as the hydrogen atom vibrates and Eq. (6-9) will not apply with complete accuracy; however, only a small error is made by assuming that the enzyme is immovable. When this assumption is made, m is the mass of a single hydrogen atom and k_h is the Hooke's law constant for the net result of all the forces acting on this adsorbed atom. As pointed out earlier, when the atom vibrates with sufficient amplitude, the force no longer varies strictly as a linear function of the displacement from the equilibrium position. Under these circumstances, the frequency of vibration in the higher energy levels differs from the frequency of vibration in the zero energy level, but this variation is not significant for the discussion at this point.

Example 6-1

A hydrogen atom is attached to the surface of an enzyme by bonding which has an elastic-force constant $k_h = 5.00 \times 10^3$ dynes cm^{-1}. Calculate the energy of vibration in the zero and the first energy level in calories per mole.

The frequency of vibration is given by Eq. (6-9). The mass of the hydrogen atom m is calculated by dividing the atomic weight M by Avogadro's number N_A:

$$m = \frac{M}{N_A} = \frac{1.008}{6.02 \times 10^{23}} = 1.674 \times 10^{-24} \text{ g}$$

$$\nu = \frac{1}{2\pi} \sqrt{\frac{k}{m}}$$

$$= \frac{1}{2 \times 3.1416} \sqrt{\frac{2.00 \times 10^5}{1.674 \times 10^{-24}}}$$

$$= 0.15915 \sqrt{11.95 \times 10^{28}}$$

$$= 0.15915 \times 3.465 \times 10^{14}$$

$$= 5.51 \times 10^{13} \text{ sec}^{-1}$$

$$\epsilon_0 = \tfrac{1}{2}h\nu = \tfrac{1}{2}(6.625) \times 10^{-27} \text{ erg sec}$$
$$\times 5.51 \times 10^{13} \text{ sec}^{-1}$$

$$= 1.82 \times 10^{-14} \text{ erg}$$

$$\epsilon_1 = 1.5h\nu = 8.65 \times 10^{-13} \text{ erg}$$

$$E_0 = 1.82 \times 10^{-14} \frac{\text{erg}}{\text{molecule}} \times 1.439$$

$$\times 10^{16} \frac{\text{cal mole}^{-1}}{\text{erg molecule}^{-1}} = 0.262 \frac{\text{kcal}}{\text{mole}}$$

$$E_1 = 2.74 \times 10^{-13} \frac{\text{erg}}{\text{molecule}} \times 1.439$$

$$\times 10^{16} \frac{\text{cal mole}^{-1}}{\text{erg molecule}^{-1}} = 0.393 \frac{\text{kcal}}{\text{mole}}$$

6-4 COMPLEXIONS AND STATES

If we could observe the vibratory motion of a hydrogen atom attached to the surface of an enzyme in a cell over a period of time, we would find the amount of the amplitude and consequently the energy of vibration frequently changing. At one moment the atom might be in the zero energy level, with the minimum of energy $\tfrac{1}{2}h\nu$; a moment later it might absorb $2h\nu$ when struck by a neighboring atom and then have $\tfrac{5}{2}h\nu$, the energy corresponding to the second energy level; a moment later, it might give up some of its energy to the water molecules in the surrounding cytoplasm and be back in

the first energy level, where $i = 1$ and the energy the atom has in its vibration is $\tfrac{3}{2}h\nu$.

An adsorbed atom can also receive energy by absorbing light. If a photon of light with energy $h\nu$ passes near the atom, it is possible for the atom to absorb the energy and move up to a higher energy level. In the same way, the atom can emit the photon of light and move down to a lower energy level. Later, in Chaps. 16 to 19 dealing with quantum mechanics, the laws which govern the absorption and emission of radiant energy will be discussed.

Now let us consider the limited artificial situation of three atoms of hydrogen, designated as atoms a, b, and c, attached side by side to the surface of an enzyme; these atoms can exchange energy with each other but not with the enzyme or with any other part of the surroundings. All the atoms are vibrating with the same frequency ν. In addition to the zero-point energy $\tfrac{1}{2}h\nu$ in each atom, there are three quanta of energy, each of amount $h\nu$, which can be shared in any way among these three atoms as long as each quantum remains intact; i.e., the quanta of energy cannot be split. Thus, there are several different patterns of sharing, shown graphically in Fig. 6-5, where the atoms are placed in different positions on the energy-level ladder.

The simplest situation is the one where, in addition to the zero-point energy, each atom has a single quantum of energy; i.e., atoms a, b, and c are all in the first energy level. Since there are only three quanta and each atom has one, there is only *one* way of making such a distribution of the energy, the type I distribution.

Now go to the other extreme and consider the situation where one atom has all three quanta and the other two have none. Atom a can be in the third energy level with atom b and atom c in the zero level. Atom b can be in the third energy level with atoms a and c in the zero level; or atom c can be in the third level and atoms a and b in the zero level. There are thus *three* ways of achieving this type II distribution.

Given three atoms and three quanta of energy, there is only one other type of distribution possible, that shown as type III in Fig. 6-5. Here atom a has two quanta, atom b has one quantum, and atom c has none; or atom a can have two quanta, atom b can have no energy, and atom c can have one quantum of energy; or there may be other permutations of the ar-

Fig. 6-5   The probability of states and complexions.

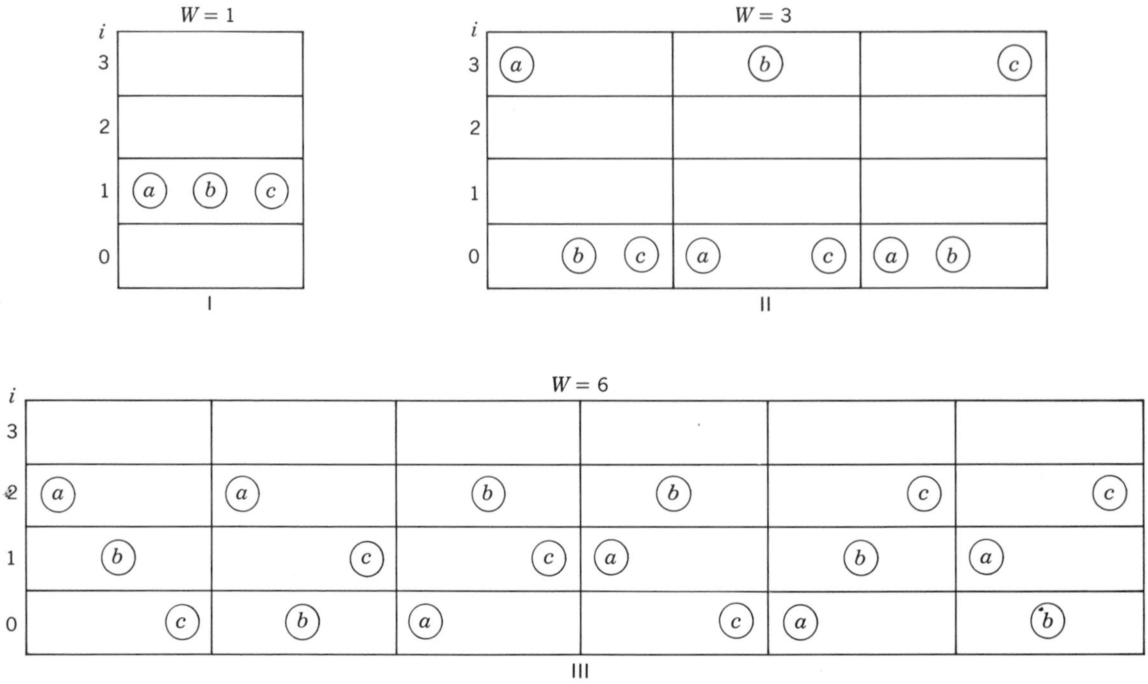

$W = 1$ (I): level 1 contains a, b, c.

$W = 3$ (II): level 3 contains a, b, c in separate arrangements; level 0 contains the other two atoms.

$W = 6$ (III): one atom in level 2, one in level 1, one in level 0, shown in six arrangements.

rangement. The diagram shows that there are *six* possibilities for this type III distribution.

The number of ways W by which any type of distribution can be formed is given by a relatively simple equation:

$$W = \frac{n!}{n_0! n_1! n_2! n_3! \cdots} \tag{6-10}$$

In this equation n is the total number of atoms, n_0 is the number of atoms in the zero energy level, n_1 is the number in the first level, and so on. The exclamation points indicate factorial numbers; thus, $0!=1$; $1!=1$; $2! = 1 \times 2$; $3! = 1 \times 2 \times 3$, etc.

To illustrate the use of Eq. (6-10), we apply it to the three different types of distribution:

$$W_I = \frac{3!}{0!3!0!0!} = \frac{1 \times 2 \times 3}{(1)(1 \times 2 \times 3)(1)(1)} = 1$$

$$W_{II} = \frac{3!}{2!0!0!1!} = \frac{1 \times 2 \times 3}{(1 \times 2)(1)(1)(1)} = 3 \tag{6-11}$$

$$W_{III} = \frac{3!}{1!1!1!0!} = \frac{1 \times 2 \times 3}{(1)(1)(1)(1)} = 6$$

It is customary to refer to each *type* of distribution as a *state*. If the three hydrogen atoms attached to the surface of the enzyme constitute the *system*, then this system can be in state I, with energy equally distributed; in state II, with one atom in level three and the other two in level zero; or in state III, with one atom in level zero, one atom in level one, and the third atom in level two. The different *ways* in which each state can be formed are called the *complexions* of that state; in other words, W is equal to the number of complexions of the state.

If we could actually see and record the distribution of energy among the three hydrogen atoms from microsecond to microsecond or even faster, we would find it constantly shifting back and forth between all the different complexions shown in Fig. 6-5. We postulate that *each complexion is equally probable*. This is in accord with a principle closely related to the equipartition of energy, and it states that if the atoms exchange quanta back and forth purely at random by collision or by radiating and absorbing photons, there is just as much chance of finding a complexion corresponding to type I as there is of finding any one of the complexions of type II or any one of the complexions of type III. But if each complexion has the same probability of appearing and *state* III has six complexions while *state* I has only one, the probability of finding a ladderlike distribution corresponding to *state* III is 6 times the probability of finding all the atoms in the same energy level, as in *state* I, because *state* III has six complexions and *state* I has only one. In other words, although each *complexion* has the same probability, the different *states* have different probabilities because some have more complexions than others.

When the system is changed from just three hydrogen atoms on the surface of a single enzyme to a system consisting of 1,000 or so hydrogen atoms on a series of enzymes, the probability of finding a ladderlike distribution of energy of the sort illustrated in type III becomes overwhelmingly greater than the probability of finding all the atoms in the same energy level, as illustrated in type I. In fact, one special ladderlike distribution of energy among the atoms is so overwhelmingly more probable than any other that we can be almost certain that in any physical experiment this distribution will be found when the system consists of 1,000 atoms or more.

Example 6-2

If three quanta of energy are distributed among four oscillators, calculate the values of W for the different possible types.

$$W_I = \frac{4!}{1!3!0!\cdots} = \frac{24}{6} = 4$$

$$W_{II} = \frac{4!}{2!0!2!} = \frac{24}{2 \times 2} = 6$$

$$W_{III} = \frac{4!}{2!1!0!1!} = \frac{24}{2} = 12$$

$$W_{IV} = \frac{4!}{3!0!0!1!} = \frac{24}{6} = 4$$

6-5   THE MAXWELL–BOLTZMANN DISTRIBUTION LAW

Since the probability of any given type of distribution depends on the number of different ways W in which

the oscillators can be distributed over the energy levels to form this given type, the most probable type of distribution will be that for which W has the maximum possible value. During the latter half of the nineteenth century there was an extended investigation of this problem, in which the leading contributors were the English physicist J. Clark Maxwell and the Austrian physicist Ludwig Boltzmann. This most probable type of distribution, now called the Maxwell-Boltzmann distribution, turns out to be the one given for the harmonic oscillator by the equation:

$$n_i = n_0 e^{-(\epsilon_i - \epsilon_0)/kT} \tag{6-12}$$

where n_i = the number of oscillators in ith energy level

n_0 = number of oscillators in the zero energy level

e = base of natural logarithms

ϵ_i = value of energy for the ith energy level

ϵ_0 = value of energy in zero level

k = Boltzmann constant ($= R/N_A$)

T = temperature at which oscillators interchange quanta of energy, °K

When $i = 0$, $\epsilon_i = \epsilon_0$, and the exponent is zero. If the exponent of e is zero, then $e^0 = 1$, so that Eq. (6-12) reduces to the identity $n_0 = n_0$. As i takes on larger values, ϵ_i also takes on larger values. Since the exponent is negative, this means that the factor $e^{-(\epsilon_i - \epsilon_0)/kT}$ becomes smaller and smaller; in other words, the number of oscillators found in each energy level continually decreases as the energy associated with the level increases. Figure 6-6 shows the distribution over the energy levels for 100 hydrogen atoms vibrating on the surface of an enzyme at normal human body temperature, 37°C, or 310°K.

In many cases the number of oscillators in the zero energy level n_0 is not known, but the total number of oscillators n_{tot} is given. It is, therefore, useful to convert Eq. (6-12) to another form expressing the number of oscillators n_i in each energy level in terms of the total number of oscillators. To get this expression, add together, or *sum*, the numbers of oscillators in all the different energy levels. This summation is expressed by the symbol $\sum_i$. In effect, this is the addition of a whole series of equations of the type shown in Eq.

Fig. 6-6   The Boltzmann distribution for 100 hydrogen atoms vibrating at the surface of an enzyme.   Temperature 37°C.
$\nu = 3 \times 10^{12}$ sec^{-1}.

n energy level

13 _____

12 _____

11 _____

10 _____

9 _____

8 _____ ◯   1

7 _____ ◯   1

6 _____ ◯◯   2

5 _____ ◯◯◯◯   4

4 _____ ◯◯◯◯◯◯   6

3 _____ ◯◯◯◯◯◯◯◯◯   9

2 _____ ◯◯◯◯◯◯◯◯◯◯◯◯◯◯◯   15

1 _____ ◯◯◯◯◯◯◯◯◯◯◯◯◯◯◯◯◯◯◯◯◯◯◯◯   24   Number of atoms

0 _____ ◯◯◯◯◯◯◯◯◯◯◯◯◯◯◯◯◯◯◯◯◯◯◯◯◯◯◯◯◯◯◯◯◯◯◯◯◯◯   38

(6-12), where the value of i ranges from zero to the highest energy level where significant numbers of molecules are to be found.   This gives

$$n_{\text{tot}} = \sum_i n_i = \sum_i n_0 e^{-(\epsilon_i - \epsilon_0)/kT} \tag{6-13}$$

Since n_0 may be regarded as a constant in this equation, it can be taken out in front of the summation sign, so that

$$n_0 = \frac{n_{\text{tot}}}{\sum_i e^{-(\epsilon_i - \epsilon_0)/kT}} \tag{6-14}$$

Inserting this expression for n_0 in Eq. (6-12) gives

$$\frac{n_i}{n_{\text{tot}}} = \frac{e^{-(\epsilon_i - \epsilon_0)/kT}}{\sum_i e^{-(\epsilon_i - \epsilon_0)/kT}} \tag{6-15}$$

This equation expresses the ratio of the number of particles in a selected level i to the total number of particles in terms of the exponentials. n_i/n_0 is the fraction of particles in a given level.   Just as an example, consider the fraction of particles in the zero level

$$\frac{n_0}{n_{\text{tot}}} = \frac{e^{-(\epsilon_0 - \epsilon_0)/kT}}{\sum_i e^{-(\epsilon_i - \epsilon_0)/kT}} = \frac{1}{\sum_i e^{-(\epsilon_i - \epsilon_0)/kT}} \tag{6-16}$$

$\epsilon_0 - \epsilon_0 = 0$, and $e^0 = 1$; the numerator on the right-hand side of the equation becomes equal to 1, and the fraction of oscillators in the zero energy level is the reciprocal of the summation that appears in the denominator.   This summation, $\sum_i e^{-(\epsilon_i - \epsilon_0)/kT}$, is so closely related to a number of the important properties of atoms and molecules that it is given a special name, the *partition function*. For purposes of brevity it is frequently denoted by the single symbol Z.

The Maxwell-Boltzmann distribution formula for systems with discrete energy levels was written in the preceding paragraphs for the specific case of a system of quantized oscillators so that each step of the development might be illustrated by a concrete example. The logic of the development is far more general, however, and applies to all aggregates of individual systems in which there are discrete energy levels and for which there are no special conditions, e.g., lack of identifiability of the systems or interaction of the systems, that restrict the permissible kinds of distributions. These special conditions will be discussed later in Chaps. 20 and 22.

There are advantages in expressing Eqs. (6-13) and (6-15) in logarithmic form when using them in calculations. In logarithmic form Eq. (6-13) becomes

$$\log \frac{n_i}{n_0} = \frac{-1}{2.303kT} (\epsilon_i - \epsilon_0) \tag{6-17}$$

and Eq. (6-15) becomes

$$\log \frac{n_i}{n_t} = \frac{-1}{2.303kT} (\epsilon_i - \epsilon_0) + \log Z \tag{6-18}$$

These last two equations are known as linear algebraic expressions. If we let $\ln (n_i/n_0) = y$, $\log (n_i/n_t) = y'$, $-1/2.303kT = a$, $\epsilon_i - \epsilon_0 = x$, and $\ln Z = b$, then Eq. (6-17) can be written as

$$y = ax \tag{6-19}$$

and Eq. (6-18) can be written as

$$y' = ax + b \tag{6-20}$$

Example 6-3

Iodine is attached to a benzene ring in the amino acid diiodotyrosine

$$\text{HO} \begin{array}{c} \text{I} \\ \bigcirc \\ \text{I} \end{array} \text{CH}_2\text{CH(NH}_2)\text{COOH}$$

which is a constituent of many proteins. The iodine atoms vibrate with a frequency of $4.59 \times$

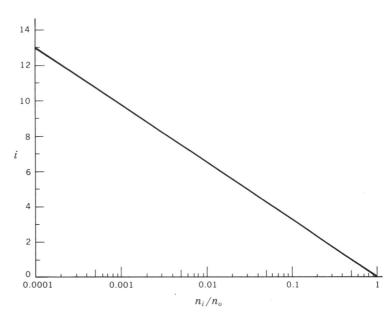

Fig. 6-7 The values of n_i/n_0 for $i = 0$ to $i = 12$ for the vibration of an iodine atom in diiodotyrosine at 37°C. $\nu = 4.59 \times 10^{12}$ sec^{-1}.

10^{12} Hz.   Calculate the value of n_{10}/n_0 at 37.00°C.

$$\epsilon_{10} - \epsilon_0 = 10h\nu = 10 \times 6.625 \times 10^{-27}$$
$$\times 4.59 \times 10^{12}$$
$$= 3.04 \times 10^{-13} \text{ erg}$$

$$37.00°C = 310.15°K$$

$$\log \frac{n_{10}}{n_0} = \frac{-1}{2.303kT}(\epsilon_i - \epsilon_0)$$

$$= \frac{-3.04 \times 10^{-13} \text{ erg}}{2.303 \times 1.38 \times 10^{-16} \times 310.15°}$$

$$= -3.08 \qquad \log \frac{n_0}{n_{10}} = 3.08$$

$$\frac{n_0}{n_{10}} = 1,202 \qquad \frac{n_{10}}{n_0} = 0.000832$$

Ans.

Figure 6-7 is the plot of log n_i/n_0 against i for an iodine atom vibrating normal to the plane of the benzene ring in diiodotyrosine. This type of vibratory motion is called the I-C bending mode because it tends to bend the bond between the iodine atom and the benzene ring. Observations of spectra show that the frequency of this motion is about 4.6×10^{12} Hz, as stated in Example 6-3.

From the graph in Fig. 6-7 one can read of the values of n_i/n_0 for each energy level up to the height where the value becomes negligibly small. The sum of these values is Z, and $n_0/n_{\text{tot}} = 1/Z$. This makes possible the calculation of n_{12}/n_{tot} from Eq. (6-18), as illustrated in Example 6-4. These two points are sufficient to determine the straight line which is the plot of log n_i/n_{tot} against i shown in Fig. 6-8.

Example 6-4

Calculate n_{12}/n_{tot} for the vibration of iodine at 37°C for which $\nu = 4.59 \times 10^{12}$ Hz.

From the graph in Fig. 6-7 the values of n_i/n_0 are found to be:

i	n_i/n_0
0	1.000
1	0.500
2	0.248
3	0.122
4	0.059
5	0.029
6	0.015
7	0.007
8	0.003
9	0.002
10	0.001
11	0.0004
12	0.0002
13	0.0001
14	0.0001

$$\sum_i \frac{n_i}{n_0} = 1.987$$

$$Z = \frac{1}{\sum_i (n_i/n_0)} = 0.5033$$

From Eq. (6-18)

$$\log \frac{n_{12}}{n_{\text{tot}}} = \frac{-(\epsilon_{12} - \epsilon_0)}{2.303kT} + \log Z$$

$$\log \frac{n_{\text{tot}}}{n_{12}} = \frac{\epsilon_{12} - \epsilon_0}{2.303kT} + \log \frac{1}{Z}$$

$$\epsilon_{12} - \epsilon_0 = 12h\nu = 12 \times 6.625 \times 10^{-27}$$
$$\times 4.59 \times 10^{12}$$
$$= 3.65 \times 10^{-13}$$

$$\log \frac{n_{\text{tot}}}{n_{12}} = \frac{3.65 \times 10^{-13}}{2.303 \times 1.38 \times 10^{-16} \times 310.15}$$
$$+ \log 1.987$$

$$= 3.70 + 0.30 = 4.00$$

$$\frac{n_{12}}{n_{\text{tot}}} = 1.00 \times 10^{-4} \qquad \textit{Ans.}$$

Since $\epsilon_i - \epsilon_0 = ih\nu$, Eq. (6-17) can be put in the alternative form

$$i = \frac{-2.303kT}{h\nu} \log \frac{n_i}{n_0} \qquad (6\text{-}21)$$

Fig. 6-8   The values of n_i/n_{tot} for $i = 0$ to $i = 12$ for the vibration of an iodine atom in diiodotyrosine at 37°C. $\nu = 4.59 \times 10^{12}$ sec^{-1}.

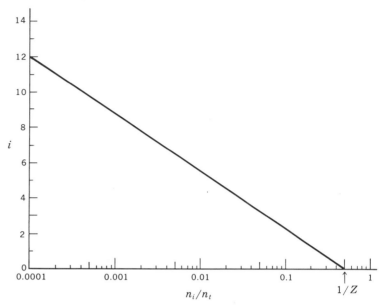

Thus, the slope of the line in Fig. 6-7 is proportional to the absolute temperature. It is therefore easy to draw lines for the values of n_i/n_0 at a number of different temperatures. Such a set of lines is shown in Fig. 6-9. Note that at 0°K all atoms are in the zero energy level. At 40°K, $n_1/n_0 = 0.01$; at 100°K, $n_1/n_0 = 0.1$, $n_2/n_0 = 0.01$, $n_3/n_0 = 0.001$, and $n_4/n_0 = 0.0001$.

Because the iodine atom is one of the heaviest elements normally found in living matter, the iodine vibrator has a relatively low frequency and an appreciable percentage of atoms are in energy levels higher than zero at body temperature. On the other hand, hydrogen atoms are so light that only a negligible percentage are above the zero energy level at body temperature when linked by a covalent bond to atoms such as carbon, nitrogen, or oxygen.

The Maxwell-Boltzmann distribution law for thermal energy is one of the most important principles in physical chemistry—indeed, in all of physical science. In our discussion above we have touched on only one of the many instances where this principle provides insight into the physical and chemical behavior of atoms and molecules. As we noted in the first chapter, every change in the state of matter, whether physical or chemical, involves a shifting in the distribution of energy. The whole life process is, in effect, a closely interwoven tapestry of energy changes. For this reason comprehension of the Maxwell-Boltzmann principle is essential both for an understanding of physical chemistry and of the physicochemical aspects of life processes.

The application of the Maxwell-Boltzmann distribution law to the velocity of molecules in the gaseous state is somewhat more complicated than the application to the harmonic oscillator. In the original development the range of velocities, and consequently of energies, was regarded as continuous instead of quantized. Instead of the number in a given energy level, the calculation gave the fraction $f_x = dn/n_t$ of the total number in the velocity range du. For gaseous molecules moving in one dimension x the expression is

$$\frac{f_x}{du} = Ae^{-(1/2)mu^2_x/kT} \tag{6-22}$$

where A is a proportionality constant. This expression closely resembles Eq. (6-12) since $\frac{1}{2}mu_x^2$ is the kinetic

Fig. 6-9 n_i/n_o **as a function of** i **at various values of** T **for the vibration of I in diiodotyrosine.** $\nu = 4.59 \times 10^{12} \text{ sec}^{-1}$.

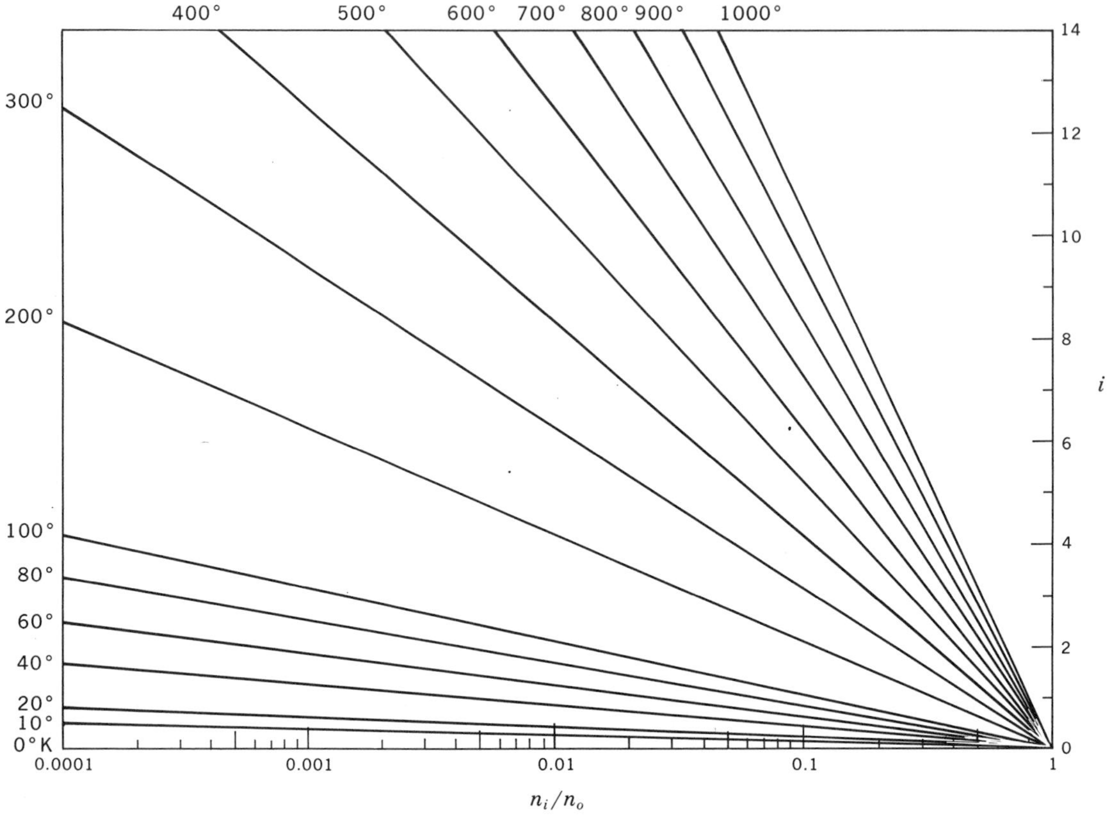

energy of the molecule. When the velocity u in three dimensions is the independent variable, the expression becomes

$$\frac{f}{du} = \frac{dn/n_{\text{tot}}}{du} = 4\pi \left(\frac{m}{2\pi kT}\right)^{3/2} e^{-(1/2)mu^2/kT} u^2 \quad (6\text{-}23)$$

A plot of f/du against u is given in Fig. 6-10 for nitrogen molecules at 0°C. As contrasted with the distribution of oscillators where the largest number have the lowest possible value of the energy, the majority of gas molecules have velocities not far from the average velocity.

6-6 THE BAROMETRIC EQUATION

One of the most interesting applications of the Maxwell-Boltzmann distribution principle is found in the variation of atmospheric pressure with height, which is great enough to cause profound physiological effects. Anyone who has visited cities with an altitude of a mile or more has experienced the feeling of shortness of breath. If the pressure is reduced, it is obvious that a lungful of air contains less oxygen at this altitude than at sea level. Consequently, when one is doing physical work that requires an increased supply of oxygen in the bloodstream, one must breathe much more rapidly in order to get the oxygen.

Consider the changes in the density of air with height in a column of air like that shown in Fig. 6-11. At a height h the downward force F_h due to the weight of the air in the column above this height is balanced by the force PA exerted by the pressure P at this height on the cross-sectional area A

$$-F_h = PA \quad (6\text{-}24)$$

Since the force exerted by the pressure is upward and the force F_h downward, the latter is given the negative sign. Consider a similar balance of the force due to the weight of the air against the force due to the pressure acting on the cross section at a height $h + dh$. This balance can be expressed

$$-(F_h - dF) = (P - dP)A \tag{6-25}$$

The downward force here is smaller by an amount dF, and the upward pressure is smaller by an amount $dP\ A$. The downward force has been decreased because the weight of the air in this cross section of height dh in Fig. 6-11 is no longer exerting force. The volume of this cross section is $A\ dh$. The weight of the air in this cross section is the density ρ multiplied by the volume, $\rho A\ dh$, and the downward force exerted by the air in this little cross section is the acceleration of gravity multiplied by the weight, $g\rho A\ dh$, which, of course, is equal to dF. Subtracting Eq. (6-24) from Eq. (6-25) gives

$$-dF = dP\ A \tag{6-26}$$

Substituting for dF, the expression in terms of the air,

$$-g\rho A\ dh = dP\ A \tag{6-27}$$

Canceling out the area A, which appears on both sides,

$$-g\rho\ dh = dP \tag{6-28}$$

The equation for an ideal gas

$$PV = nRT \tag{6-29}$$

provides an expression for ρ since this quantity is the number of grams of air per unit volume, which is in turn equal to the number of moles n in the little cross section multiplied by the molecular weight of air M and divided by the volume of the little cross section V. Thus

$$\rho = \frac{nM}{V} = \frac{MP}{RT} \tag{6-30}$$

Substituting this value for ρ in Eq. (6-28) and rearranging, we find

$$\frac{dP}{P} = -\frac{gM}{RT}\ dh \tag{6-31}$$

Integrating this equation between sea level h_0, where the pressure is P_0, and the height h, where the pressure is P,

$$\int_{P_0}^{P} \frac{dP}{P} = -\frac{gM}{RT} \int_{h_0}^{h} dh \tag{6-32}$$

so that

$$\ln \frac{P}{P_0} = -\frac{gM}{RT} h \tag{6-33}$$

where h_0 is set equal to zero since height is the distance above sea level. This equation can be put in exponential form

$$\frac{P}{P_0} = e^{-(gM/RT)h} \tag{6-34}$$

Think of the column of air as resembling an energy-level ladder, as shown in Fig. 6-12. Every molecule in the air acquires potential energy as it rises to a higher level. Thus the number of molecules of air found in a cross section at any moment is analogous to the num-

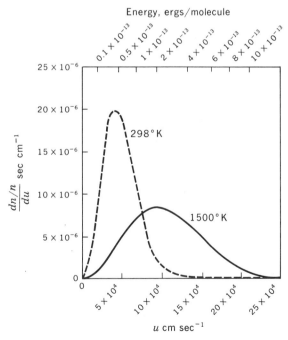

Fig. 6-10 The distribution of the speeds of N_2 molecules. (From G. M. Barrow, "Physical Chemistry," 2d ed., McGraw-Hill Book Company, New York, 1966.)

Fig. 6-11   Variation of pressure with altitude.

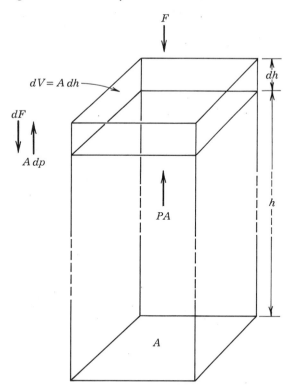

ber of molecules at that energy level. The value of the energy per mole at a particular energy level i is given by

$$E_i = gMh_i \tag{6-35}$$

The energy per molecule ϵ_i is given by

$$\epsilon_i = g \frac{M}{N_A} h_i \tag{6-36}$$

Thus, the number of moles in the ith energy level n_i divided by the number of moles in the zero energy level n_0 (or at sea level) is given by

$$\frac{n_i}{n_0} = e^{-(\epsilon_i - \epsilon_0)/kT} \tag{6-37}$$

Among the other effects produced by altitude is the reduction of the temperature at which water boils. Water boils when its vapor pressure becomes equal to the pressure of the atmosphere above it. At sea level, under normal conditions, this pressure is 1 atm; thus, water boils at the normal boiling point of 100°C. However, in a city like Denver the pressure of the atmosphere is sufficiently reduced for the water to attain a vapor pressure equal to that of the Denver atmosphere at a substantially lower temperature. Thus, since the temperature of boiling water is lower, it takes considerably longer to cook an egg in boiling water in Denver than it does in New York, since the rate of cooking depends on the temperature.

6-7   THE STATISTICAL TEMPERATURE SCALE

In Boltzmann's original derivation of the equation for the statistical distribution of energy, he merely showed that the fraction of energy lying in a given energy level varies as the total amount of energy in the assembly of oscillators changed. In the barometric equation (6-37) the quantity appearing in the denominator of the exponent is proportional to the temperature determined and defined by the ideal-gas equation. This proves that the temperature scale based on the ideal-gas equation is also a temperature scale in accord with the statistical meaning of temperature based on the Maxwell-Boltzmann distribution law.

There are certain restrictions on the use of the Maxwell-Boltzmann equation in situations like the distribution of energy among electrons where quantum relationships affect the number of electrons found in various energy levels. These special cases will be discussed in later chapters dealing with the more advanced aspects of the quantum theory.

6-8   THE EINSTEIN THEORY OF HEAT CAPACITY

We turn now from the consideration of the way in which vibrational energy is distributed among a set of oscillators at a given fixed temperature and consider how much energy is required to raise the temperature of this same set of oscillators by 1°, the heat capacity at constant volume. In the following discussion the set consists of 1 mole (Avogadro's number, 6.02×10^{23}) of oscillators, so that the heat capacity is denoted by c_v. Think of these oscillators vibrating independently of each other on the surface of an enzyme. Assume that the enzyme's surface acts as a completely rigid wall so that the only thermal motion is the vibration of the hydrogen atoms themselves toward and away from the wall in a direction at a right angle to the wall.

In Sec. 6-5, with the help of the Boltzmann equation, we calculated the fraction of the vibrating hydrogen atoms on the surface of the enzyme that would be in the zero level, the first vibrational energy level, the second vibrational level, and so on. It turned out that at body temperature (37°C) about 38 percent of the hydrogen atoms are in the zero energy level, 23 percent in the first energy level, 15 percent in the second energy level, and so on, for the strength of binding that was postulated. The Maxwell-Boltzmann equation (6-15) shows that as the temperature increases, more and more of the hydrogen atoms are in higher energy levels. The amount of energy absorbed when the temperature of the mole of oscillators is raised by 1° is the heat capacity per mole at constant volume c_v.

When it first became possible, about 60 years ago, to make accurate heat-capacity measurement from room temperature to temperatures near 0°K, observations showed that the heat capacity of metals fell off very rapidly as the temperature was lowered. Einstein came forward in 1907 with the first simple explanation of this effect. On the basis of the Maxwell-Boltzmann equation and the postulation of quantized energy levels, he showed that 1 mole of oscillators attached to a rigid surface, such as the idealized enzyme surface, and vibrating only toward and away from that surface should have a heat capacity given by

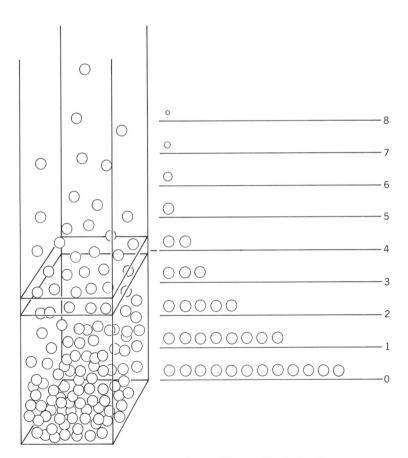

Fig. 6-12   Energy levels corresponding to different altitudes for air.

$$c_v = \frac{Ru^2 e^u}{(e^u - 1)^2} \tag{6-38}$$

$$u = \frac{h\nu}{kT} = \frac{\theta}{T} \tag{6-38a}$$

where c_v = heat capacity per mole
  R = ideal-gas constant
  h = Planck's constant
  ν = frequency of vibration
  T = temperature, °K
  k = Boltzmann's constant

The derivation of this equation is given in Appendix 4.

To simplify the equation, the quantity $h\nu/k$ is frequently designated by a single symbol θ. Since h and k are universal constants, the value of θ depends only on the frequency ν with which the oscillator vibrates. When necessary to avoid ambiguity the subscript ein will be used to identify the value of θ appropriate for the Einstein equation (6-38). The exponent of e must be dimensionless, and thus it is clear that θ must have the dimensions of temperature since $u = \theta/T$. For this reason, θ is frequently called the *characteristic temperature*. Since the value of c_v depends only on the value of u, it is possible to compute c_v as a function of u, as shown in Table 6-1. A plot of the values in this table is shown in Fig. 6-13. If $\theta = 100°$, the temperature scale for the graph is that shown at the bottom of the figure. c_v at 0°K is zero and rises slowly, reaching a value of about 0.9 R when $T = \theta$; c_v approaches R at higher temperatures in accord with the experimental observations discussed in Chap. 4.

In his original discussion, Einstein used his equation not to calculate the heat capacity of independent oscillators but of atoms in metal crystals. These atoms do not act as independent oscillators but have coupled vibrations; as a result, at low temperature the observed heat capacity is considerably higher than that calculated from Eq. (6-38). However, as an explanation of the lowering of heat capacity at low temperatures the equation was a marked advance in the theory of thermal energy.

Today the importance of the Einstein heat-capacity equation lies not in its use in predicting the heat capacity of crystals but in its application to calculations of the heat capacity associated with the internal vibration of molecules; in dozens of examples the values calculated from the Einstein equation check with observed values very closely.

6-9   THE DEBYE THEORY OF HEAT CAPACITY

Shortly after Einstein proposed the equation for the heat capacity of oscillators, Peter Debye, a Dutch physicist, pointed out that the difference between the results calculated with Einstein's equation and the results experimentally observed in crystals could be explained by assuming that in crystals not all atoms vibrate with a single frequency but vibrate over a range of frequencies, which can be determined in a manner similar to the calculation of the overtones of a violin string. The theory of overtones is important not only for the thermal vibrations in crystals but also for the vibrations of the de Broglie waves that form the essential basis for the interpretation of atomic and molecular behavior with the help of theories of quantum mechanics. For this reason, it is useful to examine in some detail the relation between thermal vibrations in a crystal and the vibrations of a violin string.

Figure 6-14 shows the pattern of vibrations of the first five modes of motion of a violin string. If the string is plucked at the center (bottom line), it sounds its fundamental tone as the whole string first arches

TABLE 6-1   The Einstein heat-capacity function

θ/T	c_v	θ/T	c_v
0	5.961	3.0	2.959
0.1	5.957	3.5	2.345
0.2	5.942	4.0	1.813
0.3	5.917	5.0	1.018
0.4	5.883	6.0	0.535
0.5	5.839	7.0	0.267
0.6	5.786	8.0	0.128
0.7	5.724	9.0	0.060
0.8	5.654	10.0	0.027
0.9	5.575	11.0	0.0120
1.0	5.489	12.0	0.0053
1.5	4.959	13.0	0.0023
2.0	4.317	14.0	0.0010
2.5	3.630		

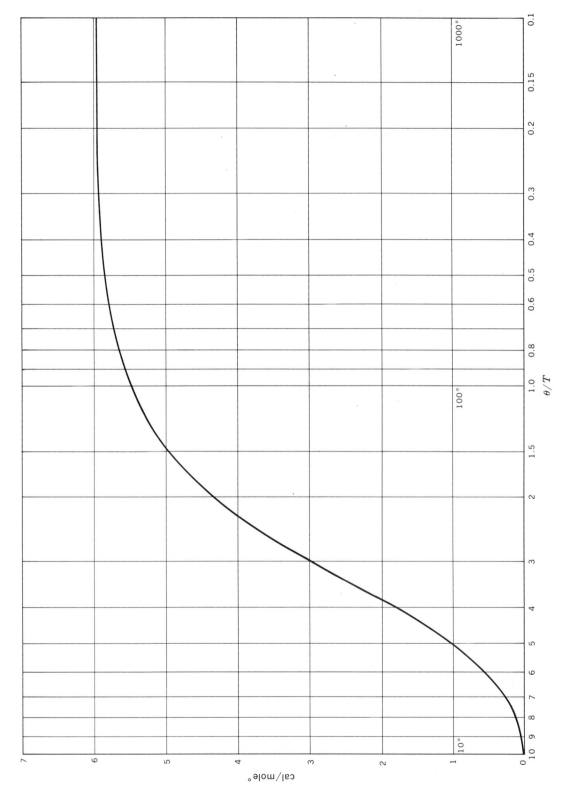

Fig. 6-13 Einstein function for heat capacity for 3 degrees of freedom. T is shown for $\theta = 100°$.

Fig. 6-14 Modes of vibration of a stretched string.

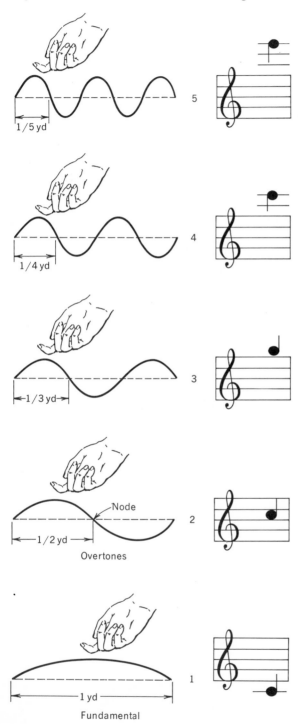

up to a maximum amplitude and then moves down, forming a concave inverted arch. Suppose that the string is of the right length and cross-sectional weight and under such tension that it sounds the note of middle C as its fundamental tone. This note has a frequency of 256 Hz on the scientific scale of pitch. (The musical scale based on A = 440 Hz is slightly different.) If the center of the string is pressed lightly and the string is plucked halfway between the center and one end, a new pattern of vibration results, as shown in line 2 of Fig. 6-14. Here the string vibrates in two parts. The two loops move alternately up and down in seesaw fashion while the string sounds the note with 512 Hz, or high C. This is, of course, one octave above the fundamental tone and has a frequency just twice as that of the fundamental. Proceeding in a similar way, as shown in line 3, one can pluck the string one-third of the way between the left peg and the center of the string and set in motion the next overtone, where there are three loops, two bowing up while one bows down. The frequency of this overtone is 3 times that of the fundamental and the note sounded is G' lying just above high C. In the next overtone (line 4) there are four loops, the frequency is 4 times that of the fundamental, and the note is high high C, two octaves above the fundamental tone. Finally, in the next overtone (line 5) with a pattern of five loops the frequency is 5 times that of the fundamental.

Debye pointed out that the thermal energy in a crystal causes it to vibrate in a way similar to the violin string. A cubic crystal is shown in Fig. 6-15. The fundamental vibration of the crystal takes place in three dimensions, as contrasted with the string, where there is vibration in only one dimension. This is because the crystal has length, breadth, and height where the string has only the one significant dimension of length as far as the pattern of vibration is concerned. In the fundamental vibration, the cube expands and contracts as a whole. Because the cube is three-dimensional, the pattern of overtones becomes somewhat more complicated than that of the string, but Debye was able to calculate the distribution of the frequencies in these overtones. He then postulated that each overtone obeys the Einstein equation and took into account the variation of frequency, from the lowest to the highest possible value ν_{max}. Thus

$$c_v = 3Rx^3 \int_0^x \frac{u^4 e^u \, du}{(e^u - 1)^2} \tag{6-39}$$

x is given by the relation

$$x = \frac{T}{\theta_{\text{deb}}} = \frac{kT}{h\nu_{\text{max}}} \tag{6-40}$$

The subscript deb is used to distinguish the value of θ appropriate for the Debye equation (6-39) from the value θ_{ein} appropriate for the Einstein equation (6-38). Thus, there is a characteristic temperature θ_{deb} that plays the same role in the Debye equation as θ_{ein} plays in the Einstein heat-capacity equation.

As in the case of the Einstein equation, one can make a table of values of the heat capacity as a function of x, as shown in Table 6-2; these values are plotted in Fig. 6-16.

At very low temperatures, the Debye equation assumes the simple form

$$c_v = \frac{12}{5}\pi^4 R \left(\frac{T}{\theta_{\text{deb}}}\right)^3 = \text{const} \times T^3 \tag{6-41}$$

Thus, at very low temperatures, the heat capacity varies simply as the cube of the absolute temperature. Figure 6-17 shows the experimentally observed heat capacities and computed curves for four different crystals; the agreement between the experimental values and values computed from the Debye equation is excellent.

In crystals where the substance consists not of individual atoms like iron or nickel but of molecules packed together in a crystal lattice, the situation is somewhat more complicated. Consider as an example a crystal composed of thymine molecules. Thymine is one of the four key constituents of the rungs in the genetic ladder found in the macromolecule of deoxyribose nucleic acid (DNA), which guides such a large portion of the biological activity inside a cell and carries the genetic message that determines the form of growth of the organism. A graphic formula of thymine is shown in Fig. 6-18, together with the formula of the simple parent molecule, benzene.

In a crystal of thymine, thermal energy is manifested in a number of different types of vibration. First of all, the thymine molecules move back and forth in translational vibration, similar to the translational vibration of atoms of iron in an iron crystal. The evidence is that the heat capacity due to this translational motion is given by the Debye formula. However, in addition to this back-and-forth translational motion, the thymine molecules also rock around a central position, a motion called *libration*.

In an ideal gas, $c_v = \frac{3}{2}R$, consisting of $\frac{1}{2}R$ for each of the 3 degrees of freedom in which there is kinetic energy. In the crystal, in addition to the kinetic energy of vibration, there is also potential energy because of the action of the forces between atoms in the crystal lattice. As a consequence $\frac{1}{2}R$ is added for each of the 3 degrees of freedom, making c_v (crystal translational vibration) $\simeq 3R$, as T becomes larger than θ_{deb}. The same considerations apply to libration, which also takes place in three dimensions; c_v (crystal librational vibration) $\simeq 3R$ as T becomes larger than θ_{deb}.

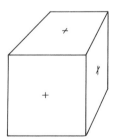

Fundamental mode:
cube expands and contracts
as a whole

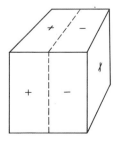

Overtone:
left half expands while
right half contracts

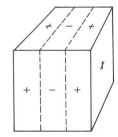

Higher overtone:
left and right thirds expand
while center third contracts

Fig. 6-15 Fundamental and two overtone modes of Debye vibrations of a cube.

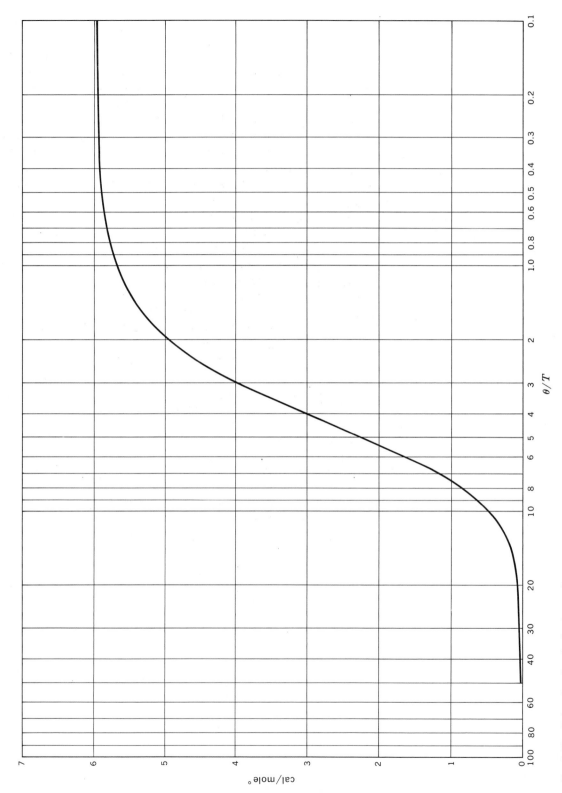

cal/mole°

θ/T

Fig. 6-16   Debye function for heat capacity.

As heat is fed into the thymine crystal under constant pressure and its temperature increases, there is an expansion of the crystal resulting in greater potential energy because of the stretching of the intermolecular forces. Thus this expansion also absorbs heat. Formulas have been devised to give approximately the heat absorbed due to this expansion. Finally, there are internal vibrations in the thymine molecule similar to the internal vibrations in simple molecules like H—H and O=O.

The total number of degrees of freedom for a thymine molecule is $3n$, where n is the number of atoms in the molecule. The formula of thymine is $C_5N_2O_2H_6$; thus there are 15 atoms in the molecule altogether and therefore $3 \times 15 = 45$ total degrees of freedom. Since six of these degrees of freedom consist of the three translational vibrations and the three librational motions, there are altogether 39 internal vibrational degrees of freedom. The best evidence indicates that all these internal degrees of freedom have heat capacities varying with temperature according to the Einstein equation. The problem is to ascertain the frequency for each of the internal degrees of freedom so that it can be used to calculate the value of θ_{ein} and thus, in turn, the contribution to the heat capacity.

Analyses of spectra have made it possible to calculate values of the frequencies in molecules of this sort. Thus, one can make quite an accurate estimate not only of the total thermal energy in a molecule like thymine but even of the energy in each of the modes of motion with which the thymine vibrates. Our knowledge of the actual chemistry of replication involving the production of duplicate DNA molecules and template molecules of RNA has not yet reached the point where we are able to ascertain exactly the role of each type of motion in replication, but it is clear that as our ability to interpret spectra improves, we shall acquire a detailed knowledge of the modes of motion of these complex molecules of biochemical interest that may ultimately enable us to understand more about the intricacies of replication.

One of the first steps in this direction has been the interpretation of the spectra of the benzene molecule, which made possible the calculation of the heat capacity of benzene and the comparison of the calculated values with the experimentally observed values.

TABLE 6-2   The Debye heat-capacity function

θ/T	c_v	θ/T	c_v
0.0	5.955	6.0	1.582
0.1	5.95	7.0	1.137
0.2	5.94	8.0	0.823
0.3	5.93	9.0	0.604
0.4	5.91	10.0	0.452
0.5	5.88	11.0	0.345
0.6	5.85	12.0	0.267
0.7	5.81	13.0	0.211
0.8	5.77	14.0	0.169
0.9	5.72	15.0	0.137
1.0	5.67	16.0	0.113
1.5	5.34	18.0	0.0796
2.0	4.92	20.0	0.0581
2.5	4.45	22.0	0.0436
3.0	3.95	24.0	0.0336
3.5	3.46	26.0	0.0264
4.0	3.00	28.0	0.0212
5.0	2.197	30.0	0.0172

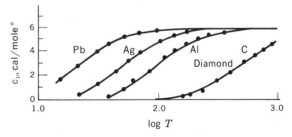

Fig. 6-17   The heat capacity observed experimentally (circles) for several elements in the crystalline state. The lines are the values calculated from the Debye heat-capacity function. Values of θ: Pb 90.3; Ag 213; Al 389; Cl 890.

These results are given in Table 6-3 and plotted in Fig. 6-19. The excellent agreement between the calculated and the observed results shows that both the Einstein and the Debye heat-capacity equations can be applied successfully to molecules of considerable complexity.

Fig. 6-18   Structural formulas and crystal vibration.

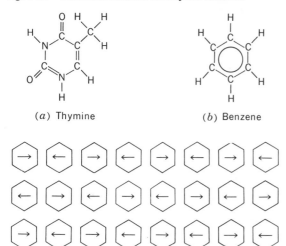

(a) Thymine   (b) Benzene

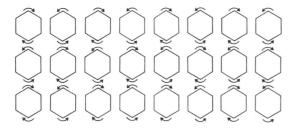

(c) Translational vibration in a benzene crystal

(d) Libration in a benzene crystal, schematic

6-10   THE HEAT CAPACITY OF LIQUIDS

The kinetic theory of gases coupled with the Einstein equation for the internal vibration in gaseous molecules provides an accurate way of calculating the heat capacity of gases and a satisfactory knowledge of the actual thermal motions of gaseous molecules. At the low end of the temperature scale the use of the Debye and Einstein equations for heat capacities together with knowledge of frequencies obtained from spectra also provides a detailed understanding of the thermal motions in crystals. But in the intermediate range, where the molecules are in the liquid state, there is still only a limited knowledge of the actual motions of the molecules. It is probable that in the liquid state the molecules undergo translational motion toward and away from each other much as they do in the crystal

state. Because the molecules are farther apart and the intermolecular forces weaker, molecules can exchange places and diffuse back and forth through the liquid far more easily than they can in the crystalline state. But the evidence is that the Debye equation can still be applied to calculate the heat capacity due to this translational vibration in the liquid state. Although the value of θ_{deb} is not known with great certainty, an error in this value makes little difference in the value of the heat capacity because in the liquid state the temperature is generally so high that c_v closely approaches the classical value of $3R$.

However, the degree to which molecules either librate or actually spin freely in the liquid state is uncertain. There is evidence that the potential barrier restricting rocking motion is sufficiently low for molecules in a number of cases actually to spin freely in the liquid state. In fact, as temperature is raised from a value just above the melting point through the liquid range to a value just below the boiling point, it seems quite likely that in most liquids the molecular motion changes from libration, at temperatures in the vicinity of the melting point, through a transition phase, in which more and more molecules spin, to a state just below the boiling point, where nearly all the molecules may be spinning freely.

For benzene it is possible to calculate theoretically the amount of heat capacity that should be due to rocking c_r and how this might be expected to change as the motion changes from rocking to free rotation. Adding together the contributions to the heat capacity due to the translational vibration, to the expansion of the molecules and then subtracting this sum from the observed heat capacity gives the residue of heat capacity presumably associated with libration (restricted rotation) and with the change of the libration to free rotation. The results of such a calculation are given in Table 6-4 for benzene and plotted in Fig. 6-20. At high temperatures, the heat capacity due to *restricted* rotation should be the same as the heat capacity due to translational vibration in three dimensions, namely, a heat capacity with a value $3R$. On the other hand, the heat capacity due to *free* rotation should be the same as for the gaseous state, where the molecules are freely rotating, namely, $\frac{3}{2}R$. Pitzer (see Table 6-4) has shown by quantum-mechanical calculations that in such a transition from rocking to free rotation, one would expect a small rise in the heat capacity above the value $3R$ and then a falling off to the value $\frac{3}{2}R$. There is at least an indication of such a kind of change in the heat capacity of liquid benzene.

Another important factor in the heat capacity of liquids is undoubtedly hydrogen bonding. Especially in liquids of biological interest hydrogen bonding must play an important role. The nature of hydrogen bonding will be discussed at greater length in the section on quantum mechanics. Here, it may be described as a force produced between neighboring molecules when a proton can partially dissociate itself from one molecule and partially associate itself with a neighboring one.

The objective of this rather brief survey of the distribution of thermal energy has been to provide a background picture of the way in which thermal energy is present in various kinds of aggregates of molecules. This makes possible a somewhat more visual and detailed appreciation of the actual nature of heat that is helpful in discussing the thermodynamic relations between heat and various other kinds of energy.

Example 6-5

θ_{ein} has the value 813° for the internal mode of vibration in the Cl—Cl molecule in which the two atoms move toward and away from each other. Calculate the contribution of this mode of motion to the heat capacity at 100, 300, 500, and 1000°K in calories per degree per mole.

Since this is a vibratory motion with 1 degree of freedom, the value is one-third of that shown in Fig. 6-13. Values of θ/T are calculated for each value of T and values of c_v read from the above figure and multiplied by $\frac{1}{3}$.

TABLE 6-3   Observed and calculated heat capacity of benzene

| | c_p, cal deg^{-1} mole^{-1} | |
T, °K	Obs[a]	Calc[b]
4	0.0195	0.018
8	0.147	0.144
15	0.920	0.908
20	1.84	1.93
30	4.24	4.45
50	8.14	8.10
70	10.16	10.09
100	11.99	12.02
120	13.24	13.24
140	14.71	14.63
160	16.26	16.22
180	18.02	17.99
200	20.02	20.06
220	22.24	22.31
240	24.75	24.86
260	27.75	27.55

[a] Data from J. E. Ahlberg, E. R. Blanchard, and W. O. Lundberg, *J. Chem. Phys.*, **5**:552–556 (1937), and H. M. Huffman, G. S. Parks, and A. C. Daniels, *J. Am. Chem. Soc.*, **52**:1547 (1930).
[b] Data from R. C. Lord, Jr., J. E. Ahlberg, and D. H. Andrews, *J. Chem. Phys.*, **5**:650 (1937).

| | θ | c_v, cal deg^{-1} mole^{-1} | | |
T, °K	$\dfrac{\theta}{T}$	3 degrees of freedom	1 degree of freedom	Ans.
100	8.13	0.12	0.04	
300	2.71	3.22	1.07	
500	1.63	4.85	1.62	
1000	0.81	5.65	1.88	

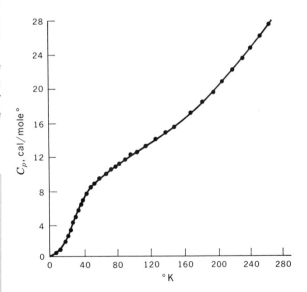

Fig. 6-19 The molal heat capacity of crystalline benzene experimentally observed (circles) and theoretically calculated (see Table 6-3).

Example 6-6

The value of θ_{deb} for benzene is 159°. Calculate the contribution made to the heat capacity by the translational vibration of the molecules in the crystal lattice at 20°K and compare with the heat capacity as calculated from the Einstein equation with $\theta_{\text{ein}} = 159°$.

Following the same procedure as in Example 6-5,

		c_v, cal deg⁻¹ mole⁻¹		
T, °K	$\dfrac{\theta}{T}$	Debye	Einstein	*Ans.*
20	7.95	0.85	0.12	

Example 6-7

Calculate the contribution to the heat capacity of benzene from translational and librational vibration of the molecules at 0°C using $\theta_{\text{deb}} = 159°$.

		c_v, cal deg⁻¹ mole⁻¹		
T, °K	$\dfrac{\theta}{T}$	3 degrees of freedom	6 degrees of freedom	*Ans.*
273	0.582	5.87	11.74	

TABLE 6-4   Heat capacity in liquid benzene due to rocking and free rotation of molecules[a]

T, °K	c_p (obs)	c_R
mp 278		
280	31.59	5.32
290	32.10	5.03
300	32.62	4.66
310	33.16	4.38
320	33.69	4.05
330	34.26	3.81
340	34.87	3.67
350	35.50	3.70
bp 353		

[a] Experimental data from G. D. Oliver, M. Eaton, and H. M. Huffman, *J. Am. Chem. Soc.*, **70**:1502 (1948); values of c_R calculated by B. L. Petterson, W. C. Roman, D. C. Rueger, and D. H. Andrews, *Abstr. Bull., Div. Phys. Chem., Am. Chem. Soc.*, **April, 1967**:179, from tables of K. S. Pitzer and W. D. Gwinn, *J. Chem. Phys.*, **10**:428 (1942).

Problems

1. If the Hooke's law constant k_h for the H_2 molecule is taken as equal to 4.60×10^5 dynes cm⁻¹, calculate the force of attraction due to the stretching of the

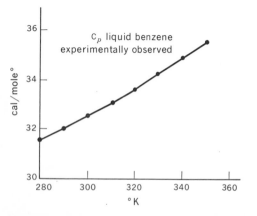

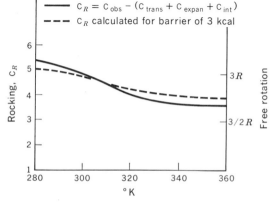

Fig. 6-20 c_p during change from libration to free rotation.

H—H bond when the bond length is increased from the equilibrium value, where $F = 0$, to a length 0.1 Å greater.

2. If the force of repulsion between two chlorine atoms is 6.60×10^{-4} dyne when the normal Cl—Cl bond is shortened by 0.200 Å, calculate the value of the Hooke's law constant, assuming that Hooke's law is obeyed.

3. The frequency of vibration ν is related to the Hooke's law constant and the molecular weight of a diatomic molecule composed of two similar atoms by

$$\nu = \frac{1}{2\pi} \left(\frac{4k_h N_A}{M} \right)^{1/2}$$

Calculate the frequency of vibration of the chlorine molecule using the value of k_h obtained in Prob. 2 and assuming $M = 70$.

4. The frequency of vibration of a hydrogen atom attached to the ring carbon in thymine found in DNA is 9.17×10^{13} vibrations per second. Calculate in calories per mole the value of the energy in each of the first six energy levels associated with the stretching of the C—H bond, assuming that Hooke's law is obeyed.

5. Calculate the number of complexions W in each of the possible states when four quanta of energy are distributed in all possible ways among three hydrogen atoms executing stretching vibrations on the surface of an enzyme.

6. Calculate the number of complexions W if two quanta of energy are distributed in all possible ways between three C—H vibrators on three neighboring molecules of thymine in DNA.

7. The net force holding an OH group on the surface of a certain enzyme is such that the Hooke's law constant k_h has the value of 5×10^5 dynes cm^{-1} when the OH group vibrates in a direction perpendicular to the enzyme surface. Assume that the mass and rigidity of the enzyme are such that the surface may be regarded as a rigid wall and that the O and H atoms are so tightly bound that they vibrate as a single mass. Calculate the frequency of vibration of the OH group.

8. Calculate the energy of vibration of such an OH group as described in Prob. 7 when it is in the first energy level.

9. Calculate the ratio of the number of molecules in the first energy level to the number in the zero level for the above oscillator at normal body temperature (37°C).

10. Make a plot of $\log(n_i/n_0)$ against ϵ_i for the above

oscillator up to the fourth energy level, at 0, 37, and 100°C.

11. From the graph in Prob. 10, calculate the value of the partition function of the above oscillator at 37°C.

12. Using the value of the partition function calculated above, calculate n_i/n_{tot} for $i = 1, 2, 3$ at 37°C for the OH oscillator described in Prob. 7.

13. Make a graph of the value of $\log(n_i/n_{tot})$ as abscissa against i as ordinate for the above oscillator at 1000°C. From the graph, make a table of the values of n_i/n_{tot} from $i = 0$ to $i = 6$. Add these fractions together and calculate the difference between this sum and unity. What does this figure represent?

14. For the mode of vibration in thymine where the ring expands and contracts as a whole, the value of θ for the Einstein heat-capacity function may be taken as 1424°. From the chart in Fig. 6-11, calculate the contribution to the heat capacity of this mode of vibration at body temperature (37°C) and at 500°C. Keep in mind that the chart gives the values for 3 degrees of freedom and that thymine has only 1 degree of freedom with this value of θ.

15. The value of θ for the Einstein heat-capacity function is 4400° for the stretching motion of hydrogen attached to the carbon in the thymine ring. To what temperature must thymine be raised if the motion of this hydrogen in stretching is to contribute 0.1 cal deg^{-1} mole^{-1} to the heat capacity?

16. The value of θ for the Einstein heat-capacity function is 4250° for the motion in thymine where the three hydrogen atoms attached to the carbon forming a methyl group stretch the CH bond, all moving out at the same time. Using the chart in Fig. 6-11, calculate the contribution to the heat capacity of this motion at 500°K.

17. Using the value of 90.3° for θ in the Debye heat-capacity function, calculate the heat capacity c_v of 1 mole of crystalline lead at 10 and at 1°K, from the chart in Fig. 6-14.

18. Using the value of 389° for the θ in the Debye heat-capacity function, calculate the heat capacity c_v of 1 mole of aluminum at 100 and at 300°K from the chart in Fig. 6-14.

19. The Debye characteristic temperature θ for crystalline neon is 63°K. At what temperature will the value of c_v for this crystal reach 0.9 of the classical value of $3R$? (Use Fig. 6-14.)

20. The Debye characteristic temperature θ for thymine may be taken as approximately 150°. Calculate the contribution to the heat capacity of 1 mole of thymine from the Debye vibrational and librational motion at body temperature, using Fig. 6-14.

REFERENCES

Gurney, R. W.: "Introduction to Statistical Mechanics," pp. 1–68, McGraw-Hill Book Company, New York, 1949.

Guggenheim, E. A.: "Boltzmann's Distribution Law," North-Holland Publishing Company, Amsterdam, 1955.

Hill, T. L.: "An Introduction to Statistical Thermodynamics," chaps. 1, 5, and 6, Addison-Wesley Publishing Company, Inc., Reading, Mass., 1960.

Davidson, N.: "Statistical Mechanics," chaps. 8, 10, and 16, McGraw-Hill Book Company, New York, 1962.

MacDonald, D. K. C.: "Introductory Statistical Mechanics for Physicists," chaps. 1 and 2, John Wiley & Sons, Inc., New York, 1963.

Andrews, F. C.: "Equilibrium Statistical Mechanics," chaps. 3–8, John Wiley & Sons, Inc., New York, 1963.

Eyring, H., D. Henderson, B. J. Stover, and E. M. Eyring: "Statistical Mechanics and Dynamics," chaps. 1 and 4, John Wiley & Sons, Inc., New York, 1964.

Knuth, E. L.: "Introduction to Statistical Thermodynamics," chaps. 8, 9, and 14, McGraw-Hill Book Company, New York, 1966.

Eyring, H. (ed.): "Physical Chemistry: An Advanced Treatise," vol. II, "Statistical Mechanics," chaps. 1 and 3, Academic Press, Inc., New York, 1967.

Rice, O. K.: "Statistical Mechanics, Thermodynamics and Kinetics," secs. 1-4, 1-5, 3-4 to 3-6, W. H. Freeman and Company, San Francisco, 1967.

Thermodynamics is that branch of science which deals with the change of energy from one form to another.   Such changes are common even in every-day life.   Bore a hole in a block of wood with a steel drill.   Immediately after the drill is withdrawn from the hole it feels warm to the touch.   Mechanical energy applied in the twisting of the drill has been converted in part into thermal energy, or heat.   The reverse process, the conversion of heat into mechanical energy, can be seen in the operation of a steam engine.   Again, consider that the thermal energy in the steam came originally from

Chapter Seven

the first law of thermodynamics

the chemical energy in the coal burned under the boiler.   The combination of oxygen from the air with the combustible material in the coal led to the conversion of the latent chemical energy in the coal into heat.   Even in the human body itself, there are countless examples of the conversion of energy.   With every heartbeat, chemical energy is being converted into mechanical energy.   Chemical energy in one form is converted into chemical energy in other forms as impulses travel along the nerves.   When air passes over the vocal chords causing them to vibrate, energy is sent out from the body in sound waves.   Thermal energy leaves the skin in the form of electromagnetic radiation in the infrared part of the spectrum.

In order to systematize the treatment of these various kinds of energy changes and to establish quantitative relations between them, a number of special concepts have been defined.   In terms of these concepts three laws of thermodynamics have been formulated which express the most general regularities observed in energy changes.

7-2   THE FIRST LAW OF THERMODYNAMICS

The first law of thermodynamics may be expressed:

In any change of state of a system, energy is never created or destroyed.

This is also known as the principle of the conservation of energy and was first discovered through observations of the changes of mechanical energy.

Consider an experiment like that illustrated in Fig. 7-1. A block of iron A and a block of wood B slide toward each other on a frictionless surface. The masses of the blocks m_A and m_B are known. The velocity of each block is measured before they collide (v_{A_1}, v_{B_1}) and after they collide (v_{A_2}, v_{B_2}). The following relation is found to hold between these quantities:

$$\tfrac{1}{2}m_A v_{A_1}{}^2 + \tfrac{1}{2}m_B v_{B_1}{}^2 = \tfrac{1}{2}m_A v_{A_2}{}^2 + \tfrac{1}{2}m_B v_{B_2}{}^2 \qquad (7\text{-}1)$$

Since the quantity $\tfrac{1}{2}mv^2$ was defined as the kinetic energy E_k, the above equation can be written

$$(E_k)_{A_1} + (E_k)_{B_1} = (E_k)_{A_2} + (E_k)_{B_2} \qquad (7\text{-}2)$$

or

$$(E_k)_1 \;=\; (E_k)_2 \qquad (7\text{-}3)$$

total kinetic   total kinetic
energy before   energy after
collision   collision

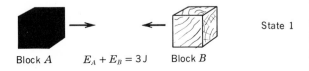

Block A $E_A + E_B = 3\,\text{J}$ Block B

State 1

Collision

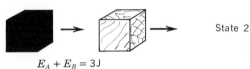

State 2

$E_A + E_B = 3\,\text{J}$

Fig. 7-1   Conservation of kinetic energy. (J = joule)

This expresses the fact that there is no change in the total kinetic energy of the system because of the change in the velocities of the bodies of the system brought about by the collision (the change of state).

If instead of sliding along on the surface, one of the blocks is tossed into the air, its velocity decreases, coming to a zero value when the block reaches the top of the trajectory and starts to fall back toward the earth. At the top of the trajectory, the kinetic energy E_k has been changed completely into potential energy E_p. This is energy the block has because of the work done against the force of gravity. When the falling block comes back to the original height from whence it was thrown, it has roughly the same kinetic energy due to its velocity downward that it had due to its velocity upward when first thrown. The potential energy has been converted back into kinetic energy again.

However, careful observation shows that the final velocity downward is slightly less than the initial velocity upward, due of course to the friction between the moving block and the air. This slows the motion of the block down in the same way that friction between the block and the surface on which it is sliding causes a decrease in velocity and a constant loss in kinetic energy. Part of the kinetic energy has been changed into thermal energy.

The history of the first law

The first suggestion of a relation between kinetic (mechanical) energy and thermal energy (heat) was made by Benjamin Thompson, a native of Woburn, Massachusetts, who left the American colonies at the time of the American Revolution, was employed by the King of Bavaria at the Munich arsenal, and was given the title of Count Rumford, the name of a town in New England where he had lived for a time. After making observations of the heat generated during the boring of cannon, Rumford suggested that the heat was proportional to the mechanical energy expended in the boring. In the following year, 1799, the English scientist Sir Humphry Davy showed that mechanical energy brought about melting when two blocks of ice were rubbed together, an effect similar to the melting produced by the absorption of heat. However, the broad principle of the conservation of energy was not clearly formulated until a number of years later.

In 1838, Julius Robert von Mayer received the degree of Doctor of Medicine from Tübingen University in Germany and shortly afterward sailed as ship's doctor on a schooner bound from Rotterdam to the Dutch East Indies. During his stay in the tropics he noticed that the venous blood of the sailors was almost as brilliantly red as the arterial blood and was told that this was typical of blood in the tropics because the body required less oxygen to maintain normal temperature. This led Mayer to a broader conception of the principle of conservation of energy, which he stated in a paper published in the *Annalen der Chemie und Pharmazie* in 1842.

Meanwhile the English scientist James Prescott Joule, a student of John Dalton, was carrying out experiments on the conversion of energy in England. He made precise measurements of the amount of heat produced by mechanical work by observing the rise of temperature of a measured amount of water when weights of known mass moved downward through measured distances under the influence of gravity (Fig. 7-2). He also carried out experiments on the conversion of electric energy into heat. His results were embodied in a paper entitled "On the Mechanical Equivalent of Heat," read before the Royal Society in 1849.

Perpetual motion of the first kind

During these years, when the quantitative basis for the principle of conservation of energy was being laid through careful scientific experiments, there was a great deal of speculation on the possibility of coupling a series of energy conversions into a cycle in such a way that energy actually could be created instead of merely being conserved. As a result of the Industrial Revolution, in the early part of the nineteenth century, there was a keen appreciation of the economic value of energy. Energy was needed to drive machines to manufacture all kinds of products. The principal source of energy to operate the factories was the chemical energy in coal. Coal had to be mined and transported to the plants that powered the factory. This was expensive.

Suppose that a combination of processes could be found in which energy could be created by machine, then such a machine could produce the power to oper-

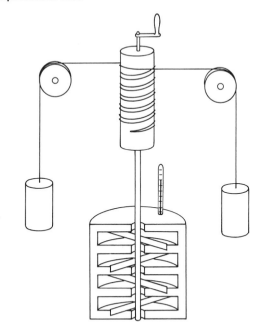

Fig. 7-2 **Joule's experiment to determine the mechanical equivalent of heat.**

ate a factory and the only cost would be its maintenance. It was conceivable that mechanical energy might be used to produce electric energy, which in turn would be changed into thermal energy and back again into mechanical energy in such abundance that there would be enough to operate a factory and also supply the initial energy needed to keep the energy-producing machine going. This imaginary device was called a *perpetual motion machine*. Hundreds of efforts were made to invent one but all without success.

Over this same period, very precise instruments had been developed for measuring mass. In chemical experiments matter in one form was changed into matter in another form, and it did not take long for chemists to be aware that in these transformations the total amount of mass never changed. Thus, the principle of conservation of mass was established. Although energy did not have quite the same tangible character as mass, the conclusion that energy also was conserved throughout periods of transformation seemed more and more reasonable. Thus, it became clear that the total amount of energy in a system is a property of the system.

The mathematical statement of the first law

As an example of how the above relationships can be put into precise mathematical form, consider a physicochemical system like that shown in Fig. 7-3. The system itself consists of 1 mole of an ideal gas confined in a cylinder equipped with a sliding piston. A slim rod leads from the piston to a platform on which a weight can be placed. There is a vacuum in the space inside the cylinder above the piston so that the upward thrust due to the pressure of the gas is balanced by the downward thrust produced by the force of gravity acting on the weight.

The device is equipped with a manometer so that the pressure of the gas can be measured. The height of the piston can be observed, and from this the volume of the gas can be calculated. There is a thermometer for observing the temperature of the gas. Precise measurements of the flow of energy into or out of the gas can be made. When the state of the gas is changed through changes in the pressure, temperature, and volume, energy is permitted to flow into or out of the gas only through certain definite channels.

The cylinder is equipped with thermal insulation (shown as heavy black lines) so that heat can flow into or out of the gas only through a metallic rod at the bottom. Thermometers inserted in this rod make it possible to calculate exactly how much heat flows into or out of the gas during a given period of time by measuring the thermal gradient. This part of the device is known as the heat channel. The amount of heat which flows in or out during a change of state is denoted by Q.

The only other way energy can enter or leave the gas is in the form of work. If the piston falls, work is performed on the gas and energy enters it. If the piston rises, energy leaves the gas as work is done in raising the weight. The amount of energy entering or leaving the system during a given change of state is designated by W.

The first law of thermodynamics asserts that the internal energy of the thermodynamic system (the gas) is a function of the state of the system. If the gas, for example, is in the standard state at 0°C under a pressure of 1 atm and with a volume of 22.4 liters mole^{-1}, the internal energy of the gas has one and only one value. Similarly, there is a unique characteristic value of the internal energy of the gas associated with each of the possible states in which the gas may be found. When the gas changes from an initial state (state 1) to a final state (state 2), the change in the internal energy of the gas $\underset{1\to2}{\Delta E}$ in passing from state 1 to state 2 always has the same value no matter by what path or in what way the change of state is brought about. The value of $\underset{1\to2}{\Delta E}$ is defined by the relation

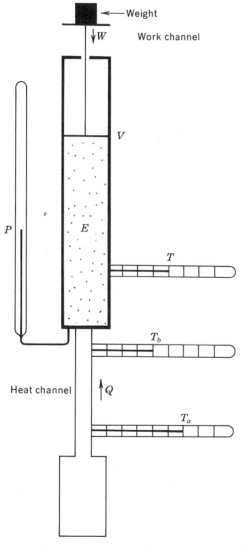

Fig. 7-3   A thermodynamic system with energy channels.

$$\underset{1\to2}{\Delta E} \equiv E_2 - E_1 \tag{7-4}$$

Fig. 7-4   A cyclic change of state.

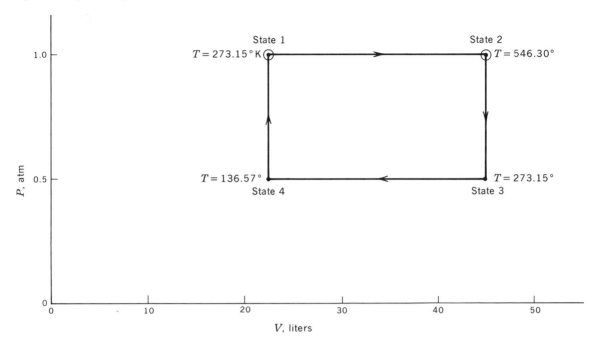

If, for example, the gas has gained energy during this change, $\underset{1\to2}{\Delta E}$ will have a positive value and it must be equal to the net algebraic sum of the amount of energy which, during this state change, has entered or left the system through the heat channel $\underset{1\to2}{Q}$ and the amount of energy which has entered or left the system through the work channel $\underset{1\to2}{W}$. Mathematically the first law of thermodynamics can be expressed

$$\underset{1\to2}{\Delta E} = \underset{1\to2}{Q} + \underset{1\to2}{W} \tag{7-5}$$

If the change of state is so small that, in the sense of the calculus, there is only an infinitesimal change in the internal energy, Eq. (7-5) can be written

$$dE = DQ + DW \tag{7-6}$$

According to the concepts of the calculus, the infinitesimal change in the energy dE is an exact differential and is therefore written using the small letter d; the expressions on the right side of the equation are not exact differentials and are therefore denoted by D. The difference between the exact differential and the other quantities can be made clear by considering the differences in their natures.

Suppose that a series of changes of state are carried out as shown in Fig. 7-4, such that at the end of the series of changes the gas is back in the original state from which it started. Such a series of changes is known as a cyclic change of state. The sum of the series of changes in the internal energy in going around such a cycle is indicated by the symbol $\oint dE$, called a circuit integral. Since at the end of the cycle, the gas is in the same state as it was at the beginning, it must have the same energy at the end as at the beginning and the algebraic sum of all the changes in energy in going around the cycle must be zero:

$$\oint dE = 0 \tag{7-7}$$

This relation is always true for an exact differential.

Any variable like E which is a true property of the system always has the circuit integral equal to zero for a cyclic change. On the other hand, for a cyclic change, the circuit integrals of DQ and DW may or may *not* be equal to zero:

$$\oint DQ \neq 0 \tag{7-8}$$

$$\oint DW \neq 0 \tag{7-9}$$

as will be made clear in the examples in the balance of the chapter.

Work and heat units

It is obvious that in Eqs. (7-5) and (7-6), Q and W must be expressed in the same units if their numerical values are to be combined algebraically to give the value of ΔE. In the majority of thermodynamic publications today, the calorie is the unit used for Q. This is also the customary unit in which to express E. The fundamental unit of W is the erg, but more convenient is the joule, which is equal to 10^7 ergs.

The first quantitative relation between the unit of thermal energy and the unit of mechanical energy was established by Joule, in whose honor the unit of mechanical work is named. While the precise experiments carried out since Joule's time have established the relationship between these units with an accuracy of about 1 part in 10,000, the International Commission on Units has recommended that the calorie be *defined* in terms of the fundamental unit of mechanical work rather than in terms of the amount of heat necessary to raise one gram of water by one degree Celsius under specified conditions, and so today this defined calorie is generally used, a unit equal to 4.184 joules.

In writing equations like (7-5) and (7-6) there is generally no indication of the conversion factor necessary to change the units of W into calories, in which Q and E will be expressed. However, in making numerical calculations the necessity of using this conversion factor must always be kept in mind.

For a change of state W can be expressed as the product of the pressure times the change in volume. For an infinitesimal change in state, the expression is

$$DW = -P\,dV \tag{7-10}$$

For a finite change in state under conditions of constant pressure

$$W_{1 \to 2} = -P(V_2 - V_1) \tag{7-11}$$

For a finite change of state where pressure is not constant

$$W_{1 \to 2} = -\int_1^2 P\,dV \tag{7-12}$$

Often pressure is expressed in units of atmospheres and volume in units of liters. In this case, the unit of work is the liter atmosphere: 1 liter atm is equal to 101.3278 absolute joules or 24.2179 defined calories.

7-3   HEAT CAPACITY

Heat capacity has been defined as the ratio between the rate at which heat enters the system and the rate of increase of temperature. If the system is heated at constant volume, the heat capacity at constant volume of 1 mole of substance c_v is defined symbolically as

$$c_v \equiv \left(\frac{DQ}{dT}\right)_v \tag{7-13}$$

The value of the energy change in the form of work DW is given in terms of pressure and the volume change by Eq. (7-10). If this expression is inserted in the expression for the first law of thermodynamics (7-6) and applied to a system of 1 mole, the relation is

$$d\mathrm{E} = DQ - P\,dv \tag{7-14}$$

For the change at constant volume $dv = 0$. If the two remaining terms in Eq. (7-14) are divided by dT, the expression becomes

$$\frac{d\mathrm{E}}{dT} = \frac{DQ}{dT} \tag{7-15}$$

This can be put in the form of partial differentials with an explicit subscript to indicate a change at constant volume

$$\left(\frac{\partial \mathrm{E}}{\partial T}\right)_v = \left(\frac{DQ}{dT}\right)_v \equiv c_v \tag{7-16}$$

This expression shows that the heat capacity at constant volume is equal to the partial derivative of the internal energy with respect to temperature at constant volume.   Recall that for an ideal monatomic gas the kinetic energy is a function of the temperature

$$E_k = \tfrac{3}{2}RT \tag{7-17}$$

The internal energy of 1 mole of an ideal gas E can be regarded as the sum of the kinetic energy and other terms, e.g., chemical energy and nuclear energy, which may be regarded as constant with respect to temperature.   This leads to the expression for the heat capacity at constant volume of an ideal gas

$$c_v = \left(\frac{\partial E}{\partial T}\right)_v = \left[\frac{d}{dT}\left(\tfrac{3}{2}RT\right)\right]_v = \tfrac{3}{2}R \tag{7-18}$$

Heat capacity at constant pressure

When heat flows into the system and a change of state takes place at constant pressure, the relation between the change in internal energy, the amount of heat, and the amount of work can be obtained by integrating Eq. (7-14) to give

$$\int_1^2 dE = \int_1^2 DQ - P\int_1^2 dV \tag{7-19}$$

Carrying out the integration from state 1 to state 2 gives

$$E_2 - E_1 = \underset{1\to2}{Q} - P(V_2 - V_1) \tag{7-20}$$

which can be rearranged

$$\underset{1\to2}{Q} = (E_2 + PV_2) - (E_1 + PV_1) \tag{7-21}$$

Recall that in an earlier chapter *enthalpy H* was defined by the relation

$$H \equiv E + PV \tag{7-22}$$

Since energy is a true property of the system according to the first law of thermodynamics, and since pressure and volume are also true properties of the system, their sum must be a true property of the system. Thus, for integration around a cyclic process there is the relation

$$\oint dH = 0 \tag{7-23}$$

Combining Eqs. (7-21) and (7-22),

$$\left(\underset{1\to2}{Q}\right)_p = H_2 - H_1 \tag{7-24}$$

Thus, the heat change $\underset{1\to2}{Q}$ when the system passes from state 1 to state 2 at constant pressure is a measure of the change in enthalpy.   Differentiating Eq. (7-22) for a change under conditions of constant pressure gives

$$(dH)_p = dE + P\,dV \tag{7-25}$$

Applying Eq. (7-14) to a change under conditions of constant pressure,

$$(dE)_p = (DQ)_p - P\,dV \tag{7-26}$$

In order to get the change with temperature, dT can be introduced on both sides in the denominator, giving

$$\left(\frac{\partial E}{\partial T}\right)_p = \left(\frac{DQ}{dT}\right)_p - P\left(\frac{\partial V}{\partial T}\right)_p \tag{7-27}$$

This can be rearranged to give

$$\left(\frac{DQ}{dT}\right)_p = \left(\frac{\partial E}{\partial T}\right)_p + P\left(\frac{\partial V}{\partial T}\right)_p$$
$$= \left[\frac{\partial(E + PV)}{\partial T}\right]_p = \left(\frac{\partial H}{\partial T}\right)_p \tag{7-28}$$

Since $(DQ/dT)_p$ is, by definition, the heat capacity at constant pressure C_p, we find for a system of 1 mole that

$$c_p = \left(\frac{\partial H}{\partial T}\right)_p \tag{7-29}$$

Since almost all the experimental observations of heat capacity are made under conditions of constant pressure, this equation is the basis for obtaining many other important thermodynamic data.

7-4   CHANGES OF STATE IN AN IDEAL GAS

In order to illustrate further applications of the first law of thermodynamics, consider some of the principal ways in which a change of state in an ideal gas can be brought about.   The nature of these changes is best illustrated by showing the initial and final states and the path of the change on a plot of pressure against volume as in Fig. 7-6.   A schematic drawing (Fig. 7-5) is also provided to emphasize the changes in the state

Fig. 7-5   Change of state at constant volume (isochoric change) of 1 mole of an ideal gas.

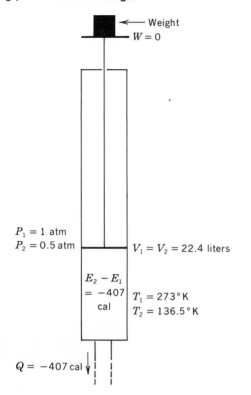

$P_1 = 1$ atm
$P_2 = 0.5$ atm

$V_1 = V_2 = 22.4$ liters

$E_2 - E_1 = -407$ cal

$T_1 = 273°K$
$T_2 = 136.5°K$

$Q = -407$ cal

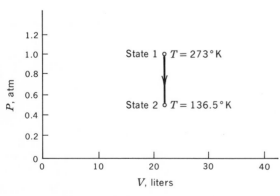

$W = 0$ (no area under change of state line)
$Q = -407$ cal

Fig. 7-6   Change of state at constant volume on the PV diagram.

variables and the amounts of energy in the form of work or heat passing into or out of the system.   In each case the system consists of 1 mole of an ideal gas and the initial state is that where $P = 1$ atm, $v = 22.4$ liters and $T = 273°K$.

Change of state at constant volume

As an example of this change of state (sometimes called an isochoric change), heat is abstracted from the gas while the volume remains constant and the temperature is reduced to one-half its initial value, as shown in Figs. 7-5 and 7-6.   Writing the first law in the form

$$dE = DQ - P\,dV \qquad (7\text{-}30)$$

it is clear that $dV = 0$ since the volume is constant. The remaining terms can then be integrated

$$\int_1^2 dE = \int_1^2 DQ \qquad (7\text{-}31)$$

giving

$$E_2 - E_1 = \underset{1 \to 2}{Q} \qquad (7\text{-}32)$$

Example 7-1

With the help of the first law of thermodynamics, calculate the heat withdrawn from the system and the change in the internal energy when the temperature of the gas is reduced to one-half its initial value at constant V.

From the ideal-gas law, the relation is first derived

$$\frac{P_2 V_2}{T_2} = \frac{P_1 V_1}{T_1}$$

$T_2 = 136.5°K$, and so $P_2 = 0.5$ atm and the line indicating the change of state on the diagram is the vertical line shown in Fig. 7-6.   The amount of heat abstracted from the gas is given by

$$\underset{1 \to 2}{Q} = C_v(T_2 - T_1) = \tfrac{3}{2}R(136.5°)$$

$$= -\tfrac{3}{2}(1.987 \text{ cal deg}^{-1})(136.5°)$$

$$= -406.8 \text{ cal}$$

According to the kinetic theory of the ideal gas,

$$E_k = \tfrac{3}{2}RT$$

Thus the change in the internal energy in passing from state 1 to state 2 is

$$(E_k)_2 - (E_k)_1 = \tfrac{3}{2}R(T_2 - T_1) = -406.8 \text{ cal}$$

From this it is clear that in such a change the energy change consists of the transfer of 406.8 cal of energy from the internal energy of the gas to the heat reservoir while there is no energy flow through the work channel.

Change of state at constant pressure

As an example of this change (sometimes called an isopiestic change), the volume of the gas is doubled while the pressure remains constant, as shown in Figs. 7-7 and 7-8. The relations between the variables can be shown by writing the first law of thermodynamics in the form

$$dE = DQ - P\,dV \tag{7-33}$$

Since the change is taking place at constant pressure, this expression can be integrated

$$\int_1^2 dE = \int_1^2 DQ - P\int_1^2 dV \tag{7-34}$$

This gives

$$E_2 - E_1 = \underset{1 \to 2}{Q} - P(V_2 - V_1) \tag{7-35}$$

Example 7-2

Calculate the change in variables and the energy flowing through the work and heat channels for the isopiestic expansion of 1 mole of an ideal monatomic gas from the initial standard state at 0°C to the final state where the volume has been doubled while the pressure remains constant.

From the equation for the ideal gas, it is clear that when the pressure remains constant and volume doubles, the temperature also must double, giving a final temperature of 546°K. The change in volume is shown in Fig. 7-7. The path of the change is shown in Fig. 7-8. The energy flowing out through the work channel is

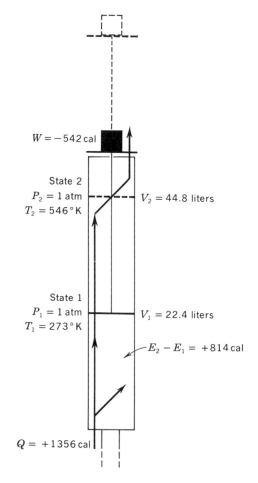

Fig. 7-7   Change of state at constant pressure (isopiestic) of 1 mole of an ideal gas.

$W = -542$ cal

State 2
$P_2 = 1$ atm
$T_2 = 546°$K
$V_2 = 44.8$ liters

State 1
$P_1 = 1$ atm
$T_1 = 273°$K
$V_1 = 22.4$ liters

$E_2 - E_1 = +814$ cal

$Q = +1356$ cal

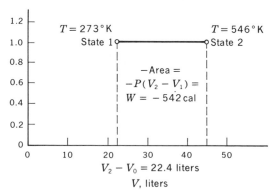

Fig. 7-8   Change of state at constant pressure on the PV diagram.

$$W_{1\to2} = 1 \text{ atm } (44.8 - 22.4 \text{ liters})$$

$$= -22.4 \text{ liter atm} \times 24.2 \text{ cal (liter atm)}^{-1}$$
$$= -542 \text{ cal}$$

According to the kinetic theory of an ideal gas, the change in internal energy is given by

$$E_2 - E_1 = \tfrac{3}{2}R(T_2 - T_1)$$
$$= \tfrac{3}{2}R(546 - 273°)$$
$$= \tfrac{3}{2}(1.987 \text{ cal deg}^{-1})(273°)$$
$$= 814 \text{ cal}$$

According to the first law of thermodynamics, the heat absorbed should be the increase in

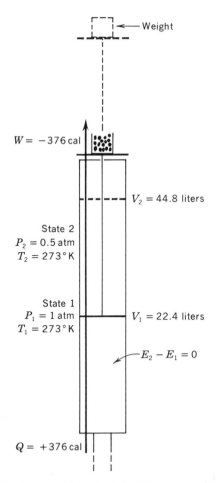

Fig. 7-9 Change of state at constant temperature (isothermal) of 1 mole of an ideal gas.

internal energy plus the amount of energy that flows out through the heat channel

$$Q_{1\to2} = (E_2 - E_1) + P(V_2 - V_1)$$
$$= 814 + 542 \text{ cal}$$
$$= 1356 \text{ cal}$$

When the experiment is performed, it is found that the system does absorb 1356 cal in the form of heat. As shown on the diagram, 814 cal of this heat goes into increasing the internal energy of the gas, while 542 cal flows straight on through the system and out through the work channel.

In Fig. 7-8 note that the area under the line that represents the change of state is equal to the pressure multiplied by the change in volume. In other words, this area is equal in magnitude and opposite in sign to the energy flow through the work channel.

Change of state at constant temperature

The change of state at constant temperature is frequently called an *isothermal* change. If volume increases as temperature remains constant, pressure must decrease. As shown in Fig. 7-9, if instead of a single block, the weight consists of a number of small steel balls, such a change can be carried out constantly, balancing the downward thrust of the weight against the upward thrust produced by the pressure of the gas. When the balls are removed one by one, the pressure is decreased and the volume increases. The change of state is shown on the PV diagram in Fig. 7-10. In order to see the relation between the variables and the energy changes, the first law of thermodynamics may again be written

$$dE = DQ - P\,dV \tag{7-36}$$

For 1 mole of an ideal gas, the relation is

$$P = \frac{RT}{V} \tag{7-37}$$

Inserting this value in Eq. (7-36) gives

$$d\mathbf{E} = DQ - RT\frac{dV}{V} \tag{7-38}$$

Isothermal

$$dE = 0$$
$$0 = DQ - P\,dV.$$
$$\therefore \ 0 = DQ$$

According to the calculus

$$\frac{d\mathbf{v}}{\mathbf{v}} = d \ln \mathbf{v} \tag{7-39}$$

Equation (7-38) can be integrated over the range state 1 to state 2

$$\int_1^2 d\mathbf{E} = \int_1^2 DQ - RT \int_1^2 d \ln \mathbf{v} \tag{7-40}$$

which gives

$$\mathbf{E}_2 - \mathbf{E}_1 = \underset{1 \to 2}{Q} - RT \ln \frac{\mathbf{v}_2}{\mathbf{v}_1} \tag{7-41}$$

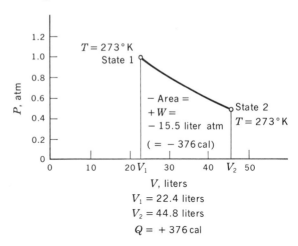

Fig. 7-10 Change of state at constant T (isothermal) on the PV diagram.

$$V_1 = 22.4 \text{ liters}$$
$$V_2 = 44.8 \text{ liters}$$
$$Q = +376 \text{ cal}$$

Example 7-3

Calculate the energy flowing out through the work channel and flowing in through the heat channel when 1 mole of an ideal gas expands isothermally from the standard state at 0°C to twice its initial volume.

According to the kinetic theory of gases, the internal energy of the gas is given by

$$\mathbf{E} = \tfrac{3}{2}RT$$

If the temperature remains constant, there is no change in the value of the internal energy so that $\mathbf{E}_2 - \mathbf{E}_1 = 0$. Thus, the heat flowing into the system $\underset{1 \to 2}{Q}$ goes right on through the system out at the top as work, as expressed by

$$\underset{1 \to 2}{Q} = RT \ln \frac{\mathbf{v}_2}{\mathbf{v}_1}$$

The relation between the natural logarithm and the common logarithm is

$$\ln \frac{\mathbf{v}_2}{\mathbf{v}_1} = 2.303 \log \frac{\mathbf{v}_2}{\mathbf{v}_1}$$

Thus the energy flowing out through the work channel is given by

$$\underset{1 \to 2}{W} = -RT\, 2.303 \log \frac{\mathbf{v}_2}{\mathbf{v}_1}$$
$$= -(1.987)(273°)(2.303)(\log 2)$$
$$= -376 \text{ cal}$$

Change of state with $Q = 0$

In this fourth type of change the heat channel is closed so that energy can enter or leave the system only by means of the work channel. This is called an *adiabatic* (no flow) change of state; the nature of the change is shown in Figs. 7-11 and 7-12. The first law of thermodynamics is written

$$dE = DQ - P\,dV \tag{7-42}$$

The term $DQ = 0$, so that the expression becomes

$$dE = -P\,dV \tag{7-43}$$

Use is now made of the relation from the ideal-gas law

$$P = \frac{RT}{V} \tag{7-44}$$

and the expression for dE obtained from the first law of thermodynamics

$$d\mathbf{E} = c_v\, dT \tag{7-45}$$

Substituting in Eq. (7-42) gives

$$c_v\, dT = -RT \frac{d\mathbf{v}}{\mathbf{v}} \tag{7-46}$$

This can be arranged and put in the logarithmic form

$$\frac{c_v}{R} d \ln T = -d \ln \mathbf{v} \tag{7-47}$$

Fig. 7-11   Adiabatic expansion of 1 mole of an ideal gas.

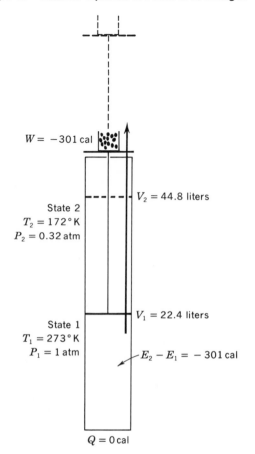

which can be integrated from state 1 to state 2

$$\frac{C_v}{R} \int_1^2 d \ln T = - \int_1^2 d \ln v \qquad (7\text{-}48)$$

giving

$$\frac{C_v}{R} \left(\ln \frac{T_2}{T_1} \right) = - \left(\ln \frac{v_2}{v_1} \right) \qquad (7\text{-}49)$$

If both logarithms are changed to the base 10,

$$\frac{C_v}{R} \log \frac{T_2}{T_1} = - \log \frac{v_2}{v_1} \qquad (7\text{-}50)$$

Example 7-4

When 1 mole of an ideal gas is in the standard state at 0°C and is expanded to double the volume under adiabatic conditions, calculate the final temperature and the amount of energy flowing into the gas through the work channel.

The final temperature can be calculated from Eq. (7-50)

$$\log \frac{T_2}{T_1} = - \frac{R}{C_v} \log \frac{v_2}{v_1}$$

$$= - \frac{R}{\frac{3}{2}R} \log 2$$

$$= - \tfrac{2}{3}(0.301) = -0.201$$

The change in internal energy of the gas is given by

$$\frac{T_1}{T_2} = 1.59$$

$$T_2 = \frac{T_1}{1.59} = 172°\text{K}$$

$$E_2 - E_1 = C_v(T_2 - T_1)$$

$$= \tfrac{3}{2}R(172 - 273°)$$

$$= - \tfrac{3}{2}(1.987)(101°) = -301 \text{ cal}$$

Put the first law of thermodynamics in the form

$$E_2 - E_1 = \underset{1 \to 2}{Q} + \underset{1 \to 2}{W}$$

Since $\underset{1 \to 2}{Q} = 0$, the energy change in the work channel is equal to the change in internal energy, which is −301 cal. This means that in this change 301 cal is abstracted from the in-

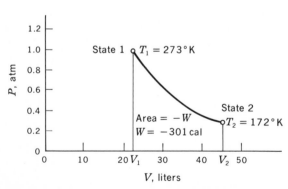

Fig. 7-12   Adiabatic expansion on the PV diagram.

ternal energy of the gas and flows out through the work channel. In Fig. 7-12 the area corresponding to this numerical value is shown.

Summary of types of state change

In Fig. 7-13 the four principal types of change of state of an ideal gas are plotted in a single diagram. In each case the work done is the area lying under the line which represents the change of state. For a change of constant volume this area is zero; no work is done. The minimum amount of work is done when the gas undergoes adiabatic change. Slightly more work is done when the change takes place at constant temperature. The maximum amount of work is done when the change takes place at constant pressure.

7-5   A THERMODYNAMIC CYCLE

In Fig. 7-14 changes of state are shown which constitute a thermodynamic cycle. Along line a, 1 mole of an ideal gas under standard conditions at 0°C is expanded adiabatically until its volume is doubled and the temperature is dropped to 172°K. Along line b the gas temperature is increased at constant volume to 273°K. Then the gas is compressed at constant temperature back to the initial state, where the pressure is 1 atm and the volume is 22.4 liters. The flow of energy through the work channel in each of these steps is given by

$$\oint DW = \underset{(a)}{-c_v(T_2 - T_1)} + \underset{(b)}{0 \text{ cal}} + \underset{(c)}{RT_3 \ln (v_1/v_3)}$$

$$(7\text{-}51)$$

Since $c_v = \frac{3}{2}R$, the numerical values can be calculated easily and are given in the expression

$$\oint DW = -301 \text{ cal} + 0 + 376 \text{ cal}$$

$$= +75 \text{ cal} \qquad (7\text{-}52)$$

The sum of the amounts of energy flowing through the heat channel is given by

$$\oint DQ = \underset{(a)}{0 \text{ cal}} + \underset{(b)}{\tfrac{3}{2}R(273° - 172°)} - \underset{(c)}{RT_3 \ln (v_1/v_3)}$$

$$(7\text{-}53)$$

Putting in the numerical values,

$$\oint DQ = 0 + 301 - 376 \text{ cal} = -75 \text{ cal} \qquad (7\text{-}54)$$

Thus, in going around the cycle, the circuit integral of the exact differentials such as dE, dV, and dT are all equal to zero, but the circuit integral of DW is $+75$ cal and the circuit integral of DQ is -75 cal. In effect, this means that 75 cal equivalent of mechanical energy has been converted into 75 cal of thermal energy. This is an illustration of the fact that the circuit integrals of W and Q are not equal to zero and that DW and DQ are not exact differentials.

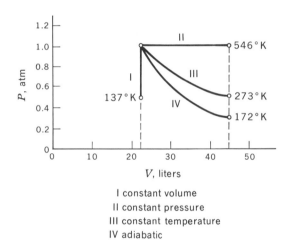

I constant volume
II constant pressure
III constant temperature
IV adiabatic

Fig. 7-13   Four principal types of change of state.

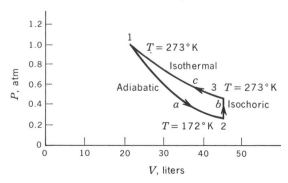

Fig. 7-14   A thermodynamic cycle.

7-6   THE EXPANSION OF NONIDEAL GASES

Next to the work of van der Waals, probably the most important early experimental investigation of nonideal gases was that carried out by Julius Thomsen, a Danish chemist, and Joule, who as we have seen, also made the first careful measurements to determine the relation between energy in the form of heat and energy in the form of work.

Joule and Thomsen performed their crucial experiment in an apparatus like that shown schematically in Fig. 7-15.   A nonideal gas was placed in the part of the cylinder shown at the left of the drawing.   Pushing the piston in forced part of this gas through a porous plug, shown at the center of the drawing, and into the cylinder on the right.   By placing thermometers on either side of the porous plug, Joule and Thomsen demonstrated that the gas underwent cooling when passing through the plug.

During the experiment, the pistons were moved at such a rate that the pressure in the left half of the cylinder was kept constant at a value P and the pressure on the right half of the cylinder was kept constant at a value P'.   When 1 mole of gas passed through the plug, the piston on the left performed an amount of work PV on the system and on the right side an amount of work $P'V'$ was done by the system on the right-hand piston.   The first law of thermodynamics applicable for this change can be written

$$E - E' = Q + W \qquad (7\text{-}55)$$

where E is the internal energy of the mole of gas before it passes through the plug and E' is the internal energy of the mole of gas after it has passed through the plug. The apparatus can be insulated so that effectively

$Q = 0$.   Since mechanical energy passes into the gas at the left in amount PV and passes out of the gas at the right in amount $P'V'$, $W = PV - P'V'$.   Therefore, Eq. (7-55) becomes

$$E' - E = PV - P'V' \qquad (7\text{-}56)$$

Rearranging,

$$E + PV = E' + P'V' \qquad (7\text{-}57)$$

Thus

$$H = H' \qquad (7\text{-}58)$$

showing that the process takes place at constant enthalpy.

By measuring how much the gas is cooled $(T' - T)$ for various values of the pressure drop $(P' - P)$, it is possible to determine the value of dT/dP at constant H.   This quantity is defined as the Joule-Thomsen coefficient μ:

$$\mu \equiv \left(\frac{\partial T}{\partial P} \right)_H \qquad (7\text{-}59)$$

At low temperatures not too far above the boiling point, μ is positive, and the gas is cooled on expansion. As temperature increases, the value of μ falls, passes through zero, and becomes negative.   When μ is negative, expansion warms the gas.   The temperature at which $\mu = 0$ is called the Joule-Thomsen inversion temperature.   This is 50°K for helium, 204.6°K for hydrogen, 621°K for nitrogen, and higher for gases with higher boiling points.

The liquefaction of gases

The cooling produced by the Joule-Thomsen expansion can be used to lower the temperature of gases to the

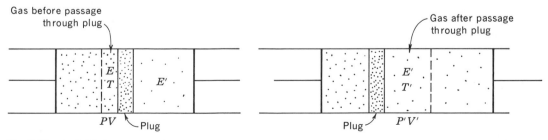

Fig. 7-15   Joule-Thomsen expansion.

point where liquefaction takes place. An apparatus for such a process is shown in Fig. 7-16. A gas like nitrogen is first compressed to a pressure of about 2,000 psi and then passed through the helical tube in the interior of the apparatus and out through an expansion valve at the bottom. The expansion at this point cools the gas, which then flows back through the space outside the helix and cools the incoming gas. The gas is then returned to the pump and recirculated. By this cyclic process the gas inside the apparatus gets colder and colder and finally condenses into liquid drops, which are collected at the bottom of the apparatus and drawn out into Dewar flasks for storage in the form of liquid.

Because of the variation of the Joule-Thomsen coefficient with temperature, the gas cools only when circulation begins at a temperature below the Joule-Thomsen inversion temperature. For this reason, in cooling to 1°K, it was first necessary to use liquid nitrogen to precool hydrogen; then the hydrogen was cooled further by a Joule-Thomsen expansion and used to precool helium; finally, helium itself was cooled by a Joule-Thomsen expansion. It was in such a cascade that helium was first liquefied by the Dutch physicist Heike Kamerlingh-Onnes at the University of Leiden in 1909. Kamerlingh-Onnes received the Nobel Prize for his pioneering work in low-temperature research, which today is frequently called cryogenics.

Figure 7-17 shows a logarithmic temperature scale with some of the points first reached by the early Joule-Thomsen expansion of gases and points attained by other processes more recently. In by far the most widely used method for the cooling of gases, the gas is made to perform actual mechanical work under adiabatic conditions. In this process the gas pushes against a piston, and the work energy comes from the kinetic energy of the gas molecules. Such an apparatus, devised by Professor S. C. Collins of the Massachusetts Institute of Technology, has revolutionized low-temperature research by producing liquid helium on a large scale by a cyclic adiabatic expansion.

The lowest temperatures ever reached have been attained with the help of intense magnetic fields, which reduce the thermal vibration in molecules with relatively high magnetic susceptibility; the heat is "squeezed" out of the atoms by the magnetic lines of

Fig. 7-16   Gas liquefier.

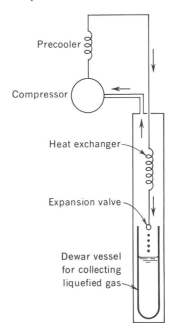

force. This method was first used by W. F. Giauque at the University of California. Employing the same principle to restrict nuclear spin, N. Kurti at Oxford University has reached temperatures within about one-millionth of a degree of 0°K. At these temperatures the heat capacity is extremely small. To warm 100 g of copper from 10^{-6} to 0.01°K requires only the amount of energy released when a common pin is dropped through a distance of 0.1 in.

In many life processes, like the replication of DNA, there are many key reactions which involve only the movement of a few atoms, so that the heat of reaction may be as small as 10^{-10} erg or less. The study of microenergy changes at extremely low temperatures may ultimately provide important knowledge for the better understanding of biochemical reactions of this sort.

7-7   THE THERMODYNAMICS OF METABOLISM

Today it is fair to say that every physicist, chemist, and biologist is in agreement that all life processes are in

Fig. 7-17   Points on temperature scale, logarithmic.

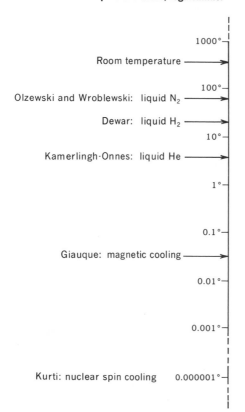

These systems are referred to as *open systems*. The human body in performing its normal life functions is an open thermodynamic system. Figure 7-18 shows on the left a laboratory thermodynamic system where in addition to the heat channel and the work channel there is a *matter channel,* whereby a measured number of moles of gas can be introduced into or taken out of the interior, where the active system, the gas, is contained. One can imagine the gas as being introduced by something like a hypodermic syringe. In this case, the first law of thermodynamics, expressing the change from the initial state to the final state, must be written with an additional term to take care of the energy introduced through this matter channel. Figure 7-18 on the right shows a human body with analogous channels. The *work channel* can be thought of as primarily the muscles, which convert the internal energy of the body into work. The *heat channel* can be thought of as the mechanisms by which the body loses heat primarily through radiation and conduction to the surrounding air. The *matter channel* can be thought of primarily as the mouth, through which matter containing energy is introduced into the human thermodynamic system. Thus, we can write for either the open laboratory thermodynamic system or the human thermodynamic system a statement of the first law of thermodynamics in the form

$$E_2 - E_1 = Q + W + M_f \qquad (7\text{-}60)$$

| change in internal energy | heat in or out | work energy in or out | energy introduced as matter or food |

It is important to keep in mind that the first law of thermodynamics applies not only to the body as a whole and to any organism as a whole but also to each individual biological cell, whether part of an organism or carrying on the life process as an independent individual.

Physiological work

Considered in detail, the work channels in a biological thermodynamic system are extremely complex. This is due to a large extent to the fact that many life processes involve *moving* material from one place to another. When such movements consist of a change of position of an arm or a leg, it is fairly obvious that movement involves work. When the arm is raised, it has a

accord with the first law of thermodynamics. No living organism, whether consisting of a single cell or of 10^{14} cells like the human body, can get the energy it needs by creating it out of nothing. Energy always flows into the organism by some means and then is converted into heat, which is radiated or conducted away from the organism, and into work, which the organism performs.

In this sense, the human body is a thermodynamic system, although admittedly a complex one. The thermodynamic systems we have been discussing in the first part of this chapter are members of a class known as *closed systems*. In these systems no material such as a gas, liquid, or solid either flows into the system during the change from the initial to the final state or flows out of the system. However, there are many thermodynamic systems of importance in physics, and biology where such a flow of matter does take place into or out of the system during the change of state.

higher potential energy in the gravitational field of the earth than when it is hanging down at the side of the body. This increase in potential energy is brought about by work performed by the muscles controlling the position of the arm. It is also fairly obvious that the blood must be kept continually circulating and that to effect this transport of the blood from one position to another work must be done to overcome forces of friction and to force the blood into certain places under pressure; this work is performed by the muscles of the heart, which produce the pumping action in this organ. It is perhaps less obvious that in the process of *growth* there is also a continual movement of molecules in solution from one place to another. This movement in many cases is very much like the expansion of a gas. Just as there is the PV factor in the expansion of a gas associated with work, in physiological solutions there is also a PV factor involved when molecules move to form a more dilute solution.

The passage of molecules in the body fluids is governed frequently by semipermeable membranes which permit some types of molecules to flow through them and restrict the flow of other molecules. As these flexible membranes move, there is also a work factor involved. It is important to keep in mind that the emphasis on the use of the ideal gas to illustrate thermodynamic relations has been justified by the desirability of proceeding from simple gaseous examples to the more complex examples encountered in the theory of solutions. Many examples of physiological work have to be regarded not only from the point of view of physical work but even electrical work, as in the operation of the nervous system.

Bioenergetics

The energy to support life on earth comes almost wholly—either directly or indirectly—from the star which is the center of our planetary system, the sun.

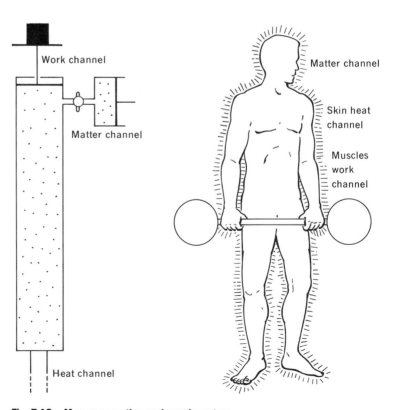

Fig. 7-18   Man as open thermodynamic system.

Today we believe that the primary source of the energy of the sun is the fusion of hydrogen atoms there into helium with the release of large amounts of nuclear energy. This maintains a temperature in the interior of the sun which may be as high as $10^{7}°C$ and a surface temperature of about 5000°C. Because the sun is so much larger than the earth, it has an enormously large radiating surface. As a result, radiant energy streams toward the earth and provides the principal source of energy that sustains life. A large part of this is visible light; there is also radiation with a wavelength a little shorter than the violet, the ultraviolet radiation which also affects many life processes.

The principal method by which solar energy is transformed into energy for use in life processes is the photosynthesis which takes place primarily in the leaves of plants. In particular, the pigment chlorophyll acts as a light trap and converts the radiant energy so that the carbon dioxide in the air and water are synthesized into products containing the solar energy as chemical energy. As an example, consider the reaction

$$6CO_2 + 6H_2 \rightarrow C_6H_{12}O_6 + 6O_2 \qquad (7\text{-}61)$$
glucose

in which the increase in enthalpy ΔH is approximately 673,000 cal mole^{-1}. When the plants that have trapped this solar energy are eaten by herbivorous animals, this chemical energy undergoes further transformation. In general, the transformation consists of oxidation processes, whereby the original solar energy trapped by the plant in the form of chemical energy is changed back into energy of work and heat and the carbon is given out by the animal in the form of carbon dioxide. It is estimated that all the plants on earth each year convert something like 17 billion tons of carbon in the form of carbon dioxide into chemical substances of which glucose may be regarded as the prototype. Our coal and petroleum reserves represent carbon that has been converted from carbon dioxide into bioproducts by plant life in the past. At the present time, the biocombustion of animal life on earth plus the combustion of coal and petroleum maintains the carbon dioxide in the air at a relatively constant level so that cycling of carbon through these various forms continues. Subsequent chapters will discuss both the macro and micro aspects of these carbon cycles: on the macro scale the energy-transforming devices like engines and furnaces and on the micro scale the energy transformations in biological cells.

Muscle energy

The discussion of the first law of thermodynamics concludes with a brief survey of the most obvious biological means by which energy is transformed into mechanical work, namely, muscle action. The commonest form of muscle is made up of a number of parallel rows of cells called myofibrils, the functional units by which the contractual machinery operates. The basic component of these muscle cells is a protein called *myosin,* a long-chain protein molecule with a molecular weight of about 350,000. Myosin molecules have the ability to catalyze the hydrolysis of adenine triphosphate (ATP) to adenine diphosphate (ADP), thereby releasing the energy that makes the muscle fiber contract and perform work.

Many mechanisms have been proposed for the process of contraction. One is a kind of folding model, in which the filaments of the myofibrils shorten. It has also been proposed that there is a sliding rather than a folding mechanism. The point to be emphasized is that there is *some mechanism* by which chemical energy brings about a mechanical change so that work is done according to the simple formula

$$W = F(x_2 - x_1) \qquad (7\text{-}62)$$

which expresses the fact that the work is equal to the force F multiplied by the distance through which it acts $x_2 - x_1$. Practically all work-producing devices, running the entire gamut from human and animal muscle to steam engines, internal combustion engines, and electric motors, end up by producing work which can be expressed by this simple equation. However, 150 years ago, when work-producing machines like the steam engine were first being built, scientists and engineers noted that it was impossible to use heat as a source of energy and then convert *all* the heat into mechanical work. There is always some heat that never is converted to work. It was the investigation of this limitation on the conversion of heat into work which led to the second great generalization of thermodynamics, to be discussed in the next chapter.

Problems

1. If 1 mole of an ideal monatomic gas at 3 atm and 300°K is expanded at constant pressure to 1.5 times the original volume, and if the pressure is produced by a 100-kg weight on top of the piston, how high is the weight raised by the expansion and what is the equivalent in calories of the work done?

2. A 50-kg weight falls through a distance of 20 cm thereby compressing an ideal gas isothermally. Calculate the amount of energy in calories that flows through the heat channel as a result.

3. If 3 moles of an ideal monatomic gas is heated at constant volume from 300 to 500°K, calculate the amount of heat flowing in through the heat channel in calories, joules, and ergs.

4. If 3 kcal of heat flows into $\frac{1}{2}$ mole of an ideal monatomic gas maintained at constant volume, what will be the final temperature of the gas if the original temperature is 250°K?

5. If 7 moles of an ideal monatomic gas expands at constant pressure from an initial state where $P = 2$ atm and $T = 250°$K to 3 times the original volume of the gas, calculate in calories the amount of energy flowing in through the heat channel, the amount of energy flowing out through the work channel, and the increase in the internal energy of the gas.

6. The force produced by the pressure of 10 moles of an ideal monatomic gas on a piston is balanced by a weight of 100 kg. If the gas is compressed at a constant pressure of 2 atm and the initial temperature is 310°K, through what distance will the weight fall, and what will be the final temperature of the gas if the volume is reduced to one-half the original volume?

7. If 3 moles of an ideal monatomic gas at 0°C and 3 atm is expanded isothermally to 10 times the original volume, calculate in calories the amount of energy that flows in through the heat channel and in liter atmospheres the amount of work which flows out through the energy channel.

8. If 1 mole of an ideal monatomic gas at 100°C and 2 atm is compressed isothermally, 2000 cal of energy flows out through the heat channel. Calculate the final volume of the gas.

9. If 5 moles of an ideal monatomic gas at an initial temperature of 310°K and an initial pressure of 1 atm is expanded adiabatically to 3 times the initial volume, calculate the final pressure and temperature of the gas.

10. A boy has an effective lung capacity of 2 liters. With his lungs filled with air at 1 atm and at body temperature (37°C), the boy holds his breath and contracts his chest cavity so that the effective volume is 1,800 cm³. Assuming that the molal heat capacity at constant volume for air is $\frac{5}{2}R$ cal deg⁻¹ mole⁻¹, calculate the temperature of the gas immediately after the contraction, assuming that there is not sufficient time for any interchange of heat between the gas and the lung tissue. Calculate the pressure of the gas in the lungs at this instant. Assuming that after a few seconds the gas has returned to body temperature while the volume remains constant, calculate the final pressure of the gas when thermal equilibrium has been attained.

11. For the experiment described in Prob. 10, calculate the increase in the internal energy of the gas in the boy's lungs in terms of calories after the adiabatic contraction has taken place. Calculate the work done by the boy's muscles in terms of joules and liter atmospheres, assuming that all the work went into the compression of the gas.

12. Taking the value of the heat capacity of air at constant volume as $\frac{5}{2}R$ cal deg⁻¹ mole⁻¹, calculate the increase in the temperature of the air when in the cylinder of an internal combustion engine 1 liter of air at 1-atm pressure and at 300°C is compressed adiabatically to one-fifth its original volume.

13. Behaving as an ideal gas, 2 liters of helium at 25°C and 1 atm is expanded first at constant pressure to a volume 4 times the original volume. The gas is then cooled to its original temperature and compressed to its original volume isothermally. Calculate the net amount of energy that has gone in or out of the work channel during this process.

14. Three cylinders each contain 1 mole of helium at 0°C and 1 atm. In the first cylinder the gas is allowed to expand at constant pressure to 10 times its original volume. In the second cylinder, the gas is allowed to expand isothermally to 10 times its original volume. In the third cylinder, the gas is allowed to expand adiabatically to 10 times its original volume. Calculate the final temperature after the expansions have taken place in the first cylinder and the third cylinder and calculate the amount of work done by the gas in each of the three cases.

15. Suppose that 1 mole of an ideal gas at 25°C and at 2-atm pressure expands reversibly at constant pressure to double its initial volume. The gas is then permitted to expand irreversibly and adiabatically against a piston where the force opposing the expansion is only one-half the original pressure and the volume now reaches 4 times the initial volume at the beginning of the experiment. Calculate the amount of energy that has flowed into the gas in the form of heat during the whole experiment, the amount of work performed by the gas in calories, and the final temperature of the gas.

16. After 1 mole of an ideal gas at 0°C and 1 atm expands irreversibly, doing no work, to a volume of 67.2 liters, it is brought back to its original volume by a process in which the pressure is maintained constant at the value reached immediately at the end of the irreversible expansion. Calculate the final temperature of the gas.

17. If 1 mole of an ideal monatomic gas at 0°C and a pressure of 1 atm is expanded adiabatically to a volume of 67.2 liters, calculate the change in the internal energy of the gas and the change in the enthalpy of the gas.

18. If 1 mole of air is heated from 25 to 50°C at constant pressure, assuming that the air behaves as an ideal gas and that the heat capacity at constant volume is $\frac{5}{2}R$ cal deg^{-1} mole^{-1}, calculate the increase in the internal energy E and in the enthalpy H during this process.

19. The same sample of air described in Prob. 18 is heated through the same range of temperature while the volume is maintained constant. Calculate the increase in internal energy E and in the enthalpy H taking place.

20. If a breath of air, in volume 1 liter, is drawn into the lungs and comes to thermal equilibrium with the body while the pressure remains constant, calculate the increase in the enthalpy of the air if the initial temperature of the air is 20°C and the pressure is 1 atm.

REFERENCES

Hargreaves, G.: "Elementary Chemical Thermodynamics," Butterworth & Co. (Publishers), Ltd., London, 1961.

Lewis, G. N., and M. Randall: "Thermodynamics," 2d ed., rev. by K. S. Pitzer and Leo Brewer, McGraw-Hill Book Company, New York, 1961.

Kirkwood, J. G., and I. Oppenheim: "Chemical Thermodynamics," McGraw-Hill Book Company, New York, 1961.

Hiebert, E. N.: "Historical Roots of the Principle of Conservation of Energy," The University of Wisconsin Press, Madison, 1962.

Mahan, B. H.: "Elementary Chemical Thermodynamics," W. A. Benjamin, Inc., New York, 1963.

Lehninger, A. L.: "Bioenergetics," W. A. Benjamin, Inc., New York, 1965.

Kestin, J.: "A Course in Thermodynamics," Blaisdell Publishing Company, Waltham, Mass., 1966.

We are all familiar with the fact that heat spontaneously flows from a hot body to a cold body when they are placed in contact while the reverse type of heat flow, from a cold body to a hot body, is never observed.   Take a pan of hot water off the stove and place it on a stone tile at room temperature.   In 10 or 15 sec enough heat flows from the pan into the tile for the pan to be a little cooler and the tile to feel warmer.   It would be surprising if after half a minute the tile became colder while the water became hotter.   No one has ever observed heat flowing from a cold to a hot body.

Chapter Eight

the second law of thermodynamics

There is a similar limitation on the direction of flow in the change of energy between the form of heat and the form of mechanical energy.   In Joule's experiment (Fig. 8-1), a falling weight turns a paddle wheel, and mechanical energy is transformed by friction into heat energy in the water, as evidenced by the increase in water temperature. Suppose the paddle wheel were placed in the water and connected to the device on which the weights are suspended.   It would be surprising if the paddle wheel turned backward, lifting the weights to a higher level, while the water became colder.   If this were to happen, one would be forced to conclude that heat was flowing out of the water and was being changed spontaneously into mechanical work, thereby lifting the weights to a greater height, where they had greater potential energy.   Such an action has never been observed to take place and we confidently believe that it never will.   From such simple experiments we conclude that there are limitations on the direction of heat flow and on the nature of processes by which heat is transformed into other forms of energy.

Fig. 8-1   Joule's experiment and its impossible antiprocess.

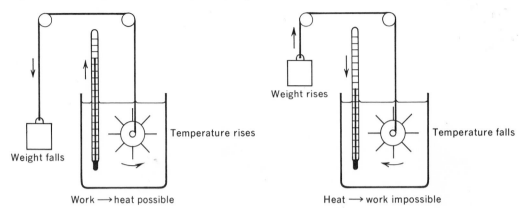

8-2   THE CARNOT CYCLE

The first clear exposition of the factors that limit the transformation of heat into other forms of energy was set forth by the French engineer-scientist Sadi Carnot at the beginning of the nineteenth century in a theoretical analysis of the transformation of heat into work by cyclic expansion and contraction of a gas. In this cycle (now called the *Carnot cycle*), the changes take place in four steps. In step I the gas is expanded isothermally; in step II it is expanded still further adiabatically; in step III it is compressed isothermally to a volume such that an adiabatic compression (step IV) restores the gas to its initial state. Each step is carried out reversibly; i.e., the force produced by the pressure of the gas on the piston is balanced by an external force so that an infinitesimal decrease in the external force causes the gas to expand and an infinitesimal increase causes the gas to contract. During each step Q and W are measured and the amount of heat converted into work is derived as a function of the temperatures of isothermal expansion and isothermal contraction. This provides the basis for deriving quantitatively the limitation on the conversion of heat into mechanical energy which is embodied in the second law of thermodynamics.

The changes of state selected as an illustration of the Carnot cycle are shown graphically in Fig. 8-2.

In the initial state (state 1), 1 mole of an ideal gas, e.g., helium, is at 0°C with $P = 1$ atm and $v = 22.4$ liters. In step I, the gas is expanded isothermally and

reversibly to $v = 44.8$ liters. For this step the first law of thermodynamics may be written:

$$Step\ I: \quad E_2 - E_1 = Q_I + W_I \tag{8-1}$$
$$ 0 \qquad +376\ cal \quad -376\ cal$$

The amount of heat absorbed is:

$$Q_I = RT \ln \frac{v_2}{v_1} = +376\ cal \tag{8-2}$$

For a process involving an ideal gas at constant temperature since there is no change in the internal energy, the work W_I must be equal numerically and opposite in sign to the heat absorbed and therefore has the value -376 cal.

In step II, the gas is expanded *adiabatically* and reversibly to state 3, in which $v_3 = 89.6$ liters. The first law of thermodynamics for this expansion may be written in the form:

$$Step\ II: \quad E_3 - E_2 = Q_{II} + W_{II} \tag{8-3}$$
$$ -301\ cal \quad 0 \quad -301\ cal$$

The numerical values are obtained with the help of the equations derived in the previous chapter for adiabatic expansion. According to Eq. (7-50),

$$\frac{c_v}{R} \log \frac{T_3}{T_2} = -\log \frac{v_3}{v_2} \tag{8-4}$$

For an ideal monatomic gas, $c_v = \frac{3}{2}R$, and so

$$\tfrac{3}{2}\log \frac{T_3}{T_2} = -\log \frac{v_3}{v_2} \tag{8-5}$$

$E_3 - E_2 = \frac{3}{2} R \Delta T$

$= \frac{3}{2}(273 - 172)R$

$= 301$

and doubling the volume reduces the temperature to $172°K$. Thus,

$$E_3 - E_2 = c_v(T_2 - T_1) = \tfrac{3}{2}R(273 - 172°)$$
$$= -301 \text{ cal} \qquad (8\text{-}6)$$

and from Eq. (8-3)

$$W_{II} = E_3 - E_2 = -301 \text{ cal} \qquad (8\text{-}7)$$

In step III the gas is compressed reversibly at the constant temperature of $172°K$ back to the volume of 44.8 liters (state 4). For this step Eq. (8-2) can be applied relating heat and volume change in an isothermal compression

$$Q_{III} = -W_{III} = RT \ln \frac{V_4}{V_3}$$
$$= (1.987)(172°)2.303 \log \frac{44.8}{89.6}$$
$$= -237 \text{ cal} \qquad (8\text{-}8)$$

Since the compression is isothermal,

$$E_4 - E_3 = 0 \qquad (8\text{-}9)$$

and for step III the first law reads:

$$\textit{Step III:} \quad \underset{0}{E_4} - \underset{-237 \text{ cal}}{E_3} = \underset{}{Q_{III}} + \underset{+237 \text{ cal}}{W_{III}} \qquad (8\text{-}10)$$

Finally, in step IV, the gas is compressed adiabatically and reversibly back to the initial volume of 22.4 liters. The equation for adiabatic compression is

$$\frac{c_v}{R} \log \frac{T_1'}{T_4} = -\log \frac{V_1}{V_4} \qquad (8\text{-}11)$$

Inserting numerical values and solving for the final temperature T_1' shows that this is the same as the initial temperature T_1. If the final volume and temperature are the same as the initial volume and temperature, the final pressure must equal the initial pressure; thus, the gas has been carried around the cycle and returned to its initial state

$$E_1 - E_4 = \frac{c_v}{R}(T_1 - T_4)$$
$$= \tfrac{3}{2}(273 - 172°)$$
$$= +301 \text{ cal} \qquad (8\text{-}12)$$

so that

$$W_{IV} = E_1 - E_4 = +301 \text{ cal} \qquad (8\text{-}13)$$

and the statement of the first law for step IV is:

$$\textit{Step IV:} \quad \underset{+301 \text{ cal}}{E_1} - \underset{0}{E_4} = \underset{301 \text{ cal}}{Q_{IV}} + \underset{}{W_{IV}} \qquad (8\text{-}14)$$

We turn now to examine the net total amount of the mechanical energy flow and of the heat flow in this cycle:

$$\oint DW = \underset{-139 \text{ cal}}{W_I} + \underset{-376 \text{ cal}}{W_{II}} + \underset{-301 \text{ cal}}{W_{III}} + \underset{+237 \text{ cal}}{W_{IV}} + {\scriptstyle +301 \text{ cal}} \qquad (8\text{-}15)$$

Therefore, 139 cal has left the system in the form of mechanical work. Similarly,

$$\oint DQ = \underset{+139 \text{ cal}}{Q_I} + \underset{+376 \text{ cal}}{Q_{II}} + \underset{0 \text{ cal}}{Q_{III}} + \underset{-237 \text{ cal}}{Q_{IV}} + {\scriptstyle 0 \text{ cal}} \qquad (8\text{-}16)$$

Therefore, 139 cal has gone into the system in the form of heat. Since, in the final state, the system has the same internal energy as in the initial state,

$$139 \text{ cal (heat at } 273°K) \rightarrow 139 \text{ cal work} \qquad (8\text{-}17)$$

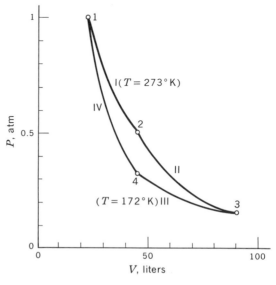

Fig. 8-2   A Carnot cycle for 1 mole of an ideal gas.

And coupled with this transformation, 237 cal of heat went into the system at 273°K and left the system at 172°K:

$$237 \text{ cal (heat at } 273°K) \rightarrow 237 \text{ cal (heat at } 172°K) \tag{8-18}$$

Thus, although heat has been changed into mechanical energy [Eq. (8-17)], it is *coupled* with a degradation of heat from 273 to 172°K, so that there is no contradiction of the statement that the reverse of Joule's experiment is never observed; i.e., the transformation of heat into work as an *isolated* process.

Carnot pointed out that there are many devices like a steam engine which will convert heat into mechanical work or energy, the kind of change which is so desirable if energy is needed to make machinery operate or to perform other processes requiring mechanical work. He also pointed out that a price must be paid for this conversion of heat into work, i.e., that other heat must be degraded into heat at a lower temperature where it is less available for doing work.

Example 8-1

Suppose that 1 mole of an ideal gas at 0°C and a pressure of 1 atm is expanded isothermally to a volume of 224 liters; it is then expanded adiabatically to a volume of 448 liters where the temperature is 172°K; it is then compressed isothermally to a volume of 44.8 liters; and, finally, it is compressed adiabatically back to the initial state at 0°C and 1 atm. Calculate the amount of heat which has been converted into work as the gas is carried through this Carnot cycle.

The heat passing into the system during the first isothermal expansion is given by

$$Q_I = -W_I = RT \ln \frac{V_2}{V_1}$$

$$= (2.303)(1.987)(273) \log 10 = 1249 \text{ cal}$$

The heat flowing out of the system in the isothermal compression is given by

$$Q_{III} = -W_{III} = RT \ln \frac{V_4}{V_3}$$

$$= (2.303)(1.987)(172)(\log 0.1)$$

$$= -787 \text{ cal}$$

Since the adiabatic expansion doubles the volume and the adiabatic compression halves the volume, the amount of work performed by the system in expanding is exactly canceled out by the amount of work performed on the system when it is compressed. Therefore, the net amount of heat which is changed into work by the cycle is given by

$$Q_{net} = 1250 - 786 \text{ cal} = 462 \text{ cal}$$

8-3   THE EFFICIENCY OF HEAT ENGINES

A heat engine takes in heat at a high temperature, transforms part of it into mechanical work, and then gives out the remainder at a lower temperature. Since mechanical work is the desired product to be obtained from the operation of the heat engine, the efficiency of the heat engine is defined as the net *output* of mechanical work W_0 divided by the heat input Q_I when the engine goes through one complete cycle. From the point of view of thermodynamics $W_0 = -W$, where the latter is the net energy passing through the work channel; so that e has a positive sign when defined as

$$e \equiv \frac{W_0}{Q_I} \tag{8-19}$$

At the end of the cycle the system returns to the same state which it had at the beginning; therefore, the internal energy at the end of the cycle is the same as it was at the beginning, and consequently, the heat taken into the system during this cycle Q_I must be exactly equal to the heat and net work given out during the cycle, $W_0 - Q_{III}$; note that Q_{III} is heat leaving the system and has a negative value.

$$Q_I = W_0 - Q_{III} \tag{8-20}$$

Thus

$$W_0 = Q_I + Q_{III} \tag{8-21}$$

When this value for W_0 is substituted in Eq. (8-19),

$$e = \frac{Q_I + Q_{III}}{Q_I} \tag{8-22}$$

Because steps I and III are isothermal,

$$\frac{T_2}{T_3} = \frac{T_1}{T_4} \tag{8-23}$$

Since T_1' (the final temperature) is equal to T_1 (the initial temperature), T_2/T_3 can be inserted in Eq. (8-11) to give

$$\frac{C_v}{R} \log \frac{T_2}{T_3} = -\log \frac{v_1}{v_4} \tag{8-24}$$

If, in Eq. (8-4), T_3/T_2 and v_3/v_2 are inverted, the equation reads

$$\frac{C_v}{R} \log \frac{T_2}{T_3} = -\log \frac{v_2}{v_3} \tag{8-25}$$

Combining these equations,

$$\frac{v_1}{v_4} = \frac{v_2}{v_3} \tag{8-26}$$

Therefore

$$\frac{v_2}{v_1} = \frac{v_3}{v_4} \tag{8-27}$$

The value of Q_I in terms of the volume change is found in Eq. (8-2), and the value for Q_{III} in terms of volume change is found in Eq. (8-8); these expressions can be substituted in Eq. (8-22) to give

$$e = \frac{RT_I \ln (v_2/v_1) + RT_{III} \ln (v_4/v_3)}{RT_I \ln (v_2/v_1)} \tag{8-28}$$

Substituting from Eq. (8-27) gives

$$e = \frac{RT_I \ln (v_2/v_1) - RT_{III} \ln (v_2/v_1)}{RT_I \ln (v_2/v_1)} \tag{8-29}$$

Canceling out terms which appear both in the numerator and the denominator reduces this to the simple expression

$$e = \frac{T_I - T_{III}}{T_I} \tag{8-30}$$

This expression shows that the efficiency of the heat engine is equal to the difference between the upper temperature at which the heat goes in and the lower temperature at which it flows out divided by the upper temperature. For the example discussed in the preceding section, the values of the temperatures may be inserted, giving

$$e = \frac{273 - 172°}{273°} = 0.37 \tag{8-31}$$

Example 8-2

Calculate the efficiency of the cycle outlined in Example 8-1 from the heat input and output.

The efficiency of the cycle is given by the amount of heat converted into work divided by the amount of heat absorbed in the isothermal expansion

$$e = \frac{Q_{net}}{Q_I} = \frac{462}{1249} = 0.37$$

8-4 ENTROPY

In addition to deriving the important expression for the efficiency of a heat engine in terms of the isothermal expansion temperature and the isothermal compression temperature, Carnot also pointed out an important relation between the ratio of the heat taken in to the temperature at which it is taken in and the heat given out to the temperature at which it is given out. This expression can be derived by writing the ratio of the heat taken in Q_I to the heat given out Q_{III} in terms of the volume changes

$$\frac{Q_I}{Q_{III}} = \frac{RT_I \ln (V_2/V_1)}{RT_{III} \ln (V_4/V_3)} \tag{8-32}$$

Again making use of the volume ratio equality in Eq. (8-27) and canceling out common terms in the numerator and the denominator, this expression becomes

$$\frac{Q_I}{Q_{III}} = \frac{-T_I}{T_{III}} \tag{8-33}$$

which can be rearranged to give

$$\frac{Q_I}{T_I} + \frac{Q_{III}}{T_{III}} = 0 \tag{8-34}$$

Recall that in steps II and IV there is no heat flow, so that Q_{II} and Q_{IV} are both equal to zero. In view of this, Eq. (8-34) is equivalent mathematically to the cyclic integral of DQ/T each increment of heat being divided by the temperature at which it enters or leaves the system

$$\oint \frac{DQ}{T} = 0 \qquad \text{reversible cycle} \tag{8-35}$$

The first law of thermodynamics rests on the experimental fact that the cyclic integral of dE is always found to be equal to zero, and the conclusion is drawn that the internal energy is therefore a true property of the system. From the experimental observations of thermodynamic cycles operating on ideal gases, an analogous relation is found for DQ/T; in any such thermodynamic cycle in which the changes are carried out reversibly, the relation expressed in Eq. (8-35) is always found to be true. This suggests that DQ/T represents an infinitesimal change in a quantity that is a true property of a system just as E is a true property. These conclusions have led to the name *entropy* for this property, denoted by the symbol S. Thus, Eq. (8-35) can be generalized to read

$$\oint dS = 0 \qquad \text{reversible cycle} \qquad (8\text{-}36)$$

The basic meaning of this property, entropy, involves other considerations far more fundamental than the value of a heat change divided by the temperature at which the heat change takes place. Fundamentally, entropy is a measure of the probability of the state of a system. The value of the heat change divided by the temperature at which it takes place is only a *measure* of the entropy change and does not in itself reveal the fundamental nature of entropy. These relations will be discussed in detail in Sec. 8-5.

Example 8-3

Calculate the entropy increase in step I of the cycle described in Example 8-1 and the entropy decrease in step III of this same cycle.
The entropy changes are

$$dS = \frac{Q_I}{T_I} = \frac{1249}{273} = 4.6 \text{ eu}$$

$$dS = \frac{Q_{III}}{T_{III}} = \frac{-787}{172} = -4.6 \text{ eu}$$

Irreversible changes

In order to see the full significance of Carnot's deductions, it is necessary to consider entropy changes in irreversible cycles. Irreversibility may be introduced into the cycle when the gas is permitted to expand and the pressure of the gas is not balanced by an equal and opposite force. This is illustrated in Fig. 8-3. In Fig. 8-3a, 1 mole of an ideal gas in the standard state at 0°C expands from an initial volume V_1 of 22.4 liters to

a final volume V_2 of 44.8 liters. As shown in the preceding chapter, the heat absorbed is

$$Q_{1 \to 2} = RT \ln \frac{V_2}{V_1} = 376 \text{ cal} \qquad (8\text{-}37)$$

The entropy change is the heat absorbed divided by the temperature at which it is absorbed.

$$S_2 - S_1 = \frac{Q_{1 \to 2}}{T} = \frac{376 \text{ cal}}{273°} = 1.38 \text{ eu} \qquad (8\text{-}38)$$

The entropy change can be expressed in terms of units which correspond to calories per degree; however, it is frequently convenient to abbreviate this unit and call it an *entropy unit* (eu).

In contrast to this reversible expansion, suppose that the weight on top of the piston is removed so that the piston rises from its initial position (corresponding to V_1) to its final position V_2 with the motion opposed by no force. Under these circumstances, no energy flows out through the work channel, no heat flows in through the heat channel, and the temperature of the gas remains constant throughout the change since its internal energy is not changed. But if entropy is a true property of the system, it must have increased by exactly the same amount during the course of this irreversible expansion as it increased when the state was changed from V_1 to V_2 by means of the reversible expansion.

In order to prove this, the gas can be returned to its initial state. Note that there is no spontaneous way by which such a return can be accomplished. The gas will not, all by itself, shrink back from a volume of 44.8 liters to a volume of 22.4 liters. As will be shown in a subsequent section, this kind of irreversible contraction of volume is never observed because it is so highly *improbable*. If one gas molecule were bouncing around in a volume of 44.8 liters, sooner or later this molecule would be found in the lower half of the volume 22.4 liters and, in effect, an irreversible contraction would have taken place. However, one might have to wait for 10 to the millionth power years before ever observing 1 g mole of gas with 6×10^{23} molecules contracting all by itself into one-half of its original volume. Since the universe is believed to be only about 10^9 years old, it is scarcely misleading to say that the irreversible contraction of an ideal gas into a volume one-half that of the initial volume is *impossible*.

If the gas is returned from a volume of 44.8 liters to the initial volume of 22.4 liters by a *reversible* compression, the entropy change is found to be

$$s_1 - s_2 = \frac{Q_{1 \to 2}}{T} = \frac{-376 \text{ cal}}{273°} = -1.38 \text{ eu} \qquad (8\text{-}39)$$

since in this compression 376 cal will leave the gas through the heat channel.

Turning now to the Carnot cycle, it is clear that if the step I of isothermal expansion takes place *irreversibly*, no work is performed and the efficiency of the cycle is reduced to zero. Even if only a part of the isothermal expansion takes place irreversibly, the work done will be less than in the reversible cycle and the efficiency will be correspondingly less. Thus, the efficiency of a reversible cycle represents the maximum possible efficiency, the maximum possible amount of work to be obtained, when the amount of heat under consideration is turned partly into mechanical work and partly into heat at a lower temperature. This relation can be represented schematically:

Reversible cycle: Q_I at T_I $\longrightarrow$ W (8-40)
                     $\searrow Q_\mathrm{III}$ at T_III

Putting in the numerical values for the example of the Carnot cycle:

376 cal in at 273°K $\xrightarrow{\text{reversible cycle}}$ 139 cal out as work
                                        $\searrow$ 237 cal out at 172°K
(8-41)

Thus, in the reversible Carnot cycle operated *forward*, 376 cal of heat enters the system at 273°K and is transformed into 139 cal which leaves the system as mechanical work and 237 cal which leaves the system as heat at 172°K. If the Carnot cycle for this particular example is operated in reverse, exactly the opposite process takes place:

376 cal out at 273°K $\xleftarrow{\text{reversible cycle}}$ 139 cal in as work
                                        $\diagdown$ 237 cal in at 172°K
(8-42)

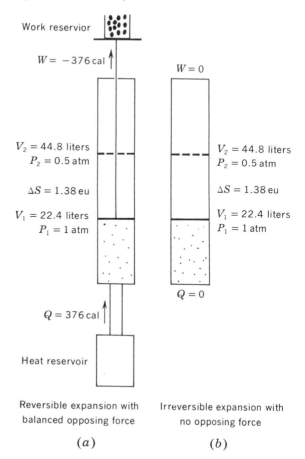

Fig. 8-3 **Isothermal expansion at 273°K.**

Work reservior

$W = -376 \text{ cal}$ ↑ $W = 0$

$V_2 = 44.8$ liters $V_2 = 44.8$ liters
$P_2 = 0.5$ atm $P_2 = 0.5$ atm

$\Delta S = 1.38$ eu $\Delta S = 1.38$ eu

$V_1 = 22.4$ liters $V_1 = 22.4$ liters
$P_1 = 1$ atm $P_1 = 1$ atm

                              $Q = 0$

$Q = 376 \text{ cal}$ ↑

Heat reservoir

Reversible expansion with   Irreversible expansion with
balanced opposing force   no opposing force

(*a*)   (*b*)

That, of course, is a refrigeration cycle where work is used to pump heat from the lower temperature (172°K) to the upper temperature (273°K).

Suppose that there is some supercycle with an efficiency of 0.50 that is used to take 376 cal of heat at 273°K and transform half of it into mechanical work so that 188 cal leaves the system as work and only 188 cal leaves the system as heat at 172°K:

376 cal in at 273°K $\xrightarrow{\text{supercycle}}$ 188 cal out as work
                                        $\searrow$ 188 cal out at 172°K
(8-43)

The reversible Carnot cycle then could be coupled to this supercycle to raise the 188 cal of heat back from 172 to 273°K. This would only require 120 cal of mechanical energy, and this refrigeration engine would deliver only 308 cal at the higher temperature:

308 cal out at 273°K ⟵ reversible cycle ⟵ 120 cal in as work
⟵ 188 cal in at 172°K

$$(8\text{-}44)$$

The operation of the supercycle forward with the reversible cycle backward would then result, after one complete cycle, in the direct conversion of 68 cal of heat into 68 cal of mechanical work:

68 cal heat in at 273°K ⟶ 68 cal out as work

$$(8\text{-}45)$$

An imaginary machine which would make it possible to convert heat completely into work or to raise heat from a lower temperature to a higher temperature without any net expenditure of work has been called *a perpetual motion machine of the second kind*. No such machine has ever been devised, and we believe that none ever will be, whether operating with ideal gases or with any other types of physical or chemical change.

Just as the first law of thermodynamics is expressed in words as the principle of the conservation of energy, these relations may be embodied in a statement called the second law of thermodynamics:

Heat is never converted completely into mechanical energy by a process in which there is no accompanying change of mechanical energy into heat or change of heat from a higher to a lower temperature.

Again it is important to keep in mind that if one could wait long enough, an ideal gas might spontaneously contract to a smaller volume; then the gas could be expanded and made to do work, and there would be a violation of the second law of thermodynamics. Thus, the second law rests on the basis that in any system consisting of large numbers of molecules the probability of such contractions taking place is overwhelmingly small.

In view of these considerations the most general statement of the second law is

$$dS \geq 0 \qquad (8\text{-}46)$$

In this expression, the equal sign applies when the change is reversible and the greater-than sign applies when the change is irreversible.

8-5   THE STATISTICAL MEANING OF ENTROPY

Volume probability

One of the most helpful ways to see the relation between the thermodynamic definition of entropy and the statistical definition in terms of probability is found in the consideration of the probable position of the molecules of a gas enclosed in a box. Of course, there is no way of directly observing the positions of individual molecules, but the kinetic theory of gases has been so successful in explaining molecular behavior, that it is possible to deduce the positions of individual molecules on the basis of this theory.

In Fig. 8-4 a single molecule is shown inside a box. In Fig. 8-4a the molecule is in the lower half of the box; in Fig. 8-4b the molecule is in the upper half; Fig. 8-4c illustrates the probability of finding the molecule in the whole box.

Neglecting the effect of gravity on the molecule, there is every reason to believe that the probability Y_a of finding the molecule in the lower part of the box is the same as the probability Y_b of finding it in the upper half, and so $Y_a = Y_b = \frac{1}{2}$. The probability Y_c of finding the molecule in the whole box is unity. The volume of the lower half of the box is denoted by V_a. The volume of the whole box is denoted by V_c. Thus, there is the relation

$$\frac{Y_c}{Y_a} = \frac{V_c}{V_a} = 2 \qquad (8\text{-}47)$$

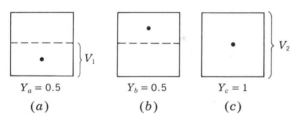

$Y_a = 0.5$ $Y_b = 0.5$ $Y_c = 1$

(a) (b) (c)

Fig. 8-4   The probability of finding a molecule either in the lower half of or in the whole box.

Consider now the probability Y_1 of finding a single molecule in the lower tenth of a box as contrasted with the probability Y_2 of finding it in the total box, as shown in Fig. 8-5a and b. Under these conditions

$$\frac{Y_2}{Y_1} = \frac{V_2}{V_1} = 10 \tag{8-48}$$

If there are two molecules in the box, the probability of finding *both* molecules in the lower tenth of the box is equal to the product of the probabilities of finding each one in the lower tenth of the box

$$\underset{\substack{\text{both molecules}}}{Y_1} = \underset{\substack{\text{molecule} \\ a}}{^aY_1} \times \underset{\substack{\text{molecule} \\ b}}{^bY_1} = 0.1 \times 0.1 = 0.1^2 = 0.01 \tag{8-49}$$

Thus

$$\underset{\substack{\text{both molecules}}}{\frac{Y_2}{Y_1}} = \underset{\substack{\text{ratio for} \\ \text{individual} \\ \text{molecule}}}{\left(\frac{Y_2}{Y_1}\right)_i^n} = 10^2 = 100 \tag{8-50}$$

Now consider 1 g mole of gas initially in the lower tenth of the volume and finally in the total volume, as shown in Fig. 8-5c and d. In this case, the number of molecules n is Avogadro's number N_A. As shown in Eq. (8-48), the value of Y_2/Y_1 for an individual molecule is equal to the ratio of the total volume to the partial volume. Thus, for 1 g mole of molecules, the quotient of the probability of finding all the gas in the whole box and the probability of finding it in the lower tenth of the box is

$$\underset{\substack{\text{g mole gas}}}{\frac{Y_2}{Y_1}} = \left(\frac{V_2}{V_1}\right)^{N_A} \tag{8-51}$$

If the logarithm of both sides of this expression is taken,

$$\ln \frac{Y_2}{Y_1} = \ln \left(\frac{V_2}{V_1}\right)^{N_A} = N_A \ln \frac{V_2}{V_1} \tag{8-52}$$

Statistical entropy ^sS is defined by the relation

$$^sS = k \ln W \tag{8-53}$$

where W is a multiplicity factor, proportional to the probability of the state of the system, and k is the

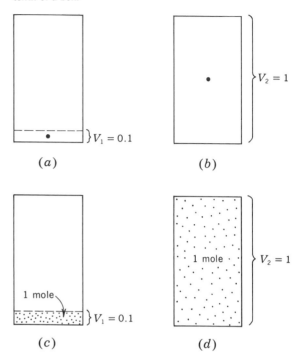

Fig. 8-5 The probability of finding molecules in the lower tenth of a box.

(a)

(b)

(c)

(d)

$V_2 = 1$

$V_1 = 0.1$

1 mole

$V_1 = 0.1$

1 mole

$V_2 = 1$

Boltzmann constant. The difference between the entropies of two states is thus a function of the ratio of the probabilities of the two states

$$^sS_2 - {}^sS_1 = k \ln \frac{W_2}{W_1} = k \ln \frac{Y_2}{Y_1} \tag{8-54}$$

Thus, the increase in entropy in passing from state 1 (gas confined in the lower tenth of the box) to state 2 (gas confined in the total box) is

$$^sS_2 - {}^sS_1 = k \ln \frac{Y_2}{Y_1} = k \ln \left(\frac{V_2}{V_1}\right)^{N_A} \tag{8-55}$$

Since $R = kN_A$

$$^sS_2 - {}^sS_1 = kN_A \ln \frac{V_2}{V_1} = R \ln \frac{V_2}{V_1} \tag{8-55a}$$

Recall that the heat $\underset{1 \to 2}{Q}$ absorbed when an ideal gas expands reversibly from volume V_1 to volume V_2 is

$$\underset{1 \to 2}{Q} = RT \ln \frac{V_2}{V_1} \tag{8-56}$$

Dividing this expression by the temperature T at which the isothermal expansion takes place gives the expression for the change in thermodynamic entropy ${}^tS$

$$\frac{Q}{T} = R \ln \frac{V_2}{V_1} = {}^tS_2 - {}^tS_1 \qquad (8\text{-}57)$$

According to the thermodynamic definition of entropy, the term on the left of this equation is equal to the change in entropy when this reversible expansion takes place. A comparison with Eq. (8-55a) shows that this is also the change in statistical entropy, and so the two can be equated

$$\underset{\text{statistical entropy}}{{}^sS_2 - {}^sS_1 = R \ln (V_2/V_1)} = \underset{\text{thermodynamic entropy}}{\frac{Q}{T} = {}^tS_2 - {}^tS_1}$$

$$(8\text{-}58)$$

Thus, it is clear that in this particular change the statistical definition of entropy is the equivalent of the thermodynamic definition of entropy. In general, this is true for all types of change of state.

Example 8-4

Calculate the ratio of the probabilities the two states of a mole of gas at 0°C and 0.1 atm and at 0°C and 1 atm.

The ratio of these probabilities is given by

$$\frac{Y_b}{Y_a} = \left(\frac{224}{22.4}\right)^{6.02 \times 10^{23}} = 10^{6.02 \times 10^{23}}$$

Entropy and complexions

In the discussion of thermal energy distribution in Chap. 6, a calculation was made of the number of complexions W corresponding to a state in which an assembly of harmonic oscillators were in thermal equilibrium with each other. Given a set of oscillators and a definite number of quanta of energy, it was found that there were a precise number W of different ways in which these quanta could be distributed among the oscillators:

$$W = \frac{N!}{N_0! \, N_1! \, N_2! \, \cdots} \qquad (8\text{-}59)$$

where N is the total number of oscillators, N_0 is the number of oscillators in the zero energy level, N_1 is the number in the first energy level, and so on.

When N is very large, there is one *type* of distribution which has a value of W so much greater than any of the others that this value may be taken as W for the state. For the majority of states discussed in the following chapters, this distribution with the maximum value of W is the Maxwell-Boltzmann distribution. Just as the probability of the state of an ideal-gas molecule is proportional to the volume [Eq. (8-48)], the probability of a state in general is proportional to or equal to the value of W. Thus by reasoning analogous to that which led to Eq. (8-55), the general statistical definition of entropy is given as

$$^sS = k \ln W \qquad (8\text{-}60)$$

In acquiring a knowledge of the relations between thermodynamic properties and physical or chemical behavior, it is helpful to keep in mind that the deductions from statistical mechanics lead to the conclusion that thermodynamic entropy is merely a logarithmic *measure* of the probability of the system. There is an advantage in using this logarithmic measure because the definitions just set forth lead to an expression of entropy in units which make it possible to relate this property directly to the quantities of heat flowing into or out of a system during a change and other thermodynamic properties of the system. But the fundamental meaning of entropy is not a quantity of heat divided by temperature but a logarithm of probability.

8-6   ENTROPY AND ORDER

Entropy of mixing

Probability and *order* are closely associated. Consider a standard pack of cards. Suppose that the pack is arranged so that all the red cards are in the upper half of the pack and all the black cards are in the lower half of the pack. With respect to color, this represents a high degree of order. Next the cards are shuffled thoroughly. The chance is extremely high (about 10^{15} to 1) that at the end of the shuffling the cards will be found to be indiscriminately mixed. It would be necessary to shuffle the pack probably about 10^{15} times before one would find all the red cards back in the top half of the pack and all the black cards in the lower half of the pack.   In other words, when a pack of cards is randomly shuffled, an end result of disorder is much more probable than an end result of order.

Suppose that 1 g mole of nitrogen molecules is in the upper part of a tube and 1 g mole of helium molecules in the lower part, as shown in Fig. 8-6, with the two gases separated by a gastight sliding partition. In the initial state when the two gases are in separate compartments, the total entropy of the system is equal to the sum of the entropies of each individual gas

$$S_1 = {}^aS_1 + {}^bS_1 = R \ln v_1 + {}^as^* + R \ln v_1 + {}^bs^*$$

(8-61)

where S_1 = total entropy

$\quad {}^aS_1$ = entropy of nitrogen

$\quad {}^bS_1$ = entropy of oxygen

$\quad {}^as^*$ = part of entropy of nitrogen not related to volume

$\quad {}^bs^*$ = part of entropy of oxygen not related to volume

When the partition is removed, the nitrogen diffuses downward and the oxygen diffuses upward, and after a short time the two gases are intimately mixed.   Assuming that the gases behave like ideal gases, the entropy of the system in the final state S_2 will be

$$S_2 = {}^aS_2 + {}^bS_2 = R \ln v_2 + {}^as^* + R \ln v_2 + {}^bs^*$$

(8-62)

The change in entropy will be

$$\Delta S = S_2 - S_1 = 2R \ln \frac{v_2}{v_1} = 2R \ln 2 \qquad (8\text{-}63)$$

The diffusion of the two gases into each other is very much like the "diffusion" of the two different colors of cards into each other when the pack with the red cards in the upper half and the black cards in the lower half is shuffled.   In both cases there is a higher entropy associated with the greater disorder found when the two species are mixed together.   Just as there is a very low probability that shuffling the cards further will ever result in the red cards returning to the upper half of the pack and the black to the lower half, there is an extremely low probability that the shuffling

Fig. 8-6   The entropy of mixing 1 mole of N_2 (open circles) and 1 mole of He (solid circles).

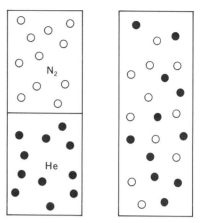

$\Delta S = 2R \ln 2 = 2.76$ eu

action of the thermal motion of the gas molecules will ever result in a situation where all the nitrogen molecules are in the upper half of the tube and all the oxygen molecules are in the lower half.

Example 8-5

Calculate the entropy of mixing of 3 moles of nitrogen and 2 moles of hydrogen behaving as ideal gases.

The entropy of mixing for the nitrogen is given by

$$S_2 - S_1 = 3R \ln \frac{V_2}{V_1}$$

$$= 3(1.987)(2.303) \log \tfrac{5}{3} = 3.04 \text{ eu}$$

and for hydrogen

$$S_2 - S_1 = 2R \ln \frac{V_2}{V_1}$$

$$= 2(1.987)2.303 \log \tfrac{5}{2} = 3.64 \text{ eu}$$

Total entropy of mixing $= 3.04 + 3.64 = 6.68$ eu

Biological order

A system like a biological cell represents a very high degree of order.   The cell contains on the average something like 10^{14} atoms grouped into molecules and macromolecules; these, in turn, are arranged with respect to each other in a pattern of extremely complex order. One of the most interesting problems connected with the chemical and physical interpretation of the life process is the search for relationships between various aspects of biological order and the thermodynamic properties of the substances within the cell. This is a topic which will be discussed at some length in many of the later chapters of this book.

8-7 // ENTROPY AND HEAT CAPACITY

The second law of thermodynamics leads directly to relations between changes in entropy and changes in heat capacity. As a basis for developing these relationships, the second law is written in the form

$$dS = \frac{DQ_{rev}}{T} \tag{8-64}$$

The subscript rev denotes that DQ is the infinitesimal amount of heat which enters or leaves the system in an infinitesimal *reversible* change. In the following discussion, the assumption will be made that all changes are reversible, and the subscript will be used only if there is any question of ambiguity.

Heat capacity at constant volume

When dT is placed in the denominator on both sides of Eq. (8-64) and the condition of constant volume is imposed, the equation becomes

$$\left(\frac{\partial S}{\partial T}\right)_v = \left(\frac{DQ}{\partial T}\right)_v \frac{1}{T} \tag{8-65}$$

Since $(DQ/\partial T)_v$ is by definition C_v,

$$\left(\frac{\partial S}{\partial T}\right)_v = \frac{C_v}{T} \tag{8-66}$$

This can be rearranged to give

$$dS = C_v \frac{dT}{T} = C_v\, d \ln T \tag{8-67}$$

Integrating this equation from state 1 to state 2,

$$\int_1^2 dS = \int_1^2 C_v\, d \ln T \tag{8-68}$$

When C_v does not vary with temperature, the integration can be carried out directly, giving

$$S_2 - S_1 = C_v \ln \frac{T_2}{T_1} \tag{8-69}$$

Heat capacity at constant pressure

For a change at constant pressure, Eq. (8-64) can be written

$$\left(\frac{\partial S}{\partial T}\right)_p = \left(\frac{DQ}{\partial T}\right)_p \frac{1}{T} \tag{8-70}$$

By definition $(DQ/\partial T)_p$ is equal to C_p. Thus,

$$\left(\frac{\partial S}{\partial T}\right)_p = \frac{C_p}{T} \tag{8-71}$$

This can be put in the form

$$dS = C_p \frac{dT}{T} \tag{8-72}$$

Integrating between state 1 and state 2,

$$S_2 - S_1 = \int_1^2 C_p\, d \ln T \tag{8-73}$$

When C_p does not vary with temperature, this equation may be written

$$S_2 - S_1 = C_p \ln \frac{T_2}{T_1} \tag{8-74}$$

The majority of chemical compounds have values of C_p that change so rapidly with temperature that Eq. (8-74) cannot be used to calculate the entropy increase with meaningful accuracy over a range of 100° or more. One of the most useful ways of calculating entropy in such cases is graphical integration. As shown in Eq. (8-73), a plot of C_p against ln T provides a curve beneath which the area corresponds to S. The area corresponding to 1 eu is the area with a height corresponding to 1 cal deg^{-1} mole^{-1} and a width corresponding to an increase of 1 in the value of the natural logarithm.

Frequently, it is more convenient to use a logarithmic scale with base 10 rather than a natural logarithm, as shown in Fig. 8-7. In that case the unit area has a width corresponding to an increase of 2.303 on the log scale. By counting squares in the area under the curve, values of S can be determined for a series of temperatures.

Fig. 8-7 The entropy of benzene at 250°K shown as the area under C_p plotted against log T.

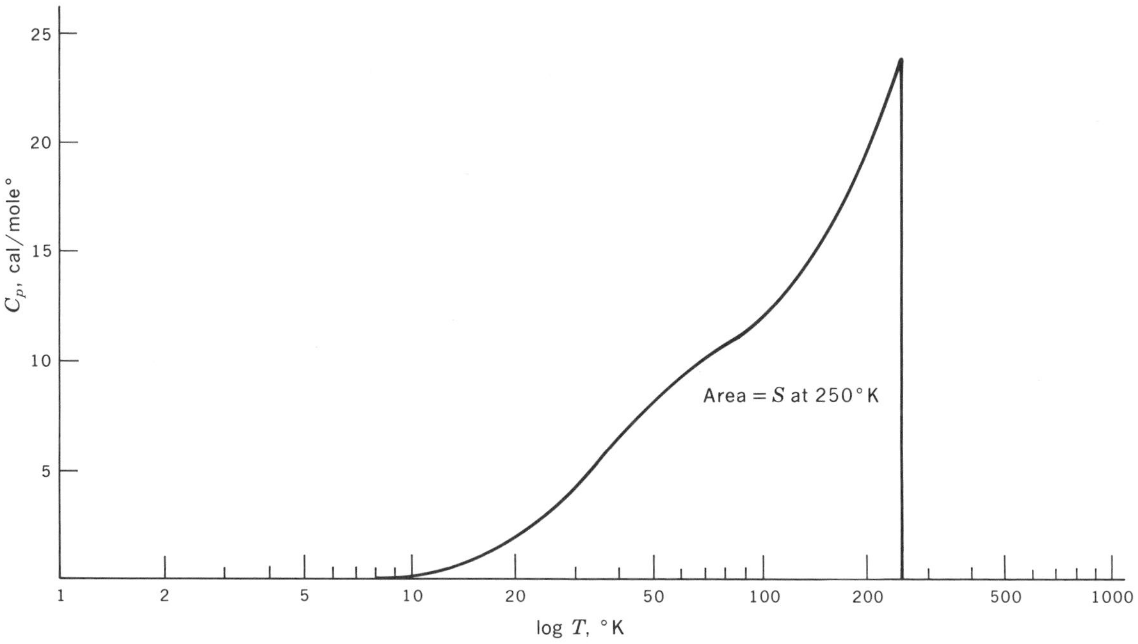

Example 8-6

If 3 moles of helium gas is heated from 0 to 100°C with pressure maintained constant at 1 atm, calculate the increase in the entropy of the gas.

The increase in entropy is given by

$$S_2 - S_1 = C_p \ln \frac{T_2}{T_1}$$

$$= 3 \times \tfrac{5}{2}(1.987)(2.303) \log \tfrac{373}{273} = 4.65 \text{ eu}$$

8-8 ENTROPY AND CHANGES OF PHASE

When there is a phase change, like the melting of ice to form water in the liquid state, heat flows into the system. Under these conditions the entropy change is given by the application of the second law of thermodynamics. For a change in which 1 mole of a substance goes from one phase to another, it is customary to express the heat change as the molal enthalpy for this particular kind of transition. For example, in fusion

$$s_l - s_c = \Delta s_f = \frac{\Delta H_f}{T_m} \tag{8-75}$$

where s_l = entropy of 1 mole of a substance in liquid state

s_c = entropy of 1 mole of a substance in crystalline state

Δs_f = entropy of fusion

ΔH_f = enthalpy of fusion

T_m = temperature of fusion or melting point, °K

In the same way for a change of phase from the liquid to the vapor state

$$s_g - s_l = \Delta s_v = \frac{\Delta H_v}{T_b} \tag{8-76}$$

where s_g = entropy of 1 mole of a substance in gaseous state

s_l = entropy of 1 mole of substance in liquid state

Δs_v = entropy of vaporization

ΔH_v = enthalpy of vaporization

T_b = boiling point or temperature of phase change °K

Since the enthalpy of fusion and the enthalpy of vaporization are both positive quantities, the entropy of fusion and the entropy of vaporization are also positive. This implies that there is a change toward a state of greater disorder in these processes.

Recall that in the crystalline state molecules are arranged in a lattice which represents a high degree of atomic or molecular order. In the liquid state molecules are arranged in a much more disorderly fashion. At the same temperature the random state is a far more probable state than the ordered state. The crystal can be thought of as analogous to the pack of cards when the cards are arranged in order, alternately red and black throughout the pack, as shown in Fig. 8-8a. The liquid state might be thought of as the pack of cards when the cards are *indiscriminately* arranged with respect to their color, as in Fig. 8-8b. While the change to greater disorder is not quite as apparent in the case of the gaseous state, the fact that larger volumes permit states of greater probability shows that the vaporization of a liquid is analogous in many ways to the expansion of a gas from a small volume to a large volume.

A number of important regularities have been observed both for the entropy of fusion and the entropy of vaporization for many molecules. They will be discussed in greater detail when the relation of thermodynamic properties to physical equilibria is surveyed.

In Fig. 8-9, values of the molal entropy of benzene are plotted against temperature. Entropy increases gradually with temperature over the crystalline range. At the melting point there is an abrupt increase in entropy due to the change from order in the crystal state to relative disorder in the liquid. The same abrupt rise is seen at the boiling point due to a similar increase in disorder in the phase change. The justification for taking S as equal to zero at $0°K$ will be discussed in the next chapter.

Example 8-7

The heat of vaporization of ethyl alcohol is 205 cal g^{-1} at 78.3°C. Calculate the entropy of vaporization.

The heat of vaporization per mole is the heat of vaporization per gram multiplied by the molecular weight, which is 46.07 g mole^{-1}. Thus, the heat of vaporization per mole is 9444 cal. Dividing this value by the boiling temperature $351°K$, we get the value of 26.9 eu for the entropy of vaporization per mole.

Problems

1. Suppose that 1 mole of water vapor (regarded as an ideal gas) at 10 atm and 400°K is expanded to twice its volume isothermally; it is then expanded adiabatically to four times the initial volume; it is compressed isothermally back to twice its initial volume; and finally, compressed adiabatically back to its initial state. Calculate the amount of heat converted into work as the mole of water vapor is put through this Carnot cycle. $C_v = 11$ cal deg^{-1} mole^{-1}.

Red Black

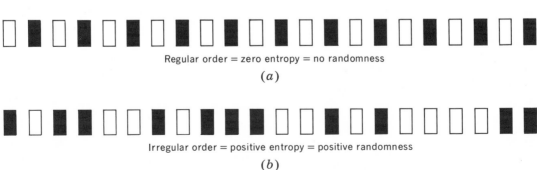

Regular order = zero entropy = no randomness

(a)

Irregular order = positive entropy = positive randomness

(b)

Fig. 8-8 The relationship between randomness and entropy illustrated by order and disorder in a sequence of red and black playing cards.

Fig. 8-9   **The entropy of 1 mole of benzene as a function of temperature.**

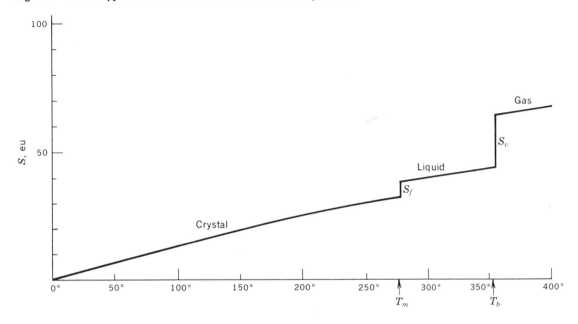

2.   Suppose that 1 mole of helium at 300°K and 1 atm is expanded adiabatically to twice its initial volume; it is then expanded isothermally to four times its original volume at the start of the experiment; it is compressed adiabatically back to double the original volume and, finally, compressed isothermally back to the original volume and temperature.   Calculate the amount of heat which has been extracted from the heat reservoir at the lower temperature and brought up to room temperature (300°K).

3.   Calculate the efficiency of the Carnot cycle described in Prob. 1.

4.   A man inhales 0.1 liter of air and allows it to come to body temperature (37°C) in his lungs at 1-atm pressure.   He then contracts his chest wall slowly enough for the air to be compressed isothermally to eight-tenths of its original volume.   Calculate the decrease in the entropy of the air and the amount of heat generated by the muscle work in this isothermal compression.

5.   A sample of air is removed from the lungs by flowing through a tube into a vacuum chamber.   If the sample is originally in the lungs at body temperature (37°C) and 1 atm and has a volume of 50 cm³, calculate the change in entropy if it goes into a chamber 1 liter in capacity without any change of temperature.

6.   Calculate the probability that the air (regarded as an ideal gas) in a room with an area of 4 by 4 m² and a height of 3 m, will spontaneously contract into the space lying between the floor and a height 0.3 m above the floor.   The temperature of the room is 25°C and the pressure is 1 atm.

7.   A tank containing 10 liters of gaseous helium at a temperature of 30°C and a pressure of 10 atm is connected by a pipe of negligible volume to a similar tank containing a similar amount of nitrogen gas under similar conditions.   Calculate the increase in entropy when the two gases are permitted to mix completely.

8.   The heat capacity at constant pressure of fluoromethane, CH_3F, is 9.0 cal deg⁻¹ mole⁻¹ at 27°C.   Assume that this substance will behave as an ideal gas and that this value of the heat capacity may be regarded as the constant value between 0 and 50°C.   Calculate the increase in the entropy of the substance when it is heated over this 50° range.

9.   The heat capacity due to the vibrations of the molecules in the crystal lattice for crystalline benzene is roughly constant at the value 11.7 cal deg⁻¹ mole⁻¹ over the range 220°K to the melting point at 278°K.   Calculate the increase in entropy due to the vibrations of the crystal lattice of 1 mole of benzene over this range of temperature.

10. Assuming that the heat capacity of liquid water is 1.00 cal deg^{-1} g^{-1}, calculate the increase in the entropy of a 300-g glass of water if a man draws it from a water cooler, where it is at a temperature of 5°C, and it is warmed in his stomach to body temperature (37°C).

11. The entropy of fusion of benzene at the melting point (278.5°K) is quoted in the literature as 8.43 cal deg^{-1} mole^{-1}. Calculate the heat of fusion of 1 mole of benzene at the melting point.

12. The entropy of vaporization of 1 mole of benzene is given in the literature as 20.85 cal mole^{-1} at the temperature of a normal boiling point (353.2°K). Calculate the heat of vaporization of benzene at the normal boiling point.

13. The heat capacity of crystalline benzene just below the melting point may be taken as 29.5 cal deg^{-1} mole^{-1}. The heat capacity of liquid benzene just above the melting point may be taken as 32.0 cal deg^{-1} mole^{-1}. Calculate the entropy increase when the temperature of 1 mole of crystalline benzene is raised from 275°K through the melting point (278.5°K) (where the substance melts) to the temperature of 285°K.

14. The value of the heat capacity at constant pressure of mercuric chloride is 24.4 cal deg^{-1} mole^{-1} over the temperature range 600 to 1000°K. The entropy at 600°K is 56.4 cal deg^{-1} mole^{-1}. Calculate the entropy per mole at 1000°K.

15. The heat capacity of titanium dichloride, TiCl$_2$, has the following values when it is heated at constant pressure:

°K	C_p
300	17.376
400	17.828
500	18.280
600	18.732

Make a plot of C_p against the logarithm of the absolute temperature and obtain a value of the increase in the entropy over this range by means of graphical integration.

16. Calculate the increase in entropy per mole when 1 mole of CO gas (regarded as a rigid molecule) is heated from 300 to 400°K.

17. A volume of oxygen gas is mixed with an equal volume of nitrogen gas at the same temperature and pressure and the entropy of mixing is found to be 1 cal deg^{-1} mole^{-1}. Assuming ideal-gas behavior, calculate the number of moles of nitrogen which were mixed with the oxygen.

18. Calculate the entropy increase when 8 moles of nitrogen is mixed with 2 moles of oxygen, both originally at the same temperature and pressure.

19. A man breathes in 200 cm^3 of air at 37°C and 1 atm, extracts from it 10 cm^3 of O$_2$, which pass into his lungs, exhales into this sample of air 5 cm^3 of CO$_2$, and breathes the sample out. Calculate the change in entropy of mixing brought about by this process. Air is 21 percent O$_2$ and 79 percent N$_2$ on a mole basis.

20. A man breathes in 300 cm^3 of air on a winter day when the temperature is -30°C. The air is warmed in his lungs at constant pressure to body temperature (37°C). Calculate the change in the entropy of the air, assuming ideal-gas behavior and that N$_2$ and O$_2$ behave as rigid diatomic molecules.

REFERENCES

Hill, T. L.: "Statistical Mechanics," McGraw-Hill Book Company, New York, 1956.

Wilson, A. H.: "Thermodynamics and Statistical Mechanics," Cambridge University Press, London, 1957.

Aston, J. G., and J. J. Fritz: "Thermodynamics and Statistical Thermodynamics," John Wiley & Sons, Inc., New York, 1959.

Fast, J. D.: "Entropy," McGraw-Hill Book Company, New York, 1963.

Bent, H. A.: "The Second Law," Oxford University Press, New York, 1965.

Knuth, E. L.: "Introduction to Statistical Thermodynamics," McGraw-Hill Book Company, New York, 1966.

Chapter Nine

the third law: free energy

9-1 THE THIRD LAW OF THERMODYNAMICS

Statistical entropy

As discussed in the preceding chapter, the entropy of a state is defined statistically by the relation

$$S = k \ln W \tag{9-1}$$

where k is the Boltzmann constant and W is the multiplicity, or the number of equivalent ways of forming the state. For an assembly of harmonic oscillators W is given by

$$W = \frac{N!}{N!\, N_1!\, N_2!\, N_3! \cdots} \tag{9-2}$$

where N is the total number of oscillators, N_0 the number of oscillators in the zero energy level, N_1 the number in the first energy level, and so on.

At 0°K there is no thermal energy in the system so that all the oscillators are in the zero energy level. Equation (9-2) takes the form

$$W = \frac{N!}{N!\, 0!\, 0!\, 0!\, 0! \cdots} = 1 \tag{9-3}$$

Since $0! = 1$, the value of W at 0°K is also equal to unity. Since the logarithm of unity is equal to zero, from Eq. (9-1) it is clear that $S = 0$ at 0°K when the above equations are applicable.

The third law of thermodynamics

During the first few decades of the twentieth century two lines of investigation led to the conclusion that for all perfectly crystalline compounds, the value of the entropy is zero at 0°K. This conclusion is now called the third law of thermodynamics. In a classic text on thermodynamics this law is stated in the following form:

If the entropy of each element in some crystalline state be taken as zero at the absolute zero of temperature, every substance has a finite positive entropy; but at the absolute zero of temperature, the entropy may become zero, and does so become in the case of perfect crystalline substances.[1]

The proof of this law is derived from a combination of experimental measurement and theoretical analysis. Experimentally, measurements have been made of heat capacity and heats of phase transitions from temperatures of the order of $1°K$ to temperatures so high that the substances are in the gaseous state. As temperature approaches $1°K$, the heat capacity becomes vanishingly small. Moreover, the variation of heat capacity with temperature obeys the equations of Einstein and Debye so closely that extrapolations to $0°K$ may be made with confidence. Thus, employing the methods discussed at the end of the preceding chapter, it is possible to calculate the increase in entropy when the temperature of a substance rises from $0°K$ to a temperature so high that the substance is in the gaseous state, behaving closely like an ideal gas.

Developments in theories of quantum statistics make possible a calculation of the statistical entropy of a substance in the gaseous state. Thus, the thermodynamic entropy and the statistical entropy can be compared.

The thermodynamic entropy is determined from the relation which has been discussed before:

$$^tS = {}^tS_0 + \int_0^T C_p \, d \ln T + \sum_i \frac{\Delta H_i}{T_i} \qquad (9\text{-}4)$$

where ^tS = thermodynamic entropy

$^tS_0$ = entropy at $0°K$

C_p = heat capacity at constant pressure

H_i = various enthalpy changes associated with phase transitions

T_i = temperatures at which phase transitions take place

It is found that the statistically calculated value of the entropy ^sS is equal numerically to the last two terms in Eq. (9-4):

$$^sS = \int_0^T C_p \, d \ln T + \sum_i \frac{\Delta H_i}{T_i} \qquad (9\text{-}5)$$

[1] G. N. Lewis and M. Randall, "Thermodynamics and the Free Energy of Chemical Substances," 1st ed., p. 448, McGraw-Hill Book Company, New York, 1923.

when the substance under consideration is in the perfect crystalline state as it is cooled toward $0°K$. The comparison of Eq. (9-4) with Eq. (9-5) shows that $^tS_0 = 0$ if $^tS = {}^sS$, as the experimental and theoretical evidence indicates.

As a specific example of this relation, the increments in the entropy of benzene over the range from $0°K$ are

$$^tS = \int_0^{T_m} C_p \, d \ln T + \frac{\Delta H_f}{T_m} + \int_{T_m}^{T_b} C_p \, d \ln T + \frac{\Delta H_v}{T_b}$$

30.89 eu   8.43 eu   2.15 eu   27.36 eu

$$^tS = 68.15 \text{ eu} \qquad ^sS = 68.14 \text{ eu} \qquad (9\text{-}6)$$

The value of ^tS is almost exactly the same as the value of ^sS, which is calculated with the help of the theorems of quantum statistics on the basis of values of the frequencies of vibration derived from spectroscopic measurements and other data.

When, as in the preceding paragraph, no special symbol is needed to make clear that extensive properties refer to the quantity of 1 g mole, the special small capital letters may not be used during the remainder of this book.

Another proof of the validity of the third law rests on experimental measurements of the entropies of two different crystalline forms of the same substance from temperatures in the neighborhood of $0°K$ up to temperatures where phase transformations change these crystals into the same form. The measurements of Eastman and McGavock combined with those of West (see References) show that the values of the heat capacity of rhombic sulfur together with the heat of transition of rhombic sulfur to monoclinic sulfur give a value for the thermodynamic entropy of monoclinic sulfur at $368.5°K$ of 9.07 eu. The heat-capacity values for monoclinic sulfur from 0 to $368.5°K$ give a value of 9.04 eu. Therefore, the two forms of sulfur must have the same value of the entropy at $0°K$; this is consistent with the conclusion from the third law that the entropies of both of these crystalline forms of sulfur are zero at $0°K$.

The measurements of Stephenson and Giauque on two crystalline modifications of phosphine show a difference of only 0.01 eu at a temperature where the two forms have been transformed into a common form.

Imperfect crystals

If, in crystallizing, molecules are not arranged in perfect crystalline order but have some randomness of orientation, S should have a value greater than zero at $0°K$. Evidence for an anomaly of this sort appeared in the measurements made by Clayton and Giauque on carbon monoxide. They had calculated that at $298.15°K$ the entropy of carbon monoxide is 47.20 eu according to quantum statistics. Extremely precise thermal measurements ranging from temperatures near $0°K$ up to $298.15°K$ showed that the value of the thermodynamic entropy is only 46.2 eu, exactly 1 eu less than the value calculated from quantum statistics. The conclusion is that the crystals of carbon monoxide have an entropy of 1.0 eu at $0°K$. Clayton and Giauque suggested that this residual entropy is due to the imperfect orientation of the molecules of CO in the crystal lattice. An ideal crystal of CO should have all the molecules oriented without any randomness; e.g., in the stylized drawing in Fig. 9-1a, all carbon atoms are on top while all oxygen atoms are on the bottom. If, instead of this perfect order, a random selection of carbon atoms is on the bottom, as shown in Fig. 9-1b, there should be residual entropy at $0°K$.

If there were a completely random orientation of the CO molecules, the residual entropy S_0^* at absolute zero should be

$$S_0^* = R \ln 2 = 1.4 \text{ eu} \qquad (9\text{-}7)$$

Clayton and Giauque also compared the experimentally determined entropy of crystalline glycerin with the entropy of a sample of glycerin which was cooled so rapidly that it remained as a supercooled liquid as temperature was reduced to about $1°K$. In this case, there appeared to be a residual entropy in the supercooled liquid glycerin of 5.6 eu.

Example 9-1

Calculate the residual entropy at absolute zero if 10 moles of carbon monoxide is cooled to this temperature and the molecules have completely random distribution in the crystal with respect to the direction of the end of the molecule containing the carbon atom.

Since there are only two possible orientations, C≡O and O≡C, the equation for the residual entropy due to random distribution is

$$nR \ln 2 = (10)(1.987)(2.303) \log 2 = 13.8 \text{ eu}$$

9-2   FREE ENERGY

Spontaneous changes of state

In order to see the relation between entropy changes and the ability of a system to change state irreversibly or spontaneously, it is helpful to specify not only the entropy change in the system itself but also the sum of the entropy changes in the system and in its surroundings. This distinction is illustrated in Fig. 9-2. In Fig. 9-2a, 1 mole of an ideal gas is shown first at the initial volume V_1. We consider the change from V_1 to twice the volume V_2 at constant pressure balanced by

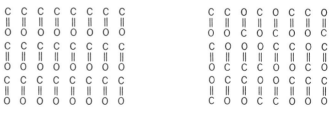

(*a*) Perfect crystal   (*b*) Crystal showing random orientation

Fig. 9-1   Regular order and residual random order in crystalline carbon monoxide (schematic).

Fig. 9-2   Entropy increase in reversible and irreversible change.

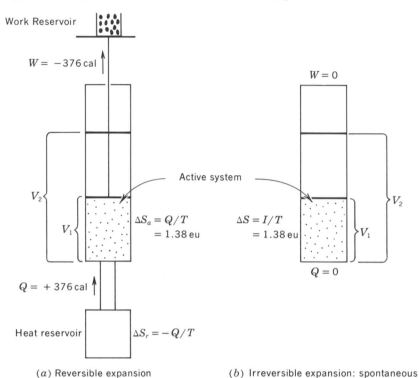

(a) Reversible expansion   (b) Irreversible expansion: spontaneous

the downward force from a weight so that the change can take place reversibly.   Under these conditions when 1 mole of an ideal gas at 0°C and a pressure of 1 atm goes from 22.4 to 44.8 liters, 376 cal passes from the heat reservoir into the gas, as calculated in Chap. 7.   If the thermodynamic system is regarded as the gas alone, there is an increase in entropy of 376 cal/273° = 1.38 eu.   However, it is important to keep in mind that this transfer of heat results in an equal change in the entropy of the heat reservoir but a change with opposite sign.   If the gas and the reservoir are thought of as combined into a total thermodynamic system, the change in entropy of this total system is zero.   These relations may be summarized:

Active system: $\Delta S = \dfrac{\Delta Q}{T} = +1.38$ eu   (9-8a)

Total system: $\Delta S = \underset{\substack{\text{active} \\ \text{system}}}{\dfrac{\Delta Q}{T}} - \underset{\text{reservoir}}{\dfrac{\Delta Q}{T}} = 0$ (9-8b)

In general, whenever a change of state takes place reversibly, the change in entropy for the *total* thermodynamic system is zero.

In contrast to this, consider the change when 1 mole of gas under the same initial conditions expands irreversibly to twice its volume, as shown in Fig. 9-2b. Here there is no force opposing the upward thrust due to the pressure of the gas.   As a consequence, no work is done, and no heat passes into the system.   Yet the entropy change is still the same, 1.38 eu.   In this case, the "amount" of irreversibility may be thought of as numerically equal to the heat measured in calories which would have passed into the system if the change had been made reversibly.   This quantity, the *irreversibility*, is denoted by the symbol I.

With this notation it is possible to write a mathematical statement of the second law of thermodynamics, using an equals sign instead of the equal-or-greater-than sign ($\geq$).   If the change of state is infinitesimal, the expression is

$$dS = \frac{DQ}{T} + \frac{DI}{T} \qquad (9\text{-}9)$$

If the change of state is finite and takes place at constant temperature, the expression is

$$\Delta S = \frac{Q}{T} + \frac{I}{T} \qquad (9\text{-}10)$$

If the change of state takes place reversibly, I is equal to zero and Eq. (9-10) takes on the familiar form $\Delta S = Q/T$; if the change of state takes place with complete irreversibility and no passage of heat, $Q = 0$ and Eq. (9-10) takes the form $\Delta S = I/T$; if there is partial irreversibility, neither Q nor I is zero and both terms appear in the equation. The second law of thermodynamics asserts that there will never be found a change where I has a negative value when the change under consideration takes place in an isolated thermodynamic system such as the total system of the gas and the heat reservoir, shown in Fig. 9-2.

Gibbs free energy

The distinguished American physicist Willard Gibbs pointed out almost 100 years ago the importance of a special combined thermodynamic function which he called the *free energy* and which is now denoted by the symbol G in his honor. Just as the internal energy E is combined with the pressure-volume product PV to form the combined function enthalpy ($H = E + PV$), these two can be combined with the negative of the product of temperature times entropy to form the function called Gibbs free energy:

$$G \equiv E + PV - TS \equiv H - TS \qquad (9\text{-}11)$$

This function is particularly important in studying changes of state taking place at constant pressure and constant temperature. Consider first an infinitesimal change of state which results in an infinitesimal change in the Gibbs free energy dG under conditions of constant pressure and temperature. Differentiating Eq. (9-11) gives

$$dG = dE + P\,dV - T\,dS \qquad (9\text{-}12)$$

The statement of the second law of thermodynamics contained in Eq. (9-9) may be transformed through multiplication by T into

$$T\,dS = DQ + DI \qquad (9\text{-}13)$$

The right-hand portion of this can be substituted for $T\,dS$ in Eq. (9-12):

$$dG = dE + P\,dV - DQ - DI \qquad (9\text{-}14)$$

The first three terms on the right-hand side are found in the statement of the first law of thermodynamics where the mechanical energy change DW is taken as $-P\,dV$:

$$dE = DQ + DW = DQ - P\,dV \qquad (9\text{-}15)$$

From this relation it is clear that in Eq. (9-14) $dE + P\,dV$ cancels out DQ, leaving the simple result

$$dG = -DI \qquad (9\text{-}16)$$

For a finite change of state under conditions of constant pressure and temperature, the finite change in the Gibbs free energy ΔG is related to the finite irreversibility by the expression

$$\Delta G = -\Delta I \qquad (9\text{-}17)$$

where ΔI is equal to the amount of heat which would have passed if the change had taken place reversibly.

In many ways ΔG is a kind of chemical *arrow* associated with a reaction. Just as an unsupported body falls in a gravitational field where the force arrow points downward, having a negative value, so a chemical reaction proceeds when the chemical arrow ΔG points downward on a G plot, i.e., has a negative value.

When all components are at 1 atm and 25°C, the reaction for the formation of water from its elements ($H_2 + \frac{1}{2}O_2 \rightarrow H_2O$) proceeds spontaneously—in fact, explosively—when the mixture is ignited. The chemical arrow ΔG has a large negative value. When the reaction is written in the reverse direction ($H_2O \rightarrow H_2 + \frac{1}{2}O_2$), the sign of the chemical arrow ΔG becomes positive; the reaction does not proceed spontaneously in this direction under these conditions. By definition a vector is a quantity that has both magnitude and significant direction. Thus it is logical to regard ΔG as a chemical vector since both its magnitude and its relation to the direction of the reaction are involved in its value. A vector is often denoted by an *arrow*, and so the term *arrow* will be used in discussing ΔG and its components.

If Eq. (9-11) is differentiated, with T constant, then

$$dG = dH - T\,dS \qquad (9\text{-}18)$$

For a finite change of state this becomes

$$\Delta G = \Delta H - T\,\Delta S \qquad (9\text{-}19)$$

Thus the chemical arrow ΔG is really the difference between two other arrows, the *enthalpy arrow* ΔH and the temperature-entropy arrow $T\,\Delta S$. Since the latter is a measure of the effect of the probability difference at temperature T, it will be called the *probability arrow*. When these two arrows are equal in magnitude and opposite in sign, the reaction will go either way reversibly, i.e., under equilibrium conditions; when the algebraic sum of the two arrows is negative, the reaction will go spontaneously (irreversibly) in the direction of the equation as written; when the algebraic sum of the two arrows is positive, the reaction will not go in the direction which the equation is written but will go in the opposite direction.

Example 9-2

At 1000°K the entropy of gaseous difluoromethane, CH_2F_2, is 77.021 cal deg^{-1} mole^{-1}; the value of the enthalpy is 14.746 kcal mole^{-1}. Calculate the value of the free energy at this temperature.

The value of the free energy is given by

$$G = H - TS = 14.746 - (1000)(77.021)$$
$$= -62.275 \text{ kcal mole}^{-1}$$

Free energy and vaporization

In order to illustrate the quantitative relation between free energy, enthalpy, and the temperature-entropy product, let us consider in more detail what happens when 1 mole of liquid water at the normal boiling point (100°C) is vaporized under a pressure of 1 atm. In this illustration the vapor is assumed to behave like an ideal gas. The equation for this change may be written

$$H_2O\ (l,\ P_1\text{ atm},\ 100°C) \rightarrow H_2O\ (g,\ P_2\text{ atm},\ 100°C)$$

state 1		state 2
$-H_1$	$+$	$H_2 = \Delta H$
$-S_1$	$+$	$S_2 = \Delta S$
$-G_1$	$+$	$G_2 = \Delta G$

$$(9\text{-}20)$$

We shall consider the vaporization of this mole of water under three different sets of conditions, as shown in Fig. 9-3: (1) under the constant pressure of 1 atm (the vapor pressure of water at this temperature); (2) under a pressure of 0.9 atm; and (3) the hypothetical vaporization under a pressure of 1.1 atm.

Case 1: Reversible vaporization at 1 atm

When the water vaporizes reversibly, the second law of thermodynamics states that

$$\Delta S = \frac{Q}{T} = \frac{\Delta H_v}{T} \qquad (9\text{-}21)$$

Thus

$$\Delta H = T\,\Delta S \qquad (9\text{-}22)$$

Consequently

$$\Delta G = \Delta H - T\,\Delta S = 0 \qquad (9\text{-}23)$$

In other words, under equilibrium conditions, where water vaporizes reversibly, the change in enthalpy is equal to the temperature multiplied by the change in

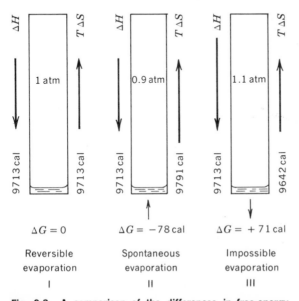

Fig. 9-3 A comparison of the differences in free-energy change between reversible and spontaneous changes of state with balance or lack of balance between the H and the $T\,\Delta S$ factors.

entropy, and consequently the free-energy change is zero.

Case 2: Irreversible vaporization at 0.9 atm

In this case 1 mole of water passes from the liquid state to the vapor state at a pressure of 0.9 atm. Since entropy is a true property of the system, the entropy change in this case is the same *as if* the change took place in two steps; the first step is the vaporization of the water reversibly to 1 mole of H_2O at 1 atm with a volume of 30.6 liters; the second step is the isothermal expansion of 1 mole of water vapor from the volume of 30.6 liters to the volume of 34.0 liters, which it has at a temperature of 373°K and a pressure of 0.9 atm. For such an isothermal expansion the entropy change is given by

$$\Delta S = R2.303 \log \frac{V_2}{V_1} = R2.303 \log \frac{34.0}{30.6} = 0.21 \text{ eu}$$
(9-24)

Thus, the total entropy change in going from 1 mole of liquid water to 1 mole of water vapor at 1 atm and then to 1 mole of vapor at 0.9 atm is $26.04 + 0.21$ eu $= 26.25$ eu. Thus, the factor $T\Delta S$ is 9791 cal for the irreversible change from liquid water to water vapor at 0.9 atm. Therefore, the change in free energy is

$$\Delta G = \Delta H - T\Delta S$$
(9-25)

−78 cal   9713 cal   −9791 cal

since ΔH is essentially independent of the pressure under which the water is evaporated when the vapor behaves as an ideal gas. Since

$$-\Delta G = \Delta I$$
(9-26)

−(−78 cal)   78 cal

the irreversibility or the drive to make the reaction go is 78 cal.

Case 3: Hypothetical vaporization at 1.1 atm

We now calculate what the free-energy change would be if it were possible for water in the liquid state at 1 atm to pass directly into the vapor state at 1.1 atm. Again, this process can be treated in two steps. The first step is the evaporation of 1 mole of water at 1 atm, where the entropy change is 26.04 eu. The second step is the compression of the water vapor from 1 atm to 1.1 atm, where the entropy change can be calculated:

$$\Delta S = R2.303 \log \frac{V_2}{V_1} = R2.303 \log \frac{27.8}{30.6}$$

$$= -0.19 \text{ eu}$$
(9-27)

From this we see that the entropy change in going from 1 mole of liquid water to 1 mole of water vapor at a pressure of 1.1 atm is 25.85 eu. Multiplying this by 373°K shows that $T\Delta S = 9642$ cal. Thus, the free-energy change for this hypothetical vaporization is given by

$$\Delta G = \Delta H - T\Delta S$$
(9-28)

+71 cal   9713 cal   −9642 cal

In this case there is a *negative* irreversibility

$$-\Delta G = \Delta I$$
(9-29)

−(+71 cal)   −71 cal

The negative value for ΔI shows that it is impossible for the water to go spontaneously from the liquid state to the gaseous state at 1.1 atm at a temperature of 373°K. It also means that the reverse process, the condensation of water vapor at a temperature of 373°K and a pressure of 1.1 atm, takes place spontaneously and irreversibly.

Free energy as the difference between enthalpy and the temperature-entropy product

The examples of vaporization under different pressures clarify the meaning of free energy. Think of the water molecules as subject to two different driving forces, or tendencies. As shown in Fig. 9-3 by the arrow pointing downward, there is a tendency for the water molecules to go from the gaseous state into the liquid state. This tendency is produced, first, by the van der Waals attraction between the molecules. This effect is measured by ΔE, the increase in energy necessary to overcome this attraction when the molecules go from liquid to gas. In the second place the pressure of 1 atm which the piston exerts on the gas also adds to the tendency of the molecules to go into the liquid state; this additional term is measured by $P\Delta V$, the work done when 1 mole of water vapor is formed and the piston is pushed up. Since $\Delta E + P\Delta V = \Delta H$, the enthalpy change, or enthalpy arrow ΔH tends to change the system from gas to liquid.

The opposing tendency, the tendency of the molecules to go from liquid to gas, is measured by the arrow, $T \Delta S$. The gaseous state is much more probable than the liquid state because in the gaseous state the molecules are scattered through a far larger volume. The higher the temperature, the faster the molecules move and the greater the effect of increased volume in furthering the passage from the small-volume state (liquid) to the large-volume state (gas). Thus, the tendency, or driving force, is measured by multiplying the entropy change (the logarithmic measure of the probability change) by the absolute temperature, giving the arrow $T \Delta S$, the probability factor.

When the pull downward ΔH is exactly balanced by the push upward $T \Delta S$, the process will go either way under an infinitesimal change of pressure; in other words, it is reversible. Under these conditions, ΔG, which is the measure of the difference between the downward tendency and the upward tendency, will be zero; the length of the vector arrow downward is exactly the same as the length of the vector arrow upward.

The free energy has the same significance for chemical processes; it always represents the difference between two arrows representing tendencies. ΔH is the net result of the enthalpy factors; $T \Delta S$ is the net result of the probability factors. As a rule, enthalpy tends to drive a reaction one way while probability tends to drive it the other way. The direction in which the reaction actually moves *spontaneously* depends upon which of these two arrows is the larger.

It is not always as easy to see the direct physical relation between these factors in a chemical process as it is in a physical process, but as we explore more deeply and in more detail the nature of many kinds of chemical processes in the remaining chapters of this book, the importance of this aspect of free energy should become clear.

Example 9-3

Calculate the free-energy change when 1 mole of benzene at the normal boiling point of 353.2°K vaporizes into the gaseous state at a pressure of 0.1 atm. The enthalpy of vaporization is 7.364 kcal mole^{-1}.

In this example, the process of vaporization may be regarded as taking place in two steps. First, there is the vaporization of 1 mole of liquid benzene to vapor at 1 atm. The change in entropy for this process is given by

$$\Delta S_v = \frac{\Delta H_v}{T_b} = \frac{7364}{353.2} = 20.85 \text{ eu}$$

The second step is the expansion of the benzene vapor to reduce its pressure to one-tenth of the original pressure of 1 atm so that the volume increases by a factor of 10. The entropy change in this process is given by

$$S_2 - S_1 = R \ln \frac{V_2}{V_1} = 1.987 \times 2.303 \log 10$$

$$= 4.58 \text{ eu}$$

The total entropy change in passing from the liquid state to vapor at a pressure of 0.1 atm is therefore the sum of these two entropy changes, which is

$$\Delta S_{\text{tot}} = 20.85 + 4.58 = 25.43 \text{ eu}$$
$$\Delta G = \Delta H - T \Delta S$$
$$= 7.364 - 353.2° \times 25.43$$
$$= -1.618 \text{ kcal} \qquad Ans.$$

9-3   CHANGE OF FREE ENERGY WITH OTHER VARIABLES

When the expression defining G [Eq. (9-11)] is differentiated,

$$dG = dE + P\,dV + V\,dP - T\,dS - S\,dT \qquad (9\text{-}30)$$

As noted in the derivation of Eq. (9-16), when $DW = -P\,dV$ the first law becomes

$$dE = DQ + DW = DQ - P\,dV \qquad (9\text{-}31)$$

The second law may be written in the form

$$dS = \frac{DQ}{T} \cdot + \frac{DI}{T} \qquad (9\text{-}32)$$

Rearranging,

$$T\,dS = DQ + DI \qquad (9\text{-}33)$$

so that

$$DQ = T\,dS - DI \qquad (9\text{-}34)$$

Substituting this expression in Eq. (9-31),

$$dE = T\,dS - DI - P\,dV$$

When the change takes place reversibly,

$$DI = 0 \qquad (9\text{-}35)$$

so that

$$dE = T\,dS - P\,dV \qquad (9\text{-}36)$$

Substituting this expression in Eq. (9-30) gives

$$dG = V\,dP - S\,dT \qquad (9\text{-}37)$$

Change of G with T

If temperature is the only variable while pressure is maintained constant, $dP = 0$ and

$$dG = -S\,dT \qquad (9\text{-}38)$$

Rearranging this into the familiar partial-differential form,

$$\left(\frac{\partial G}{\partial T}\right)_p = -S \qquad (9\text{-}39)$$

When a substance is heated at constant pressure, the expression for the increase in free energy is

$$G_2 - G_1 = \int_1^2 S\,dT \qquad (9\text{-}40)$$

where G_1 is the value of G at T_1 and G_2 is the value of G at T_2. If $T_2 - T_1$ is sufficiently small for S to be effectively constant, then

$$G_2 - G_1 = S\int_1^2 dT = S(T_2 - T_1) \qquad (9\text{-}41)$$

Example 9-4

Calculate the change in the free energy of 1 mole of acetylene when it is heated from 500 to 600°K. The average entropy over this range is 55 cal deg^{-1} mole^{-1}.

The rate of change of free energy with temperature at constant pressure is equal to the negative of the entropy [Eq. (9-39)]. If we regard the entropy as constant over this temperature interval, the change in free energy is therefore given by

$$\Delta G = -S\,\Delta T = -55 \times 100 = -5.5 \text{ kcal}$$

Change of G with P

In a similar manner, if pressure is the variable while temperature is constant so that $dT = 0$, then from Eq. (9-37)

$$dG = V\,dP \qquad (9\text{-}42)$$

Rearranging this to the familiar partial-differential form gives

$$\left(\frac{\partial G}{\partial P}\right)_T = V \qquad (9\text{-}43)$$

The meaning of these relations will be explored further in the next chapter, where phase changes are considered in the perspective of free-energy relations.

If the substance under consideration is a liquid, so that V is essentially constant, then

$$G_2 - G_1 = V\int_1^2 dP = V(P_2 - P_1) \qquad (9\text{-}44)$$

If the substance behaves like an ideal gas, so that $V = RT/P$, then

$$G_2 - G_1 = RT\int_1^2 dP = RT \ln \frac{P_2}{P_1} \qquad (9\text{-}45)$$

Example 9-5

Calculate the change in free energy when 1 mole of helium gas at 300°K is expanded from 1 to 0.1 atm, behaving as an ideal gas.

The rate of change of free energy with pressure under conditions of constant temperature is equal to the volume. Thus, the change in free energy of the helium is

$$\int_1^2 dG = \int_1^2 V\,dP = \int_1^2 RT \frac{dP}{P} = RT \ln \frac{P_2}{P_1}$$

$$= (1.987)(300)(2.303) \log \frac{0.1}{1}$$

$$= -1373 \text{ cal mole}^{-1}$$

9-4   CALCULATION OF FREE ENERGY FROM ENTHALPY AND ENTROPY

Since free energy is the difference between the enthalpy arrow ΔH and the arrow that is a measure of the effectiveness of probability as a driving force $T\,\Delta S$, it is instructive to see how free energy can be calculated from values of H and S. The general equation giving the value of H from absolute zero up to a selected temperature T_a is

$$H = H_0^\circ + \int_0^{T_m} C_p \, dT + H_f + \int_{T_m}^{T_b} C_p \, dT + H_v + \int_{T_b}^{T_a} C_p \, dT \qquad (9\text{-}46)$$

$$\underset{\substack{\text{crystal} \\ \text{range}}}{} \qquad \underset{\substack{\text{liquid} \\ \text{range}}}{} \qquad \underset{\substack{\text{gas} \\ \text{range}}}{}$$

Figure 9-4 has a graph of the values of the enthalpy of benzene from 0 to 400°K. Because there is more energy in 1 mole of benzene at 0°K than there is in six carbon atoms of graphite and three hydrogen molecules in the standard state at absolute zero, the enthalpy of benzene at absolute zero H_0° has a positive value. Above absolute zero, the enthalpy rises as heat is put into the crystal until the melting point is reached. Then, at constant temperature, the enthalpy increases by the amount H_f when the benzene melts and forms the liquid state. Then the enthalpy again increases over the liquid range up to the boiling point. Here again, at constant temperature, the enthalpy increases by the amount H_v when the benzene is vaporized, and, finally, the enthalpy increases as heat is put into the gas, raising its temperature up to 400°K.

In a similar way, the expression for S can be written

$$\left(\frac{\partial G}{\partial T}\right)_p = -S \qquad (9\text{-}49)$$

Thus, if S is always positive, the value of G always falls with increasing temperature. Since $G = H - TS$, this means that as the values of H and TS both rise with temperature, the value of TS must rise more rapidly than that of H.

Figure 9-5 shows the values of the free energy of benzene plotted from 0 to 400°K. Note that at 0°K the value of the entropy is zero and therefore the slope is zero. In other words, the graph at 0°K is horizontal; i.e., it is tangential to the horizontal line going through the value of the free energy at 0°K. This value is, of course, just the value of H_0° since the value of the entropy at this temperature is zero. Experimentally the value H_0° is obtained in the following way. By conven-

$$S = S_0^\circ + \int_0^{T_m} C_p \, d\ln T + S_f + \int_{T_m}^{T_b} C_p \ln dT + S_v + \int_{T_b}^{T_a} C_p \ln dT \qquad (9\text{-}47)$$

$$\underset{\substack{\text{crystal} \\ \text{range}}}{} \qquad \underset{\substack{\text{liquid} \\ \text{range}}}{} \qquad \underset{\substack{\text{gas} \\ \text{range}}}{}$$

Since benzene is normally in the form of essentially perfect crystals at 0°K, $S_0^\circ = 0$. The entropy increases in a manner very similar to enthalpy with increasing temperature.

Figure 9-4 also shows a graph of the probability factor TS obtained by multiplying at each temperature the value of S by T. The value of G at any temperature is just the difference between the value of the graph for H and the value of the graph for TS.

According to the third law, the value of the entropy at 0°K is always zero or positive. According to Eq. (8-71),

$$\frac{\partial S}{\partial T} = \frac{C_p}{T} \qquad (9\text{-}48)$$

C_p must always be positive; heat must flow into the system when the system changes to a higher temperature. Therefore, if entropy has a value of zero or greater than zero at 0°K and always increases with increasing temperature, S is always positive.

According to Eq. (9-39),

tion, H_0° for each chemical element in its standard crystalline state at 0°K is set equal to zero. Measurements of C_p and of any enthalpy changes ΔH_{ph} due to phase changes for the elements thus make possible the calculation of the sum of the enthalpies H_i° of the quantities necessary to form the compound at 298.15°K (25°C), the standard temperature at which enthalpies are tabulated. Thus

$$\sum_i n_i H_i^\circ = \int_0^{298.15^\circ} \sum_i n_i (C_p)_i \, dT + \sum_i n_i (\Delta H_{\text{ph}})_i \qquad (9\text{-}50)$$

where n_i is the number of moles of each element needed to form the compound. The values of C_p for the compound are measured together with any heats of phase change so that the enthalpy change from 0°K to 25°C (ΔH°) for the compound is known. The enthalpy of formation ΔH_f° of the compound at 25° is either measured or calculated. Then H_0° is given by

$$H_0^\circ = \Delta H_f^\circ - \Delta H^\circ + \sum_i n_i H_i^\circ \qquad (9\text{-}51)$$

Fig. 9-4 H and $T \Delta S$ for benzene showing G as $H - T \Delta S$.

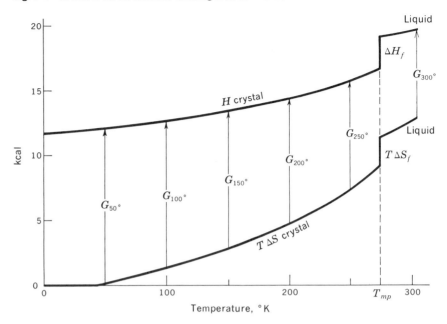

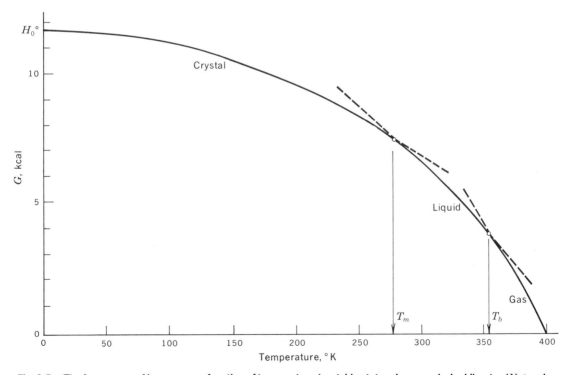

Fig. 9-5 The free energy of benzene as a function of temperature (unstable states shown as dashed lines). (*Note:* when G is plotted to scale as in this diagram, the discontinuities in slope at T_m and T_b are not easily seen.)

For practical calculations H_0° is seldom used, tables being based on another convention that H° for an element is zero at 25°C. However, to see the basic significance of H, TS, and G it is more helpful to show plots based on the convention that H_0° for elements is zero at 0°K. As temperature increases above absolute zero, the value of the entropy gets higher and higher and therefore the slope of the free-energy curve gets more and more negative. As the melting point is reached, there is a discontinuous change in slope but not any discontinuous change in the value of G as is the case of H and S; that is, as the benzene melts, there is no change in free energy. In other words, since the heat of fusion H_f is precisely equal to the product of the temperature of fusion T_f by the change in entropy on fusion S_f when the fusion takes place reversibly, there is no change in free energy in passing from the crystalline to the liquid state. This should be obvious because the two states are in equilibrium with each other at this point and therefore under equilibrium conditions a change of physical state should have no free-energy change associated with it.

As the temperature rises above the melting point, the free energy falls still faster since the entropy is larger and the free-energy curve must therefore have an even greater negative slope. At the boiling point there is again a discontinuous change of slope but not any discontinuity in the value of G. The liquid benzene is in equilibrium with benzene vapor during the process of vaporization, and therefore there is no change in free energy when the state changes reversibly from liquid to vapor. Graphs of this sort will be examined in considerably more detail when physical equilibrium is discussed in the perspective of free-energy change in Chap. 10.

9-5 STATISTICAL THERMODYNAMICS

Statistical entropy defined strictly in terms of probability is linked directly to thermodynamic entropy, which is merely a logarithmic measure (the heat change divided by the temperature) of the change in probability. Since energy has the same meaning in quantum statistics as it has in thermodynamics, we expect to find the equations of statistics which deal

with energy linked to similar equations in thermodynamics. This relationship between quantities like energy, enthalpy, and entropy based on statistical considerations and the same quantities based on thermodynamic considerations emerges most clearly from the experimental proofs of the validity of the third law of thermodynamics discussed earlier in this chapter. We now examine in a little more detail some of these relationships.

The Sackur-Tetrode equation

As the implications of the quantum theory were more and more clearly understood during the early part of the twentieth century, it became possible to develop statistical equations for the value of the entropy of matter under a number of different conditions. We consider first the equation for the entropy of the ideal monatomic gas developed by Sackur and Tetrode, where only the factors involved with the translational motion of the molecules are included. The general form of this equation is

$$S = 2.303R(\tfrac{3}{2} \log M + \tfrac{5}{2} \log T - \log P - 0.5055)$$

$$(9-52)$$

where M is the molecular weight of the gas and R is the gas constant.

In the discussion of the Debye heat-capacity equation we noted that the waves of thermal vibration form a spectrum running all the way from the longest wave, where just 1 wavelength extends from the left-hand face of the crystal to the right-hand face of the crystal, to the shortest significant waves, where the wavelength is twice the distance between two atoms in the crystal if the crystal is monatomic. Recall that the energy levels were quantized by the boundary condition that the wavelength had to be an integral fraction of this distance between the two crystal faces, a fact that made it possible to calculate the heat capacity with the help of the Debye function.

In a similar way, in modern quantum mechanics each molecule has associated with it a wave of the kind first proposed by the French physicist de Broglie. If a gas is in a box, these waves are quantized by the condition that there must always be an integral number of half wavelengths between the two faces of the box. This is the quantum-mechanical basis of the Sackur-

Tetrode equation. It was with the use of an equation similar to this that the entropy of benzene in the gaseous state was calculated statistically, giving a value that checked almost exactly with the value of entropy based on experimental measurements of the heat capacity and calculated by the thermodynamic equations presented in the previous sections. This statistical basis of entropy is a very useful perspective in which to view a number of the problems of biochemistry and physiology.

Statistical functions

A development of the relations embodied in the Boltzmann equation leads to a direct relation between the various thermodynamic quantities for a group of oscillators and the partition function Z. Thus the energy E of 1 mole of oscillators is

$$E = \frac{RT^2}{Z}\left(\frac{\partial Z}{\partial T}\right)_v \qquad (9\text{-}53)$$

where $Z = \sum_i e^{-\epsilon_i/kT}$. In a similar way it is possible to show that the entropy for 1 mole of oscillators is

$$S = R \ln Z + \frac{RT}{Z}\left(\frac{\partial Z}{\partial T}\right)_v \qquad (9\text{-}54)$$

Since the effect of variation of pressure on oscillators is normally negligible in almost all physicochemical changes, there is an advantage in using a second definition of free energy which omits the PV factor. This defines the so-called Helmholtz free energy A, given by

$$A \equiv E - TS \qquad (9\text{-}55)$$

Combining the equation for the energy E and the equation for the entropy S, the Helmholtz free energy A is given by

$$A = -RT \ln Z \qquad (9\text{-}56)$$

This again represents the difference between the energy arrow (in this case, E) which makes the reaction move in one direction and the probability arrow TS which makes the reaction move in the opposite direction.

In Sec. 6-5 the equation for the Boltzmann distribution for an assembly of oscillators was derived:

$$\frac{n_i}{n_{tot}} = \frac{e^{-(\epsilon_i-\epsilon_0)/kT}}{\sum_i e^{-(\epsilon_i-\epsilon_0)/kT}} \qquad (9\text{-}57)$$

where n_i is the number of oscillators in the ith state and n_{tot} is the total number of oscillators. In logarithmic form this is

$$\log \frac{n_i}{n_{tot}} = \frac{-\epsilon_i - \epsilon_0}{2.303kT} - \log Z \qquad (9\text{-}58)$$

where $Z = \sum_i e^{-(\epsilon_i-\epsilon_0)/kT}$. This equation shows the linear relation between $\log (n_i/n_0)$ and ϵ_i at constant temperature. A plot of the first quantity against the second gives the straight line shown in Fig. 9-6.

If Eq. (9-58) is put in the form

$$\epsilon_i - \epsilon_0 = -2.303kT \log \frac{n_i}{n_0} - 2.303kT \log Z \qquad (9\text{-}59)$$

this is in the algebraic form

$$y = ax + b \qquad (9\text{-}60)$$

from which it is clear that the intercept of the graph on the axis where $\log (n_i/n_0) = 0$ is just the value of A per molecule (A^*) if the ordinate is $\epsilon_i - \epsilon_0$, the energy per molecule in the ith level.

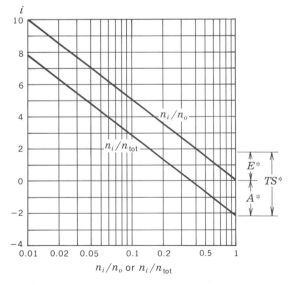

Fig. 9-6 Approximate Boltzmann distribution for adenine in DNA over energy levels for intermolecular thermal vibration ($\bar{\nu} = 100 \text{ cm}^{-1}$). Values for A^*, E^*, and TS^* are also shown.

The value of the average energy per molecule E^* can be calculated from

$$E^* = \frac{\sum_i n_i \epsilon_i}{\sum_i n_i} \tag{9-61}$$

where the values of n_i are read from the line corresponding to n_i/n_{tot}. Since $TS^* = E^* - A^*$, this quantity can be read directly from the graph.

Example 9-6

Calculate the entropy of 1 mole of helium gas at body temperature (37°C).

The value for the entropy of a monatomic gas is given by the Sakur-Tetrode equation and for helium will be

$$S = 2.303T(\tfrac{3}{2} \log M + \tfrac{5}{2} \log T$$
$$- \log P - 0.5055)$$
$$= (2.303)(1.987)(\tfrac{3}{2} \log 4.00$$
$$+ \tfrac{5}{2} \log 310 - \log 1 - 0.5055)$$
$$= 12.12 \text{ eu}$$

9-6   FREE ENERGY AS BIODRIVING FORCE

The concept of free energy provides the means for seeing the quantitative relationship between biochemical energy and the driving force which makes biochemical reactions go forward. We have seen that the free-energy change ΔG for a reaction is always equal to the difference between the enthalpy arrow ΔH and the probability arrow $T \Delta S$

$$\Delta G = \Delta H - T \Delta S \tag{9-62}$$

The enthalpy change ΔH for the reaction is, of course, just the difference between the sum of the enthalpies of the products of the reaction and the sum of the enthalpies of the reactants. As a rule, all these quantities depend primarily on the energy content of the various substances that are the reactants and the products; the enthalpies of these substances vary very little with concentration. By way of contrast, the probability factors determining the $T \Delta S$ product are highly dependent on the concentration of the substances involved. Thus the tendency for the reaction to go ir-

reversibly (the net free-energy arrow of the reaction) varies with the concentration of the reactants and products because these concentrations influence the probability factors.

As we have seen in the previous sections, where the relationship between free energy and the tendency for a reaction to go irreversibly has been discussed, there must be a drop in the value of the free energy if the reaction is to proceed spontaneously. Just as a stone left unsupported in midair always *falls* spontaneously and never rises spontaneously, so the free energy always falls when a reaction goes forward spontaneously.

The values of the free energies of a number of important compounds in biochemical changes have been measured, and tables of these values provide a most significant guide to understanding the dynamics of the life process. Of course, the change of free energy or chemical arrow does not alone determine the actual rate at which the reaction takes place. As we shall see shortly in our study of reaction kinetics, other kinetic factors play a prominent role in determining how fast a reaction actually goes. As an example from inorganic chemistry, consider the reaction of hydrogen gas with oxygen gas to form water; there is a large decrease in the free energy, a large chemical arrow tending to make the reaction go. But it is easy to demonstrate that a mixture of hydrogen and oxygen gas can be left for long periods of time without any appreciable reaction taking place; the reaction does not go because there are rate barriers which, apart from thermodynamic considerations, slow the reaction rate down almost to zero. These factors will be discussed in the chapter on chemical kinetics. But we must remember that if there were no thermodynamic chemical arrow to make the reaction go, it would never proceed at all, no matter how favorable the kinetic factors might be.

Problems

1. In the crystalline state the molecule ($N{\equiv}N{=}O$) can be oriented either with a nitrogen atom at one end or an oxygen atom at one end. Calculate the residual entropy due to completely random orientation of 20 moles of N_2O in the crystalline form at 0°K.

2. When gaseous NO is condensed into the crystal, the molecules dimerize into the form N_2O_2, where the two molecules form a planar rectangle with oxygen atoms at diagonally opposite corners.

$$\begin{array}{c} N\!-\!O \\ | \quad | \\ O\!-\!N \end{array}$$

When this substance is crystallized with relative speed, the squares may orient in two different ways. Calculate the residual entropy due to this random orientation in 10 moles of the crystal of this substance at 0°K.

3. At 500°K the entropy of crystalline boron in the beta-rhombohedral form is 3.257 cal deg^{-1} mole^{-1}. The enthalpy of this same substance is 1032 cal mole^{-1}. Calculate the free energy at this temperature.

4. At 2000°K the entropy of liquid aluminum is 22.358 cal deg^{-1} mole^{-1}. The enthalpy is 13,007 cal mole^{-1}. Calculate the value of free energy at this temperature.

5. At 1000°K the enthalpy of aluminum monofluoride, AlF, regarded as an ideal gas, is 8057 cal mole^{-1}. The free energy is $-53,429$ cal mole^{-1}. Calculate the entropy at this temperature.

6. The normal boiling point of aluminum metal is 2740°K. The enthalpy of vaporization is 70.7 kcal mole^{-1}. Calculate the change in free energy if 1 mole of aluminum metal vaporizes into a space where it attains a pressure of 0.2 atm at the temperature of the normal boiling point, assuming that the vapor is an ideal gas.

7. The mean value of the entropy of aluminum vapor over the range between the normal boiling point (2740°K) and 3000°K is 50.7 cal mole^{-1}. Calculate the change in free energy when 1 mole of liquid aluminum is vaporized at the normal boiling point and the vapor is heated to 3000°K.

8. At 298.1°K, carbon (graphite) has a value of the entropy equal to 5.81 joules deg^{-1} g atom^{-1} and a value of the enthalpy of 1,052 joules g atom^{-1}. At the same temperature, carbon (diamond) has a value of the entropy of 2.28 joules deg^{-1} g atom^{-1} and a value of the enthalpy of 1,806 joules g atom^{-1}. Calculate the change in free energy when 168.1 g of carbon in the form of diamond changes to graphite. Which form of carbon is the most stable at this temperature and 1-atm pressure?

9. Using the Sakur-Tetrode equation (9-52), calculate the entropy of argon gas at the pressure of 1 atm and at the following temperatures: 100, 200, 300, and 400°K.

10. Using the Sakur-Tetrode equation (9-52), calculate the entropy of argon at 200°K and at the following values of the pressure: 0.1, 0.5, 1, 1.5, and 2.0 atm.

11. The value of the enthalpy of argon gas at 1 atm expressed in joules per gram atom is 3,035 at 100°K and 5,240 at 200°K. Using the values of the entropy calculated in Prob. 9, calculate the values of the free energy for this substance at 1 atm and 100 and 200°K.

12. The value of the Gibbs free energy of nitrogen gas at 1 atm is -2775 cal mole^{-1} at 200°K and -3865 cal mole^{-1} at 250°K. Assuming that the rate of change with temperature is linear over this range, calculate the value of the entropy at 225°K.

13. The average value of the entropy of oxygen gas at 1 atm between 100 and 150°K is 20.3 cal deg^{-1} mole^{-1}. Calculate the change in the Gibbs free energy over this temperature interval.

14. The values of the free energy of nitrogen over the range 100 to 298°K at 1 atm are:

T, °K	G, cal deg^{-1} mole^{-1}
100	-732
150	-1718
200	-2775
250	-3865
298	-4954

Assuming that nitrogen behaves like an ideal gas, calculate the values of G at 0.1 and 10 atm and plot all three sets of values at these five temperatures.

15. On semilog paper, make a plot of the entropy of argon gas over the range 0.1 to 1000°K at 0.1, 1, and 10 atm, assuming that argon will remain in the gaseous state and will behave as an ideal gas. Use the data from Prob. 9 to get points for the line at 1 atm and the Sackur-Tetrode equation to calculate the other points.

16. On semilog paper, make a plot of the entropies of helium, neon, argon, and xenon over the temperature range 100 to 1000°K, all at 1 atm, using the Sackur-Tetrode equation to calculate the points.

17. If argon acts like an ideal gas as temperature is lowered toward 0°K, calculate the temperature at which the value of the entropy becomes zero, using the Sackur-Tetrode equation.

18. Extrapolate the graphs plotted in Prob. 16 to low temperature and estimate the temperatures at which the values of the entropy would become zero for helium and xenon if these substances continue to behave like ideal gases.

19. Using the Sackur-Tetrode equation, calculate the pressure necessary to reduce the entropy to zero at 100°K if argon behaves like an ideal gas under indefinitely increased pressure. *Note:* Under the conditions of extremely high pressure in very dense stars, astrophysicists regard entropy as essentially zero.

20. Using the Sackur-Tetrode equation, calculate the pressure necessary to reduce the entropy of helium to zero at 1°K if helium behaves like ideal gas as pressure is indefinitely increased.

REFERENCES

Nernst, W.: *Nachr. Kgl. Ges. Wiss. Göttingen Math.- Phys. Kl.,* **1906**:1.

Planck, M.: "Thermodynamik," 3d ed., p. 279, Veit and Co., Leipzig, 1911.

Sackur, O.: *Ann. Physik,* (4)**36**:598 (1911); **40**:67 (1913).

Tetrode, H.: *Ann. Physik,* (4)**38**:434 (1912).

Gibson, G. E., and W. F. Giauque: *J. Am. Chem. Soc.,* **45**:93 (1923).

Clayton, J. O., and W. F. Giauque: *J. Am. Chem. Soc.,* **54**:2610 (1932).

Eastman, E. D., and W. C. McGavock: *J. Am. Chem. Soc.,* **59**:145 (1937).

Stephenson, C. C., and W. F. Giauque: *J. Chem. Phys.,* **5**:149 (1937).

West, E. D.: *J. Am. Chem. Soc.,* **81**:29 (1959).

Lewis, G. N., and M. Randall: "Thermodynamics," rev. 2d ed., by K. S. Pitzer and L. Brewer, chap. 3, p. 128, McGraw-Hill Book Company, New York, 1961.

Fast, J. D.: "Entropy," McGraw-Hill Book Company, New York, 1963.

Stull, D. R., E. F. Westrum, and G. C. Sinke: "Chemical Thermodynamics of Organic Compounds," John Wiley & Sons, Inc., New York, 1970.

Chapter Ten

physical equilibrium

10-1 THE NATURE OF PHASES

In their classic text, "Thermodynamics and the Free Energy of Chemical Substances," Lewis and Randall remark that in all thermodynamics there is no concept more fundamental than the idea of *equilibrium,* and so, as we turn from the presentation of the theoretical nature of the laws of thermodynamics to their applications to physical and chemical reactions, it may be helpful to review once more the nature of equilibrium.

Equilibrium

The definition of equilibrium is closely tied to the definition of a physicochemical state.   When fixed external conditions completely determine the physicochemical state of a portion of matter, its properties do not change with time, but there are two distinct patterns for this condition of constancy: (1) if the intensive properties are not uniform throughout this portion of matter, it is in a *steady state;* (2) if the intensive properties are uniform throughout the entire volume, the matter is in a true *equilibrium state.*   To provide an example of a *steady state,* a tube containing liquid water can be placed so that one end is in contact with a heat reservoir at a relatively high temperature and the other end is in contact with another heat reservoir at a relatively low temperature.   If the temperatures of the reservoirs are maintained constant, there will be no change in the properties of the tube of water with time but heat will flow continually in at the hotter end of the tube and out at the colder end of the tube.   The temperature of the tube is not uniform; therefore, it is a physicochemical system in a steady state.   On the other hand, if the surroundings of the tube are all at the same temperature and pressure and the water itself has uniform temperature and pressure throughout, it is in a true equilibrium state.

The concept of physicochemical equilibrium is closely related to the concept of mechanical equilibrium.   When a mechanical system is subjected momentarily to a force that tends to change its state and then returns to its initial state when the force is removed, it is said to be in a true equilibrium state. For example, imagine a ball lying at the bottom of a concave bowl. If the ball is subjected to a slight force so that it rolls away from the bottom of the bowl upward along the concave surface, it will return to the bottom of the bowl when the force is removed. This is a true equilibrium state. If instead of lying on such a concave surface the ball lies on a horizontal plane surface, a slight force will move it to a new position on the plane; but when the force is removed, the ball does not return to its original position. The ball is in a quasi-equilibrium state. As a third possibility, the ball may be balanced at the point of maximum height on the surface of an inverted bowl; if the ball is subjected to a slight horizontal force, it starts rolling down the surface and continues to roll even when the force is removed. Under these conditions, the ball is in a nonequilibrium state.

In the same way, a portion of matter can exist in a state with no change of its properties with time; it is physicochemically *at rest*. If a slight disturbing force is applied producing a small change in state, and if when the force is removed, the portion of matter returns to its original state, it is said to be in a state of true equilibrium.

In this chapter the discussion is concerned primarily with physicochemical systems in which two or more portions of the same kind of chemical substance are in different physicochemical states but have the kind of physical contact which permits an exchange of molecules between the different states. These different states are designated as the different *phases* of the system. As discussed in Chap. 4, water can exist in the gaseous phase, in the liquid phase, and in a number of different crystalline phases. At the triple-point temperature of 273.16°K, under a pressure of 4.58 mm Hg, the three phases of water can exist side by side in contact with each other under conditions where molecules of H_2O can pass back and forth between the different phases. Under these conditions there is no change in the physical properties of the system with time. If the pressure were made momentarily slightly higher, an appreciable number of the molecules would pass from the less dense phases (the

vapor and the crystalline phase) to the state of maximum density (the liquid phase). Thus, the relative amount of water in the different phases would alter, and the pressure would fall to the triple-point pressure, at which the system would again be in true equilibrium.

10-2 THE THERMODYNAMICS OF PHASE CHANGES

When two different phases of a substance are in physical contact, they are in equilibrium with each other if the free energy per mole of the first phase G_1 is equal to the free energy per mole of the second phase G_2

$$G_1 = G_2 \qquad (10\text{-}1)$$

Since by definition

$$G \equiv H - Ts \qquad (10\text{-}2)$$

from Eqs. (10-1) and (10-2) it follows that

$$H_1 - Ts_1 = H_2 - Ts_2 \qquad (10\text{-}3)$$

For a change of 1 mole of a substance from phase 1 to phase 2 under equilibrium conditions, these relations may be expressed in terms of the delta notation

$$\Delta G = G_2 - G_1 = 0 \qquad (10\text{-}4)$$

In general, ΔH and Δs are not equal to zero

$$\Delta H = H_2 - H_1 \neq 0 \qquad (10\text{-}5)$$
$$T \Delta s = T(s_2 - s_1) \neq 0 \qquad (10\text{-}6)$$

As an illustration of the application of these equations, consider the changes in phase as 1 mole of crystalline benzene at 0°K is warmed, melts, and changes to liquid at 279°K and finally at 353°K passes to the vapor phase under a pressure of 1 atm. These changes in free energy are shown graphically in Fig. 10-1. The numerical values of the free energy per mole in this figure are based on the convention that the value of the enthalpy for all the chemical elements in the standard state at 0°K is equal to zero. As pointed out in Chap. 5, this convention has a limited usefulness. Since values of H at 0°K must be determined by extrapolation from low-temperature heat-capacity values, they do not provide too firm a thermodynamic basis for extended calculations. Therefore, almost

Fig. 10-1   The free energy of benzene as a function of temperature (unstable states shown as dashed lines).

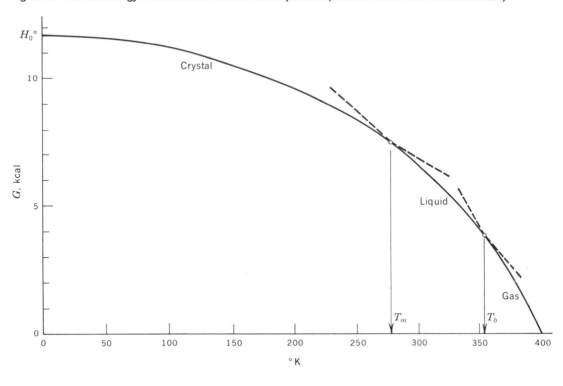

all tables of thermodynamic quantities are based on the alternative convention that the value of the enthalpy for all the chemical elements in the standard state at 25°C is equal to zero.   While the latter definition provides a firmer basis for precise calculations, the former convention is more helpful in clarifying a number of the theoretical relations.

Melting

Consider first the relationship between the free energy per mole for the crystalline phase of benzene G_c and the free energy per mole in the liquid phase G_l.   The normal melting point is defined as that temperature at which, under 1-atm pressure, the crystalline phase and the liquid phase will be in equilibrium with each other so that

$$G_c = G_l \tag{10-7}$$

By definition,

$$G_c = H_c - TS_c \tag{10-8}$$

and

$$G_l = H_l - TS_l \tag{10-9}$$

so that Eq. (10-7) can be written

$$H_c - TS_c = H_l - TS_l \tag{10-10}$$

In order to complete the set of equations related to this phase change, the equations in delta notation are included:

$$\Delta H = H_l - H_c \tag{10-11}$$

$$\Delta S = S_l - S_c \tag{10-12}$$

$$\Delta G = G_l - G_c \tag{10-13}$$

To show more clearly the relations between the free energy of the crystalline state and the free energy of the liquid state, the portion of the curves in Fig. 10-1 in the neighborhood of the melting point have been enlarged and are shown in Fig. 10-2.   Note first of all the intersection at the melting point of the curve for the free energy of the crystal G_c and the curve for the free

Fig. 10-2   Free-energy relations near the melting point of benzene, 278.7°K (scale approximate).

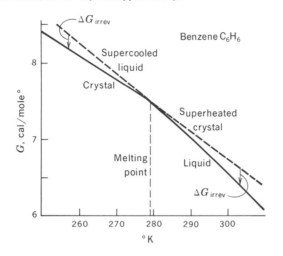

energy of the liquid G_l. This intersection is just another way of portraying the relation expressed in Eq. (10-7).

The relationship between the rate of change of free energy with temperature and entropy was derived in Chap. 8:

$$\left(\frac{\partial G}{\partial T}\right)_p = -s \tag{10-14}$$

Since the entropy of the liquid phase is greater than the entropy of the crystalline phase, the slope of the curve for G_l on this plot is steeper than the slope of G_c.

Unstable phases

On the scale shown in Fig. 10-2, the lines for G_c and G_l are essentially straight lines, so that it is possible to extrapolate both of them beyond the melting point; these extrapolations are shown in the figure as dashed lines. The extrapolation of the line for G_l to temperatures below the melting point signifies the free energy which the substance has when in the liquid state below the melting point. Physically this state can be produced by cooling a liquid below the melting point in a quiescent condition with no crystals present. This state is called the *supercooled liquid.*

The dashed line representing the extrapolation of the line for G_c to temperatures above the melting point represents the free energy of superheated crystals.

There is considerable difficulty in raising the temperature of a substance in the crystalline state above the melting point and maintaining the substance there as a crystal. Molecules at the surface in general rearrange to form the liquid phase just as soon as the temperature rises above that of the melting point.

In passing from the supercooled liquid state to the crystalline state at a temperature below the melting point, the point on the diagram representing the state of the system *falls* in going from the dashed curve to the continuous curve. This change of state can be expressed by

$$C_6H_6(l) \rightarrow C_6H_6(c) \tag{10-15}$$

In accordance with the convention for a change of state, the change in a property is set equal to the value of that property in the final state minus the value of the property in the initial state; thus, the equation for the change in free energy is written

$$-G_l + G_c = \Delta G \tag{10-16}$$

As may be seen on the diagram, the free energy for the supercooled liquid is greater than the free energy for the crystal at any temperature below the melting point, and consequently the value of ΔG is negative.

In exactly the same way, the change from the superheated crystal to the liquid state above the melting point may be written

$$C_6H_6(c) \rightarrow C_6H_6(l) \tag{10-17}$$

and the corresponding change in free energy is

$$-G_c + G_l = \Delta G \tag{10-18}$$

As may be seen from the diagram, above the temperature of the melting point, the free energy in the crystalline state is higher than the free energy in the liquid state, and consequently the value of ΔG for this change is also negative.

Irreversibility

These changes illustrate the principle that was discussed in the previous chapter, namely, the relation between the value for the free-energy change and the tendency of the system to make a change from one phase to another in a spontaneous and irreversible way. When the free energy of one phase is equal to the free energy of the other phase, the change from one to the other is such that $\Delta G = 0$ and the change takes place under equilibrium conditions, i.e., *reversibly.* When ΔG is a negative quantity, the change takes place spontaneously and irreversibly. This is true for the change from the supercooled liquid to the crystalline state or from the superheated crystalline state to the liquid state. When ΔG equals a positive quantity, it is impossible to have the change taking place spontaneously and irreversibly. In other words, below the melting point the crystal never melts spontaneously and forms the supercooled liquid state because such a change would have a positive value of ΔG. In the same way, above the melting point a liquid never changes spontaneously to form a superheated crystal because this change also would have a positive value of ΔG.

Enthalpy-probability balance

Finally, we should recall that ΔG represents the balance between two factors which influence the direction of change of phase. This is expressed by the equation

$$\Delta G = \Delta H - T \Delta S \qquad (10\text{-}19)$$
$$\underset{\text{factor}}{\text{enthalpy}} \quad \underset{\text{factor}}{\text{probability}}$$

The enthalpy factor is, in turn, the sum of two factors

$$\Delta H = \Delta E + P \Delta V \qquad (10\text{-}20)$$
$$\underset{\text{factor}}{\text{energy}} \quad \underset{\text{factor}}{\text{expansion}}$$

Although ΔG is equal to zero at the melting point, neither of the two factors of which it is composed is individually equal to zero. Whether the change takes place below the melting point, at the melting point, or above the melting point, the system always gains energy in passing from the crystalline state to the liquid state because the binding is always tighter in the crystalline state and energy must be introduced into the system to loosen this binding and permit the molecules to pass to the less-ordered array found in the liquid state. Therefore, $\Delta E \neq 0$. This difference in packing and in order is illustrated schematically in Fig. 10-3. Since the normal liquid occupies greater volume than the crystalline state, the factor $P \Delta V$ is also positive for the change from the crystalline to the liquid state. The sum of these two factors is ΔH, the enthalpy factor, which is thus positive for the change from the crystal-

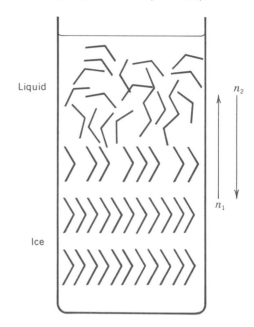

Fig. 10-3 Equilibrium between ice and water at 0°C (schematic). n_1 (number of molecules ice to water per second) $= n_2$ (number of molecules water to ice per second).

line to the liquid state; consequently, ΔH is negative for the change from the liquid to the crystalline state. Since ΔH and ΔG have the same sign in Eq. (10-19), a negative value for ΔH tends to make ΔG have a negative value. Thus, the positive ΔH factor favors the tendency for the system to change spontaneously from the liquid to the crystalline state. This may be thought of as due to the forces of attraction which act between molecules and tend to pull the molecules as close together as possible. Also, the force of the atmosphere tends to compress the system, so that the $P \Delta V$ factor also favors the change from the liquid state to the crystalline state when the latter is denser than the former.

The other portion of ΔG consists of the term $-T \Delta S$. As indicated schematically in Fig. 10-3, the liquid state is less orderly than the crystalline state. Just as a pack of playing cards after shuffling is in a less orderly (more probable) state than the same pack with all the red cards on top and all the black cards on the bottom, the liquid state is less orderly (more probable) than the crystalline state. Since entropy is the logarithmic measure of probability, the liquid state has a larger entropy per mole than the crystalline state. Therefore, Δs has a positive value; and since this appears with a negative sign in the expression for ΔG, a positive value of Δs contributes negatively to the value of ΔG. Thus, the probability factor tends to make the system go spontaneously and irreversibly from the crystalline state to the liquid state, a tendency exactly opposite to that caused by the enthalpy factor.

From this point of view, it is clear that at the melting point the enthalpy factor tending to bring the system from the liquid state to the crystalline state is exactly balanced by the probability factor tending to bring the system from the crystalline state to the liquid state, so that the two factors cancel each other out. ΔG is equal to zero, and the change takes place reversibly. Below the melting point, primarily because T is smaller, the probability factor $T \Delta s$ becomes smaller than the enthalpy factor ΔH, and the influence of the former dominates, causing the system to go irreversibly from the liquid state to the crystalline state. Above the melting point, the situation is reversed; primarily because the value of T is now larger, the value of $T \Delta s$ becomes larger than ΔH and therefore the probability factor dominates and the system tends to go irreversibly from the crystalline state to the liquid state.

Example 10-1

The heat of fusion of benzene is 2348 cal $mole^{-1}$; the melting point is $278.7°K$. Calculate the entropy of fusion.

$$\Delta s_f = \frac{\Delta H_f}{T_m} = \frac{2350 \text{ cal mole}^{-1}}{278.7°K}$$

$$= 8.43 \text{ eu mole}^{-1} \qquad Ans.$$

Example 10-2

The molar free energy of crystalline benzene at the melting point is approximately 7500 cal $mole^{-1}$, and the molar entropy is 30.89 eu. Calculate the molar free energy for the supercooled liquid and for crystalline benzene at $0°C$ and the decrease in free energy when 1 mole passes from the former state to the latter, assuming no significant change in the entropies of these two states over this temperature range.

$$G_l(273.2°K) = G_l(278.7°K)$$
$$+ 39.32 \text{ eu } (278.7 - 273.2°)$$
$$= 7500 + 39.32(5.5°)$$
$$= 7716 \text{ cal mole}^{-1}$$
$$G_c = 7500 + 30.89(5.5°)$$
$$= 7670 \text{ cal mole}^{-1}$$
$$\Delta G = G_c - G_l = 46 \text{ cal mole}^{-1} \qquad Ans.$$

Vaporization

Figure 10-4 shows on an enlarged scale the portion of the curve of free energy against temperature for benzene in the neighborhood of the normal boiling point. The thermodynamic relations for this phase change are almost exactly the same as those just discussed for free energy in the neighborhood of the freezing point. The extrapolation of the line for the free energy of the liquid below the temperature of the normal boiling point gives the curve shown as a dashed line for the free energy of supercooled vapor. Physically this state can be produced by cooling the vapor slowly under conditions where no droplets of the liquid are present. In a similar way, the extrapolation of the line for the free energy of the liquid to temperatures above that of the normal boiling point gives the dashed line which represents the free energy of the superheated liquid. Physically, many liquids can be heated above the boiling point and maintained in this superheated liquid state with very little vaporization.

Just as in the case of the change from crystalline to liquid state, the free-energy change in going from the liquid state to the vapor state can be broken down into the two factors, the enthalpy factor ΔH_v and the probability factor $T \Delta s_v$.

In order to make a substance pass from the liquid state to the vapor state it is necessary to introduce energy into the system to overcome the forces of attraction between the molecules. For this reason, for the change expressed by the equation

$$C_6H_6(l) \rightarrow C_6H_6(g) \qquad (10\text{-}21)$$

the value of ΔE_v is always positive. Since the volume per mole of the vapor is always greater than the volume per mole of the liquid, the factor $P \Delta v$ is also always

Fig. 10-4   Free-energy relations near the boiling point of benzene, 353.3°K.

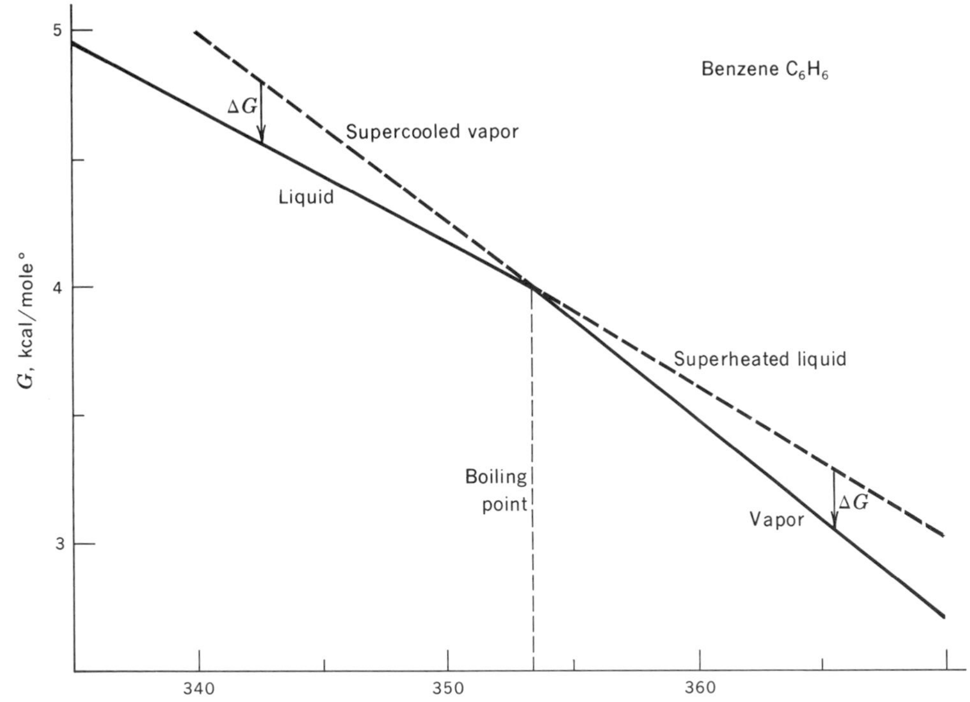

positive.   Thus, there is always a positive value for $\Delta H = \Delta E + P \Delta V$.

The value of Δs for the change of state in vaporization is also always positive.   Recall the discussion of the relation between entropy and volume for a gas in Chap. 8.   If 1 mole of benzene vapor is confined in a closed volume, the probability of finding the molecules uniformly distributed throughout the volume is enormously greater than the probability of finding all the molecules clustered together in one small corner of the volume.   In the same way, the probability of finding the molecules evenly distributed as vapor throughout the volume is enormously greater than the probability of finding them all clustered together in a small pocket of liquid in one corner of the volume.   Since the factor $T \Delta s$ appears with a negative sign in the expression for the free energy, a positive value of Δs tends to make a negative value for ΔG.

Thus, the effect of the energy and the external atmosphere may be thought of as producing a tendency (the enthalpy factor) to make the system go from the vapor to the liquid state; on the other hand, the probability factor tends to make the system go from the less probable liquid state to the far more probable vapor state.   At the normal boiling point these two factors exactly balance each other so that the free-energy change $\Delta G = 0$ and the change can take place either way reversibly.   As the system changes to temperatures below that of the normal boiling point, the lower value of T in the factor $T \Delta s$ makes this factor smaller and ΔH dominates in determining the sign of ΔG.   In other words, the forces of attraction between the molecules plus the external pressure tend to force the molecules together, carrying them from the vapor phase to the liquid phase.   Thus, below the normal boiling point under pressure of 1 atm the vapor tends to go spontaneously and irreversibly from the vapor state to the liquid state.

On the other hand, above the temperature of the normal boiling point, primarily because T becomes greater in the factor $T \Delta s$, the probability factor dominates, and the push to send the molecules out into the vapor state is greater than the pull tending to bring them together from the vapor into the liquid state; consequently, the system goes spontaneously and irreversibly from the liquid state to the vapor state under a pressure of 1 atm at temperatures above the normal boiling point.

Example 10-3

The molar heat of vaporization of C_6H_6 is 7.34 kcal at the normal boiling point 80.1°C. Calculate the entropy of vaporization.

$$\Delta s_v = \frac{\Delta H_v}{T_m} = \frac{7340 \text{ cal}}{273.2 + 80.1°}$$
$$= 20.8 \text{ eu} \qquad Ans.$$

Example 10-4

Calculate the decrease in the free energy of benzene when 1 mole of superheated vapor at 86.0°C condenses. Use the data from the previous problem.

$$G_l - G_v = -(T - T_b)\,\Delta s_v = -5.9° \times 20.8 \text{ eu}$$
$$= -123 \text{ cal} \qquad Ans.$$

10-3   THE ENTROPY OF PHASE CHANGES

While the values of the *enthalpy* per mole associated with phase changes vary over a wide range, depending on the size and force field of the molecules involved, the values of the *entropy* per mole associated with phase changes such as fusion and vaporization lie within relatively narrow ranges.

Entropy of fusion

The value of the entropy of fusion reflects the change from a state of relatively high order in the crystal to a state of relative randomness in the liquid. In order to understand in some detail the connection between entropy and order, recall the equation discussed in Chap. 6 connecting entropy s and the number of ways W in which a molar state can be realized:

$$s = k \ln W \tag{10-22}$$

where k is the Boltzmann constant.

W can be thought of as the product of several factors:

$$W = W_1 \times W_2 \times W_3 \times \cdots \tag{10-23}$$

For example, in the case of the crystal, W_1 may be taken as the measure of crystal order. In the case of carbon monoxide, discussed at some length in Chap. 9, W_1 has the value of unity in a perfect crystal, where all the molecules are aligned in the same way. In a crystal containing 1 mole of CO and formed under conditions of cooling so that the molecules are aligned randomly with regard to the carbon and the oxygen ends, $W_1 = 2N_A$, where $N_A = 6.02 \times 10^{23}$. Thus, the increment in entropy s_r due to the randomness is given by the expression

$$s_r = k \ln W_1 = k \ln 2^{6 \times 23} = R \ln 2 \tag{10-24}$$

W_2 is the multiplicity due to the vibration and libration in the crystal lattice, and W_3 is the multiplicity due to internal vibration in the molecules. In order to simplify the discussion, the equations in the remainder of this section will be put in the form applicable to 1 mole of the substance so that Avogadro's number instead of occurring as an exponent on the characteristic of the logarithm will appear as multiplying k, making the factor in front of the logarithm equal to the gas constant R.

Turning now to the consideration of the liquid state, the value of W_1 per molecule will be considerably larger than the value of unity found in the perfect crystal, because the molecules in the liquid state are randomly oriented. While the degree of this randomness varies considerably from one substance to another, the average value can be estimated from the fact that the quasi rotation of molecules in the liquid state tends to make them behave as if they were spherical in shape, unless the basic shape of the molecule differs greatly from a compact form. D. H. Andrews, R. L. Coleman, J. Feuer, and R. A. Schuck have pointed out that roughly spherical molecules tend to come together in hexagonal close packing. Assuming that each molecule lies in a cage made up of the spherical fields of force of the neighboring molecules in roughly hexagonal packing, the interior surface of the cage contains 14 potential-energy minima. An unsymmetrical molecule oriented with its head and tail resting in opposite potential minima may spin about this primary axis through a series of potential minima and may rest momentarily in any one of six orientations. Thus, according to this model, there are $14 \times 6 = 84$ orientations with roughly equal probability. In other words, for the liquid, the part of the entropy s_0 due to the different orientations of a molecule is given by

$$s_0 = R \ln 84 = 8.8 \text{ eu} \tag{10-25}$$

Because of the less dense packing in the liquid state, the molecules are able to vibrate more freely with a lower frequency. If the elastic-force constant controlling this vibration is reduced by 50 percent, there will be an increase in volume and also an entropy increase s_v of 4.2 eu according to the Debye equation for heat capacity. Assuming this to be the case, the entropy of fusion Δs_f for simple unsymmetrical molecules is given by

$$\Delta s_f = s_0 + s_v = 8.8 + 4.2 \text{ eu} = 13.0 \text{ eu} \tag{10-26}$$

If a molecule is symmetrical, the number of different orientations in the liquid state is reduced in proportion to the symmetry. For example, the benzene molecule has a symmetry number σ with a value of 12. Think of a benzene molecule as lying flat on a plane. When the molecule is rotated around an axis normal to the plane through an angle of 360°, six equivalent positions for the molecule will be found. In other words, each time the molecule is rotated through 60° it assumes the exact configuration it had before rotation, since the carbon and hydrogen atoms are indistinguishable from each other. In addition to these six orientations, there are six others which may be found when the molecule is turned over so that its opposite face is now in contact with the plane. Thus, there are in all 12 equivalent positions found when the benzene molecule is put through all possible rotations. In other words, for benzene, the value in the liquid state of W_1 is only one-twelfth the value of W_1 for a completely unsymmetrical molecule. Thus, to compare the entropy of fusion of benzene with the normative value derived in Eq. (10-26), a term s_σ must be added, as given by

$$s_\sigma = R \ln \sigma = R \ln 12 = 4.9 \text{ eu} \tag{10-27}$$

In this way we arrive at a value for the normalized entropy of fusion Δs_{fn} given by the expression

$$\Delta s_{fn} = \Delta s_{\text{exper}} + R \ln \sigma \tag{10-28}$$

where Δs_{exper} is the experimentally observed entropy of fusion determined by dividing the experimentally measured enthalpy of fusion ΔH_f by the temperature of the melting point T_m expressed in degrees Kelvin. Inserting the appropriate values in Eq. (10-28), we find

$$\Delta S_{fn} = 8.4 + 4.9 \text{ eu} = 13.4 \text{ eu} \tag{10-29}$$

In view of the approximate nature of the assumption about the value of W_1 in the hexagonal-close-packed order in the liquid state and the approximate nature of the calculation of s_v, the agreement of the figure 13.4 eu for benzene with the calculated value of 13.0 eu may be taken as an indication of the correctness of this picture of the dependence of the entropy of fusion on the change in order. In Table 10-1 values of the entropy of fusion are listed for a few compounds together with values of σ to indicate the varieties of compounds for which this approach may be used to calculate Δs_f. In 1968 A. Bondi (see References) published an extensive survey of entropies of fusion, discussing many other factors which influence crystal-liquid equilibria in cases more complex than the above example. The effect of the reduction of molecular symmetry on the equilibrium between crystalline and liquid phases is shown in Fig. 10-5.

TABLE 10-1   Entropies of fusion

Compound	σ	Δs_f, eu	$R \ln \sigma$, eu	Δs_{fn}, eu
CH_3CH_3	6	7.6	3.6	11.2
$CH_3CH_2CH_3$	2	9.9	1.4	11.3
CH_2CH_2	4	7.7	2.8	10.5
CH_3SH	3	9.4	2.2	11.6
$HCHO$	2	11.5	1.4	12.9
$CFCl_3$	3	10.1	2.2	12.3
$CHBr_3$	3	9.4	2.2	11.6
CH_2I_2	2	10.3	1.8	12.1
CH_3Cl	3	8.8	2.2	11.0
C_6H_6	12	8.4	4.9	13.4
$AsCl_3$	3	9.4	2.2	11.6
$AsBr_3$	3	9.2	2.2	11.4

Fig. 10-5   Partial orientation and lowered entropy caused by hydrogen bonding (schematic).

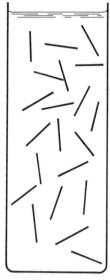 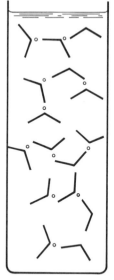

Completely random orientation in normal liquid.

Partial orientation in water caused by hydrogen bonding (o indicates bonding hydrogen atom. Note that molecules are ordered so that a number of H atoms are placed near O atoms on other molecules).

Example 10-5

The molar heat of fusion of chloroform, $CHCl_3$, is 2200 cal mole^{-1}. The melting point is $-63.49°$. Calculate the normalized entropy of fusion for $\sigma = 3$.

The experimental entropy of fusion is

$$\Delta s_f = \frac{\Delta H_f}{T_m} = \frac{2200 \text{ cal}}{209.7°} = 10.5 \text{ eu}$$

The normalizing factor to correct for symmetry is

$$s\sigma = R \ln \sigma = 1.987 \times 2.303 \log 3 = 2.2 \text{ eu}$$

The normalized entropy of fusion is

$$\Delta s_{fn} = 10.5 + 2.2 \text{ eu} = 12.7 \text{ eu} \qquad Ans.$$

Example 10-6

Estimate the molar entropy of fusion of HCHO from the symmetry rule.

The value of σ for HCHO is 2.

$$R \ln 2 = 1.4 \text{ eu}$$

$$13 - 1.4 \text{ eu} = 11.6 \text{ eu (estimated value)}$$

$$Ans.$$

The experimentally observed value is 11.5 eu.

Entropy of vaporization

In the last quarter of the nineteenth century, several scientists pointed out that the values of the entropy of vaporization of many liquids lay in the neighborhood of 21 eu. Today this generalization is known as Trouton's rule after the scientist who showed most conclusively the existence of this regularity. While no complete theoretical explanation of this value of 21 eu has been forthcoming, the clustering of experimental values around it indicates that for a wide variety of substances the change in molecular order passing from the liquid state to the vapor state is about the same. In Table 10-2 values of Δs_v are given for several substances.

In many liquids, in addition to the usual van der Waals forces attracting molecules to one another, additional forces tend to orient the molecules and make the randomness in the liquid less than it would be if only van der Waals forces were operating. Water is an example of such a liquid. As shown in Fig. 10-5, the hydrogen atoms of the water molecules tend to orient themselves so that a hydrogen atom lies between the oxygen atom to which it was originally completely attached and the oxygen atom of another water molecule. This is due to a force of attraction between the hydrogen atom and a pair of unshared electrons on the oxygen atom in the neighboring water molecule. This phenomenon, known as *hydrogen bonding,* will be discussed in greater detail in Chap. 13. If the molecules are ordered by the operation of forces like those produced by hydrogen bonding, there is less randomness in the liquid state and consequently the liquid has a lower entropy. If the liquid has lower entropy, the increase in entropy in going from the liquid to the vapor state Δs_v will be greater (Fig. 10-6). A survey of the values of Δs_v for about 200 compounds shows an average value of 21.2 cal deg^{-1} mole^{-1} for liquids where there is little or no evidence of hydrogen bonding. On the other hand, for liquids with hyrogen bonding the values of Δs_v are considerably higher. Values of Δs_v for both types of liquids are given in Table 10-2.

Vapor pressure

One of the most important aspects of physical equilibrium is illustrated by the pressure of vapor in physical equilibrium with the liquid state; this is called the *vapor pressure* of the liquid. In Chap. 4, the equation for the vapor pressure was stated without derivation. It is now possible to use the principles of thermodynamics to derive this equation.

Consider first the free energy of a phase as a function both of pressure and temperature. If pressure is changed by the amount dP and temperature is changed by the amount dT, the change in free energy dG will be given by

$$dG = \left(\frac{\partial G}{\partial P}\right)_T dP + \left(\frac{\partial G}{\partial T}\right)_p dT \tag{10-30}$$

Inserting the values for the partial differentials of G derived from the previous chapter,

$$dG = v\, dP - s\, dT \tag{10-31}$$

For the liquid state, this equation becomes

$$dG_l = v_l\, dP - s_l\, dT \tag{10-32}$$

For the vapor state, the expression is:

$$dG_v = v_v\, dP - s_v\, dT \tag{10-33}$$

If the quantities in both of these equations are defined as referring to 1 mole of the substance, then when liquid and vapor are in equilibrium, any change in the free energy of the liquid must equal the change in free energy of the vapor if the changes are such as to maintain equilibrium between the two phases

$$dG_l = dG_v \tag{10-34}$$

Combining this expression with Eqs. (10-32) and (10-33),

$$v_l\, dP - s_l\, dT = v_v\, dP - s_v\, dT \tag{10-35}$$

which can be rearranged to give

TABLE 10-2   Entropies of vaporization

Compound	s_v, eu	Compound	s_v, eu
AsBr$_3$	20.0	CH$_3$NO$_2$	30.7
AsCl$_3$	18.6	CH$_3$OH	26.5
AsF$_3$	22.5	CH$_3$SH	21.0
AsH$_3$	19.8	CH$_3$NH$_2$	23.1
BCl$_3$	20.0	CH$_3$CH$_2$OH	26.2
BF$_3$	25.0	CH$_3$CH$_2$NH$_2$	23.0
CBr$_4$	22.6	H$_2$O	26.0
CCl$_4$	20.5	H$_2$S	21.0
CF$_4$	20.7	D$_2$O	34.2
CFCl$_3$	20.1	NH$_3$	23.3
CF$_2$Cl$_2$	20.0	N$_2$H$_4$	25.1
CHCl$_3$	21.0	N$_2$O	21.4
CH$_2$O	23.0	PBr$_3$	20.8
CH$_3$Br	20.7	PCl$_3$	20.9
CH$_3$Cl	20.7	PF$_4$	19.9
CH$_3$F	21.7	PF$_5$	21.7
CH$_3$I	21.2	SiF$_4$	24.4

$$\frac{dP}{dT} = \frac{s_v - s_l}{v_v - v_l} = \frac{\Delta s_v}{\Delta v_v} \tag{10-36}$$

For purposes of approximate calculation, the value of v_l may be taken as negligible compared with v_v, so that the expression becomes

$$\frac{dP}{dT} = \frac{\Delta s_v}{v_v} \tag{10-37}$$

If the vapor behaves as an ideal gas, then

$$v_v = \frac{RT}{P} \tag{10-38}$$

The entropy of vaporization can be expressed in terms of the enthalpy of vaporization and the temperature at which the vaporization takes place

$$\Delta s_v = \frac{\Delta H_v}{T} \tag{10-39}$$

Equation (10-37) then becomes

$$\frac{dP}{P} = \frac{\Delta H_v}{R} \frac{dT}{T^2} \tag{10-40}$$

Applying the rules of the calculus, this expression becomes

Fig. 10-6 The effect of molecular orientation in liquids lowers S_L and raises ΔS_v.

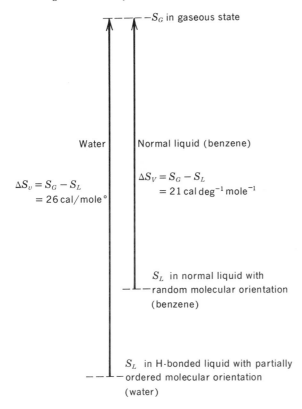

regarded as the vapor pressure P at some temperature $T = T_2$ at a value different from that of the normal boling point. The expression then becomes

$$\ln P = -\frac{\Delta H_v}{R}\left(\frac{1}{T} - \frac{1}{T_b}\right) \tag{10-44}$$

which gives the vapor pressure P as a function of the temperature T with the value of the enthalpy of vaporization ΔH_v and the temperature of the normal boiling point T_b appearing as constants. This expression can be rearranged to give

$$R \ln P = -\Delta H_v \frac{1}{T} + \frac{\Delta H_v}{T_b} \tag{10-45}$$

Since the last term in this equation is just the entropy of vaporization at the normal boiling point Δs_v, the equation can be put in the form

$$\left.\begin{aligned} R \ln P &= -\Delta H_v \frac{1}{T} + \Delta s_v \\ y &= ax + b \end{aligned}\right\} \tag{10-46}$$

Equation (10-46) is in the form of an algebraic equation for a straight line, where a is the slope of the line and b is the intercept of the line with the axis where $x = 0$. Such a plot is shown in Fig. 10-7. Above the critical temperature there is no distinction between the liquid and the vapor state; consequently, the vapor pressure of a liquid is a significant quantity only in the range below the critical temperature where the liquid phase and the vapor phase can exist as distinct phases in equilibrium with each other. Although the line in Fig. 10-7 representing $R \ln P$ has no physical meaning at temperatures above the critical temperature, the point where this line intersects the right-hand vertical axis is significant. When this line is extrapolated to the axis where $1/T$ is equal to zero (theoretically where the temperature is infinitely high), the intercept is the value of the entropy of vaporization at the normal boiling point; this can be proved by inserting the value of zero for $1/T$ in Eq. (10-46).

According to Trouton's rule, all liquids which have normal randomness in the orientation of molecules have the same entropy of vaporization, 21.2 eu. The lines representing the value of $R \ln P$ for these liquids, therefore, all intersect the axis of infinite temperature

$$d \ln P = -\frac{\Delta H_v}{R}\left(d\,\frac{1}{T}\right) \tag{10-41}$$

The assumption now is made that the variation of ΔH_v with temperature can be regarded as negligible over the range of temperature for which this equation is to be applied. Integrating between T_1, at which the pressure is P_1, and T_2, at which the pressure is P_2, the expression becomes

$$\int_{P_2}^{P_1} d \ln P = -\frac{\Delta H_v}{R}\int_{T_2}^{T_1} d\,\frac{1}{T} \tag{10-42}$$

Carrying out the integration,

$$\ln \frac{P_2}{P_1} = -\frac{\Delta H_v}{R}\left(\frac{1}{T_2} - \frac{1}{T_1}\right) \tag{10-43}$$

For many purposes it is convenient to let $P_1 = 1$ atm so that $T_1 = T_b$, the temperature of the normal boiling point expressed in degrees Kelvin. Then P_2 may be at a common focus. This point is designated as the Trouton focus and is shown in Fig. 10-8.

Liquids like water and ammonia have abnormally low randomness of molecular orientation and therefore abnormally high entropies of vaporization. Thus, the vapor-pressure lines for such liquids when extrapolated do not pass through the Trouton focus but intersect the axis at a point considerably higher.

The abnormally low randomness or abnormally high order of molecular arrangement in liquid water is important for many problems in biology. According to the data presented in Table 4-2, the interior of a biological cell is filled with material that is about 90 percent water. Moreover, the substances which make up the remaining 10 percent of the cell's fluid are molecules like monosaccharides, proteins, and amino acids, which also tend to form hydrogen bonds. There is general agreement among cytologists that in the semifluid interior of the cell there is greater order than in a normal liquid like benzene where hydrogen bonding is effectively absent. Since growth takes place to a considerable extent through the process of transferring molecules from this semifluid interior of the cell to cell walls and other more ordered molecular aggregates, it is particularly important to get as much information as possible about ordering in the liquid state.

Another point of interest in cytology is the connection between molecular symmetry and molecular entropy. As an example, in the process of cell growth and replication, molecules pass from the cytoplasm to form macromolecules like DNA and RNA. There is every reason to believe that this process is controlled by the balancing of an enthalpy factor ΔH and a probability factor $T \Delta s$; the probability factor, in turn, is significantly affected both by the molecular ordering in the phase from which the molecules come and the phase into which the molecules go in the growth process; the symmetry and internal configurations in the molecules also affect the entropy.

Example 10-7

The vapor pressure of CF_4 is 100 mm at $-150.7°C$, and its normal boiling point is $-127.7°C$. Calculate the enthalpy of vaporization.

$$\Delta H_v = -\left(\frac{1}{T} - \frac{1}{T_b}\right)^{-1} R \ln P$$

$$= -(0.0081632 - 0.0068728)^{-1}$$

$$\times 4.576 \log \tfrac{100}{760}$$

$$= 0.0012904^{-1} \times 4.576 \log 7.60$$

$$= 3828 \times 0.8808 = 3373 \text{ cal mole}^{-1}$$

Ans.

Example 10-8

The normal boiling point of CH_3SH is 7.6°C. Using Trouton's rule, estimate the enthalpy of vaporization.

$$\Delta H_v = 21.2 \times T_b$$

$$= 21.2 \text{ cal deg}^{-1} \text{ mole}^{-1} \times 280.8$$

$$= 5.95 \text{ kcal mole}^{-1} \qquad Ans.$$

Note: The observed enthalpy of vaporization is 5.9 kcal mole^{-1}.

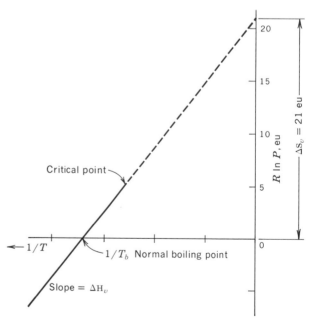

Fig. 10-7   The linear plot of $R \ln P$ against $1/T$ for the vapor pressure P of a normal liquid. (*Note:* $1/T$ increases toward the left.)

Fig. 10-8   Plot of $R \ln P$ against $1/T$ showing Trouton focus for normal liquid vapor pressure P and abnormal intersection point for water caused by molecular orientation.

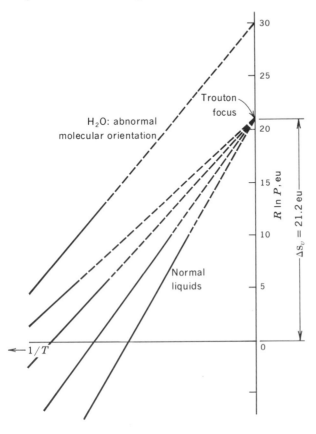

10-4   THE EFFECT OF PRESSURE ON PHASE CHANGES

Pressure and melting temperature

In the analysis of the various factors that affect the free energy, the relation between free energy and pressure was

$$\left(\frac{\partial G}{\partial P}\right)_T = v \qquad (10\text{-}47)$$

This equation shows that the free energy of the crystal is raised by the application of pressure, the rate of raising by increasing pressure at constant temperature being equal to the molar volume. Figure 10-9 shows the graphs of the free energy of benzene in the crystal-

line state plotted against temperature for pressures of 1 and 5 atm.

The free energy of the liquid is also raised by the application of the pressure, but because the volume occupied by 1 mole of liquid is so much larger in most cases than the volume occupied by 1 mole of crystals, the effect of applying pressure is greater; in other words, the distance between the free-energy curve for the liquid at 5 atm and at 1 atm normally is greater than the distance between the free-energy curve for the crystal at 5 atm and at 1 atm. Thus, the point of intersection of the free-energy curve for the crystalline state and for the liquid state normally is displaced to higher temperatures by pressure.

For water the effect is exactly the opposite. The volume of 1 mole of water in the crystalline state is larger than the volume of 1 mole of water in the liquid state (Fig. 10-10). This phenomenon is the basis of skating on ice. When the force produced by the weight of the body is applied over the relatively small area of the surface of the bottom of a skate, a significantly high pressure is produced on the ice. This pressure reduces the melting point of the ice, so that the crystals immediately subjected to the pressure melt and form a thin film of water between the surface of the skate and the surface of the ice, reducing the friction. If the ice were sufficiently cold, the pressure would not be enough to melt the ice and skating would be difficult, if not impossible.

Metallic bismuth exhibits the same anomaly as water. In theory, it might be possible to skate on the surface of metallic bismuth if the metal were held slightly below its melting point at 271°C.

Pressure and boiling point

Equation (10-47) applies equally well to the change of free energy with pressure for a gas. However, the molar volume for a gas is approximately 1,000 times greater than the molar volume for a liquid or a crystal. Consequently, the effect of pressure on the boiling point is very much greater than the corresponding effect on the melting point. Figure 10-11 is a plot of the free energy of the liquid and the gas in the neighborhood of the boiling point. The effect of pressure on the free energy of the gas is so much greater than the effect on the liquid that the latter may be regarded as negligible.

Example 10-9

The normal density of CH_3SH at 20°C is 0.868. Calculate the increase in the value of the free energy due to an increase in pressure of 10 atm, assuming that the change in density is negligible over this pressure range.

$$\left(\frac{\Delta_G}{\Delta P}\right)_T = v$$

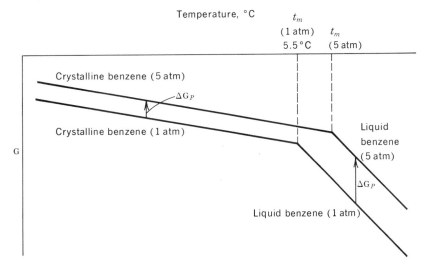

Fig. 10-9 The raising of the melting point of crystalline benzene when subjected to pressure (not to scale).

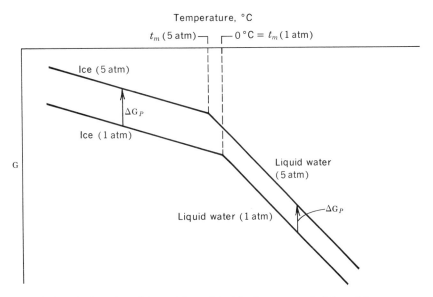

Fig. 10-10 The lowering of the melting point of ice by pressure (not to scale).

Fig. 10-11 The elevation of the boiling point by pressure (not to scale).

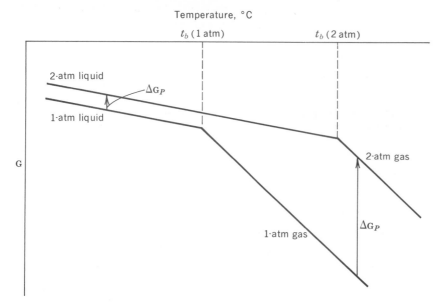

$$v = \frac{48.10 \text{ g mole}^{-1}}{0.868 \text{ g cm}^{-3}}$$

$$= 55.41 \text{ cm}^3 \text{ mole}^{-1}$$

$$\Delta G = v\,\Delta P = 55.41 \times 10$$

$$= 554.1 \text{ cm}^3 \text{ atm mole}^{-1}$$

$$= 554.1 \times 0.0242 \text{ cal cm}^{-3} \text{ atm}^{-1}$$

$$= 13.4 \text{ cal mole}^{-1} \qquad Ans.$$

10-5   PHASE DIAGRAMS

Water

One of the most helpful aids to the study of the application of thermodynamics to physical equilibria is the plot of corresponding values of pressure and temperature at which two phases are in equilibrium. Such a plot for water is shown in Fig. 10-12. Since the primary objective of this figure is to show the relative positions of the lines separating the different phases, the diagram is not drawn to scale. For example, in the right-hand part of the figure, the line is drawn for the values of the pressure and temperature at which the liquid phase and the vapor phase are in equilibrium.

This shows the vapor pressure as a function of temperature. Since the vapor pressure is a logarithmic function of temperature, it would be almost impossible to show in detail the relations around the melting point and at the same time include values of the pressure up to the boiling point on any diagram of reasonable size. Qualitatively speaking, any point on the diagram below the line corresponds to the system in the vapor state, where the pressure is less than the vapor pressure. Any point on the diagram above the line indicates that the system is completely in the liquid phase since the pressure here is higher than the vapor pressure.

At 0.01°C, the vapor phase, liquid phase, and crystal phase can all exist in equilibrium with each other. This is the *triple point* of the system. The extrapolation of the liquid-vapor line to temperatures below the triple-point temperature corresponds to the vapor pressure of the supercooled liquid. The vapor pressure of the crystal is shown by a line lying slightly below that for the supercooled liquid. Note that the area to the right of this line corresponds to a state where the system is all in the vapor phase and the area to the left of this line corresponds to a state where the system normally is all in the crystal phase.

At temperatures below the triple-point temperature

and at pressures above the triple-point pressure, the liquid phase and the crystal phase can be in equilibrium with each other. The corresponding equilibrium pressures and temperatures are shown by the line that slants upward and to the left from the triple point. This line separates the area on the diagram corresponding to the system completely in the liquid phase and the area corresponding to the system completely in the crystal phase.

Figure 10-13 is the phase diagram for water at relatively high pressures. The area shown in Fig. 10-12 would be too small to be seen on the scale used in Fig. 10-13, since the former figure covers a range of about 2 atm and this corresponds to a distance considerably less than 0.001 in. in Fig. 10-13. The liquid-crystal line of Fig. 10-12 can be seen in Fig. 10-13 in extended form slanting upward and to the left from the point on the zero pressure axis at 0°C. In Fig. 10-13, normal ice is designated as *ice I*. When the pressure rises to about 2,000 atm, a new form of crystalline ice appears called *ice III*. At this pressure ice I and ice III can exist in contact and in equilibrium with each other. This is

an example of still another type of physical equilibrium, an equilibrium between two crystalline phases. The diagram also shows the line for the equilibrium between normal ice (ice I) and ice II as well as the equilibrium between ice II and ice V and the equilibrium between ice III and ice V. While ice I is less dense than liquid water, the other crystalline forms of water (ice) encountered at the higher pressures are more dense than liquid water at these pressures. Because of this, an increase in pressure shifts the equilibrium to higher temperatures in contrast with the effect of pressure on the equilibrium between liquid water and ice I where the equilibrium is shifted to lower temperatures with rising pressure. Under sufficient pressure, liquid water will even be in equilibrium with ice in the form of ice VII at a temperature corresponding to the boiling point of water.

Sulfur

The phase diagram for sulfur is shown in Fig. 10-14. Like the phase diagram for water, this is not drawn to scale, so that phenomena at relatively low and rela-

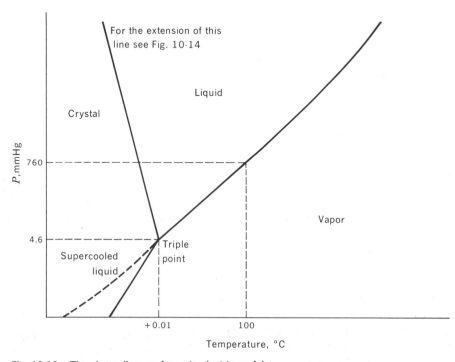

Fig. 10-12   The phase diagram for water (not to scale).

Fig. 10-13   The phase diagram for water under high pressure. (After G. M. Barrow, "Physical Chemistry," McGraw-Hill Book Company, New York, 1966.)

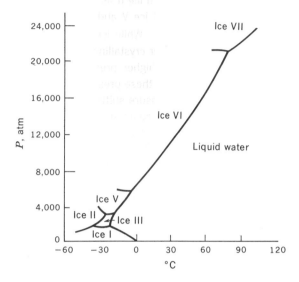

and there is another triple point where monoclinic sulfur, liquid sulfur, and sulfur vapor are all in equilibrium with each other.

By cooling liquid sulfur below this point, it can be put into the supercooled phase (supercooled with respect to monoclinic sulfur), and the vapor pressure can be measured to the left of the point at 118.75°C; this is shown as a dashed line extended to the left down to the true triple point at 112.8°C.

By going to pressures higher than the vapor pressure, the pressure-temperature points can be determined for the equilibrium between rhombic and monoclinic sulfur as well as the pressure-temperature points for the equilibrium between rhombic sulfur and liquid and between monoclinic sulfur and liquid. These are the three lines extending upward from the vapor-pressure curves and coming to a common intersection

tively high pressures can be included on the same diagram. At the lower left the line indicates the temperature and pressure at which rhombic crystals of sulfur S_α are in equilibrium with sulfur vapor. As the system is heated to higher temperatures, there is a transition temperature (95.5°C) at which rhombic sulfur and crystalline sulfur in the monoclinic form are in equilibrium with each other. This point is thus a triple point where two crystalline forms of sulfur and also sulfur vapor are all in equilibrium. It is possible to heat rhombic sulfur above this point as the change to monoclinic sulfur takes place very slowly. Thus, the line at the lower left, which represents the vapor pressure of rhombic sulfur, can be extended to temperatures above the triple point to the temperature where rhombic sulfur melts (112.8°C). This is the normal triple-point temperature of sulfur, where crystalline sulfur in the rhombic form, liquid sulfur, and sulfur vapor are all in equilibrium. However, if the rhombic sulfur goes through the phase transition at 95.5°C and changes into monoclinic sulfur (S_β), the vapor-pressure curve of the monoclinic sulfur can be determined; this is shown as the heavy dark line extending to the right and upward from the pseudo triple point at 95.5°C. At 118.75°C, the monoclinic sulfur melts,

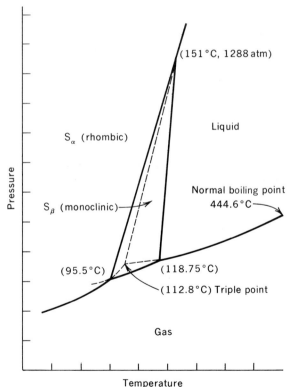

Fig. 10-14   The phase diagram of sulfur (not to scale). (After T. Moeller, "Inorganic Chemistry," John Wiley & Sons, Inc., New York, 1952.)

at 151°C and 1,288 atm. Above this pressure, rhombic sulfur is in equilibrium with liquid sulfur.

The many factors influencing equilibria between crystals and other crystalline forms or liquid forms will be discussed in greater detail in the chapter on the crystalline state.

Problems

1. For chloroform, $CHCl_3$, the enthalpy of vaporization is 6962 cal mole^{-1} at the temperature of the normal boiling point, 61.5°C. Calculate the entropy of vaporization.

2. For ethyl alcohol, C_2H_5OH, the enthalpy of vaporization is 9384 cal mole^{-1} at the normal boiling point (78.3°C). For ethyl bromide, C_2H_5Br, the enthalpy of vaporization is 7008 cal mole^{-1} at the normal boiling point (38.4°C). Calculate the entropy of vaporization for each of these compounds. In which compound does the entropy of vaporization indicate the presence of hydrogen bonding in the liquid state?

3. For methyl chloride, CH_3Cl, the enthalpy of vaporization is 5120 cal mole^{-1} at the normal boiling point (−23.8°C). For methyl alcohol, CH_3OH, the enthalpy of vaporization is 8400 cal mole^{-1} at the normal boiling point of 64.7°C. Calculate the entropy of vaporization for each of these compounds at the normal boiling points and indicate which compound shows the most evidence of hydrogen bonding.

4. For ethane, C_2H_6, the enthalpy of fusion is 683 cal mole^{-1} at the normal melting point (−183°C). Calculate the observed entropy of fusion. Assuming that this compound has a symmetry number of 6, calculate the normalized entropy of fusion.

5. For normal pentane, C_5H_{12}, the enthalpy of fusion is 2009 cal mole^{-1} at the melting point (−129.7°C). The difference between the entropy of vaporization and the Trouton constant indicates that the randomness in the liquid state lies below the normal randomness by 1.1 eu. The various configurations of the CH_2 groups in the liquid state add an excess entropy of 4.4 eu mole^{-1}. Assume that for this molecule the symmetry number is equal to 2 and calculate the normalized entropy of fusion.

6. For urea, NH_2CONH_2, the enthalpy of fusion is 3611 cal mole^{-1} at the normal melting point (132.7°C). Assuming that this compound has a symmetry number of 2, calculate the normalized entropy of fusion.

7. For diethyl ether, $(C_2H_5)_2O$, the enthalpy of vaporization is 6608 cal mole^{-1} at the normal boiling point (34.6°C). Calculate the driving force in terms of the free-energy change when 1 mole of supercooled vapor at a pressure of 1 atm and a temperature 5° below the boiling point condenses to form the liquid state.

8. For methyl alcohol, commonly called *wood alcohol,* CH_3OH, the enthalpy of vaporization is 8400 cal mole^{-1} at the normal boiling point (64.7°C). Calculate the driving force or change in free energy when 1 mole of superheated liquid vaporizes at a temperature 5° above the normal boiling point under a pressure of 1 atm.

9. Ethyl acetate has an enthalpy of vaporization of 7800 cal mole^{-1} at the temperature of the normal boiling point (77.2°C). What is the change in free energy when 1 mole of this substance as supercooled vapor 15° below the normal boiling point condenses to form the liquid state?

10. Urea, NH_2CONH_2, has an entropy of fusion of 8.9 eu at the normal melting point (132.7°C). Calculate the free-energy change when 1 mole of this substance in the supercooled liquid state passes to the crystalline state at 8° below the normal melting point.

11. Diethyl ether, $(C_2H_5)_2O$, has an entropy of fusion of 11.1 eu at the normal freezing point (−116.3°C). Calculate the change in free energy when 1 mole of this substance goes from the supercooled liquid state to the crystalline state at 15° below the melting point.

12. The straight-chain hydrocarbon $C_{43}H_{88}$ is reported to have an entropy of fusion of 96 eu at its normal melting point (85°C). If 1 mole of this substance in the crystalline state were suddenly heated 5° above the normal melting point and melting was delayed because of the great length of the molecules, what would the free-energy change be when 1 mole of this substance passed from the superheated crystalline state to the liquid state?

13. The temperatures and corresponding vapor pressures of four halogen derivatives of carbon are as follows:

P, atm	0.1316	1
CCl_2F_2	−68.6°C	−29.8°C
CCl_3F	−23.0	+23.7
CCl_4	+23.0	+76.7
CF_4	−150.7	−127.7

Make a plot of $R \ln P$ against the reciprocal of the corresponding temperatures in degrees Kelvin. Extrapolate the vapor-pressure lines so obtained to the point where $1/T = 0$. Measure the height on the ordinate axis for each of these intersections in terms of entropy units.

14. In the table below are given values of the vapor pressure of water and ammonia together with the corresponding values of the temperatures. Make plots similar to those in Prob. 13 and note that the entropies of vaporization lie considerably higher, as indicated by the intersection of the extrapolated vapor-pressure curves with the ordinate axis.

P, atm	0.1316	1
NH_3	−68.4°C	−33.6°C
H_2O	+51.6	+100.0

15. The vapor pressure of formaldehyde, HCHO, is 100 mm Hg at −70.6°C and 760 mm Hg at −19.5°C. Calculate the heat of vaporization and entropy of vaporization at the normal boiling point.

16. The normal boiling point of methyl iodide is 42.4°C. Using Trouton's rule, calculate the vapor pressure at 0°C.

17. The vapor pressure of crystalline hydrogen sulfide is 10 mm Hg at −116.3°C and 100 mm Hg at −91.6°C. Calculate the enthalpy of sublimation.

18. The vapor pressure of liquid hydrogen sulfide is 400 mm at −71.8°C and 760 mm Hg at −60.4. Calculate the enthalpy of vaporization.

19. From the data from Probs. 17 and 18, calculate the enthalpy of fusion and the entropy of fusion at the melting point −85.5°C.

20. The density of liquid methanol, CH_3OH, is 0.7928 at 20°C. Calculate the increase in free energy when the pressure changes from 1 to 10 atm, assuming no change in the density over this pressure range.

21. Ethyl alcohol has a normal boiling point of 78.4°C. At 400-mm pressure, this liquid boils at 63.5°C. At what temperature will the liquid boil when it is under a pressure of 2 atm?

REFERENCES

Marsh, J. S.: "Principles of Phase Diagrams," McGraw-Hill Book Company, New York, 1935.

Timmermans, J.: "Physico-chemical Constants of Pure Organic Compounds," Elsevier Publishing Company, Amsterdam, 1950.

Smoluchowski, R., J. E. Mayer, and W. A. Weyl (eds.): "Phase Transformations in Solids," John Wiley & Sons, Inc., New York, 1951.

Timmermans, J.: "Les Constants physiques des composés organiques cristallisés," Masson et Cie, Paris, 1953.

Rossini, F. D., B. J. Mair, and Anton J. Streiff: "Hydrocarbons from Petroleum," Reinhold Publishing Corporation, New York, 1953.

Fine, M. E.: "Phase Transformations in Condensed Systems," The Macmillan Company, New York, 1964.

Brout, R. H.: "Phase Transitions," W. A. Benjamin, Inc., New York, 1965.

Reid, R. C., and T. K. Sherwood: "The Properties of Gases and Liquids," 2d ed., McGraw-Hill Book Company, New York, 1966.

Andrews, D. H., R. L. Coleman, J. Feuer, and R. A. Schuck: A Regularity in Entropies of Fusion, *Abstr. Bull., Div. Phys. Chem., Am. Chem. Soc.,* **April, 1967:** 178.

Bondi, A.: "Molecular Crystals, Liquids and Glasses," John Wiley & Sons, Inc., New York, 1968.

Chapter Eleven

solutions

11-1 THE NATURE OF SOLUTIONS

To understand the role which solutions play in life processes, it is essential first of all to perceive the difference between physical mixtures and chemical solutions. Imagine sand ground so fine that the individual particles cannot be distinguished by the unaided human eye. If 100 cm³ of this powdered sand is stirred into 500 cm³ of water, a milky liquid is formed. This liquid can be poured from one beaker to another and appears to be completely homogeneous; the sand seems to have disappeared completely into the water, or, in the language of physical chemistry, it appears to be *dissolved* in the water.

But a study of a drop of this liquid with a high-power microscope shows this conclusion to be erroneous. Through the microscope, individual grains of sand can be seen suspended in the liquid, each surrounded by clear water. The diameter of one of these particles of sand can be estimated at about 10^{-4} cm. If the sand is pure quartz, its chemical composition is SiO_2. Knowing the approximate size of the atoms of silicon and the atoms of oxygen, it can be calculated that in each grain of sand there are more than 10^{12} atoms of each of these elements. Thus, on the atomic scale, there is a volume of space where billions of atoms of silicon and oxygen are grouped together in the way that gives them the properties associated with quartz; on the other hand, in the space between the grains of sand there are billions of hydrogen and oxygen atoms associated in the form of H_2O molecules in the way that gives them the characteristic properties of liquid water. The grains of sand and the water constitute a *physical mixture*. This mixture is characterized by the fact that at a degree of fineness (10^{-4} cm) still far from approaching atomic dimensions (10^{-8} cm) there are relatively large volumes with the characteristic properties of one of the components of the mixture and other large volumes with the characteristic properties of the other component of the mixture. This situation is illustrated schematically in Fig. 11-1.

Now consider what happens when the same amount of finely powdered table salt, NaCl, is mixed with 500 cm³ of water. The grains of salt

Fig. 11-1   Enlarged views of a physical mixture of sand and water.

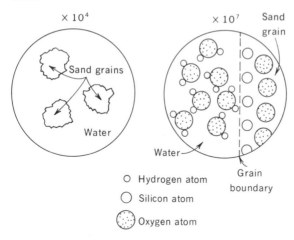

disappear in the water and form a clear solution. Like the mixture of finely powdered sand and water, this liquid appears homogeneous to the unaided eye. But when a drop of this fluid is compared with a drop of the sand-water mixture under a microscope (Fig. 11-2), there is a striking difference; no matter how high the optical magnification, no grains can be seen. All the evidence indicates homogeneity down to the atomic level; each Na^+ is surrounded by H_2O molecules, and each Cl^- is surrounded by H_2O molecules. The properties of a portion of this combination of ingredients have taken on new values, differing both from those of one of the pure ingredients H_2O and from the other pure ingredient $NaCl$; this liquid is therefore called a true *solution*.

In general, the component which crystallizes out when its concentration is sufficiently increased is called the *solute*, and the other component is called the *solvent*. In the case just discussed, $NaCl$ is the solute, and H_2O is the solvent. For a solution like that of ethyl alcohol in water at room temperature where neither component crystallizes, the component present in the larger amount is usually called the solvent, but in a solution made up of equal parts of two components there is no distinction between solute and solvent.

There is also some ambiguity in distinguishing true chemical solutions from physical mixtures when the molecules of the solute are very large or when the solid particles in the physical mixture are very small. When

the dimensions of the molecules of the solute or of the particles are of the order of magnitude lying between 10^{-6} and 10^{-3} cm, the liquids usually are called *colloidal solutions,* which will be discussed in detail in Chap. 23.

11-2   SOLUTION THERMODYNAMICS

Concentration units

In order to relate the properties of solutions to each other thermodynamically, it is necessary to define units of concentration. The total number of moles present in the solution is designated by the symbol n_{tot}; the number of moles of component 1 is designated by the symbol n_1, and the number of moles of component 2 by the symbol n_2. As an example, consider a solution of methanol (component 1) in water (component 2), where the numerical values are

$$n_{tot} \quad = \quad n_1 \quad + \quad n_2 \tag{11-1}$$

10 moles   0.2 mole   9.8 moles

For thermodynamic relations, the most useful unit of concentration is the *mole fraction* X_i, defined as the number of moles of the component divided by the total number of moles; the subscript refers to the component. For the example described above, the values of the mole fraction are defined by the equations

$$X_1 = \frac{n_1}{n_1 + n_2} = \frac{0.2 \text{ mole}}{10 \text{ moles}} = 0.02 \tag{11-2}$$

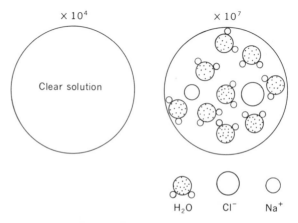

Fig. 11-2   Enlarged views of a true solution ($NaCl$ in H_2O).

$$X_2 = \frac{n_2}{n_1 + n_2} = \frac{9.8 \text{ moles}}{10 \text{ moles}} = 0.98 \qquad (11\text{-}3)$$

Thus

$$X_1 + X_2 = 1 \qquad (11\text{-}4)$$

In very dilute solutions, where n_1 is very much less than n_2, there is the approximate relation

$$\frac{n_1}{n_2} \simeq \frac{n_1}{n_1 + n_2} = X_1 \qquad (11\text{-}5)$$

since

$$n_1 + n_2 \simeq n_2 \qquad (11\text{-}6)$$

Another useful concentration unit is *molarity,* usually designated by M and defined by the equation

$$M_1 \text{ moles liter}^{-1} = n_1 \text{ moles} \frac{1,000 \text{ cm}^3 \text{ liter}^{-1}}{V \text{ cm}^3} \qquad (11\text{-}7)$$

where n_1 is the number of moles of component 1 in the solution and V is the volume of the solution expressed in cubic centimeters.

Still another concentration unit is *weight molarity* (sometimes called *molality*) usually designated by m_W and defined by the equation

$$(m_W)_1 \text{ moles kg}^{-1} \text{ solvent} = n_1 \frac{1,000 \text{ g kg}^{-1}}{W_2 \text{ g solvent}} \qquad (11\text{-}8)$$

where W_2 is the weight of the solvent in grams.

Example 11-1

A solution is composed of 75 g of ethanol (1) and 25 g of methanol (2) dissolved in 900 g of water (3). Calculate the mole fractions.

$$n_1 = \frac{75 \text{ g}}{46.07 \text{ g mole}^{-1}} = 1.63 \text{ moles}$$

$$n_2 = \frac{25 \text{ g}}{32.04 \text{ g mole}^{-1}} = 0.78 \text{ mole}$$

$$n_3 = \frac{900 \text{ g}}{18.02 \text{ g mole}^{-1}} = \underline{49.94 \text{ moles}}$$

$$n_{\text{tot}} = 52.35 \text{ moles}$$

$$X_1 = \frac{1.63}{52.35} = 0.031$$

$$X_2 = \frac{0.78}{52.35} = 0.015$$

$$X_3 = \frac{49.94}{52.35} = 0.954$$

Example 11-2

The volume of the solution described in Example 11-1 is 1,020 cm³. Calculate the molarity for components 1 and 2.

$$M_1 = 1.63 \frac{1,000}{1,020} = 1.59 \text{ moles liter}^{-1}$$

$$M_2 = 0.78 \frac{1,000}{1,020} = 0.76 \text{ mole liter}^{-1}$$

Example 11-3

Calculate the molality for components 1 and 2 in Example 11-1.

$$(m_W)_1 = 1.63 \frac{1,000}{900} = 1.81 \text{ moles kg}^{-1}$$

$$(m_W)_2 = 0.78 \frac{1,000}{900} = 0.87 \text{ mole kg}^{-1}$$

Partial molal properties

When two or more substances are mixed together to form a solution, the properties of each component of the solution depend upon its concentration in the solution. The relationship between volume and concentration provides one of the simpler examples of this type of dependence. The simplest situation is found when each of the two types of molecules making up the solution occupies the same volume, i.e., has the same effective radius.

In order to see how individual molecular volumes affect the total volume of the solution, consider what happens when two kinds of small shot are mixed together. Suppose that we have 1,000 small iron shot and 1,000 small brass shot, each spherical and with the same radius. Suppose that the size is such that 20 of these shot when closely packed together occupy a volume of 1 cm³. If the 500 brass shot are poured into a graduated cylinder, they will occupy a volume of 500 cm³; if the 500 iron shot are then poured on top of the brass shot, they will fill up the cylinder to the

1,000-cm³ mark. Since all the shot are the same size, there will be no change in the total volume if the shot are shaken and thoroughly mixed together. If 20 more brass shot are then added, the total volume will be increased by 1 cm³ no matter whether these brass shot lie on top or are shaken and mixed into the total collection.

Suppose now that there are other brass shot which are so small in diameter that 100 of them closely packed occupy a volume of 1 cm³. If 500 cm³ of these shot are mixed with 500 cm³ of the iron shot, the total volume will not be 1,000 cm³ but some smaller figure because the small brass shot can fit into the interstices between the iron shot. In this case, the total volume of the mixture of the two kinds of shot will depend on the concentration of each. If the mixture consists almost entirely of the small brass shot, adding 1 cm³ of additional brass shot will increase the total volume of 1 cm³; but if the total mixture consists almost entirely of iron shot, adding 1 cm³ of brass shot to the total mixture may increase the total volume by an amount less than 0.01 cm³.

A solution made up of molecules of ethanol and water in many ways resembles the mixture made up of the large iron shot and the small brass shot. The formula for ethanol is C_2H_5OH; the water molecule, H_2O,

is considerably smaller in volume. If the ethanol molecules and the water molecules had identical volume, and there were no interaction between the molecules to change the volume, 500 cm³ of ethanol added to 500 cm³ of water should give a solution with a volume of 1,000 cm³. In such a case the solution would be called an *ideal* solution with respect to volume. In an analogous way, 500 cm³ of helium at 1 atm and 0°C mixed with 500 cm³ of argon also at 1 atm and 0°C occupies a total volume of 1,000 cm³; these gases form an ideal mixture or gaseous "solution." Since the water molecules and ethanol molecules do *not* have identical volumes, and since they do interact, their solution actually has a smaller volume than that predicted on the basis of ideal behavior.

Table 11-1 lists extremely precise measurements made at the National Bureau of Standards on the observed volumes of aqueous ethanol solutions at different values of X_1, the mole fraction of ethanol. The volume of 1 mole of an ideal solution (v_{ideal}) is calculated from the relation

$$v_{ideal} = X_1 v_1^\circ + X_2 v_2^\circ \qquad (11\text{-}9)$$

This shows the volume per mole that the solution would have if it were ideal. Here v_1° is the volume per mole of pure ethanol, and v_2° is the volume per mole of pure water. This relation can be represented graphically as shown in Fig. 11-3a. The part of the total volume occupied by the molecules of ethanol is $X_1 v_1^\circ$, and the value of this quantity is shown in the diagram. There is also a line showing the value of $X_2 v_2^\circ$. As can be seen from the table, the observed volume of 1 mole of the solution is somewhat smaller than the calculated ideal volume; the difference between these is shown in the last column of the table and plotted in Fig. 11-3b. This difference reaches a maximum at $X_1 = 0.425$.

For such a nonideal solution a quantity called the *partial molal* volume, $(\partial V/\partial n_i)_{T,P,X_i}$, is defined for each component i by the equation

$$dV = \left(\frac{\partial V}{\partial n_i}\right)_{T,P,X_i} dn \qquad (11\text{-}10)$$

The partial molal volume also is frequently denoted by the shorter symbol $\bar{v}_i$. When Eq. (11-10) is integrated

TABLE 11-1   Ideal molar volume $\overline{v}_{ideal}$ and observed partial molar volume $\overline{v}_{obs}$ as a function of mole fraction X_i for a soluton of ethanol (1) and water (2)[a]

X_i	$\overline{v}_{ideal}$, cm³	$\overline{v}_{obs}$, cm³	$\overline{v}_{ideal} - v_{obs}$, cm³
0.0000	18.0681	18.0681	0.0000
0.0416	19.7593	19.5664	0.1929
0.0890	21.6850	21.2272	0.4578
0.1435	23.8978	23.1859	0.7119
0.2068	26.4668	25.5686	0.8982
0.2811	29.4856	28.4682	1.0174
0.3697	33.0838	32.0041	1.0797
0.4771	37.4458	36.3680	1.0778
0.6100	42.8429	41.8643	0.9786
0.7787	49.6962	48.9932	0.7029
1.0000	58.6829	58.6829	0.0000

[a] *Natl. Bur. Std. U.S., Circ.* 19: "Standard Density and Volumetric Tables," 6th ed., Oct., 1924.

at constant T, P, and X_i from $n_i = 0$ to $n_i = 1$, with $(\partial V / \partial n_i)_{T,P,X_i}$ constant, the expression becomes[1]

$$\int_0^{V^\dagger} dV = \left(\frac{\partial V}{\partial n_i} \right)_{T,P,X_i} \int_0^1 dn_i \qquad (11\text{-}11)$$

Since

$$\int_0^1 dn_i = 1 \qquad (11\text{-}12a)$$

and

$$\int_0^{V^\dagger} dV = \mathbf{v}_i^* \qquad (11\text{-}12b)$$

then

$$\mathbf{v}_i^* = \left(\frac{\partial V}{\partial n_i} \right)_{T,P,X_i} \equiv \bar{\mathbf{v}}_i \qquad (11\text{-}13)$$

Since the above integration of dn_i from zero to the value of 1 mole is the physical equivalent of adding 1 mole of component i to a solution sufficiently large for the concentration to remain effectively constant during this addition, $\mathbf{v}_i^*$ is the increase in volume of the solution due to this addition; this shows the physical meaning of the *partial molal volume*. In other words, this quantity is the increase in the volume of the solution when 1 mole of component 1 is added while temperature, pressure, and concentration are maintained constant.

There are a number of ways of obtaining values for $\bar{\mathbf{v}}_i$. These are discussed at some length in Lewis and Randall (see References). For example, it is possible to have a large amount of solution in a volumetric flask with a small calibrated neck; $\frac{1}{10}$ mole of a component can be added to this solution without appreciably changing the concentration; and the increment in volume can be determined. Multiplying this increment by 10 gives the increment in volume per mole at this concentration for this component.

The value of the partial molal volume can also be obtained by plotting the volume of 1,000 g of the solvent containing various numbers of moles of the solute. The slope of the tangent to the curve corresponds to the partial molal volume. In Fig. 11-4 such a plot is

[1] The symbol $\mathbf{v}_i^*$ denotes that portion of the volume of the solution effectively occupied by 1 mole of the solute at the concentration under consideration.

Fig. 11-3 (a) The variation of molar volume $\mathbf{v}_{\text{ideal}}$ with mole fraction X, for an ideal solution of ethanol and water; (b) deviation of observed molar volume from the ideal molar volume $\mathbf{v}_{\text{ideal}}$ for a solution of ethanol and water.

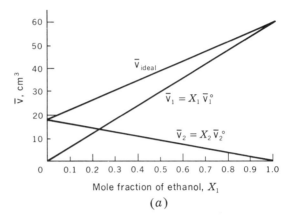

(a)

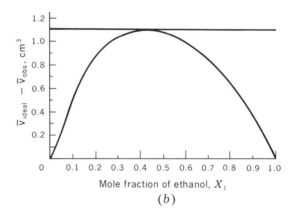

(b)

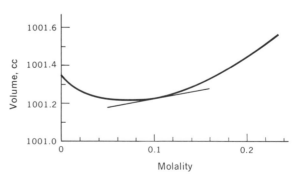

Fig. 11-4 The volume of 1,000 g of H_2O containing varying amounts of $MgSO_4$. [After T. Kohlrausch and W. Hallwachs, *Ann. Physik.*, (3)53:14 (1894).]

shown for a solution of magnesium sulfate (component 1) in water at 18°C. A line is drawn tangent to the curve at a value of the molality of 0.1; this line has a slope corresponding to an increase 0.090 cm³ for an increment of 0.10 mole. The value of $\bar{v}_1$ is thus 0.90 cm³ mole⁻¹.

Gibbs-Duhem equation

The total volume of a two-component solution can be expressed

$$V = \bar{v}_1 n_1 + \bar{v}_2 n_2 \qquad (11\text{-}14)$$

The total differential of this expression is

$$dV = \bar{v}_1\,dn_1 + n_1\,d\bar{v}_1 + \bar{v}_2\,dn_2 + n_2\,d\bar{v}_2 \qquad (11\text{-}15)$$

On the other hand, when a small increment dn_1 and another small increment dn_2 are added to a solution, the total increment in volume is

$$dV = \bar{v}_1\,dn_1 + \bar{v}_2\,dn_2 \qquad (11\text{-}16)$$

Subtracting Eq. (11-16) from Eq. (11-15) gives

$$0 = n_1\,d\bar{v}_1 + n_2\,d\bar{v}_2 \qquad (11\text{-}17)$$

This is called the Gibbs-Duhem equation; it can also be put in the form

$$\frac{d\bar{v}_1}{d\bar{v}_2} = \frac{-n_2}{n_1} \qquad (11\text{-}18)$$

This equation shows that the rate at which one partial molal volume varies with respect to the other is equal to the negative of the inverse ratio of the number of moles making up the solution.

Open thermodynamic system

The thermodynamic systems considered up to this point have had a channel by which energy in the form of work can enter or leave and another channel by which energy in the form of heat can enter or leave; these were called *closed* thermodynamic systems. In studying the thermodynamic properties of solutions, it is helpful to consider systems with an additional channel, called a *matter channel,* by which varying amounts of different kinds of matter may be introduced into, or withdrawn from, the system; such a system is called an *open* thermodynamic system (Fig. 11-5). As in the discussion of volume changes, partial molal values can

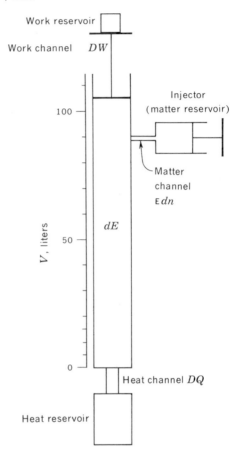

Fig. 11-5 Open thermodynamic system. $dE = DQ + DW + \varepsilon\,dn$.

be obtained for any of the extensive variables of the system, being the change in that variable when 1 mole of a component is introduced into the system under conditions where the volume of the system is sufficiently large to ensure that introduction of this component does not appreciably change its concentration. Thus, the partial molal enthalpy $\bar{H}_1$ of component 1 is defined by the equation

$$dH = \left(\frac{\partial H}{\partial n_1}\right)_{T,P,X_1} dn_1 = \bar{H}_1\,dn_1 \qquad (11\text{-}19)$$

where H is the total enthalpy of the system. Other extensive thermodynamic properties such as entropy and free energy are defined in a similar manner and will be discussed in some detail in the later part of this chapter.

11-3   IDEAL SOLUTIONS

In the previous section a two-component ideal solution was likened to a mixture of brass shot and iron shot each having the same radius so that the total volume remained the same whether the two kinds of shot were separated or mixed together.   An ideal solution can also be compared to the mixture of two ideal gases which have no effect on each other when mixed.   According to Dalton's law of partial pressures, in such a mixture of ideal gases

$$P_{tot} = P_1 + P_2 \tag{11-20}$$

where P_1 and P_2 are the partial pressures of each of the components.   The expression for the total pressure and for the two partial pressures can also be written in the form of the usual ideal-gas equations

$$P_{tot} V = n_{tot} RT \tag{11-21a}$$

$$P_1 V = n_1 RT \tag{11-21b}$$

$$P_2 V = n_2 RT \tag{11-21c}$$

where V = volume of container enclosing gas mixture
   R = gas constant
   T = temperature of gas mixture, °K
   n_{tot} = total number of moles

Equation (11-21b) divided by Eq. (11-21a) becomes

$$\frac{P_1}{P_{tot}} = \frac{n_1}{n_{tot}} \tag{11-22}$$

Recall that n_1/n_{tot} is, by definition, the quantity called the mole fraction X_1, so that

$$\frac{P_1}{P_{tot}} = X_1 \tag{11-23}$$

This equation can be rearranged

$$P_1 = X_1 P_{tot} \tag{11-24}$$

Note that P_{tot} is the same pressure that would be found in the container if all the molecules in the container were molecules of component 1, a fact that can be expressed

$$P_{tot} = P_1^\circ \tag{11-25}$$

where P_1° denotes the pressure that would be found in the box if all the molecules there were component 1 molecules.   Consequently,

$$P_1 = X_1 P_1^\circ \tag{11-26}$$

In an ideal solution, there is an analogous relation where P_1 is the *vapor pressure* of component 1, X_1 is the mole fraction of component 1 in the solution, and P_1° is the vapor pressure of pure component 1.   This important relation is known as Raoult's law.   A similar relationship holds for component 2.

$$P_2 = X_2 P_2^\circ \tag{11-27}$$

Raoult's law can be interpreted in terms of the physical meaning of vapor pressure.   For example, if 1 mole of 2-methylheptane were placed in a 1-liter container at a temperature of 35.6°C, and if there were no forces of attraction between the molecules so that this substance acted as an ideal gas, the pressure in this container would be 25.3 atm.   Because of the attraction between the molecules of this substance, however, nearly all the substance will be found in the liquid state, occupying a volume of only 146 cm³, while a small part will be in the gaseous state above the liquid at a pressure of 100 mm, the vapor pressure of 2-methylhexane at this temperature.   These contrasting states are shown schematically in Fig. 11-6.   Suppose now that

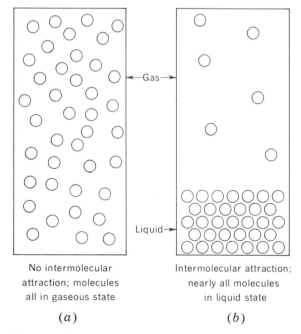

No intermolecular attraction; molecules all in gaseous state

Intermolecular attraction; nearly all molecules in liquid state

(a)

(b)

Fig. 11-6   Dependence of state on intermolecular attraction.

half the 2-methylhexane is removed and replaced by 3-methylhexane. The molecule of this substance closely resembles the molecule of 2-methylhexane since the only difference is the shift of the methyl group from the 2 position to the 3 position. Both substances behave essentially as ideal gases. If both were in the container under the hypothetical conditions where there were no forces of attraction between the molecules, the pressure in the box would not be changed appreciably by substituting one set of molecules for the other.

The vapor pressure of 3-methylhexane at the temperature under consideration (35.6°C) is 106 mm. This indicates that the attraction between the molecules of one substance is about the same as the attraction between the molecules of the other substance and, presumably, the same as the attraction of a molecule of one substance for a molecule of the other substance. In the gaseous state as ideal gases the partial pressure of each substance is proportional to its concentration expressed as mole fraction. Consequently, it is logical to expect that the partial vapor pressure of each substance over a solution will also be proportional to its concentration in the solution expressed as mole fraction. This is the relation expressed by Eqs. (11-26) and (11-27). Thus, the total pressure over the solution will be

$$P_{\text{tot}} = P_1 + P_2 = X_1P_1^\circ + X_2P_2^\circ \qquad (11\text{-}28)$$

In practice, Raoult's law is applicable to many solutions where the vapor pressures of the pure components are considerably different from each other. In Fig. 11-7 the partial vapor pressures are shown for each of the components in a solution consisting of benzene and toluene, together with the dashed line showing the total pressure.

Figure 11-8 is a schematic cross section of a portion of a solution of toluene and benzene together with toluene and benzene vapor. In the liquid state the mole fraction of toluene X_1 is 0.8, and the mole fraction of benzene X_2 is 0.2. Because of the higher vapor pressure of benzene, the composition in the vapor $^gX_1 = 0.6$ and $^gX_2 = 0.4$ where the symbol ^gX is used to denote the mole fraction in the gaseous state.

Example 11-4

At 20°C benzene has a vapor pressure of 74 mm Hg, and toluene has a vapor pressure of 22 mm Hg. The two substances form a nearly ideal solution at this temperature. Calculate the partial pressure of each component when $X=0.50$ and also calculate the total pressure at this composition.

The Raoult's law equations for this system are

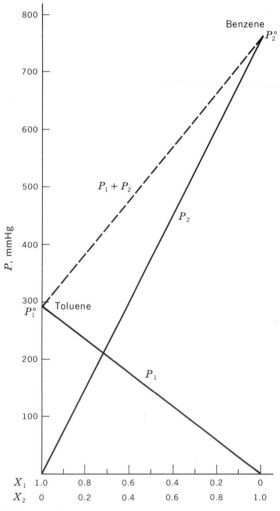

Fig. 11-7 Raoult's law graphs for the vapor pressures in ideal solution of benzene and toluene at 80.1°C.

$$P_a = P_a^\circ X_a = 22 \times 0.5 = 11 \text{ mm Hg}$$
$$P_b = P_b^\circ X_b = 74 \times 0.5 = 37 \text{ mm Hg}$$
$$P_{tot} = P_a + P_b = 11 + 37 = 48 \text{ mm Hg}$$

Example 11-5

If 2 moles of benzene and 2 moles of toluene are mixed together to form an ideal solution at 40°C, calculate the change in free energy for the system when this mixing takes place.

We assume that there will be no change in the enthalpy and that the effective volume for each constituent will be doubled. The entropy change is given by

$$\Delta S^M = nR \ln 2 = 5.51 \text{ eu}$$

Therefore, the free-energy change will be given by the expression:

$$\Delta G^M = -TS = -1725 \text{ cal}$$

Entropy and concentration

Two kinds of molecules normally form an ideal solution when mixed together if the attraction of each for its own kind is approximately the same as the attraction of one kind for the other. Since enthalpy depends primarily on the attractive forces operating on the molecules, there is very little change (if any) in enthalpy when molecules pass from a pure liquid into an ideal solution; i.e., there is no change of enthalpy with *concentration* in an ideal solution. If two different liquids form an ideal solution with each other, such a solution can be thought of as being formed by mixing portions of the two liquids together just as one can mix two ideal gases together to form an ideal-gas mixture. When ideal gases are mixed, entropy increases; the same kind of increase takes place when two liquids are mixed to form an ideal solution

$$\Delta S^M = -Rn_1 \ln \frac{n_1}{n_1 + n_2} - Rn_2 \ln \frac{n_2}{n_1 + n_2} \quad (11\text{-}29)$$

where ΔS^M is the increase in entropy due to the mixing and n_1 and n_2 are the numbers of moles respectively of components 1 and 2. The portion of the entropy of mixing per mole of component 1 is given by

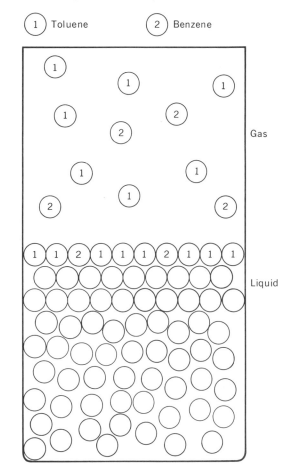

Fig. 11-8 Schematic cross section of liquid where $X_1 = 0.8$ and $X_2 = 0.2$. Because of the higher vapor pressure of benzene, the composition in the vapor is $^gX_1 = 0.6$ and $^gX_2 = 0.4$.

(1) Toluene   (2) Benzene

$$\Delta \bar{s}_1 = \bar{s}_1^\circ - s_1^\circ = -R \ln \frac{n_1}{n_1 + n_2} = -R \ln X_1 \quad (11\text{-}30)$$

A similar expression holds for component 2

$$\Delta \bar{s}_2 = \bar{s}_2 - s_2^\circ = -R \ln \frac{n_2}{n_1 + n_2} = -R \ln X_2 \quad (11\text{-}31)$$

The total entropy of mixing per mole of solution is

$$\Delta S^M = -X_1 R \ln X_1 - X_2 R \ln X_2 \quad (11\text{-}32)$$

so that

$$\Delta S^M = X_1 \Delta \bar{s}_1 + X_2 \Delta \bar{s}_2 \quad (11\text{-}33)$$

These three quantities are plotted in Fig. 11-9.

Fig. 11-9   The effect of mixing in an ideal solution on the total entropy of the solution ΔS^M, on the partial molal entropy of component 1, Δs_1, and on the partial molal entropy of component 2, Δs_2.

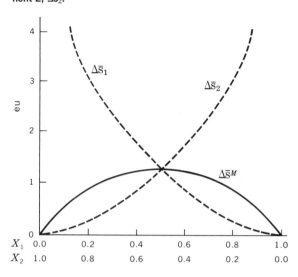

Consider first the partial molal entropy of component 1, $\Delta \bar{s}_1$, due to mixing. In a liquid consisting wholly of pure component 1, this quantity is zero as there is no disorder due to mixing in a pure liquid. As component 2 is added, the partial molal entropy due to the presence of component 1 in the solution increases, as shown by the rising curve as the value of $\Delta \bar{s}_1$ gets less in passing to the left. As the mole fraction X_1 approaches zero, the partial molal entropy goes to infinity, but the contribution to the entropy of the solution due to the entropy of component 1 is $X_1 \Delta \bar{s}_1$. Thus, although $\Delta \bar{s}_1$ approaches infinity, X_1 approaches zero even more rapidly. Thus the curve for the total entropy of mixing ΔS^M becomes zero both at the point where $X_1 = 0$ and where $X_1 = 1$.

Free energy and concentration

In line with the definition of free energy itself, the partial molal free energy is defined as the difference between the partial molal enthalpy and the partial molal entropy multiplied by the absolute temperature

$$\bar{g} \equiv \bar{h} - T\bar{s} \tag{11-34}$$

Thus for component 1 the change in the partial molal free energy $\Delta \bar{g}_1$ due to the formation of the solution is

$$\Delta \bar{g}_1 = \Delta \bar{h}_1 - T \Delta \bar{s}_1 \tag{11-35}$$

For an ideal solution $\Delta \bar{h}_1 = 0$. Substituting from Eq. (11-30), the expression for the change in free energy due to solution is

$$\Delta \bar{g}_1 = RT \ln X_1 \tag{11-36}$$

A similar equation is found for component 2

$$\Delta \bar{g}_2 = RT \ln X_2 \tag{11-37}$$

Thus the total change in free energy of the system due to the formation of the solution is

$$\Delta G^M = X_1 RT \ln X_1 + X_2 RT \ln X_2 \tag{11-38}$$

where ΔG^M is the change in free energy for the formation of 1 mole of solution in which the mole fractions of the components are X_1 and X_2. A plot of ΔG^M, $\Delta \bar{g}_1$, and $\Delta \bar{g}_2$ is shown in Fig. 11-10. This plot has the same character as the plot of ΔS^M since Eq. (11-38) closely resembles Eq. (11-33). Thus, for an ideal solution, the entropy change due to the formation of the solution has a maximum positive value for ΔS^M at $X_1 = 0.5$ and $X_2 = 0.5$, while the value of ΔG^M has a minimum value at the same point on the graph.

Substances are frequently precipitated out of solution by raising the concentration and put into solution by lowering the concentration. Substances come out of solution if the partial mole free energy in the solution is greater than the molal free energy of the substance in the crystalline state. Substances go into solution if the partial molal free energy in the solution is less than the molal free energy in the crystalline state. When substances are precipitated out, or brought into, solution through concentration changes, the action basically consists in changing the value of the partial molal entropy, which, in turn, changes the value of the partial molal free energy and thus brings about change of state in the substance.

In biological growth processes, the basic action is usually the change of a substance from a state that resembles a *solution* state, with greater randomness and higher entropy, to a *solid* state that resembles the crystalline state with a lower randomness and lower entropy. Growth is brought about by achieving the correct imbalance between the enthalpy and the entropy factor so that there is a negative change in the free energy when the process takes place. The enthalpy factor depends primarily on the structure of the molecule itself in relation to the chemical nature of the solvent molecules and is not easily changed. On the other hand, the entropy factor depends to a large extent on the concentration so that by controlling the concentration, the entropy factor is controlled and, in turn, the free-energy factor is controlled. For this reason, the dependence of both entropy and free energy on concentration is of the greatest importance in the study of biological processes.

Example 11-6

Calculate the increase in entropy due to mixing when 0.1 mole of benzene (1) is mixed with 0.9 mole of toluene (2) to form an ideal solution.

$$\Delta S^M = -0.1R \ln 0.1 - 0.9R \ln 0.9$$

$$= +(0.1)(1.987)(2.303) \log \frac{1}{0.1}$$

$$+(0.9)(1.987)(2.303) \log \frac{1}{0.9}$$

$$= (0.4576)(1) + (4.118)(0.0457)$$

$$= 0.4576 + 0.1882 = 0.6458 \text{ eu} \quad Ans.$$

Example 11-7

Calculate the decrease in free energy due to mixing in Example 11-6 if the temperature is 37°C.

$$\Delta G^M = \Delta H^M - T \Delta S^M = 0 - (310.15)(0.6458)$$
$$= -200.3 \text{ cal mole}^{-1} \quad Ans.$$

11-4   NONIDEAL SOLUTIONS

While there are many pairs of substances like benzene and toluene that resemble each other closely enough to form ideal solutions, the majority of all the possible pairs of different substances either form solutions which deviate considerably from the ideal or fail to be soluble in each other to any appreciable extent. In order to provide effective ways for applying thermodynamics to nonideal solutions, G. N. Lewis suggested modified forms of the equations for ideal solutions.

Fugacity and activity

One of the most basic equations valid both for ideal and nonideal systems expresses the relation between the change of free energy with pressure and volume

$$\left(\frac{\partial G}{\partial P}\right)_T = V \tag{11-39}$$

With the understanding that the temperature is constant, this equation can be written in the simpler notation of the calculus

$$dG = V\,dP \tag{11-40}$$

When this equation is applied to an ideal gas, where $PV = nRT$, it can be put in the form

$$dG = \frac{nRT}{P}\,dP \tag{11-41}$$

or in the equivalent logarithmic form:

$$dG = nRT\,d \ln P \tag{11-42}$$

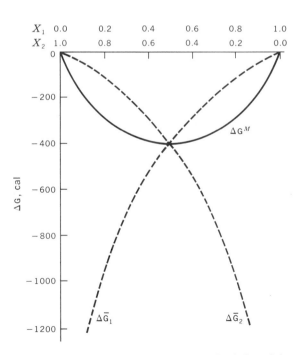

Fig. 11-10 The effect of mixing in an ideal solution of the total free energy per mole of the solution ΔG^M, on the partial molal free energy of component 1, $\Delta \bar{G}_1$, and on the partial molal free energy of component 2, $\Delta \bar{G}_2$.

In order to express the change in free energy when a system is changed from state 1 to state 2 at constant temperature this equation can be put in integral form

$$\int_1^2 dG = nRT \int_1^2 d \ln P \tag{11-43}$$

When the integration is carried out, this relation becomes

$$G_2 - G_1 = nRT \ln \frac{P_2}{P_1} \tag{11-44}$$

When the gas does not exhibit ideal behavior, this relation might be modified by adding correction terms to G. As a more convenient approach, G. N. Lewis suggested the use of a new quantity called *fugacity* as a substitute for the pressure. Fugacity f is defined so that the following relation is valid:

$$G_2 - G_1 = nRT \ln \frac{f_2}{f_1} \tag{11-45}$$

When applied to 1 mole of substance, this expression can be put in the form

$$\ln f_2 = \frac{G_2}{RT} + \left(\ln f_1 - \frac{G_1}{RT} \right) \tag{11-46}$$

State 1 can always be selected to provide a pressure so low that the gas behaves like an ideal gas, and therefore

$$P_1 = f_1 \tag{11-47}$$

Inserting this value for f_1 in Eq. (11-46) gives

$$\ln f_2 = \frac{G_2}{RT} + \lim_{P_1 \to 0} \left(\ln P_1 - \frac{G_1}{RT} \right) \tag{11-48}$$

This is the defining equation for fugacity. The second term on the right is a constant when the temperature is constant; thus, Eq. (11-48) can be differentiated

$$\left(\frac{\partial \ln f}{\partial G} \right)_T = \frac{1}{RT} \tag{11-49}$$

Because of the relation

$$\left(\frac{\partial G}{\partial P} \right)_T = v \tag{11-50}$$

there results

$$\left(\frac{\partial \ln f}{\partial P} \right)_T = \frac{v}{RT} \tag{11-51}$$

With the substitution of fugacity for pressure, all the thermodynamic equations applicable to ideal gases also apply to real gases.

Activity

In applying thermodynamics to nonideal solutions, Lewis proposed the use of the property called *activity*, which bears a relation to *concentration* comparable to the relation between fugacity and pressure. *Activity* is defined by relating concentration to partial molal free energy.

Recall the relation between free energy and pressure for an ideal gas

$$G_2 - G_1 = RT \ln \frac{P_2}{P_1} \tag{11-52}$$

Let P_2 correspond to the vapor pressure of a component in a solution at concentration X_2, and let P_1 correspond to the vapor pressure of the same component at concentration X_1. Since this component in the vapor state with free energy G_2 is in equilibrium with the same component in the liquid state with partial molal free energy $\bar{G}_2$, these two values of the free energy must be equal. The same relation holds between G_1 and $\bar{G}_1$ so that

$$G_1 = \bar{G}_1 \qquad G_2 = \bar{G}_2 \tag{11-53}$$

According to Raoult's law, obeyed by ideal solutions,

$$P_1 = P° X_1 \qquad P_2 = P° X_2 \tag{11-54}$$

Dividing the second equation by the first gives

$$\frac{P_2}{P_1} = \frac{X_2}{X_1} \tag{11-55}$$

Using Eqs. (11-53) and (11-55), Eq. (11-52) can be changed to

$$\bar{G}_2 - \bar{G}_1 = RT \ln \frac{X_2}{X_1} \tag{11-56}$$

Activity a is defined in the same manner as fugacity in terms of a change from a concentration so dilute that the solution behaves ideally to a state where the solution may be nonideal

$$a_2 = \frac{\bar{G}_2}{RT} + \lim_{X_1 \to 0} \left(\ln X_1 - \frac{\bar{G}_1}{RT} \right) \tag{11-57}$$

since at sufficiently high dilution, where $X_1 \to 0$, $X_1 = a_1$. When a is substituted for X, all the equations for ideal solutions can be applied to nonideal solutions.

The *activity coefficient* γ (gamma) is defined by the equation

$$\gamma \equiv \frac{a}{X} \tag{11-58}$$

Positive Raoult deviations

For solutions the deviations from ideal behavior fall into two classes: there are the solutions where the vapor pressure is higher than that calculated on the basis of Raoult's law, in other words, *positive* deviations, and there are the solutions where the vapor pressure is lower than that calculated from Raoult's law, in other words, *negative* deviations. Examples of positive deviations are shown in Fig. 11-11 for the solution system where component 1 is carbon disulfide, CS_2, and component 2 is acetone, CH_3COCH_3. The line labeled P_1 is the graph of the partial vapor pressure of carbon disulfide as its mole fraction is varied from 0 to 1. The point on the right-hand ordinate axis labeled P_1° denotes the vapor pressure of pure carbon disulfide. The dashed line connecting this point with the zero point on the left-hand axis denotes the vapor pressure carbon disulfide would have at the various values of X_1 if the behavior of the solution were ideal. In the same way, the line labeled P_2 denotes the observed partial vapor pressure for acetone over the same range of concentration. The point on the left-hand ordinate axis labeled P_2° denotes the vapor pressure of pure acetone. The dashed line connecting this point with the zero point at the right side of the diagram shows the partial vapor pressure acetone would have if the solution behaved ideally.

The curved line labeled $P_1 + P_2$ shows the observed total vapor pressure above the solution over this concentration range. The dashed line connecting P_2° and P_1° shows the total vapor pressure the solution would have over this concentration range if it behaved ideally.

There are several features of the curve for P_1 which should be noted. At the left side of the diagram, where very small amounts of CS_2 are present in almost pure acetone, the partial pressure of the carbon disulfide is roughly 3 times greater than it would be if the

Fig. 11-11 The observed partial pressures (solid lines) of carbon disulfide P_1 and of acetone P_2 combined in solution. The dashed lines show the corresponding pressures in ideal solutions. (After "Solubility of Nonelectrolytes," 3d ed., by J. H. Hildebrand and R. L. Scott. Copyright 1950 by Reinhold Publishing Corporation, New York. By permission of Van Nostrand Reinhold Company.)

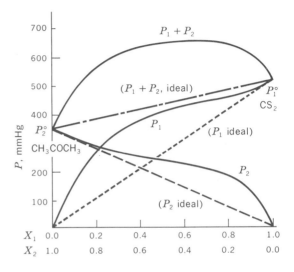

solution were ideal. The acetone molecules exert a much stronger force of attraction toward each other than they do toward the molecules of carbon disulfide. As a result, the CS_2 molecules are subjected to a force tending to move them out of the solution, superimposed on the normal tendency to escape from the solution. Consequently, the observed vapor pressure for CS_2 is higher than that calculated on the basis of ideal-solution behavior; the deviation from Raoult's law is positive. In the same way, at the right-hand side of the diagram, the vapor pressure of the small amount of acetone in almost pure carbon disulfide is larger than the vapor pressure calculated on the basis of Raoult's law. Contrasting with this, note that the graph of P_2 at the left side of the diagram becomes tangent to the dashed line for the ideal vapor pressure as X_2 approaches the value of unity. In other words, as a little carbon disulfide is added to pure acetone, the vapor pressure of the pure acetone is reduced in accord with Raoult's law. The same type of behavior is observed when a little acetone is introduced into pure carbon disulfide, as shown by the tangential approach of the graph of P_1 to the dashed line indicating the partial vapor pressure of CS_2 when Raoult's law is obeyed.

Negative Raoult deviations

Figure 11-12 shows the vapor-pressure lines for the solution system where chloroform, $CHCl_3$, is component 1 and acetone is component 2. In this system, the deviations from Raoult's law are negative. The graph of the partial vapor pressure P_1 of chloroform lies below the dashed line indicating the ideal partial vapor pressure. The same is true for the partial vapor pressure P_2 of acetone. Consequently, the graph of the total vapor pressure above the solution $(P_1 + P_2)$ lies below the line joining the point denoting the vapor pressure of pure acetone $P_2°$ and the point denoting the vapor pressure of pure chloroform $P_1°$ and showing the ideal total vapor pressure.

The same characteristics are observed for the graphs of partial vapor pressure that were seen in the case of positive Raoult deviations.

Henry's law

An extreme case of deviation from ideality is found in solution systems where one component is a liquid but the other component at the temperature of the system is a gas in its pure state. In 1803 the English chemist William Henry made a study of the dependence of solubility on pressure for the gaseous component in such systems. In such a case, the pressure of the gas above the solution in which it is dissolved may be regarded as the partial vapor pressure for the gaseous component. Henry found that instead of obeying Raoult's law $(P_1 = P_1°X_1)$, the relation between the partial vapor pressure and the mole fraction is given by

$$P_1 = {}^hkX_1 \qquad (11\text{-}59)$$

where ${}^hk$ is a constant different from the Raoult's law constant $P_1°$. In effect, Henry's law is an expression of the fact that the observed vapor pressure for the component present in very low concentrations can be approximated by a straight line having a slope different from that of the graph corresponding to Raoult's law. In Fig. 11-13 these relations are illustrated for the solution system carbon disulfide–acetone, which was discussed in the section dealing with positive deviations from Raoult's law. The dashed line is shown connecting the zero point at the left with the point at the right $P_1°$ denoting the pressure of pure carbon disulfide. Above this is the graph of the observed partial vapor pressure of carbon disulfide. Finally, the dashed line tangent to the curve of the observed partial vapor pressure at very low concentrations of carbon disulfide is the graph of Henry's law. The extrapolation of this line to the right-hand axis shows the vapor pressures which would be observed if Henry's law were obeyed over the entire range up to $X_1 = 1$. As can be seen from Eq. (11-59), when $X_1 = 1$, $P_1 = {}^hk$. In other words, ${}^hk$ is the value of the vapor pressure which pure carbon disulfide would have if the vapor pressure obeyed Henry's law instead of Raoult's law. Thus, Henry's law can be put into a form more closely resembling Raoult's law if the Henry's law constant ${}^hk$ is denoted by the symbol $P_1{}^h$

$$P_1 = P_1{}^hX_1 \qquad (11\text{-}60)$$

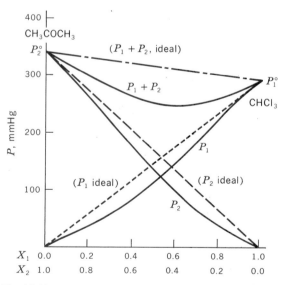

Fig. 11-12 The partial vapor pressures in solution of chloroform P_1 and acetone P_2. (After "The Solubility of Nonelectrolytes," 3d ed., by J. H. Hildebrand and R. L. Scott. Copyright 1950 by Reinhold Publishing Corporation, New York. By permission of Van Nostrand Reinhold Company.)

In Fig. 11-14, the position of the graph for P_1 (observed) shows that at very low concentrations the partial vapor pressure observes Henry's law; then as concentration is increased, the graph deviates from

Henry's law and finally at high concentrations approaches tangentially the line for Raoult's law.

Equation (11-60) shows that X_1, the mole fraction of the gas in solution, is directly proportional to the pressure of the gas above the solution P_1 since the term $P_1{}^h$ is a constant. This relationship is emphasized in the verbal statement of Henry's law:

The mole fraction of a gas in solution, at a fixed temperature, is directly proportional to the pressure of the gas above the solution.

Table 11-2 gives values of the amount of gas in cubic centimeters dissolved in 1 liter of water when the gas is under 1 atm of pressure and the water is at 0°C. In some cases, the attraction between the molecules of the gas and the molecules of the solvent is far greater than the attraction between the molecules of the gas for each other or even the attraction of the molecules of the solvent for each other; there may be such close association that one can speak of a compound being formed between the molecules of the gas and the molecules of the liquid solvent. This is true of ammonia dissolved in water. As seen in Table 11-2, the amount of ammonia dissolved is enormously greater than that of any of the other gases. Carbon dioxide also has an abnormally high solubility, a fact of considerable biological importance.

The solubility of gases in liquids can be expressed in various sets of units. One of the more useful is the *solubility coefficient* $\mathbb{S}$, defined as the ratio between the volume in the gaseous state of the solute and the volume of the solvent. If, in the expression of Henry's law, the mole fraction X_1 is replaced by the equivalent in terms of number of moles, Henry's law can be written

$$P_1 = P_1{}^h \frac{n_1}{n_1 + n_2} \qquad (11\text{-}61)$$

When n_1 is very small, this can be replaced by the approximate relation

$$P_1 = P_1{}^h \frac{n_1}{n_2} \qquad (11\text{-}62)$$

Thus, for a given amount of solvent ($n_2 = $ const), n_1 is proportional to P_1. If the number of moles of component 1, n_1, is measured in the gaseous state, the volume is given by

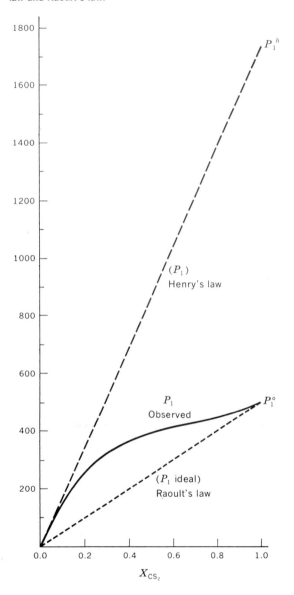

Fig. 11-13 The observed values of the partial pressure of CS_2 in solution with acetone showing the graphs of Henry's law and Raoult's law.

$$V_1 = n_1 \frac{RT}{P_1} \qquad (11\text{-}63)$$

when the gas behaves ideally. Substituting the value of n_1 from Eq. (11-62) in Eq. (11-63) results in

$$V_1 = \frac{n_2 RT}{P_1{}^h} \qquad (11\text{-}64)$$

Fig. 11-14 The observed values of the partial pressure of CS_2 in solution with $CHCL_3$, showing the graphs of Henry's law and of Raoult's law.

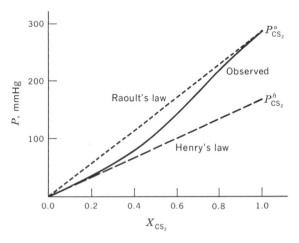

(11-65), the expression for the solubility constant becomes

$$S = \frac{1}{P_1{}^h} \frac{RT}{v_2} \tag{11-66}$$

Some values of S are given in Table 11-3.

Another quantity used to express the solubility of a gas is the *Bunsen absorption coefficient* usually designated by the symbol α and defined as the ratio of the volume (reduced to standard conditions) of the gas dissolved at 1 atm to the volume of the solvent. In symbolic form this definition is written

$$\alpha \equiv \frac{V_1^\circ}{V_2} \tag{11-67}$$

where the superscript denotes standard conditions. Because of the relation

$$\frac{V_1}{V_1^\circ} = \frac{T_1}{273.15^\circ} \tag{11-68}$$

the relation between S and α is

$$S = \alpha \frac{T_1}{273.15^\circ} \tag{11-69}$$

Values of α are given in Table 11-4.

Example 11-8

For the system chloroform-acetone, estimate from the diagram the Henry's law constants.

Extrapolate the tangent to each curve; the intercepts show that $P_a{}^h = 150$ mm Hg and that $P_b{}^h = 175$ mm Hg, where a denotes chloroform.

Example 11-9

Using the values for the Henry's law constants, calculate the vapor pressure of chloroform when its mole fraction is equal to 0.2 and the vapor pressure of acetone when its mole fraction is equal to 0.1.

The equations for Henry's law are

$$P_a = P_a{}^h X_a = 150 \times 0.2 = 30 \text{ mm Hg}$$
$$P_b = P_b{}^h X_b = 175 \times 0.1 = 18 \text{ mm Hg}$$

Example 11-10

Calculate the volume of carbon dioxide that will dissolve in 1 liter of water at 37°C.

TABLE 11-2 Amount of gas at $P = 1$ atm dissolved in 1 liter of H_2O at 0°C[a]

Gas	Amount, cm³
He	14.87
H_2	21
N_2	23.9
CO	35.4
O_2	48.9
CO_2	1,713
NH_3	1.3×10^6

[a] From J. P. Amsden, "Physical Chemistry for Premedical Students," 2d ed. Copyright 1950. McGraw-Hill Book Company. Used by permission.

Thus, the volume of the gas dissolved V_1 is independent of the pressure and depends only on the amount of solvent n_2, Henry's law constant $P_1{}^h$, the gas constant R, and the temperature T in degrees Kelvin. The definition of the solubility coefficient can be written

$$S = \frac{V_1}{V_2} = \frac{V_1}{n_2 v_2} \tag{11-65}$$

where v_2 is the volume per mole of the solvent. If the value of n_2 derived from Eq. (11-64) is inserted in

According to the values of the solubility coefficient in Table 11-2, there is a decrease of 158 cm^3 in the amount of CO_2 dissolving in water as temperature is changed from 30 to 40°C. Therefore, the decrease in the value in going from 30 to 37°C will be $158 \times 0.7 = 111$ cm^3. Subtracting this from the solubility coefficient at 30°C gives a solubility coefficient of 655 at 37°C. Therefore, the volume of CO_2 dissolved in 1 liter of water at this temperature will be 655 cm^3.

Gas solubility in biological processes

The solubility of oxygen and carbon dioxide in the bloodstream plays a most important role in a number of biological processes. Oxygen from the blood provides a source of energy when it is transported to various parts of the body and undergoes combination with other substances through oxidation processes. Carbon dioxide is formed, as well as the many products of this "combustion," and must ultimately be eliminated from the bloodstream. If it were necessary to get the supply of oxygen to all the places it is needed merely through the normal solubility of oxygen and water, body chemistry would be far different from what it is. Actually, there is a large negative deviation from Raoult's law when oxygen goes into solution in the blood. Because of this negative deviation, the vapor pressure of oxygen in the blood is far less than would be expected if Raoult's law were obeyed and the solubility is correspondingly greater. The reason is that the blood contains *hemoglobin,* which effectively forms a compound with oxygen, thus increasing its solubility. This compound formation can be prevented by the presence of other gases, e.g., carbon monoxide, CO, and hydrogen cyanide, HCN. In the presence of these gases, the negative Raoult deviation is greatly reduced and the organism quickly dies from lack of oxygen.

Dissolved carbon dioxide affects the mechanism of sexual reproduction of hydra. The hydra is a tiny water animal commonly found in ponds and equipped with tentacles surrounding its mouth, by which it catches water fleas and similar organisms for food. In fresh water, where the concentration of CO_2 is low, the hydra flourishes by asexual reproduction, in which a bud appears, grows into a side tube, puts out ten-

TABLE 11-3	Solubility coefficient S for CO_2[a]
t, °C	S
0	1.713
10	1.238
20	0.943
30	0.766
40	0.608

[a] From J. P. Amsden, "Physical Chemistry for Premedical Students," 2d ed. Copyright 1950. McGraw-Hill Book Company. Used by permission.

TABLE 11-4	The Bunsen absorption coefficient[a]			
t, °C	CO	CO_2	O_2	N_2
0	0.0354	1.713	0.0489	0.0239
10	0.0282	1.194	0.0390	0.0196
20	0.0232	0.878	0.0310	0.0164
30	0.0200	0.665	0.0261	0.0138
40	0.0178	0.530	0.0231	0.0118

[a] From J. P. Amsden, "Physical Chemistry for Premedical Students," 2d ed. Copyright 1950. McGraw-Hill Book Company. Used by permission.

tacles, pinches off, and begins life on its own. By way of contrast, in stagnant water, where there is a high content of dissolved CO_2 gas, the hydra develops differentiated reproductive organs and reproduces sexually. The chemical mechanism of this effect is not known.

11-5 FRACTIONAL DISTILLATION

Because in almost all solutions formed from liquid components the vapor pressures of the components differ from each other at any temperature, one component can be separated out by evaporation, a process frequently called *fractional distillation.* The underlying reason for the possibility of fractional distillation can be seen by examining the diagram for a typical system obeying Raoult's law. Figure 11-15 shows the total vapor pressure plotted against composition in

Fig. 11-15 **The vapor pressure of a benzene-toluene mixture** at 20°C plotted against the mole fraction of benzene in the liquid and against the mole fraction of benzene in the vapor state. [Data from R. Bell and T. Wright, *J. Phys. Chem.*, 31:1884 (1927).]

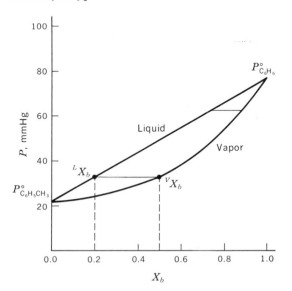

mole fractions for the binary system toluene-benzene. As noted in the section on Raoult's law, the graph of the total vapor pressure of the system is a straight line connecting the point on the left which shows the vapor pressure of pure toluene and the point on the right which shows the vapor pressure of pure benzene at the temperature under consideration. Denoting toluene by the subscript t and benzene by the subscript b, the expression for Raoult's law can be written

$$P_t = P_t^\circ X_t \qquad P_b = P_b^\circ X_b \qquad (11\text{-}70)$$

The equations for the composition of the vapor are

$$^vX_t = \frac{P_t}{P_{\text{tot}}} \qquad ^vX_b = \frac{P_b}{P_{\text{tot}}} \qquad (11\text{-}71)$$

where ^vX_t and ^vX_b respectively denote the mole fraction of toluene and benzene in the vapor. Using these equations, it is possible to plot the graph of the total vapor pressure of the solution as a function not of the composition of the solution but of the composition of the vapor. This graph is the lower curved line in Fig. 11-15 labeled vapor.

Consider a solution where $^vX_t = 0.5$ and $^vX_b = 0.5$.

The point on the vapor curve corresponding to this composition of solution lies above 0.5 on the scale of X_b. The total pressure corresponds to 34 mm Hg. The composition of the liquid which has this vapor pressure is given by the point on the liquid curve at the height corresponding to 34 mm Hg. This point lies at 0.2 on the scale of X_b. Thus, a solution in which the mole fraction of benzene is 0.2 is in equilibrium with vapor where the mole fraction of benzene is 0.5.

When the benzene-toluene mixture is distilled, far more benzene passes from the liquid into the vapor than toluene. Consequently, the vapor becomes richer in benzene, and the liquid left behind becomes richer in toluene. By condensing the vapor and re-evaporating it several times, almost complete separation can be effected. An apparatus for such a process is shown in Fig. 11-16.

For a pure liquid, the temperature at which the total vapor pressure of a solution becomes equal to 1 atm is the normal boiling point of the solution. If the normal boiling point of a solution is plotted against the composition of the solution, the result is the so-called *boiling-point diagram;* this type of diagram is plotted in Fig. 11-17 for the system toluene-benzene.

Azeotropes

If a solution does not show ideal behavior, the boiling-point diagram becomes somewhat more complicated than that shown in Fig. 11-17. For example, the system acetone-chloroform shows negative Raoult deviations (Fig. 11-12), and the boiling point takes the form shown in Fig. 11-18. Consider a liquid where the mole fraction for chloroform is 0.4. This solution has a normal boiling point of 62.4°C. At this temperature the mole fraction of chloroform in the vapor is only 0.3. Therefore, as the liquid boils, it loses more acetone than chloroform and gets richer in chloroform. As a result, the temperature of boiling rises and the composition of the liquid shifts to the right until it reaches the value 0.68. As shown by the diagram, the composition of the vapor and the liquid is the same when the liquid boils at this temperature, the maximum on both the liquid and the vapor curve. Therefore, as the liquid boils at this temperature, there is no change in its composition and the boiling point remains constant. A solution which boils at a constant temperature without change of composition is called *azeotrope*. The components of an azeotrope cannot be separated from each other by simple fractional distillation.

11-6 PARTIAL MISCIBILITY

Having seen the various degrees to which solutions deviate from the ideal in their behavior, it is logical to ask about the nature of any limit on the possible degree of deviation. The answer to this question is to be found by considering the relations between the values of the partial molal free energy and the concentration. Consider a solution made up of n_1 moles of component 1 and n_2 moles of component 2. If dn_1 moles are added to the solution,

$$d\bar{G}_1 = \left(\frac{\partial \bar{G}_1}{\partial n_1}\right)_{T,P} dn_1 \qquad (11\text{-}72)$$

The composition of the solution can be changed in an equivalent way by adding simultaneously dn_1 and dn_2 such that $dn_1/dn_2 = n_1/n_2$. This leaves the composition of the solution unchanged, and therefore there is no change in either $\bar{G}_1$ or $\bar{G}_2$. Then dn_2 moles can be removed from the solution by distillation, and the composition will be the same as when dn_1 mole was added as in the first process considered above. The change in $\bar{G}_1$ is given by

$$d\bar{G}_1 = \frac{\partial \bar{G}_1}{\partial n_2} dn_2 \qquad (11\text{-}73)$$

Since a solution of the same composition is arrived at by either of these alternative routes,

$$dn_1\left(\frac{\partial \bar{G}_1}{\partial n_1}\right)_{P,T,X_1} = -dn_2\left(\frac{\partial \bar{G}_1}{\partial n_2}\right)_{P,T,X_1} \qquad (11\text{-}74)$$

But from the calculus, it is clear that

$$\left(\frac{\partial \bar{G}_1}{\partial n_2}\right)_T = \left(\frac{\partial^2 G}{\partial n_1 \, \partial n_2}\right)_T = \left(\frac{\partial \bar{G}_2}{\partial n_1}\right)_T \qquad (11\text{-}75)$$

Combining Eqs. (11-72), (11-74), and (11-75) gives

$$n_1\left(\frac{\partial \bar{G}_1}{\partial n_1}\right)_T + n_2\left(\frac{\partial \bar{G}_2}{\partial n_1}\right)_T = 0 \qquad (11\text{-}76)$$

A similar expression can be found with respect to a change of n_2

$$n_1\left(\frac{\partial \bar{G}_1}{\partial n_2}\right)_T + n_2\left(\frac{\partial \bar{G}_2}{\partial n_2}\right)_T = 0 \qquad (11\text{-}77)$$

If the independent variable is omitted, the equations can be written

$$n_1 \, d\bar{G}_1 + n_2 \, d\bar{G}_2 = 0 \qquad (11\text{-}78)$$

This equation was first derived by Willard Gibbs but is frequently attributed either to P. Duhem or to M. Margules. By dividing both terms by $n_1 + n_2$, the equation becomes

$$X_1 \, d\bar{G}_1 + X_2 \, d\bar{G}_2 = 0 \qquad (11\text{-}79)$$

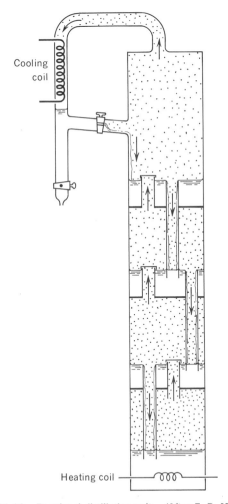

Fig. 11-16 Fractional distillation unit. (After E. B. Millard, "Physical Chemistry for Colleges," McGraw-Hill Book Company, New York, 1953.)

Fig. 11-17   The boiling-point diagram for the system toluene-benzene (almost ideal).

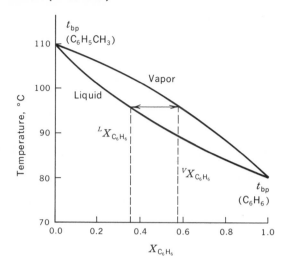

From the relation

$$\overline{G} - G° = RT \ln a \qquad (11\text{-}80)$$

it follows that

$$d\overline{G} = RT\, d \ln a \qquad (11\text{-}81)$$

Consequently, Eq. (11-79) can be written

$$X_1\, d \ln a_1 + X_2\, d \ln a_2 = 0 \qquad (11\text{-}82)$$

In a binary solution there is always the relation

$$X_1 = -X_2 \qquad (11\text{-}83)$$

so that

$$dX_1 = -dX_2 \qquad (11\text{-}84)$$

and

$$d \ln X_1 = -d \ln X_2 \qquad (11\text{-}85)$$

This leads to

$$\frac{\partial \ln a_1}{\partial \ln X_1} = \frac{\partial \ln a_2}{\partial \ln X_2} \qquad (11\text{-}86)$$

If Raoult's law is valid over the whole range of composition for the vapor pressure of one component of a binary solution, it must hold for the other component. When Raoult's law is valid, a_1 equals X_1 and, consequently,

$$\frac{\partial \ln a_1}{\partial \ln X_1} = 1 \qquad (11\text{-}87)$$

and this, according to Eq. (11-86), results in

$$\frac{\partial \ln a_2}{\partial \ln X_2} = 1 \qquad (11\text{-}88)$$

showing that Raoult's law also holds for the vapor pressure of component 2.

When Henry's law holds for the vapor pressure of component 2,

$$a_2 = P_2{}^h X_2 \qquad (11\text{-}89)$$

Consequently,

$$\frac{\partial \ln a_2}{\partial \ln X_2} = 1 \qquad (11\text{-}90)$$

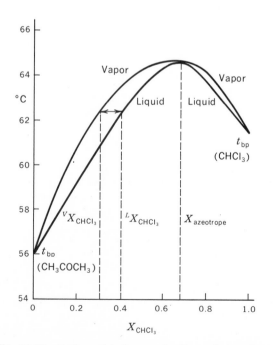

Fig. 11-18   Boiling-point diagram for the system acetone-chloroform.   (Data from "International Critical Tables," McGraw-Hill Book Company, New York, 1926–1930.)

Then, according to Eq. (11-86), Eq. (11-87) must be valid, and Raoult's law is approached as an approximation by the vapor pressure of component 2.

Because of the relations demonstrated above between the natural logarithm of the activity and the natural logarithm of the mole fraction for the two components, Margules suggested that the activity coefficient ($\gamma = a/X$) might be expressed

$$\ln \gamma_1 = -JX_2^2 \qquad \log \gamma_1 = -J'X_2^2 \qquad (11\text{-}91)$$

$$\ln \gamma_2 = -JX_1^2 \qquad \log \gamma_2 = -J'X_1^2 \qquad (11\text{-}92)$$

When $J = 0$, Raoult's law is obeyed. As J takes on larger and larger positive values, there are larger and larger positive deviations from Raoult's law. In Fig. 11-19 graphs are shown for the changes of vapor pressure with composition for $J' = 0.8$ and $J' = 1.6$. Assuming ideal behavior for the gas, $\gamma = (P_1)_{obs}/(P_1)_{ideal}$.

At the larger values for J, two points in the range of composition have the same partial vapor pressure for component 1. On the graph for $J' = 1.6$ this is true for $X_1 = 0.04$ and $X_1 = 0.92$. When this situation is found, the system separates into two liquid phases, one having the composition corresponding to the point on the left of the diagram and the other having the composition corresponding to the point on the right of the diagram. When two liquids are not completely miscible but form two liquid phases when shaken together, the phenomenon is known as *partial miscibility*.

The change from one phase to two phases discussed above is very similar to the change from one phase to two phases when a gaseous system represented by a point on the PV diagram just above the critical point is cooled just below the critical temperature. Both for the appearance of the two liquid phases of a solution and for the appearance of the two phases (liquid and gas) in an originally gaseous system, the isotherm develops the characteristic S shape which indicates that two different states of the system have the same molal free energy.

It is instructive to consider this phenomenon in solutions as temperature is varied, with a corresponding variation in the value of J. For example, a mixture of phenol, C_6H_5OH, and water shows partial miscibility over a considerable range of temperature. In Fig. 11-20 the miscibility curve is plotted with the weight percent of phenol as the abscissa and the temperature as the ordinate. At any given temperature, a mixture of the composition shown on the left branch of the curve is in equilibrium with the mixture of the com-

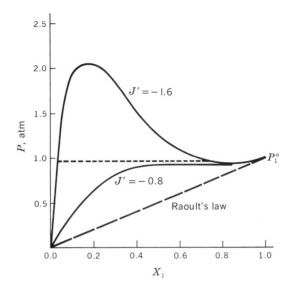

Fig. 11-19 The graph of deviations from Raoult's law following Margules's equation: $\ln \gamma_1 = -J'X_2^2$, $\gamma_1 = (P_1)_{obs}/(P_1)_{ideal}$.

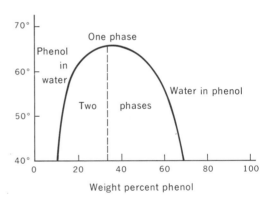

Fig. 11-20 Two-phase system phenol-water, showing zone of partial miscibility. (After J. P. Amsden, "Physical Chemistry for Premedical Students," McGraw-Hill Book Company, New York, 1959.)

position shown on the right branch of the curve at the same temperature. As temperature rises, the liquids become more and more miscible, and finally, at a temperature of 65.9°C, the boundary between the two liquid phases disappears and the liquids are completely miscible.

It is frequently possible to separate impurities from a liquid by shaking it with another liquid with which it is only partially miscible when a second liquid can be found in which the impurities are more soluble than in the original liquid.   Thus, the impurities are transferred from one liquid phase to the other as the mixture is shaken, and the original liquid is purified.

Partial miscibility generally exists between liquids like water on the one hand, where there is a possibility for strong hydrogen bonding involving either of the hydrogen atoms, and liquids like phenol, where in addition to the presence of a group, —OH, which makes possible hydrogen bonding to H_2O, there is also a group, C_6H_5—, which has little or no affinity for the water molecules.   Many of the substances transported by blood are organic molecules which have radicals consisting of carbon atoms surrounded by hydrogen atoms which by themselves would have little if any solubility in water.   However, most of these radicals are attached to more highly polar groups like NH_2 or OH, which have a strong affinity for water, so that it is possible to get small amounts of these molecules into solution, amounts sufficient for them to be transported to the various places in the organism where they are needed for biochemical processes.   Thus, the phenom-

enon of limited solubility and the related phenomena of partial miscibility are important for understanding the nature of many biological processes involving such kinds of molecules.

11-7   LIQUID-SOLID EQUILIBRIA

Especially in biological processes, there are many important examples of physical equilibrium between a solution and one of its components in the solid or crystalline phase.   We consider first a system consisting of two pure substances which are completely miscible in the liquid state forming an essentially ideal solution which is in equilibrium with one or more of these components in the crystalline state.   An example of such a binary system is the solution consisting of o-dinitrobenzene, $o\text{-}C_6H_4(NO_2)_2$,   and m-dinitrobenzene, $m\text{-}C_6H_4(NO_2)_2$, which may be in equilibrium with either or both of the components in the crystalline phase. For purposes of brevity, the first of these substances will be referred to simply as ortho and the second as

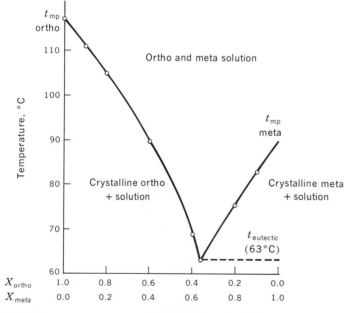

Fig. 11-21   The composition in mole fraction and the temperature at which crystalline o-dinitrobenzene (*left*) and crystalline m-dinitrobenzene (*right*) are in equilibrium with a solution of ortho and meta.

Fig. 11-22   The binary freezing-point system: o- and m-dinitrobenzene.

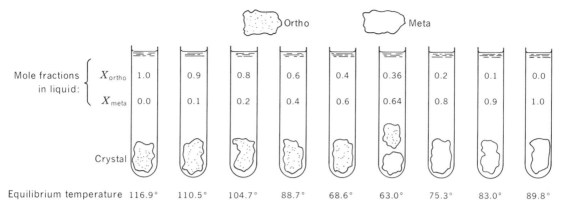

Mole fractions in liquid:	X_{ortho}	1.0	0.9	0.8	0.6	0.4	0.36	0.2	0.1	0.0
	X_{meta}	0.0	0.1	0.2	0.4	0.6	0.64	0.8	0.9	1.0

Equilibrium temperature 116.9°   110.5°   104.7°   88.7°   68.6°   63.0°   75.3°   83.0°   89.8°

meta; the mole fraction of the first will be denoted by the symbol X_o and of the second by the symbol X_m.

Figure 11-21 shows the temperatures at which crystalline ortho and crystalline meta are in equilibrium with a liquid where composition varies from $X_o = 1$ to $X_o = 0$. In order to make clear the physical meaning of the different points on this diagram, Fig. 11-22 shows a series of test tubes containing liquids of varying mole fractions of the two components and also crystals of one or both of the two components, the tubes corresponding to points on the diagram in Fig. 11-21. Pure ortho melts at 116.9°C, so that a liquid with the composition $X_o = 1$ is in equilibrium with crystalline ortho at this temperature, as shown in the tube on the left. When meta is added to the solution, so that X_o is reduced to 0.9, the temperature of the equilibrium between the solution and crystalline ortho falls to 110.5°C. As more and more meta is added, the temperature of equilibrium continues to fall, until at 63.0°C the solution is in equilibrium with both crystalline ortho and crystalline meta. Then, as still more meta is added, the solution no longer is in equilibrium with crystalline ortho but only with crystalline meta. When the mole fraction of meta rises to the point where $X_m = 1$, the pure liquid meta is in equilibrium with crystalline meta at 89.8°C.

This relation of the temperature of equilibrium between liquid and crystal to the composition of the liquid can be shown on a diagram like that in Fig. 11-23, where free energy G is plotted against temperature t. At the far right is the intersection of the free-energy curve for crystalline ortho and liquid ortho, at the temperature of 116.9°C. In the same way that the addition of a second component lowers the vapor pressure, it also lowers the partial molal free energy of the first component in the solution. When the mole fraction of ortho falls to 0.8, the curve for the partial molal free energy of ortho is displaced downward, so that it now intersects the curve for crystalline ortho at a temperature of 104.7°C. If X_o drops to 0.6, the intersection falls to 88.7°C.

The relation between X_o and the temperature of liquid-crystal equilibrium in degrees Kelvin can be expressed by an equation that is strictly analogous to the equation for the variation of vapor pressure with temperature (for derivation see Appendix):

$$\log X_o = -\frac{\Delta° H_f}{2.303R}\left(\frac{1}{T} - \frac{1}{°T_f}\right) \qquad (11\text{-}93)$$

where $\Delta° H_f$ is the heat of fusion of o-dinitrobenzene and $°T_f$ is the melting point. Just as a plot of the logarithm of the vapor pressure against reciprocal temperature gives a straight line, the plot of the logarithm of the mole fraction against the reciprocal temperature for equilibrium between the solution and the crystalline phase is also a straight line. In Fig. 11-24 the linear graphs for equilibrium with crystalline ortho and for equilibrium with crystalline meta are shown calculated from Eq. (11-93) and the similar equation applicable to meta. Experimentally observed temperatures of equilibrium are also plotted, showing how closely this system obeys the thermodynamic equation for an ideal

Fig. 11-23   The free energy of o-dinitrobenzene in the crystalline state, in pure liquid, and in solution where $X_o = 0.8$ and where $X_o = 0.6$ (G values estimated).

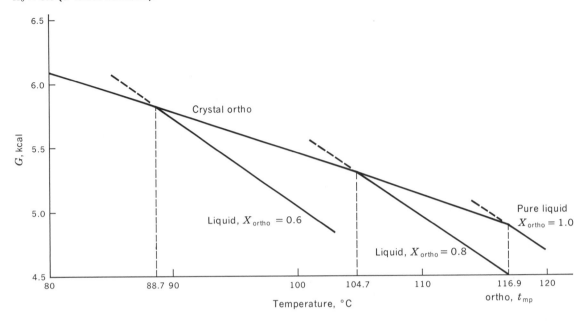

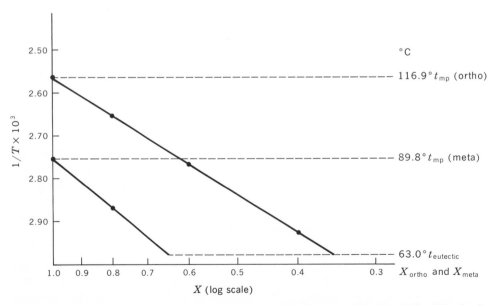

Fig. 11-24   The binary system o-dinitrobenzene and m-dinitrobenzene.   Experimentally determined values are shown as small solid circles.   The straight lines show values calculated from Eq. 11-93.

solution. The point at which the solution is in equilibrium simultaneously with both crystalline ortho and crystalline meta is called the *eutectic point.*

Time-temperature curves

Phase diagrams of the type shown in Fig. 11-21 are generally obtained by preparing a series of solutions of the compounds to be studied at different concentrations and extracting heat from these solutions at a uniform rate while observing the time rate of change of temperature. An apparatus for this purpose (Fig. 11-25) consists of a series of vessels, shown in cross section. The outer vessel is a typical double-walled, cylindrical Dewar flask E with a strip on both sides which is left unsilvered so that the interior of the flask can be observed. The top is closed by two cork disks C_1 and C_2 which support the various tubes leading into the interior of the apparatus. The Dewar is filled about half full with a light paraffin oil. A U-shaped glass tube D goes to the bottom of the Dewar and is wound with a coil of nichrome wire through which an electric current can be passed to heat the interior of the Dewar; a stream of cold air can also be circulated through this tube for cooling. Through the central cork, a glass tube A leads down to the center of the Dewar. From the bottom of this tube is suspended a small test tube B which holds the sample of the solution to be studied. Within this tube there is a still smaller tube M containing one junction of a thermocouple, the wires of which are shown protruding at the top T_1. With the help of this thermocouple junction the temperature of the solution can be determined accurately and the tube can be used for stirring the solution as required. Another thermocouple T_2 provides a means for measuring the temperature difference between the outer surface of the sample tube B and the Dewar fluid. A third thermocouple T_3 measures the temperature of the Dewar fluid. The sample tube is surrounded by an air jacket inside the shield tube S supported from the cork C_3, and can be heated by passing hot air through the tube H.

From a knowledge of the melting point of a substance, one can predict roughly the shape of the curves in Fig. 11-21. The first step in determining the precise shape of these curves is the determination of accurate values of the freezing temperatures of the components

Fig. 11-25 **Apparatus for obtaining time-temperature curves for the plotting of phase diagrams.**

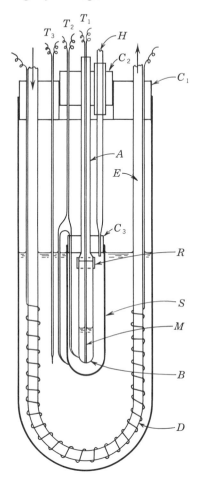

in their pure state. For example, a sample of pure ortho can be placed in the sample tube and the temperature of the Dewar liquid can be lowered to a point about 20° below the freezing point of the sample. By controlling the temperature of the Dewar fluid, a constant temperature gradient can be maintained between the sample and its surroundings so that heat is extracted from the sample at a constant rate. Readings of the temperature of the sample taken at equally spaced time intervals provide data for a plot like that shown in Fig. 11-26. At the beginning of the experiment the sample is completely liquid; the temperature falls at a constant rate, so that the graph of the dia-

Fig. 11-26   Time-temperature curve for pure o-dinitrobenzene.

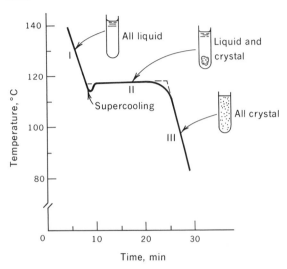

gram at the point (I) is a straight line. The temperature continues to fall until the liquid is a degree or two below the freezing point in the supercooled state. Crystallization then can be induced either by scratching the side of the sample tube with the thermocouple tube or by inserting a small bit of crystal into the liquid through the tube containing the thermocouple T_1. As shown on the diagram, temperature then rises because of the heat liberated by the process of crystallization. Crystallization continues at a steady rate (II) such that the heat produced by crystallization per unit of time exactly equals the heat withdrawn through external cooling from the system per unit of time. Meanwhile, the rate of extraction of heat from the system is maintained constant by keeping the temperature difference between the sample and its surroundings at a constant value. As the sample becomes almost completely crystalline, the crystallization process takes place at a slower rate so that it cannot compensate completely for the heat withdrawn from the system. The temperature falls, and, finally, when the system is completely crystalline, the graph of temperature against time once more becomes a straight line (III); the slope of this line is somewhat steeper than the slope of the initial portion (I) because the heat capacity of the crystal is less than that of the liquid.

If there were no supercooling, the ideal curve would follow the path shown by the dotted lines instead of showing the slight dip due to supercooling at the beginning of crystallization and the rounded shoulder between parts II and III. In theory, for a pure liquid, part II will be precisely horizontal because the temperature remains constant as long as there is true equilibrium between the crystalline phase and the liquid phase. Freezing points can be determined with an accuracy of 0.001 to 0.0001°C without too great difficulty; but unless the liquid is carefully freed from any dissolved gases, there will be a slight lowering of the freezing point due to the presence of the gaseous components. In the careful studies made by H. G. Leopold and H. W. Foote to determine with precision the reference point for the centigrade scale temperature at 0°C, it was found that if liquid water is exposed to the atmosphere at a pressure of 1 atm, the gases dissolved in the water lower the temperature appreciably. This phenomenon is in large part responsible for the difference between the absolute temperature for 0°C, which is 273.15°K, and the temperature of the triple point for water which is 273.16°C.

Figure 11-27 shows a similar time-temperature curve observed with a sample of liquid which has the composition $X_o = 0.8$ and $X_m = 0.2$. If the size of the sample is the same as that used in determining the curve shown in Fig. 11-26, the slope of the curve in section I will be the same as in the previous figure, since the heat capacity of meta is only slightly different from the heat capacity of ortho. When the temperature reaches that of the true equilibrium between crystalline ortho and liquid with the composition $X_o = 0.8$, then if there is no supercooling, the graph shows an abrupt break; but instead of displaying a horizontal portion like II in Fig. 11-26, the graph immediately starts to fall. As ortho begins to crystallize out, the mole fraction of ortho begins to decrease. Consequently, the point on the corresponding phase diagram in Fig. 11-28, representing the state of the system, begins to fall along the graph to lower temperatures. If there is supercooling, the graph of time against temperature dips below the equilibrium temperature. When crystallization is induced, the temperature rises momentarily and then starts falling again on the line that is characteristic of the changing composition of the liquid. For this reason, it is necessary to extrap-

olate back to the intersection with the time-temperature curve for the liquid range (dotted line) in order to
obtain the true temperature for the equilibrium of crystalline ortho with solution at composition $X_o = 0.8$.

Eutectics

As shown on Fig. 11-28, the curve representing equilibrium between the liquid phase and crystalline ortho
intersects the curve representing equilibrium between
the liquid phase and crystalline meta at the temperature of 63.0°C and the composition $X_o = 0.36$. This
temperature is called the *eutectic temperature,* and
the composition is called the *eutectic composition.*
The intersection is called the *eutectic* of the system.
In theory, as the temperature of a system is reduced
with a constant rate of extraction of heat, both crystalline phases will start to separate out simultaneously
at the eutectic temperature. If the system were cooled
slowly enough, one might expect the rate of separation
of ortho and the rate of separation of meta to be the
same so that the composition of the system would not
change during this dual crystallization. If this were
the case, temperature should remain constant during
this crystallization since there is no change in composition (III). In practice, many phenomena associated
with supercooling are found as heat is extracted from
systems at the eutectic point so that the actual time-
temperature curves are not ideal in shape. When the
system is completely crystalline, the temperature falls
once more at a rate determined by the rate of extraction of heat and the heat capacity of the system (IV).

Example 11-11

In the ideal solution formed with o- and m-dini-
trobenzene, calculate the lowering of the temperature when sufficient m-dinitrobenzene is
added to o-dinitrobenzene to make the mole
fraction of the latter equal to 0.8; the enthalpy
of fusion of o-dinitrobenzene is 5460 cal mole^{-1}.

The formula relating mole fraction and temperature is

$$\log X_1 = \frac{-\Delta H_f}{2.303R}\left(\frac{1}{T} - \frac{1}{T_f}\right)$$

The lowering for the melting point is given in reciprocal degrees absolute by the formula

$$\frac{1}{T_f} - \frac{1}{T} = \frac{\log X}{\Delta H_f/2.303R}$$

Putting in numerical values,

$$\frac{1}{T_f} - \frac{1}{T} = \frac{-0.0969}{5,460/4.58} = 0.0000813$$

Adding this quantity to $1/T_f$ gives

$$0.0000813 + 0.0025634 = 0.0026447$$

This is the value of $1/T$, where T is the temperature in degrees Kelvin when liquid of the composition $X_1 = 0.8$ is in equilibrium with crystalline ortho. This temperature is 104.9°C.
Subtracting it from the melting temperature of
pure ortho (116.9°) shows that this mole fraction of ortho theoretically should produce a
freezing-point lowering of 12.0°C.

Compound formation

Occasionally the two components of a system unite to
form a compound of definite chemical composition

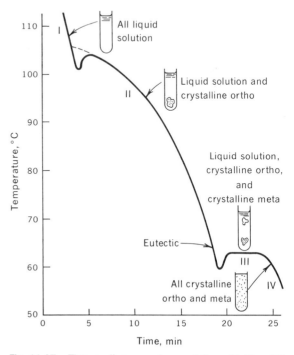

Fig. 11-27 Time-cooling curve for a solution with $X_o = 0.8$ and $X_m = 0.2$.

Fig. 11-28   Path on binary diagram followed by substance in cooling shown in Figs. 11-22 and 11-27.

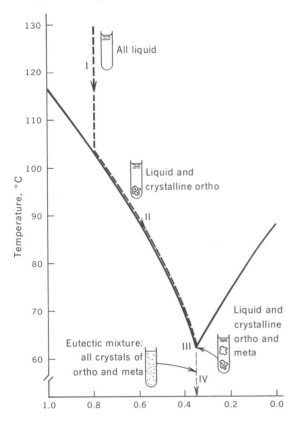

stances, and phase diagrams have been recorded where whole series of compounds appear.

Solid solutions

In the discussion in the previous paragraphs the assumption has been made that when one of the components appears in the crystalline phase, the crystals are made up solely of this component; they are pure crystals. In contrast, there are many phase systems where the two components resemble each other closely enough so that the molecules of one may partially replace the molecules of the other in the crystal lattice. Crystals which contain variable amounts of more than one component are called *solid solutions*.

In the system bismuth-tin, bismuth metal can dissolve to a limited extent in the crystals of tin, replacing the atoms of tin in the crystal lattice. The phase diagram for this system is shown in Fig. 11-30. Consider a liquid for which $X_{Bi} = 0.20$. As this system in the liquid state is cooled and drops to the temperature of 190°C, crystals begin to separate out. In a normal binary phase system these would be crystals of pure tin; but because bismuth is soluble in tin, these crystals have the composition denoted by the point where the horizontal dashed line drawn from point I on the liquid curve intersects the curve on the left marked "solid." At each temperature, this solid curve indicates the composition of crystals which will be in equilibrium with the liquid having the composition indicated by the liquid curve at this temperature.

Such a system in many ways resembles the two-phase liquid system in which a phenol containing a little water is in equilibrium with a water phase containing a little phenol. Here a crystalline phase of tin containing a little bismuth is in equilibrium with a liquid phase containing tin and a much higher concentration of bismuth. A knowledge of the behavior of solid solutions is helpful in understanding many of the equilibria between solid phases and liquid phases found in biological systems. The structural pattern of a typical solid solution is shown in Fig. 11-31.

Binary systems are found occasionally where the mutual solubility of components in the crystalline phase is so complete that there is no eutectic (Fig. 11-32). For example, at 1000°C a liquid solution with composition $X_{Au} = 0.22$ will be in equilibrium with

which separates out in the crystalline phase, e.g., the system with magnesium and zinc metals as the two components. In this system, a compound is formed with the formula $MgZn_2$, and the phase diagram can be separated into two parts, as shown in Fig. 11-29. At the left of the vertical dashed line is the normal binary phase diagram for the system, where one component is pure Mg and the other component is the compound $MgZn_2$. On the right side of the diagram is again a normal binary system, where the two components are $MgZn_2$ and pure Zn. There are two eutectic points. The maximum of the curve between the two eutectics lies above the point on the zero axis for $X_{Zn} = 0.66$ which corresponds to the compound $MgZn_2$. The temperature of the maximum of the curve at this point is the melting point of this compound.

Compound formation occurs between many sub-crystals which consist of a solid solution with composition $X_{Au} = 0.31$ and $X_{Ag} = 0.69$.

The binary system CuZn shows evidence of the formation of many different compounds and solid solutions between these two elements. The result is the complicated phase diagram shown in Fig. 11-33. The data for such a system are usually obtained by a process called *thermal analysis;* a series of liquids of different composition are cooled and time-temperature curves are plotted; from the position of the breaks in these curves the data for the phase diagram can be obtained.

The phase rule

Starting in 1875, Gibbs published a remarkable series of papers under the general title On the Equilibrium of Heterogeneous Substances, in which he established the fundamental mathematical relations for heterogeneous equilibrium. There is a special relation of general importance called the *Gibbs phase rule,* which he derived by considering the relation between the degrees of freedom of a system f, the number of phases p, and the number of components c.

In describing the motion of a molecule, the number of degrees of freedom is the minimum number of coordinates which must be assigned specified values in order to determine precisely the position of the molecule; in describing the state of a system, the number of degrees of freedom is equal to the number of independently variable intensive properties. For a system containing p phases and c components, the state is determined by specifying the temperature, the pressure, and the amounts of each component in each phase. Thus, the total number of variables is $pc + 2$. The symbol ^jn_i will be used to denote the number of moles of component i in phase j. In the same way, the mole fraction of each component in each phase is denoted by the symbol ^jX_i, giving the relation

$$^jX_i = \frac{^jn_i}{^jn_1 + {^jn_2} + \cdots {^jn_c}} \tag{11-94}$$

When j retains a fixed value,

$$^jX_1 + {^jX_2} + \cdots + {^jX_c} = 1 \tag{11-95}$$

Therefore, in each phase the last mole fraction in this series ^jX_c need not be specified, and for each phase

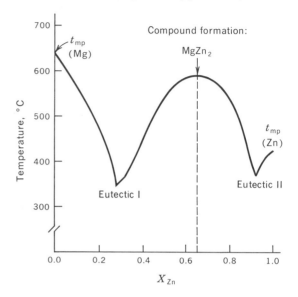

Fig. 11-29 The binary system magnesium-zinc compound formation and two eutectics. (From Max Hansen, "The Constitution of Binary Alloys," p. 928. Copyright © 1958. McGraw-Hill Book Company. Used by permission.)

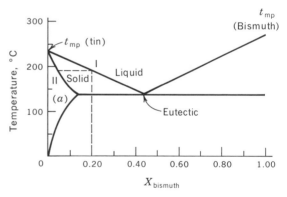

Fig. 11-30 The binary system tin-bismuth showing the solid solution of bismuth in tin in area (*a*). (From Max Hansen, "The Constitution of Binary Alloys," p. 337, McGraw-Hill Book Company, New York, 1968.)

only $c - 1$ mole fractions must be fixed. Since there are p phases, this means that the total number of mole fractions to be fixed is $p(c - 1)$. In addition to these independent variables, there are the independent variables of pressure and temperature, so that the total number of independent variables is $p(c - 1) + 2$.

Fig. 11-31   Bismuth atoms (dotted circles) replacing tin atoms (open circles) in random fashion to form a solid solution (schematic).

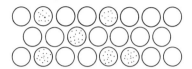

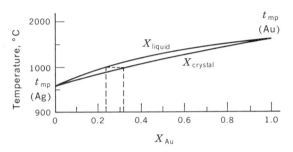

Fig. 11-32   The binary system silver-gold showing solid solution over the entire range.   (From Max Hansen, "The Constitution of Binary Alloys," p. 11, McGraw-Hill Book Company, New York, 1958.)

At equilibrium, the partial molal free energy of each component in each phase must be equal to its partial molal free energy in every other phase, a relation expressed by the equations

$$^1\overline{G}_1 = {}^2\overline{G}_1 = {}^3\overline{G}_1 = \cdots = {}^p\overline{G}_1$$
$$^1\overline{G}_2 = {}^2\overline{G}_2 = {}^3\overline{G}_2 = \cdots = {}^p\overline{G}_2$$
$$\text{------------------------}$$
$$^1\overline{G}_c = {}^2\overline{G}_c = {}^3\overline{G}_c = \cdots = {}^p\overline{G}_c$$

$$(11\text{-}96)$$

The number of independent variables is therefore decreased by an amount equal to the number of equal signs in this set of equations, which is $c(p - 1)$.   Since the number of degrees of freedom f equals the total number of variables minus the number of restraints,

$$f = p(c - 1) + 2 - c(p - 1)$$
$$= c - p + 2 \qquad (11\text{-}97)$$

This is the symbolic expression of the Gibbs phase rule.

When the phase rule is applied to a system consisting only of a single pure substance, $c = 1$.   If there is only one phase, e.g., the gaseous phase, $p = 1$.   In a cylinder completely filled with water vapor, there are two intensive variables, temperature and pressure.   Applying the phase rule,

$$f = c - p + 2$$
$$2 = 1 - 1 + 2 \qquad (11\text{-}98)$$

Both temperature and pressure can be varied independently; on a PV diagram any point in the plane where there is only one phase is a permissible state for the system.

If the temperature is lowered so that liquid water appears, there are two phases and the phase rule has the values

$$f = c - p + 2$$
$$1 = 1 - 2 + 2 \qquad (11\text{-}99)$$

Either temperature or pressure may be varied, but when one of these is fixed, the other is fixed and no longer independent.

If temperature is lowered so that ice also appears, there are three phases, the system is at the triple point, and the expression for the phase rule is

$$f = c - p + 2$$
$$0 = 1 - 3 + 2 \qquad (11\text{-}100)$$

Both temperature and pressure are fixed if the system is at the triple point, where three phases are in simultaneous equilibrium.

Consider the two-component system consisting of both o- and m-dinitrobenzene.   In the vapor state the expression for the phase rule is

$$f = c - p + 2$$
$$3 = 2 - 1 + 2 \qquad (11\text{-}101)$$

Temperature, pressure, and concentration all may be varied independently.

If the temperature is lowered and the system is condensed completely into the liquid state, the same equation applies.   If the behavior of this liquid is considered under the usual conditions where the pressure is constant at 1 atm, there is one less variable and the phase rule reads

$$f = c - p + 1$$
$$2 = 2 - 1 + 1 \qquad (11\text{-}102)$$

In other words, where the system is completely liquid in the area above the two composition lines in Fig. 11-21 the point representing the state of the system may range anywhere, with both temperature and composition being variable.

If the temperature is lowered so that crystalline ortho is present in equilibrium with the liquid, there are two phases and the phase rule is

$$f = c - p + 1$$
$$1 = 2 - 2 + 1$$

(11-103)

Now the point representing the system can move anywhere up or down on the temperature-composition line; there is 1 degree of freedom; but both temperature and composition cannot be varied independently. If the temperature is lowered still further to the eutectic temperature, there will be three phases, the solution in equilibrium with both ortho and meta crystals. At this point the expression for the phase rule is

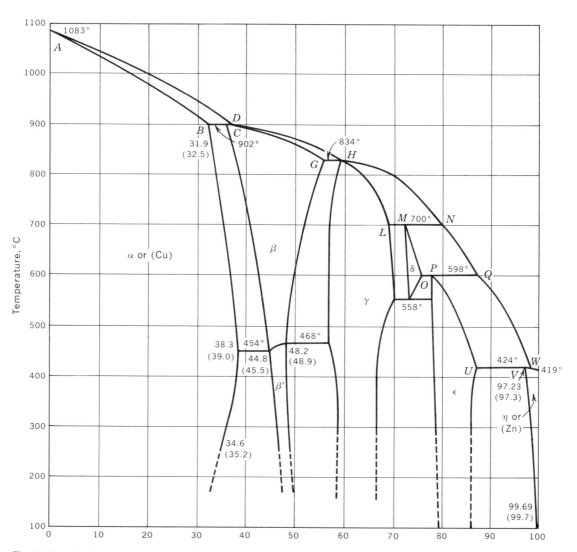

Fig. 11-33   The binary system copper-zinc showing compound formation and solid solutions.   (From Max Hansen, "The Constitution of Binary Alloys," p. 650, McGraw-Hill Book Company, New York, 1958.)

$$f = c - p + 1$$
$$0 = 2 - 3 + 1 \tag{11-104}$$

There are no degrees of freedom; temperature, pressure, and composition are all fixed.

11-8   MOLECULAR-WEIGHT DETERMINATION

When a crystalline substance is dissolved in a liquid and the vapor pressure is lowered in accordance with Raoult's law, the boiling point of the liquid is raised and the freezing point is lowered.   Observations of the boiling-point and freezing-point changes provide ways of determining the molecular weight of the dissolved substance.

Boiling-point elevation

Figure 11-34 is a plot of $R \ln P^\circ$ against $1/T$, where P° is the vapor pressure of pure benzene and the range of temperature covers the 10° above the normal boiling

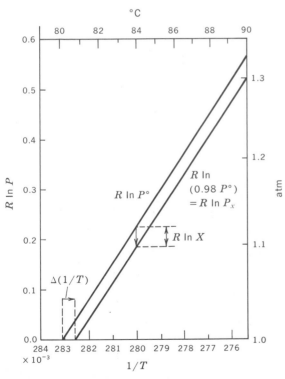

Fig. 11-34   Boiling-point elevation T for benzene, $X = 0.98$. $\Delta T = (1/T_x)^{-1} - (1/T^\circ)^{-1}$.

point of benzene (80.103°C).   If a crystalline substance with negligible vapor pressure is dissolved in benzene to form an ideal solution and the value of the mole fraction of benzene X_b is reduced from the value of 1 for pure benzene to the value 0.98, then according to Raoult's law the vapor pressure of the solution is given by

$$P = P^\circ X_b = (1 \text{ atm})(0.98) \tag{11-105}$$

The ratio of the vapor pressure of the solution P_1 at one temperature and the vapor pressure P_2 at another temperature is given by the usual vapor-pressure equation

$$R \ln \frac{P_2}{P_1} = -\Delta H_v \left(\frac{1}{T_2} - \frac{1}{T_1} \right) \tag{11-106}$$

where ΔH_v is the enthalpy of vaporization; for an ideal solution, this is the same as the enthalpy of vaporization for the pure liquid.   Let T_1 be the temperature in degrees Kelvin of the normal boiling point of *pure benzene* and T_2 the temperature at which the vapor pressure of the *solution* is equal to 1 atm; then Eq. (11-106) can be written

$$R \ln \frac{1}{P^\circ X_b} = -\Delta H_v \left(\frac{1}{T_2} - \frac{1}{T_1} \right) \tag{11-107}$$

where X_b is the mole fraction of benzene in the solution and P° is the vapor pressure of pure benzene at T_2. Since the difference between T_1 and T_2 is relatively small compared with their absolute values, the expression on the right-hand side of Eq. (11-107) may be rewritten

$$R \ln \frac{P^\circ X_b}{1} \simeq -\Delta H_v \frac{T_2 - T_1}{T_1^2} \tag{11-108}$$

Since $T_2 - T_1$ is small, $P^\circ = 1$; using the abbreviation ΔT for the elevation of the boiling point ($T_2 - T_1$), the equation can be put in the form

$$R \ln X_b = -\Delta H_v \frac{\Delta T}{T_1^2} \tag{11-109}$$

Where $X_b < \frac{1}{2}$, $\ln X_b$ can be expressed in the form of a series

$$\ln X_b = (X_b - 1) - \tfrac{1}{2}(X_b - 1)^2 + \tfrac{1}{3}(X_b - 1)^3 + \cdots \tag{11-110}$$

Since the value of X_b lies so close to unity and $X_b = 1 - X_a$, Eq. (11-109) can be written

$$\Delta T = \frac{RT_1^2}{\Delta H_v} X_a \tag{11-111}$$

Recall that the definition of X_a is given by

$$X_a = \frac{n_a}{n_a + n_b} \simeq \frac{n_a}{n_b} \simeq \frac{W_a/M_a}{W_b/M_b} \tag{11-112}$$

The molality of the solute in the solvent is given by

$$m_W = \frac{1,000 \, n_a}{W_b} = \frac{1,000 \, W_a}{M_a W_b} \tag{11-113}$$

Combining these last two expressions with Eq. (11-111) gives

$$\Delta T_B = \frac{RT_1^2}{\Delta H_v} \frac{W_a}{W_b} \frac{M_b}{M_a}$$

$$= \left(\frac{RT_1^2}{\Delta H_v} \frac{M_b}{1,000} \right) m_W = K_B m_W \tag{11-114}$$

The factor appearing on the right in parentheses is the constant characteristic of the solvent and is called the *molal boiling-point-elevation constant.* Values of this constant for a number of solvents are listed in Table 11-5. Knowing the value of ΔT and the value of the constant, the molality m_W can be determined immediately, and from the weight of the sample and the weight of the solvent in which it was dissolved, the molecular weight can be calculated with the help of Eq. (11-113).

Example 11-12

If 0.1 mole of diphenyl ether is dissolved in 500 g of benzene, calculate the boiling-point elevation.

The boiling-point elevation is

$$\Delta T_b = K_b m_w = 2.53 \times 0.1 \frac{1,000}{500} = 0.506°C$$

Example 11-13

If 18.016 g of glucose is dissolved in 100 g of water and the boiling-point elevation is found to be 0.51°C, calculate the molecular weight of glucose.

The molecular weight is

$$M = 1,000 \frac{K_b}{\Delta T_b} \frac{W_{\text{solute}}}{W_{\text{solv}}}$$

$$= 1,000 \frac{0.51}{0.51} \frac{18.016}{100} = 180$$

TABLE 11-5   Boiling-point-elevation constants

	t, °C	K_b, °C per unit of molality
Acetic acid	118.1	3.07
Acetone	56.5	1.71
Benzene	80.1	2.53
Bromobenzene	155.0	6.26
Carbon disulfide	46.3	2.34
Carbon tetrachloride	76.8	5.03
Chloroform	61.3	3.63
Ethyl alcohol	78.4	1.22
Ethyl ether	34.6	2.02
Water	100.0	0.51

Freezing-point depression

The depression of the freezing point of a solvent is related to the molality of the solute dissolved in it by an expression analogous to that for the elevation of the boiling point. The mole fraction of the solvent X_b is related to the freezing point of the pure solvent T_1 and to the freezing point of the solution T_2 by

$$R \ln X_b = -\Delta H_f \left(\frac{1}{T_2} - \frac{1}{T_1} \right) \tag{11-115}$$

Since $\Delta T = T_1 - T_2$, and since this value is small compared with the absolute values T_1 and T_2, Eq. (11-115) can be put in the form

$$R \ln X_b = \Delta H_f \frac{\Delta T_f}{T_1^2} \tag{11-116}$$

Using the series approximation for X_b and transforming to X_a, this can be put in the form

$$\Delta T_f = \frac{RT_1^2}{\Delta H_f} X_a \tag{11-117}$$

Expressing X_a in terms of weight molality m_W, the above equation becomes

$$\Delta T_f = \left(\frac{RT_1^2}{\Delta H_f} \frac{M_b}{1,000} \right) m_W = K_f m_W \tag{11-118}$$

The expression in the parentheses is known as the *molal freezing-point-depression constant,* and values of this constant are given in Table 11-6. From the

TABLE 11-6   Freezing-point-depression constants

	t, °C	K_f, °C per unit of molality
Acetic acid	16.6	3.90
Benzene	5.5	5.12
Benzophenone	49.0	9.80
Biphenyl	70.0	8.00
Bromoform	7.8	14.4
Cyclohexane	6.5	20
Camphor	173	40
Formic acid	8.40	2.77
Naphthalene	80.22	6.9
Phenol	40.90	7.00
Water	0.00	1.86

values in this table the value of the molality can be determined directly from a knowledge of the freezing-point depression, and from the latter the molecular weight is easily calculated.

Example 11-14

If 0.1 mole of naphthalene is dissolved in 1,000 g of benzene, calculate the freezing-point depression.

The freezing-point depression is

$$\Delta T_f = K_f m_W = 5.12 \times 0.1 = 0.512°C$$

Example 11-15

If 3.60 g of glucose is dissolved in 100 g of water and the freezing-point depression is found to be 0.372°C, calculate the molecular weight of glucose.

The molecular weight is

$$M = 1,000 \, \frac{K_f}{\Delta T_f} \frac{W_{\text{solute}}}{W_{\text{solv}}}$$

$$= 1,000 \, \frac{1.86}{0.372} \frac{3.60}{100} = 180$$

Problems

1.   The vapor pressure of pure $SnCl_4$ is 50 mm Hg at the temperature where the vapor pressure of pure CCl_4 is 213 mm Hg. These two substances form an essentially ideal solution. Calculate the vapor pressure of each component and the total vapor pressure over a solution where the mole fraction of CCl_4 is 0.6.

2.   The observed total vapor pressure above a solution of CCl_4 and $SnCl_4$ is 100 mm Hg. Using the data in Prob. 1, calculate the partial pressure of CCl_4 and of $SnCl_4$, assuming that the temperature is the same as that of the solution in Prob 1.

3.   At a certain composition the partial pressures of CCl_4 and $SnCl_4$ are equal. Calculate the total pressure over the solution of this composition and the partial pressure of each of the components, and calculate the composition of the solution, assuming that the temperature and the vapor pressures of the pure components are the same as in Prob. 1.

4.   If values of the vapor pressure of chloroform are measured over the range $X_1 = 0$ to $X_1 = 0.3$ when chloroform is dissolved in acetone, it is found that Henry's law applies and that the value of the Henry's law constant is 175 mm Hg. Calculate the value of the vapor pressure when $X_1 = 0.225$.

5.   For a solution of acetone in chloroform, it is found that the value of the Henry's law constant is 150 mm Hg when the solution is at a temperature of 35.17°C. Calculate the value of the vapor pressure of acetone when the mole fraction is 0.12.

6.   Assuming that Henry's law is applicable over sufficient range of composition to make the calculation valid, compute the composition at which the Henry's law pressure of chloroform equals the Henry's law pressure of acetone at 35.17°C using the data in the previous problems.

7.   Calculate the entropy of mixing when 2 moles of CCl_4 is mixed with 10 moles of $SnCl_4$.

8.   Calculate the entropy of mixing when 20 g of benzene forms an ideal solution with 80 g of toluene.

9.   Using the answers obtained in Prob. 3, calculate the entropy of mixing when the appropriate amount of CCl_4 is mixed with the appropriate amount of $SnCl_4$ so that there is 1 mole of solution and the vapor pressure of $SnCl_4$ is equal to the vapor pressure of CCl_4.

10.   Calculate the lowering of free energy due to the mixing of the components described in Prob. 7 at 30°C.

11.   Calculate the lowering of the free energy caused by the mixing of the components described in Prob. 9 at 37°C.

12. m-Nitrotoluene and p-nitrotoluene form an approximately ideal solution when mixed together. The melting point of m-nitrotoluene is 16°C. The enthalpy of fusion is 3270 cal mole^{-1}. Calculate the tempera-

ture at which the solution will be in equilibrium with crystals of m-nitrotoluene when the equilibrium temperature is reduced by the addition of sufficient p-nitrotoluene to reduce the value of the concentration of the m-nitrotoluene to $X = 0.95$.

13. o-Dinitrobenzene and p-dinitrobenzene form a nearly ideal solution when mixed together. When sufficient o-dinitrobenzene is added to p-dinitrobenzene to reduce the mole fraction of the latter to 0.793, it is found that the melting point is reduced from 173.5°C (pure para) to 159.9°C. From this information, calculate the enthalpy of fusion of p-dinitrobenzene.

14. Calculate the entropy of fusion for pure p-dinitrobenzene from the information obtained in Prob. 13.

15. From the graph in Fig. 11-21, calculate the amount by which the partial molal free energy of o-dinitrobenzene must be reduced in order for liquid ortho in solution in m-dinitrobenzene to be in equilibrium with crystalline ortho at 80.0°C.

16. Estimate the activity coefficient for CS_2 dissolved in acetone (Fig. 11-11) at the concentration where its partial pressure is 100 mm Hg.

17. Calculate the value of the Margules coefficient J for CS_2 dissolved in acetone from the data in Prob. 16.

18. Calculate the activity of a component for $X = 0.10$ in a solution in which the Margules coefficient $J = -0.5$.

19. From Fig. 11-17, estimate the composition of the vapor in equilibrium with a liquid mixture of benzene and toluene which has a normal boiling point at 100°C.

20. Using the data in Table 11-5, calculate the boiling-point elevation when 10 g of diphenyl ether, $C_6H_5OC_6H_5$, is dissolved in 100 g of benzene, C_6H_6.

21. When 20.7 g of biphenyl is dissolved in 150 g of benzene, the boiling point of the solution is 82.4°C. Calculate the approximate molecular weight of biphenyl.

22. If 3.00 g of phenol, C_6H_5OH, is dissolved in 120 g of bromoform, $CHBr_3$, calculate the freezing point of the mixture.

23. Suppose that 3.5644 g of anthracene is dissolved in 100 g of benzene and the mixture is found to freeze at 4.476°C; assuming that the freezing point of pure benzene is 5.500°C, calculate the molecular weight of anthracene.

REFERENCES

Leopold, H. G., and H. W. Foote: The Effect of Dissolved Gas on Freezing Points, *Am. J. Sci.*, **11**:42–46 (1926).

Hildebrand, J. H., and R. L. Scott: "The Solubility of Nonelectrolytes," 3d ed., Reinhold Publishing Corporation, New York, 1950.

Timmermans, J.: "Physico-chemical Constants of Binary Systems," Elsevier Publishing Company, Amsterdam, 1950.

Timmermans, J.: "Les Constantes physiques des composés organiques Cristallisés," Masson et Cie, Paris, 1953.

Prigogine, I.: "The Molecular Theory of Solutions," Interscience Publishers, Inc., New York, 1957.

Hala, E., J. Pick, V. Fried, and O. Vilim: "Vapour-Liquid Equilibrium," Pergamon Press, New York, 1958.

Rowlinson, J. S.: "Liquids and Liquid Mixtures," Academic Press, Inc., New York, 1959.

Lewis, G. N., and M. Randall: "Thermodynamics," 2d ed., rev. by K. S. Pitzer and L. Brewer, McGraw-Hill Book Company, New York, 1961.

Hildebrand, J. H., and R. L. Scott: "Regular Solutions," Prentice-Hall, Inc., Englewood Cliffs, N.J., 1962.

In a paper[1] entitled The Evolution of the Concept of Chemical Equilibrium from 1775 to 1923, Maurice W. Lindauer traces the history of concepts of chemical change. He points out that over 2,000 years ago the Greek philosopher-scientist Hippocrates proposed that chemical action is the result of a kinship between reacting substances, and that in the thirteenth century the alchemist Geber arranged several metals in a primitive affinity series according to their reactions with sulfur, mercury, and oxygen. The concept of affinity developed very gradually during the next 500 years and continued to retain its unidirectional character.

Chapter Twelve

chemical equilibrium

For chemists realized that two substances with an affinity for each other might react under suitable circumstances to yield a product but failed to recognize that the products under different circumstances might recombine and yield the original reactants.

In 1777, the German chemist C. F. Wenzel published a paper in which he attempted to estimate chemical affinities from the rate at which different metals were dissolved by acids. He observed that this rate was influenced by the quantity of acid as well as by its nature. The idea of the reversibility of a reaction was first suggested in 1799 by C. L. Berthollet, one of Napoleon's advisers, who read a paper before the National Institute of Egypt, pointing out that chemical combinations are influenced both by affinity and by quantity of reactants. Berthollet had observed the large deposits of sodium carbonate on the shores of the Natron Lakes in Egypt and recognized that they were due to the reaction

$$CaCO_3 + 2NaCl \rightarrow CaCl_2 + Na_2CO_3 \qquad (12\text{-}1)$$

which is the reverse of the familiar reaction

$$Na_2CO_3 + CaCl_2 \rightarrow 2NaCl + CaCO_3 \qquad (12\text{-}2)$$

resulting in the precipitation of calcium carbonate.

[1] *J. Chem. Educ.,* **39**:384 (1962).

In this paper Berthollet's ideas were still in the embryonic stage, but he continued to call attention to the effect of quantity on chemical reactions in later publications, and this factor was discussed from time to time by other prominent chemists during the following 50 years.

The first comprehensive statement of the law of chemical equilibrium (law of mass action) appeared in 1864 in a paper by C. N. Guldberg and P. Waage, of the University of Christiania in Norway. They introduced the term active mass, which is essentially the same as the modern term concentration. For the general reaction

$$A + B \rightleftharpoons C + D \qquad (12\text{-}3)$$

where the concentrations of the various substances are respectively [A], [B], [C], and [D], the condition of equilibrium is

$$K = \frac{[C][D]}{[A][B]} \qquad (12\text{-}4)$$

where K is a constant.

This proposal was expressed in precise quantitative form by the Dutch physical chemist van't Hoff in 1884; and through his work and the contemporary work of Willard Gibbs, the law of mass action was put on a firm thermodynamic basis. For the reaction where a moles of A combine with b moles of B to form c moles of C and d moles of D

$$a\text{A} + b\text{B} \rightleftharpoons c\text{C} + d\text{D} \qquad (12\text{-}5)$$

the equilibrium constant K is given by the expression

$$K = \frac{[C]^c[D]^d}{[A]^a[B]^b} \qquad (12\text{-}6)$$

12-1   THE THEORY OF CHEMICAL EQUILIBRIUM

Equilibrium constant and free energy

In order to develop a thermodynamic theory of equilibrium it is helpful to have a way of expressing precisely the extent to which a reaction proceeds. Consider a reaction in solution for which the chemical equation is Eq. (12-5). The terms on the left are called the *reactants,* and the terms on the right the *products.*

Denote the number of moles of A taking part in the reaction by the symbol n_A and the moles of the other constituents by similar symbols. Then the relation follows:

$$\frac{n_A}{a} = \frac{n_B}{b} = \frac{n_C}{c} = \frac{n_D}{d} = \lambda \qquad (12\text{-}7)$$

where λ denotes the amount by which the reaction proceeds and is called the *progress variable.* From the definition of activity in Eq. (11-57), the relation follows

$$\bar{G}_A - G_A^\circ = RT \ln a_A \qquad (12\text{-}8)$$

where the activity a_A is the concentration variable which can be substituted for concentration [A] as used in Eq. (12-6). For the purpose of making the derivation as simple as possible, let us assume that the substances form ideal solutions so that the activity a_A equals the concentration of this component [A]. Then each component has an equation similar to Eq. (12-8) as follows:

$$\bar{G}_A = G_A^\circ + RT \ln [A] \qquad (12\text{-}9)$$
$$\bar{G}_B = G_B^\circ + RT \ln [B] \qquad (12\text{-}10)$$
$$\bar{G}_C = G_C^\circ + RT \ln [C] \qquad (12\text{-}11)$$
$$\bar{G}_D = G_D^\circ + RT \ln [D] \qquad (12\text{-}12)$$

When the reaction proceeds to the extent of 1 mole ($\lambda = 1$) under equilibrium conditions, the change in free energy ΔG for the system is

$$\Delta G = c\bar{G}_C + d\bar{G}_D - a\bar{G}_A - b\bar{G}_B \qquad (12\text{-}13)$$

The change in free energy under standard conditions ΔG° is given by the expression

$$\Delta G^\circ = c\bar{G}_C^\circ + d\bar{G}_D^\circ - a\bar{G}_A^\circ - b\bar{G}_B^\circ \qquad (12\text{-}14)$$

Combining these equations with Eqs. (12-9) to (12-12) gives

$$\Delta G = \Delta G^\circ + RT \{c \ln [C]$$
$$+ d \ln [D] - a \ln [A] - b \ln [B]\} \qquad (12\text{-}15)$$

When the reaction takes place under equilibrium conditions, $\Delta G = 0$. If the term in braces in the last equation is combined into a single term, the relation becomes

$$\Delta G^\circ = -RT \ln \frac{[C]^c[D]^d}{[A]^a[B]^b} \qquad (12\text{-}16)$$

which can be written

$$\frac{[C]^c[D]^d}{[A]^a[B]^b} = e^{-\Delta G^\circ/RT} \tag{12-17}$$

Since at constant temperature the term on the right is a constant and not dependent on concentration, it can be denoted by the symbol K and the equation can then be written

$$K = \frac{[C]^c[D]^d}{[A]^a[B]^b} \tag{12-18}$$

The constant K is called the *equilibrium constant*. A comparison of Eq. (12-18) with Eq. (12-6) shows that the two are identical. Thus, the law of mass action expressed in Eq. (12-6) is shown to be derivable from the laws of thermodynamics. Another useful way of expressing this relation is

$$\Delta G^\circ = -RT \ln K \tag{12-19}$$

These equations also apply to reactions in mixtures of ideal gases.

Equilibrium and entropy

In view of the definition of free energy, Eq. (12-15) can be written

$$\Delta H - T\,\Delta S = \Delta H^\circ - T\,\Delta S^\circ$$
$$+ RT(c \ln [C] + d \ln [D] - a \ln [A] - b \ln [B]) \tag{12-20}$$

In an ideal solution, the change in enthalpy will be independent of concentration, so that

$$\Delta H = \Delta H^\circ \tag{12-21}$$

Equation (12-20) can then be put in the form

$$\Delta S - \Delta S^\circ$$
$$= -R(c \ln [C] + d \ln [D] - a \ln [A] - b \ln [B]) \tag{12-22}$$

The difference between the entropy of 1 mole of component A in an ideal solution at concentration [A] and in solution under standard conditions is

$$s - s^\circ = -R \ln [A] \tag{12-23}$$

Thus, the term $-aR \ln [A]$ is just the entropy change when a moles of A are changed from *standard concen-*

tration, [A] $= 1$, to the concentration which appears in the expression for the equilibrium constant. Thus, when the system is changed from the state where all the components are at standard concentration to the state where all the components are at concentrations resulting in equilibrium, *the equilibrium is brought about solely by the change in the entropies of the components due to the change in their concentrations.* This fact can be emphasized by writing the general relation

$$\Delta G \quad - \quad \Delta G^\circ \quad = \quad -T\left(-R \ln \frac{[C]^c[D]^d}{[A]^a[B]^b}\right) \tag{12-24}$$

change in free change in free change in entropy in passing
energy under energy under from standard state to
equilibrium standard equilibrium state
conditions conditions
(equal to 0)

12-2   EXAMPLES OF EQUILIBRIA

Gaseous equilibria

When a substance A is in the gaseous state, its concentration (A) in units of moles per liter is equal to the inverse of the molal volume v in units of liters per mole.[1] If the substance behaves as an ideal gas, there is the relation

$$(A) = \frac{1}{v_A} = \frac{P_A}{RT} \tag{12-25}$$

When these values are substituted for the concentration values in Eq. (12-18), we find the expression

$$K = \frac{(P_C/RT)^c(P_D/RT)^d}{(P_A/RT)^a(P_B/RT)^b} \tag{12-26}$$

On factoring terms, this becomes

$$K = \left[\frac{P_C^c P_D^d}{P_A^a P_B^b}\right][(RT)^{c+d-a-b}]^{-1} \tag{12-27}$$

At constant temperature the expression in the brackets is equal to a constant, generally denoted by the symbol K_P or the equilibrium constant when the intensive variable is pressure. When there is a need to denote the nature of the equilibrium constant unambiguously, this

[1] When the distinction between the gas and liquid phases is significant, concentration in the gas phase will be denoted by parentheses, instead of brackets.

quantity is frequently written K_C when the intensive property is concentration. Thus,

$$K_P = \frac{P_C{}^c P_D{}^d}{P_A{}^a P_B{}^b} \qquad (12\text{-}28)$$

Because of the relation

$$G - G° = RT \ln P \qquad (12\text{-}29)$$

for an ideal gas, there is also the relation for an *equilibrium* for ideal gases[1]

$$\Delta G° = -RT \ln K_P \qquad (12\text{-}30)$$

When the gas does not behave ideally, the free energy can be expressed in terms of the fugacity

$$G - G° = RT \ln f \qquad (12\text{-}31)$$

and the equilibrium constant will be

$$K_f = \frac{f_C{}^c f_D{}^d}{f_A{}^a f_B{}^b} \qquad (12\text{-}32)$$

The relation between the equilibrium constant expressed in terms of concentration and the equilibrium constant expressed in terms of pressure is

$$K_P = K_C (RT)^{\sum_i x_i} \qquad (12\text{-}33)$$

where

$$\sum_i x_i = (c + d) - (a + b) \qquad (12\text{-}34)$$

For any reaction in general $\sum_i x_i$ is always equal to the algebraic sum of the stoichiometric coefficients for the reactants taken as negative and the stoichiometric coefficients for the products taken as positive.

The formation of water

The chemical equation for the formation of water from its elements is

$$H_2 + \tfrac{1}{2}O_2 \rightleftarrows H_2O \qquad (12\text{-}35)$$

Consider first this reaction over a temperature range where the components are all in the gaseous state and each one has a partial pressure of 1 atm (stand-

---

[1] If concentration is the variable, $\Delta G°$ refers to the change between reactants at a concentration of unity and products each at a concentration of unity; if pressure is the variable, then $\Delta G°$ refers to the change between reactants at a pressure of 1 atm and products each at a pressure of 1 atm.

ard state). The changes in the different thermodynamic variables can then be expressed

$$-(H_{H_2}° + \tfrac{1}{2}H_{O_2}°) + H_{H_2O}° = \Delta H° \qquad (12\text{-}36)$$

$$-(S_{H_2}° + \tfrac{1}{2}S_{O_2}°) + S_{H_2O}° = \Delta S° \qquad (12\text{-}37)$$

$$-(G_{H_2}° + \tfrac{1}{2}G_{O_2}°) + G_{H_2O}° = \Delta G° \qquad (12\text{-}38)$$

$\Delta G°$ is the change in free energy when 1 mole of hydrogen combines with $\tfrac{1}{2}$ mole of oxygen to form 1 mole of water while each component is at a partial pressure of 1 atm. This quantity is frequently called the *standard free energy of formation* $\Delta G_f°$ since it is the free-energy change for the formation of water from its elements under standard conditions.

The free-energy change is related to the enthalpy change $\Delta H°$ and the entropy change $\Delta S°$ by the expression

$$\Delta G° = \Delta H° - T \Delta S° \qquad (12\text{-}39)$$

The values of each of these quantities for selected temperatures are given in Table 12-1, and the values of the entropy of formation are given in Table 12-2.

Note that the value of enthalpy of formation $\Delta H_f°$ changes very little over the range from 500 to 5000°K. Roughly speaking, the energy necessary to pull the hydrogen and oxygen molecules apart subtracted from the energy given out when the atoms recombine to form water is not affected very much by a change in temperature. In the same way, the difference between the entropy per mole of the water molecules on the one hand and the sum of the entropy per mole for hydrogen and for $\tfrac{1}{2}$ mole of oxygen ΔS_f changes relatively little over this temperature range, as shown also in Table 12-2. But because the probability factor $T \Delta S$ is the *product* of the entropy change ΔS and T, this part of the free-energy change rises from about 6 kcal at 500°K to over 70 kcal at 5000°K. Thus, at 500°K, $\Delta G_f°$ is -52 kcal, but at 5000°K it is $+10$ kcal. Thus, when all components are at 1 atm and the system is at 500°K, there is such a large decrease in free energy that the reaction goes explosively. On the other hand, at 5000°K, the reaction goes in the reverse direction, water decomposing.

This point is illustrated in Fig. 12-1, in which is plotted a curve for the sum of the free energies of the reactants $G_{H_2 + (1/2)O_2}°$ and the curve for the free energy of the product $G_{H_2O}°$. The curve for the free energy

TABLE 12-1   Thermodynamic variables for the formation of gaseous water from its elements[a]

T, °K	ΔH_f°, kcal mole^{-1}	$T\,\Delta S_f^\circ$, kcal mole^{-1}	ΔG_f°, kcal mole^{-1}	log K_P
500	−58.277	+5.918	−52.359	22.885
1000	−59.246	+13.209	−46.037	10.061
1500	−59.824	+20.532	−39.292	5.724
2000	−60.150	+27.757	−32.393	3.540
3000	−60.530	+42.101	−18.429	1.342
4000	−60.910	+56.571	−4.339	0.237
4307	−61.062	+61.062	0	0
5000	−61.465	+70.924	+9.862	−0.431

[a] D. R. Stull (ed.), "JANAF Tables," Dow Chemical Company, Midland, Mich., 1963.

per mole equivalent of the reactants at low temperatures lies above the curve for the free energy per mole of the products; in passing from one to the other there is a large drop in the value of the free energy, and consequently the reaction proceeds spontaneously with violence. As temperature rises, the difference between the two curves gets less and less, and finally the curves intersect at 4307°K. At this temperature, there is no difference between the free energy for the reactants and for the products, and $\Delta G° = 0$; in other words, when the constituents are all at a partial pressure of 1 atm, the reaction proceeds reversibly under equilibrium conditions at this temperature. As temperature rises higher, the curve for the reactants lies below the curve for the product; water decomposes spontaneously when the partial pressure of all the reactants is at 1 atm.

Consider now the change brought about at a temperature of 400°K when the pressure of the reactants is lowered and the pressure of the product is raised. In Fig. 12-2 the curve of the free energy for the reactants under standard conditions is shown at the top of figure $G_{H_2+(1/2)O_2}^\circ$. At the bottom of the figure is shown the curve for the free energy of the product $G_{H_2O}^\circ$. If the pressure of the hydrogen is reduced to 10^{-9} atm and the oxygen to 10^{-20} atm, the curve for the free energy of the reactants drops down to the middle of the diagram, as shown. In reverse fashion, if the pressure of H_2O is raised to 1.735×10^{10} atm, the curve of the free energy of the product rises to the middle of the diagram, going up by 19.8 kcal.

TABLE 12-2   Values of ΔS_f° for H_2O[a]

T, °K	ΔS_f°, eu
1000	13.209
2000	13.879
3000	14.034
4000	14.143
5000	14.185

[a] D. R. Stull (ed.), "JANAF Tables," Dow Chemical Company, Midland, Mich., 1963.

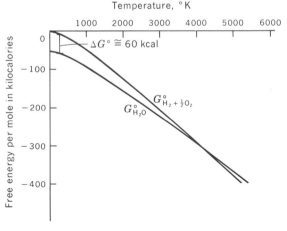

Fig. 12-1   The free energy per mole of $H_2 + \frac{1}{2}O_2$ and of H_2O in the standard state at 1 atm.

Fig. 12-2   Establishing equilibrium at 400°K by raising pressure of H_2O and lowering pressure of H_2 and O_2.

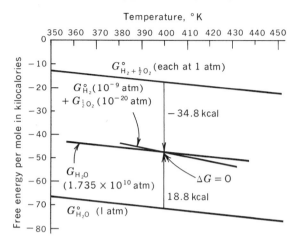

Now the curves under these conditions intersect at 400°K. At this point $\Delta G = 0$. The ratio of these equilibrium pressures is thus equal to the equilibrium constant

$$K_P = \frac{P_{H_2O}}{P_{H_2}P_{O_2}^{1/2}} = \frac{1.735 \times 10^{10}}{(10^{-9})(10^{-10})^{1/2}}$$

$$= 1.735 \times 10^{24} \text{ atm}^{-1/2} \qquad (12\text{-}40)$$

and the latter is related to the free energy by the expression

$$\Delta G° = -RT \ln K = -53.52 \text{ kcal} \qquad (12\text{-}41)$$

If the pressure of the hydrogen were raised by a factor of 10, the pressure of oxygen would have to be reduced by a factor of 100 in order to maintain equilibrium

$$K_P = \frac{1.735 \times 10^{10}}{(10^{-8})(10^{-12})^{1/2}} = 1.735 \times 10^{24} \text{ atm}^{-1}$$

$$(12\text{-}42)$$

In conclusion, it is important to note that when this reaction takes place at low temperatures, the enthalpy factor predominates; there is a strong tendency for the water to form because the probability factor $T \Delta S_f°$ is relatively small compared with the enthalpy change. As temperature increases, this increases the "weight" of the probability; in other

words, because the probability change $\Delta S_f°$ is multiplied by the temperature in the expression for the free energy, this probability factor counts more and more in the balance determining the free energy.

The probability factor favors the dissociation of the water into the elements primarily because the atoms involved are scattered more widely when in the form of the elemental molecules H_2 and O_2 than they are when combined to form H_2O; just as a gas in a scattered state (occupying a large volume) is in a more probable state than when condensed (occupying a smaller volume), so the state is more probable where the atoms are more widely scattered. Here again concentration affects equilibrium because it affects probability.

In studying the thermodynamics of evaporation, a diagram was drawn like that in Fig. 12-3a, showing that the ΔH factor tends to pull the molecules together (condensation) while the $T \Delta S$ factor tends to drive the molecules apart (vaporization). Equilibrium is found when these two factors exactly balance each other so that their difference, the free energy ΔG, is zero. In exactly the same way there is a balance of the same factors in a chemical equilibrium. If the reactants and product are portrayed separately, as in Fig. 12-3b, instead of all mixed in together as they really are, one can indicate the ΔH and the $T \Delta S$ factors by arrows which balance exactly at equilibrium. The ΔH arrow is, of course, the difference between H for 1 mole of water and H for 1 mole of hydrogen plus $\frac{1}{2}$ mole oxygen; ΔS is likewise the difference between S for the product and S for the reactants. The length and direction of the arrow for ΔG determines whether a reaction will proceed spontaneously from right to left, will be balanced in equilibrium, or will proceed spontaneously from left to right.

The synthesis of ammonia

Ammonia can be produced from nitrogen and hydrogen according to the equation

$$\tfrac{1}{2}N_2 + \tfrac{3}{2}H_2 \rightleftarrows NH_3 \qquad (12\text{-}43)$$

where both reactants and the product are in the gaseous state. When the pressures are measured in atmospheres, the expression for the equilibrium constant is

Fig. 12-3   A comparison of (a) physical and (b) chemical equilibrium.

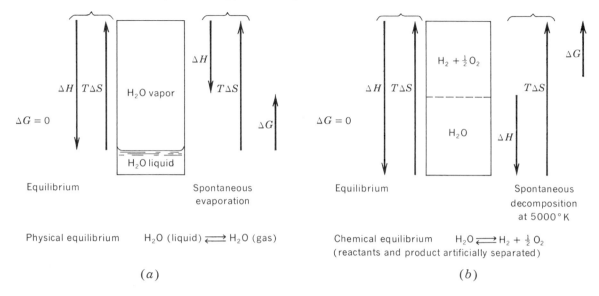

Physical equilibrium H_2O (liquid) $\rightleftharpoons$ H_2O (gas)

(a)

Chemical equilibrium $H_2O \rightleftharpoons H_2 + \frac{1}{2} O_2$
(reactants and product artificially separated)

(b)

$$K_P = \frac{P_{NH_3}}{P_{N_2}^{1/2} P_{H_2}^{3/2}} = 1.28 \times 10^{-2} \text{ atm}^{-1} \quad (12\text{-}44)$$

The numerical value of the equilibrium constant is that found at 400°C. If the reaction is written

$$N_2 + 3H_2 = 2NH_3 \quad (12\text{-}45)$$

the expression for the equilibrium constant is

$$K_P = \frac{P_{NH_3}^2}{P_{N_2} P_{H_2}^3} = 1.64 \times 10^{-4} \text{ atm}^{-2} \quad (12\text{-}46)$$

Since 2 moles of ammonia are formed in the reaction as written, $\Delta G°$ for this reaction will be twice as large as that for Eq. (12-43), where only 1 mole is formed. Consequently, the value of K_P is altered in view of the relation

$$\Delta G° = -RT \ln K_P \quad (12\text{-}47)$$

In using values of K_P or $\Delta G°$ care must be taken to specify the exact chemical equation with which these quantities are associated.

When equilibrium is maintained at higher values of the pressure, there is deviation from the ideal-gas laws and consequently a deviation in the value of K. As shown in Table 12-3, there is a trend toward higher values of K_P at higher total pressures.

TABLE 12-3   Values of $K_P = \dfrac{P_{NH_3}}{P_{H_2}^{3/2} P_{N_2}^{1/2}}$ at various pressures[a]

t, °C	K_P, atm^{-1}		
	At 10 atm	At 100 atm	At 1,000 atm
400	0.0129	0.0137	
450	0.00659	0.00725	0.02328
500	0.00381	0.00402	

[a] Data from A. T. Larson and R. L. Dodge, *J. Am. Chem. Soc.*, **45**:1925 (1923).

Example 12-1

Assuming all constituents to be ideal gases, what will be the pressure of hydrogen in equilibrium with water vapor at 0.1 atm and oxygen at 0.01 atm at a temperature of 400°K?

$$K_P = \frac{P_{H_2O}}{P_{H_2} P_{O_2}^{1/2}} = \frac{0.1}{P_{H_2}(0.01)^{1/2}}$$

$$= 1.735 \times 10^{24} \text{ atm}^{-1/2}$$

$$P_{H_2O} = (1.735 \times 10^{24})^{-1} \times (0.1)(0.1^{-1})$$

$$= 5.76 \times 10^{-25} \text{ atm}$$

Example 12-2

Assuming all constituents to be ideal gases, what will be the partial pressures of N_2 and H_2 in equilibrium with NH_3 at 1 atm at 400°C, if $P_{N_2} = P_{H_2} = x$?

$$K_P = \frac{P_{NH_3}}{x^{1/2}x^{3/2}} = \frac{1}{x^2} = 1.28 \times 10^{-2} \text{ atm}^{-1}$$

$$x = (1.28 \times 10^{-2})^{-1/2} = 8.8 \text{ atm}$$

Equilibria in solution

The equation for the reaction between acetic acid and ethyl alcohol to form ethyl acetate and water can be written

$$CH_3COOH + C_2H_5OH \rightleftharpoons CH_3COOC_2H_5 + H_2O \tag{12-48}$$

Expressing the concentration in moles per liter of solution, the equilibrium constant is written

$$K_C = \frac{[CH_3COOC_2H_5]}{[CH_3COOH][C_2H_5OH]} = 4.0 \text{ (mole liter}^{-1})^{-1} \tag{12-49}$$

The numerical value is the one found at 25°C. In writing the expression for an equilibrium constant for a reaction where water is formed, it is customary to omit the concentration of water when the equation is applied to conditions where the concentrations of the other components are so small that the concentration of water is effectively constant.

Another similar reaction is the uniting of glycine and alanine by a peptide bond with the elimination of water

$$CH_3CH(NH_2)COOH + NH_2CH_2COOH \rightarrow$$
Ala   Gly

$$CH_3CH(NH_2)CONHCH_2COOH + H_2O \tag{12-50}$$
Ala—Gly

The expression for the equilibrium constant at 25°C is

$$K_C = \frac{[Ala—Gly]}{[Ala][Gly]} = 1.07 \times 10^3 \text{ (mole liter}^{-1})^{-1} \tag{12-51}$$

Example 12-3

At 25°C, what will be the concentration of ethyl acetate in equilibrium with 0.1 m acetic acid and 0.1 m ethyl alcohol, assuming an ideal solution?

$$K_C = \frac{[CH_3COOC_2H_5]}{[CH_3COOH][C_2H_5OH]} = \frac{x}{[0.1][0.1]}$$

$$= 4 \text{ (mole liter}^{-1})^{-1}$$

$$x = 4 \times 10^{-2} \text{ mole liter}^{-1}$$

Example 12-4

At 25°C, what will be the concentration of alanine in equilibrium with 0.1 m alanylglycine and 1 m glycine, assuming an ideal solution?

$$K_C = \frac{[Ala—Gly]}{[Ala][Gly]} = \frac{[0.1]}{[x][1]}$$

$$= 1.07 \times 10^3 \text{ (mole liter}^{-1})^{-1}$$

$$x = [0.1][1.07]^{-1} 10^{-3}$$

$$= 9.35 \times 10^{-5} \text{ mole liter}^{-1}$$

Heterogeneous equilibria

Calcium carbonate in the solid state is in equilibrium with solid calcium oxide when exposed to the proper pressure of carbon dioxide gas according to the equation

$$CaCO_3(s) \rightleftharpoons CaO(s) + CO_2(g) \tag{12-52}$$

The equilibrium constant is given by the expression:

$$K_P = \frac{[CaO(s)]P_{CO_2}}{[CaCO_3(s)]} = P_{CO_2} \tag{12-53}$$

Because $CaCO_3$ and CaO are in the solid state, their concentrations are fixed and can be assigned the value of unity. Thus the equilibrium constant is just the value of the pressure of CO_2 above the mixture of the two solids. In this respect it resembles a vapor pressure. Actually, each of the solids also has a small vapor pressure but so small as to be negligible compared with the pressure of CO_2 even though this value is extremely small at room temperature; at 25°C, $K_P = P_{CO_2} = 1.175 \times 10^{-23}$ atm, explaining the stability of the large natural deposits of $CaCO_3$. The decomposition pressures of CO_2 for other carbonates are given by similar equations. Table 12-4 lists the temperatures at which the decomposition pressures of various carbonates reach a value of 1 atm.

A similar equation also applies when one of the constituents in an equilibrium reaction is in the liquid state. For example, NOBr decomposes in the following way

$$2NOBr \rightleftharpoons 2NO + Br_2$$

When the partial pressure of bromine is sufficiently high, the bromine condenses. Without bromine condensation the equilibrium expression at 25°C is

$$K_P = \frac{P_{Br_2}P_{NO}^2}{P_{NOBr}^2} = 1.0 \times 10^{-2} \text{ atm} \qquad (12\text{-}54)$$

At 25°C the vapor pressure of liquid bromine $P_{Br_2(l)}$ is 0.282 atm, so that for bromine in the liquid state, this expression becomes

$$K_P = \frac{P_{Br_2(l)}P_{NO}^2}{P_{NOBr}^2} = \frac{[0.282 \text{ atm}] P_{NO}^2}{P_{NOBr}^2} \qquad (12\text{-}55)$$

Thus, an alternative expression for the equilibrium constant can be written:

$$K_P^* = \frac{P_{NO}^2}{P_{NOBr}^2} = \frac{K_P}{[0.282 \text{ atm}]} = 3.5 \times 10^{-2} \text{ atm}^{-1}$$

$$\qquad (12\text{-}56)$$

where the term in brackets is the constant pressure of the bromine vapor, the vapor pressure of the liquid bromine at this temperature.

If the vaporization of a substance is written in the form of an equation similar to a chemical equation, the vapor pressure can be regarded as an equilibrium constant. For example, writing

$$H_2O(l) \rightleftharpoons H_2O(g) \qquad (12\text{-}57a)$$

then

$$K_P = \frac{P_{H_2O}}{[H_2O(l)]} = P_{H_2O} \qquad (12\text{-}57b)$$

where the concentration term for liquid water $[H_2O(l)]$ is a constant at constant temperature and may be set equal to unity.

In the same way the solubility of a substance may be treated as an equilibrium. For example, o-dinitro-

TABLE 12-4   Temperatures at which the pressure of CO_2 above various carbonates reaches a value of 1 atm

Compound	t, °C
Ag_2CO_3	275
$BaCO_3$	1360
$CaCO_3$	900
$CdCO_3$	350
$FeCO_3$	400
Li_2CO_3	1270
$MgCO_3$	400
$MnCO_3$	327
$PbCO_3$	340
$SrCO_3$	1280

benzene in solution in liquid m-dinitrobenzene and with solid o-dinitrobenzene exhibits the relation

$$o\text{-}DNB(s) \rightleftharpoons o\text{-}DNB(soln) \qquad (12\text{-}58a)$$

and the equilibrium-constant expression is

$$K_X = \frac{X_{o\text{-}DNB}}{X_{o\text{-}DNB}(s)} \qquad (12\text{-}58b)$$

Since the mole fraction of the o-DNB in the solid state is fixed, $X_{o\text{-}DNB}(s)$ can be set equal to unity. When $X_{o\text{-}DNB}$ becomes equal to unity at the melting temperature of o-DNB, K_X becomes equal to unity also.

Example 12-5

Calculate the value of the partial pressure of NO in equilibrium with NOBr (1 atm) and Br_2 at 25°C when (a) liquid Br_2 is present and (b) when $P_{Br_2} = 0.1$ atm.

(a) $\quad K_P^* = \dfrac{P_{NO}^2}{P_{NOBr}^2} = \dfrac{P_{NO}^2}{(1 \text{ atm})^2}$

$\qquad = 3.5 \times 10^{-2}$

$\quad P_{NO} = (3.5 \times 10^{-2})^{1/2} = 0.19$ atm

(b) $\quad K_P^* = \dfrac{P_{NO}^2 P_{Br_2}}{P_{NOBr}^2} = \dfrac{P_{NO}^2 (0.1 \text{ atm})}{(1 \text{ atm})^2}$

$\qquad = 1.0 \times 10^{-2}$ atm

$\quad P_{NO} = (0.010 \text{ atm}^2)^{1/2} = 0.10$ atm

Example 12-6

Using the data from Fig. 11-21, Calculate K_X for Eq. (12-58) at 105°C. At this temperature $X_{o\text{-DNB}} = 0.8$, so that $K_X = X_{o\text{-DNB}} = 0.8$.

Calculation of K

Because of the close relationship between the equilibrium constant K and the standard free energy $\Delta G°$, the former can be calculated easily from the latter. There are many tables of values of $\Delta G°$, especially at 25°C, that are of great value in making such calculations. In particular, the values of the standard free energy of formation $\Delta G_f°$ can be combined so that values of K can be calculated for reactions between compounds other than those of formation from the elements.

For example, at 25°C, the following standard free energies of formation from the elements are found:

$$C(s) + O_2(g) = CO_2(g) \qquad \Delta G_f° = -94,260 \text{ cal} \tag{12-59}$$

$$C(s) + \tfrac{1}{2}O_2(g) = CO(g) \qquad \Delta G_f° = -32,778 \text{ cal} \tag{12-60}$$

If Eq. (12-60) is subtracted from Eq. (12-59), the expression is:

$$CO(g) + \tfrac{1}{2}O_2(g) = CO_2(g) \qquad \Delta G° = -61,482 \text{ cal} \tag{12-61}$$

From the relation between K_P and $\Delta G°$,

$$\Delta G° = -RT \ln K_P \tag{12-62}$$

the value of K_P for this reaction can be obtained by inserting numerical values and solving for K_P

$$-61,482 = -1.987 \times 298.15 \ln K_P$$
$$K_P = 1.166 \times 10^{45} \text{ atm}^{-1/2} \tag{12-63}$$

12-3 THE CHANGE OF K WITH T, P, V, AND c

Variation of K with T

Since the value of K for vaporization is the value of the vapor pressure, as shown in Eq. (12-58), it is clear that the formula for the variation of K with temperature must be similar to the formula for the variation of vapor pressure with temperature. This can be demonstrated using the relation

$$\Delta G° = -RT \ln K \tag{12-64}$$

Differentiating with respect to temperature at constant pressure, we obtain

$$\frac{\partial \Delta G°}{\partial T} = -RT \frac{\partial \ln K}{\partial T} - R \ln K \tag{12-65}$$

Since the left side of the equation is just $-\Delta S°$, we get

$$-\Delta S° = -RT \frac{\partial \ln K}{\partial T} - R \ln K \tag{12-66}$$

Multiplying by T gives

$$-T \Delta S° = -RT^2 \frac{\partial \ln K}{\partial T} - RT \ln K \tag{12-67}$$

Using Eq. (12-64), this can be transformed to

$$-T \Delta S° = -RT^2 \frac{\partial \ln K}{\partial T} + \Delta G° \tag{12-68}$$

Since

$$\Delta G° + T \Delta S° = \Delta H° \tag{12-69}$$

then

$$\frac{\partial \ln K}{\partial T} = \frac{\Delta H°}{RT^2} \tag{12-70}$$

Since $d(1/T) = -(1/T^2)\, dT$, Eq. (12-70) can be put in the form

$$d \ln K = \frac{-\Delta H°}{R} d\frac{1}{T} \tag{12-71}$$

This can be integrated between T_0 and T, at which K has the values K_0 and K, assuming that $\Delta H°$ is constant over this range

$$\ln \frac{K}{K_0} = -\frac{\Delta H°}{R} \left(\frac{1}{T} - \frac{1}{T_0} \right) \tag{12-72}$$

If T_0 is taken as the temperature at which $K_0 = 1$, then

$$\ln K = -\frac{\Delta H°}{R} \left(\frac{1}{T} - \frac{1}{T_0} \right) \tag{12-73}$$

This is exactly analogous to the vapor-pressure equation

$$\ln P = -\frac{\Delta H_v}{R} \left(\frac{1}{T} - \frac{1}{T_b} \right) \tag{12-74}$$

where T_b is the temperature at which $P = 1$ atm. Equation (12-73) can be put in the form

$$\log K = \frac{-\Delta H^\circ}{2.303\ R}\left(\frac{1}{T} - \frac{1}{T_0}\right) \qquad (12\text{-}75)$$

Thus a plot of $\log K$ against $1/T$ is linear if the system behaves ideally and ΔH° does not vary with temperature. A plot of $R \ln K$ against $1/T$ is shown in Fig. 12-4 for the reaction

$$H_2 + S = H_2S \qquad (12\text{-}76)$$

based on the data given in Table 12-5.

Example 12-7

From the data in Table 12-5 calculate the value of ΔH° for the reaction $H_2 + S \rightleftharpoons H_2S$.

Using the formula for the variation of $R \ln K$ with $1/T$, the value of ΔH° is

$$\Delta H^\circ = \frac{R \ln K_2 - R \ln K_1}{1/T_1 - 1/T_2}$$

$$= \frac{1.177 - 9.272}{0.0009775 - 0.0005999}$$

$$= \frac{-8.095}{3.776 \times 10^{-4}} = -21.43 \text{ kcal} \qquad Ans.$$

Free-energy function

In 1930 the American physical chemist W. F. Giauque pointed out the advantages in tabulating values of $(G - H_0^\circ)/T$ at various temperatures in order to make possible the calculation of values of G by temperature interpolation. This combination is called the *free-energy function* and varies much less rapidly with temperature than G, so that interpolation is much more accurate. H_0° is the enthalpy under standard conditions at $0^\circ K$ and is calculated from measurements of heat capacity between 0 and $298.15^\circ K$, as discussed in Chap. 7.

For any reaction

$$aA + bB = cC + dD \qquad (12\text{-}77)$$

$$\Delta \frac{G^\circ - H_0^\circ}{T} = c\left(\frac{G^\circ - H_0^\circ}{T}\right)_C + d\left(\frac{G^\circ - H_0^\circ}{T}\right)_D$$

$$-a\left(\frac{G^\circ - H_0^\circ}{T}\right)_A - b\left(\frac{G^\circ - H_0^\circ}{T}\right)_B \qquad (12\text{-}78)$$

Fig. 12-4 Equilibrium between H_2, S, and H_2S. (Original data from G. Preuner and W. Schupp, *Z. Phys. Chem.*, **68**:157 (1909), and from M. Randall and F. R. Bichowsky, *J. Rm. Chem. Soc.*, **40**:368 (1918); calculations and plot from G. N. Lewis and M. Randall, "Thermodynamics," 2d ed., rev. by K. S. Pitzer and L. Brewer, McGraw-Hill Book Company, New York, 1961.)

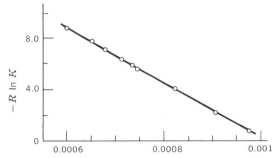

TABLE 12-5   Equilibrium for the formation of H_2S from the elements[a]

T, $^\circ K$	$R \ln K$
1023	9.272
1218	5.975
1362	4.130
1473	2.944
1667	1.177

[a] From G. N. Lewis and M. Randall, "Thermodynamics," 2d ed., rev. by K. S. Pitzer and L. Brewer. Copyright 1961. McGraw-Hill Book Company. Used by permission.

Also

$$\Delta \frac{G^\circ - H_0^\circ}{T} = \frac{\Delta G^\circ}{T} - \frac{\Delta H_0^\circ}{T} \qquad (12\text{-}79)$$

and

$$\ln K = -\frac{\Delta G^\circ}{RT} = -\frac{\Delta H_0^\circ}{RT} + \frac{1}{R}\Delta\frac{G^\circ - H_0^\circ}{T} \qquad (12\text{-}80)$$

Example 12-8

Given the values of the free-energy functions and ΔH_0°, find the value of K_P for the formation of HCl(g) at 25°C.

$(\Delta H_0^\circ)_f$ for HCl is $-22,019$ cal.

$$\frac{(\Delta H_0^\circ)_f}{RT} = \frac{-22,019}{1.987 \times 298.15^\circ}$$

$$= -37.168$$

$$\frac{1}{R}\Delta\frac{G^\circ - H_0^\circ}{T} = \frac{1}{R}\left(\underset{\text{HCl}}{37.72} - \underset{\frac{1}{2}\text{Cl}_2}{\frac{45.93}{2}} - \underset{\frac{1}{2}\text{H}_2}{\frac{24.42}{2}}\right)$$

$$= \frac{2.54}{1.987} = 1.28$$

$$\ln K_P = -(-37.17 + 1.28) = +35.89$$

$$\log K_P = \frac{+35.89}{2.303} = +15.584$$

$$K_P = 3.84 \times 10^{15}$$

Variation of K with P

If all the components taking part in a chemical equilibrium are ideal gases, a change in total pressure does not affect the value of K_P as long as this change in pressure is brought about in such a way that the ratios of the partial pressures are not affected. Consider the reaction

$$a\text{A} + b\text{B} = c\text{C} + d\text{D} \tag{12-81}$$

for which

$$K_P = \frac{P_C{}^c P_D{}^d}{P_A{}^a P_B{}^b} \tag{12-82}$$

The relations of the partial pressures to the total pressure P are

$$P_A = PX_A \qquad P_B = PX_B \tag{12-83}$$

and so forth, where

$$X_A = \frac{n_A}{n_A + n_B + n_C + n_D} \tag{12-84}$$

and so forth. If n_i moles of an inert gas are introduced into the reaction chamber, then

$$X_A' = \frac{n_A}{n_A + n_B + n_C + n_D + n_i} \tag{12-85}$$

and

$$\frac{X_A'}{X_A} = \frac{n_A + n_B + n_C + n_D}{n_A + n_B + n_C + n_D + n_i} = r \tag{12-86}$$

The values of X_B, X_C, and X_D are also reduced by the same ratio r.

The original total pressure is related to volume and temperature by the equation

$$PV = (n_A + n_B + n_C + n_D)RT \tag{12-87}$$

After the introduction of the inert gas the total pressure P' is

$$P'V = (n_A + n_B + n_C + n_D + n_i)RT \tag{12-88}$$

so that

$$\frac{P'}{P} = \frac{1}{r} \tag{12-89}$$

Thus the values of the partial pressures P_A, P_B, P_C, P_D remain unchanged

$$P_A' = P'X_A' = P\left(\frac{1}{r}\right)X_A r = PX_A = P_A \tag{12-90}$$

If the gases do not behave ideally, the values of the fugacities may be dependent on the total pressure, and the value of K_P will be affected.

Even when the gases behave ideally, the value of K_X is dependent on the total pressure, since

$$K_X = \frac{(X_C)^c(X_D)^d}{(X_A)^a(X_B)^b} \tag{12-91}$$

and the equilibrium constant for the reaction K_X after the introduction of the inert gas is given by the expression

$$K_X' = \frac{(X_C')^c(X_D')^d}{(X_A')^a(X_B')^b}$$

$$= \frac{(rX_C)^c(rX_D)^d}{(rX_A)^a(rX_B)^b} \tag{12-92}$$

$$= K_X r^{\Sigma_i j_i}$$

where

$$\Sigma_i j_i = c + d - a + b \tag{12-93}$$

Variation of K with V

If the volume of a reaction chamber is changed, the values of K are in general affected. Suppose that the initial volume V is changed to a new volume

$(V' = sV)$. Then each partial pressure is changed from $P_A = n_A RT/V$ to a new value

$$P_A' = \frac{n_A RT}{V'} = \frac{n_A RT}{Vs} \qquad (12\text{-}94)$$

so that

$$P_A' = P_A s^{-1} \qquad (12\text{-}95)$$

Consequently,

$$K_P' = \frac{(P_C')^c(P_D')^d}{(P_A')^a(P_B')^b} = \frac{(s^{-1}P_C)^c(s^{-1}P_D)^d}{(s^{-1}P_A)^a(s^{-1}P_B)^b} \qquad (12\text{-}96)$$

$$K_P' = s^{-\Sigma_i j_i} K_P \qquad (12\text{-}97)$$

Variation of K with concentration

The addition of solvent to an ideal solution affects the equilibrium in exactly the same way as the increase in volume for an ideal gaseous system. In a solution the concentration [A] of component A in moles per liter is given by the expression

$$[A] = \frac{n_A}{V_s} \qquad (12\text{-}98)$$

where V_s is the volume of the solution. Suppose that the volume is increased from this initial value to a new value $(V_s' = V_s s)$. Then

$$[A]' = [A]s^{-1} \qquad (12\text{-}99)$$

and

$$K_C' = K_C s^{-\Sigma_i j_i} \qquad (12\text{-}100)$$

just as in the case for the change of volume of a gaseous equilibrium system.

If the solution is not ideal, K_C will also be affected by the changes in the activities due to the new pattern of interaction between solutes and solvents at the new concentration.

Le Châtelier's principle

Writing in the year 1888, the French physical chemist Henry Louis Le Châtelier summarized the results of the study of the effect of changes of thermodynamic variables on equilibria in a conclusion which today is called *Le Châtelier's principle:*

If a change of variable shifts the equilibrium, the shift will be in a direction that tends to restore the variable to its original value.

Consider an equilibrium like that between hydrogen, oxygen, and water at 400°

$$H_2 + \tfrac{1}{2}O_2 \rightleftarrows H_2O \qquad \Delta H° = -52{,}042 \text{ cal} \qquad (12\text{-}101)$$

Heat is evolved when H_2 and O_2 combine to form H_2O. At this temperature

$$K_P = \frac{P_{H_2O}}{P_{H_2}P_{O_2}^{1/2}} = 1.73 \times 10^{29} \text{ atm}^{-1/2} \qquad (12\text{-}102)$$

If the temperature is raised to 500°K, then

$$K_P = 7.67 \times 10^{22} \text{ atm}^{-1/2} \qquad (12\text{-}103)$$

There is now less water and more H_2 and O_2; water has dissociated due to the increase in temperature; the process of dissociation absorbs heat; the absorption of heat *tends* to lower the temperature, moving this variable (which was originally changed) back toward its original value.

Consider the effect of changing pressure on a physical system like liquid water in equilibrium with its vapor

$$H_2O(l) \rightleftarrows H_2O(g) \qquad (12\text{-}104)$$

If the pressure is raised, water condenses, thus *tending* to lower the pressure and restore the value of the variable (which was originally changed) to its initial value.

Finally consider the equilibrium in solution:

$$(Ala) + (Gly) \rightleftarrows (Ala\text{—}Gly) \qquad (12\text{-}105)$$

for which

$$K_C = \frac{(Ala\text{—}Gly)}{(Ala)(Gly)} \qquad (12\text{-}106)$$

If the concentration of Ala is increased, it tends to shift the system in the forward direction indicated by Eq. (12-105); such a shift tends to decrease the concentration of Ala, moving this variable back toward the initial value which it had before the change was made.

Summary

The equation for the approximate variation of K_P with T for an ideal gas [Eq. (12-72)] provides a helpful insight into the nature of equilibrium when put in the form

$$\ln \frac{K_2}{K_1} = -\frac{\Delta H^\circ}{R}\left(\frac{1}{T_2} - \frac{1}{T_1}\right) \qquad (12\text{-}107)$$

where K_2 is the equilibrium constant at T_2 and K_1 is the equilibrium constant at T_1. This equation can be rearranged

$$-(R \ln K_2 - R \ln K_1) = \frac{\Delta H^\circ}{T_2} - \frac{\Delta H^\circ}{T_1}$$

$$= \Delta S_2 - \Delta S_1 \qquad (12\text{-}108)$$

where ΔS_2 and ΔS_1 are the changes in entropy when the reaction takes place *under equilibrium conditions* at T_2 and T_1, respectively.

This expression can be compared with a similar one derived from the relation between K, ΔG, and ΔG° at these same two temperatures

$$\Delta G_1 - \Delta G_1^\circ = RT_1 \ln K_1$$
$$\Delta G_2 - \Delta G_2^\circ = RT_2 \ln K_2 \qquad (12\text{-}109)$$

Although ΔG_1 and ΔG_2 are each equal to zero since since they are the free-energy changes at equilibrium, they can be left in the expression and expanded

$$(\Delta H_1 - T_1 \Delta S_1) - (\Delta H_1^\circ - T_1 \Delta S_1^\circ) = RT_1 \ln K_1 \qquad (12\text{-}110)$$

$$(\Delta H_2 - T_2 \Delta S_2) - (\Delta H_2^\circ - T_2 \Delta S_2^\circ) = RT_2 \ln K_2$$

If the gases behave ideally, $\Delta H_1 = \Delta H_1^\circ$ and $\Delta H_2 = \Delta H_2^\circ$, so that T_1 and T_2 can be eliminated, giving

$$\Delta S_1 - \Delta S_1^\circ = -R \ln K_1$$
$$\Delta S_2 - \Delta S_2^\circ = -R \ln K_2 \qquad (12\text{-}111)$$

Subtracting the second from the first gives

$$\Delta S_2 - \Delta S_1 - (\Delta S_2^\circ - \Delta S_1^\circ) = -(R \ln K_2 - R \ln K_1) \qquad (12\text{-}112)$$

As pointed out earlier, the entropy for nearly all substances varies much more slowly with temperature than the free energy; also, the reactants on the one hand and the products on the other are apt to have about the same increase in entropy over a given temperature range *when in the standard state*. For this reason the difference between the sum of the entropies of the reactants and the sum of the entropies of the products all in the standard state ($\Delta S_2^\circ - \Delta S_1^\circ$) is almost negligible compared with the difference between

the entropies in the equilibrium state ($\Delta S_2 - \Delta S_1$). For the formation of water from its elements, the term $\Delta S_2^\circ - \Delta S_1^\circ$ contributes only enough to make about a 2 percent change in the value of log K_P over the range 500 to 1000°K. Therefore, this term may be eliminated, leaving the expression in the form

$$\underline{\Delta S_2 - \Delta S_1} = -(R \ln K_2 - R \ln K_1) \qquad (12\text{-}113)$$

entropy changes
under equilibrium
conditions

This equation is identical with Eq. (12-108), which is just the familiar equation relating $R \ln K$ to $1/T$.

The terms on the right are also merely changes in entropy. For the formation of water from its elements,

$$H_2 + \tfrac{1}{2}O_2 \rightleftarrows H_2O \qquad (12\text{-}114)$$

we have

$$-R \ln K_2 = -R \ln \frac{P_{H_2O}}{P_{H_2}P_{O_2}^{1/2}}$$

$$= -R \ln P_{H_2O} - (-R \ln P_{H_2}) - (-R \ln P_{O_2}) \qquad (12\text{-}115)$$

Recall that the change in entropy in passing from the standard state, where P_{H_2O} is 1 atm, to the state with the pressure P_{H_2O} is

$$S - S^\circ = -R \ln P_{H_2O} \qquad (12\text{-}116)$$

This change *adjusts* the system from the standard state to the equilibrium state. The term $-R \ln K$ is the *equilibrium adjustment entropy*. Because the standard-state difference between the reactants' entropy and the products' entropy varies so little with temperature, the *equilibrium-state difference* change with temperature $\Delta S_2 - \Delta S_1$ is just the change in the *equilibrium adjustment entropy*, $-R \ln (K_2/K_1)$.

Recall now that when temperature is raised, the change in free energy under equilibrium conditions ΔG must be maintained at zero value. For a system of ideal gases, where $\Delta H = \Delta H^\circ$ is relatively invariant with temperature, this means that ΔS must change in inverse proportion to temperature to maintain $\Delta G = 0$

$$0 = \Delta G = \Delta H^\circ - T \Delta S \qquad (12\text{-}117)$$

This simple relationship is the basis not only of the equation relating log K to $1/T$ but also of the equations relating the logarithm of the vapor pressure and of the solubility mole fraction to $1/T$ also. All equilibria, whether chemical or physical, are governed by this simple relation to temperature insofar as they approach ideal behavior. This provides a basic pattern of relationships to which varying degrees of nonideal behavior can be compared in order to understand the underlying reasons for nonideality better.

In constructing a chemical ensemble like a biological organism, nature is faced with the problem of bringing about a very large number of reactions between different substances with varying degrees of chemical affinity $\Delta G°$. These reactions must proceed in a controlled and coordinated pattern, not explosively. To a considerable degree this is attained by maintaining the system in a state close to equilibrium conditions. This state is reached by adjusting the entropies of the component through variations of their concentrations in a manner appropriate to the temperature [Eq. (12-72)] for the optimum functioning of the organism.

Problems

1. At a temperature of 327°C, an equilibrium mixture of N_2, H_2, and NH_3 is formed. If the partial pressure of N_2 is 10 atm and the partial pressure of H_2 is 15 atm, calculate the partial pressure of NH_3 in this mixture and the total pressure of the system. The equilibrium constant at this temperature is equal to 4.42×10^{-2} atm^{-1}.

2. At 1600°K potassium hydroxide, KOH, can be formed from potassium vapor and oxygen and hydrogen gas, according to the equation

$$K + \tfrac{1}{2}O_2 + \tfrac{1}{2}H_2 = KOH$$

The equilibrium constant for this reaction is 6.3×10^5 atm^{-2}. If, in an equilibrium mixture the partial pressure of oxygen is 1 atm and the partial pressure of hydrogen is 1 atm, what partial pressure of K will be needed to maintain a partial pressure of KOH at 1 atm?

3. Lithium fluoride, LiF, is formed from lithium vapor and fluorine gas according to the equation

$$Li + \tfrac{1}{2}F_2 = LiF$$

At a temperature of 2500°K, it is found that in an equilibrium mixture the partial pressures of the constituents will be 10^{-2} atm for F_2, 10^{-4} atm for Li, and

10^2 atm for LiF. Calculate the equilibrium constant for this reaction.

4. Calculate from the data obtained in Prob. 3 the change in free energy when 1 mole of lithium combines with $\tfrac{1}{2}$ mole of F_2 to form 1 mole of LiF when the partial pressure of each of the constituents is 1 atm at 2500°K.

5. The change in free energy when 1 mole of Hg combines with 1 mole of F_2 to form 1 mole of HgF_2 is -6.3 kcal. At 700°K under standard conditions calculate the equilibrium constant for this reaction at this temperature when the partial pressures are expressed in atmospheres.

6. Gaseous iodine monochloride, ICl, is formed from gaseous Cl_2

$$\tfrac{1}{2}I_2 + \tfrac{1}{2}Cl_2 = ICl$$

In an equilibrium mixture at 500°K the partial pressure of I_2 equals that of Cl_2, and the partial pressure of ICl is 1 atm. The standard free energy of formation at this temperature for this reaction is -4.043 kcal mole^{-1}. Calculate the total pressure of the system under these conditions.

7. From 500 to 1500°K the standard enthalpy of formation for the reaction in Prob. 6 is -3.36 kcal mole^{-1}. When the partial pressures are expressed in atmospheres, the value of log K_P at 1000°K is 1.035. Calculate the value of log K_P at 1500°K.

8. For the formation reaction of mercuric chloride, $HgCl_2$, the value of log K_P is 5.555 at 1000°K when the partial pressures of the components are expressed in atmospheres. At 1500°K, the value of log K_P is 2.000. Calculate the average standard enthalpy of formation over this range.

9. For the formation of ethyl acetate and water from ethyl alcohol and acetic acid at room temperature, the equilibrium constant is approximately equal to 4. Calculate the standard free-energy change under these conditions.

10. In an equilibrium mixture of ethyl alcohol, acetic acid, ethyl acetate, and water, the concentrations of ethyl alcohol and acetic acid are equal. If the concentration of ethyl acetate is 0.4 m when the reaction is in equilibruim at room temperature, what will be the concentration of ethyl alcohol if $K_C = 4$?

11. In an equilibrium mixture of $CaCO_3$, CaO, and CO_2, the partial pressure of CO_2 is observed to be 3.871 atm at a temperature of 1000°C. Calculate the free-energy change under standard conditions for the reaction at this temperature.

12. At a temperature of 25°C, the compound NOBr dissociates according to the reaction

2NOBr = 2NO + Br₂

In an equilibrium mixture where the pressure is sufficiently high to form liquid Br_2, the constant K_P for the heterogeneous reaction is equal to 3.5×10^{-2}. If the partial pressure of NOBr is 10 atm, what will be the partial pressure of NO?

13. If hydrogen at 1-atm pressure and benzene at 1-atm pressure are placed in a flask at 385°C and allowed to come to chemical equilibrium according to the reaction

$$C_6H_6(g) + 3H_2(g) \rightleftharpoons C_6H_{12}(g)$$

calculate the pressure of C_6H_{12} when equilibrium is attained. The value of K_P is approximately 10^{-3} atm^{-3}. *Note:* When the amount of product formed is small, the change in the pressures of the reactants may be neglected.

14. Calculate the value of $\Delta G°$ in Prob. 13.

15. PCl₅ decomposes according to the equation

$$PCl_5(g) \rightleftharpoons PCl_3(g) + Cl_2(g)$$

If 1 mole of PCl₅ is placed in a 10-liter flask at 250°C and at equilibrium 0.47 mole of Cl_2 is present, calculate the value of K_C.

16. Calculate the value of $\Delta G°$ in Prob. 15.

17. The compound adenosine triphosphate (ATP) hydrolyzes at 38°C to form adenosine diphosphate (ADP) and PO(OH)₃ according to the reaction

A(P)(P)(P) + H₂O → A(P)(P) + PO(OH)₃

with $\Delta G° = -10,000$ cal. *Note:* (P) denotes the phosphate radical —PO(OH)₂. Calculate K_C for this reaction using moles per liter as the concentration unit.

18. Adenosine phosphate, A(P), hydrolyzes at 38°C according to the equation

A(P) + H₂O → A + HO(P)

with $\Delta G° = 2000$ cal. Calculate the value of K_C using moles per liter as the concentration unit.

19. If the concentration of A(P) is 0.01 mole per liter, calculate the concentrations of A and HO(P) present at equilibrium.

20. If the concentration of HO(P) is 0.001 mole per liter and is equal to the concentration of A, calculate the concentration of A(P) in equilibrium with this mixture.

REFERENCES

Coull, J., and E. B. Stuart: "Equilibrium Thermodynamics," pp. 411–447, John Wiley & Sons, Inc., New York, 1964.

Patton, A. R.: "Biochemical Energetics and Kinetics," pp. 1–24, 32–41, W. B. Saunders Company, Philadelphia, 1965.

Barrow, G. M.: "Physical Chemistry," 2d ed., pp. 211–262, McGraw-Hill Book Company, New York, 1966.

Daniels, F., and R. A. Alberty: "Physical Chemistry," 3d ed., pp. 190–240, John Wiley & Sons, Inc., New York, 1966.

Fairley, J. L., and G. L. Kilgour: "Essentials of Biological Chemistry," 2d ed., pp. 133–218, Reinhold Publishing Corporation, New York, 1966.

Gucker, S. T., and R. L. Seifert: "Physical Chemistry," pp. 394–414, 735–780, W. W. Norton & Company, Inc., New York, 1966.

Hamill, W. H., R. R. Williams, Jr., and C. MacKay: "Principles of Physical Chemistry," 2d ed., pp. 196–213, Prentice-Hall, Inc., Englewood Cliffs, N.J., 1966.

Baldwin, E.: "Dynamic Aspects of Biochemistry," 5th ed., pp. 42–60, Cambridge University Press, London, 1967.

Wyatt, P. A. H.: "Energy and Entropy in Chemistry," The Macmillan Company, New York, 1967.

Klotz, I. M.: "Energy Changes in Biochemical Reactions," pp. 14–71, Academic Press, Inc., New York, 1967.

13-1 THE NATURE OF AQUEOUS IONS

When dissolved in water, many substances are found not as individual solute molecules surrounded by solvent but in a dissociated condition as electrically charged *molecular* fragments called *ions*. Sometimes this dissociation, or ionization, process is partial, sometimes practically complete. The cause of ionization is to be found in the nature and properties of water.

The properties of water

Water is both the most abundant and the most important of all liquids. The earth's oceans consist of water with a small amount of dissolved solute; since the oceans cover the greater part of the surface of the earth, water is as abundant as air. Water is important because it is the primary fluid of life. The human body is more than 70 percent water; the fluids which circulate within it are largely water; even in the interior of the 10^{14} biological cells of which a human being is composed, aqueous solutions constitute almost all the media in which the most basic biochemical reactions take place.

These facts are particularly striking because water is one of nature's most peculiar substances. Consider, for example, the properties of the hydrides of the elements that constitute group VI in the periodic table of chemical elements. As pointed out by the American chemist Linus Pauling, if one starts at the bottom of the periodic table and makes a plot of the melting points of some of the hydrides of the elements of group VI, the data for the compounds H_2Te, H_2Fe, and H_2S show the expected trend; yet the curve extrapolated to H_2O predicts a melting point for water of $-100°C$, some $100°$ below the observed melting point at $0°C$. A similar graph of the boiling points of the same compounds, when extrapolated, predicts a boiling point for H_2O at $-80°C$, or $180°$ below the observed boiling point at $100°C$. Other properties are equally unusual, as can be seen from Table 13-1.

During the last few decades evidence has accu-

Chapter Thirteen

ionic solutions

TABLE 13-1   A comparison of the properties of water with those of other liquids

	c_p at 0°C, cal g^{-1}	t_m, °C	H_f, cal g^{-1}	t_b, °C	H_v, cal g^{-1}
H_2O	1.007	0.00	79.7	100	539.5
H_2SO_4	0.34^b	10.5	24.0	330	122
C_2H_5OH	0.535	−117.3	24.9	78.5	204
C_6H_6	0.39^a	+5.5	30.3	80.1	94
CCl_4	0.198	−22.8	4.2	76.8	46

^{a} At 6°C.
^{b} At 11°C.

mulated showing that the properties of water are abnormal because water molecules attract one another with a force that is far stronger than the normal van der Waals force of attraction between ordinary molecules. This extra force is due largely to the phenomenon of hydrogen bonding, mentioned in Chap. 4. While bound primarily to one oxygen atom, a hydrogen nucleus can approach close to an unshared pair of electrons on another oxygen atom; then there is a kind of sharing of the hydrogen nucleus between the two oxygen atoms producing a bondlike force of attraction. About 6000 cal of energy is required to break this *hydrogen bond* and separate the water molecules so joined. This energy may be compared with energy of only 300 cal required to separate two molecules of CH_4 held together by van der Waals forces. On the other hand, the energy required to break a normal covalent O—H bond is about 111,000 cal.

In a water molecule, each hydrogen nucleus lies at a distance of 0.96 Å from the oxygen nucleus, and the two O—H bonds are separated by an angle of 104.5°. The electron cloud is distributed about the three nuclei in a way that leaves a net residual positive charge on the side of the molecule toward the two hydrogen nuclei and a net residual negative charge on the side toward the oxygen nucleus. This results in a dipole moment, as shown in Fig. 13-1. Because of the dipole moment, the water molecules orient themselves as shown in Fig. 13-2 when water is placed between two oppositely charged metallic plates; this action is in accord with Le Châtelier's principle, as it tends to reduce the effect of the strain

produced by the charge on the plates by lowering the electric potential difference between the two plates.

In ice, water molecules are grouped in the ordered lattice as shown in the cross-sectional drawing in Fig. 13-3a. In liquid water there is still some residual order, as shown in Fig. 13-3b, although the molecules can move about much more freely than is possible in the crystalline state. The kind of order induced in liquid water by hydrogen bonding is shown diagrammatically in Fig. 13-4.

The dissociation of water

Water itself can dissociate according to the equation

$$H_2O \rightleftharpoons H^+ + OH^- \tag{13-1}$$

The value of the equilibrium constant (K_w^*) is

$$K_w^* = a_{H^+} \frac{a_{OH^-}}{a_{H_2O}} \tag{13-2a}$$

In this chapter the activity will be defined in terms of concentration units, with $a° = 1\ m$. For dilute aqueous solutions a_{H_2O} does not change appreciably with concentration and is set equal to unity by convention; the value of K_w is defined by the expression

$$K_w \equiv a_{H^+} \times a_{OH^-} \simeq [H^+][OH^-] = K_w' \tag{13-2b}$$

where [H$^+$] and [OH$^-$] are the concentrations of these ions. Because of the definition of activity the product (or ratio) of *activities* in the mass-action law is a true constant denoted by K, with appropriate subscripts if needed. The product (or ratio) of analogous concentrations is not, in general, strictly constant except for very dilute solutions and is therefore denoted by a different symbol K', with appropriate subscripts as needed. Equation (13-2b) illustrates this usage with respect to the hydrogen ion.

The hydrogen ion itself consists of a bare proton, and there is some evidence that it tends to associate with another undissociated water molecule. For this reason, the dissociation is frequently represented in the literature by an equation

$$2H_2O \rightleftharpoons H_3O^+ + OH^- \qquad (13\text{-}3)$$

in order to emphasize the fact that the proton does not wander freely through the liquid. However, the actual arrangement of atoms and molecules is certainly more complex than might be deduced from Eq. (13-3); and for this reason, in this chapter the equations will be written in the simplest possible form, where H^+ is used to denote the hydrogen ion. It is important to keep in mind that the chemical behavior of water and of the substances dissolved in it in the form of ions must always be interpreted in a way that takes into account the clustering of water molecules around charged particles.

The interaction of ions and water

Many substances scarcely dissolve at all in solvents with little or no molecular dipole moment but dissolve freely in water, to form ions about which the water molecules orient themselves. Thermodynamic considerations explain this difference in behavior. Consider first the formation of a solution of sodium chloride in benzene. To change 1 g mole of Na^+Cl^-, in the form of the usual crystal lattice, to Na^+ and Cl^- as individual ions so widely separated in the gaseous state that they exert no force on each other, the amount of energy needed is 186 kcal $mole^{-1}$. About the same amount of energy would be required to transform 1 mole of the crystal into individual ions in solution in a solvent like benzene at a dilution so great that the ions would not exert forces on each other. If the ions in solution were in equilibrium with the crystal lattice according to the equation

$$NaCl(c) \rightleftharpoons Na^+ + Cl^- \qquad (13\text{-}4)$$

in benzene
solution

the free-energy difference between the two states would be zero and the equation for the free energy

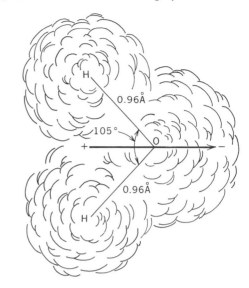

Fig. 13-1   Water molecule showing dipole moment.

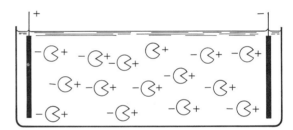

Fig. 13-2   Water molecules oriented between two charged plates.

in terms of enthalpy and the temperature-entropy factor would have numerical values approximately equal to those given in the following expression:

$$\Delta G \;=\; \Delta H \;-\; T\,\Delta S \qquad (13\text{-}5)$$
$$\;\;0 \qquad +186\text{ kcal} \qquad -186\text{ kcal}$$

If the entropy increase ΔS is regarded as being due to the dilution, it is equal to $\Delta H/T$ and is related to the concentration by the expression

$$\frac{\Delta H}{T} = \Delta S \simeq -R \ln m \qquad (13\text{-}6)$$

From this, m can be calculated:

$$m = -\text{antilog}\,\frac{-\Delta S}{R} = 10^{-136} \qquad (13\text{-}7)$$

Fig. 13-3   The arrangement of molecules in (a) ice and (b) in liquid water.

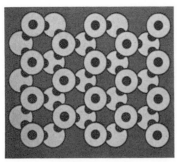

(a) Ice structure

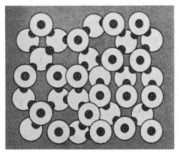

(b) Water structure

```
          . .
      H : O :
          . .
        H       H
  . .         . .
H : O : H : O : H : O :
  . .         . .
    H    H    H
  . .  . .  . .
  : O : H : O : H : O : H
  . .         . .
    H    H
```

Fig. 13-4   Natural orientation of water molecules under the influence of hydrogen bonding. (After M. J. Sienko and R. A. Plane, "Chemistry," McGraw-Hill Book Company, New York, 1957.)

This expression shows the almost infinitesimal solubility of NaCl in benzene.

By way of contrast, when crystalline sodium chloride is dissolved in water, the water molecules around the Na$^+$ orient themselves so that the negative end of the dipole points toward the ion and the water molecules around the Cl$^-$ orient themselves so that the positive side of the dipole points toward the ion. The amount

by which the energy of the system is lowered due to this process is called the *energy of hydration*. There is no direct way of measuring this energy, but it can be estimated at least with an accuracy of about 10 percent. The sum of the energy of hydration of the Na$^+$ and Cl$^-$ is about 185 kcal, so that the energy necessary to bring about a separation of the ions (186 kcal) is roughly balanced by the lowering of the energy due to hydration (185 kcal). Thus, for the process

$$NaCl(c) \rightleftarrows Na^+(aq) + Cl^-(aq) \qquad (13\text{-}8)$$

the thermodynamic expression is approximately

$$\Delta G = \Delta H - T\Delta S \qquad (13\text{-}9)$$
$$\quad 0 \qquad +1\text{ kcal} \quad -1\text{ kcal}$$

so that according to this rough calculation the temperature-entropy factor is $+1$ kcal and the value of ΔS is given by the expression

$$\frac{\Delta H}{T} = \Delta S = +R \ln m \simeq +2 \text{ cal deg}^{-1} \text{ mole}^{-1}$$
$$(13\text{-}10)$$

In making an accurate calculation many other factors must be taken into consideration. This rough calculation, comparing the solution of NaCl in benzene with the solution in water, is presented only to show the qualitative differences. These approximate enthalpy relations are shown diagrammatically in Fig. 13-5. Figure 13-6 shows schematically the action of water molecules in separating Na$^+$Cl$^-$ into hydrated ions. This is, in general, the pattern of the process by which crystalline substances dissolve in water to form ionic solutions.

The theory of Arrhenius

Some of the early experiments on the conduction of electricity through solutions strongly suggested the presence of ions. If two metallic plates are connected by wires to form a circuit including a battery and an ammeter to show the passage of electric current, insertion of the plates into various solutions immediately discloses a striking difference. When the plates are immersed in a solution of a crystalline substance like sugar in water, there is no evidence for the passage of any appreciable amount of current through the solution. On the other hand, when the same two plates are inserted in an aqueous solution of NaCl, a measurable current flows.

Another kind of evidence for the existence of ions came from the amount by which a substance like sodium chloride lowers the freezing point of water as compared with the lowering produced by a substance like sugar. As noted in Table 11-6, the freezing point of water is depressed by 1.86°C when the concentration of sugar in the water is 1 m. The same concentration of sodium chloride lowers the freezing point by almost twice this amount. The Swedish chemist and physicist Arrhenius therefore suggested that the sugar goes into solution with one molecule of sugar surrounded by water molecules but the sodium chloride dissociates to form the two ions Na^+ and Cl^-. Thus, for 1 mole of sodium chloride dissolved, there are 2 moles of particles formed, half of them positively charged and half of them negatively charged, and the result is that the freezing point is lowered by approximately twice the amount found for the solution of sugar. Some substances caused a lowering of the freezing point of a magnitude suggesting only partial dissociation into ions; other substances caused a lowering suggesting that the mole of solute produced considerably more than twice the number of charged particles in solution, as was found with sodium chloride. Consequently, many types of measurements were made in the effort to determine the extent to which dissociation takes place and the extent of the hydration of the ions.

13-2   EXPERIMENTAL STUDIES OF IONIC SOLUTIONS

Substances which dissolve in water to form ions are called *electrolytes,* and their aqueous solutions are called *ionic* or *electrolytic* solutions.

Conductance

One of the first quantitative methods for studying the nature of electrolytes in solution was based on the measurement of the electric conductance of such solutions. The Arrhenius theory provided the first basis for the interpretation of the results.

Many years earlier there had been extensive studies

Fig. 13-5   Enthalpy relations between NaCl(s), gaseous ions, and hydrated ions. (Numerical values are only approximate.)

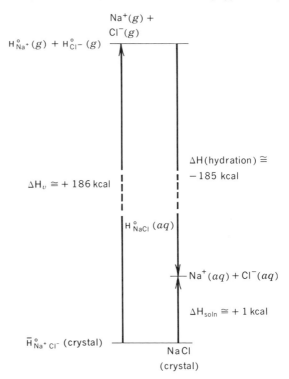

Fig. 13-6   The action of water in separating Na^+ and Cl^-.

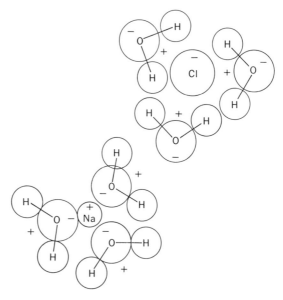

Fig. 13-7   Vessel for conduction measurement, 1 cm³ in volume with metallic electrodes on left and right faces, each 1 cm² in area.

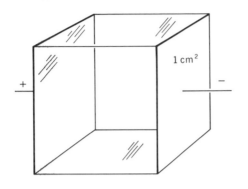

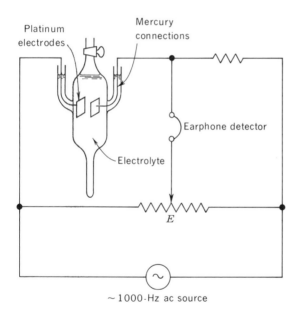

Fig. 13-8   Apparatus for measuring the conductivity of solutions.

of the electric conductance of metals. These gave rise to Ohm's law, which can be written

$$\mathcal{E} = I \times r \qquad (13\text{-}11)$$

10 volts   1 amp   10 ohms
electric   current   resistance
potential

According to this law, when a current I flows through a wire having resistance r, the voltage drop between the end of the wire where the current enters and the end of the wire where it leaves is $\mathcal{E}$. The same law can be stated in terms of electric conductance C, which is the reciprocal of the resistance r. In a number of textbooks the unit of conductance is called the *mho* (ohm spelled backward). Thus, Ohm's law can be written

$$I = \mathcal{E} \times C \qquad (13\text{-}12)$$

The conductance of 1 cm³ of solution is known as the *specific conductance* of that solution and is generally designated by the Greek letter kappa κ. If it were possible to arrange a cube of square cross section with electrodes as shown in Fig. 13-7, the specific conductance could be measured directly, since the electrode on each face would have an area of 1 cm² and the cube would contain 1 cm³ of solution. As a matter of practical convenience, it is easier to use a conductance cell of the type shown in Fig. 13-8. By measuring the conductance C of such a cell when filled with a solution of known specific conductance κ, the cell constant c_0 can be determined from the relation:

$$\kappa = c_0 C \qquad (13\text{-}13)$$

In order to relate conductance to concentration it is necessary to define the concentration unit for ionic solutions. With solutes which do not dissociate to form ions, the concentration is 1 m $(m = 1)$ when 1 mole (6.02×10^{23} molecules) of solute is present in 1 liter of solvent. For ionic solutions, the concentration is said to be 1 N $(N = 1)$ when there is 1 g $equiv$ of ions present in 1 liter of solvent: the gram equivalent is that amount of solute which when completely dissociated into ions will produce 6.02×10^{23} positive charges and a similar number of negative charges. The correspondence between gram mole and gram equivalent is illustrated with examples in Table 13-2.

The *equivalent conductance* of a solution is defined as the conductance between two electrodes one centimeter apart when current flows from one to the other through a solution containing precisely one gram equivalent of the solute. If the solution in the cell shown in Fig. 13-9 contains 1 g equivalent of solute, the conductance of this cell is the equivalent conductance of the solution. The relation between equivalent conductance Λ and specific conductance κ is:

$$\Lambda = \kappa V_e \qquad (13\text{-}14)$$

where V_e is the volume of the solution in cubic centimeters just sufficient to contain precisely 1 gram equivalent of the solute. When conduction takes place, the positive ions move toward the negatively charged electrode and the negative ions move toward the positively charged electrode (Fig. 13-10). In a very dilute solution, an ion travels without being subject to any influence from other ions that may be present. Under these conditions, the solution is said to be effectively at infinite dilution or zero concentration and its equivalent conductance Λ_0 can be expressed as the sum of the conductance of the positive ion, or *cation*, λ_0^+ and the conductance of the negative ion, or *anion*, λ_0^-:

$$\Lambda_0 = \lambda_0^+ + \lambda_0^- \qquad (13\text{-}15)$$

One of the pioneers in the measurement of conductances was the German physical chemist F. Kohlrausch, who found that the plot of λ against the square root of concentration $\sqrt{N}$ gives almost linear graphs. Some of these are shown in Fig. 13-11. By extrapolating these graphs to $N = 0$, values of Λ_0 can be found. For weak electrolytes like acetic acid, such extrapolation is almost impossible, as can be seen from the graph. Since the ions do not influence each other at sufficiently low concentration, the value of Λ_0 can be expressed as the sum of other terms

$$(\Lambda_0)_{\text{HAc}} = (\Lambda_0)_{\text{NaAc}} + (\Lambda_0)_{\text{HCl}} - (\Lambda_0)_{\text{NaCl}} \qquad (13\text{-}16)$$

where the abbreviation Ac is used for the acetate radical, $-\text{OOCCH}_3$. Since the terms on the right of this expression are all for strong electrolytes, they can be determined by the extrapolation method; their algebraic sum gives the value of the equivalent conductance for acetic acid at zero concentration.

Mobility and transference numbers

As an ion moves through the solution under the influence of an electric potential, the action is much like that when an object falls slowly through a liquid under the influence of a gravitational potential. In the latter case, a downward force acts on the mass of

TABLE 13-2   Relation between the mole and the gram equivalent[a]

	1 mole, g	1 g equiv, g
Ion:		
H_3O^+	19.02	19.02
Na^+	23.00	23.00
Ca^{++}	40.08	20.04
Al^{3+}	26.97	8.99
OH^-	17.01	17.01
Cl^-	35.46	35.46
SO_4^{--}	96.06	48.03
PO_4^{3-}	94.98	31.66
Molecule:		
HCl	36.47	36.47
Na_2SO_4	142.06	71.03
$AlCl_3$	133.35	44.45
$Al_2(SO_4)_3$	342.12	57.02

[a] After J. P. Amsden, "Physical Chemistry for Premedical Students," 2d ed. Copyright 1950. McGraw-Hill Book Company. Used by permission.

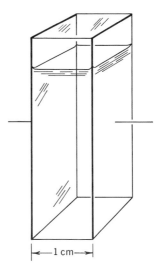

Fig. 13-9   A glass cell, with electrodes 1 cm apart, filled with an electrolytic solution containing exactly 1 equiv. The conduction is then equal to Λ.

the object because of the field of gravity. When a lead ball falls through a column of water, its speed of fall depends on the weight of the ball, its volume, and

Fig. 13-10   Schematic picture of positive ions moving toward the negative electrode while negative ions move toward the positive electrode.

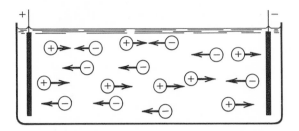

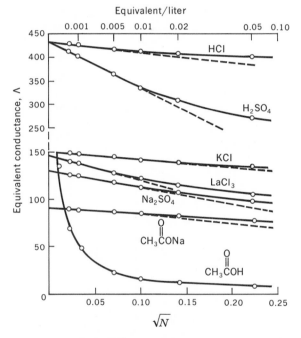

Fig. 13-11   The equivalent conductance Λ plotted against $\sqrt{N}$ for aqueous electrolytic solutions at 25°C. (After G. M. Barrow, "Physical Chemistry," 2d ed., McGraw-Hill Book Company, New York, 1966.)

the viscosity of the liquid through which it moves. In the same way, an ion moving through an electric field travels at a speed dependent on its charge since the charge determines the force acting on the ion; the speed is also dependent on the volume of the ion, on the viscosity of the solution, and on the strength of the electric field. If the lead ball has several plastic

balls fastened around it, it moves through the liquid more slowly because of the increased resistance to flow. In the same way, ions with water molecules attached to them move through the solution more slowly.

Measurements of the speed of motion of ions under different conditions provide a way of estimating the amount of hydration of the ions. The speed of motion can be measured if the ion has some characteristic property such as color or a high refractive index. As an ion cloud moves forward along a tube, there will be a change in the color or in refractive index at the forward edge of the cloud and the change in position of the edge per unit of time provides a measure of the speed. The speed of travel of the ion in centimeters per second under a potential drop of 1 volt cm^{-1} is called the mobility U. In a solution containing pairs of positive and negative ions, if the speed of travel and the concentration of one of the ions is known, it is possible to determine the fraction of the total conductance due to that ion. This fraction is called the *tranference number t*. The ionic conductance is related to the equivalent conductance by the equations

$$t_0{}^+ = \frac{U_0{}^+}{U_0{}^+ + U^-} \qquad \lambda_0{}^+ = t_0^+ \Lambda_0$$

$$\text{(13-17)}$$

$$t_0{}^- = \frac{U_0{}^-}{U_0{}^+ + U_0{}^-} \qquad \lambda_0{}^- = t_0^- \Lambda_0$$

Values of ionic conductances and mobilities (Table 13-3) provide information which is helpful in determining the degree of hydration of an ion. The hydration, in turn, affects both the enthalpy and entropy of the ion and thus plays a prominent role in determining equilibrium constants and chemical behavior.

Osmotic pressure

Quantitative measurements of the amounts of dissociation into ions in solution were first obtained through the determination of values of *osmotic pressure* in 1885 by van't Hoff. This quantity is a special kind of pressure produced in a solution which is in contact with another solution through the medium of a *semipermeable membrane,* a membrane which permits passage of (is permeable to) certain kinds of particles in the solution but does not permit the passage of other kinds of particles present. The particles may be either uncharged molecules or charged ions. A cell for the measurement of osmotic pressure is shown schematically in Fig. 13-12. The left side of the cell is filled with pure water. The water molecules are shown as small v marks. The right side of the cell is filled with a solution of electrolytes. Each ion is shown with an outer covering of water molecules attached to it by the force of attraction between the charge on the ion and the dipoles of the water molecules. Across the center of this cell is a barrier, which may be made from a porcelain disk, having channels just large enough to permit the passage of water molecules back and forth but small enough so that the ions with their coatings of attached water molecules cannot pass through.

At equilibrium the partial molal free energy of the water on the left side $\bar{G}°$ must equal the partial molal free energy of the water on the right side of the solution $\bar{G}$. In other words, equal numbers of water molecules must pass from left to right and from right to left per second. If the total pressure on the left and right were the same, $\bar{G}°$ would be greater than $\bar{G}$ since the latter is related to the former by the equation

$$\bar{G} = \bar{G}° + RT \ln X_1 \qquad (13\text{-}18)$$

where X_1 is the mole fraction of the water. Since X_1 is less than unity, the logarithm is negative, making the term on the far right correspond to a negative quantity. However, if the total pressure on the right side of the cell is increased by having the height of the column of solution on the right greater than the height of the column of pure water on the left, the rate at which the free energy of the water on the right is raised by increasing the pressure is

$$\left(\frac{\partial \bar{G}}{\partial P}\right)_T = \bar{V}_1 \qquad (13\text{-}19)$$

Integrating this equation from unit pressure to the pressure P_{eq} which establishes equilibrium and assuming that $\bar{V}_2$ is constant, we get

$$\bar{G} - \bar{G}° = \int_1^{P_{eq}} \bar{V}_1 \, dP = \bar{V}_1(P_{eq} - 1) \qquad (13\text{-}20)$$

Since the partial molal volume of the water $\bar{V}_1$ does not change appreciably with pressure, the integration can

TABLE 13-3   Properties of ions at infinite dilution (25°C)

Ion	Mobility, cm sec^{-1} (volt/cm^{-1})$^{-1}$	Equivalent conductance, mhos
H$^+$	363×10^{-5}	349.8
Li$^+$	40.1	38.7
Na$^+$	52.0	50.2
K$^+$	76.2	73.5
Mg^{++}	55	53.1
Ca^{++}	61.6	59.5
OH$^-$	205	197.6
Cl$^-$	79.0	76.3
Br$^-$	81.2	78.3
I$^-$	79.6	76.8
NO$_3^-$	74.0	71.4
SO$_4^{--}$	82.7	79.8

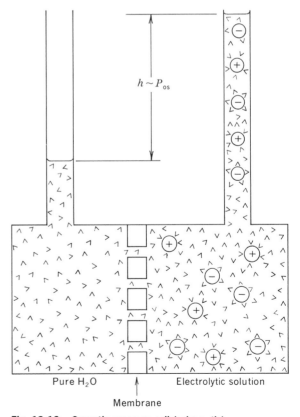

$h \sim P_{os}$

Pure H$_2$O

Electrolytic solution

Membrane

Fig. 13-12   Osmotic pressure cell (schematic).

be carried out as shown by the term on the right. Thus

$$-RT \ln X_1 = \bar{V}_1 \Delta P = P_{os}\bar{V}_1 \qquad (13\text{-}21)$$

where ΔP is the increment in pressure necessary to establish equilibrium, i.e., the osmotic pressure P_{os}.

The mole fraction of water X_1 is equal to $1 - X_2$ where X_2 is the mole fraction of the solute, and when X_2 is much less than 1, $\ln(1 - X_2) \simeq -X_2$ and $X_2 = n_2/(n_1 + n_2) \simeq n_2/n_1$; also under these conditions $\bar{V}_1 \simeq v_1$ and $n_1\bar{V}_1 \simeq V$, the total volume of the solution. When these values are substituted in Eq. (13-21), the expression is

$$P_{os}V = n_2RT \qquad (13\text{-}22)$$

Thus the osmotic pressure due to the solute is related to the total volume V of the solution containing the solute n_2, to the number of moles of the solute, and to the temperature T by an equation exactly like the ideal-gas law. In dilute solution, where the solute molecules or ions wander through the solvent like gas molecules in space, they produce the osmotic pressure almost exactly like gas molecules producing ordinary pressure. Thus it is not surprising that so many of the thermodynamic equations for solutions are almost identical with the analogous equations for gases. This is the justification for the statement made in Chap. 1 that a thorough understanding of the behavior of gases provides an essential basis for the understanding of the somewhat more complex behavior of solutions.

When the solute on the right side in Fig. 13-12 does not dissociate, the values of the osmotic pressure experimentally measured are those predicted on the basis of this theory. However, electrolytes exhibit pressures much higher than those predicted for ideal behavior, as shown by the data in Table 13-4.

As a measure of this deviation from ideal behavior, van't Hoff proposed the definition of a factor i known as the *van't Hoff factor*

$$i = \frac{\text{observed value of colligative property}}{\text{normal value of colligative property}}$$

Denoting by α the fraction of the molecules of solute which have dissociated to form ions, the undissociated fraction will be equal to $1 - \alpha$. The number of ions formed when one molecule of solute dissociates is denoted by n. Then the number of moles of solute particles after the dissociation of 1 mole will be $n\alpha + (1 - \alpha)$ or $1 + (n - 1)\alpha$. The relation between i, n, and α can be expressed

$$i = \frac{\text{no. of particles after dissociation}}{\text{no. of particles before dissociation}}$$
$$= \frac{1 + (n - 1)\alpha}{1} \qquad (13\text{-}23)$$

Rearranging this expression,

$$\alpha = \frac{i - 1}{n - 1} \qquad (13\text{-}24)$$

TABLE 13-4 Calculated and observed values of osmotic pressure at 0°C and corresponding values of the van't Hoff coefficient i

	Concentration					
	0.00 m		0.001 m		0.01 m	
	P_{os}, atm	i	P_{os}, atm	i	P_{os}, atm	i
Ideal undissociated solute	0	1	0.0224	1	0.224	1
Sugar, $C_{12}H_{22}O_{11}$	0	1	0.0224	1	0.224	1
KCl	0	2	0.0442	1.97	0.435	1.94
$BaCl_2$	0	3	0.0648	2.85	0.610	2.72

Freezing and boiling points

Many measurements have been made of the values of i calculated from measurements of freezing-point depression and boiling-point elevation produced by varying amounts of solutes in different kinds of solvents. When measurements with smaller and smaller concentrations are extrapolated to zero concentration, the value of i approaches the value of n, the number of ions formed from each solute molecule when it is completely dissociated. In Table 13-5 these latter values are listed on the lines corresponding to zero value for the concentration. In Table 13-6 values of α corresponding to observed values of i are listed for several substances.

TABLE 13-5   Values of the van't Hoff factor i calculated from freezing-point-depression data

	0.00 m	0.01 m	0.1 m	1.0 m
NaCl	2	1.94	1.87	1.81
HCl	2	1.94	1.89	2.12
$CuSO_4$	2	1.45	1.12	0.93
$MgSO_4$	2	1.53	1.21	1.09
$CaCl_2$	3	2.70	2.60	3.03
H_2SO_4	3	2.46	2.12	2.17
$Pb(NO_3)_2$	3	2.63	2.13	1.31
$K_3Fe(CN)_6$	4	3.36	2.85	
$FeCl_3$	4	3.72	3.22	4.40

Example 13-1

The specific conductance of a certain solution is 0.3568 mho. When placed in a cell, the conductance is 0.02681 mho. Calculate the cell constant c_0.

According to Eq. (13-13),

$$c_0 = \frac{\kappa}{C} = \frac{0.3588 \text{ mho}}{0.02681 \text{ mho}} = 13.38$$

Example 13-2

The equivalent conductance Λ_0 at $N = 0$ has the values

NaCl	KCl	NaAc
126.45	149.86	91.0

Calculate the value of Λ_0 for KAc.

$(\Lambda_0)_{\text{KAc}} = (\Lambda_0)_{\text{KCl}} - (\Lambda_0)_{\text{NaCl}} + (\Lambda_0)_{\text{NaAc}}$

$114.41 = 149.86 - 126.45 + 91.0$ *Ans.*

TABLE 13-6   Values of the van't Hoff factor i and the degree of dissociation α

Substance	Concentration, moles liter^{-1}	i	α
Sucrose	0.05	1.0	0.0
	0.01	1.0	0.0
	0.00	1.0	0.0
$MgSO_4$	0.05	1.3	0.3
	0.01	1.5	0.5
	0.00	2.0	1.0
NaCl	0.05	1.9	0.9
	0.01	1.9	0.9
	0.00	2.0	1.0
$K_3Fe(CN)_6$	0.05	3.1	0.7
	0.01	3.4	0.8
	0.00	4.0	1.0
$MgCl_2$	0.05	2.7	0.8
	0.01	2.9	0.9
	0.00	3.0	1.0
HCl	0.05	1.9	0.9
	0.01	1.9	0.9
	0.00	2.0	1.0
$FeCl_3$	0.05	3.4	0.8
	0.01	3.7	0.9
	0.00	4.0	1.0

Example 13-3

At a concentration of 0.04 mole liter^{-1} a solution of $MgSO_4$ has $i = 1.4$. Calculate the degree of dissociation according to the van't Hoff equation.

Since $MgSO_4$ dissociates to give Mg^{++} and SO_4^{--}, $n = 2$.

$$\alpha = \frac{i - 1}{n - 1} = \frac{0.4}{1} = 0.4 \qquad Ans.$$

Example 13-4

When $m = 0.11$, a solution of $BaCl_2$ at 0°C has an osmotic pressure of 0.671 atm. If there were no dissociation, the osmotic pressure would be 0.246. Calculate the amount of dissociation according to the van't Hoff equation.

$$i = \frac{0.669}{0.246} = 2.72$$

Since $BaCl_2 \rightarrow Ba^{++} + 2Cl^-$, $n = 3$.

$$\alpha = \frac{i - 1}{n - 1} = \frac{1.72}{2} = 0.86 \qquad Ans.$$

13-3   THE THERMODYNAMICS OF IONIC SOLUTIONS

The Ostwald dilution law

As an extension of the research of van't Hoff and Arrhenius, the German physical chemist W. F. Ostwald pointed out that the factor α could be inserted in the mass-action law and provide a theoretical relation for the variation of the degree of dissociation with concentration.

Consider the dissociation of the compound AB into two ions according to the equation

$$AB \rightleftharpoons A^+ + B^- \qquad (13\text{-}25)$$

$$m(1 - \alpha) \quad m\alpha \quad m\alpha$$

If the concentration of AB is m moles liter^{-1} before dissociation, then after the fraction α of the compound is dissociated, the concentrations of the undissociated part and of the two ions will be given by the terms appearing under the chemical symbols in Eq. (13-25). The mass-action law then takes the form

$$K = \frac{[A^+][B^-]}{[AB]} = \frac{m\alpha \times m\alpha}{m(1 - \alpha)} = \frac{m\alpha^2}{1 - \alpha} \qquad (13\text{-}26)$$

This is the Ostwald dilution law. According to the Arrhenius theory,

$$\alpha = \frac{\Lambda}{\Lambda_0} \qquad (13\text{-}27)$$

Substituting for α in Eq. (13-26) gives

$$K = \frac{\Lambda^2 m}{\Lambda_0(\Lambda_0 - \Lambda)} \qquad (13\text{-}28)$$

When measurements are made of Λ at different concentrations and values of Λ_0 obtained by extrapolation, the values of K calculated from Eq. (13-28) are constant for varying concentrations in the case of electrolytes which must be at extremely small concentrations in order to dissociate in large amounts; these are called *weak electrolytes*. On the other hand, the values of K are not constant as concentration is varied when the electrolytes are almost completely dissociated even at high concentrations; these electrolytes are called *strong electrolytes*. An example of a test of Ostwald's law is given in Table 13-7.

As more data accumulated, the defects in the Arrhenius theory became more apparent. Not only did the Ostwald dilution law break down, but the values of α for strong electrolytes obtained from conductance ratios were not in agreement with those obtained from measurements of osmotic pressure, freezing-point depression, and boiling-point elevation.

One of the strongest arguments for the Arrhenius theory was the similarity of heats of neutralization of strong acids and bases. If these substances are completely dissociated, one might expect that in the reaction

$$H^+ + OH^- \rightleftharpoons H_2O \qquad (13\text{-}29)$$

TABLE 13-7   Values of the equilibrium constant for the dissociation of acetic acid at various concentrations (25°C)[a]

m	$K_\Lambda \times 10^5$[b]	$K \times 10^5$[c]
0.0000280	1.76	1.75
0.000218	1.77	1.75
0.00241	1.79	1.75
0.0200	1.81	1.74
0.1000	1.79	1.70
0.2000	1.75	1.65

[a] Data from D. A. MacInnes and T. Shedlowsky, *J. Am. Chem. Soc.*, **54**:1429 (1932).
[b] Directly calculated from measurements of the equivalent conductance.
[c] Corrected for interionic attraction and variations in ionic mobility.

the ions H^+ and OH^- would unite with the evolution of the same amount of energy, irrespective of the parent ions to which they were attached in the undissociated state. Careful measurements of the heats of neutralization showed that there were variations which could not be accounted for on the basis of the Arrhenius theory.

Almost as soon as these discrepancies appeared, the Dutch chemist J. A. W. van Laar pointed out that one might expect the strong electrostatic forces present in an ionic solution to influence the properties of the ions. This problem was also discussed by the English physicist S. R. Milner in 1912. The first detailed analysis of the situation was made in 1923 by Debye and the German physicist E. Hückel, who started with the assumption that in strong electrolytes there is complete dissociation into ions and that the observed deviations from the Arrhenius theory are due to the electrostatic interactions of the ions.

Ionic activities

The best approach to this problem is provided by the use of the concept of activity proposed by G. N. Lewis, the quantity which has the dimensions of a concentration but which is related to the thermodynamic free energy by the simple equation

$$\bar{G} = \bar{G}_+^\circ + RT \ln a \tag{13-30}$$

In order to simplify the notation, the partial molal free-energy $\bar{G}$ frequently is denoted by the symbol μ and called the *chemical potential*. If a is expressed in molarity units, then as the concentration becomes smaller and smaller, the values of a approach the values of the concentration m. The ratio of a to m is called the activity coefficient γ.

Denoting the activity of the cation by the symbol a_+ and the activity of the anion by the symbol a_-, a quantity called the mean activity $a_\pm$ is defined by

$$(a_+ a_-)^{1/2} \equiv a_\pm \tag{13-31}$$

for an electrolyte like NaCl with singly charged ions. When the ions have multiple charges, exponents are introduced into the definition of the mean activity. For example, if the salt dissociates in solution according to the equation

$$C_{\nu_+} A_{\nu_-} \rightarrow \nu_+ C + \nu_- A \tag{13-32}$$

where ν_+ is the number of ions of plus charge and ν_- of negative charge, then the mean activity is defined by the equation

$$a_\pm \equiv (a_+^{\nu_+} a_-^{\nu_-})^{1/\nu} \tag{13-33}$$

where

$$\nu \equiv \nu_+ + \nu_- \tag{13-34}$$

For the dissociation equation

$$Al_2(SO_4)_3 \rightleftarrows 2Al^{3+} + 3SO_4^{--} \tag{13-35}$$

then

$$a_\pm \equiv (a_{Al}^2 \, a_{SO_4}^3)^{1/5} \tag{13-36}$$

Individual ionic activity coefficients are defined as

$$\gamma_+ \equiv \frac{a_+}{m_+} \qquad \gamma_- \equiv \frac{a_-}{m_-} \tag{13-37}$$

and the mean activity coefficient is defined

$$\gamma_\pm \equiv (\gamma_+^{\nu_+} \gamma_-^{\nu_-})^{1/\nu} \equiv \frac{a_\pm}{(m_+^{\nu_+} m_-^{\nu_-})^{1/\nu}}$$

$$= \frac{a_\pm}{m_\pm} \tag{13-38}$$

A comparison of measured thermodynamic quantities with the concentrations makes possible the calculation of the values of $a_\pm$ and $\gamma_\pm$. These methods are discussed in detail in works listed in the References. A quantity called the ionic strength is defined as

$$I \equiv \frac{1}{2} \sum_i m_i z_i^2 \tag{13-39}$$

where z_i is the number of charges on the ith ion. In order to define the standard state for a strong electrolyte, the assumption is made that there is complete dissociation. The thermodynamic equation of equilibrium takes the form for HCl

$$K = \frac{a_+ a_-}{a_2} \tag{13-40}$$

where K is the equilibrium constant and a_2 is the activity of the solute. At infinite dilution the activity of each ion is equal to its molality. Values of the activity and associated thermodynamic quantities for hydrochloric acid are given in Table 13-8.

TABLE 13-8   Activity in aqueous HCl at 25°C[a]

m	$\gamma_{\pm}$	$a_{\pm}$	a_{HCl}	$\bar{G}_{HCl} - \bar{G}^{\circ}_{HCl}$, cal
0.0005	0.975	0.000488	2.4×10^{-7}	-9030
0.001	0.965	0.000965	9.3×10^{-7}	-8230
0.005	0.928	0.00464	2.15×10^{-5}	-6370
0.01	0.904	0.00904	8.17×10^{-5}	-5580
0.05	0.830	0.0415	1.72×10^{-3}	-3771
0.1	0.796	0.0796	6.34×10^{-3}	-2999
0.5	0.757	0.3785	1.43×10^{-1}	-1151
1.0	0.809	0.809	6.55×10^{-1}	-251
5.0	2.38	11.19	1.42×10^{2}	$+2935$
10.0	10.44	104.4	1.09×10^{4}	$+5508$

[a] From G. N. Lewis and M. Randall, "Chemical Thermodynamics," 2d ed., rev. by K. S. Pitzer and L. Brewer.   Copyright 1961.   McGraw-Hill Book Company. Used by permission.

The Debye-Hückel theory

As the concentration of an electrolytic solute is lowered toward zero, the distribution of ions becomes completely random because they are too far apart to be subject to any electrostatic attraction or repulsion; consequently, the activity coefficient becomes unity. As concentration increases, the ions get closer together and the electrostatic force between the ions is given by Coulomb's law

$$F = \frac{e_1 e_2}{Dr^2} \tag{13-41}$$

where e_1, e_2 = charges on interacting ions
$\qquad D$ = dielectric constant
$\qquad r$ = distance between ions

Under the influence of this force, a negative ion tends to attract a cloud of positive ions around it, and vice versa.   This reduces the activity coefficient of the electrolyte; the higher the charges on the ions, the greater the effect.   The tendency to form the ionic cloud is partially offset by the tendency of the cloud to disperse because of thermal energy.   Just as a swarm of gaseous molecules placed in a vacuum immediately disperses because of the thermal motions of the molecules, the ion cloud tends to expand and disperse.   The concentration m_+ of positive ions at a distance r from another positive ion in the solution is given by

$$m_+ = me^{-e_1 \phi_r / kT} \tag{13-42}$$

where m = concentration of ions averaged throughout total solution
$\qquad e_1$ = charge on a single electron
$\qquad \phi_r$ = electric potential at a distance r from central ion
$\qquad k$ = Boltzmann constant

The concentration of the negative ions m_- at a distance r from the central ion is

$$m_- = me^{+e_1 \phi_r / kT} \tag{13-43}$$

Note that the change in concentration with distance is analogous to the change in the pressure of the earth's atmosphere with height.   Molecules at a greater height above the earth's surface have a greater *potential energy* and therefore a higher value of H because of the effect of the earth's gravitational field.   To maintain equilibrium with molecules at the earth's surface, the *free energy* G must remain constant; if H increases, Ts must also increase since $G = H - Ts$; $s \sim R \ln v \sim -R \ln P = -R \ln C$; that is, the concentration C must decrease.   In the same way the effective concentration of negative ions must decrease in moving *away* from a positive ion, while the effective concentration of positive ions must decrease in moving toward a positive ion, as in each case the value of H is increased.

For dilute solutions, the logarithm of the activity coefficient γ_i for the ion species i is given by

$$\ln \gamma_i = \frac{-e_1^3 z_i^2}{(DkT)^{3/2}} \sqrt{\frac{2\pi 6.06 \times 10^{23} I}{1,000}} \qquad (13\text{-}44)$$

where z_i is the number of charges on the ion and I is the ionic strength of the solution as defined earlier. At $298.15°K$ in water

$$\log \gamma_i = -0.509 z_i^2 \sqrt{I} \qquad (13\text{-}45)$$

$$\log \gamma_{\pm} = -0.509 z_+ z_- \sqrt{I} \qquad (13\text{-}46)$$

This provides a limiting law for the value of the activity coefficient at low concentrations much as the ideal-gas law is a limiting law for the relation between P, V, and T at low pressures. This law holds well up to an ionic strength of 0.01, but even at this concentration there may be large deviations if for any pair of oppositely charged ions the product of the charge numbers becomes greater than 4. In Fig. 13-13, the observed activity coefficients for sodium chloride and zinc sulfate are compared with the Debye-Hückel limiting law.

The enthalpy of ions

In their textbook on thermodynamics, Lewis and Randall remark: "Because of the immense confusion existing in the literature regarding various sorts of heat of solution, heat of dilution and the like, we shall enter with some minuteness into the problem of the partial molal enthalpy." In an introductory textbook of physical chemistry it is not possible to devote as much space to this important problem, but a summary of the thermodynamic approach will be helpful in understanding the nature of many of the ionic equilibria which play such a basic role in biochemistry.

In discussing the enthalpy of ions, the first step is the selection of the standard reference states. For the solvent (component 1), pure liquid water is the reference state, and its molal heat content at any temperature is denoted by H_1°. This is identically equal to $\overline{H}_1^\circ$, which denotes the partial molal enthalpy of water in any aqueous solution where the concentration of the solute approaches zero.

As a rule, the solute (component 2) does not exist as a pure liquid at temperatures of primary interest in chemistry. For this reason, the standard state for

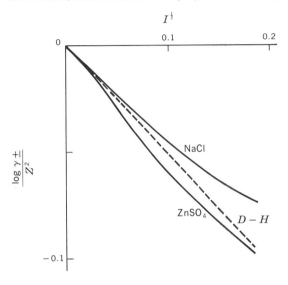

Fig. 13-13 Deviations from the Debye-Hückel limiting law (D—H) for NaCl and ZnSO$_4$. (From G. N. Lewis and M. Randall, "Chemical Thermodynamics," 2d ed., rev. by K. S. Pitzer and L. Brewer, McGraw-Hill Book Company, New York, 1961.)

the solute is the somewhat paradoxical state of infinite dilution. The thermodynamic values of the variables for this state are the partial molal quantities. The partial molal enthalpy of the solute in this standard state is denoted by $\overline{H}_2^\circ$. For many purposes it is helpful to consider relative enthalpies, which are defined for solvent and solute by the expressions

$$\overline{L}_1 = \overline{H}_1 - \overline{H}_1^\circ \qquad (13\text{-}47)$$

$$\overline{L}_2 = \overline{H}_2 - \overline{H}_2^\circ \qquad (13\text{-}48)$$

Therefore, in the standard state $\overline{L}_1 = 0$ and $\overline{L}_2 = 0$. The values of these quantities for solutions of sodium chloride are given in Fig. 13-14.

Especially with concentrated solutions, a sharp distinction must be drawn between the *total* or *integral* heat of solution and the *partial molal* (differential) heat of solution. The former is the amount of heat absorbed or evolved, measured with a calorimeter when 1 mole of solute is dissolved in pure water. The value of this quantity depends on the amount of water used, i.e., on the concentration of the resultant solution. By way of contrast, the *partial molal* heat of solution is the heat absorbed or evolved when 1

Fig. 13-14   Partial molal enthalpies of NaCl and H₂O. (From G. N. Lewis and M. Randall, "Chemical Thermodynamics," 2d ed., rev. by K. S. Pitzer and L. Brewer, McGraw-Hill Book Company, New York, 1961.)

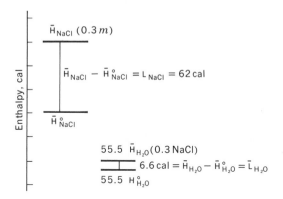

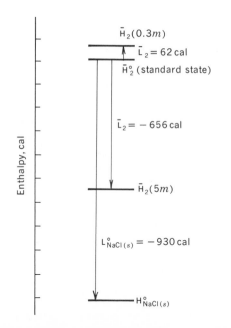

Fig. 13-15   Relative enthalpies of solution. (From G. N. Lewis and M. Randall, "Chemical Thermodynamics," 2d ed., rev. by K. S. Pitzer and L. Brewer, McGraw-Hill Book Company, New York, 1961.)

mole of solute is added to a quantity of *solution* so large that the addition of the 1 mole of solute does not appreciably change the concentration of the solution. When sulfuric acid is added to water, the total

and partial heats of solution differ by 350 cal at 0.5 m and by 2000 cal at 5.0 m.

In order to express these relations mathematically, δq denotes the amount of heat measured and is negative if heat is evolved from the system and positive if heat is absorbed by the system when the salt goes into solution. dn_2 is the amount of the solute in the crystalline state which is added, expressed in moles, so that $\delta q/dn_2$ is the partial heat of the solution of the salt per mole. The molal enthalpy of the solid salt is denoted by $H_2(s)$, so that the enthalpy of the crystalline solute added is $H_2(s)\,dn_2$. Then the relation between the partial heat of solution, the change in enthalpy, and the change in relative enthalpy can be expressed

$$\frac{\delta q}{dn_2} = \bar{H}_2 - \bar{H}_2(s) = \bar{L}_2 - L_2(s) \tag{13-49}$$

If crystalline solute is dissolved in such a large amount of pure solvent that the resulting solution has a concentration still effectively zero, then $\bar{L}_2 = 0$ and $\delta q/dn_2 = -L_2(s)$, as shown in Fig. 13-15.

When 1 mole of water is added to a fairly concentrated salt solution, appreciable quantities of heat may be either evolved or absorbed by the system. This is referred to as the *heat of solution of water*. This heat is defined by equations strictly analogous to those for the solute

$$\frac{\delta q}{dn_1} = \bar{L}_1 \tag{13-50}$$

Methods for determining partial molal enthalpies are discussed in many of the textbooks of thermodynamics listed at the end of the chapter.

Partial molal entropies

There are no experimental methods for measuring *partial molal entropies* directly, but since these quantities are related to partial molal enthalpies and partial molal free energies by the expression

$$\bar{G} = \bar{H} - T\bar{S} \tag{13-51}$$

the difference between the latter quantities divided by the absolute temperature provides values of $\bar{S}$. The partial molal free energy is frequently determined from measurements of the partial vapor pressure and also from measurements of electromotive force, which will be discussed in the next chapter.

In view of the relation

$$\bar{G} = \bar{G}° + RT \ln a \qquad (13\text{-}52)$$

the partial molal entropy can be related to the activity

$$\bar{S} - \bar{S}° = \frac{\bar{L}}{T} - R \ln a \qquad (13\text{-}53)$$

The partial molal entropy in the standard state is defined in the usual way by the expression

$$\bar{S}° \equiv \lim_{m \to 0} (\bar{S}_m + R \ln m) \qquad (13\text{-}54)$$

where $\bar{S}_m$ is the partial molal entropy of the solute at concentration m.

Partial molal free energy

The *partial molal free energy* is defined by the equation

$$\bar{G}° \equiv \lim_{m \to 0} (\bar{G}_m + RT \ln m) \qquad (13\text{-}55)$$

The most useful values are those corresponding to the partial molal free energy of the formation of the ion from its elements. It is not possible to calculate absolute values of the thermodynamic quantities for individual ions, since in any physical or chemical reaction measurable in the laboratory with current experimental techniques there are always equivalent numbers of positive and negative ions. For this reason it is customary to assign for the most common ion, H$^+$, a value of 0 to $\bar{S}°$, to $\bar{C}_p°$, and to the enthalpy of formation $\Delta \bar{H}_f°$, and the free energy of formation, $\Delta \bar{G}_f°$, from H$_2$ in the standard state. Using this convention, values of the thermodynamic data for aqueous ions at 298.15°K are listed in Table 13-9.

Example 13-5

At 298.15°K the value of $\Delta \bar{H}°$ of solution KCl is 4.11 kcal mole^{-1}, for $m = 4.82$ and $\gamma_\pm = 0.588$. The entropy of crystalline KCl is 19.76 cal deg^{-1} mole^{-1}. Calculate the combined entropy of the two ions.

$$\Delta \bar{G}° = -2RT \ln \gamma m$$
$$= -1.23 \text{ kcal mole}^{-1}$$

$$\Delta \bar{S}° = \frac{4110 + 1230}{298.15°}$$
$$= 17.9 \text{ cal deg}^{-1} \text{ mole}^{-1}$$
$$\bar{S}_+° + \bar{S}_-° = \Delta \bar{S}° + S°(s) = 19.76 + 18.0$$
$$= 37.7 \text{ cal deg}^{-1}$$

From Table 13-9

$$\bar{S}_{K^+}° + \bar{S}_{Cl^-}° = 24.5 + 13.2$$
$$= 37.7 \text{ cal deg}^{-1} \text{ mole}^{-1}$$

Example 13-6

From measurements of heat capacity it is found that $\bar{S}_{Cd}°(s) = 12.3$ and $\bar{S}_{H_2}°(g) = 31.23$ cal deg^{-1} mole^{-1} at 298.15°K. When Cd is dissolved in very dilute H$_2$SO$_4$,

$$Cd + 2H^+ \rightleftarrows Cd^{++} + H_2(g)$$
$$\Delta H = -16,700 \text{ cal}$$

and

$$\Delta G° = 18,580 \text{ cal}$$

Calculate $\bar{S}_{Cd^{++}}$.

$$\Delta S° = \frac{-16,700 + 18,580}{298.2} = 6.30 \text{ cal deg}^{-1}$$

Since $\bar{S}_{H_2} = 0$,

$$\Delta S° = \bar{S}_{Cd^{++}} + \bar{S}_{H_2} - \bar{S}_{Cd}$$
$$= \bar{S}_{Cd^{++}} + 12.3 - 31.23$$
$$\bar{S}_{Cd^{++}} = 6.30 - 18.93$$
$$= -12.6 \text{ cal deg}^{-1} \text{ mole}^{-1}$$

13-4   IONIC EQUILIBRIA

The role of H$^+$

Not only because water dissociates to form H$^+$ but because so many other compounds either dissociate to form H$^+$ or form ions which can unite with H$^+$, the hydrogen ion plays a key role in aqueous solution chemistry. The concentration of this quantity, therefore, appears in many mass-action equations for equilibrium constants and affects the free energy because of the relation

$$\Delta G = \Delta G° - RT \ln K \qquad (13\text{-}56)$$

TABLE 13-9   Thermodynamic properties of aqueous ions at 25°C[a]

Ion	$\bar{S}°$, cal deg^{-1} mole^{-1}	$\Delta \bar{H}_f°$, kcal mole^{-1}	$\Delta \bar{G}°$, kcal mole^{-1}
H^+	0	0	0
OH^-	−2.52	−54.96	−37.59
F^-	−2.3	−78.66	−66.08
Cl^-	+13.2	−40.02	−31.35
Br^-	+19.29	−28.90	−24.57
I^-	+26.14	−13.37	−12.35
S^{--}	−4.	+7.8	+20.6
HS^-	+15.0	−4.10	3.00
SO_4^{--}	+4.1	−216.90	−177.34
HSO_4^-	+30.52	−211.70	−179.94
NH_4^+	+26.97	−31.74	−19.00
NO_3^-	+35.0	−49.37	−26.43
PO_4^{3-}	−52.0	−306.9	−245.1
HPO_4^{--}	−8.6	−310.4	−261.5
$H_2PO_4^-$	+21.3	−311.3	−271.3
$HCOO^-$	+21.9	−98.0	−80.0
HCO_3^-	+22.7	−165.18	−140.31
CO_3^{--}	−12.7	−161.63	−126.22
CH_3COO^-	+20.8	−116.84	−89.02
$C_2O_4^{--}$	+10.6	−195.7	−159.4
CN^-	+28.2	+36.1	+39.6
Sn^{++}	−5.9	−2.39	−6.27
Tl^+	+30.4	+1.38	−7.75
Zn^{++}	−25.45	−36.43	−35.18
Cu^{++}	−23.6	+15.39	+15.53
Cd^{++}	−14.6	−17.30	−18.58
Ag^+	+17.67	+25.31	+18.43
Fe^{3+}	−70.1	−11.4	−2.53
Fe^{++}	−27.1	−21.0	−20.30
Al^{3+}	−74.9	−125.4	−115.
Mg^{++}	−28.2	−110.41	−108.99
Ca^{++}	−13.2	−129.77	−132.18
Sr^{++}	−9.4	−130.38	−133.2
Ba^{++}	+3.	−128.67	−134.0
Li^+	+3.4	−66.55	−70.22
Na^+	+14.4	−57.28	−62.59
K^+	+24.5	−60.04	−67.46
Rb^+	+28.7	−59.4	−67.65
Cs^+	+31.8	−62.6	−70.8

[a] From G. N. Lewis and M. Randall, "Chemical Thermodynamics," 2d ed., rev. by K. S. Pitzer and L. Brewer. Copyright 1961. McGraw-Hill Book Company. Used by permission.

The values of (H^+) commonly encountered range over magnitudes differing in ratio by 10^{15}. As a matter of convenience the Danish physical chemist S. P. Sorensen proposed a logarithmic measure of hydrogen-ion concentration denoted by the symbol pH and basically defined by the equation

$$pH \equiv -\log (H^+) \tag{13-57}$$

A more precise working definition of pH based on electromotive force measurements will be described in Chap. 14.

In discussing ionic equilibria, the basic equations are theoretically applicable only when concentration factors are expressed in terms of activities. For example, (H^+) and (OH^-) are strictly the activities of these ions, and this usage is followed in the balance of this chapter. Of course, in dilute solutions the activity coefficient approaches unity, so that equations expressed in terms of activity a frequently apply with considerable accuracy to experimental data expressed in terms of concentration $[c]$. Whenever there is need for a distinction between activities and concentrations, the former will be denoted by parentheses (H^+) and the latter by brackets $[H^+]$. Values of the equilibrium constant based on activities will be denoted by K and those based on concentration data only by K'.

Other logarithmic measures, similar to pH, are

$$pOH = -\log (OH^-) \tag{13-58}$$

$$pK = -\log K \tag{13-59}$$

For the reaction involving the dissociation of water into ions

$$H_2O \rightleftharpoons H^+ + OH^- \tag{13-60}$$

the mass-action expression is

$$K_w = (H^+)(OH^-) \tag{13-61}$$
$$1.0 \times 10^{-14} = (10^{-7})(10^{-7})$$

where the term (H_2O) normally occurring in the denominator on the right-hand side is set equal to unity and omitted since it is essentially a constant in almost all situations of interest. In the p notation the equation can be written

$$pK_w = pH + pOH \tag{13-62}$$
$$14 = 7 + 7$$

Again, because the concentrations of interest nearly all involve negative exponents, the negative sign was introduced into the definitions in Eqs. (13-57) to (13-59) to avoid continually writing negative signs in the p notation.

Solubility product

For a crystalline electrolyte like $CaCO_3$ in equilibrium with its saturated solution, the equation can be written

$$CaCO_3(s) \rightleftharpoons Ca^{++} + CO_3^{--} \tag{13-63}$$

and the mass-action expression is

$$K = \frac{(Ca^{++})(CO_3^{--})}{(CaCO_3)(s)} \tag{13-64}$$

Because the term in the denominator is a constant, it is equal to unity and the expression is written

$$K_{sp} = (Ca^{++})(CO_3^{--}) \tag{13-65}$$

The equilibrium constant has the subscript sp denoting solubility product, a term implying that at any given temperature for a saturated solution in equilibrium with the crystalline state, the product of the concentration of the cation by the concentration of the anion will be a constant if the solution is ideal. For a number of relatively insoluble electrolytes, this condition is fulfilled. Values of K_{sp} at 25°C for a number of electrolytes are given in Table 13-10.

The solubility product is related to the free-energy change by the equation

$$\Delta G° = -RT \ln K_{sp} \tag{13-66}$$

The variation of K_{sp} with temperature is discussed in Chap. 14 and illustrated in Table 13-11.

Example 13-7

In a solution of $CaCO_3$ at 15°C in the form of calcite, the value of (Ca^{++}) is raised to 10^{-2} m by the addition of $CaCl_2$. To what value will (CO_3^{--}) be reduced?

$$K_{sp} = 0.99 \times 10^{-8} = (10^{-2})(CO_3^{--})$$
$$(CO_3^{--}) = 0.99 \times 10^{-6} \ m \qquad Ans.$$

Example 13-8

In a solution at 18°C in equilibrium with crystalline BaF_2, the value of $Ba^{++} = 0.042$ m and the value of $F^- = 0.020$ m. Calculate the value of K_{sp}.

$$K_{sp} = (Ba^{++})(F^-)^2 = (0.042)(0.020^2)$$
$$= 1.68 \times 10^{-5} \ m^2 \qquad Ans.$$

TABLE 13-10   Values of K_{sp} at 25.00°C

AgCl	1.62×10^{-10}
Al(OH)$_3$	3.7×10^{-15}
BaCO$_3$	8.1×10^{-9}
BaIO$_3$	6.5×10^{-10}
BaSO$_4$	1.08×10^{-10}
CaCO$_3$ (calcite)	8.7×10^{-9}
CaC$_2$O$_4$	2.50×10^{-9}
CuC$_2$O$_4$	2.87×10^{-8}
FeC$_2$O$_4$	2.1×10^{-7}
LiCO$_3$	1.7×10^{-8}
SrCO$_3$	1.6×10^{-9}

TABLE 13-11   Values of K_{sp} at various temperatures

	t, °C	K_{sp}
Al(OH)$_3$	18	1.1×10^{-15}
	25	3.7×10^{-15}
BaF$_2$	10.0	1.60×10^{-5}
	18.0	1.70×10^{-5}
	25.0	1.73×10^{-5}
BaSO$_4$	15	8.2×10^{-11}
	25	1.08×10^{-10}
	50	1.98×10^{-10}

Example 13-9

For Ag$_2$(C$_2$O$_4$) the value of $K_{sp} = 1.3 \times 10^{-11}$. Calculate the concentration of the ion C$_2$O$_4^{--}$ in a solution in equilibrium with crystalline Ag$_2$(C$_2$O$_4$) when the concentration of Ag$^+$ is 1.00×10^{-3} m.

The formula for the solubility product is

$$K_{sp} = (Ag^+)^2(C_2O_4^{--}) = 1.3 \times 10^{-11}$$

The formula for the concentration of C$_2$O$_4^{--}$ is

$$(C_2O_4^{--}) = \frac{1.3 \times 10^{-11}}{(1.00 \times 10^{-3})^2} = 1.3 \times 10^{-5}\ m$$

Ans.

Acids and bases

The dissociation of an acid (HA) can be written

$$HA = H^+ + A^- \tag{13-67}$$

The mass-action expression is:

$$K_a = \frac{(H^+)(A^-)}{(HA)} \tag{13-68}$$

The subscript a is attached to the equilibrium constant to denote that it applies to this particular type of dissociation. An acid like H$_3$PO$_4$ can undergo stepwise dissociation; each dissociation has its own equilibrium constant. A number of dissociation constants are listed in Table 13-12. Acids where $K_a > 1$ are called *strong acids,* and those for which $K_a < 1$ are called *weak acids.*

Similar relations apply to bases

$$BOH = B^+ + OH^- \tag{13-69}$$

$$K_b = \frac{(B^+)(OH^-)}{(BOH)} \tag{13-70}$$

Some values of K_b are listed in Table 13-13. For ideal solutions, values of K_a and K_b vary with temperature in exactly the same way as K_{sp}.

Indicators

Because the value of the concentration of hydrogen ion plays such an important role in the chemistry of so many solutions, special methods have been devised for its quick determination. One of the most important of these is the employment of substances called *indicators.* In general, an indicator is a substance which in its undissociated form has one color and in its dissociated form as an ion has another color. The general expression for the dissociation of an indicator HInd may be written

$$HInd = H^+ + Ind^- \tag{13-71}$$

The equilibrium expression for this is

$$K_{Ind} = \frac{(H^+)(Ind^-)}{(HInd)} \tag{13-72}$$

If the hydrogen-ion concentration is high, the concentration of the indicator ion is decreased and the

TABLE 13-12   Values of pK_a' at 25°C for some acids in aqueous solution

pK_a'	Acid	pK_a'	Acid
1.3	HOOCCOOH	5.7	$HOOC(CH_2)_2COO^-$
1.6	H_5IO_6	6.4	H_2CO_3
1.8	H_3PO_3	6.4	$Citric^{--}$
1.9	H_2SO_3	6.5	$HCrO_4$
1.9	HSO_4^-	6.7	$H_2PO_3^-$
2.0	$HClO_2$	7.0	$H_2As_5O_4^-$
2.0	H_3PO_2	7.0	H_2S
2.1	H_3PO_4	7.1	$H_2N_2O_2$
2.2	$H_4P_2O_6$	7.1	$H_2SO_3^-$
2.3	$H_3As_5O_4$	7.2	$HClO$
2.4	Cyanobutyric	7.2	H_2PO_4
2.5	H_2SeO_3	7.3	$H_2P_2O_6^{--}$
2.5	Quinolinic	8.0	$N_2H_3^+$
2.5	Cyanoacetic	8.4	HIO_6^-
2.7	Bromoacetic	8.4	HSO_4^-
2.8	$H_3P_2O_6^-$	8.7	$HBrO$
2.8	$CH_2(COOH)_2$	9.2	H_3AsO_3
2.9	$ClCH_2COOH$	9.2	H_3BO_3
3.1	$HOC(CH_2)_2(COOH)_3$	9.2	NH_4^+
3.2	HF	9.3	HCN
3.3	HNO_2	9.9	$(CH_3)_3NH^+$
3.7	HCOOH	10.0	HIO
3.7	HOCN	10.0	C_6H_5OH
3.9	$CH_3CHOHCOOH$	10.0	$HP_2O_6^{3-}$
4.2	$(CH_2)_2(COOH)_2$	10.3	HCO_3^-
4.3	$^-OOCCOOH$	10.7	$CH_3NH_3^+$
4.3	$(CH_2)_3(COOH)_2$	11.0	$H_2N_2O_2^-$
4.7	CH_3COOH	11.0	$(CH_3)_2NH_2^+$
4.7	$HOOC(CH_2)_5COOH$	11.0	HSe^-
4.8	$CH_3(CH_2)_2COOH$	11.0	HTe^-
4.8	$Citric^-$	11.5	$H_2A_5O_4^{--}$
4.9	CH_3CH_2COOH	11.8	H_2O_2
5.1	Malic	11.8	HPO_4^{--}
5.3	$HOOC(CH_2)_3COO^-$	12.9	HS^-
5.7	$HOOC(CH_2)COO^-$	15.0	$H_3IO_6^{--}$

indicator is shifted from the ionic form to the undissociated form. Suppose that the ionic form is colored yellow and that the undissociated form is colored red; this is true of the indicator called *methyl red*. At the point where the majority of the molecules change from the ionic form to the undissociated form there is a sharp and distinct change in color. As an example, at pH < 4.2 in a solution of HCl, the presence of the hydrogen ion forces the methyl red into the undissociated form. As a base like NaOH is added, the value of pH changes sharply when the acid is neutralized and free base is present. This is shown in Fig. 13-16. Methyl red indicator changes color and is yellow when pH > 6.3.

In the same way when acid is added to a basic solution, the ratio of the undissociated to the dissociated form of the indicator changes sharply at the point where (H^+) equals the value of K_{Ind}

TABLE 13-13   Values of pK_b' and pK_a' of bases at 25°C in aqueous solution

pK_b'	Base	pK_a'
2.9	$(C_2H_5)_2NH$	11.1
3.3	$(C_2H_5)_2NH_2$	10.7
3.3	$(CH_3)(CH_2)_2NH_2$	10.7
3.4	$Al(OH)_3$	9.3
4.7	Pyridine	5.2
8.8	$C_6H_5NH_2$	4.6

$$\frac{K_{\text{Ind}}}{(H^+)} = \frac{(\text{Ind}^-)}{(\text{Ind})} \tag{13-73}$$

This equation is obtained from Eq. (13-72) merely by shifting the terms (H^+) from the right side to the left side of the equation. When $(H^+) = K_{\text{Ind}}$, then $(\text{Ind}^-) = (\text{Ind})$; the indicator is half in the ionic form and half in the undissociated form. When the hydrogen-ion concentration rises beyond this point, the quantity on the right-hand side becomes a smaller and smaller fraction; the majority of the indicator molecules have shifted to the undissociated form, Ind. When the value of (H^+) equals that of K_{Ind}, there is a sharp change in the color of the indicator. The value of pH corresponding to the sharp change of color for a number of indicators is given in Table 13-14.

Example 13-10

A certain solution has a hydrogen-ion concentration of 4.00×10^{-3} m. Calculate the value of pH for this solution.

The formula relating pH to (H^+) is

$$\begin{aligned} pH &= -\log(H^+) = -\log(4 \times 10^{-3}) \\ &= -\log 4 - \log(10^{-3}) \\ &= -0.602 + 3 = +2.398 \qquad \textit{Ans.} \end{aligned}$$

Example 13-11

For the indicator thymol blue, the value of pH is 2.0 when half the indicator is in the un-ionized form. Calculate the percentage of the indicator in this form in the solution specified in Example 13-10.

The equation relating the pH of the solution to the pH at the midpoint of the indicator and the fraction in the undissociated state is

$$pH = -\log K_{\text{eq}} - \log\frac{(\text{HInd})}{(\text{Ind}^-)}$$

$$= pH_{\text{mdpt}} - \log\frac{(\text{HInd})}{(\text{Ind}^-)}$$

$$\log\frac{(\text{HInd})}{(\text{Ind}^-)} = pH_{\text{mdpt}} - pH = 2.0 - 2.398$$

$$= -0.398$$

$$\log\frac{\text{Ind}^-}{\text{HInd}} = 0.398 \qquad \frac{\text{Ind}^-}{\text{HInd}} = 2.5 = \frac{1 - x_{\text{HInd}}}{x_{\text{HInd}}}$$

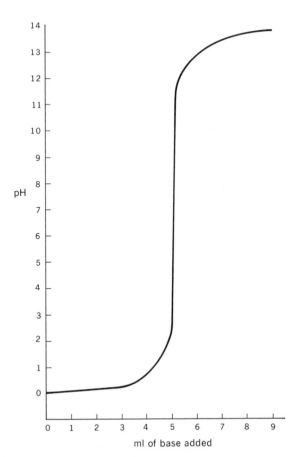

Fig. 13-16   Titration of HCl against a basic solution.

where

$$\text{Fraction of HInd} = x_{\text{HInd}} = \frac{1}{3.5} = 0.29$$

Ans.

Ampholytes

Many of the molecules in the various body fluids can dissociate to produce either hydrogen ions or hydroxyl ions. Molecules of this type are called *ampholytes*. The amino acids are among the most important ampholytes in biochemistry. One of the simplest examples is glycine, which has the formula in the neutral state H_2NCH_2COOH and dissociates to form hydrogen ions according to the equation

$$H_2NCH_2COOH = H_2NCH_2COO^- + H^+ \tag{13-74}$$

The equilibrium expression for this reaction is

$$K_a = \frac{(H^+)(H_2NCH_2COO^-)}{(H_2NCH_2COOH)} \tag{13-75}$$

The equation for the basic reaction of glycine is

$$H_2NCH_2COOH + H^+ = {}^+H_3NCH_2COOH \tag{13-76}$$

The equilibrium expression for this reaction is

$$K_b = \frac{({}^+H_3NCH_2COOH)}{(H^+)(H_2NCH_2COOH)} \tag{13-77}$$

The ionized forms of glycine can change to a doubly ionized form by the reaction

$$\begin{aligned}{}^+H_3NCH_2COOH &= {}^+H_3NCH_2COO^- + H^+ \\ H_2NCH_2COO^- + H^+ &= {}^+H_3NCH_2COO^-\end{aligned} \tag{13-78}$$

with the corresponding equilibrium expression

$$K_{ba} = \frac{({}^+H_3NCH_2COO^-)(H^+)}{({}^+H_3NCH_2COOH)} \tag{13-79}$$

$$K_{ab} = \frac{({}^+H_3NCH_2COO^-)}{(H_2NCH_2COO^-)(H^+)} \tag{13-80}$$

Such a compound can become doubly ionized by directly transferring a hydrogen ion from one end to the other by the reaction:

$$H_2NCH_2COOH = {}^+H_3NCH_2COO^- \tag{13-81}$$

The equilibrium expression for this reaction is

TABLE 13-14   pH values for some indicators at the midpoint HInd/Ind$^-$ of color change (25°C)

Substance	pH	Color change with increasing acidity
Thymol blue	2.0	Red → yellow
2,4-Dinitrophenol	3.5	Colorless → yellow
Bromcresol green	4.8	Yellow → blue
Methyl red	5.3	Red → yellow
Phenol red	7.6	Yellow → red
Phenolphthalein	9.2	Colorless → red
Nitramine	11.9	Colorless → orange

$$K_d = \frac{K_{ba}}{K_b} = \frac{K_{ab}}{K_a} = \frac{({}^+H_3NCH_2COO^-)}{(H_2NCH_2COOH)} \tag{13-82}$$

The neutral molecule of glycine is relatively insoluble in water. The charged water molecules tend to orient with respect to each other and to be attracted to each other by dipole attraction and hydrogen bonding, so that the attraction of the water molecules for each other is far stronger than the attraction of the water molecules toward neutral glycine. On the other hand, if the glycine molecule has a positive charge at one end and a negative charge at the other, making it a doubly charged ion, the water molecules tend to cluster around each end, turning their positive charges toward the negative end of the glycine and their negative charges toward the positive end of the glycine. This reduces the free energy and makes the glycine relatively soluble. All the experimental evidence points to the fact that in aqueous solution glycine is present largely as the doubly charged ion.

Buffers

When a solution contains a weak acid and a corresponding anion or a weak base and the corresponding cation, one can add more of an acid or a base and find that the value of pH is not changed nearly as much as if the weak acid or the weak base were not present. Because weak acids and weak bases reduce the change otherwise expected, they are called *buffers*. To show this buffering action, consider what happens when two drops of 1 *m* NaOH solution are added to 1 liter of a solution of HCl containing 10^{-5} mole of solute. The two drops of sodium hydroxide solution contain about 1.1×10^{-4} mole of solute. The reaction can be written

$$\text{NaOH} + \text{HCl} = \text{NaCl} + \text{H}_2\text{O} \qquad (13\text{-}83)$$

$\quad$ 10^{-5} mole $\quad$ 10^{-5} mole $\qquad$ 10^{-5} mole

This neutralizes all the acid and leaves behind 1.0×10^{-4} mole of NaOH, which shifts the pH from the original value of 5.0 up to 10.0.

Now consider what happens if the same amount of sodium hydroxide solution (1.1×10^{-4} mole) is added to a solution which contains 1.8000 moles of sodium acetate and 1.0000 mole of acetic acid in 1 liter of water. The value of K_a for acetic acid is 1.8×10^{-5}. The equilibrium expression for the relation between the hydrogen-ion concentration, the acid-ion concentration, and the undissociated acetic acid is

$$K_a = \frac{(\text{H}^+)(\text{Ac}^-)}{(\text{HAc})} = \frac{(10^{-5})(1.8)}{(1)} = 1.8 \times 10^{-5}$$

$$(13\text{-}84)$$

The presence of the sodium acetate prevents the acetic acid from dissociating to any appreciable extent. The sodium acetate being essentially completely dissociated provides a concentration of Ac^- of 1.8 moles liter^{-1}. Thus, as the numerical values in Eq. (13-84) show, the concentration of hydrogen ion must be 10^{-5} mole liter^{-1}, giving a value of pH of 5.0. The 1.1×10^{-4} mole of sodium hydroxide added reacts with the acetic acid to form sodium acetate

0 00011 mole NaOH + 0.00011 mole HAc

$\qquad$ = 0.00011 mole NaAc + H_2O $\quad$ (13-85)

The equilibrium expression (13-84) now becomes

$$K_a = \frac{(\text{H}^+)(\text{Ac}^-)}{(\text{HAc})} = \frac{(10^{-5})(1.80011)}{(0.99989)} = 1.8 \times 10^{-5}$$

$$(13\text{-}86)$$

The concentration of the hydrogen ion is scarcely changed at all, and the value of pH remains essentially constant.

To summarize, the addition of the two drops of sodium hydroxide solution without the buffer change the concentration of H^+ by a factor of 100,000; with

the buffer, the concentration is not changed appreciably.

The importance of buffering action in biochemistry will be discussed in later sections of this chapter.

Example 13-12

A solution at 25°C is made up of 1 mole of NaAc and 1 mole of HAc dissolved in 1 liter of H_2O. To this is added 0.02 mole of NaOH. Calculate the value of pH before and after the NaOH is added.

The formula expressing the equilibrium before the addition of NaOH is

$$K_a = \frac{(\text{H}^+)(\text{Ac}^-)}{(\text{HAc})} = 1.8 \times 10^{-5}$$

$$(\text{H}^+) = \frac{(1.8 \times 10^{-5})(1)}{(1)}$$

The value for pH before the addition of the NaOH is

$$\text{pH}_{\text{init}} = -\log(1.8) - \log 10^{-5}$$
$$= 5 - 0.255 = 4.745$$

After the addition of the NaOH, the equilibrium expression is

$$K_a = \frac{(\text{H}^+)(1.02)}{(0.98)} = 1.8 \times 10^{-5}$$

$$(\text{H}^+) = 1.73 \times 10^{-5} \ m$$

Thus, the final value of pH is

$$\text{pH}_{\text{final}} = -\log 1.73 - \log 10^{-5}$$
$$= 5 - 0.238 = 4.762 \qquad Ans.$$

Hydrolysis

When the salt of a strong base and a weak acid like sodium acetate is dissolved in water, there is a tendency for some of the acetate ions to unite with the water, forming undissociated acetic acid according to the equation

$$\text{NaAc} + \text{HOH} = \text{Na}^+ + \text{HAc} + \text{OH}^- \qquad (13\text{-}87)$$

The reaction can be written more simply as the reaction of the acetate ion with water after the former has been formed through the dissolving of the NaAc

$$\text{Ac}^- + \text{HOH} = \text{HAc} + \text{OH}^- \qquad (13\text{-}88)$$

The equilibrium expression for this is

$$K_{hydr} = \frac{(HAc)(OH^-)}{(Ac^-)} \tag{13-89}$$

The subscript hydr is added to the equilibrium constant because this process is called *hydrolysis*. The expression in (13-89) can be rearranged to give

$$K_{hydr} = \frac{(HAc)(OH^-)}{(Ac^-)} = \frac{(OH^-)}{(Ac^-)/(HAc)}$$

$$= \frac{(H^+)(OH^-)}{(H^+)(Ac^-)/(HAc)}$$

$$= \frac{K_w}{K_a} \tag{13-90}$$

In a similar manner it can be shown that the hydrolysis of a weak base takes place according to the same pattern. For the base, K_{hydr} is equal to K_w/K_b.

Complex ions

The evidence from heats of hydration points to the fact that in aqueous solution most ions are bound very tightly to the surrounding water molecules. The strength of this binding is so great that one is almost justified in regarding the ion and the water as forming an actual compound. When the binding is strong and the geometry of the situation favors a distinct integral number of water molecules, these "compounds" may be written with definite chemical formulas. For example, there is evidence that the silver ion over a large range of concentration in solution forms the compound $Ag(H_2O)_2^+$. When NH_3 is added to such a solution of a silver ion, the NH_3 can replace the H_2O, and an ion with the formula $Ag(NH_3)_2^+$ can be formed.

Such an ion is called a *complex ion*. There are many examples in the literature of such ions: $Cu(NH_3)(H_2O)_3^{++}, Cu(NH_3)_2(H_2O)_2^{++}, Cu(NH_3)_3(H_2O)^{++}, Cu(NH_3)_4^{++}, Hg(NH_3)_2^{++}$. Compounds of the type just listed decompose at low concentrations of NH_3, so that a chemical equation can be written and an expression for the equilibrium constant can be formulated. Examples are

I. $Ag^+ + NH_3 = Ag(NH_3)^+$ (13-91)

$$K_I = \frac{(Ag(NH_3)^+)}{(Ag^+)(NH_3)} \tag{13-92}$$

II. $Ag(NH_3)^+ + NH_3 = Ag(NH_3)_2^+$ (13-93)

$$K_{II} = \frac{(Ag(NH_3)_2^+)}{(Ag(NH_3)^+)(NH_3)} \tag{13-94}$$

III. $Ag^+ + 2NH_3 = Ag(NH_3)_2^+$ (13-95)

$$K_{III} = K_I K_{II} = \frac{(Ag(NH_3)_2^+)}{(Ag^+)(NH_3)^2} \tag{13-96}$$

13-5   THE IONIC THERMODYNAMICS OF BLOOD

There are endless examples of the importance of ions in the bloodstream; thus, the ion HCO_3^+ acts as a buffer in maintaining a constant pH in the bloodstream. Again, not only must the right ions be present in the blood, but also if a solution is injected into the blood, it is essential that the concentration of ions in the solution be such that the injection does not disturb bloodstream concentrations.

Sodium ions have an effect on the heart which can stop its beating. This can be counteracted by the presence of calcium ions; but these, in turn, produce an overcontraction of the heart muscle, and the effect of the calcium ions can be counteracted to a certain extent by potassium ions. Other ions of physiological importance are those of copper, iron, magnesium, manganese, and the halogens as well as phosphate ions.

Particularly in mammals, maintenance of life depends on the delivery of oxygen to the various parts of the living tissue and the removal of the carbon dioxide which is one of the important products of the combustion reactions due to the oxygen. Oxygen is introduced by way of the lungs, and the rate of respiration is, in turn, controlled by the concentration of CO_2, O_2, and H^+ in the bloodstream. Since respiration itself has an influence on these quantities, this is an example of chemical cybernetic action where factor A influences factor B, factor B influences factor C, and factor C, in turn, influences factor A. The details of these reactions will be discussed in Chap. 27. The oxygen in the blood is transported largely by the formation of a chemical compound between the oxygen and the hemoglobin (Hb) in the blood. The increasing concentration of CO_2 in the blood decreases the tendency of O_2 to combine with the hemoglobin.

In all such actions as these, the various thermodynamic factors ultimately determine the direction and the extent of the reactions. The enthalpy factor favors the union of the oxygen with the hemoglobin because the compound formation lowers the enthalpy of the system; this action is opposed by the entropy factor. The changing concentration of O_2 in the blood changes the entropy. When the concentration of oxygen in the blood falls below a certain level, the entropy increase on dissociation of the Hb—O bond is so great as to outweigh the enthalpy effect.

The hydration of the ions with the water molecules in the bloodstream acts in the direction of providing a free-energy decrease to make the ions more soluble. This is particularly important in the case of the more complex ions like the amino acids, which would be relatively insoluble in the blood if the clustering of the water molecules around the charged parts of the ions did not produce an increased enthalpy factor ΔH to increase solubility. On the other hand, the orientation of the water molecules in the blood due to the presence of the ion reduces the entropy ΔS, and this has an effect $T \Delta S$ opposing that of the energy factor.

In addition, a number of interaction factors like the buffering action and the solubility-product action also affect the concentration of the various ions in the bloodstream. When ions leave the blood to become incorporated into more oriented aggregates like the ribosomes in the individual cells or even the cytoplasmic fluid within the cell, these thermodynamic factors are still at work. The probability factor helps transport ions from a region of higher concentration to a region of lower concentration or even remove ions where it is necessary to overcome an energy of attachment of these ions.

Body acid-base balance

By and large, the biochemical reactions which constitute the metabolic processes in the body tend to contribute large amounts of H^+ ions to the body fluids. The fluids present outside of the body cells generally have a value of pH in the range between 7.35 and 7.5. By way of contrast, there are some types of cell where the fluid within the cell may have a pH as high as 8.0 or a value considerably lower than 7.0. Values as low as 5.0 have been reported in the prostate gland. It is clear that many acids must be transported from the point where they are formed in the various body cells to the places where they are excreted from the body.

The respiratory mechanism participates to a high degree in the regulation of the *acid-base balance*. The CO_2 can diffuse readily from the blood through the membranes in the lungs and into the air. A slight increase in the hydrogen-ion concentration in the blood stimulates the removal of excess CO_2 from the extracellular fluids to the expired air. A decrease in the concentration of CO_2 or in the hydrogen-ion concentration in the blood acts in the opposite direction and cuts down the rate at which CO_2 is removed. The excretion of various substances in the urine also tends to regulate the acid-base balance. The normal value of pH in the urine is about 6.0. The difference between this and the pH of 7.4 in the plasma represents acid which is removed by the renal action. Another mechanism which maintains the acid-base balance is the exchange of H^+ and K^+ ions across the membranes surrounding the cells in the body. This will be discussed further in Chap. 23.

Cytoplasm

In the matrix fluid of the cytoplasm there is every reason to believe that there will be, first of all, an increased orientation due to the hydrogen bonding between the water molecules. With the presence in this fluid of many complex ions with multiple charges, the possibilities for an even greater degree of order certainly exist. All these factors may affect the entropy of the various substances that are present in the matrix fluid of the cytoplasm. The more these ions are ordered through such influences, the lower will be the value of ΔS when they move toward the various sites in the cell where they ultimately join still more ordered aggregates of atoms as part of the process of growth. Thus, ordering in the cytoplasm increases the free-energy lowering, which promotes growth.

With regard to the energy factors of the reactions taking place in the matrix fluid of the cytoplasm, one can conclude that there will be influences at work similar to those found in the solution of ions in familiar aqueous solutions. There is every reason to believe that the water molecules in the matrix fluid of the cytoplasm cluster around charges present, whether they are on monatomic ions like K^+ or on the charged groups like $-NH_3^+$ or $-CO_2^-$. The amino acids and the proteins are present as amphoteric ions in the matrix fluid of the cytoplasm. The energy factor reduces the value of ΔH when the molecule changes from the neutral form to the ionized form because of the clustering of the H_2O dipoles around the charge sites. It is therefore interesting to ask whether such an amino acid or molecule in leaving the matrix fluid of the cytoplasm and joining a more ordered group of molecules in the more solid structure of the cell wall or in one of the other more organized parts of the cell will undergo an increase or decrease in the enthalpy. If the amino acid were linked to the other molecules in the cell wall only by van der Waals forces, one would expect an enthalpy increase in such a process. In fact, thinking of the values of the enthalpy of hydration, one might expect that the enthalpy factor would oppose such a passage from the matrix fluid of the cytoplasm to the cell wall. One would also expect the entropy to favor the transport of molecules from the more ordered state in the walls or in the mitochondria to the less ordered state in the matrix fluid of the cytoplasm, as discussed in more detail in Sec. 14-5.

In light of this, it seems clear that there must be more than mere van der Waals forces operating to bring the molecules out of the matrix fluid of the cytoplasm and into the more ordered aggregates during the process of growth. Of course, when molecules join to form a macromolecule like DNA, chemical bonds are formed along the length of the molecule and the two strands of the molecule are joined by relatively strong hydrogen bonds. This tends to create an energy factor that might move the molecules from the matrix fluid of the cytoplasm to the DNA with a tendency sufficiently great to overcome the opposing tendency of the entropy factor. At the present time, the data are insufficient to permit the calculation either of the changes in enthalpy or the changes in entropy in any precise way. At best, we can only survey the biochemical thermodynamics of the cell in a very broad perspective. Nevertheless, the principles of thermodynamics can provide impor-

tant outlines that enable us to see the tapestry of biochemical reactions in the cell with a growing clarity and in a perspective from which a clearer understanding of the interrelationships may soon emerge.

Problems

1. At 25°C, a 0.001 N solution of KCl has a specific conductance κ of 1.47×10^{-4}. Calculate the equivalent conductance at this concentration.
2. At 25°C, a 0.005 N solution of KCl has an equivalent conductance Λ of 143 mhos. Calculate the specific conductance at this temperature.
3. At 25°C, the mobility in centimeters per second has been reported to be 7.62×10^{-4} for K^+ and 7.9×10^{-4} for Cl^- when these ions are in solutions effectively at infinite dilution. Calculate the transport numbers for these ions if the equivalent conductance under these conditions for K^+ is 73.5 mhos; calculate the equivalent conductance for Cl^- and for the solution of KCl in water at infinite dilution.
4. At 25°C and infinite dilution, the equivalent conductance for $(NH_4)_2SO_4$ is 227.3 mhos in aqueous solution. The respective mobilities for the ions are 7.6×10^{-4} and 8.3×10^{-4} cm sec^{-1}. Calculate the equivalent conductances for each ion.
5. The freezing-point-depression constant for water is 1.86 deg mole^{-1} liter^{-1}. At a concentration of 0.01 m, $Pb(NO_3)_2$ depresses the freezing point by 4.90×10^{-2} deg. Calculate the van't Hoff i factor for this substance at this concentration.
6. A 0.001 N solution of HCl freezes at -0.0037°C. Calculate the value of the van't Hoff factor i.
7. Calculate the value of the degree of dissociation α of the substances for which the van't Hoff i factor was calculated in Probs. 5 and 6.
8. Measurements of conductance of aqueous solutions of HCl indicate that the value of the equivalent conductance in a solution of 0.001 m is 421.36 mhos and at infinite dilution is 426.16 mhos. Assuming that the dissociation constant α is equal to the ratio of the equivalent conductance at a concentration to the equivalent conductance in an infinitely dilute solution, calculate the value of α and compare with the value calculated in Prob. 7.
9. For BaF_2 at 18°C, the value of K_{sp} is 1.7×10^{-6}. If the concentration of F^- is 2.3×10^{-3} m, what will be the concentration of the Ba^{++} ions when this solution is in equilibrium with crystalline BaF_2 at this temperature?

10. For a solution of $CaSO_4$ at $10°C$, it is found that the concentration of the Ca^{++} ion is $2 \times 10^{-2}\ m$ and the concentration of the SO_4^{--} ion is $9.7 \times 10^{-3}\ m$ in a solution that is in equilibrium with crystalline calcium sulfate. Calculate the value of K_{sp} for this solution.

11. The normal value of pH for blood is 7.4. Calculate the concentration of H^+ in moles per liter in blood.

12. The value of pH in urine can vary from 4.5 to 8.2. Calculate the number of moles of H^+ secreted in 1 liter of urine at the extreme limit of alkalinity and at the extreme limit of acidity.

13. For nitrous acid, HNO_2, at $25°C$ the value of K_a is 4×10^{-4}. If this substance is present in a solution where pH = 4.0, calculate the fraction of the HNO_2 molecules that will dissociate into ions.

14. For a solution of ammonium hydroxide, NH_4OH, the value of K_b is 1.8×10^{-5} at $25°C$. If the value of pH is 4.5 in the solution where ammonium hydroxide is present, calculate the concentration of NH_4^+ ions in this solution.

15. The indicator phenolphthalein is approximately half dissociated into the ionic form when the value of pH is 9.0. Calculate the percentage in the ionic form at pH of 9.5.

16. The indicator phenol red is half in the ionic form when the value of the pH is 7.2. If the ratio of the undissociated form to the ionic form is 1.5, calculate the value of pH.

17. Calculate the value of the partial molal entropy of NaBr in aqueous solution in the standard state.

18. Calculate the value of the partial molal entropy in the standard state for $Al_2(SO_4)_3$.

19. Calculate the value of $\Delta \bar{G}_f^\circ$ and $\Delta \bar{H}_f^\circ$ for $Al_2(SO_4)_3$.

20. Calculate the average value of the heat of solution of $BaSO_4$ to form Ba^{++} and SO_4^{--} in the standard state over the range 18 to $50°C$.

REFERENCES

Gurney, R. W.: "Ionic Processes in Solution," McGraw-Hill Book Company, New York, 1953.

Fuoss, R. M.: "Electrolytic Conductance," Interscience Publishers, Inc., New York, 1959.

Hamer, W. H. (ed.): "The Structure of Electrolytic Solutions," John Wiley & Sons, Inc., New York, 1959.

Davis, K. S., and J. A. Day: "Water: The Mirror of Science," Doubleday Anchor Books, New York, 1961.

Lewis, G. N., and M. Randall: "Thermodynamics," 2d ed., rev. by K. S. Pitzer and L. Brewer, McGraw-Hill Book Company, New York, 1961.

Monk, C. B.: "Electrolytic Dissociation," Academic Press, Inc., New York, 1961.

Vanderwerf, C. A.: "Acids, Bases and the Chemistry of the Covalent Bond," Reinhold Publishing Corporation, New York, 1961.

Cantarow, A., and B. Schepartz: "Biochemistry," 3d ed., W. B. Saunders Company, Philadelphia, 1962.

Christensen, H. N.: "Body Fluids and the Acid-Base Balance," W. B. Saunders Company, Philadelphia, 1964.

Scheraga, H. A.: Behavior of Ions in Relation to Biochemical and Biophysical Systems, *Ann. N.Y. Acad. Sci.*, **125** (2):249–772 (1965).

Conway B. E.: Electrolyte Solutions: Solvation and Structural Aspects, *Ann. Rev. Phys. Chem.*, **17**:481 (1966).

Dreisbach, D.: "Liquids and Solutions," Houghton Mifflin Company, Boston, 1966.

Kirkwood, J. G.: "Theory of Solutions," Gordon and Breach, Science Publishers, Inc., New York, 1968.

Eisenberg, D., and W. Kauzmann: "The Structure and Properties of Water," Oxford University Press, New York, 1969.

In electrolytic solutions many reactions take place in which electrons are added to or subtracted from the solutes. For example, when an electron is added, a ferric ion, Fe^{3+}, is changed to a ferrous ion, Fe^{++}. This reaction can be written $Fe^{3+} + e^- \rightarrow Fe^{++}$. When it proceeds from left to right, it is regarded as a *reduction* reaction by analogy with the change from ferric oxide, Fe_2O_3, to ferrous oxide, FeO, where the latter is in a less oxidized, or more reduced, state.

In many oxidation or reduction reactions the tendency of atoms or molecules in solution to gain or lose electrons can be measured directly with the

Chapter Fourteen

electrochemistry

help of electric circuits where contacts are made with the solutions through metal bars or wires, called electrodes, which are inserted directly into the reacting solutions and provide *sources* or *outlets* into which electrons flow from the reacting species. The devices in which such measurements are made are called *electromotive force,* emf, *cells.*

The presence of both positive and negative ions homogeneously distributed in the solutions in emf cells results in a kind of paradoxical state of matter which permits many kinds of chemical reactions that are practically impossible between un-ionized molecules. When any appreciable numbers of charged molecules are formed, all having the same sign of charge, large electrostatic forces of repulsion immediately come into action and prevent the formation of any further charged molecules of this sign being formed. Thus chemical reactions involving the presence of a single kind of charge do not take place in significant amounts. On the other hand, when both positively and negatively charged species are present and uniformly mixed together, there are no large electrostatic forces due to clusters of ions all having the same sign. Yet individual ions can move freely through the clusters of homogeneously mixed positive and negative ions; they can either give up their charges or add more; by exchanging electrons syntheses can be carried out, energy can be transferred, molecular structure

Fig. 14-1   The Daniell emf cell.

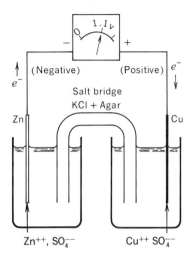

can be rearranged, and many other kinds of chemical action can be achieved that are impossible in a less flexible nonionic environment.

The tendency of a chemical reaction in an emf cell to proceed depends upon the free-energy change of the reaction. The electric potential difference between the electrodes of the cell is directly proportional to this free-energy change. Thus, the study of reactions by electrochemical methods provides valuable information from which the basic thermodynamic data for chemical reactions can be obtained.

14-1   ELECTROMOTIVE FORCE CELLS

There are many kinds of emf cells, a number of which will be discussed in this chapter. One of the simpler types, called a *Daniell cell,* is shown schematically in Fig. 14-1. On the left is a beaker containing an aqueous solution of $ZnSO_4$, for which $a = 1\ m$. Into this is inserted a zinc metal bar, which serves as an electrode. On the right is a similar beaker filled with an aqueous solution of $CuSO_4$, also at unit activity, and containing a copper metal electrode. An inverted U tube connects the two beakers; this tube, filled with an aqueous jelly of agar containing potassium chloride, is called a *salt bridge.* The whole apparatus is at a temperature of 25°C.

Electrically conducting wires connect the two electrodes to a voltmeter. In precision measurements a more sensitive potential measuring device called a *potentiometer* is used; this will be described in a later section.

When the electric circuit is completed, the voltmeter indicates a voltage of $+1.10$ volts and a small electric current flows. According to the conventions of electrical engineering, this *current* comes out of the electrode on the right (called the cathode) and flows through the wires and the voltmeter from right to left, entering the apparatus again through the electrode on the left (called the anode). The negatively charged *electrons* flow in the *opposite* direction as shown in the figure. The flow of ions in the solutions completes the circuit. On the right copper ions move to the cathode, remove electrons, and are deposited as metallic copper. Potassium ions flow from left to right and chloride ions from right to left through the salt bridge. Atoms of metallic zinc pass from the electrode on the left into the solution as Zn^{++} ions, leaving behind electrons which flow up the electrode and into the wire to the voltmeter. Electrons also flow down the wire, into the right-hand electrode and neutralize the Cu^{++} ions, which deposit on the electrode as metallic copper.

The chemical reaction which takes place in the right half of the cell can be written

$$Cu^{++}(aq) + 2e^- \rightarrow Cu(s) \tag{14-1}$$

In the left half the reaction is

$$Zn(s) \rightarrow Zn^{++}(aq) + 2e^- \tag{14-2}$$

Added together, these expressions give the total cell reaction

$$Cu^{++}(aq) + Zn(s) \rightarrow Cu(s) + Zn^{++}(aq) \tag{14-3}$$

The reason the reactions proceed in this direction, and not in the reverse direction, can be seen by considering the nature of the process in each half of the cell. As shown schematically in Fig. 14-2, an atom of zinc on the surface of the zinc electrode on the left side of the cell is in contact with water molecules. There is a tendency for this zinc atom to leave two electrons in the electrode and move into the solution as a Zn^{++} ion, about which water molecules cluster, oriented in a way that reduces the potential energy due to the electric forces. This results in a certain "pressure" of electrons on the electrode, a pressure represented in the diagram by the length of the arrow X. At the right side there is a similar pressure of electrons on the copper electrode, a pressure which is represented by the length of the arrow Y. Since the pressure X on the left is greater than the pressure Y on the right, there is a net pressure in the circuit represented by the length of the arrow $X - Y$ which causes the electrons to flow through the wire from left to right. The measure of this net pressure is the potential difference between the two electrodes ($+1.10$ volts). As a result $Zn(s)$ atoms move from the anode and enter the solution as Zn^{++} ions, while Cu^{++} ions move out of the solution and deposit as $Cu(s)$ on the cathode. Thus there is a *reduction* taking place at the right electrode and an *oxidation* at the left electrode.

The measurement of cell potential

The flow of electrons is measured in units of charge per second. The unit of charge is the *coulomb;* and when one coulomb per second passes through the circuit, the current is one *ampere*. If a length of wire has an electric resistance of one ohm, a current of one ampere flowing through the wire produces an electrical potential of one *volt* between the ends of the wire.

The potential of an emf cell is denoted by $\mathcal{E}$; it is defined as the voltage measured between the electrodes when no current is flowing. A schematic diagram of the circuit for measuring potential is shown in Fig. 14-3. A battery causes a current of known value to flow through a calibrated resistance R, called a potentiometer. The emf cell is connected in series with a galvanometer; one end of the cell-galvanometer circuit is connected to the left side of the calibrated resistance and the other end to an arm which can make contact at calibrated points along the resistance R. The potential from the emf cell tends to make a current flow in the direction indicated by the arrow C on the diagram through the cell and through the galvanometer; but the potential from the battery tends to make the current flow through the galvanometer and emf cell as shown by the arrow B in the

Fig. 14-2 Daniell cell showing relative electron "pressure" of Zn^{++} ions X and Cu^{++} ions Y and net pressure in external circuit $X—Y$.

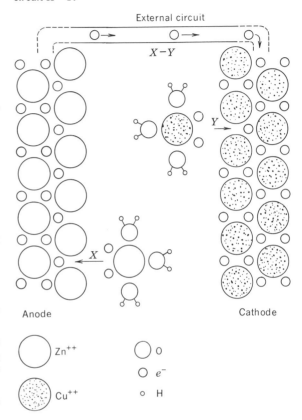

Fig. 14-3 A potentiometer circuit (simplified) for measuring the potential of an emf cell by balancing its potential against that of a working battery through a galvanometer G.

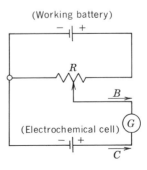

opposite direction. The position of the contact arm is adjusted until the potential from it is balanced as closely as possible against the potential from the emf cell, so that effectively no current flows through the cell. From the calibration of the resistance, the potential $\mathcal{E}$ is determined.

In practice, the potentiometer is far more complicated than shown in the diagram; the most accurate potentiometers can measure voltages with a precision of 10^{-7} volt. Usually such accuracy is not required. The potential of the cell depends not only on the kinds of metals in the electrodes and the nature and concentration of the solutions but also on such factors as strains in the metals of the electrode and the manner of establishing contact between the two solutions. Two different cells carefully constructed to be as similar to each other as possible may show differences in potential of 10^{-2} volt or more. The reproducibility of the results determines the amount of precision which will be of significance in the electrical measurement.

14-2   CELL REACTIONS

The changes in the enthalpy ΔH, entropy Δs, and free energy ΔG in the Daniell cell can be expressed in the usual form, where the values correspond to the amount of change which results when 1 atom of zinc forms 1 mole of zinc ions and 1 mole of copper ions deposits out in the form of 1 g atom of copper:

$$-(H_{Zn(s)} + \overline{H}_{Cu^{++}}) + (\overline{H}_{Zn^{++}} + H_{Cu(s)}) = \Delta H \quad (14\text{-}4)$$

$$-(s_{Zn(s)} + \overline{S}_{Cu^{++}}) + (\overline{S}_{Zn^{++}} + s_{Cu(s)}) = \Delta s \quad (14\text{-}5)$$

$$-(G_{Zn(s)} + \overline{G}_{Cu^{++}}) + (\overline{G}_{Zn^{++}} + G_{Cu(s)}) = \Delta G \quad (14\text{-}6)$$

reactants   products

The faraday

A special unit is used to measure the amount of electric charge which flows through the wire from the right electrode to the left electrode when 1 *g equiv* of ions is formed at the surface of the left electrode and 1 *g equiv* of ions leaves the solution and deposits as metal on the surface of the right electrode. This is called the *faraday* $\mathcal{F}$ in honor of the British physical chemist who did so much of the pioneering research

in electrochemistry. $\mathcal{F}$ is equal to 9.64870×10^4 absolute coulombs per equivalent. On both the zinc ion and the copper ion the number of charges n is equal to 2; thus, when the reaction proceeds to the extent where 1 atom of zinc goes into solution in the form of ions and 1 g atom of copper is deposited out from ions in solution, the amount of charge flowing through the electrical circuit is 2 $\mathcal{F}$. In general, when 1 g atom of ions is formed, the amount of charge flowing is $n\mathcal{F}$.

Electrical work and free energy

If a wire of low electric resistance is used to connect the left electrode to the right electrode directly, the reaction takes place in an *irreversible* manner much like that of a gas expanding into a vacuum. The potential of the cell is only partly opposed by the drop in potential due to the resistance of the connecting wire, so that the reaction proceeds with relative rapidity and ions are formed and move into the solution much like the molecules of a gas diffusing irreversibly into an an evacuated space.

The cell reaction can be made to take place *reversibly* by balancing the cell potential against another source of potential which is only infinitesimally less than the cell potential. As the reaction proceeds reversibly, the amount of heat entering or leaving the cell ΔH can be measured. As the reaction forms 1 mole of ions, an amount of charge $n\mathcal{F}$ is moved against the balancing potential $\mathcal{E}_b$ which differs only infinitesimally from the cell potential; thus, the amount of work performed by the cell is $n\mathcal{F}\mathcal{E}$; this amount of energy *leaves* the system.

In order to relate this electrical work to the free energy, we apply the first and second laws of thermodynamics to the cell reaction. The first law is stated

$$dE = DQ + DW \quad (14\text{-}7)$$

The cell differs from the typical thermodynamic systems considered in earlier chapters because it has *two* channels by which energy can enter or leave it in the form of work DW. First, there is the familiar *expansion* channel, where the work performed against external pressure P when the volume of the solution changes by the amount dV is $P\, dV$. In addition to this, the presence of the electrodes provides an *electrical-work* channel; as stated above, when electrons are moved through an electric potential, the work performed is the product of the amount of charge multiplied by the potential through which the charge is moved. For an infinitesimal amount of reaction in the cell, this work can be expressed as $\mathscr{E}\,dQ$ where dQ is the infinitesimal amount of charge moved. Thus, the total amount of energy leaving the system DW through the work channel is

$$DW = -P\,dV - \mathscr{E}\,dQ \qquad (14\text{-}8)$$

The second law of thermodynamics can be expressed in the form

$$T\,dS = DQ \qquad (14\text{-}9)$$

when the reaction proceeds reversibly. Equations (14-7) to (14-9) can be combined to give

$$dE = T\,dS - P\,dV - \mathscr{E}\,dQ \qquad (14\text{-}10)$$

which can be rearranged as

$$-\mathscr{E}\,dQ = dE + P\,dV - T\,dS \qquad (14\text{-}11)$$

The free energy per mole is defined by the equation

$$\mathrm{G} \equiv \mathrm{E} + P\mathrm{v} - T\mathrm{s} \qquad (14\text{-}12)$$

Consequently, the total differential dG is given by

$$dG = dE + P\,dV + V\,dP - T\,dS - S\,dT \qquad (14\text{-}13)$$

Almost all cell reactions take place under conditions of constant pressure ($dP = 0$) and constant temperature ($dT = 0$), so that Eq. (14-13) becomes

$$dG = dE + P\,dV - T\,dS \qquad (14\text{-}14)$$

Comparing Eqs. (14-14) and (14-11), we see that

$$dG = -\mathscr{E}\,dQ \qquad (14\text{-}15)$$

When the reaction changes 1 g atom of zinc into 1 g atom of zinc ions and 1 g atom of copper ions into 1 g atom of copper, as shown in Eq. (14-3), the expression for the change in free energy is

$$\Delta G = -\mathscr{E}\,\Delta Q = -n\mathscr{F}\mathscr{E} \qquad (14\text{-}16)$$

change in   electrical
free energy   work

Thus, the change in free energy is equal to the negative of the electrical work performed, and a measurement of the cell potential $\mathscr{E}$ provides a direct measure of the free-energy change of the reaction. Of course, to be numerically equal, the change in free energy and the electrical work must be expressed in the same units. As written, the electrical work is in the units of volt-equivalent-faradays. One of these units is equal to 2.3060×10^4 cal.

As will be discussed in Sec. 14-4, the value of $\mathscr{E}$ depends on the concentrations of the ions. When the concentrations correspond to unit activity, the reaction takes place in the standard state and the equation relating free energy and electromotive force is written

$$\Delta\mathrm{G}^\circ = -n \times \mathscr{F} \times \mathscr{E}^\circ$$

−50.7 kcal mole⁻¹   −2 equiv mole⁻¹   23.060 cal volt⁻¹ equiv⁻¹   1.10 volt

$$(14\text{-}17)$$

Since the value of $\mathscr{E}^\circ$ for the Daniell cell at 25°C is $+1.10$ volt, the measurement of $\mathscr{E}^\circ$ makes possible the calculation of the value of $\Delta\mathrm{G}^\circ$, as shown above. From the calorimetric measurement of the heat given out when this reaction takes place, $\Delta\mathrm{H}^\circ = -51.8$ kcal. From the relation

$$\Delta\mathrm{G}^\circ = \Delta\mathrm{H}^\circ - T \times \Delta S^\circ$$

−50.7 kcal mole⁻¹   −51.8 kcal mole⁻¹   298.15°   −3.66 cal deg⁻¹ mole⁻¹

$$(14\text{-}18)$$

the value of ΔS° can be calculated, as indicated above.

As shown at the beginning of this section, the cell reaction can be thought of as taking place in two parts: the reaction at the surface of the zinc electrode and the reaction at the surface of the copper electrode. In each of these reactions there is a contribution to the free energy which, in turn, is made up of two parts: (1) the change in enthalpy and (2) the temperature multiplied by the change in entropy.

In order to remove a zinc atom from its position in the crystal lattice in the metallic electrode, a great deal of energy must be introduced into the system. However, when this zinc ion goes into the water, the water molecules surrounding it are oriented and attracted to it by the force between the charge on the ion and the dipoles in the water molecules. In this process of hydration, a great deal of energy leaves the system. The process of hydration also reduces the entropy by partially orienting the H_2O molecules in a more regular order; but there is an increase in entropy because the zinc ions in solution have far less order than in the crystal lattice. Opposite kinds of changes in enthalpy and entropy take place at the copper electrode. From the complex balance of all these changes results the net free-energy change. As shown in Eq. (14-17) for the case of the Daniell cell operating under standard conditions at 25°C, this change is -50.7 kcal mole^{-1} and tends to make the reaction in Eq. (14-3) proceed from left to right. The nature of these factors affecting the free energy and the cell potential will be examined in more detail after some of the more technical features of cells have been described.

14-3   CELL CONVENTIONS AND TECHNIQUES

The hydrogen electrode

One of the most useful types of electrode for studying ionic solutions is made of platinum over which hydrogen gas bubbles at a pressure of 1 atm. This is called the standard *hydrogen electrode*. In Fig. 14-4 such an electrode is shown coupled with a copper electrode of the type used in the Daniell cell. The chemical reaction at the hydrogen electrode is

$$H_2(g, 1 \text{ atm}) \rightarrow 2H^+(aq, a = 1\ m) + 2e^- \qquad (14\text{-}19)$$

The reaction at the copper electrode is

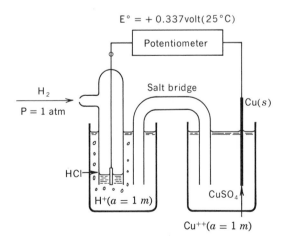

$E° = +0.337 \text{volt}(25°C)$

Fig. 14-4 $Cu^{++}(a = 1\ m)/Cu(s)$ **electrode coupled to a standard hydrogen half-cell.**

$$Cu^{++}(aq, a = 1\ m) + 2e^- \rightarrow Cu(s) \qquad (14\text{-}20)$$

Adding these two equations together gives

$$H_2(g, 1 \text{ atm}) + Cu^{++}(aq, a = 1\ m) \rightarrow$$
$$2H^+(aq, a = 1\ m) + Cu(s) \quad (14\text{-}21)$$

The value of $\mathcal{E}°$ is 0.34 volt.

Half-cells

Since the cell reaction consists of two parts, the cell may be regarded as composed of two *half-cells;* but although the potential of the whole cell can be measured with relative ease, it is never possible to measure the potential of a single half-cell. Suppose that the right half of the cell in Fig. 14-4 were removed. In order to measure the potential of the remaining part of the cell, some electrical connection would have to be made to the electrolyte of the left half-cell. Any wire introduced here would in itself constitute an electrode and therefore form a second half-cell.

Although there is no possibility of measuring the potential of a single half-cell experimentally, half-cell potentials are so useful in making calculations that a convention has been adopted setting the potential of the hydrogen half-cell equal to zero when it is operated in the standard state with hydrogen ion at unity activity and the pressure of the hydrogen gas at 1 atm. When the copper half-cell is coupled with the hydrogen half-cell at 25°C, the total potential is found to be $+0.34$ volt. Therefore, this is taken as the potential of the copper half-cell, and this relation is expressed

$$\mathcal{E}°_{\text{tot}} \quad = \quad \mathcal{E}°_{\text{Cu}} \quad - \quad \mathcal{E}°_{\text{H}} \qquad (14\text{-}22)$$
$$\text{+0.34 volt} \quad \text{+0.34 volt} \quad \text{0 volt}$$

Standard reduction potentials

The potentials of hundreds of half-cells have been measured and tabulated. The classic work on inorganic emf cells is Latimer's,[1] and that on organic and biochemical emf cells is Clark's (see References). A number of the values of $\mathcal{E}°$ for half-cells with reduction reactions are listed in Tables 14-1 and 14-2.

[1] This book was given the title "Oxidation Potentials" as Latimer wrote half-cell reactions as oxidation reactions rather than as reduction reactions; the difference in convention is discussed further in the following section.

TABLE 14-1   Values of standard reduction potentials $\mathcal{E}°$ for half-cells at 25°C[a]

Based on convention Pt/H$_2$(1 atm), H$^+$($a = 1\ m$)//M$^+$/M(s) gives tabulated $\mathcal{E}°$[a]

Half-cell reaction	Electrode	$\mathcal{E}°$, volts
$F_2(g) + 2e^- = 2F^-$	$F^-\mid F_2(g)$, Pt	+2.866
$O_3 + 2H^+ + 2e^- = H_2O + O_2$	$H^+\mid O_3, O_2$, Pt	+2.07
$Ag^{++} + e^- = Ag^+$	$Ag^{++}, Ag^+\mid$ Pt	+1.98
$PbO_2 + SO_4^{--} + 4H^+ + 2e^- =$ $PbSO_4 + 2H_2O$	$H^+, SO_4^{--}\mid PbO_2, PbSO_4$, Pb	+1.685
$Ce^{4+} + e^- = Ce^{3+}$	$Ce^{4+}, Ce^{3+}\mid$ Pt	+1.61
$Cl_2 + 2e^- = 2Cl^-$	$Cl^-, Cl_2\mid$ Pt	+1.3595
$Tl^{3+} + 2e^- = Tl^+$	$Tl^{3+}, Tl^+\mid$ Pt	+1.25
$4H^+ + O_2 + 4e^- = 2H_2O$	$H^+\mid O_2$, Pt	+1.229
$Br_2(l) + 2e^- = 2Br^-$	$Br^-\mid Br_2(l)$, Pt	+1.0652
$V(OH)_4^+ + 2H^+ + e^- = VO^{++} + 3H_2O$	$H^+, V(OH)_4^+, VO_2^{++}\mid$ Pt	+1.00
$AuCl_4^- + 3e^- = Au + 4Cl^-$	$AuCl_4^-, Cl^-\mid$ Au	+1.00
$2Hg^{++} + 2e^- = Hg_2^{++}$	$Hg^{++}\mid$ Hg	+0.906
$Ag^+ + e^- = Ag$	$Ag^+\mid$ Ag	+0.7991
$Hg_2^{++} + 2e^- = 2Hg$	$Hg_2^{++}\mid$ Hg	+0.792
$Fe^{3+} + e^- = Fe^{++}$	$Fe^{3+}, Fe^{++}\mid$ Pt	+0.771
$C_6H_4O_2 + 2H^+ + 2e^- = C_6H_4(OH)_2$	$C_6H_4O_2, C_6H_4(OH)_2, H^+\mid$ Pt	+0.699
$HgSO_4 + 2e^- = 2Hg + SO_4^{--}$	$SO_4^{--}\mid HgSO_4, Hg(l)$	+0.6153
$I_3^- + 2e^- = 3I^-$	$I_3^-, I^-\mid I_2$	+0.536
$MB + 2H^+ + 2e^- = MBH_2$	$MBH_2, MB, H^+\mid$ Pt	+0.53
$Cu^+ + e^- = Cu$	$Cu^+\mid$ Cu	+0.521
$O_2 + 2H_2O + 4e^- = 4OH^-$	$OH^-\mid O_2$, Pt	+0.4011
$VO^{++} + 2H^+ + e^- = V^{3+} + H_2O$	$H^+, VO^{++}, V^{3+}\mid$ Pt	+0.361
$Fe(CN)_6^{3-} + e^- = Fe(CN)_6^{4-}$	$Fe(CN)_6^{3-}, Fe(CN)_6^{4-}\mid$ Pt	+0.36
$Cu^{++} + 2e^- = Cu$	$Cu^{++}\mid$ Cu	+0.337
$UO_2^{++} + 4H^+ + 2e^- = U^{4+} + 2H_2O$	$U^{4+}, UO_2^{++}, H^+\mid$ Pt	+0.334
$Hg_2Cl_2 + 2e^- = 2Hg + 2Cl^-$	$Hg_2Cl_2, Cl^-\mid Hg(l)$	+0.2678
$AgCl + e^- = Ag + Cl^-$	$Cl^-\mid$ AgCl, Ag	+0.2224
$Sn^{4+} + 2e^- = Sn^{++}$	$Sn^{4+}, Sn^{++}\mid$ Pt	+0.154
$Cu^{++} + e^- = Cu^+$	$Cu^{++}, Cu^+\mid$ Pt	+0.153
$CuCl + e^- = Cu + Cl^-$	$Cl^-\mid$ CuCl, Cu	+0.137
$AgBr + e^- = Ag + Br^-$	$Br^-\mid$ AgBr, Ag	+0.0713
$2H^+ + 2e^- = H_2$	$H^+\mid H_2$, Pt	0.0000
$Hg_2I_2 + 2e^- = 2Hg + 2I^-$	$I^-\mid Hg_2I_2, Hg(l)$	−0.0405
$Pb^{++} + 2e^- = Pb$	$Pb^{++}\mid$ Pb	−0.126
$Sn^{++} + 2e^- = Sn$	$Sn^{++}\mid$ Sn	−0.136
$AgI + e^- = Ag + I^-$	$I^-\mid$ AgI, Pt	−0.1518
$CuI + e^- = Cu + I^-$	$I^-\mid$ CuI, Cu	−0.185
$Ni^{++} + e^- - Ni$	$Ni^{++}\mid$ Ni	−0.250
$V^{3+} + e^- = V^{++}$	$V^{3+}, V^{++}\mid$ Pt	−0.255
$Co^{++} + 2e^- = Co$	$Co^{++}\mid$ Co	−0.277
$Tl^+ + e^- = Tl$	$Tl^+\mid$ Tl	−0.336
$PbSO_4 + 2e^- = SO_4^{--} + Pb$	$SO_4^{--}\mid PbSO_4$, Pb	−0.359
$Ti^{3+} + e^- = Ti^{++}$	$Ti^{++}, Ti^{3+}\mid$ Pt	−0.369
$Cd^{++} + 2e^- = Cd$	$Cd^{++}\mid$ Cd	−0.403

TABLE 14-1   Values of standard reduction potentials $\mathscr{E}°$ for half-cells at 25°C[a] (Continued)

Based on convention $Pt/H_2(1 \text{ atm})$, $H^+(a = 1\ m)//M^+/M(s)$ gives tabulated $\mathscr{E}°$[a]

Half-cell reaction	Electrode	$\mathscr{E}°$, volts
$Cr^{3+} + e^- = Cr^{++}$	$Cr^{++}, Cr^{3+} \mid Pt$	−0.41
$Eu^{3+} + e^- = Eu^{++}$	$Eu^{++}, Eu^{3+} \mid Pt$	−0.43
$Fe^{++} + 2e^- = Fe$	$Fe^{++} \mid Fe$	−0.440
$U^{4+} + e^- = U^{3+}$	$U^{3+}, U^{4+} \mid Pt$	−0.61
$Zn^{++} + 2e^- = Zn$	$Zn^{++} \mid Zn$	−0.763
$Mn^{++} + 2e^- = Mn$	$Mn^{++} \mid Mn$	−1.180
$Al^{3+} + 3e^- = Al$	$Al^{3+} \mid Al$	−1.662
$U^{3+} + 3e^- = U$	$U^{3+} \mid U$	−1.80
$Th^{4+} + 4e^- = Th$	$Th^{4+} \mid Th$	−1.90
$Mg^{++} + 2e^- = Mg$	$Mg^{++} \mid Mg$	−2.363
$Na^+ + e^- = Na$	$Na^+ \mid Na$	−2.714
$Ca^{++} + 2e^- = Ca$	$Ca^{++} \mid Ca$	−2.866
$Sr^{++} + 2e^- = Sr$	$Sr^{++} \mid Sr$	−2.888
$Ba^{++} + 2e^- = Ba$	$Ba^{++} \mid Ba$	−2.905
$K^+ + e^- = K$	$K^+ \mid K$	−2.925
$Li^+ + e^- = Li$	$Li^+ \mid Li$	−3.045
$H_2O^- + H_2O + 2e^- = 3OH^-$	$OH^-, H_2O^- \mid Pt$	+0.88
$O_2 + 2H_2O + 4e^- = 4OH^-$	$OH^- \mid O_2, Pt$	−0.401
$S^{--} = S + 2e^-$	$S^{--} \mid S, Pt$	−0.447
$PbCO_3 + 2e^- = Ni + 2OH^-$	$CO_3^{--} \mid PbCO_3, Pb$	−0.506
$Ni(OH)_2 + 2e^- = Ni + 2OH^-$	$OH^- \mid Ni(OH)_2, Ni$	−0.72
$2H_2O + 2e^- = H_2 + 2OH^-$	$OH^- \mid H_2, Pt$	−0.828
$SO_4^{--} + H_2O + 2e^- = SO_3^{--} + 2OH^-$	$OH^-, SO_4^{--}, SO_3^{--} \mid Pt$	−0.93
$ZnO_2^{--} + 2H_2O + 2e^- = Zn + 4OH^-$	$OH^-, ZnO_2^{--} \mid Zn$	−1.216
$HPO_3^{--} + 2e^- = H_2PO_2^- + 3OH^-$	$OH^-, HPO_3^{--}, H_2PO_2^- \mid Pt$	−1.57
$Ca(OH)_2 + 2e^- = 2OH^- + Ca$	$OH^- \mid Ca(OH)_2, Ca \mid Pt$	−3.03

[a] Data from W. M. Latimer, "Oxidation Potentials," 2d ed., Prentice-Hall, Inc., Englewood Cliffs, N.J., 1952, and A. J. de Bethune, T. S. Licht, and N. Swedeman, *J. Electrochem. Soc.*, **106**:616 (1959).

Cell conventions

In order to specify unambiguously the nature and properties of cells, a number of conventions are widely used in the chemical literature. These are summarized in the following rules:

1. In order to designate the composition of an emf cell, an expression is written in a special form to show the nature of the electrodes, the electrolytes, and the means (such as a salt bridge) by which they are joined. For example, the Daniell cell in Fig. 14-1 is denoted by

$$Zn(s) \mid Zn^{++}(aq, a = 1\ m) \mid KCl(aq) \mid$$
$$Cu^{++}(aq, a = 1\ m) \mid Cu(s) \quad (14\text{-}23)$$

The symbol appearing at the left, $Zn(s)$, shows that the electrode on the left is made of solid copper. The bar indicates a change of phase. The next symbol, reading from left to right, shows that the active ion in this electrolytic aqueous solution is Zn^{++} at an activity of $1\ m$. The ion not entering into the chemical reaction, in this case SO_4^{--}, is not shown. The symbol $\mid KCl(aq) \mid$ denotes a salt bridge with

TABLE 14-2   Standard reduction potentials at 25°C of half-cells of biochemical interest where $pH = 7$

Half-cell reaction	$\mathscr{E}_0'$, volts
Ferredoxin (ox) $\rightleftarrows$ ferredoxin (red)	-0.432[a]
Nicotinamide-adenine-dinucleotide phosphate$^+$ $\rightleftarrows$ nicotinamide-adenine-dinucleotide phosphate (red)	-0.324[a]
Nicotinamide-adenine-dinucleotide$^+$ $\rightleftarrows$ nicotinamide-adenine-dinucleotide (red)	-0.320[b]
$CH_3COCH_3 \rightleftarrows CH_3CHOHCH_3$	-0.296[a]
1,3-diphosphoglycene^{4-} $\rightleftarrows$ glyceraldehyde 3-phosphate $+ HPO_4^{--}$	-0.286[a]
$CH_3CHO \rightleftarrows CH_3CH_2OH$	-0.197[a]
Dihydraoxyacetone phosphate^{--} $\rightleftarrows$ L-glycerol 1-phosphate^{--}	-0.192[a]
Pyruvate $\rightleftarrows$ L-lactate$^-$	-0.190[a]
Oxalacetate^{--} $\rightleftarrows$ L-malate^{--}	-0.166[a]
Flavoprotein (ox) $\rightleftarrows$ flavoprotein (red)	-0.060[a]
Coenzyme Q (ox) $\rightleftarrows$ coenzyme Q (red)	$+0.110$[b]
Cytochrome C (ox) $\rightleftarrows$ cytochrome C (red)	$+0.260$[b]
Cytochrome oxidase (ox) $\rightleftarrows$ cytochrome oxidase (red)	$+0.550$[b]
$2H^+ + \tfrac{1}{2}O_2 \rightleftarrows H_2O$	$+0.82$[b]

[a] K. Burton, *Ergeb. Physiol. Biol. Chem. Exp. Pharmakol.*, **49**:212 (1957).
[b] B. Harrow and A. Mazur, "Textbook of Biochemistry," 9th ed., W. B. Saunders Company, Philadelphia, 1966.

KCl as the conduction electrolyte. This bridge is in contact on the right with Cu^{++} in aqueous solution at $a = 1\ m$. Finally, the electrode on the far right is solid copper metal, $Cu(s)$.

2. Each cell can be regarded as composed of two half-cells connected together by the means shown at the center of the designation, such as $|KCl(aq)|$. The half-cell on the right has the standard designation and cell reaction

$$Cu^{++}(aq, a = 1\ m) \mid Cu(s)$$
$$Cu^{++}(aq, a = 1\ m) + 2e^- = Cu(s) \qquad (14\text{-}24)$$

The half-cell on the left, regarded as an *independent* half-cell and not as part of the Daniell cell, has the standard designation and cell reaction

$$Zn^{++}(aq, a = 1\ m) \mid Zn(s)$$
$$Zn^{++}(aq, a = 1\ m) + 2e^- = Zn(s) \qquad (14\text{-}25)$$

Note that in the designation of the whole cell, the half-cell on the right appears in the form of its standard designation but the half-cell on the left appears with the order of its parts reversed.

3. *The standard potential* of a half-cell is taken as the potential of the whole cell formed when this half-cell under standard conditions at 25°C makes up the right-hand portion, the left-hand portion consisting of a standard hydrogen half-cell, the pressure of the hydrogen being equal to 1 atm, and the activity of H^+ being equal to $1\ m$. For example, in cell notation

$$Pt, H_2(1\ atm) \mid H^+(aq, a = 1\ m) \mid\mid$$
$$Cu^{++}(aq, a = 1\ m) \mid Cu(s)$$
$$\mathscr{E}° = +0.34\ volt \qquad (14\text{-}26)$$

(*Note:* The designation of the nature of the bridge is frequently omitted, with the understanding that an arrangement is used which either will make no contribution to the potential of the whole cell or for which the contribution is known and corrected for.)

Thus, the standard notation for the right half-cell and its potential is

$$Cu^{++}(aq, a = 1\ m)\ |\ Cu(s)\quad \mathcal{E}° = +0.34\ \text{volt}$$

$$(14\text{-}27)$$

All values of $\mathcal{E}°$ correspond to 25°C, to metals in the solid state (s), to unit activity for ionic species $(a = 1\ m)$, and to unit fugacity for gas pressures $(f = 1\ \text{atm})$, unless otherwise noted.

In the same way the standard potential for the half-cell on the left of the Daniell cell in Fig. 14-1 is

$$Zn^{++}(aq, a = 1\ m)\ |\ Zn(s)\quad \mathcal{E}° = -0.76\ \text{volt}$$

$$(14\text{-}28)$$

Note that in the total cell notation for the Daniell cell, this half-cell is denoted in the reverse order.

Some standard half-cell potentials at 25°C are listed in Table 14-1.

4. In line with the above conventions, the left half-cell reaction, as it appears in the total cell designation, is always an oxidation; the right half-cell reaction is always a reduction. The half-cell reactions in the standard table of values of $\mathcal{E}°$ are always reductions. (*Note:* In Latimer's classic work on oxidation-reduction potentials and in some other texts, the half-cell reactions are written as oxidations, and the values of $\mathcal{E}°$ have the opposite sign from those in Table 14-1. In nearly all the biochemical literature, the half-cell reactions are written as reductions, but many values of the potentials refer to the standard state, in which the activity of H^+ is $10^{-7}\ m$; such values are generally designated by the symbols $\mathcal{E}'$, $\mathcal{E}'_0$, E' or E'_0.)

5. *The combination of standard half-cell potentials* is effected by taking as listed the potential of the half-cell on the right side of the cell to be formed and subtracting from it the listed value of the half-cell to be put on the left side. For example, the Daniell cell standard potential with the copper electrode on the right and the zinc electrode on the left is calculated thus

$$
\begin{array}{ll}
Cu^{++}\ |\ Cu & \mathcal{E}° = +0.34\ \text{volt} \\
-(Zn^{++}\ |\ Zn) & -(\mathcal{E}° = -0.76\ \text{volt}) \\
\hline
Zn\ |\ Zn^{++}\ ||\ Cu^{++}\ |\ Cu & \mathcal{E}° = +1.10\ \text{volts}
\end{array}
$$

$$(14\text{-}29)$$

The sign of the potential so obtained is the polarity of the right-hand electrode of the total cell. Note that this convention is in line with that for the determination of the free-energy change in a reaction from the values of the free energy for the individual components; the free energy of the components on the left of the equal sign (the reactants) is subtracted from the free energy of the components on the right (the products).

6. To obtain the chemical reaction for a cell, the left half-cell reaction must be subtracted from the right half-cell reaction with the usual algebraic convention that the signs of all components in the part being subtracted are negative and that the sign changes when any part is shifted from the left side of the equals sign to the right side or vice versa. For example, to get the equation for the Daniell cell:

$$
\begin{array}{ll}
\textit{Right side:} & Cu^{++} + 2e^- = Cu \\
\textit{Left side:} & -(Zn^{++} + 2e^- = Zn) \\
\hline
\textit{Cell reaction:} & Zn + Cu^{++} = Zn^{++} + Cu
\end{array}
$$

$$(14\text{-}30)$$

7. When the value of $\mathcal{E}°$ has the positive sign, the value of $\Delta G°$ is negative and the reaction proceeds spontaneously as written according to rule 6. For example,

$$Zn|Zn^{++}||Cu^{++}|Cu \quad \mathcal{E}° = +1.10\ \text{volts}$$

$$Zn + Cu^{++} \xrightarrow[\text{spontaneously}]{}$$

$$Zn^{++} + Cu \quad \Delta G° = -50.73\ \text{kcal mole}^{-1}$$

$$(14\text{-}31)$$

The Weston standard cell

In order to calibrate potentiometers so that the readings of the resistance contacts correspond to known potentials, it is necessary to have a potential source with a value that is precisely known to serve as a reference standard. One of the most useful potential standards is the *Weston standard cell* (Fig. 14-5), which consists of two half-cells joined together to form an H configuration. A platinum wire is sealed through the bottom of each leg of the H to provide contacts with the electrodes. The negative electrode on the left is an amalgam of mercury and 12.5 percent cadmium by weight. The electrolyte of the cell consists of a saturated solution of hydrated cadmium sulfate, $CdSO_4 \cdot \frac{8}{3}H_2O$, in which there are large crystals of this compound. The positive electrode is a pool of pure mercury covered by a paste formed by grinding mercury with mercurous sulfate, Hg_2SO_4. Above the paste and in contact with it is a saturated solution of $CdSO_4$. The openings at the top of each vertical portion of the H are sealed to prevent the evaporation of water.

In the conventional notation, the cell is

Hg, 12.5%Cd | Cd$^+$(aq satd 3CdSO$_4$ · 8H$_2$O) ||

$\qquad$ Hg$_2^+$(aq satd Hg$_2$SO$_4$) | Hg(l)   (14-32)

The variation with temperature t of the potential of the Weston saturated cell is given by

$$\mathcal{E} = 1.018410 - 4.93 \times 10^{-5}(t - 25°)$$
$$- 8.0 \times 10^{-7}(t - 25°) + 1 \times 10^{-8}(t - 25°)^3$$
$$(14\text{-}33)$$

Thus, at 25°C the potential of the cell is 1.018410, and the temperature coefficient of the potential is -0.0493 mv deg^{-1}.

Standard cells are often prepared in which the solution does not have the concentration of cadmium sulfate corresponding to saturation at 25°C but a concentration corresponding to saturation at 4°C. Thus, the solution is not saturated at 25°C. Since such a cell has a temperature coefficient of potential which is practically zero, it is therefore convenient under circumstances where the temperature may vary slightly. However, the potential is not as reproducible or as invariant with time as that of a saturated cell. Such a cell can be used only if the accuracy of the measurement may vary by as much as 0.01 percent.

If the potential of a Weston cell is to be maintained constant and at a maximum, the measuring current establishing the balance of the potential against the unknown potential must be less than 10^{-9} amp.

Fig. 14-5   The Weston standard cell. $\mathcal{E} = 0.7768$ volt (25°C).

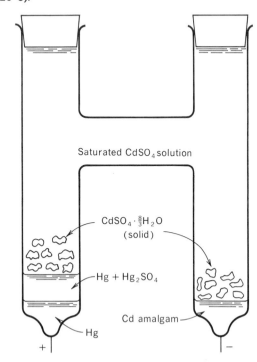

Saturated CdSO$_4$ solution

CdSO$_4 \cdot \frac{8}{3}$H$_2$O (solid)

Hg + Hg$_2$SO$_4$

Cd amalgam

Hg

+

−

The calomel electrode

An electrode can be formed from a portion of metal in contact with an insoluble salt of the metal, which is, in turn, in contact with a solution containing the anion of the salt. The most frequently used electrode of this kind is the *calomel electrode*, in which metallic mercury is in contact with calomel, Hg_2Cl_2; the latter, in turn, is in contact with a solution containing chloride ions. Such an electrode is shown in Fig. 14-6 and is denoted by

Cl$^-$(aq) | Hg$_2$Cl$_2$(s), Hg(l)
$\frac{1}{2}$Hg$_2$Cl$_2$ + e^- = Hg + Cl$^-$ (14-34)

The saturated calomel electrode is prepared by grinding up Hg_2Cl_2 with crystalline KCl. The concentration of Cl$^-$ is established by the equilibrium at saturation with the crystalline KCl.

The glass electrode

As an alternative to the electrode made with a platinum wire and bubbling hydrogen gas for measuring the

Fig. 14-6 Saturated calomel half-cell coupled to a standard copper electrode. $Hg(l) | Hg, Hg_2Cl_2$ paste $| Hg_2Cl_2(aq\ satd) \| CuSO_4(aq) | Cu(s)$.

$$-(E_1^\circ)\, 0.2676v + (E_2^\circ)\, 0.337\, v = (E^\circ)\, 0.069\, v$$

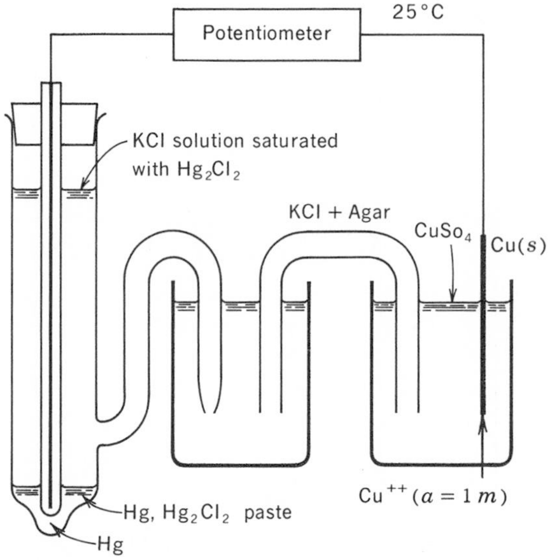

concentration of H$^+$, an electrode can be constructed in which equilibrium with H$^+$ ions in solution is established through a glass diaphragm so thin that ions can migrate through it. Such an electrode is shown in Fig. 14-7 combined with a calomel electrode to form a device for measuring pH. The glass electrode operates in both oxidizing and reducing media and has the advantage of being unaffected by proteins and sulfur compounds which cause malfunctioning of platinum electrodes.

As shown in Fig. 14-7, there is a thin glass diaphragm at the bottom of the electrode. The lower surface of the diaphragm is in contact with the solution to be studied, and the upper surface is in contact with a dilute solution of KCl and acetic acid in which is immersed a platinum wire coated with Ag and AgCl. The potential of this electrode varies with the concentration of H$^+$ in the same ways as that of the hydrogen-gas electrode. The glass electrode can be coupled with the calomel electrode, which has a constant potential; the total cell potential thus provides a measure of the concentration of H$^+$. The total cell is denoted by the expression

$$Ag, AgCl(s) \mid KCl(aq), CH_3COOH \mid M \mid H^-(aq) \mid KCl(aq) \mid$$
$$Cl^-, KCl(satd,\ aq) \mid Hg_2Cl_2(s), Hg(l) \qquad (14\text{-}35)$$

where M denotes the diaphragm which acts as a semipermeable membrane.

The glass-electrode potential varies by 0.0591 volt per pH unit at 25°C; the potentiometer is coupled with the electrodes shown in the figure and is generally graduated directly in terms of pH. The glass diaphragm itself produces a small potential sometimes called the asymmetry potential; for this reason, such a device must be calibrated against a solution where the pH has been determined by other means. The total device is frequently called a pH meter.

Inert-metal electrodes

An electrode can be formed from an inert metal in contact with a solution containing two different oxidation states of an ion. As an example, consider the electrode formed by a platinum wire in contact with a solution containing Fe^{++} and Fe^{3+} where the ions are at the respective concentrations m_2 and m_3. The notation for the electrode and the electrode reaction is

$$Fe^{++}(aq, m_2), Fe^{3+}(aq, m_3) \mid Pt$$
$$Fe^{3+} + e^- = Fe^{++} \qquad (14\text{-}36)$$

Another electrode of this type is the hydroquinone-quinone, $QH_2 \cdot Q$, Q electrode. The cell notation and equation are

$$H^+(m, aq), Q, QH_2 \mid Pt$$

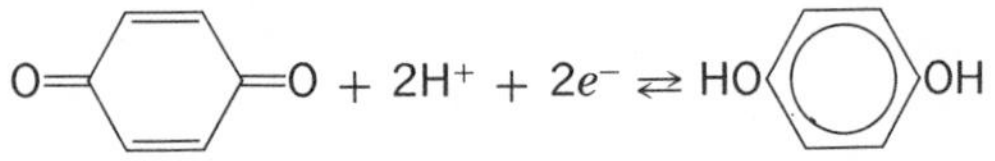

$$O=\!\!\!\left\langle\;\bigcirc\;\right\rangle\!\!\!=O + 2H^+ + 2e^- \rightleftarrows HO\!\!\left\langle\;\bigcirc\;\right\rangle\!\!OH$$

$$(14\text{-}37)$$

This electrode is commonly known as the *quinhydrone electrode* from the name of the crystalline compound $QH_2 \cdot Q$ which can be added to water to form the electrolyte.

Junction potentials

When two different electrodes are coupled together by forming a direct contact between the electrolyte of one and the electrolyte of the other, a component of the cell potential is contributed by the junction. At the point where the two electrolytes meet, there is diffusion of the ions across the junction. As a rule, ions diffuse at different rates, and this gives rise to a potential. This component of the cell potential must be eliminated or corrected in order to obtain precise significant values of the electrode potentials. The junction potential can be minimized by the use of a salt bridge consisting of a potassium chloride solution with gelatin added to reduce the rates of diffusion. The mobility of the potassium ions is about the same as that of the chloride ions, so that the contribution to the total potential from diffusion is very small. If potassium chloride cannot be used because of a chemical reaction with one of the electrolytes in the half-cells, e.g., with a solution of silver nitrate, a salt bridge of ammonium nitrate can be substituted.

Example 14-1

Calculate the total voltage from a cell which on the left has a half-cell with the reaction $Mg^{++} + 2e^- = Mg(s)$ and on the right has the half-cell for which the reaction is $Cu^{++} + 2e^- = Cu(s)$.

Looking in Table 14-1, we find that for the half-cell on the left the potential is -2.37 volts and for the half-cell on the right the potential is $+0.34$. In order to get the potential for the total cell, the potential on the left must be subtracted from the potential on the right, giving

$$0.34 \text{ volt} - (-2.37 \text{ volts}) = +2.71 \text{ volts}$$
$$Ans.$$

Example 14-2

Calculate the potential for a cell which has on the left the half-cell with the equation $Cu^{++} + e^- = Cu^+$ and on the right the half-cell with the equation $Fe^{3+} + e^- = Fe^{++}$.

Looking in Table 14-1, we find the value of $+0.521$ for the half-cell on the left and $+0.77$ for the half-cell on the right. Thus, subtracting the value of the half-cell on the left from the value of the half-cell on the right, we get

$$+0.77 \text{ volt} - (+0.52 \text{ volt}) = +0.25 \text{ volt}$$
$$Ans.$$

Example 14-3

Calculate the free energy for the half-cell where the reaction is $Cd^{++} + 2e^- = Cd(s)$.

Looking in Table 14-1, we find that the potential for this cell is -0.40 volt. To get the answer, we insert numerical values in the equation

$$\Delta G = -n\mathscr{F}\mathscr{E} = -(2)(23.060)(-0.40)$$
$$= +18.4 \text{ kcal} \qquad Ans.$$

Example 14-4

From the value for $\mathscr{E}°$ for (a) the reaction $Cu^{++} + e^- = Cu^+$ and for (b) the reaction $Cu^{++} + 2e^- = Cu(s)$, calculate the value of $\mathscr{E}°$ for (c) the reaction $Cu^+ + e^- = Cu(s)$.

As previously pointed out, it is not correct to take the numerical difference between the value of $\mathscr{E}°$ for the two given reactions in order

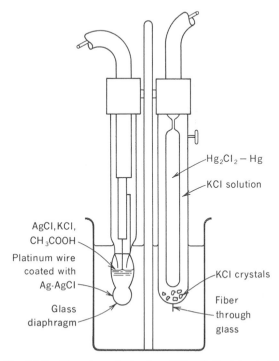

Fig. 14-7 The glass electrode. (After F. Daniels, J. W. Williams, P. Bender, R. A. Alberty, and C. D. Cornwell, "Experimental Physical Chemistry," 6th ed., McGraw-Hill Book Company, New York, 1962.)

$Hg_2Cl_2 - Hg$

KCl solution

AgCl, KCl, CH_3COOH

Platinum wire coated with Ag-AgCl

KCl crystals

Glass diaphragm

Fiber through glass

to get the value of $\mathscr{E}°$ asked for. The simple subtraction gives the value of $+0.184$ volt, which is incorrect. In order to get the correct answer it is necessary to calculate the values of $\Delta G°$ for each of the given reactions and to take the difference between the values of $\Delta G°$ in order to get the answer as follows:

$$\Delta G_a° = -n_a \mathscr{F} \mathscr{E}_a° = -1(23.060)(+0.153)$$
$$= -3.528 \text{ kcal}$$
$$\Delta G_b° = -n_b \mathscr{F} \mathscr{E}_b° = -2(23.060)(+0.337)$$
$$= 15.542 \text{ kcal}$$
$$\Delta G_c° = \Delta G_b° - \Delta_a° = -15.542 - (-3.528 \text{ kcal})$$
$$= -12.014 \text{ kcal}$$
$$\mathscr{E}_c° = -\Delta G_c° \mid n_c \mathscr{F} = -\frac{-12.014}{(1)(23.060)}$$
$$= +0.521 \text{ volt} \qquad Ans.$$

Note that the correct answer differs from the answer obtained by the incorrect procedure of subtracting $\mathscr{E}_a°$ from $\mathscr{E}_b°$ directly.

14-4   CELL EQUILIBRIA

$\mathscr{E}°$ and the equilibrium constant

As explained in Sec. 14-2, there is a relation between the free-energy change under standard conditions $\Delta G°$ and the standard electromotive force $\mathscr{E}°$ of a cell in which this reaction takes place. In a Daniell cell, where the zinc ions and the copper ions are both at unit activity at 25°C, $\mathscr{E}° = +1.100$ volts. The reaction in the Daniell cell is

$$Zn(s) + Cu^{++}(aq) \rightleftharpoons Zn^{++}(aq) + Cu(s) \qquad (14\text{-}38)$$

If the reaction is allowed to proceed reversibly so that 1 g mole of metallic zinc is converted into zinc ions and copper ions are converted into 1 g mole of copper, $\Delta G° = -50.7$ kcal mole^{-1}. The relation between this quantity and the emf is

$$\Delta G° = -n\mathscr{F}\mathscr{E}° \qquad (14\text{-}39)$$

The standard change in free energy is related to the equilibrium constant for this reaction by the expression

$$\Delta G° = -RT \ln K \qquad (14\text{-}40)$$

where the equilibrium constant is the quotient of the activities for which the reaction proceeds under equilibrium conditions

$$K = \left(\frac{a_{Zn^{++}}}{a_{Cu^{++}}}\right)_{eq} \qquad (14\text{-}41)$$

Thus, the emf is related to the free-energy change under standard conditions by the expression

$$\mathscr{E}° = \frac{RT}{n\mathscr{F}} \ln K \qquad (14\text{-}42)$$

Consider now the expression for the free-energy change ΔG when the reaction proceeds not under equilibrium conditions but under conditions where the values of the activities may be set at any amounts which do not correspond to equilibrium conditions. This quotient of activities Q is

$$Q = \frac{a_{Zn^{++}}}{a_{Cu^{++}}} \qquad (14\text{-}43)$$

where the activity terms can take on any values. The general expression relating free-energy change and activity quotient is

$$(\Delta G)_b - (\Delta G)_a = RT \ln Q_b - RT \ln Q_a \qquad (14\text{-}44)$$

where the subscript a refers to a change when the activity coefficients have the value Q_a and the subscript b refers to a change where the activity coefficients have a quotient equal to Q_b. When the change a corresponds to standard conditions and the change b corresponds to equilibrium conditions, this expression becomes

$$\Delta G_{eq} - \Delta G° = RT \ln K - RT \ln 1 \qquad (14\text{-}45)$$

Note that under equilibrium conditions $\Delta G_{eq} = 0$ and $Q_b = K$; under standard conditions $\Delta G_a = \Delta G°$ and $Q_a = 1$. If change a takes place at standard conditions but change b takes place under some arbitrarily selected nonequilibrium condition, then

$$\Delta G - \Delta G° = RT \ln Q \qquad (14\text{-}46)$$

since $RT \ln Q_a = 0$. The relation between the free-energy change under nonequilibrium conditions ΔG and the corresponding emf $\mathscr{E}$ is

$$\Delta G = -n\mathscr{F}\mathscr{E} \qquad (14\text{-}47)$$

Thus, Eq. (14-46) can be written

$$-n\mathfrak{F}\mathscr{E} + n\mathfrak{F}\mathscr{E}° = RT \ln Q \qquad (14\text{-}48)$$

This can be rearranged to express the emf $\mathscr{E}$ as a function of $\mathscr{E}°$ and the activity coefficient Q:

$$\mathscr{E} = \mathscr{E}° - \frac{RT}{n\mathfrak{F}} \ln Q \qquad (14\text{-}49)$$

In the standard hydrogen cell the only concentration variable is the activity of the hydrogen ion; consequently Eq. (14-49) becomes in this case

$$\mathscr{E} = \mathscr{E}° - \frac{RT}{n\mathfrak{F}} \ln a_{H^+} \qquad (14\text{-}50)$$

$\mathscr{E}$ and pH

In the preceding chapter the discussion of pH was based on the definition proposed by Sorensen in 1909

$$pH \equiv -\log C_{H^+} \qquad (14\text{-}51)$$

where C_{H^+} is the molar concentration. During the last 50 years, improvements both in experiments and theory have made a different definition desirable.

At the present time the most accurate and convenient way of determining pH is by means of emf. As previously pointed out, there are many factors which make the relation between *concentration* and other thermodynamic quantities extremely complex. For this reason, G. N. Lewis suggested that thermodynamic relations might better be expressed in terms of the quantity *activity,* analogous to concentration but defined in such a way that it has relatively simple relations with the other principal thermodynamic variables. Thus, it is *activity* and not *concentration* that bears a direct relation to emf. With this in mind a new definition of pH has been adopted based on measurements with a cell composed of a hydrogen electrode and a calomel electrode where the concentration of KCl is $0.1\ N$. When in this cell the value of a_{H^+} is equal to unity, the emf has the value 0.3358 volt. This quantity is denoted by the symbol $\mathscr{E}_{ref}$. The relation between the emf and hydrogen-ion activity can be written

$$\mathscr{E} = \mathscr{E}_{ref} - \frac{RT}{\mathfrak{F}} \ln a_{H^+} \qquad (14\text{-}52)$$

Using the functional relationship expressed in Eq. (14-51), this expression can be changed to

$$\mathscr{E} = \mathscr{E}_{ref} + 2.303 \frac{RT}{\mathfrak{F}} pH \qquad (14\text{-}53)$$

which can be rearranged to give

$$pH \equiv \frac{\mathscr{E} - \mathscr{E}_{ref}}{2.303\ RT/\mathfrak{F}} \qquad (14\text{-}54)$$

This equation serves as the new definition of pH, based on an emf standard.

Concentration cells

In deriving the basic thermodynamic equations for emf cells in the preceding sections, a relationship was established between emf and the activities of the various ionic constituents which are, in turn, functions of concentration. A cell reaction can be written in the generalized form

$$aA + bB = cC + dD \qquad (14\text{-}55)$$

where the terms may represent both the ionized and the un-ionized constituents in solution.

When the reaction takes place under standard conditions of unit activity,

$$\Delta G° = \Delta H° - T\ \Delta s° \qquad (14\text{-}56)$$

When the reaction takes place with arbitrarily selected activities equal to $a_A{}^a$, $a_B{}^b$, $a_C{}^c$, and $a_D{}^d$, the free-energy change is

$$\Delta G = \Delta H - T\ \Delta s \qquad (14\text{-}57)$$

It is a necessary consequence of the definition of activity that it is a function of concentration satisfying the equation

$$\Delta G - \Delta G° = T\left(R \ln \frac{a_C{}^c a_D{}^d}{a_A{}^a a_B{}^b}\right) \qquad (14\text{-}58)$$

Subtracting Eq. (14-56) from Eq. (14-57),

$$\Delta G - \Delta G° = (\Delta H - \Delta H°) - T(\Delta s - \Delta s°) \qquad (14\text{-}59)$$

In an ideal solution, $\Delta H - \Delta H° = 0$. A comparison of Eqs. (14-58) and (14-59) shows that for such a solution

$$\Delta s - \Delta s° = R \ln \frac{a_C{}^c a_D{}^d}{a_A{}^a a_B{}^b} \qquad (14\text{-}60)$$

Thus, for such a solution, the term in parentheses in Eq. (14-58) represents the difference between the entropy change when the reaction takes place with all constituents at unit activity and the entropy change when the reaction takes place with the constituents at the activities shown. In other words, the term in parentheses is the entropy change brought about by changing from unit activities to the arbitrarily selected set of activities shown between the parentheses.

As previously discussed, the relations between free energy and emf are

$$\Delta_G = -n\mathfrak{F}\mathcal{E} \tag{14-61}$$

$$\Delta_{G}° = -n\mathfrak{F}\mathcal{E}° \tag{14-62}$$

Combining these expressions with Eq. (14-58) gives

$$\mathcal{E} = \mathcal{E}° - \frac{RT}{n\mathfrak{F}} \ln \frac{a_C{}^c a_D{}^d}{a_A{}^a a_B{}^b} \tag{14-63}$$

This is the general equation relating emf and the activities of the constituents in solution in the cell. At 25°C, with appropriate numerical values inserted, this becomes

$$\mathcal{E} = \mathcal{E}° - \frac{(1.987)(298.15)(2.303)}{n(23.060)} \log \frac{a_C{}^c a_D{}^d}{a_A{}^a a_B{}^b} \tag{14-64}$$

which reduces to the following expression, often referred to as the Nernst equation, in honor of Walther Nernst, the German physical chemist:

$$\mathcal{E} = \mathcal{E}° - \frac{0.0591}{n} \log \frac{a_C{}^c a_D{}^d}{a_A{}^a a_B{}^b} \tag{14-65}$$

An emf cell can be constructed by combining two half-cells which have the same electrodes and the same electrolytes but with the latter at different concentrations. For example, consider a cell with a zinc half-cell on the left in which $a_{Zn^+} = 2\ m$, and on the right a zinc half-cell with the electrolyte at unit activity. The cell designation is

$$Zn(s) \mid Zn^{++}(a = 2\ m) \parallel Zn^{++}(a = 1\ m) \mid Zn(s) \tag{14-66}$$

The cell equation is

$$Zn^{++}(a = 1\ m) = Zn^{++}(a = 2\ m) \tag{14-67}$$

At 25°C, the emf is given by the expression

$$\mathcal{E} = \mathcal{E}° - \frac{0.0591}{2} \log \frac{a_{Zn^{++}}}{1}$$

$$= 0.763 - \frac{0.0591}{2} \log 2 \tag{14-68}$$

$$= 0.763 - 0.009 = 0.754 \text{ volt}$$

At the higher activity of zinc ions in the left portion of the cell, the entropy of these ions is reduced below that at standard activity. When the reaction proceeds, the equation for the left half-cell is

$$Zn(s) = Zn^{++}(a = 2\ m) \tag{14-69}$$

The free-energy change is

$$\Delta G = \Delta H - T\,\Delta S \tag{14-70}$$

ΔS is the difference between the entropy of the zinc in solution as ions and in the crystalline form

$$\Delta s = \bar{s}_{ion} - s_{solid} \tag{14-71}$$

If $\bar{s}_{ion}$ is reduced, Δs is reduced; thus, Δ_G has a less negative value, and the value of $\mathcal{E}$ is smaller.

Example 14-5

Calculate the value of $\mathcal{E}$ for the half-cell in which the reaction $Cu^{++}(0.1\ m) + 2e^- = Cu(s)$ takes place at 25°C.

At this temperature the value can be obtained from the equation

$$\mathcal{E} = \mathcal{E}° - \frac{0.059}{n} \log [Cu^{++}]^{-1}$$

$$= 0.34 + \frac{0.059}{2} \log 0.1$$

$$= +0.34 \text{ volt} - 0.029(-1.000)$$

$$= 0.31 \text{ volt} \qquad Ans.$$

Example 14-6

Calculate the potential for the cell which has on the left side the reaction $Fe^{++}(a = 1\ m) = Fe^{3+}(a = 0.1\ m) + e^-$ and on the right side the reaction $Cr^{3+}(a = 0.2\ m) + e^- = Cr^{++}(a = 1\ m)$ when both reactions take place at 25°C.

For the half-cell on the left side, the potential is given by

$$\mathscr{E}_{\text{left}} = \mathscr{E}° - \frac{0.059}{1} \log [Fe^{3+}]^{-1}$$

$$= +0.77 - 0.059 \log [0.1]^{-1}$$

$$= +0.77 - 0.059 = +0.71 \text{ volt}$$

For the half-cell on the right side, the equation is

$$\mathscr{E}_{\text{right}} = \mathscr{E}° - \frac{0.059}{1} \log [Cr^{3+}]^{-1}$$

$$= -0.41 - 0.059 \log 0.2^{-1}$$

$$= -0.41 - 0.059(0.699)$$

$$= -0.41 - 0.04 = -0.45 \text{ volt}$$

$$\mathscr{E}_{\text{cell}} = \mathscr{E}_{\text{right}} - \mathscr{E}_{\text{left}} = -1.16 \text{ volt} \qquad Ans.$$

Redox couples

In Table 14-1 there are a number of examples of half-cells in which electrons enter an atom in a positively ionized form, thereby changing it into an ion with a lower positive charge. For example, Fe^{3+} changes into Fe^{++} according to the reaction

$$Fe^{3+} + e^- = Fe^{++} \tag{14-72}$$

The half-cell designation is

$$Fe^{++}, Fe^{3+} \mid Pt \qquad \mathscr{E}° = +0.771 \text{ volt} \tag{14-73}$$

Such a pair of ions is known as a *redox couple*. The term redox is namely a shorter designation for reduction-oxidation. When the above reaction goes from left to right, Fe^{3+} is reduced to Fe^{++}; when the reaction proceeds from right to left, Fe^{++} is oxidized to Fe^{3+}.

The potential of such a cell, frequently called a *redox potential,* varies with concentration as shown in Eq. (14-63). Considering such a cell at 25°C and writing the activity of the oxidized form as [ox] and of the reduced form as [red], the equation becomes

$$\mathscr{E} = \mathscr{E}° - \frac{0.0591}{n} \log \frac{[\text{red}]}{[\text{ox}]} \tag{14-74}$$

The potential of the half-cell is equal to $\mathscr{E}°$ not only when the two species of ions are at unit activity but also whenever [ox] = [red]. A plot of the fraction in the reduced form against the voltage is shown for two couples in Fig. 14-8.

Fig. 14-8   The potential of redox couples plotted against the fraction x in the reduced form.

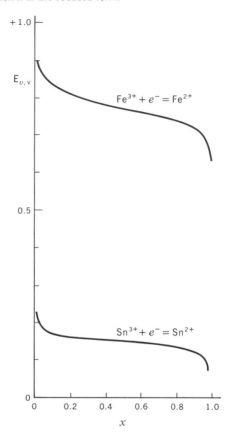

Redox indicators

In a number of substances the color of the oxidized form differs from the color of the reduced form. Thus, color and the potential of the couple are related. An interesting example is methylene blue, which is colorless in the reduced form and blue in the oxidized form. The structural formulas are

MBH₂, methylene white (reduced form)

$$\begin{bmatrix} \overset{+}{C}H_3 \\ \vdots \\ N \\ \vdots \\ CH_3 \end{bmatrix} \cdots \text{(structure)} \cdots \begin{matrix} CH_3 \\ N \\ H \quad CH_3 \end{matrix}^{+}$$

MB, methylene blue (oxidized form)

The reaction for the half-cell is abbreviated

$$MBH_2 = MB - 2H^+ + 2e^- \qquad (14\text{-}75)$$

The equation relating emf and the activities is

$$\mathscr{E} = \mathscr{E}° - \frac{0.059}{2} \log \frac{[MBH_2]}{[MB][H^+]^2} \qquad (14\text{-}76)$$

which can be expanded to

$$\mathscr{E} = \mathscr{E}° - \frac{0.059}{2} \log \frac{[MBH_2]}{[MB]} + 0.059 \log [H^+]$$

$$= \mathscr{E}° - \frac{0.059}{2} \log \frac{[MBH_2]}{[MB]} - 0.059 \text{ pH} \qquad (14\text{-}77)$$

If the potential is plotted against the fraction in oxidized form for a given value of pH, a curve is obtained similar to that shown in Fig. 14-8. Increasing values of pH shift the curve to higher positions on the graph without changing its shape (Fig. 14-9).

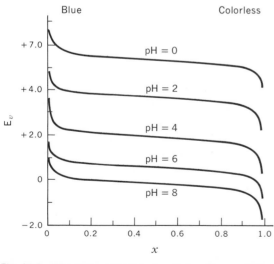

Fig. 14-9 The redox potential of methylene blue-methylene white plotted against the fraction x in reduced form at various values of pH at 30°C.

If a small amount of indicator like methylene blue is placed in a solution with a redox potential, the relative amounts of the two forms of the indicator adjust so that the redox potential of the indicator in solution matches the redox potential of the solution. For example, if a solution contains ions forming a redox couple with the ion in the reduced form at twice the activity of the ion in the oxidized form, this establishes a redox potential given by Eq. (14-74), where [red]/[ox] = 2. The ratio of the activity of the reduced form to the activity of the oxidized form of the indicator then adjusts so that the redox potential of the indicator matches the redox potential of the solution and the color of the indicator is thus a measure of the redox potential of the original solution. When the redox potential of the indicator depends on pH, as it does with methylene blue, the pH of the solution must be known in order that the potential of the indicator can be read from the color change according to the curve corresponding to that value of pH, as in Fig. 14-9. The useful ranges for several redox indicators are shown in Table 14-3.

Example 14-7

A small amount of methylene blue is added to a solution in which pH = 6.0. The color indicates that half of the indicator is in the reduced form. What is the redox potential of the solution?

Referring to Fig. 14-9, the curve for pH = 6 crosses the ordinate for 0.50 at a potential of +0.55 volt, so that the latter is the redox potential of the solution.

$\mathscr{E}$ and K_{sp}

In some emf cells the chemical reaction involves the solution of a relatively insoluble salt. In many such cases the equilibrium constant calculated from $\mathscr{E}°$ is the solubility product K_{sp}.

$$Ag \,|\, Ag^+ \,||\, Br^- \,|\, AgBr(s), \, Ag(s) \qquad (14\text{-}78)$$

The reactions and the values of the emf for the half-cells and the total cell are

Right: $AgBr(s) + e^- = Ag(s) + Br^- \quad \mathscr{E}_{right}$

$$= +0.0713 \qquad (14\text{-}79)$$

TABLE 14-3   Range of redox potential in volts for several redox indicators in solutions where $pH = 0$

Indicator	Color	Approximate voltage
Methylene blue	None	0.23
	Blue	0.43
1-Naphthol-2-sulfonic	None	0.57
acid indophenol	Blue-green	0.75
Diphenylamine	None	0.65
sulfonic acid	Violet-red	0.83
o-Phenanthroline	Red	0.95
ferrous complex	Blue	1.14
o,o'-Diphenylamine	None	1.15
dicarbonic acid	Violet-blue	1.34
Tripyridylruthenium	None	1.21
dichloride	Yellow	1.42

Left: $Ag^+ + e^- = Ag(s)\ \mathcal{E}°_{left}$

$$= +0.7991 \qquad (14\text{-}80)$$

$$\overline{Ag Br(s) = Ag^+ + Br^-\ \mathcal{E}°}$$

$$= -0.7278 \qquad (14\text{-}81)$$

Subtracting Eq. (14-80) from Eq. (14-79) gives the total cell reaction. The emf of the total cell is obtained in the same way. The equilibrium constant for the cell reaction is

$$K = a_{Ag^+} a_{Br^-} = K_{sp} \qquad (14\text{-}82)$$

since the AgBr is in the crystalline form and does not appear in the activity quotient. The relation between the solubility product and the emf at 25°C is

$$\mathcal{E} = \mathcal{E}° - \frac{0.05915}{n} \log(a_{Ag^+} a_{Br^-}) \qquad (14\text{-}83)$$

Inserting the measured numerical values for the emf $\mathcal{E}$ of the cell when the activities correspond to the equilibrium with solid AgBr and the standard electromotive force $\mathcal{E}°$ corresponding to the presence of each ionic species at unit activity gives

$$0 = -0.7278 - \frac{0.05915}{1} \log K_{sp} \qquad (14\text{-}84)$$

When this equation is solved, it is found that $K_{sp} = 4.8 \times 10^{-13}$.

The temperature dependence of $\mathcal{E}°$

Because of the direct relation between $\mathcal{E}°$ and the change in free energy ΔG, the temperature dependence

of the former at constant pressure can be determined from the latter

$$\left(\frac{\partial \Delta G}{\partial T}\right)_P = -\Delta s = -n\mathcal{F}\left(\frac{\partial \mathcal{E}}{\partial T}\right)_P \qquad (14\text{-}85)$$

From the relation

$$\Delta G = \Delta H - T\,\Delta s \qquad (14\text{-}86)$$

the change in entropy can be expressed as a function of ΔG and ΔH:

$$\Delta s = -\frac{\Delta G - \Delta H}{T} \qquad (14\text{-}87)$$

Because $\Delta G = -n\mathcal{F}\mathcal{E}$, the above expression can be written

$$\Delta s = \frac{n\mathcal{F}\mathcal{E} + \Delta H}{T} \qquad (14\text{-}88)$$

which can be combined with Eq. (14-85) to give the expression known as the Gibbs-Helmholtz equation

$$\mathcal{E} + \frac{\Delta H}{n\mathcal{F}} = T\left(\frac{\partial \mathcal{E}}{\partial T}\right)_P \qquad (14\text{-}89)$$

With this equation, measurements of $\mathcal{E}$ and of $(\partial \mathcal{E}/\partial T)_P$ can be used to calculate the value of ΔH. As a rule, values of ΔH obtained in this way are found to be more accurate than any obtained by calorimetric measurements.

Example 14-8

In the cell

$$Pb(s), Pb(Cl_2) \mid Cl^- \parallel Cl^- \mid Hg(l), Hg_2Cl_2(s)$$

the chemical reaction is

$$Pb(s) + Hg_2Cl_2(s) = PbCl_2(s) + 2Hg(l)$$

At 25°C, $\mathcal{E}° = 0.5357$ volt, and $(\partial \mathcal{E}/\partial T)_P = 1.45 \times 10^{-4}$ volt deg^{-1}. Calculate ΔH.

$$
\begin{aligned}
\Delta H &= n\mathcal{F}\left[T\left(\frac{\partial \mathcal{E}}{\partial T}\right)_P - \mathcal{E} \right] \\
&= 2 \times 23062.3\,(298.15 \times 0.000145 \\
&\qquad\qquad\qquad\qquad\qquad - 0.5359) \\
&= -22.724 \text{ kcal}
\end{aligned}
$$

Example 14-9

At 25°C the value of $\mathcal{E}°$ is 3.21 volts for the cell in which the reaction is $Ca + Cu^{++} = Ca^{++} + Cu$. The enthalpy change in this reaction is -145.16 kcal. Calculate the rate at which $\mathcal{E}°$ changes with temperature.

The answer is obtained by inserting the numerical value in the equation

$$
\begin{aligned}
\mathcal{E}° + \frac{\Delta H}{n\mathcal{F}} &= T\frac{d\mathcal{E}}{dT} \\
\frac{d\mathcal{E}}{dT} &= \frac{\mathcal{E}° + \Delta H/n\mathcal{F}}{T} \\
&= \frac{3.21 + 145.16/46.12}{298°} \\
&= \frac{3.21 + 3.15}{298} \\
&= 0.021 \text{ volt deg}^{-1}
\end{aligned}
$$

The enthalpy of ions

In a review of the literature pertaining to electrolyte solutions, published in 1966, B. E. Conway, of the University of Ottawa, remarks that a notable feature of recent research on electrolytic solutions is the attention now being given to the quantitative behavior and properties of ions and polyions in relation to biochemical and biophysical systems and the role of water. The great majority of chemical reactions in biochemical systems take place in aqueous solution. Each ion in aqueous solution is to a greater or lesser

degree hydrated; the enthalpy of the ion is strongly influenced by the amount of hydration. The process of hydration also tends to orient water molecules in the vicinity of the ion; orientation reduces the entropy of the system. The enthalpy and entropy changes associated with hydration strongly affect the free energy of the ion, and the value of the free energy is a major factor in determining the role which the ion plays in the various biochemical reactions. Therefore, data on the enthalpy, the entropy, and the free energy of ions provide a useful body of knowledge to aid in achieving a deeper understanding of the nature of many biochemical reactions. Either as directly observed or as indirectly calculated, emf potentials play a key role both in the acquisition of these thermodynamic data and in their interpretation. The remainder of this chapter is devoted to these aspects of emf potentials and related data and their use in interpreting the nature of biochemical reactions.

For such interpretations, the most informative values of enthalpy change are those for the passage of ions from the gaseous state to the hydrated state in solution. The chemical reactions for two such changes are

$$Na^+(g) + Cl^-(g) = Na^+(aq) + Cl^-(aq)$$
$$\Delta H_{hydr} = -185 \text{ kcal mole}^{-1} \quad (14\text{-}90)$$

$$H^+(g) + Cl^-(g) = H^+(aq) + Cl^-(aq)$$
$$\Delta H_{hydr} = -351 \text{ kcal mole}^{-1} \quad (14\text{-}91)$$

If, instead of going into aqueous solution, the ions from the gaseous state come together to form a crystal lattice, the process is

$$Na^+(g) + Cl^-(g) = NaCl(s)$$
$$\Delta H_c° = -186.2 \text{ kcal mole}^{-1} \quad (14\text{-}92)$$

As seen by a comparison with Eq. (14-90), the enthalpy change for the formation of crystalline sodium chloride from ions in the gaseous state differs from the enthalpy change for the formation of a sodium chloride aqueous solution in the standard state by only a small amount. As the ions move together under the influence of the electric forces of attraction, the energy lost from the system is just about the same whether a crystal lattice is formed or a solution of hydrated ions is formed. Values of the enthalpy changes both for the formation of crystals and for the formation of aqueous solutions in the standard state are listed in Table 14-4 for a number of pairs of ions.

When an aqueous solution is formed according to the process expressed in Eq. (14-90), part of the enthalpy change is due to the hydration of Na^+ and part to Cl^-; but there is considerable difficulty in calculating the amount of enthalpy change to be ascribed to each ion.   One of the best assignments is that made by R. M. Noyes (see References), who uses the method first developed by Max Born but assumes that the dielectric constant around an ion is much closer to unity than to the macroscopic dielectric constant of water.   For the rare-gas type of cations like Na^+, the effective dielectric constant is assumed to be a function of the ionic radius and nearly independent of the ionic charge.   Noyes concludes that a single charge is more than sufficient to orient and polarize the surrounding water molecules and that one or two additional charges make almost no further

TABLE 14-5   Standard enthalpy of hydration for individual ions at 25°C

	$\Delta H°$, kcal mole^{-1}	
Ion	Morris[a] (av)	Noyes[b]
H^+	−263	−270
Li^+	−127	−130
Na^+	−99	−104
K^+	−79	−84
Rb^+	−72	−78
Cs^+	−68	−70
Mg^{++}		−474
Ca^{++}		−395
Sr^{++}		−354
Ba^{++}		−326
Cu^{++}		−516
Zn^{++}		−503
Fe^{++}		−473
Co^{++}		−505
Ni^{++}		−520
Mn^{++}		−455
Cr^{++}		−457
Fe^{3+}		−1038
Cr^{3+}		−1044
Al^{3+}		−1135
La^{3+}		−806
F^-	−121	
Cl^-	−88	
Br^-	−81	
I^-	−71	

[a] D. F. C. Morris, *Struct. Bonding (Berlin)*, **4**:77 (1968).
[b] R. M. Noyes, *J. Am. Chem. Soc.*, **84**:513 (1962); **86**:97 (1964).

change.   Values of the enthalpies of individual ions based on Noyes' calculations are given in Table 14-5 and plotted in Fig. 14-10 together with values from a more recent calculation by D. F. C. Morris.

The entropy of ions

As we have seen, solute molecules in dilute aqueous solution behave very much like the molecules of an ideal gas.   Osmotic pressure can be calculated on this basis, and the change of entropy with concentration obeys the formula applicable to an ideal gas.

The absolute entropy of an ideal gas can be cal-

TABLE 14-4   Standard enthalpy of hydration of ion pairs at 25°C

	$\Delta H°^c$, kcal mole^{-1}	$\Delta H°^{hydr}$, kcal mole^{-1}
HF		−381
HCl		−351
HBr		−344
HI		−334
LiF	−247	−240
LiCl	−201	−209
NaF	−216	−215
NaCl	−186	−182
NaBr	−176	−176
NaI	−163	−164
KF	−194	−196
KCl	−167	−163
KBr	−160	−155
KI	−150	−145
RbF	−183	−189
AgCl	−218	−203
$SrCl_2$	−504	−517
AlF_3	−1467	−1473
$AlCl_3$	−219	−1380

Note: The comparable enthalpy of the ions in the crystalline state is listed in the first column to show resemblances and differences.

Fig. 14-10   The enthalpy of hydration of ions plotted against the ionic radius.   [From R. M. Noyes, *J. Am. Chem. Soc.*, 84:513 (1962); D. F. C. Morris, *Struct. Bonding* (*Berlin*), 4:77 (1968).]

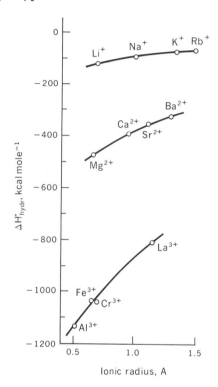

Ionic radius, A

$$\bar{s}^\circ = \tfrac{3}{2}R \ln M + 37 - 270\frac{z}{r^2} \qquad (14\text{-}94)$$

where $\bar{s}^\circ$ is the conventional entropy based upon $\bar{s}^\circ_{H^+} = 0$ and r is the effective radius in angstrom units, which is 1.0 Å greater than the crystal radius for negative ions and 2.0 Å greater than that for positive ions.   Values of $\bar{s}^\circ$ are plotted in Fig. 14-11.

The effect of the ionic charge in orienting water molecules is shown in Fig. 14-12.   The smaller the ion, the closer the water molecules can approach it and the greater the effect of orientation.   The larger the charge on the ion, the greater the number of water molecules in its vicinity which have their orientation affected.   Calculations have been made to show very roughly the number of water molecules oriented by different ions.   This is called the *hydration number,* and values are shown in Table 14-6.

Coupled with the effect of the charge will be the effect of hydrogen bonding in producing orientation of the water molecules surrounding an ion.   This is illustrated schematically in Fig. 14-13.

The entropy of hydration of ions can be calculated on the same basis as the enthalpy of hydration.   For example, when Na^+ and Cl^- in the gaseous state are changed to hydrated ions, the expression is

$$Na^+(g) + Cl^-(g) = Na^+(aq) + Cl^-(aq)$$

$$\Delta S^\circ_{\text{hydr}} = -44.3 \text{ eu mole}^{-1} \quad (14\text{-}95)$$

Values of the entropy of hydration calculated by Noyes are listed in Table 14-7 together with the partial molal entropy of the ions based on the convention that $\bar{s}^\circ = 0$ for H^+.

culated from statistical considerations by the Sackur-Tetrode equation.   In 1951, R. E. Powell and W. M. Latimer proposed a modification of the Sackur-Tetrode equation for molecules in aqueous solution

$$\bar{S}^\circ = \tfrac{3}{2}k \ln M + s_{\text{int}} + 10 - 0.22 \text{ v}^\circ \qquad (14\text{-}93)$$

where $\bar{S}^\circ$ = standard partial molal entropy of solute per mole

R = gas constant

M = molecular weight

s_{int} = internal entropy per mole

v° = molal volume in pure liquid state

When the solute particle is an electrically charged ion, the presence of the charge affects the orientation of the surrounding molecules of water drastically. Powell and Latimer found an empirical equation for the ion entropies

The free energy of ions

The free energy of hydration for individual ions can be calculated by the procedure of Born as modified by Noyes.   For the change from the ion in the gaseous state to hydrated ions, values are found as shown in the following example:

$$Na^+(g) \rightarrow Na^+(aq) \qquad \Delta G_{\text{hydr}} = -97 \text{ kcal mole}^{-1}$$
$$Cl^-(g) \rightarrow Cl^-(aq) \qquad \Delta G_{\text{hydr}} = -75 \text{ kcal mole}^{-1}$$

$$Na^+(g) + Cl^-(g) \rightarrow Na^+(aq) + Cl^-(aq)$$
$$\Delta G_{\text{hydr}} = -172 \text{ kcal mole}^{-1}$$

$$(14\text{-}96)$$

The free energy of the formation of the ion from its element or elements in the standard state at 25°C can be calculated on the basis of the convention that the free energy of formation of H^+ under standard conditions at 25°C is equal to zero. Values of the free energy of hydration and of the standard free energy of formation are listed in Table 14-8.

Example 14-10

Calculate the value of ΔH_{hydr} of AlF_3 from the values of ΔH°_{hydr} for the individual ions and compare with the value listed for AlF_3.

$$\underset{-1135}{\Delta H^\circ_{hydr}(Al^{3+})} + \underset{3(-121)}{3\Delta H^\circ_{hydr}(F^-)} = \underset{-1498\ kcal}{\Delta H^\circ_{hydr}(AlF_3)}$$

The listed value is -1473 kcal.   *Ans.*

Example 14-11

From the values of $\Delta \bar{G}^\circ_f$ for the individual ions, calculate the value of $\mathcal{E}°$ at 25°C for the half-cell Pt | Fe^{++}, Fe^{3+}.

$$\underset{-20.30}{\Delta \bar{G}^\circ_f(Fe^{++})} - \underset{-2.53}{\Delta \bar{G}^\circ_f(Fe^{3+})} = \underset{-17.77\ kcal\ mole^{-1}}{\Delta G°}$$

$$\mathcal{E}° = \frac{-\Delta G}{n} = \frac{17.77}{1 \times 23.06} = +0.771\ volt \qquad Ans.$$

The experimentally determined value is $+0.771$ volt.

14-5   BIOCHEMICAL OXIDATION-REDUCTION POTENTIALS

In a living organism there are a great many chemical reactions taking place which individually result in an *increase* of the free energy of the system. Since a chemical reaction takes place spontaneously only when it produces a *decrease* in free energy, these biochemical reactions can occur only when coupled with other reactions in such a way that the *net* result is a free-energy decrease. Oxidation-reduction reactions are by far the most important source of this coupling energy, which can force other reactions to proceed in the direction opposite to that which they would spontaneously pursue. In other words, oxidation-reduction couples convey the driving force which makes the life process go forward.

As an example, consider the equilibrium between

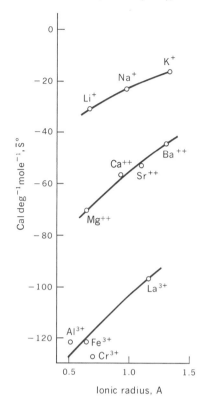

Fig. 14-11   The entropy of hydration of ions. [From R. M. Noyes, *J. Am. Chem. Soc.*, 84:513(1962); 86:97 (1964).]

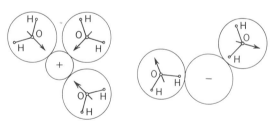

Fig. 14-12   The orientation of water molecules about ions. (After K. B. Harvey and G. R. Porter, "Physical Inorganic Chemistry," Addison-Wesley Publishing Company, Inc., Reading, Mass., 1963.)

two amino acids and the compound they form when linked together, as in a protein chain

$$CH_3CH(NH_2)COOH + CH_2(NH_2)COOH \rightleftarrows$$

Ala   Gly

$$CH_3CH(NH_2)CO—HNCH_2COOH + H_2O \quad (14\text{-}97)$$

Ala Gly

TABLE 14-6   Hydration number[a]

Ion	No.	Ion	No.
Li⁺	5 ± 1	Br⁻	1 ± 1
Na⁺	5 ± 1	I⁻	1 ± 1
K⁺	4 ± 2	Mg⁺⁺	15 ± 2
Rb⁺	3 ± 1	Ba⁺⁺	12 ± 4
F⁻	4 ± 1	Al³⁺	26 ± 5
Cl⁻	1 ± 1		

[a] After K. B. Harvey and G. B. Porter, "Physical Inorganic Chemistry," Addison-Wesley Publishing Company, Inc., Reading, Mass., 1963.

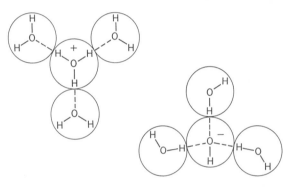

Fig. 14-13   Water molecules oriented by hydrogen bonding. (After K. B. Harvey and G. B. Porter, "Physical Inorganic Chemistry," Addison-Wesley Publishing Company, Reading, Mass., 1963.)

The formation of the *peptide* bond (—) together with the formation of H_2O brings about an increase in the standard free energy $\Delta G°$ of about 5.5 kcal. Accordingly, under the conditions commonly found in a living cell, the compound alanine-glycine tends to react with water and break up into the two amino acids, alanine and glycine. In order to make the reaction proceed in the opposite direction from left to right, a driving force must be applied. This is the direction in which the reaction must go when a protein is synthesized as a step in the life process. Oxidation-reduction reactions provide the energy necessary to produce this driving force.

In the physical world there are many ways of transporting energy to places where it is needed to make processes proceed. Energy can be transported mechanically, e.g., in the levers and gears of a machine, or electrically in metallic wires. But in living organisms, where energy must be delivered to specific sites on molecules and molecular complexes, neither of these methods is suitable; the only possible way is by the transportation of chemical energy contained in molecules that have the specific structure to ensure that the energy will go to the exact place in the reacting molecules where it is needed.

Oxidation-reduction couples are such energy carriers. Take a chromous ion, Cr^{++}, as an example: this ion has a tendency to give up an electron and become a chromic ion, Cr^{3+}, according to the equation

$$Cr^{++} \rightarrow Cr^{3+} + e^- \qquad \mathscr{E}° = 0.41 \text{ volt} \qquad (14\text{-}98)$$

The value of $\mathscr{E}°$ at 25°C for this reaction is +0.41 volt; this is a measure of a kind of "electron pressure," the pressure which the electron exerts in trying to leave the Cr^{++} ion. A similar reaction for vanadium is

$$V^{++} \rightarrow V^{3+} + e^- \qquad \mathscr{E}° = 0.26 \text{ volt} \qquad (14\text{-}99)$$

Thus, if a Cr^{++} ion collides with a V^{3+} ion, the electron pressure of the former can force the latter to accept an electron; because the electron pressure of Cl^{++} is greater than that of V^{++}, energy is transferred by this process.

For many oxidation-reduction couples found in biochemical systems, the chemical reaction takes place with the formation of H^+ ions. Consequently, the activity of H^+, that is, the value of pH, plays a role in determining the value of $\mathscr{E}°$ for the couple. The value of $\mathscr{E}°$ corresponds to unit activity for all the components of the reaction involving concentration. In almost all biochemical systems the activity of H^+ is not unity (pH = 0), corresponding to an acidic solution, but is equal to or close to $10^{-7} m$ (pH = 7), corresponding to an essentially neutral solution. For this reason, in studying biochemical oxidation-reduction couples, it is more helpful to have tabulated the values of $\mathscr{E}$ corresponding to $10^{-7} m$ as the activity for the hydrogen ion. To designate this, the symbol $\mathscr{E}'_0$ is customarily used. The value of the free-energy change under these conditions is designated by a similar symbol so that the relation is

$$\Delta G'_0 = -n\mathscr{F}\mathscr{E}'_0 \qquad (14\text{-}100)$$

The value of $\mathcal{E}_0'$ is related to the value of $\mathcal{E}°$ by the expression

$$\mathcal{E}_0' = \mathcal{E}° + \frac{RT}{n\mathcal{F}} \ln (H^+)^x = \mathcal{E}° + \frac{RT}{n\mathcal{F}} \ln (10^{-7})^x$$

$$(14\text{-}101)$$

where x is the appropriate power for the couple in question. For example, consider the half-cell in which the reaction is

$$2H^+ + \tfrac{1}{2}O_2 + 2e^- = H_2O \qquad (14\text{-}102)$$

For this reaction, $x = 2$, so that at 25°C Eq. (14-101) takes the form

$$\mathcal{E}_0' = \mathcal{E}° - \frac{0.05916}{2} \log (10^{-7})^2$$

$$= +1.2292 - \frac{0.05916}{2}(-7)(2)$$

$$= +1.2292 + 0.4141 = +0.8151 \text{ volt} \quad (14\text{-}103)$$

TABLE 14-8 $\Delta G^°_{hydr}$ **and** $\Delta \overline{G}^°_f$ **for aqueous ions at 25°C**[a]

Ion	$\Delta G^°_{hydr}$, kcal mole^{-1}	$\Delta \overline{G}^°_f$, kcal mole^{-1}
H^+	-259	0
OH^-		-37.59
F^-	-103	-66.08
Cl^-	-75	-31.35
Br^-	-68	-24.57
I^-	-59	-12.35
Li^+	-121	-70.22
Na^+	-97	-62.59
K^+	-79	-67.46
Rb^+	-74	-67.65
Cs^+	-67	-70.8
Mg^{++}	-453	-108.99
Ca^{++}	-378	-132.18
Sr^{++}	-338	-133.2
Ba^{++}	-313	-13.40
Cu^{++}	-495	$+15.53$
Zn^{++}	-482	-35.18
Fe^{++}	-451	-20.30
Fe^{3+}	-1002	-2.53
Al^{3+}	-1099	-115

[a] From R. M. Noyes, *J. Am. Chem. Soc.*, **84**:513 (1962); **86**:97 (1964).

TABLE 14-7 $\Delta S^°_{hydr}$ **and** $\overline{S}°$ **for aqueous ions at 25°C**[a]

Ion	$\Delta S^°_{hydr}$, kcal mole^{-1}	$\overline{S}°$, cal deg^{-1} mole^{-1}
H^+	-29.3	0
OH^-		-2.52
F^-	-34.1	-2.3
Cl^-	-20.3	$+13.2$
Br^-	-16.3	$+19.29$
I^-	-11.3	$+26.14$
Li^+	-31.7	$+3.4$
Na^+	-23.8	$+14.4$
K^+	-15.7	$+24.5$
Rb^+	-12.9	$+28.7$
Cs^+	-12.2	$+31.8$
Mg^{++}	-70.3	-28.2
Ca^{++}	-56.8	-13.2
Sr^{++}	-53.1	-9.4
Ba^{++}	-44.8	$+3$
Cu^{++}	-72.2	-23.6
Zn^{++}	-70.5	-25.45
Fe^{++}	-76.7	-27.1
Fe^{3+}	-121	-70.1
Al^{3+}	-121.1	-74.9

[a] From R. M. Noyes, *J. Am. Chem. Soc.*, **84**:513 (1962); **86**:97 (1964).

In Fig. 14-14 the vertical scale shows values of $\mathcal{E}_0'$ in volts; the horizontal lines located at various heights with respect to this scale show the values of $\mathcal{E}_0'$ for a number of oxidation-reduction couples of biochemical importance. Just as the Cr^{++}–Cr^{3+} couple oxidizes the V^{++}–V^{3+} couple, any couple in column I will oxidize any couple lying at a position below it and will be reduced by any couple lying above it when the concentrations of the components of the two couples are all unity or when the ratio of concentrations for each couple are the same. For example, the reaction will proceed from left to right in the expression

Cresyl blue (ox) + methylene blue (red)
= cresyl blue (red) + methylene blue (ox)   (14-104)

Of course, because of the concentration effect on the value of $\mathcal{E}'$, it would be possible to make the reaction proceed from right to left by altering the ratios of the concentrations. In column II are some examples of

Fig. 14-14   Redox potentials for a series of indicators, carrier, and enzyme couples. (After B. Harrow and A. Mazur, ''Textbook of Biochemistry, 9th ed., W. B. Saunders Company, Philadelphia, 1966, as adapted from Stephenson, ''Biochemical Metabolism,'' Longmans, Green & Co., Ltd., London, 1939.)

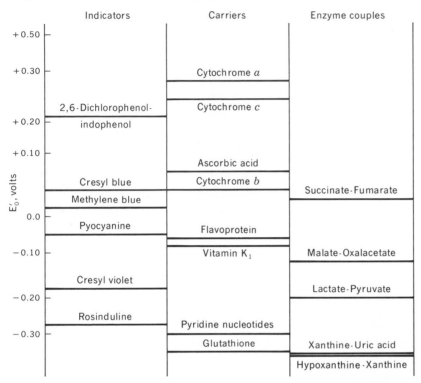

biochemical carriers of energy which behave in a manner similar to the indicator compounds listed in column I.

In column III are listed couples which undergo reactions only in the presence of appropriate enzymes. For example, if the methylene-blue couple were mixed with the hypoxanthine-xanthine couple which lies below it on the $\mathscr{E}_0'$ scale, the former would not oxidize the latter. However, if the enzyme xanthine dehydrogenase is added, the transfer of hydrogen to the dye takes place at a fairly rapid rate. In other words, there is a *tendency* for a couple in column III to be oxidized by any other couple lying above it on the $\mathscr{E}_0'$ scale, but the *rate* of reaction may be negligibly small unless other substances of the appropriate type called enzyme *catalysts* are present. This catalytic action will be discussed in the next chapter, dealing with chemical kinetics.

There are even instances where two couples can be present in the same solution each with its own appropriate enzyme but the reaction will not take place unless there is a *carrier* substance present. For example, hypoxanthine will not reduce fumaric acid when the two are mixed together even though each of these is in the presence of its catalytic enzyme; these substances can be made to interact, however, in the presence of cresyl violet, which serves as an intermediary or carrier between the two couple systems. The role of the carrier can be shown by writing the equations

Hypoxanthine + cresyl violet (ox)

$\qquad \rightarrow$ xanthine + cresyl violet (red)   (14-105)

Fumarate + cresyl violet (red)

$\qquad \rightarrow$ succinate + cresyl violet (ox)   (14-106)

The substances listed in column II function as carriers between biochemical couples.

Mitochondrial couples

In the interior of biological cells are found certain subcellular bodies called *mitochondria*. Within the mitochondria reactions take place which are important sources of energy to keep the life processes within the cell moving forward. In the course of these processes hydrogen atoms are removed from a number of the molecules within the mitochondrion and are joined with oxygen through a sequential oxidation-reduction process. The hydrogen and electron transfers take place in a stepwise series of reactions constituting a chain. In this way the chemical energy derived from the oxidation can be fed in appropriate small amounts to the molecules which require it as the life process goes forward.

The enzymes directly responsible for the oxidation are in the interior of the mitochondrion. The mitochondrial membrane is a substance made of lipoproteins, molecules with a structure partly resembling fats and partly proteins. On its surface is an organized array of other macromolecules, called coenzymes, which play the part of carriers similar to the action of cresyl violet. These include nicotinamide adenine dinucleotide (NAD), flavin adenine dinucleotide (FAD), flavin mononucleotide (FMN), and the cytochromes labeled respectively a, b, c, and a_3. The oxidized and reduced states of these substances are shown in Fig. 14-15 together with the value of $\mathcal{E}_0'$ for each couple. In the mitochondrion these couples make it possible to link the oxidation of any particular compound to the reduction of a specific member of the sequence of different molecules which participate in the chain of reactions by which molecular oxygen is consumed, carbon dioxide is liberated, and the energy so derived is transferred and becomes internal free energy in the energy-carrier class of compounds of which adenosine triphosphate (ATP) is the most important member.

The open structure of ATP is shown in Fig. 14-16a; the structure of the ATP complex with Mg^{++} is shown in Fig. 14-16b; and the structure of adenosine diphosphate is shown in Fig. 14-16c. ATP is found in biological cells in a concentration ranging between 0.001

and 0.005 m. At pH 7.0 all three of its phosphate groups are completely ionized. This ion tends to form complexes with divalent metallic ions found in the cell such as Mg^{++} and Ca^{++}, and the Mg^{++} complex is the normal state. The structure of this complex is important because in the energy-transfer reaction, the ATP must coincide geometrically with the underlying structure of the enzyme which controls the reactions; this aspect of the kinetics of the reaction will be discussed in detail in the following chapter.

The energy carried by ATP is released by enzymatic hydrolysis according to the equation

$$ATP^{4-} + H_2O \rightarrow ADP^{3-} + HPO_4^{--} + H^+ \qquad (14\text{-}107)$$

with the formation of adenosine diphosphate ions, phosphate ions, and hydrogen ions. At pH 7.0 and 25°C, $\Delta G_0' = -7$ kcal under standard conditions where each product is at unit thermodynamic activity. As pointed out previously, in the living biological cell the concentrations of ATP, ADP, and phosphate ions are much lower than unit activity, and complexes are formed with Mg^{++}. It is possible that under these conditions the free energy of hydrolysis may be as high as 12 kcal mole^{-1}.

The free-energy change for the oxidation of 1 mole of glucose is

$$\alpha\text{-D-Glucose} + 6O_2 \rightarrow 6CO_2 + 6H_2O$$
$$\Delta G° = -600 \text{ kcal} \qquad (14\text{-}108)$$

In order to use this large quantity of free energy efficiently, the oxidation must be carried out not in one step, as shown above, but in a number of much smaller steps which provide small bundles of energy that can be transported to different molecular sites in the organism where needed. With the help of the couples shown in Fig. 14-15 linked with appropriate enzymes, the energy derived from stepwise oxidation of compounds like glucose is transferred to compounds like ATP by reactions which are similar to the reverse of Eq. (14-107) and is distributed where needed. This illustrates the prime importance of oxidation-reduction couples in biological systems.

Example 14-12
From the value of $\Delta G°$ at 25°C estimated for the change of ATP to ADP and P, calculate the

Fig. 14-15 Ordered series of biochemical redox reactions in mito-chondria. (Data from B. Harrow and A. Mazur, "Textbook of Biochemistry," 9th ed., W. B. Saunders Company, Philadelphia, 1966.)

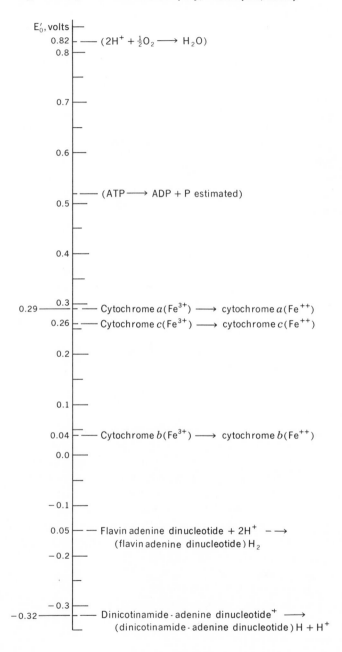

E'_0, volts

0.82 — $(2H^+ + \frac{1}{2}O_2 \longrightarrow H_2O)$

0.8

0.7

0.6

0.5 — $(ATP \longrightarrow ADP + P$ estimated$)$

0.4

0.29 — 0.3 — Cytochrome $a(Fe^{3+}) \longrightarrow$ cytochrome $a(Fe^{++})$

0.26 — Cytochrome $c(Fe^{3+}) \longrightarrow$ cytochrome $c(Fe^{++})$

0.2

0.1

0.04 — Cytochrome $b(Fe^{3+}) \longrightarrow$ cytochrome $b(Fe^{++})$

0.0

−0.1

0.05 — Flavin adenine dinucleotide $+ 2H^+ - \longrightarrow$
(flavin adenine dinucleotide) H_2

−0.2

−0.3

−0.32 — Dinicotinamide · adenine dinucleotide$^+ \longrightarrow$
(dinicotinamide · adenine dinucleotide) $H + H^+$

value of $\mathscr{E}_0'$ for a hypothetical emf cell in which this reaction takes place at $[ATP^{4+}]=[ADP^{3+}]=[HPO_4{}^{++}] = 10^{-3}\ m$ and $n = 1$.

$$\Delta G_0' = \Delta G^\circ + RT \ln \frac{[HPO_4{}^{++}][ADP^{3+}]}{[ATP^{4+}]}$$

$$= -7.0 \text{ kcal mole}^{-1} + 1.987 \text{ cal deg}^{-1}$$

$$\text{mole}^{-1} \times 298.15^\circ \ln \frac{(10^{-3})(10^{-3})}{10^{-3}}$$

$$= -7.0 + 0.592 \times 2.303 \times -3$$

$$= -7.000 - 4.1 = -11.1 \text{ kcal mole}^{-1}$$

$$\mathscr{E}_0' = \frac{-\Delta G_0'}{n\mathscr{F}} = -\frac{-11.1}{23.1} = 0.48 \text{ volt} \quad Ans.$$

Example 14-13

From the value of ΔG° at 25°C for the oxidation of 1 mole of glucose, calculate the number of moles of ATP which can be synthesized from 1 mole of glucose by reversing the reaction in Example 14-12 if all the free energy from the glucose oxidation is used in the synthesis at 25°C under standard conditions.

$$\frac{\Delta G^\circ(\text{glucose})}{\Delta G^\circ(\text{ATP})} = \frac{600 \text{ kcal}}{7 \text{ kcal mole}^{-1}} = 86 \text{ moles}$$

$$Ans.$$

Example 14-14

How many moles of cytochrome $c(Fe^{3+})$ can be formed from cytochrome $c(Fe^{++})$ with the free energy derived from the oxidation of 1 mole of glucose?

For the reaction

$$\text{Cytochrome } c(Fe^{++}) = \text{cytochrome } c(Fe^{3+}) + e^-$$

$$\mathscr{E}^\circ = -0.26 \text{ volt}$$
$$\Delta G^\circ = -n\mathscr{F}\mathscr{E}^\circ = -1 \times 23.1 \times -0.26$$
$$= 6 \text{ kcal}$$

$$\frac{\Delta G^\circ(\text{glucose})}{\Delta G^\circ(\text{cyt})} = \frac{600 \text{ kcal}}{6 \text{ kcal mole}^{-1}}$$

$$= 100 \text{ moles} \quad Ans.$$

Fig. 14-16 Structures (schematic in two dimensions) for adenosine triphosphate (ATP): (a) in extended ionized form; (b) complexed with Mg^{++}. Adenosine diphosphate (ADP) is shown in (c).

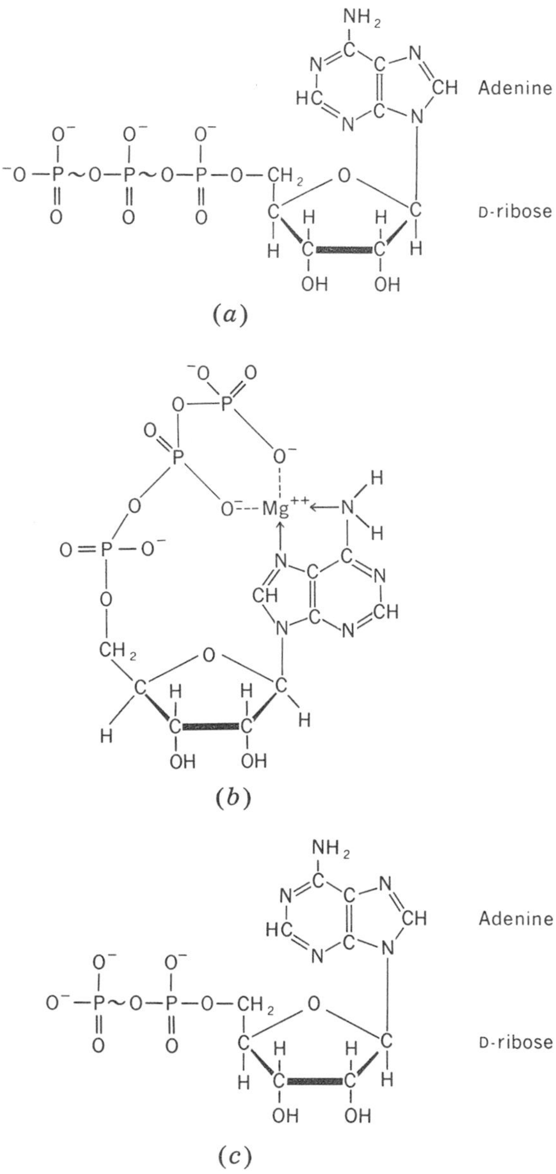

(a)

(b)

(c)

Problems

1. Calculate the potential of a cell at 25°C where the left half-cell has the reaction $Ba^{++} + 2e^- = Ba(s)$ and the right half-cell has the reaction $F_2 + 2e^- = 2F^-$.

2. Calculate the potential at 25°C for a cell where the half-cell on the left has the equation $Al^{3+} + 3e^- = Al(s)$ and the half-cell on the right has the equation $Cl_2 + 2e^- = 2Cl^-$.

3. Calculate the free-energy change for the half-cell reaction at 25°C

$$Cd^{++} + 2e^- = Cd(s).$$

4. Calculate the free-energy change for the half-cell reaction at 25°C

$$U^{3+} + 3e^- = U(s).$$

5. Calculate the free-energy change $\Delta G°$ at 25°C for the emf cell where the left half has the reaction $Ni^{++} + 2e^- = Ni(s)$ and the right half has the reaction $Hg^{++} + 2e^- = Hg$.

6. Calculate the value of $\Delta G°$ at 25°C for the cell where the left half has the reaction $Th^{4+} + 4e^- = Th(s)$ and the right half has the reaction $Ag^+ + e^- = Ag(s)$.

7. Calculate the net free-energy change when cytochrome $a(Fe^{3+})$ is reduced to cytochrome $a(Fe^{++})$ and thereby oxidizes coenzyme Q from the reduced to the oxidized state.

8. What redox reaction in the list of those in mitochondria (Fig. 14-14) will supply the minimum amount of free energy necessary to change L-malate^{--} to oxalacetate^{--}? Calculate the net free-energy change.

9. If a compound requires 3.5 kcal of free energy for its reduction, what is the $\mathcal{E}_0'$ of its redox reaction, assuming that $n = 1$? What reaction in the list in Fig. 14-15 will provide the minimum free energy required?

10. Calculate the difference in free energy between the redox reaction for cytochrome b and cytochrome c.

11. If the free energy of the change from ATP to ADP is used to oxidize cytochrome c in the reduced form to the oxidized form, how many moles of the latter will be oxidized?

12. At 25°C, a half-cell is found to have the potential of -0.300 volt. The equation for the cell is $V^{3+} + e^- = V^{++}$. If the activity of V^{++} has the value of 0.85 m, what is the value of the activity of V^{3+}?

13. At 25°C, the potential is found to be 0.8167 volt for the cell where the reaction is $Ag^+ + e^- = Ag(s)$. Calculate the value of the concentration of Ag^+.

14. For a cell with the reaction $\frac{1}{2}H_2(g) + AgCl(s) = Ag(s) + H^+(aq) + Cl^-(aq)$, the value of $\mathcal{E}°$ is 0.3341 volt, and $\Delta H° = -9391$ cal at 25°C. Calculate the value of $\partial E/\partial T$.

15. A cell is constructed where $n = 1$, $\mathcal{E}° = +0.331$,

and the entropy change for the whole cell under standard conditions at 25°C is $+15.3$ eu. Calculate the value of the enthalpy change and the value of $d\mathcal{E}/dT$ for this cell.

16. The indicator diphenylamine sulfonic acid changes from deep violet red at $\mathcal{E} = 0.93$ to colorless at $\mathcal{E} = 0.75$. Calculate the ratio of the value of [red]/[ox] from full color to the value at which the solution is colorless at 25°C ($n = 1$).

17. At 30°C, the value of $\mathcal{E}$ is found to be 3.7 volt for a solution of methylene blue where [red] = [ox]. What is the value of pH in this solution?

18. When the value of pH is 8.0, the value of $\mathcal{E}$ is -0.020 for the redox couple methylene blue. In a solution where the ratio of the colored form to the colorless form is the same as above, the potential is found to be 0.00 volt. Estimate the value of pH in this solution.

19. In a solution where the color of methylene blue is determined against a set of standard colors, it is found that the ratio [ox]/[red] = 3 at 30°C and pH = 7.0. Estimate the redox potential.

20. In a solution at 30°C, the color of the methylene blue indicates that the fraction in the reduced form is 0.55. The redox potential of the solution is $+0.60$ volt. Estimate the value of pH.

REFERENCES

Powell, R. E., and W. M. Latimer: An Application of the Sackur-Tetrode Equation to Ions, *J. Chem. Phys.*, **19**:1139 (1951).

Conway, B. E.: "Electrochemical Data," Elsevier Publishing Company, Amsterdam, 1952.

Latimer, W. M.: "Oxidation Potentials," 2d ed., Prentice-Hall, Inc., Englewood Cliffs, N.J., 1959.

Clark, W. M.: "Oxidation-Reduction Potentials of Organic Systems," The Williams & Wilkins Company, Baltimore, 1960.

Nachmansohn, D. (ed.): "Molecular Biology," pp. 37–47, Academic Press, Inc., New York, 1960.

Ives, D. J. G., and G. J. Janz: "Reference Electrodes," Academic Press, Inc., New York, 1961.

MacInnes, D. A.: "The Principles of Electrochemistry," Dover Publications, Inc., New York, 1961.

Daniels, F., J. W. Williams, P. Bender, R. A. Alberty, and C. D. Cornwell: "Experimental Physical Chemistry," 6th ed., pp. 183–211, McGraw-Hill Book Company, New York, 1962.

Karlson, P.: "Introduction to Modern Biochemistry," pp. 185–209, trans. by Charles H. Doering, Academic Press, Inc., New York, 1963.

Bates, R. G.: "Determination of pH: Theory and Practice," John Wiley & Sons, Inc., New York, 1963.

Noyes, R. M.: The Entropy of Hydration of Ions, *J. Am. Chem. Soc.*, **84:**513–522 (1962); **86:**97 (1964).

Conway, B. E.: Electrolyte Solutions: Solvation and Structural Aspects, *Ann. Rev. Phys. Chem.*, **17:**481 (1966).

Fairley, J. L., and G. L. Kilgour: "Essentials of Biological Chemistry," 2d ed., Reinhold Publishing Corporation, New York, 1966.

Harrow, B., and A. Mazur: "Textbook of Biochemistry," 9th ed., pp. 167–198, W. B. Saunders Company, Philadelphia, 1966.

Hampel, C. A. (ed.): "The Encyclopedia of Electrochemistry," Reinhold Publishing Corporation, New York, 1967.

Morris, D. F. C.: Ionic Radii and Enthalpy of Hydration, *Struct. Bonding (Berlin)*, **4:**63–82 (1968).

In the preceding chapters physical and chemical processes have been discussed principally from the point of view of thermodynamics rather than kinetics. Strictly speaking, the laws of pure thermodynamics apply only to systems that are in true equilibrium. Of course, this equilibrium is thought of not as a static process like the equilibrium when the two arms of a chemical balance are perfectly matched in weight but as a dynamic equilibrium in which the rate of conversion of the components on the left side of the equals sign in a chemical equation into the components on the right side is exactly the same as the rate of conversion in the opposite direction from right to left. In thermodynamics, the concept

Chapter Fifteen

chemical kinetics

of the quantity *free-energy* change for the reaction is derived, a quantity which is equal to zero when the system is in equilibrium. The fact is also established that if the free-energy change is negative, there is a *driving force* which tends to act on the system in such a direction that the components on the left of the equals sign, called the *reactants,* are converted into the components on the right of the equals sign, called the *products;* under these conditions the rate at which the *forward* reaction takes place (from left to right) is greater than the rate at which the *backward* reaction takes place from right to left. However, in thermodynamics, no attempt is made to find out *how much* the rate of reaction increases as the difference between the free energy of the reactants and the free energy of the products increases. The factor of *time* does not enter in to pure thermodynamics.

Another important aspect of thermodynamics as contrasted with kinetics is the lack of concern with mechanism. The laws of thermodynamics are completely independent of the mechanism by which the reaction takes place. The statement is frequently made that from the point of view of thermodynamics, the reactants flow into one end of a "black box"; there they are mysteriously changed into the products which flow out of the other end of the box; thermodynamics is not concerned with, and cannot tell anything about, the mechanism by which the change takes place within the box. In fact, there may be several paths by which such a change can take place, and each may have its own distinct mechanism; but the change of the state variables as the system passes from state I (the reactants) to state II (the products) is independent of the path by which the change takes place; i.e., is independent of mechanism. The whole structure of thermodynamics is founded on this concept. By way of contrast, in chemical kinetics, the rate at which a system passes from state I to state II is highly dependent on the path by which the change takes place. For this reason, chemical kinetics is greatly concerned with the nature of the mechanism by which the change takes place.

Consider the reaction by which hydrogen and oxygen in the form of gases change into water, also in the form of a gas. At 400°K (127°C) the reaction and its free-energy change are

$$H_2(g) + \tfrac{1}{2}O_2(g) \rightarrow H_2O(g) \qquad \Delta G° = -53.517 \text{ kcal} \tag{15-1}$$

The value of the equilibrium constant K_p is 1.74×10^{29} atm$^{-1/2}$. From this, it is evident that there is a large driving force favoring the progress of the reaction from left to right under standard conditions. Yet when gaseous hydrogen and oxygen are mixed together at this temperature, the reaction forming water proceeds at an almost infinitesimally slow rate. However, if a spark is introduced into the mixture, the path by which the reaction proceeds is changed and the formation of water takes place so rapidly that the mixture explodes.

The same considerations apply to changes of physical state. The vapor pressure of water at 25°C is 23.756 mm Hg. If liquid water is placed in a closed container where the pressure of water vapor in the space above the liquid water is only 20 mm Hg, the water evaporates with moderate rapidity, changing from the liquid to the vapor state by the process

$$H_2O(l) \rightarrow H_2O(g) \qquad \Delta G° = -103 \text{ cal} \tag{15-2}$$

But if stearic acid is added so that a unimolecular film is formed over the surface of the water, the rate of evaporation is greatly reduced because the mecha-

nism of the process of evaporation changes; yet the change in free energy in going from the liquid to the vapor state still has the value -103 cal.

15-1   RATES OF CHANGES OF PHYSICAL STATE

The factors affecting rates of change of state can be described most simply by considering a change of physical state like that shown schematically in Fig. 15-1, which depicts a large reservoir of gas at sea level connected to a tubular column rising to a height of a little more than 10,000 m (10^6 cm). Near the top of the column there is a hole 1 cm^2 in area through which the gas diffuses into an evacuated space. The pressure in the reservoir at the bottom of the column is 1 atm, and the temperature throughout reservoir and column is 25°C.

With the help of the kinetic theory of gases, it is possible to calculate the rate at which gas diffuses through the hole. Since the purpose of the analysis of this physical change is to lay the basis for the study of rates of chemical change, rate will be expressed in terms of moles per second.

The pressure P of the gas at the level of the hole is considerably less than the pressure P_0 of the gas at ground level because of the action of the force of gravity. Imagine first the artificial situation shown in Fig. 15-2, where on the left side of the hole a tube 1 cm^2 in cross section a contains gas molecules at the same density as those in the column in Fig. 15-1 but with all molecules moving to the right toward the hole with the average velocity u, which is the same as that of the molecules of the column of gas in Fig. 15-1 at the level of the hole. Then, the number of cubic centimeters of gas moving per second through the hole is Au. Actually, in a typical column of gas the molecules are moving in all possible directions in space. Only half of them have a component of motion toward the hole. Of this half, some move directly toward the hole, others at an angle, and still others vertically. As shown in the kinetic theory of gases, the average component of motion of the molecules toward the hole is one-half of the average velocity. The product of two factors each equal to one-half is one-quarter. Therefore, the rate of diffusion of the gas through the unit-area hole in terms of cubic centimeters per second is $1/4\,u$. The rate r_g in grams per second is $\frac{1}{4}\rho u$, where ρ is the density of the gas in grams per cubic centimeter. The rate r in units of moles per second is $\frac{1}{4}u/\text{v}$, where v is the volume of 1 mole of gas in cubic centimeters per mole.

According to the kinetic theory of gases, the average velocity u is related to the rms velocity U_{rms} by the equation

$$\bar{u} = 0.921\,U_{\text{rms}}$$
$$= \frac{0.921\,\sqrt{u_1^2 + u_2^2 + u_3^2 + \cdots + u_N^2}}{N} \quad (15\text{-}3)$$

where u_1, u_2, u_3, $\ldots$, u_N are the velocities of the individual molecules and N is the total number of molecules. U_{rms} is related to the kinetic energy E_k per mole of the gas by the equation

$$\text{E}_k = \tfrac{1}{2}MU_{\text{rms}}^2 = \tfrac{3}{2}R_eT \quad (15\text{-}4)$$

In the above equation M stands for the molecular weight of the gas. As shown above, E_k is also equal to $\frac{3}{2}$ multiplied by the gas constant R_eT, where the gas constant R_e is expressed in ergs per degree per mole. Thus, it follows that

$$\bar{u} = 0.921\,\sqrt{\frac{3R_eT}{M}} \quad (15\text{-}5)$$

The rate r_c in units of grams per second for diffusion through an aperture of cross-section area a is

$$r_c = \frac{a}{4}\rho\bar{u} = \frac{a}{4}\frac{M}{\text{v}}\bar{u} \quad (15\text{-}6)$$

The rate r in units of moles per second is

$$r = \frac{ar_c}{M} = \frac{a}{4}\frac{\bar{u}}{\text{v}} \quad (15\text{-}7)$$

According to the ideal-gas law,

$$\text{v} = \frac{R_cT}{P} \quad (15\text{-}8)$$

where R_c is the gas constant expressed in cubic centimeters per atmosphere per mole. From this it follows that

$$r = \frac{a}{4}\frac{\bar{u}}{R_cT/P} = \frac{a}{4}\frac{\bar{u}}{R_cT}P \quad (15\text{-}9)$$

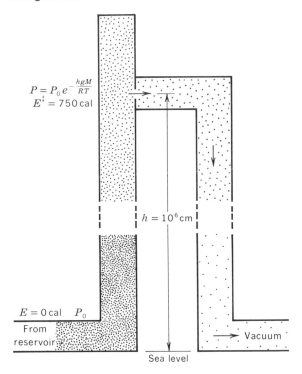

Fig. 15-1   Diffusion of a gas against a gravitational field and through a hole.

$$P = P_0 e^{-\frac{hgM}{RT}}$$
$$E^\ddagger = 750\,\text{cal}$$

$$h = 10^6\,\text{cm}$$

$E = 0\,\text{cal}$ P_0

From reservoir

Vacuum

Sea level

Combining this with Eq. (15-5),

$$r = \frac{a}{4}\frac{0.921}{R_cT}\sqrt{\frac{3R_eT}{MP}} \quad (15\text{-}10)$$

Combining the numerical values of the constants leads to

$$r = \left(\frac{44.31\,a}{\sqrt{MT}}\right)P \quad (15\text{-}11)$$

For oxygen gas at a temperature of 25°C, the constant term in parentheses is equal to 0.4536 mole sec^{-1} atm^{-1}.

In Chap. 6 the variation of pressure with altitude was studied, and Eq. (6-34) was derived, namely,

$$\frac{P}{P_0} = e^{-(Mgh/R_eT)} \quad (15\text{-}12)$$

where P = pressure at a height of h cm, atm
P_0 = pressure at level where $h = 0$, atm

Fig. 15-2   Kinetics of gaseous diffusion.

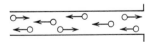

(*a*) Artificial velocity distribution:
all molecules moving in a parallel stream
normal to hole, all moving to the right.

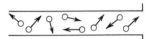

(*b*) Artificial velocity distribution:
all molecules move in parallel streams normal to
hole, half toward the right and half toward the
left. One-half of molecules pass through hole.

(*c*) Random velocity distribution:
the equivalent of one-fourth of the molecules
in the cylindrical space normal to the hole
gets through the hole.

e = basis of natural logarithm
M = molecular weight, g
R_e = gas constant, ergs deg^{-1} mole^{-1}
T = temperature, °K
g = acceleration of gravity under standard
conditions = 980.665 cm sec^{-1}

For the problem under consideration, $M = 32.00$ g mole^{-1} and $h = 10^6$ cm; the product Mgh is the increase in the average energy of 1 mole of gas when it is moved upward against the force of gravity to the height h. This is equal to 3.138×10^{10} ergs, or 3.138×10^3 joules. Since 1 joule is equal to 0.239 defined calorie, the increase in energy per mole of the gas is 750 cal. In accordance with chemical usage, the symbol $E^{\ddagger}$ will be used to denote this increase in energy due to the height. Using this notation and combining Eqs. (15-11) and (15-12) gives

$$r = \left(\frac{44.31}{\sqrt{MT}}\right) e^{-E^{\ddagger}/RT} P_0 \tag{15-13}$$

When considering the variation of the rate of diffusion of the gas through the hole at height h as a function of the pressure at ground level ($h = 0$), Eq. (15-13) can be written in the very simple form

$$r = kP_0 \tag{15-14}$$

where k is an abbreviation for all the terms which precede P_0 in Eq. (15-13) and is called the *reaction-rate constant*. In many analyses of reaction-rate problems, the terms preceding the exponential factor in Eq. (15-13) are denoted by the symbol A, which in this case is defined by the equation

$$A \equiv \frac{44.31\,a}{\sqrt{MT}} \tag{15-15}$$

A is sometimes called the *preexponential factor*. In this example, the dimensions of A are moles per second per atmosphere. The actual reaction in this case is the diffusion of the gas through the hole and takes place as a result of a series of individual *events*, the passage of individual molecules through the hole. Whether any individual molecule undergoes this reaction is a question of probability. The extremely complex chain of events leading to the reaction is made up of the large number of collisions of the gas molecules with each other and with the walls of the vessel. As can be seen from Eq. (15-15), the probability of the passage of a gas molecule through the opening is proportional to the size of the opening a. Roughly speaking, the probability of passage through the hole in any given unit of time is smaller if the molecule is heavier, since heavier molecules move more slowly; this is shown by the factor $\sqrt{M}$ in the denominator in Eq. (15-15). Under conditions where the pressure remains the same, an increase in temperature causes a decrease in density of the gas and thus decreases the probability that any molecule will pass through the hole in any given unit of time, as shown by $\sqrt{T}$ in the denominator of Eq. (15-15). Since A is in effect a probability factor, it is possible to define a special kind of entropy term denoted by the symbol $S^{\ddagger \circ}$ by means of the equation

$$S^{\ddagger \circ} \equiv R \ln \frac{A}{A^{*}} \tag{15-16}$$

which is similar in form to the equation relating the change in entropy of a gas when the probability of the state of the gas is increased by the change in volume

$$S_{II} - S_I = R \ln \frac{Y_{II}}{Y_I} \tag{15-17}$$

where S_I and S_{II} are the values of the entropy in state I and state II and the ratio Y_{II}/Y_I is equal to the ratio of the probabilities in these two states, which, in turn, is equal to the ratio of the volumes V_{II}/V_I. A^* has the numerical value of unity and the same dimensions as A. Note that the zero point of the scale for $S^{\ddagger\circ}$ is fixed by the relation that the rate r must be unity in the units chosen for measurement of rate when $E^{\ddagger} = 0$, $S^{\ddagger\circ} = 0$, and $P = 1$ in the units chosen for the measurement of pressure.

Using the definition expressed in Eq. (15-16),

$$\ln \frac{A}{A^*} = \frac{S^{\ddagger\circ}}{R} \tag{15-18}$$

Thus, k can be expressed

$$k = A^* e^{-E^{\ddagger}/RT} e^{S^{\ddagger\circ}/R} \tag{15-19}$$

The two exponential terms can be combined

$$k = A^* e^{-(E^{\ddagger} - TS^{\ddagger})/RT} \tag{15-20}$$

When the reaction is thought of in terms of the disappearance per second of a certain number of moles from the system, the energy change due to the contraction of the system should be taken into account, in order to express the precise change in the total energy of the system. In thermodynamics, *enthalpy* is used for this purpose and is defined by the equation

$$H \equiv E + PV \tag{15-21}$$

The change in enthalpy per mole is then given by

$$\Delta H = \Delta E + \Delta(Pv) \tag{15-22}$$

In reaction kinetics, the change referred to by the symbol Δ is the change from the normal ground state to the state of higher energy, or *activated state,* in which the molecules are capable of undergoing the reaction. In the example under consideration, this corresponds to the change from the molecules of the oxygen at ground level to the state at the high altitude, where the molecules with the higher energy Mgh can pass through the hole. Thus, $E^{\ddagger}$ is called the *activation energy* and is the increase in energy per mole of the molecules when they move from the ground state to the activated state. There is a similar increase in

activation enthalpy H, which can be defined in terms of Eq. (15-22) and written in reaction-rate symbolism as

$$H^{\ddagger} = E^{\ddagger} + \Delta(Pv)^{\ddagger} \tag{15-23}$$

In most chemical reactions, the term $\Delta(Pv)^{\ddagger}$ is so small as to be negligible. In the physical reaction under consideration here, oxygen may be regarded as an ideal gas so that when the oxygen passes from the ground state to the activated state, the increase in volume per mole is exactly compensated for by the decrease in pressure and the term $\Delta(Pv)^{\ddagger}$ is effectively equal to zero. Under these circumstances, Eq. (15-20) can be written

$$k = A^* e^{-(H^{\ddagger} - TS^{\ddagger})/RT} \tag{15-24}$$

There is an advantage in using $H^{\ddagger}$ instead of $E^{\ddagger}$ even though the two may be effectively equivalent. Shortly, we shall want to contrast the rate equations for a pair of reactions which proceed in opposite directions and which correspond to chemical *equilibrium* when the two rates are equal. When $H^{\ddagger}$ is used instead of $E^{\ddagger}$, the rate equations combine to give the familiar thermodynamic equations.

Again, in a manner analogous to thermodynamics, a quantity $G^{\ddagger\circ}$, called the *free energy of activation* is defined by the expression

$$G^{\ddagger\circ} \equiv H^{\ddagger} - TS^{\ddagger\circ} \tag{15-25}$$

Thus, Eq. (15-26) may be put in the form

$$k = A^* e^{-G^{\ddagger\circ}/RT} \tag{15-26}$$

or in the alternative form

$$G^{\ddagger\circ} = -RT \ln \frac{k}{A^*} \tag{15-27}$$

These equations are all closely analogous to equations relating similar quantities in thermodynamics.

Opposed rates

Consider now an apparatus like that shown in Fig. 15-3, which is similar to that in Fig. 15-1 but has two reservoirs of gas so that there can be diffusion in both directions through the hole at the top. The reservoir at the left is at ground level and contains oxygen at a pressure denoted by the symbol P_f atm.

Fig. 15-3   Opposed diffusion streams of gas in a gravitational field.

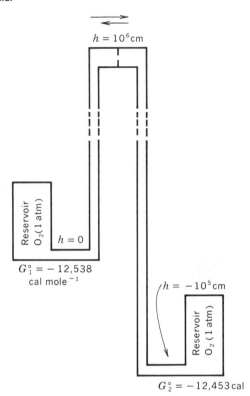

$G_1^\circ = -12,538$
cal mole^{-1}

$G_2^\circ = -12,453$ cal

The reservoir at the right is located in a well 10^5 cm below ground level so that $r = -10^5$ cm. The gas in this reservoir is maintained at a pressure of P_b atm. The temperature of the entire system is kept at 25°C.

Under these circumstances, molecules from the reservoir on the left eventually pass through the hole at the top from left to right and go to the reservoir on the right. The equation expressing such a change is

$$O_2(P_f, h = 0) \xrightarrow{f} O_2(P_b, h = -10^5 \text{ cm}) \qquad (15\text{-}28)$$

This change is called the *forward* reaction, and quantities associated with it will be designated by a subscript f.

Molecules from the reservoir on the right ascend the column and in smaller numbers pass from the right to the left through the hole. This change of state is called the *back* reaction and is expressed by the equation

$$O_2(P_b, h = -10^5 \text{ cm}) \xrightarrow{b} O_2(P_f, h = 0) \qquad (15\text{-}29)$$

or, in alternative form with the direction of the equation reversed,

$$O_2(P_f, h = 0) \xleftarrow{b} O_2(P_b, h = -10^5 \text{ cm}) \qquad (15\text{-}30)$$

Equation (15-29) can be subtracted from Eq. (15-28), or Eq. (15-30) can be added to Eq. (15-28), and the result is

$$O_2(P_f, h = 0) \underset{b}{\overset{f}{\rightleftharpoons}} O_2(P_b, h = -10^5 \text{ cm}) \qquad (15\text{-}31)$$

The rate r_f at which the forward reaction takes place is given by

$$r_f = k_f P_f = A^* e^{-G_f^{\ddagger\circ}/RT} P_f \qquad (15\text{-}32)$$

The rate r_b at which the back reaction takes place is given by

$$r_b = k_b P_b = A^* e^{-G_b^{\ddagger\circ}/RT} P_b \qquad (15\text{-}33)$$

$G_f^{\ddagger\circ}$ and $G_b^{\ddagger\circ}$ are the free energies of activation when the reaction takes place under standard conditions, i.e., when $P_f = P_b = 1$.

The difference between the two rates is given by the expression

$$r_f - r_b = k_f P_f - k_b P_b = A^*(e^{-G_b^{\ddagger\circ}} P_f - e^{-G_b^{\ddagger\circ}}) P_b \qquad (15\text{-}34)$$

Now, suppose that P_f and P_b are adjusted so that $r_f = r_b$ and equilibrium conditions are established. Then

$$k_f P_f = k_b P_b \qquad (15\text{-}35)$$

By definition, the equilibrium constant K is equal to the ratio P_f/P_b

$$K \equiv \frac{P_f}{P_b} \qquad (15\text{-}36)$$

Rearranging Eq. (15-35),

$$\frac{k_f}{k_b} = \frac{P_f}{P_b} \qquad (15\text{-}37)$$

Thus, the important conclusion is reached that the equilibrium constant is equal to the rate constant for the forward reaction divided by the rate constant for the back reaction

$$K = \frac{k_f}{k_b} \qquad (15\text{-}38)$$

When the expanded expressions for k_f and k_b are used to replace these quantities in the ratio, then

$$K = \frac{e^{-G_f^{\ddagger\circ}/RT}}{e^{-G_b^{\ddagger\circ}/RT}} \qquad (15\text{-}39)$$

Combining exponents,

$$K = e^{-(G_f^{\ddagger\circ}-G_b^{\ddagger\circ})/RT} \qquad (15\text{-}40)$$

The relation between K and the standard free-energy change was derived in Chap. 12

$$\Delta G^\circ = -RT \ln K \qquad (15\text{-}41)$$

Rearranging this expression in exponential form,

$$K = e^{-\Delta G^\circ/RT} \qquad (15\text{-}42)$$

Combining Eqs. (15-40) and (15-42),

$$G_f^{\ddagger\circ} - G_b^{\ddagger\circ} = \Delta G^\circ \qquad (15\text{-}43)$$

In other words, the change in free energy when the reaction takes place under standard conditions ΔG° is equal to the difference between the *standard* free energies of activation for the forward and backward reactions.

When the reaction takes place under standard conditions,

$$S_f^{\ddagger\circ} - S_b^{\ddagger\circ} = \Delta S^\circ \qquad (15\text{-}44a)$$

where $S_f^{\ddagger\circ}$ and $S_b^{\ddagger\circ}$ are the entropies of activation under standard conditions. When the gas is ideal, $H^\ddagger$ is not affected by the pressure and under all circumstances

$$H_f^\ddagger - H_b^\ddagger = \Delta H^\circ \qquad (15\text{-}44b)$$

Consider now the case where P_f and P_b have values which are neither unity nor result in equilibrium. When the components of the reactions are not in the standard state, pressure terms must enter into the rate expressions

$$r_f = k_f P_f \qquad (15\text{-}45a)$$
$$r_b = k_b P_b \qquad (15\text{-}45b)$$

The values of the entropy of activation and of the free energy of activation under nonstandard conditions are denoted by symbols without the superscript circle, $S^\ddagger$ and $G^\ddagger$. The relations are

$$S_f^\ddagger = S_f^{\ddagger\circ} + R \ln P_f \qquad (15\text{-}46a)$$
$$S_b^\ddagger = S_b^{\ddagger\circ} + R \ln P_b \qquad (15\text{-}46b)$$

$$G_f^\ddagger = H_f^\ddagger - TS_f^\ddagger = G_f^{\ddagger\circ} - RT \ln P_f \qquad (15\text{-}47a)$$
$$G_b^\ddagger = H_b^\ddagger - TS_b^\ddagger = G_b^{\ddagger\circ} - RT \ln P_b \qquad (15\text{-}47b)$$

Under nonstandard conditions

$$r_f = A^* e^{-G_f^\ddagger/RT} = k_f P_f \qquad (15\text{-}48a)$$
$$r_b = A^* e^{-G_b^\ddagger/RT} = k_b P_b \qquad (15\text{-}48b)$$

where A^* is the rate at which the reaction takes place when $G^\ddagger = 0$. The zero of the scales for $G^\ddagger$ and $S^\ddagger$ usually are selected so that A^* is the unit rate, such as 1 mole sec^{-1}.

The meaning of these important equations can be shown graphically by drawing *reaction profiles*. The profile for the *free energy* of the reaction defined in Eq. (15-31) with $P_f = P_b = 1$ atm is shown in Fig. 15-4. For the forward reaction, the initial state I has

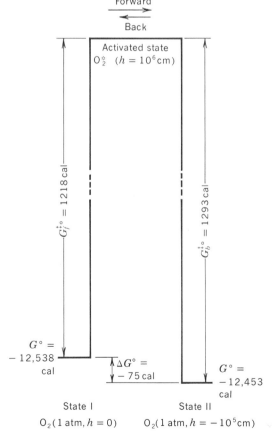

Fig. 15-4 Free-energy profile for O_2 diffusion in a gravitational field under standard conditions at 25°C.

Fig. 15-5 Enthalpy profile for O_2 diffusion in a gravitational field under standard conditions at 25°C.

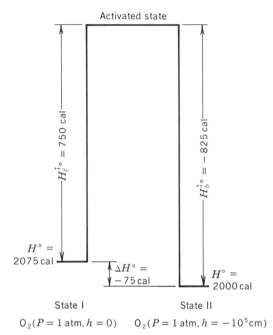

Activated state

$H_f^{\ddagger\circ} = 750\ cal$

$H_b^{\ddagger\circ} = -825\ cal$

$H^\circ = 2075\ cal$

$\Delta H^\circ = -75\ cal$

$H^\circ = 2000\ cal$

State I   State II

$O_2(P = 1\ atm, h = 0)$ $O_2(P = 1\ atm, h = -10^5 cm)$

a value of the free energy shown by the horizontal line (lower left) marked $O_2(1\ atm,\ h = 0)$. This is the state in the reservoir at the lower left in Fig. 15-3. In the forward reaction the gas rises up through the gravitational field until a few of the molecules attain the free energy in the activated state (O_2^*) with the value $G_f^{\ddagger\circ} = 1218\ cal$. The gas then diffuses through the hole, and molecules pass down the column at the right, losing the standard free energy of activation $G_b^{\ddagger\circ} = 1293\ cal$. At the end of the reaction, the gas is in state II, with the standard pressure of 1 atm and a value of $h = -10^5$ cm. The difference between the height representing $G_f^{\ddagger\circ}$ and $G_b^{\ddagger\circ}$ is the height representing ΔG°, the standard free-energy change for the reaction, which has the value -75 cal.

The *enthalpy parts* of the two values of the free energies of activation are shown schematically as in Fig. 15-5. The level for state I is the same as in the previous figure. Here, however, the height in the activated state corresponds to a value of $H_f^{\ddagger} = 750$ cal. The level for H in state II is below that of state I because the reservoir is at a depth of 10^5 cm below

the ground level of state I. Thus, the height of the enthalpy change on the right-hand side is $H_b^{\ddagger} = +825$ cal. The difference between $H_f^{\ddagger}$ and $H_b^{\ddagger}$ is $\Delta H^\circ = -75$ cal, the standard enthalpy change for the reaction.

The other part of the standard free energy of activation is made up of the negative of the product of T and the standard entropy of activation $S_f^{\ddagger\circ}$, as shown in Fig. 15-6. The horizontal line at the lower left corresponds to the temperature-entropy product for oxygen gas at a temperature of 25°C, $P = 1$ atm, and $h = 0$. The value of $-TS_f^{\ddagger\circ}$ is plotted vertically. The reason for the choice of a negative value will be explained in connection with Fig. 15-7. The height at the left corresponds to the value of this product for the forward reaction, 468 cal. Since the same probability factors, such as the size of the hole, the molecular weight, and the temperature, appear in exactly the same way for the backward reaction as they do for the forward reaction, $-TS_b^{\ddagger\circ}$ has the same value as for the forward reaction.

The free energy of activation for the forward reaction can be expressed as the sum of the enthalpy term and the negative temperature-entropy product, as shown in the equation

$$G_f^{\ddagger\circ} = H_f^{\ddagger} + (-TS_f^{\ddagger\circ}) \tag{15-49}$$

1218 cal   750 cal   478 cal

A similar equation holds for the back reaction

$$G_b^{\ddagger\circ} = H_b^{\ddagger} + (-TS_b^{\ddagger\circ}) \tag{15-50}$$

1293 cal   825 cal   468 cal

The probability factor A has a value which is generally a positive fraction less than unity; consequently, the value of $S^{\ddagger\circ}$ is generally negative since it is the logarithm of a positive fraction less than unity. As shown by the numerical values under Eqs. (15-49) and (15-50), the factors $-TS_f^{\ddagger\circ}$ and $-TS_b^{\ddagger\circ}$ have the positive value of 468 cal in this particular example. Thus, as shown in Fig. 15-7, the activation enthalpy and the activation temperature-entropy factor add together to produce the value of the activation free energy.

In the literature, these heights are frequently referred to as barriers over which (in terms of energy) the molecules must pass in order to react. Formerly, most of the discussion was concerned with the barrier representing the standard energy of activation. This, of course, is more precisely represented as an *enthalpy* barrier as in Fig. 15-5. Somewhat later, it was realized that the real barrier the molecules have to cross is the *free-energy barrier,* shown in Fig. 15-7, which is really the two barriers, temperature-entropy and enthalpy, placed one on top of the other. This visual representation is frequently helpful in understanding the factors which govern the rate of the reaction. In order to present this relationship in as simple and vivid form as possible, it is helpful to regard the free energy of activation as made up of the *sum* of the enthalpy of activation and the *negative* temperature-entropy activation factor, which numerically is almost always a positive quantity.

The actual numerical value of the *enthalpy* of activation depends, of course, on the units in which it is expressed, which may be calories, kilocalories, or any other useful energy unit. The numerical value of the *entropy* of activation depends not only on the choice of the energy part of the unit, such as the calories in calories per degree per mole, but also on the choice of the unit in which the *rate* is expressed. In this example, r_f has been expressed by the equation

$$r_f = \frac{44.31}{\sqrt{MT}}\, ae^{-H_f{}^{\ddagger\circ}/RT}P_0$$

$$= Ae^{-H^{\ddagger\circ}/RT}P_0 \tag{15-51}$$

As can be seen from the derivation in the earlier part of this section, the units of r are moles sec^{-1} cm^{-2} atm^{-1}. When $H^{\ddagger} = 0$, $P_0 = 1$ atm, and $a = 1$ cm^2, the rate is numerically equal to $44.31/\sqrt{MT}$. The zero of the scale for $S^{\ddagger\circ}$ must be selected so that under the standard rate conditions described in the previous sentence, the relation

$$r_f^\circ = e^{S^{\ddagger\circ}/RT} = \frac{44.31}{\sqrt{MT}} \tag{15-52}$$

holds. r_f° is rate under standard conditions where the enthalpy of activation is zero, the aperture is 1 cm^2, and the pressure is 1 atm. Thus, the absolute value of $S_f^{\ddagger\circ}$ is given by

$$S_f^{\ddagger\circ} = RT \ln r^\circ = RT \ln \frac{44.31}{\sqrt{MT}} \tag{15-53}$$

Since the zero point for the scale for free energy of activation depends on the zero point chosen for the entropy of activation, the former also depends on the units in which the rate of reaction is measured. These relations will be illustrated by a number of examples in the following section.

When the reaction takes place not under standard conditions but under *equilibrium* conditions, where the rate forward r_f is equal to the rate backward r_b, the profile for the reaction takes on a somewhat different form, as shown in Fig. 15-8. At equilibrium $r_f = r_b$; therefore, the following relation must hold:

$$r_f = e^{-G_f{}^{\ddagger\circ}}P_f = e^{-G_b{}^{\ddagger\circ}}P_b = r_b \tag{15-54}$$

From Fig. 15-7, it is clear that $G_b^{\ddagger\circ}$ is greater than $G_f^{\ddagger\circ}$; therefore, P_b^e must be enough greater than P_f^e to compensate for this difference; the superscript e denotes that the values of the pressure are those at equilibrium. The activation free energy for the backward reaction is greater than the activation free energy for the forward reaction because the activation enthalpy for the backward reaction is greater than the activation enthalpy for the forward reaction. This difference is removed by making the total activation entropy $S_b^{\ddagger e}$ for the back reaction less than the total activation entropy for the forward reaction $S_f^{\ddagger e}$, a result achieved by increasing P_b. This new value for the activation entropy for the back reaction is designated by $S_b^{\ddagger e}$. This relationship can be expressed in the form of an equation

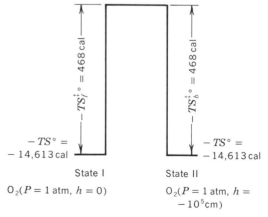

Fig. 15-6 Temperature-entropy profile for O$_2$ diffusion in a gravitational field under standard conditions at 25°C.

$$G_f^{\ddagger e} = H_f^{\ddagger\circ} - TS_f^{\ddagger e} = H_b^{\ddagger\circ} - TS_b^{\ddagger e} = G_b^{\ddagger e} \quad (15\text{-}55)$$

1218 cal   750 cal   468 cal   825 cal   393 cal   1218 cal

where

$$S_f^{\ddagger e} = S_f^{\ddagger\circ} + R \ln P_f^e \quad (15\text{-}55a)$$
$$S_b^{\ddagger e} = S_b^{\ddagger\circ} + R \ln P_b^e \quad (15\text{-}55b)$$

Example 15-1

Calculate the value of $H_f^{\ddagger}$ and $H_b^{\ddagger}$ in kilocalories when nitrogen flows from a reservoir at ground level, through a hole at a height of 10^5 m, and down into a reservoir at a height of 10^4 m. Assume that g does not vary with height.

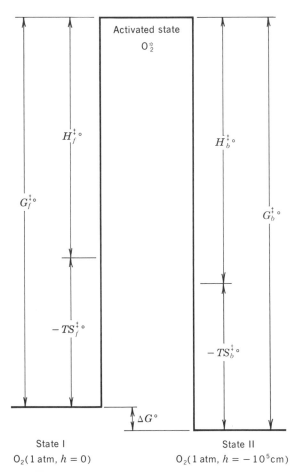

State I
$O_2(1 \text{ atm}, h = 0)$

State II
$O_2(1 \text{ atm}, h = -10^5\text{cm})$

Fig. 15-7 Free-energy profile for O_2 diffusion under standard conditions at 25°C in a gravitational field showing $G^{\ddagger\circ}$ as composed of the sum of $H^{\ddagger\circ}$ and $-TS^{\ddagger\circ}$.

$$H_f^{\ddagger} = Mgh = 28.02 \text{ g mole}^{-1} \times 981 \text{ cm sec}^{-2}$$
$$\times 10^7 \text{ cm} = 2.749 \times 10^{11} \text{ ergs}$$
$$2.749 \times 10^{11} \text{ ergs} \times 0.2390$$
$$\times 10^{-7} \text{ cal erg}^{-1} = 6.57 \text{ kcal}$$
$$H_b^{\ddagger} = Mgh = 28.02 \text{ g mole}^{-1} \times 981 \text{ cm sec}^{-2}$$
$$\times 9 \times 10^6 \text{ cm}$$
$$= 2.474 \times 10^{11} \text{ ergs} = 5.91 \text{ kcal} \quad \textit{Ans.}$$

Example 15-2

In the previous example the hole has a cross section of 0.1 cm^2, $P_f = 10$ atm, $P_b = 0.1$ atm, and $T = 25°C$.   Calculate r_f and r_b.

$$r_f = \frac{44.31}{\sqrt{MT}} e^{-H_f\ddagger/RT} P_f$$

$$= \frac{44.31}{\sqrt{28.02 \times 298.15}} e^{-6600/(1.987 \times 298.15)}$$
$$\times 10 \text{ atm}$$

$$= 0.4848 \times 1.452 \times 10^{-5} \times 10$$
$$= 7.039 \times 10^{-5} \text{ mole sec}^{-1}$$

$$r_b = \frac{44.31}{\sqrt{MT}} e^{-H_b\ddagger/RT} P_b$$

$$= \frac{44.31}{\sqrt{28.02 \times 298.15}} e^{-5880/(1.987 \times 298.15)}$$
$$\times 0.1 \text{ atm}$$

$$= 0.4848 \times 4.89 \times 10^{-5} \times 0.1 \text{ atm}$$
$$= 2.37 \times 10^{-6} \text{ mole sec}^{-1} \quad \textit{Ans.}$$

Example 15-3

Calculate the values of $-TS_f^{\ddagger\circ}$, $-TS_b^{\ddagger\circ}$, $G_f^{\ddagger\circ}$, and $G_b^{\ddagger}$ and $-TS_f^{\ddagger}$, $-TS_b^{\ddagger}$, $G_f^{\ddagger}$, and $G_b^{\ddagger}$ in the preceding example.

$$-TS_f^{\ddagger\circ} = -TS_b^{\ddagger\circ} = -TR \ln \frac{44.31}{\sqrt{MT}}$$
$$= -298.15 \times 1.987 \times 2.303$$
$$\times (-0.3146)$$
$$= 429 \text{ cal mole}^{-1}$$
$$G_f^{\ddagger\circ} = H_f^{\ddagger} - TS_f^{\ddagger\circ} = 6600 + 429$$
$$= 7029 \text{ cal mole}^{-1}$$
$$G_b^{\ddagger\circ} = H_b^{\ddagger} - TS_b^{\ddagger\circ} = 5880 + 429$$
$$= 6309 \text{ cal mole}^{-1}$$
$$-TS_f^{\ddagger} = -TS_f^{\ddagger\circ} - RT \ln P_f = 429 - 1364$$
$$= -935 \text{ cal mole}^{-1}$$

$$G_f^{\ddagger} = H_f^{\ddagger} + (-TS_f^{\ddagger}) = 6600 - 935$$
$$= 5665 \text{ cal mole}^{-1}$$
$$-TS_b^{\ddagger} = -TS_b^{\ddagger\,°} - RT \ln P_b = 429 + 1364$$
$$= 1793 \text{ cal mole}^{-1}$$
$$G_b^{\ddagger} = H_b^{\ddagger} + (-TS_b^{\ddagger}) = 5880 + 1793$$
$$= 7673 \text{ cal mole}^{-1}$$

15-2 RATES OF CHEMICAL REACTIONS

The rates of chemical reactions are governed by the same laws as the rates of physical reactions, but the mechanisms are more difficult to ascertain and depict. In a physical reaction like the diffusion of a gas through a hole, the "reactant," the gas at the initial pressure, is on one side of the hole and the "product," the gas at the final pressure, is on the other side of the hole, so that they are physically separated. In a chemical reaction like the combination of gaseous hydrogen and iodine to form hydrogen iodide, reactants and products are all mixed up in the same space. For the diffusion of the gas, the components of the probability factor A are easy to visualize and can be calculated with precision. The cross section of the hole has a specific value. The velocity of the molecules can be visualized and calculated. On the other hand, in the chemical reaction, where hydrogen molecules collide with iodine molecules, a direct calculation of the probability factor is extremely difficult. Only a rough estimate can be made of the cross section of the molecules effective in collision. There is a great deal of evidence that the probability of reaction depends on the orientation of the molecules with respect to each other when they collide. But in spite of these difficulties, much progress has been made in determining the factors affecting the rates of chemical reaction.

The fundamental equation

In studying the kinetics of a chemical reaction, it is necessary, first of all, to specify clearly the reaction under consideration. As for the physical reaction like the diffusion of a gas through a hole, a *forward* reaction and a *back* reaction can be written, which combined with the appropriate algebraic signs give

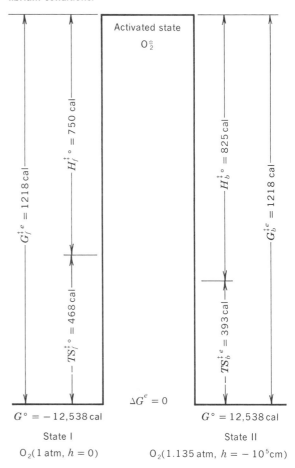

Fig. 15-8 Free-energy profile for the diffusion of O_2 in a gravitational field at 25°C under pressures that produce equilibrium conditions.

the total chemical equation. For gaseous hydrogen and iodine combining to form hydrogen iodide, the equations are

$$H_2(g) + I_2(g) \xrightarrow{f} 2HI(g) \tag{15-56a}$$
$$2HI(g) \xrightarrow{b} H_2(g) + I_2(g) \tag{15-56b}$$
$$H_2(g) + I_2(g) \underset{b}{\overset{f}{\rightleftharpoons}} 2HI(g) \tag{15-56c}$$

In order to define rate, it is necessary to adopt specific units. In this case the most basic unit is moles per second, so that when the reaction is proceeding at the rate of 1 mole sec^{-1}, 1 mole of gaseous hydrogen is combining with 1 mole of gaseous iodine per second.

As another example, consider the isomerization of methyl isocyanide to methyl cyanide according to the reaction

$$CH_3NC(g) \rightarrow CH_3CN(g) \qquad (15\text{-}57)$$

As contrasted with the reaction expressed by Eq. (15-56a), only one species of molecule appears on the left side of the arrow.

Reaction order

The *order* of a reaction is defined as the sum of the exponents of the concentration terms in the expression for the reaction rate. As an example, the reaction rate for the change expressed in Eq. (15-57) is

$$r_f = k_f P_{CH_3CN} = k(CH_3CN) \qquad (15\text{-}58)$$

When the discussion is confined to a reaction considered in only one direction, the subscripts f and b can be omitted. In order to avoid the use of subscripts denoting the components, it is also customary to write (CH_3CN) to denote the partial or total pressure of the reacting component instead of the more elaborate notation in which this pressure is denoted by P_{CH_3CN}. In Eq. (15-58) there is only one component, with the implied exponent of unity. Therefore, this reaction is called a *first-order* reaction. By way of contrast, the rate expression for the reaction denoted by Eq. (15-56a) is

$$r_f = k_f(H_2)(I_2) \qquad (15\text{-}59)$$

Here, there are two concentration terms, each with an implied exponent of unity, so that the sum of the exponents is equal to 2, and this is called a *second-order* reaction. The rate expression for the reaction denoted by Eq. (15-56b) is

$$r_b = k_b(HI)^2 \qquad (15\text{-}60)$$

Here there is only one concentration term appearing (HI), but since it has the exponent 2, this is called a second-order reaction also. When the quantities in parentheses in these last two equations have the values that result in chemical equilibrium, $r_f = r_b$ and

$$k_f(H_2)(I_2) = k_b(HI)^2 \qquad (15\text{-}61)$$

Rearranging, as in the preceding section, we get

$$K = \frac{k_f}{k_b} = \frac{(HI)^2}{(H_2)(I_2)} \qquad (15\text{-}62)$$

where K is the equilibrium constant.

For a generalized reaction expression

$$aA + bB + \cdots = xX + yY + \cdots \qquad (15\text{-}63)$$

the rate for the forward reaction is given by the usual expression when the mechanism of the reaction is a simple, normal pattern

$$r_f = k_f(A)^a (B)^b \cdots \qquad (15\text{-}64)$$

The order of the reaction is the sum of the exponents

$$Order = a + b + \cdots \qquad (15\text{-}65)$$

The same relations hold for the back reaction

$$r_b = k_b (X)^x (Y)^y \cdots \qquad (15\text{-}66)$$
$$Order = x + y + \cdots \qquad (15\text{-}67)$$

Sometimes the rate is independent of any concentration or pressure factor and then the reaction has *zero order*.

Frequently, the rate does not depend on the concentration or pressure factors in the simple way illustrated above. For example, the chemical equation for the decomposition of nitrogen pentoxide is generally written

$$2N_2O_5 \rightarrow 4NO_2 + O_2 \qquad (15\text{-}68)$$

The rate of this decomposition is found to obey the equation

$$r_f = \frac{-d(N_2O_5)}{dt} = k_f(N_2O_5) \qquad (15\text{-}69)$$

where the term $-d(N_2O_5)/dt$ denotes the time rate of change of the partial pressure of N_2O_5 and is thus a measure of the rate of the reaction. As the reaction proceeds forward, (N_2O_5) takes on smaller and smaller values. If the rate is to be regarded as a positive quantity when the reaction goes forward, it is necessary to insert a negative sign, since in the calculus the derivative of a decreasing quantity has a negative value. When (N_2O_5) denotes pressure measured in atmospheres, the dimension of k will be sec^{-1} and the dimensions of the rate will be atm sec^{-1}.

Molecularity

From the point of view of chemical kinetics and reaction mechanisms, many chemical reactions do not take place in as simple a manner as might be judged from the way in which they are written. There are frequently a number of individual steps which can be expressed as chemical reactions involving intermediate products; all the equations for the steps added together then give the overall expression for the reaction.

In much of the earlier literature of chemical kinetics, reactions of the first, second, and third orders were called unimolecular, bimolecular, and trimolecular. Today, the concept of *molecularity* of a reaction is used to denote the molecular mechanism by which the reaction proceeds. For example, the reaction expressed by Eq. (15-56b) is believed to take place when two HI molecules collide with sufficient kinetic energy and the proper mutual orientation to rearrange and form H_2 and I_2. Such an elementary process involving two molecules is called a *bimolecular reaction*. The term molecularity should be used only to describe individual elementary reactions. *Reaction order* refers to the rate equation based on experimental data; *molecularity* applies to the mechanism of the reaction.

Rate measurements

By far the simplest situation in which to make a measurement of a chemical reaction rate is one where only a single type of reaction is taking place in a single direction. In gaseous diffusion, the gas diffuses in only one direction through the hole if the molecules are promptly pumped off as soon as they get through the hole. In chemical reactions, the same result can be accomplished by removing the products of the reaction either by physical means, e.g., condensation, or by chemical means as soon as they are formed. Reactions can also be studied under conditions of temperature and pressure such that the rate of the back reaction is negligibly small compared to the rate of the forward reaction, thus ensuring that the measurement depends effectively only on a process proceeding in a single direction.

In order to make a measurement of a rate, it is necessary to find some measurable property the changes of which constitute a measure of the rate.

For the diffusion of a gas, the change in the pressure of the gas in the chamber out of which it is diffusing constitutes a measure of the rate of the reaction. In the case of the decomposition of a gas, the change in pressure in the chamber containing the gas can give a measure of the rate of the reaction if the reaction procedure is started with the pure gas, if the decomposition yields either more molecules or fewer than were originally present, and if measurements are made before any appreciable quantities of the products accumulate to start an effective back reaction. Changes in the value of pH are frequently used to study rates of reaction in solution.

A number of biochemical reactions can be followed by measuring the intensity of a light beam having a frequency absorbed by one of the reactants or one of the products. When β-methylaspartate is deaminated by the enzyme β-methylaspartase, only the reaction product, mesaconate, absorbs light in the vicinity of 2400 Å. The change in the amount of absorption of light of this wavelength thus provides a measure of the rate of the reaction.

Zero-order reactions

The deamination of β-methylaspartate was studied by Williams and Williams,[1] who found that the concentration of the reactant could be varied over a tenfold range without any alteration of the reaction rate. Therefore, the exponent of the concentration term is effectively zero, and this reaction has *zero order*.

In a nuclear reaction, like the radioactive decomposition of radon, the rate of decomposition in terms of moles per second for a given mass of material is independent of the pressure under which this material is placed. In this sense, the reaction may be regarded as zero order. However, in a container of constant volume at constant temperature, the higher the pressure, the more radon is present. The more radon there is, the more decomposition there will be in terms of atoms per second in this container. From this point of view, the reaction is not a zero-order reaction.

[1] V. R. Williams and H. B. Williams, "Basic Physical Chemistry for the Life Sciences," p. 225, W. H. Freeman and Company, San Francisco, 1967.

First-order reactions

The general rate equation for a first-order reaction can be written

$$r = -\frac{d(A)}{dt} = k(A) \tag{15-70}$$

where (A) is a pressure or concentration term for component A. The reaction itself might be written

$$A \rightarrow B + C \tag{15-71}$$

If for A we substitute radon gas, ^{222}Rn, for B polonium, ^{218}Po, and for C helium, ^{4}He, the equation reads

$$^{222}Rn \rightarrow {}^{218}Po + {}^{4}He \tag{15-72}$$

and the rate equation is

$$r = -\frac{d(^{222}Rn)}{dt} = k(^{222}Rn) \tag{15-73}$$

The rate of change of the pressure of radon gas is proportional to the pressure of the gas in a container at constant volume and temperature. The general expression, Eq. (15-70), can be rearranged

$$-\frac{d(A)}{(A)} \equiv -d \ln (A) = k \, dt \tag{15-74}$$

This can be integrated to show the dependence of the amount of change from an initial state of pressure or concentration $(A)_1$ to a final state $(A)_2$

$$-\int_1^2 d \ln (A) = k \int_1^2 dt \tag{15-75a}$$

$$-\ln \frac{(A)_2}{(A)_1} = k(t_2 - t_1) \tag{15-75b}$$

Frequently, the initial pressure or concentration is expressed by the symbol $(A)_0$, t_1 is also set equal to zero, and the term (A) is used to denote the pressure or concentration at any subsequent time t. Then Eq. (15-75b) can be put in the form:

$$\ln \frac{(A)}{(A)_0} = -kt \tag{15-75c}$$

or in the equivalent exponential form

$$(A) = (A)_0 \, e^{-kt} \tag{15-75d}$$

The most useful expression for numerical calculations is the logarithmic form

$$\log (A) = -\frac{kt}{2.303} + \log (A)_0 \tag{15-75e}$$

When the term log (A) is plotted against the time, the result is a straight line with a slope equal to $-k/2.303$. After a certain lapse of time, $t_{1/2}$, the value of (A) will be reduced to one-half of its initial value $(A)_0$. This is shown by inserting the values in Eq. (15-75e)

$$\log \frac{(A)}{(A)_0} = \log \tfrac{1}{2} = -0.30103 = -\frac{kt_{1/2}}{2.303} \tag{15-76}$$

This special elapsed time is called the *half-life* of the reaction, because during this time half of the original material disappears. The relation between the half-life and the reaction rate constant k is

$$t_{1/2} = \frac{0.693}{k} \tag{15-77a}$$

$$k = \frac{0.693}{t_{1/2}} \tag{15-77b}$$

The half-life for $^{222}Rn = 3.825$ days. Examples of semilogarithmic plots for several first-order reactions are given in Fig. 15-9. Values of k are listed in Table 15-1.

Second-order reactions

A reaction is classified as second order if the sum of the exponents of the pressure or concentration terms in the rate equation is equal to 2. Consider the general chemical reaction of the form

$$2A \rightarrow B + C \tag{15-78}$$

for which the rate expression is

$$r = -\frac{d(A)}{dt} = k(A)^2 \tag{15-79}$$

This can be rearranged to the form

$$\frac{dA}{(A)^2} = -k \, dt \tag{15-80}$$

and then integrated between an initial concentration $(A)_0$ and a concentration (A)

$$\int_{(A)_0}^{(A)} \frac{d(A)}{(A)^2} = -k \int_{t_0}^{t} dt \tag{15-81}$$

which gives the expression

$$\frac{1}{(A)} - \frac{1}{(A)_0} = kt \tag{15-82}$$

For such a reaction, a plot of the reciprocal of the pressure or concentration term (A) against time is a straight line. When at $t_{1/2}$, $(A) = \frac{1}{2}(A)_0$, then

$$t_{1/2} = \frac{1}{(A)_0 k} \tag{15-83}$$

Another type of chemical reaction which may be second order is

$$A + B \rightarrow C \tag{15-84}$$

The rate equation may be written in the form

$$\frac{dx}{dt} = k[(A)_0 - x][(B)_0 - x] \tag{15-85}$$

where x represents the number of moles of A which have reacted with B. The differential equation can then be put in the form

$$\frac{dx}{[(A)_0 - x][(B)_0 - x]} = k\,dt \tag{15-86}$$

This expression on integration gives

$$\frac{1}{(A)_0 - (B)_0} \ln \frac{(B)_0[(A)_0 - x]}{(A)_0[(B)_0 - x]} = kt \tag{15-87}$$

Figure 15-10 shows graphs for several second-order reactions. Values of k are listed in Table 15-2.

Higher-order reactions

Reactions of third order in the gas phase are extremely rare, but a number of reactions in solution have rate equations which follow the form for the third order. If the general reaction is written

$$A + B + C \rightarrow D + E \tag{15-88}$$

then, when the initial concentration of all three components have the same value $(A)_0$, the equation for the relation between concentration and time is

$$\frac{1}{2}\left[\frac{1}{(A)^2} - \frac{1}{(A)_0{}^2}\right] = kt \tag{15-89}$$

15-3   RATE AND TEMPERATURE

When the concept of the rate of a chemical reaction began to take a clear and quantitative form, the importance of the variation of this rate with temperature

Fig. 15-9   Semilogarithmic graphs of first-order reactions. [(a) from J. H. Raley, F. F. Rust, and W. E. Vaughan, *J. Am. Chem. Soc.*, 70:88, 1336, 2767 (1948); (b) from L. C. Bateman, E. D. Hughes, and C. K. Ingold, *J. Am. Chem. Soc.*, 1940:960; (c) from R. A. Ogg, *J. Chem. Phys.*, 15:337 (1947); 18:572 (1950); F. Daniels and E. H. Johnston, *J. Am. Chem. Soc.*, 43:53 (1921).]

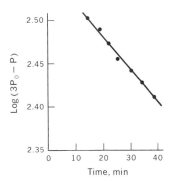

(a) $(CH_3)_3COOC(CH_3)_3 \rightarrow 2(CH_3)_2CO + C_2H_6$

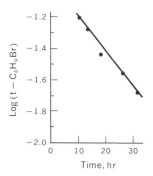

(b) $(CH_3)_3CBr + H_2O \rightarrow (CH_3)_3COH + HBr$

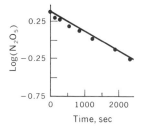

(c) A first-order complex reaction:
$$N_2O_5 \rightarrow NO_2 + NO_3$$
$$NO_2 + NO_3 \rightarrow NO_2 + O_2 + NO$$
$$NO + N_2O_5 \rightarrow 3NO_2$$

TABLE 15-1   Values of k for first-order gas reactions

Reaction	k, sec^{-1}	t, °C
$N_2O_5 \rightarrow NO_3 + NO_2$	2.7×10^{-1}	25
$(CH_3)_3COOC(CH_3)_3 \rightarrow 2(CH_3)_2CO + C_2H_6$	1.4×10^{-4}	147.2
$(CH_3)_3Br + H_2O \rightarrow (CH_3)_3$	1.4×10^{-5}	25
$\begin{array}{c} CH_2\!-\!CH \\ \mid \quad \parallel \\ CH_2\!-\!CH \end{array} \rightarrow CH_2\!=\!CH\!-\!CH\!=\!CH_2$	1.8×10^{-11}	25
$C_2H_5Br \rightarrow C_2H_4 + HBr$	8.8×10^{-27}	25
$\begin{array}{c} CH_2 \\ \diagup \quad \diagdown \\ CH_2\!-\!\!-\!CH_2 \end{array} \rightarrow CH_3\!-\!CH\!=\!CH_2$	3.4×10^{-33}	25

was recognized almost immediately. The fundamental basis for reaction-rate theory was first established by Arrhenius. As a point of departure he took the van't Hoff equation for the temperature coefficient of the equilibrium constant

$$\frac{d \ln K}{dT} = \frac{\Delta H}{RT^2} \tag{15-90}$$

and the relation between the equilibrium constant

$$K = \frac{k_f}{k_b} \tag{15-91}$$

and pointed out that a reasonable equation for the variation of the rate constant with temperature would be

$$\frac{d \ln k}{dT} = \frac{H^\ddagger}{RT^2} \tag{15-92}$$

$H^\ddagger$ denotes the increase in enthalpy per mole necessary to raise the reactants to a state of activation in which they have sufficient energy to rearrange atoms according to the equation of the reaction. In his original presentation Arrhenius used energy E instead of enthalpy H and referred to the term $H^\ddagger$ appearing in Eq. (15-92) as *activation energy*. He assumed that this quantity was independent of temperature and integrated Eq. (15-92) to give

$$\int_1^2 d \ln k = \frac{H^\ddagger}{R} \int_1^2 \frac{dT}{T^2} = -\frac{H^\ddagger}{R} \int_1^2 d\frac{1}{T} \tag{15-93a}$$

$$\ln \frac{k_2}{k_1} = -\frac{H^\ddagger}{R}\left(\frac{1}{T_2} - \frac{1}{T_1}\right) \tag{15-93b}$$

Letting $k_1 = 1$ at temperature T_1 and letting $k_2 = k$ at T, so that they are unrestricted variables, then

$$\ln k = -\left(\frac{H^\ddagger}{R}\frac{1}{T} - \frac{H^\ddagger}{RT_1}\right) \tag{15-94}$$

Since T_1 is the value of the temperature in degrees Kelvin at which $k = 1$, the choice of the units in which k is measured determines the value of T_1. Arrhenius denoted the constant term $H^\ddagger/RT_1$ by $\ln A$, so that Eq. (15-94) takes the form

$$\ln k = -\frac{H^\ddagger}{RT} + \ln A \tag{15-94a}$$

and can be put in the exponential form

$$k = Ae^{-H^\ddagger/RT} \tag{15-94b}$$

The term A is also called the *frequency factor* as well as *preexponential factor*. The equation is now generally called the *Arrhenius equation*.

The physical significance of this equation can be seen by comparing it with the equation derived from the kinetic theory of gases for the reaction where an ideal gas diffuses through a hole, Eq. (15-13). For this process

$$k = \frac{44.31a}{\sqrt{MT}}\, e^{-H^\ddagger/RT} = Ae^{-H^\ddagger/RT} \tag{15-95}$$

From the derivation of this equation, the conclusion is drawn that the preexponential factor A is really a probability factor, denoting the probability that the molecules of the gas will pass through the hole. The larger the cross section a of the hole, the higher the probability the molecules will pass through. The larger the molecular weight M, the more slowly the molecules will move and the lower the probability that they will pass through the hole in a given unit of time. The higher the temperature, the lower the density of the gas per unit volume and, therefore, the smaller the chance of molecules passing through the hole in a given unit of time. For diffusion of an ideal gas this equation shows that A does depend on the inverse of the square root of the temperature, but this variation with temperature is small compared with the exponential variation with temperature in the second half of the equation. Therefore, one can postulate a kind of ideal reaction in which neither A nor $H^{\ddagger}$ varies with temperature. This ideal reaction is very much like the ideal-gas and the ideal thermodynamic system for which, by the use of activity and fugacity, the equations are cast in an extremely simple form relating free energy to the factors which depend on concentration.

Since $H^{\ddagger}/T_1$ has the dimensions of entropy, it is logical to denote this quantity by the symbol $S^{\ddagger\circ}$, the entropy change when 1 mole of the substance under standard conditions goes to the activated state at the temperature where the rate constant k becomes equal to unity. Thus, Eq. (15-94a) can be put in the form

$$\ln k = -\frac{H^{\ddagger}}{RT} + S^{\ddagger\circ} = -\frac{H^{\ddagger} - TS^{\ddagger\circ}}{RT} = G^{\ddagger\circ}$$

(15-96)

and the Arrhenius equation can be written to show the simple relation between the rate constant and the standard activation free energy

Fig. 15-10   Semilogarithmic plots of second-order reactions. [(a) from I. Dostrovsky and E. D. Hughes, *J. Am. Chem. Soc.*, 68:157 (1946); (b) from F. Daniels and D. A. Alberty, "Physical Chemistry," 3d ed., John Wiley & Sons, Inc., New York, 1966.]

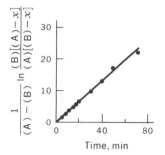

(a) $r = 5.5 \times 10^{-3} (A)(B)$
$A = (CH_3)_2(CH_2Br)CH$
$B = NaOCH_2CH_3$

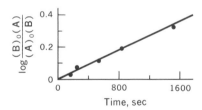

(b) $r = 0.107 (A)(B)$
$A = (OH^-)$
$B = CH_3COOCH_2CH_3$

TABLE 15-2   Values of k for second-order gas reactions

Reaction	k, moles liter^{-1} sec^{-1}	t, °C
$H + D_2 \rightarrow HD + D$	3×10^5	25
$CH_3 + H_2 \rightarrow CH_4 + H$	1.5×10^1	25
$CH_3CH_2O^- + (CH_3)_2CHCH_2Br \rightarrow$		
$\quad (CH_3)_2CHCH_2OCH_2CH_3 + Br^-$	6×10^{-3}	95
$H_2 + I_2 \rightarrow 2HI$	1.7×10^{-18}	25
$HI + HI \rightarrow H_2 + I_2$	2.4×10^{-21}	25

$$k = k_0 e^{-G^{\ddagger \circ}/RT} \tag{15-97}$$

where k_0 is a constant with the numerical value of unity and the dimensions of k. Another useful form of Eq. (15-96) is

$$R \ln k = -H^{\ddagger} \frac{1}{T} + S^{\ddagger \circ} \tag{15-98}$$

which algebraically is equivalent to

$$y = bx + a \tag{15-98a}$$

where $b = -H^{\ddagger}$ and $a = S^{\ddagger \circ}$. Thus, when experimental measurements are made of the variation of the rate constant k with temperature and $R \ln k$ is plotted against $1/T$, the ideal rate theory predicts that the graph will be a straight line with the slope equal to $-H^{\ddagger}$. At the point where $1/T = 0$, the intercept of the graph with this vertical axis will be equal to $S^{\ddagger \circ}$. On the graph, $G^{\ddagger \circ}$ is equal to the vertical distance from the graph to the horizontal axis at which $R \ln k = 0$. These relations are illustrated in Fig. 15-11a for the forward reaction and in Fig. 15-11b for the back reaction of the equation

$$\mathrm{H_2 + I_2 \underset{b}{\overset{f}{\rightleftharpoons}} 2HI} \tag{15-99}$$

In Fig. 15-11c both graphs are drawn. At any temperature, the vertical distance between the graphs is $R \ln K$, where K is the equilibrium constant. Since $K = k_f/k_b$, $R \ln K = R \ln k_f - R \ln k_b$, proving the above relation. The experimentally determined values of k_G for reaction (15-99) are shown in Fig. 15-11d, plotted as log k_G against $1/T$.

Using $1/T$ as the variable denoted by the abscissa, the zero of this coordinate axis corresponds to the point at which $T = \infty$, the boundary of the temperature scale which is the theoretical limit complementary to the theoretical limit at which $T = 0$, the absolute zero of the temperature scale. Both these theoretical limits are unattainable experimentally but are important theoretically. Just as the lower limit determines the values of the absolute entropy, the upper limit at infinite temperature gives the value of the entropy of activation.

The zero for the $1/T$ scale is placed at the right, rather than on the left side as is customary, in order to make the comparison between these graphs and graphs with a T scale easier. In this way on both sets of graphs an increase in temperature is denoted by a movement to the right. In Fig. 15-11e, there is a sheaf of lines showing the slope for various values of $H^{\ddagger}$ in kilocalories.

In Fig. 15-12a, b, and c the reaction profiles corresponding to the graphs in Fig. 15-11 are shown.

The meaning of $H^{\ddagger}$

In a *physical* reaction like the one described in Sec. 15-1, only those molecules can undergo the reaction (diffuse through the hole) which have a value of the kinetic energy sufficiently high to permit them to move upward against the force of gravity to the height of the hole. In nearly all *chemical* reactions, atoms must move apart against the force of the chemical bonds between atoms. As discussed in Chap. 6, the force of a chemical bond may be thought of as similar to the force produced by a spiral spring. When the bond is shortened below its normal length, the atoms move close together, the spring is compressed, and there is a force tending to move the atoms apart. When the atoms move further apart than the normal length of the bond, a force of attraction is exerted tending to pull the atoms back together again. In the normal state, the atoms at each end of such a bond are vibrating back and forth with both kinetic and potential energy. In order to undergo the atomic rearrangement which constitutes the reaction, it is generally necessary for the distance separating the atoms to be increased; in other words, the bonds must be stretched.

In a whole mole of substance, various numbers of molecules are in different energy levels with respect to this vibration which stretches the bond. The distribution of the numbers of molecules over these energy levels follows the customary Maxwell-Boltzmann equation. Suppose that it is necessary for the molecule to be in the tenth energy level in order to have the atoms far enough apart for the reaction to take place. The fraction of the molecules in this tenth energy level will be given by

$$\frac{n_{10}}{n_{\text{tot}}} = \frac{e^{-E_{10}/RT}}{\sum\limits_i e^{-E_i/RT}} \tag{15-100}$$

where n_{10} = number in tenth energy level

n_{tot} = total number of molecules

E_{10} = energy of molecule in tenth energy level, cal mole^{-1} or kcal mole^{-1}

The expression in the denominator with the summation sign is the usual *partition function,* omitting any factors which might be introduced by different values of the multiplicity for energy levels. This function may be regarded as a constant with respect to a comparison of the fraction of molecules in any energy level with the height of that energy level. Thus, the value of E_{10} is analogous to the value of $H^{\ddagger}$ in the example of the diffusion of molecules through a hole in a gravitational field. In other words, there is a theoretical barrier height for a chemical reaction resembling the actual, physical barrier in the physical reaction, and $H^{\ddagger}$ is the height of this barrier measured in terms of energy units.

There are very few cases in which knowledge of the details of atomic and molecular structure is sufficient for $H^{\ddagger}$ to be calculated with the help of quantum mechanics. However, as the analysis in the previous section shows, an experimental study of the variation of reaction rate with temperature provides values of $H^{\ddagger}$ sufficiently accurate to serve as the basis for a number of useful calculations. In the remainder of this chapter and in other chapters in this book dealing with the kinetic aspects of physical and chemical reactions, $H^{\ddagger}$ will be thought of as a measure of the height of the energy barrier which must be crossed if the reaction is to proceed.

The meaning of $S^{\ddagger\circ}$

Again referring back to the example of the physical reaction in which a molecule rises in a gravitational field and diffuses through a hole, it can be seen that $S^{\ddagger\circ}$ is a quantity sufficiently related to thermodynamic entropy for the name "activation entropy" to be justified. On the other hand, this activation entropy differs in a number of respects from thermodynamic entropy, and the distinction between the two is important in analyzing the various relations between thermodynamics and kinetics.

As mentioned earlier, the preexponential factor A is essentially a probability factor; and if $S^{\ddagger\circ} = R \ln (A/A^{*})$, *then* there is the same relation between the activation entropy and probability as is found for thermodynamic entropy and probability.

In the physical reaction, A increases as the size of the hole increases and decreases as the molecular weight and temperature increase. $S^{\ddagger\circ}$ undergoes a corresponding logarithmic increase as the area of the hole increases and a logarithmic decrease with respect to the increase of $\sqrt{M}$ and $\sqrt{T}$.

In the chemical reaction, the effective cross-sectional areas of reacting molecules correspond roughly with the cross-sectional area of the hole in the physical reaction. This analogy is even more exact in the phenomenon of catalysis, which will be discussed in a following section. Other probability factors governing the size of A and thus influencing logarithmically the magnitude of $S^{\ddagger\circ}$ are connected with the quantum-mechanical aspects of the reaction. These will be discussed in some detail in the chapters on quantum mechanics. At this point, it is helpful to think of $S^{\ddagger\circ}$ as a measure of the influence of these probability factors on the reaction. When $S^{\ddagger\circ}$ is multiplied by T, the result expressed in calories or kilocalories provides direct means by which the relative effects of the energy barrier $H^{\ddagger}$ on the one hand and the probability factors $TS^{\ddagger\circ}$ on the other hand can be compared *in the same units,* thus providing a direct insight into the nature of the factors which control the rate of the reaction.

The analysis of the dependence of free energy on temperature showed that given a set of concentration conditions, the change with temperature of the free energy is due preponderantly to the change in the TS term. Just as the ideal conditions are visualized in which ΔH remains constant with temperature for a given reaction, so can one visualize an ideal situation in which ΔS may also be thought of as essentially invariant with temperature. The change, then, in the value of ΔG for the reaction is due to the fact that $\Delta G = \Delta H - T \Delta S$, and it is the presence of the T in this expression which is the predominant cause of the variation of ΔG with temperature.

A strictly analogous situation exists with respect to reaction rates. Thinking of the rate as given by the equation

$$r = r_0 e^{-G^{\ddagger\circ}/RT} = r_0 e^{-(H^{\ddagger}-TS^{\ddagger\circ})/RT} \qquad (15\text{-}101)$$

it is the presence of the T multiplying the $S^{\ddagger\circ}$ plus the presence of T in the denominator of the exponent

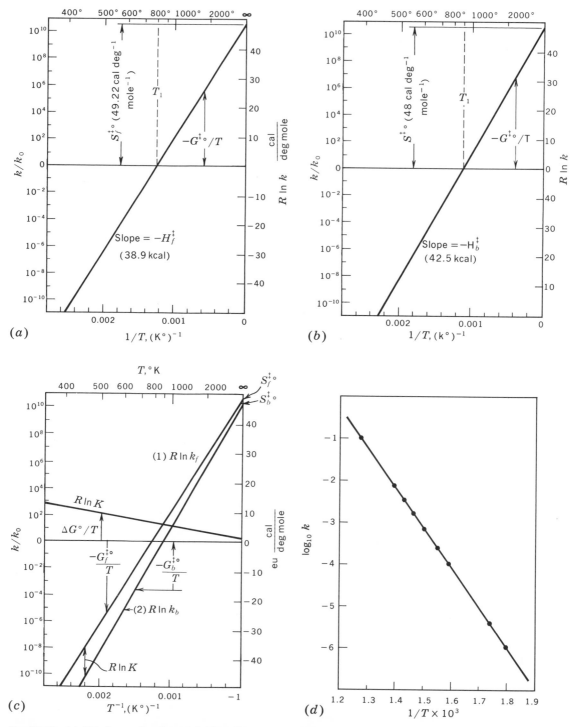

Fig. 15-11 (a) $R \ln k$ as a function of $1/T$ for $H_2 + I_2 \longrightarrow 2HI$; (b) $R \ln k$ as a function of $1/T$ for $2HI \longrightarrow H_2 + I_2$; (c) $R \ln k$ plotted for (1) $H_2 + I_2 \longrightarrow 2HI$, (2) $2HI \longrightarrow H_2 + I_2$ and $R \ln k$ for $H_2 + I_2 \rightleftharpoons 2HI$; (d) experimental data for the reaction $2HI \longrightarrow H_2 + I_2$. [(d) from M. Bodenstein, *Z. Physik. Chem.*, **29**:295 (1899).]

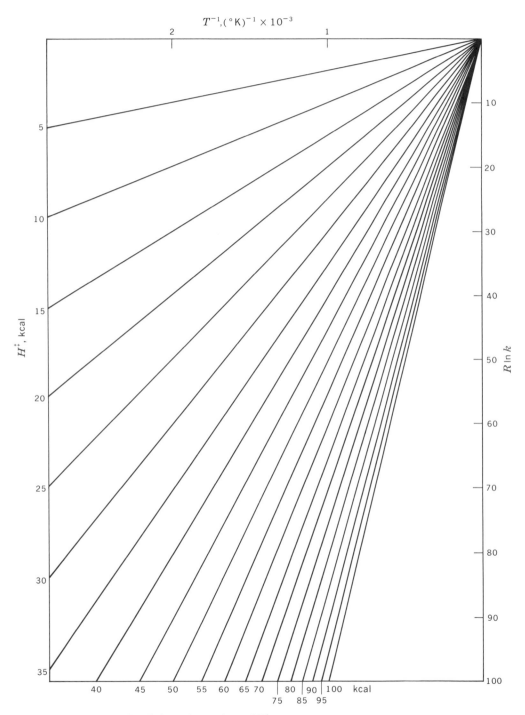

Fig. 15-11(e)   Slope of $R \ln k$ for various values of $H^{\ddagger}$.

Fig. 15-12   Reaction profiles for $H_2 + I_2 \rightleftharpoons 2HI$.

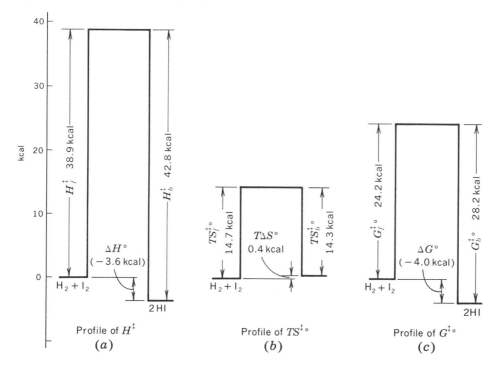

Profile of $H^\ddagger$
(a)

Profile of $TS^{\ddagger\circ}$
(b)

Profile of $G^{\ddagger\circ}$
(c)

that governs the variation of the rate with temperature.

Values of $H^\ddagger$, $TS^{\ddagger\circ}$, and $G^{\ddagger\circ}$ for a number of reactions are listed in Table 15-3, and corresponding graphs are plotted in Fig. 15-13. As $1/T$ approaches zero, the lines converge toward a common focal point at about 45 cal deg^{-1}; this indicates that there is a common value of $S^{\ddagger\circ}$ for many reactions, similar to the Trouton value for ΔS_v of 22 cal deg^{-1} and the common value for the normalized ΔS_f of 12 cal deg^{-1}. The broader significance of these relations will be examined in considerably more detail in connection with the specific examples of reaction rates discussed in the following sections.

Example 15-4

From the data in Table 15-3, calculate the value of k at 25°C for the reaction $NO_2 + CO \rightarrow NO + CO_2$ in (moles liter^{-1})$^{-1}$ sec^{-1}.

$$RT \ln k = -G^{\ddagger\circ} = -17.8 \text{ kcal mole}^{-1}$$

$$\log k = \frac{-17.8}{2.303 \times 1.987 \times 298.15} = -13.05$$

$$\frac{1}{k} = 1.3 \times 10^{13}$$

$$k = 8 \times 10^{-14} \text{ (moles liter}^{-1}\text{)}^{-1} \text{ sec}^{-1} \quad Ans.$$

Example 15-5

Assuming that $H^\ddagger$ and $S^{\ddagger\circ}$ do not vary with temperature, calculate the value of k for $H_2 + I_2 = 2HI$ at 500°K, using the data in Table 15-3.

$TS^{\ddagger\circ}$ at 298.15°K is 14.7 kcal, according to Daniels and Alberty.   At 500°K

$$TS^{\ddagger\circ} = 14.7 \times \frac{500}{298.15} = 24.7 \text{ kcal}$$

$$G^{\ddagger\circ}_{500} = H^\ddagger - TS^{\ddagger\circ}_{500} = 38.9 - 24.7$$
$$= 14.2 \text{ kcal}$$

$$RT \ln k = -G^{\ddagger\circ} = -14.2 \text{ kcal}$$

$$\log k = \frac{-14,200}{2.303 \times 1.987 \times 298.15}$$
$$= -10.41$$

TABLE 15-3   Values of $H^{\ddagger}$, $TS^{\ddagger\circ}$, and $G^{\ddagger\circ}$ at 25°C[a]

No.	Reaction	$H^{\ddagger}$, kcal mole^{-1}	$TS^{\ddagger\circ}$, kcal mole^{-1}	$G^{\ddagger\circ}$, kcal mole^{-1}
1[b]	$NO + O_3 \rightarrow NO_2 + O_2$	2.3	12.3	-10.0
2[c]	$NO + O_3 \rightarrow NO_2 + O_2$	2.5	12.1	-9.6
3[b]	$H + D_2 \rightarrow HD + H$	6.5	14.0	-7.5
4[c]	$H + C_2H_6 \rightarrow H_2 + C_2H_5$	6.8	12.9	-6.1
5[c]	$NO + NO_2Cl \rightarrow NOCl + NO_2$	6.9	12.1	-5.2
6[c]	$NO + O_3 \rightarrow NO_3 + O$	7.0	13.3	-6.3
7[c]	$CH_3 + C_6H_5CH_3 \rightarrow CH_4 + C_6H_5CH_2$	7.0	9.5	-2.5
8[c]	$CH_3 + \text{iso-}C_4H_{10} \rightarrow CH_4 + C_4H_9$	7.6	9.5	-1.9
9[c]	$CF_3 + C_2H_6 \rightarrow CF_3H + C_2H_5$	7.7	11.4	-3.7
10[c]	$CH_3 + n\text{-}C_5H_{12} \rightarrow CH_4 + C_5H_{11}$	8.1	10.9	-2.8
11[c]	$F_2 + ClO_2 \rightarrow FClO_2 + F$	8.5	10.2	-1.7
12[c]	$H + H_2 \rightarrow H_2 + H$	8.8	15.0	-5.9
13[c]	$CD_3 + C_6H_6 \rightarrow CD_3H + C_6H_5$	9.2	10.1	-0.9
14[c]	$CF_3 + CH_4 \rightarrow CF_3H + CH_3$	9.5	10.9	-1.4
15[c]	$CH_3 + CH_3COCH_3 \rightarrow CH_4 + CH_2COCH_3$	9.7	11.7	-2.0
16[c]	$CH_3 + H_2 \rightarrow CH_4 + H$	10.0	12.7	-2.7
17[c]	$CD_3 + C_2H_6 \rightarrow CD_3H + C_2H_5$	10.4	11.3	-1.3
18[c]	$NO_2 + F_2 \rightarrow NO_2F + F$	10.4	12.5	-2.1
19[c]	$CH_3 + C_2H_6 \rightarrow CH_4 + C_2H_5$	11.2	10.6	$+0.2$
20[c]	$CD_3 + CD_3COCD_3 \rightarrow CD_4 + CD_2COCD_3$	11.3	12.0	-0.7
21[c]	$H + CH_4 \rightarrow H_2 + CH_3$	12.0	13.6	-1.6
22[c]	$CD_3 + CH_4 \rightarrow CD_3H + CD_3$	14.0	10.9	$+3.1$
23[c]	$Br + H_2 \rightarrow HBr + H$	11.3	12.0	-0.7
24[b]	$N_2O_5 \rightarrow NO_3 + NO_2$	18.5	19.1	-0.6
25[c]	$NO + Cl_2 \rightarrow NOCl + Cl$	20.3	13.0	$+7.3$
26[c]	$2NOCl \rightarrow 2NO + Cl_2$	24.5	13.6	$+10.9$
27[b]	$N_2O_5 \rightarrow (N_2O_4 \rightleftarrows 2NO_2) + \frac{1}{2}O_2$	24.7	27.1	-2.4
28[c]	$2NO_2 \rightarrow 2NO + O_2$	26.6	9.1	$+17.5$
29[c]	$NO_2 + CO \rightarrow NO + CO_2$	31.6	13.8	$+17.8$
30[b]	Cyclobutene $\rightarrow$ 1,3-butadiene	32.5	19.5	$+13.0$
31a[b]	$H_2 + I_2 \rightarrow 2HI$	38.9	14.7	$+24.2$
31b[d]	$H_2 + I_2 \rightarrow 2HI$	40.7	15.0	$+25.7$
32a[b]	$2HI \rightarrow H_2 + I_2$	42.5	14.4	$+28.1$
32b[d]	$2HI \rightarrow H_2 + I_2$	43.7	13.9	$+29.8$
33[b]	$C_2H_5Br \rightarrow C_2H_4 + HBr$	53.9	20.0	$+33.9$
34[b]	Cyclopropane $\rightarrow$ propylene	65.0	22.4	$+42.6$

[a] The zero point of the $S^{\ddagger\circ}$ and $G^{\ddagger\circ}$ scale is selected so that for first-order reactions the dimensions of k are sec^{-1} and for second-order reactions are (mole liter^{-1})$^{-1}$ sec^{-1}.

[b] F. Daniels and R. A. Alberty, "Physical Chemistry," 3d ed., p. 344, John Wiley & Sons, Inc., New York, 1966.

[c] K. J. Laidler, "Chemical Kinetics," 2d ed., pp. 125, 127, McGraw-Hill Book Company, New York, 1965.

[d] I. Amdur and G. G. Hammes, "Chemical Kinetics," p. 75, McGraw-Hill Book Company, New York, 1966.

$$\frac{1}{k} = 2.6 \times 10^{-10}$$

$$k = 3.8 \times 10^{-11} \text{ (moles liter}^{-1})^{-1} \text{ sec}^{-1}$$

Example 15-6

Correct the value of k in Example 15-5 to take into account the variation of A proportional to $\sqrt{T}$.

$$k = k \text{ (uncorrected)} \sqrt{\frac{T}{298.15}}$$

$$= 3.8 \times 10^{-11} \sqrt{\frac{500}{298.15}} = 4.9 \times 10^{-11}$$

15-4   TYPES OF COUPLED REACTIONS

In the previous discussion of reaction rates, attention has been focused on the simplest situation, where a single reaction takes place in only one direction. The presumption is made that the products of the reaction can be removed so that no back reaction is proceeding to any appreciable extent. In this section, several more complicated situations are analyzed which are frequently met in experimental studies of reaction rates. These include opposed reactions, which are simultaneously proceeding both forward and backward but at different rates; reactions which take place concurrently; reactions which take place sequentially; and chain reactions.

Opposed reactions

Suppose that a forward and reverse reaction can take place between two molecular species X and Y according to the equation

$$X \underset{b}{\overset{f}{\rightleftharpoons}} Y \qquad (15\text{-}102)$$

Suppose that at the initial point in time, when $t = 0$, the concentration of $X = X_0$ and the concentration of $Y = 0$. The concentration of X decreases with time according to the equation

$$\frac{d(X)}{dt} = -k_f(X) \qquad (15\text{-}103)$$

As soon as the concentration of Y becomes appreciable, the back reaction proceeds according to the expression

$$\frac{d(Y)}{dt} = -k_b(Y) \qquad (15\text{-}104)$$

Under the initial conditions specified, the relation

$$(Y) = (X)_0 - (X) \qquad (15\text{-}105)$$

holds. This value can be inserted in Eq. (15-104), and since both these reactions take place simultaneously, the total rate of change in the concentration of X is given by

$$\frac{d[(X)_0 - (X)]}{dt} = -\frac{d(X)}{dt} = -k_b[(X)_0 - (X)] \qquad (15\text{-}106)$$

The net change in (X) is given by

$$\frac{d(X)}{dt} = -k_f(X) + k_b[(X)_0 - (X)] \qquad (15\text{-}107)$$

At equilibrium, the rate forward is equal to the rate back, so that

$$k_f(X)_{eq} = k_b(Y)_{eq} = k_b[(X)_0 - (X)_{eq}] \qquad (15\text{-}108)$$

The relation can then be expressed entirely in terms of k_b

$$\frac{d(X)}{dt} = k_b \frac{(X)_0}{(X)_{eq}} [(X)_{eq} - (X)] \qquad (15\text{-}109)$$

This expression can be integrated between the initial state $(X)_0$ and the state where the concentration is (X)

$$\int_{(X)_0}^{(X)} \frac{d(X)}{(X)_{eq} - (X)} = k_b \frac{(X)_0}{(X)_{eq}} \int_0^t dt \qquad (15\text{-}110)$$

This leads to the expression

$$\ln \frac{(X)_0 - (X)_{eq}}{(X) - (X)_{eq}} = k_b \frac{(X)_0}{(X)_{eq}} t \qquad (15\text{-}111)$$

Thus, a plot of log $[(X)_{eq} - (X)]$ against time is linear. The value of k_b can be calculated from the slope and the value of the ratio of the initial concentration of X to the equilibrium concentration of X. Using this information in Eq. (15-108) permits calculating the value of k_f.

Detailed balancing

Considerations of this sort lead to a relationship which is sometimes regarded as an integral part of the so-called zero law of thermodynamics. Suppose

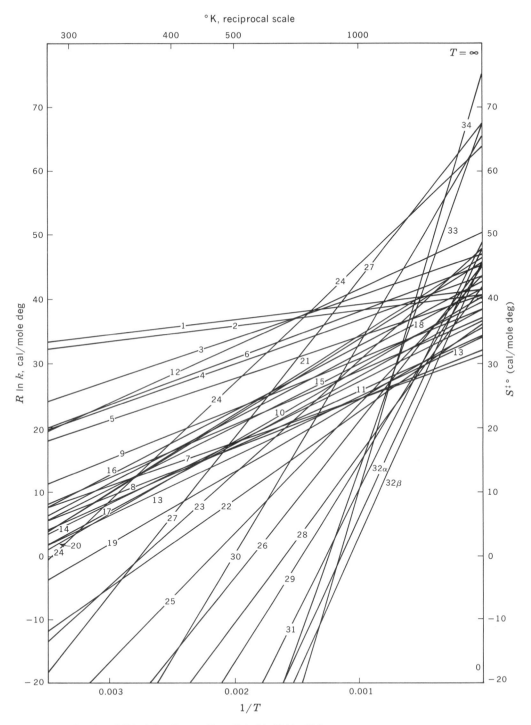

°K, reciprocal scale

Fig. 15-13 Graphs of $R \ln k$ for the reactions listed in Table 15-3.

that there are three molecular species X, Y, Z and that mutual equilibria are established:

$$(X) \xrightleftharpoons[b]{f} (Y)$$

$$(Y) \xrightleftharpoons[b]{f} (Z) \qquad (15\text{-}112)$$

$$(X) \xrightleftharpoons[b]{f} (Z)$$

In such a situation, the forward and reverse rates in each equilibrium must be equal to each other. It is not possible to have a situation like that shown in the diagram below, where the net result is a conversion of X into Y, Y into Z, and Z into X around a circular path

X → Y
  ↖ ↙
   Z

Simultaneous reactions

Suppose that X can be converted either into Y or into Z, according to the reactions:

$$X \xrightarrow{k_1} Y \qquad X \xrightarrow{k_2} Z \qquad (15\text{-}113)$$

The rate of disappearance of X is given by the expression:

$$-\frac{d(X)}{dt} = k_1(X) + k_2(X) = (k_1 + k_2)(X) \qquad (15\text{-}114)$$

As discussed in the section dealing with first-order reactions, this expression can be put into the integral form

$$(X) = (X)_0 e^{-(k_1+k_2)t} \qquad (15\text{-}115)$$

The rate of increase of the concentration of Y is given by the expression

$$\frac{d(Y)}{dt} = k_1(X) = k_1(X)_0 e^{-(k_1+k_2)t} \qquad (15\text{-}116)$$

which, in a similar way, can be put in the integral form:

$$(Y) = \frac{-k_1(X)_0}{k_1 + k_2} e^{-(k_1+k_2)t} + \frac{k_1(X)_0}{k_1 + k_2}$$

$$= \frac{k_1}{k_1 + k_2}(X)_0(1 - e^{-(k_1+k_2)t}) \qquad (15\text{-}117)$$

In other words, when the reactions run parallel under conditions such that the passage of time no longer

influences the fraction of X being converted into Y, this fraction y is given by the expression

$$y = \frac{k_1}{k_1 + k_2} \qquad (15\text{-}118)$$

Similarly, the fraction z of X being converted into Z is

$$z = \frac{k_2}{k_1 + k_2} \qquad (15\text{-}119)$$

and the concentration of Z is given by the expression

$$(Z) = \frac{k_2}{k_1 + k_2}(X)_0(1 - e^{-(k_1+k_2)t}) \qquad (15\text{-}120)$$

Sequential reactions

Suppose that instead of occurring simultaneously, a pair of reactions occurs sequentially according to the expression

$$X \xrightarrow{k_1} Y \xrightarrow{k_2} Z \qquad (15\text{-}121)$$

Then, the rate of change of the concentrations of the three components is given by the expression

$$\frac{d(X)}{dt} = -k_1(X) \qquad (15\text{-}122a)$$

$$\frac{d(Y)}{dt} = k_1(X) - k_2(Y) \qquad (15\text{-}122b)$$

$$\frac{d(Z)}{dt} = k_2(Y) \qquad (15\text{-}122c)$$

The usual initial conditions are assumed where, at $t = 0$, the concentrations of X, Y, and Z are $(X)_0$, $(Y)_0 = 0$, and $(Z)_0 = 0$. The integrated form for (X) is

$$(X) = (X)_0 e^{-k_1 t} \qquad (15\text{-}123)$$

This can be inserted into Eq. (15-122b) to give

$$\frac{d(X)}{dt} = k_1(X)_0 e^{-k_1 t} - k_2(Y) \qquad (15\text{-}124)$$

The value of (Y) is given by

$$(Y) = \frac{k_1}{k_2 - k_1}(X)_0(e^{-k_1 t} - e^{-k_2 t}) \qquad (15\text{-}125)$$

The total concentration at all times is equal to the original concentration

$$(X_0) = (X) + (Y) + (Z) \qquad (15\text{-}126)$$

and so the concentration of (Z) is given by

$(Z) = (X)_0 - (X) - (Y)$

$$= (X)_0\left[1 + \frac{1}{k_1 - k_2}(k_2e^{-k_1t} - k_1e^{-k_2t})\right] \quad (15\text{-}127)$$

The steady state

Sequences of consecutive reactions form an important part of the total reaction pattern in many portions of living matter. Frequently, a state is reached in which the rates of the members of such a sequence of reactions do not change with time. In such a case, the portion of the system consisting of the sequence of reactions is said to be in a *steady state*.

In order to see the essential nature of the steady state, consider first a series of physical reactions like those in which a gas flows through a series of small holes as shown in Fig. 15-14. To make the mechanism as simple as possible, we assume constant temperature and relate the rate of flow in moles per second to the cross-sectional area of the hole a, to the pressure P at the entrance to the hole, and to the rate of flow b through the hole of unit cross section at unit pressure. In these terms the rate of flow from left to right r_f is given by

$$r_{f_1} = a_1bP_1 \quad (15\text{-}128)$$

The rate of flow in the reverse direction r_b is given by

$$r_{b_1} = a_1bP_2 \quad (15\text{-}129)$$

where P_2 is the pressure on the right side of the hole. The net rate of flow r_1 is just the difference between the flow forward and the flow backward

$$r_1 = r_{f_1} - r_{b_1} = a_1b(P_1 - P_2) \quad (15\text{-}130)$$

When gas flows through a sequential series of three holes and the steady state is attained,

$$r_1 = r_2 = r_3 \quad (15\text{-}131)$$

Suppose that a vacuum is maintained on the right side of the third hole so that $P_4 = 0$. Since P_1 is maintained at a preselected value, there are two equations which can be solved simultaneously to find the values of the two unknowns P_2 and P_3:

$$a_1b(P_1 - P_2) = a_2b(P_2 - P_3) \quad (15\text{-}132)$$

$$a_2b(P_2 - P_3) = a_3b(P_3 - P_4) \quad (15\text{-}133)$$

A comparable chemical case is found when there is a series of first-order reactions such as

$$W \underset{}{\overset{(1)}{\rightleftharpoons}} X \underset{}{\overset{(2)}{\rightleftharpoons}} Y \underset{}{\overset{(3)}{\rightleftharpoons}} Z \quad (15\text{-}134)$$

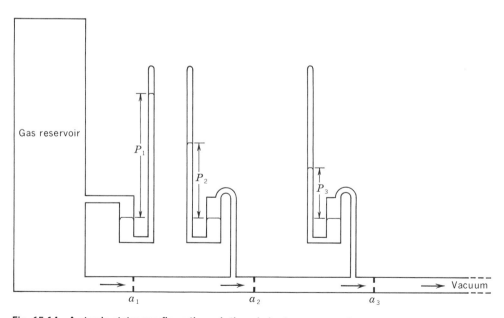

Fig. 15-14 A steady state; gas flows through three holes in sequence at a constant rate. $r_1 = r_2 = r_3$.

The forward reaction rates are given by the equations

$$r_{f_1} = k_{f_1}(W)$$
$$r_{f_2} = k_{f_2}(X) \qquad (15\text{-}135)$$
$$r_{f_3} = k_{f_3}(Y)$$

The back reaction rates are given by the equations

$$r_{b_1} = k_{b_1}(X)$$
$$r_{b_2} = k_{b_2}(Y) \qquad (15\text{-}136)$$
$$r_{b_3} = k_{b_3}(Z)$$

The three net rates are

$$r_1 = k_{f_1}(W) - k_{b_1}(X)$$
$$r_2 = k_{f_2}(X) - k_{b_2}(Y) \qquad (15\text{-}137)$$
$$r_3 = k_{f_3}(Y) - K_{b_3}(Z)$$

If (W) is maintained at a constant fixed value $(W)_0$ and the final product Z is removed so that $(Z) = 0$, the concentrations of the intermediate products are determined by the two simultaneous equations

$$k_{f_1}(W)_0 - k_{b_1}(X) = k_{f_2}(X) - k_{b_2}(Y) \qquad (15\text{-}138a)$$
$$k_{f_2}(X) - k_{b_2}(Y) = k_{f_3}(Y) - k_{b_3}(Z) \qquad (15\text{-}138b)$$

From the second of these two equations, (Y) can be expressed as a function of (X)

$$(Y) = \frac{k_{f_2}(X)}{k_{b_2} + k_{f_3}} \qquad (15\text{-}139)$$

Inserting this expression for (Y) in Eq. (15-138a) gives

$$(X) = (W)_0 \frac{k_{f_1}}{k_{f_2} + k_{b_1} - [k_{b_2}k_{f_2}/(k_{b_2} + k_{f_3})]} \qquad (15\text{-}140)$$

Thus, the rate at which the reaction proceeds in the steady state is given by

$$r_1 = r_2 = r_3$$
$$= (W)_0 k_{f_1} \left[1 - \frac{k_{b_1}}{k_{f_2} + k_{b_1} - [k_{b_2}k_{f_2}/(k_{b_2} + k_{f_3})]} \right]$$
$$(15\text{-}141)$$

An examination of Eq. (15-141) shows the effect of having any one single member of the sequence with a small forward reaction rate. If k_{f_1} is very small, the rate in the steady state is small. If k_{f_2} is very small, the second term in the expression within the brackets

approaches unity so that the total expression within the brackets becomes small and the steady-state rate is small. If k_{f_3} is very small, the second term of the expression within the brackets again approaches unity and the total rate is small.

The pattern of relations found in the reactions by which DNA and RNA are synthesized provides a striking example of the importance of rate relations in parallel and sequential reactions. Both the values of $H^{\ddagger}$ and $S^{\ddagger\circ}$ play a key role in these reactions. The essence of the genetic process is the transmission of information and its application to guide other chemical reactions. Information and entropy are directly related, the change in one being merely the negative of the change in the other. For this reason, $S^{\ddagger\circ}$ plays a particularly significant role in a number of important biochemical reactions. This will be discussed in detail in Chaps. 24 to 27.

Chain reactions

Many reactions proceed relatively slowly when the reactants combine and rearrange to form the products by what may be called the normal mechanism associated with the reaction; however, in many cases a parallel but different path can be followed in going from the reactants to the products where the rate of the reaction is much faster and a sequence or chain of reactions is followed which frequently involve free radicals.

Consider the combination of hydrogen and chlorine in the gaseous state at room temperature to give hydrogen chloride according to the reaction

$$H_2 + Cl_2 \rightleftarrows 2HCl \qquad (15\text{-}142)$$

In the dark and in relatively clean containers this reaction proceeds so slowly that the product can hardly be detected even after long periods of time. On the other hand, when a beam of light falls on the mixture of the gases, the reaction proceeds quickly. It is believed that the energy from the light dissociates the chlorine molecule to form two chlorine atoms

$$Cl_2 \xrightarrow{h\nu} Cl + Cl \qquad (15\text{-}143)$$

One of these atoms then reacts with a molecule of hydrogen

$$Cl + H_2 \rightarrow HCl + H \qquad (15\text{-}144)$$

and the free atom of hydrogen thus formed reacts with a molecule of chlorine

$$H + Cl_2 \rightarrow Cl + HCl \qquad (15\text{-}145)$$

The free chlorine atom thus formed reacts again with more hydrogen according to Eq. (15-144); and a chain, or sequence, of reactions is initiated in which free chlorine atoms are alternately formed. The reaction rate constants for these reactions are many orders of magnitude higher than the reaction rate constant for the normal path of the reaction shown in Eq. (15-142).

Since the system in the form of HCl has a lower energy than in the form of H_2 and Cl_2, the reaction is energetically self-sustaining and the chain is broken only when the free atom reacts with another molecule on the wall of the vessel or in the presence of a third body, such as another molecule, which can remove the activation energy and break the chain.

Sometimes there is sufficient energy in such a reaction for more than one free radical to be formed, and the chain branches and proliferates. In such a case, an explosion may result. There is an analogy between such a relationship and the infection of a biological cell by a virus, in which a single virus particle enters the cell, reacts with the cell contents, and produces dozens of new virus particles.

Example 15-7

In a first-order reaction the value of $(X)_0$ is 10 atm, the value of $(X)_{eq}$ is 0.32 atm, and k_b has the value 0.054 sec^{-1}. Calculate the value of (X) after 6 sec.

$$-k_b \frac{(X)_0}{(X)_{eq}} = 0.054 \text{ sec}^{-1} \frac{10 \text{ atm}}{3.2 \text{ atm}} \times 6 = 1.01$$

$$\log \frac{(X)_0 - (X)_{eq}}{(X) - (X)_{eq}} = \frac{1.01}{2.303} = 0.44$$

$$\log \frac{10 - 3.2}{(X) - 3.2} = 0.44$$

$$\frac{10 - 3.2}{(X) - 3.2} = 2.76$$

$$(X) = \frac{10 - 3.2}{2.76} + 3.2 = 5.66 \text{ atm} \qquad Ans.$$

Example 15-8

Calculate the value of (Z) at $t = 9.0$ sec in a reaction where $X \xrightarrow{k_1} Y \xrightarrow{k_2} Z$ and $k_1 = 0.065$ sec^{-1}, $k_2 = 0.055$ sec^{-1}, and $(X)_0 = 5.66$ atm.

$$(Z) = (X)_0 \left[1 + \frac{1}{k_1 - k_2}(k_2 e^{-k_1 t} - k_1 e^{-k_2 t}) \right]$$

$$= 5.66 \text{ atm} \left[1 + \frac{1}{0.065 - 0.055} \right.$$
$$\left. (0.055 e^{-0.065 \times 9.0} - 0.065 e^{-0.055 \times 9.0}) \right]$$

$$e^{-0.065 \times 9.0} = e^{-0.585} = 0.557$$

$$e^{-0.055 \times 9.0} = e^{-0.495} = 0.610$$

$$(Z) = 5.66 \left[1 + 100(0.0306 - 0.0397) \right]$$

$$= 5.66 (1 - 0.91) = 5.66 \times 0.09$$

$$= 0.5 \text{ atm} \qquad Ans.$$

Example 15-9

Calculate the steady-state rate in a sequence of reactions

$$W \rightarrow X \rightarrow Y \rightarrow Z$$

where $(W)_0$ is constant at 0.681 mole liter^{-1}, $(Z) = 0$ at all times, and the reaction rate constants have the following values:

$$k_{f_1} = 0.056 \text{ sec}^{-1}$$

$$k_{b_1} = 0.035 \text{ sec}^{-1} \qquad k_{f_2} = 0.0015 \text{ sec}^{-1}$$

$$k_{b_2} = 0.0017 \text{ sec}^{-1} \qquad k_{f_3} = 0.095 \text{ sec}^{-1}$$

$$r = (W)_0 k_{f_1} \left\{ 1 - \frac{k_{b_1}}{k_{f_2} + k_{b_1} - [k_{f_2} k_{b_2}/(k_{b_2} + k_{f_3})]} \right\}$$

$$= (0.681)(0.056) \left[1 - \right.$$

$$\left. \frac{0.035}{0.0015 + 0.035 - (0.0015 \times 0.0017)/(0.0017 + 0.095)} \right]$$

$$= (0.038) \left(1 - \frac{0.035}{0.0365 - 0.00002} \right)$$

$$= (0.038)(1 - 0.960) = 0.038 \times 0.04$$

$$= 0.0015 \text{ mole sec}^{-1} \text{ liter}^{-1}$$

15-5 CATALYSIS

There are many reactions for which an alternative path is provided by a substance called a *catalyst*.

When the reaction follows this alternative path, either the value of $H^{\ddagger}$ or $S^{\ddagger}$ or the values of both are altered in such a way that the free-energy barrier for the reaction is lowered and the reaction proceeds at a much more rapid rate. When a substance increases the rate of a reaction but is not itself changed in the process, it is called a catalyst and the shift in mechanism is called *catalytic action*.

Heterogeneous catalysis

When the catalyst is a solid substance and the catalytic action takes place on its surface, the process is called *heterogeneous catalysis*. As an example of a reaction of this type, consider the process by which nitrogen and hydrogen combine to form ammonia

$$N_2 + 3H_2 \rightarrow 2NH_3 \qquad (15\text{-}146)$$

According to Le Châtelier's principle, as temperature gets higher, there is less tendency for the nitrogen and hydrogen to combine and more tendency for the ammonia to dissociate. In order to have the reaction take place at lower temperatures, where ΔG° has a reasonably large negative value and where there is a large value of k, iron is used as a catalyst.

Sometimes impurities in the gases are adsorbed on the catalyst surface and prevent the molecules of the reactants from coming in contact with it and reacting. Impurities with this effect are called *catalytic poisons*. Substances can sometimes be added to the catalyst, in this case, traces of vanadium and related metals, which slow down the poisoning of the catalyst. Such substances are called *promoters*.

If a catalyst were to affect the equilibrium constant of the reaction while not reacting itself, the second law of thermodynamics would be violated. Then the forward reaction with the catalyst could be coupled with the back reaction without the catalyst, and heat could be raised from a lower to a high temperature or the system could be brought spontaneously from a state of higher entropy to a state of lower entropy.

The action of the catalyst is illustrated in Fig. 15-15. For example, the catalyst may lower the value of $H^{\ddagger}$, thus lowering the energy barrier. In the example of the gas diffusing upward in a gravitational field and through a hole, a "catalyst" would provide a new hole at a lower altitude. In such a situation the energy

required to move molecules of gas up to the hole level is reduced; in other words, $H^{\ddagger}$ is reduced.

If the catalytic action takes place on the surface of a solid, the larger the surface provided, the more rapidly the reaction proceeds. For the gas diffusing upward in a gravitational field and through a hole, enlarging the cross-sectional area of the hole is analogous to enlarging the area of the catalyst on which the chemical reaction can proceed. This increases the value of $S^{\ddagger\circ}$, and since $G^{\ddagger\circ} = H^{\ddagger} - TS^{\ddagger\circ}$, the effect of increasing $S^{\ddagger\circ}$ is to decrease $G^{\ddagger\circ}$. Since the rate is a negative function of $G^{\ddagger\circ}$,

$$r = r_0 e^{-G^{\ddagger\circ}/RT} \qquad (15\text{-}147)$$

a decrease in $G^{\ddagger\circ}$ increases the rate.

Heterogeneous catalysis will be discussed in more detail in Chap. 23 dealing with surface chemistry.

Homogeneous catalysis

There are many reactions in solution in which a substance like hydrogen ion or hydroxyl ion causes a rapid increase in the rate. For example, methyl acetate hydrolyzes very slowly according to the reaction

$$CH_3COOCH_3 + H_2O \rightarrow CH_3COOH + CH_3OH$$
$$(15\text{-}148)$$

but, in the presence of H^+ the reaction can proceed much more rapidly by the alternative path

$$CH_3COOCH_3 + H^+ \rightarrow CH_3{}^+C\!\!\overset{OH}{\underset{}{\diagup}}\!\!O\!\!-\!\!CH_3$$

$$CH_3{}^+C\!\!\overset{OH}{\underset{}{\diagup}}\!\!O\!\!-\!\!CH_3 \rightarrow CH_3\!\!-\!\!\overset{OH}{\underset{\underset{H\ \ \ H}{O^+}}{\overset{|}{\underset{|}{C}}}}\!\!-\!\!O\!\!-\!\!CH_3 \rightarrow$$

$$CH_3COOH + CH_3OH + H^+$$
$$(15\text{-}149)$$

Enzyme catalysis

In a large number of biochemical reactions, rates are affected by the presence of substances called *enzymes*, which themselves do not undergo change and which for this reason can be regarded as catalysts. The substance on which the enzyme exerts its catalytic action is called the *substrate*.

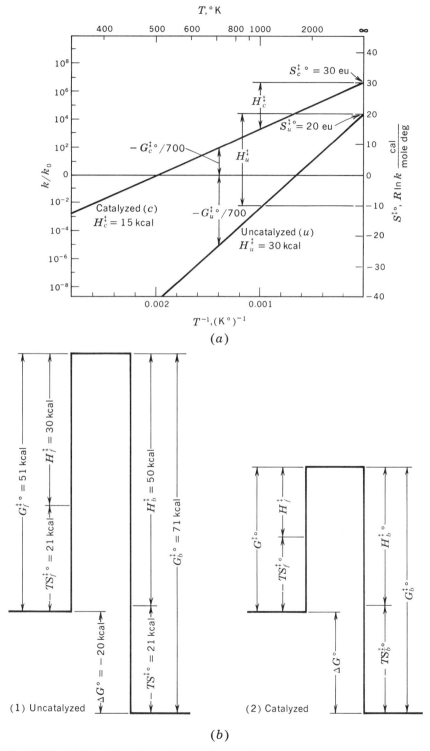

Fig. 15-15 (a) The effect of a catalyst on the rate of a reaction at different temperatures; (b) the effect of a catalyst on the reaction profile.

The enzyme is a complex molecule, organic in character, and having a molecular weight very roughly of the order of 10^3 or higher. The constituent atoms are generally linked together in such a way that the enzyme presents to the surrounding solution an active zone which roughly resembles part of the active surface of a catalyst. On this surface there are active sites where molecules of the substrate are adsorbed and the catalytic action takes place. Because of the complexity of the linkage and the variation in number of sites, there is difficulty in defining a significant molecular weight for the enzyme.

Thousands of different enzyme reactions have been studied, and there are certainly many different kinds of enzyme action, considering the variety of substances which will interact with enzymes. The number of active sites available affects the value of $S^{\ddagger\circ}$ for any reaction. The enzyme also lowers the value of $H^{\ddagger}$, and there is considerable difficulty in determining in detail just how these actions take place. The effect of the enzyme on $H^{\ddagger}$ appears to be intimately linked with the relation of the geometry of the enzyme surface to the geometry of the substrate. The rather crude analogy of a lock and key has been suggested, visualizing a substrate molecule as a key which must be of the proper shape to fit the contour of the active site (the lock) in order to have the reaction take place.

One of the most important recent developments in biochemistry has been the elucidation of the mechanism of the synthesis of DNA (deoxyribonucleic acid). There is evidence that in the biological cell, just before replication, the DNA molecule consists of two strands of long-chain molecules which in chemical composition and geometry are related to each other and are linked face to face by hydrogen bonding. During the process of replication the two molecules separate, each serving as a kind of enzymatic template for the synthesis of a duplicate DNA molecule, which consists of a long series of sequential reactions. In each step the value of $H^{\ddagger}$ is relatively low if the reaction proceeds with the reactant consisting of a molecule of the type needed at that particular point to replicate the chain, but if another molecule of a different type tries to enter the chain, the value of $H^{\ddagger}$ is so very high that the chances of the reaction's

occurring are extremely small. Thus, the *information* in the chain as expressed in the sequence guides the synthesis of the new duplicate molecule. In this kind of enzymatic catalysis, variations in both $H^{\ddagger}$ and $S^{\ddagger}$ are dominant factors in the control.

Such a complex biochemical reaction involves the laws of surface chemistry, the geometry of macromolecules, the relation between entropy and information, and the principles of that area of chemical theory lying at the boundary between thermodynamics and chemical kinetics and sometimes called *nonequilibrium thermodynamics*. For this reason, a detailed discussion of enzyme kinetics will be postponed until after discussion of these background topics.

Example 15-10

A catalyst increases the rate of a first-order reaction from $r = 1.5 \times 10^{-2}$ sec^{-1} to $r_c = 4.6$ sec^{-1} at 260°C. Calculate the decrease in $H^{\ddagger}$, assuming that $S^{\ddagger\circ}$ is not affected by the catalyst.

$$r = r_0 e^{-G^{\ddagger\circ}/RT}$$

$$r_c = r_0 e^{-G_c^{\ddagger\circ}/RT}$$

$$RT \ln \frac{r_c}{r} = G^{\ddagger\circ} - G_c^{\ddagger\circ} = H^{\ddagger} - H_c^{\ddagger}$$
$$- T(S^{\ddagger\circ} - S_c^{\ddagger\circ})$$

Assuming $S^{\ddagger\circ}$ is not changed by the catalytic action,

$$H^{\ddagger} - H_c^{\ddagger} = RT \ln \frac{r_c}{r}$$

$$= (2.303)(1.987)(533°) \log \frac{4.6}{0.015}$$

$$= (2439 \text{ cal})(\log 307)$$

$$= (2.439 \text{ kcal})(2.487)$$

$$= 6.06 \text{ kcal} \qquad Ans.$$

Example 15-11

A first-order reaction rate is measured with the results shown in the table.

The pressure was maintained at 1 atm in all the experiments. Assuming that $H^{\ddagger}$ and $S^{\ddagger\circ}$

	Without catalyst k, sec^{-1}	With catalyst (k_c), sec^{-1}
$T_1 = 473°$K	1.76×10^{-2}	6.10
$T_2 = 573°$K	0.804	88.1

do not vary with temperature, calculate the value of $H^{\ddagger}$ and $S^{\ddagger°}$ with and without catalyst.

Denoting the values with catalyst by subscript c, the following relations are found:

$$R \ln \frac{k_2}{k_1} = -H^{\ddagger}\left(\frac{1}{T_2} - \frac{1}{T_1}\right)$$

$$(1.987 \times 10^{-3})(2.303) \log \frac{0.804}{0.0176}$$

$$= -H^{\ddagger}(0.001745 - 0.002114)$$

$$H^{\ddagger} = \frac{(4.5757 \times 10^{-3})(1.6608)}{0.0003689} = 20.6 \text{ kcal}$$

$$R \ln \frac{(k_2)_c}{(k_1)_c} = -H_c^{\ddagger}\left(\frac{1}{T_2} - \frac{1}{T_1}\right)$$

$$(1.987 \times 10^{-3})(2.303) \log \frac{88.1}{6.10}$$

$$= -H_c^{\ddagger}(0.0003689)$$

$$H_c^{\ddagger} = \frac{(4.5757 \times 10^{-3})(1.1595)}{0.0003689} = 14.4 \text{ kcal}$$

$$r = k = e^{-(H^{\ddagger} - TS^{\ddagger°})/RT}$$

$$RT \ln k = -H^{\ddagger} + TS^{\ddagger°}$$

$$S^{\ddagger°} = R \ln k + \frac{H^{\ddagger}}{T}$$

$$= 1.987 \ln 0.0176 + \frac{20,600}{473}$$

$$= (1.987)(2.303)(\log 1.76 - 2) + 43.5$$

$$= (4.5757)(-1.755) + 43.5$$

$$= 35.5 \text{ eu}$$

$$S^{\ddagger°} = R \ln k_c + \frac{H_c^{\ddagger}}{T}$$

$$= 4.5757 \log 6.10 + \frac{14,400}{473}$$

$$= (4.5757)(0.7853) + 30.4$$

$$= 34.0 \text{ eu} \quad Ans.$$

Problems

1. The isotope of radon with atomic weight 209 has a half-life of 31 min. Calculate the value of k for its decomposition in min^{-1}.

2. An isotope of thorium, ^{228}Th, has a half-life of 1.90 years. Calculate the value of k for the decomposition of this isotope in sec^{-1}.

3. The most abundant isotope of thorium, ^{232}Th, has a half-life of 1.39×10^{10} years. Calculate the amount of time required to reduce the weight of 1 g of this isotope by 1 μg (10^{-6} g).

4. The half-life of ^{226}Ra is 1,620 years. Calculate the value of k for its decomposition in years^{-1}.

5. A sample of ^{226}Ra contains 1 g mole. After 1,000 years, how many atoms are left?

6. The half-life of ^{224}Ac is 2.9 hr. After decomposing for 4 hr, 1 mg is left of the original sample. What was the initial weight of ^{224}Ac?

7. The half-life for a unimolecular hydrolysis of ethyl acetate is 143 min, and if after 160 min 0.01 m HAc has been produced, what has the original concentration of EtOAc?

8. At a certain value of pH, the value of k for the unimolecular hydrolysis of ethyl acetate is 5.61×10^{-3} min. Calculate the half-life for this reaction.

9. At a certain concentration of hydrogen ion the half-life for the unimolecular hydrolysis of ethyl acetate is 125 min; calculate how long it will take to reduce the concentration from 0.0080 to 0.0020 m at this value of pH.

10. A solution of ethyl acetate at 30°C undergoes a hydrolysis reaction where $k = 0.00502$. At 50°C, $k = 0.0193$. Calculate the value of the activation enthalpy.

11. Calculate $\Delta S^{\ddagger°}$ for the reaction in Prob. 10.

12. Calculate $\Delta G^{\ddagger°}$ for the reaction in Prob. 10.

13. Cyclopropane decomposes by a unimolecular reaction to form propylene. When A is expressed in sec^{-1}, the value of A is 1.5×10^{15} and the value $H_f^{\ddagger}$ is 65.0 kcal mole^{-1}. Calculate the value of k_f at 900°K. If the pressure of cyclopropane in a vessel is 1.00 atm, calculate the pressure after 10 sec has elapsed when the gas is kept at a temperature of 900°K.

14. Cyclobutene decomposes to form 1,3-butadiene by a unimolecular reaction for which the value of $H_f^{\ddagger}$ is 32.5 kcal mole^{-1}. If the temperature at which the reaction takes place is reduced from 200 to 100°C, calculate the fraction by which k_f is reduced.

15. The value of k for the first-order decomposition of cyclobutene, CH_2—CH=CH—CH_2, to 1,3-butadiene, CH_2=CH—CH=CH_2, is given by the formula $1.2 \times 10^{13} e^{-32500/RT}$ sec^{-1}. Calculate the value of k at 350°C and the amount by which temperature must be reduced to cut the rate in half.

16. The value of k for the first-order reaction $C_2H_5Br \rightarrow C_2H_4 + HBr$ is given by the formula $2.6 \times 10^{13} e^{-53900/RT}$ sec^{-1}. Calculate the value of $G^{\ddagger \circ}$ at 200 and 300°C.

17. If the introduction of a catalyst into the reaction described in Prob. 16 increases the value of $\Delta S_f^{\ddagger \circ}$ by 20 percent and cuts the value of $\Delta H_f^{\ddagger}$ by 25 percent, calculate the value of $G^{\ddagger \circ}$ at 300°C in the presence of a catalyst.

18. Draw the energy profile, the entropy factor $-T \Delta S_f^{\ddagger}$, and the free-energy profile for the reaction described in Prob. 16.

19. At what temperature will the value of k for the decomposition of C_2H_5Br to C_2H_4 and HBr become equal to 1 sec^{-1}?

20. At what temperature will the value of k for the decomposition of (a) cyclobutane and (b) cyclopropane be equal to 1 sec^{-1}?

REFERENCES

Johnson, F. H., H. Eyring, and M. J. Polissar: "The Kinetics of Molecular Biology," John Wiley & Sons, Inc., New York, 1954.

Semenov, N. N.: "Some Problems of Chemical Kinetics and Reactivity," vol. 2, trans. by J. E. S. Bradley, Pergamon Press, New York, 1959.

Slater, N. B.: "Theory of Unimolecular Reactions," Cornell University Press, Ithaca, N.Y., 1959.

Benson, S. W.: "The Foundations of Chemical Kinetics," McGraw-Hill Book Company, New York, 1960.

Melander, L.: "Isotopes Effects on Reaction Rates," The Ronald Press Company, New York, 1960.

Frost, A. A., and R. G. Pearson: "Kinetics and Mechanism," 2d ed., John Wiley & Sons, Inc., New York, 1961.

Taylor, H. S.: Fifty Years of Chemical Kinetics, *Ann. Rev. Phys. Chem.,* **13**:1 (1962).

Bak, T. A.: "Contributions to the Theory of Chemical Kinetics," W. A. Benjamin, Inc., New York, 1963.

Eyring, H., and E. M. Eyring: "Modern Chemical Kinetics," Reinhold Publishing Corporation, New York, 1963.

Leffler, J. E., and E. Grunwald: "Rates and Equilibria of Organic Reactions," John Wiley & Sons, Inc., New York, 1963.

Balandin, A. A., et al.: "Catalysis and Chemical Kinetics," Academic Press, Inc., New York, 1964.

Laidler, K. J.: "Chemical Kinetics," 2d ed., McGraw-Hill Book Company, New York, 1965.

Patton, A. R.: "Biochemical Energetics and Kinetics," W. B. Saunders Company, Philadelphia, 1965.

Walter, C.: "Steady-state Applications in Enzyme Kinetics," The Ronald Press Company, New York, 1965.

Bray, H. G., and K. White: "Kinetics and Thermodynamics in Biochemistry," Academic Press, Inc., New York, 1966.

Amdur, I., and G. G. Hammes: "Chemical Kinetics," McGraw-Hill Book Company, New York, 1966.

Vetter, K. J.: "Electrochemical Kinetics," Academic Press Inc., New York, 1967.

Williams, V. R., and H. B. Williams: "Basic Physical Chemistry for the Life Sciences," W. H. Freeman and Company, San Francisco, 1967.

Horiuti, J., and T. Nakamura: On the Theory of Heterogeneous Catalysis, *Advan. Catalysis,* **17**:1967.

Dence, J. B., H. B. Gray, and G. S. Hammond: "Chemical Dynamics," W. A. Benjamin, Inc., New York, 1968.

Benson, S. W.: "Thermochemical Kinetics," John Wiley & Sons, Inc., New York, 1968.

Boudart, M.: "Kinetics of Chemical Processes," Prentice-Hall, Inc., Englewood Cliffs, N.J., 1968.

Rideal, E. K.: "Concepts in Catalysis," Academic Press, Inc., New York, 1968.

16-1 THE QUANTUM REVOLUTION

In the preceding 15 chapters physical chemistry has been presented in two contrasting perspectives. Part of the time attention has been focused on the study of the observable aspects of this branch of science, the quantities that can be measured directly like mass, volume, energy, enthalpy, temperature, pressure, and concentration; and part of the time the discussion has involved the micro mechanisms which cannot be observed directly, such as the details of atomic collisions, the orientation of molecules in crystals and liquids, and the kinetic-energy and entropy barriers.   Of course, the nature of these

Chapter Sixteen

quantum theory

micro mechanisms must be inferred from the observables, since atoms, molecules, and energy quanta never can be seen directly.

We now turn to take an even more penetrating look into atomic and molecular mechanisms.   Instead of treating the atom as a solid elastic particle like that postulated in the kinetic theory of gases, it becomes necessary in the more advanced parts of physical chemistry to explore the details of atomic structure in terms of electrons and protons.   It is even essential to look more carefully at the nature of the electrons and protons themselves and to examine how they influence each other at a distance. Strangely enough, this most fundamental aspect of science shows that electrons and protons act on each other as if they were spread out and interpenetrating, like *waves* instead of being separate *particles*.   Thus, there is revealed the complementary aspect of the two perspectives—the *wave,* which emphasizes the continuity, and the *particle,* which emphasizes the discontinuity, of nature.

Complementarity

Philosophers have been aware of these two complementary aspects of nature for many centuries.   Over 2,000 years ago in ancient Greece, Leucippus and his pupil Democritus concluded on purely philosophical grounds that matter was not indefinitely continuous and it could not be sliced finer and finer without limit. They proposed the concept of the ultimately uncutable (*atomos*) particle. But shortly after this Aristotle, trained under the great Greek philosopher Plato, founded a school of philosophy which emphasized continuity; and this school dominated human thinking for almost 2,000 years. When, just a few hundred years ago, men finally turned from speculation to experimentation, the idea of particles as the fundamental building blocks of nature was revived. Dalton founded the kinetic theory of gases on atomic concepts, and Newton tried to explain the laws of optics in terms of particles of light. And so, in the eighteenth century, there was considerable awareness of the grounds for viewing nature in the particle perspective.

In the nineteenth century the pendulum seemed to swing back toward the alternative perspective of continuity. The discovery of the interference phenomena in the transmission of light led to the electromagnetic theory of light based on the idea of continuous waves. There were even some scientists who proposed that the concept of the atom should be abandoned. But, with the coming of the twentieth century, more and more evidence accumulated stressing that the concepts of continuity and discontinuity were not mutually exclusive but complementary. Planck and Einstein showed that light has not only a wave aspect but also a particle aspect, thus laying the cornerstone for the foundation of the quantum theory. And 20 years later, de Broglie proposed that the fundamental particles, the electron and the proton, had a wave aspect. This hypothesis was almost immediately verified by the experiments of Davisson and Germer and of Bragg and dozens of others.

During the last half century, as biochemistry has played a more and more prominent role in the life sciences, there has been a tendency to view life processes primarily as particulate, as movement of atoms, as exchange of quanta of energy. This is understandable. The structural formula of a biochemical molecule is written in terms of symbols that emphasize the stringing together of particulate atoms. Biochemical reactions are portrayed in terms of the exchange of these particulate atoms. In fact, the knowledge of particulate structure has advanced far ahead of the knowledge of the holistic aspect of

these molecules. Thus, the evidence that "the whole is merely equal to the sum of the parts" has overshadowed the indications that in certain aspects "the whole is far greater than the sum of the parts" such as the quantum-mechanical interaction in molecules like DNA.

The use of quantum theory to interpret biochemical reactions is still in its infancy, but in evaluating this new approach, it is essential to keep in mind that in quantum theory the aspect of the whole has something far beyond the sum of the content of the individual parts.

The biochemist A. Szent-Györgyi (Nobel Laureate in Medicine, 1937, and now at the Marine Biological Laboratories, Woods Hole, Massachusetts) has repeatedly emphasized this. One of the foremost pioneers in the development of the mathematical structure of the quantum theory, Hermann Weyl, of the Institute for Advanced Study at Princeton, New Jersey, has written that in quantum theory: "the whole is always more, is capable of a much greater variety of wave states, than the combination of its parts." In the same context he says later: "In this very radical sense, quantum physics supports the doctrine that the whole is more than the combination of its parts." In commenting on the use of quantum theory for the interpretation of life processes, Weyl[1] concludes:

Even the atomic physical processes have very little similarity with the gross macroscopic action of a machine. Every atom is already a whole of quite definite structure; its organization is the foundation for possible organizations and structures of the utmost complexity . . . there is no reason to see why the theoretical symbolic construction should come to a halt before the facts of life and of psyche. It may well be that the sciences concerned have not yet reached the required level but this limitation is neither fundamental nor permanent. . . . The fact that in Nature all is woven into one whole, that space, matter, gravitation, the forces arising from the electromagnetic field, the animate and inanimate are indissolubly connected, strongly supports the belief in

[1] Hermann Weyl, "Philosophy of Mathematics and Natural Science," p. 214, Princeton University Press, Princeton, N.J., 1949. Quoted by permission.

the unity of Nature and hence in the unity of the scientific method. There are no reasons to distrust it.

In the coming pages we shall examine a few of the aspects of the quantum theory and the basic physics on which it rests, in the hope that this will provide an insight which will enable us to see life processes not merely as a congeries of particles but as a unity of dynamism which reveals not only in the atom, in the molecule, in the macromolecule, and in the cell but also in the total organism a significant holistic aspect.

Waves

While individual single waves, e.g., a tidal or shock wave, play a significant role in some fields of science, a series of waves in regular sequence, like sound and light waves, is far more important in the greater part of physics and chemistry.

Mathematically speaking, the simplest types of wave sequences are those defined by a sine or cosine function. Consider the vertical displacement of the central portion of a stretched elastic cord which is so long that there are no boundary effects observable because of the fixed position of its ends. When motionless, the cord displays the pattern of a straight line; but if it is moved by an arm attached to a wheel rotating at constant angular velocity, then sequences

of sinusoidal waves travel away from the arm toward the left and toward the right along the cord, as shown in Fig. 16-1. The important properties of the wave are summarized in Table 16-1 and illustrated in Fig. 16-2.

Consider the train of waves which moves to the right when generated as shown in Fig. 16-2; the crest of any individual wave travels at a constant velocity, u cm sec^{-1}. The distance between two adjoining wave crests is defined as the wavelength λ. The number of waves per centimeter is called the wave number σ or $\bar{\nu}$. At any fixed point, the string rises and falls a certain number of times per second, called the *frequency* ν. The length of time of one complete vibration is the period τ. The maximum amount by which the string is displaced at any time as the wave train passes by is called the amplitude A.

These definitions lead to the following relations:

$$\lambda \times \nu = u \tag{16-1}$$
$$\text{cm} \quad \text{sec}^{-1} \quad \text{cm sec}^{-1}$$

$$\sigma = \bar{\nu} = 1/\lambda \tag{16-2}$$
$$\text{cm}^{-1} \quad \text{cm}^{-1} \quad \text{1/cm}$$

$$\tau = 1/\nu \tag{16-3}$$
$$\text{sec} \quad \text{1/sec}^{-1}$$

The same basic concepts apply to many different types of wave trains. For example, a series of water

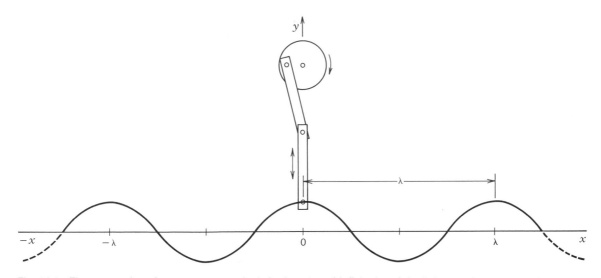

Fig. 16-1 The propagation of waves on a stretched elastic string of infinite length by linkage with rotatory motion.

TABLE 16-1	Properties of waves	
	Symbol	Usual dimensions
Wavelength	λ	cm
Wave number	σ or $\bar{\nu}$	cm^{-1}
Period	τ	sec
Frequency	ν	sec^{-1} (Hz)
Velocity	u or c	cm sec^{-1}
Amplitude	A	cm

waves may closely resemble the waves on the vibrating string when considered in cross section parallel to the direction of travel. Sound waves consist of alternate compression and rarefaction zones, adjoining zones of similar character being separated by a distance which is the wavelength λ; but the direction of the amplitude displacement is parallel to the direction of wave travel, rather than normal to it as in the case of the string. Light waves resemble the waves of the string, but the amplitude corresponds to an electrodynamic form of potential energy. In quantum mechanics, waves represent mathematical relations associated with charge distribution and the probability of atomic events. Table 16-2 lists values and properties of electromagnetic waves.

The vibrating string

The mathematical nature of the waves on a vibrating string can be deduced by considering the forces operating in a small segment of the string (Fig. 16-3) as the part drawn as a heavy black line. At the left end of the segment, there is a force pulling to the left and denoted by the length of the line marked T. The component of the force in the y direction of the coordinate system is denoted by the line marked T_y and is related to T by:

$$T_y = -T \sin \phi_0 \simeq -T\left(\frac{dy}{dx}\right)_{x_0} \qquad (16\text{-}4)$$

Since T_y decreases as ϕ_0 or $(dy/dx)_{x_0}$ increase, the negative sign appears on the right.

At the right end of the segment there is a similar relation

$$T'_y = T \sin (\phi_0 + d\phi) \simeq T\left(\frac{dy}{dx}\right)_{x_0+dx}$$

$$\simeq T\left[\left(\frac{dy}{dx}\right)_{x_0} + \left(\frac{d^2y}{dx^2}\right)_{x_0} dx\right] \qquad (16\text{-}5)$$

The resultant force $(T_s = T_y + T'_y)$ acting on the segment (when the amplitude of vibration is small) is given by

$$T_s = T\frac{d^2y}{dx^2}dx \qquad (16\text{-}6)$$

In order to simplify the problem as much as possible, the assumption is made that the string has a uniform cross section, that the weight per unit length is ρ, and that T and ρ do not vary with x. This force multiplied by the mass of the segment of the string $\rho\,dx$ is equal to the acceleration of the string in the y direction

$$\underset{\text{force}}{T\frac{\partial^2y}{\partial x^2}} = \underset{\text{mass} \times \text{acceleration}}{\rho\frac{\partial^2y}{\partial t^2}} \qquad (16\text{-}7)$$

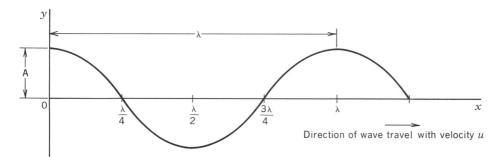

Fig. 16-2   Instantaneous view of sine wave traveling from left to right.

TABLE 16-2   The electromagnetic spectrum[a]

Radiation	Wavelength λ			σ, cm^{-1}	$\nu = c/\tau$, sec^{-1}	Ergs particle^{-1}	$E = h\nu$ electric acceleration units	kcal mole^{-1}
	cm	μ	Å					
Radio	10^4			10^{-4}	3.3×10^6	2×10^{-20}		
Microwave	1	10^4		1	3.3×10^{10}	2×10^{-16}		2.86×10^{-3}
	10^{-1}	10^3		10	3.3×10^{11}	2×10^{-15}		2.86×10^{-2}
Far infrared	10^{-2}	10^2		10^2	3.3×10^{12}	2×10^{-14}		2.86×10^{-1}
	10^{-3}	10		10^3	3.3×10^{13}	2×10^{-13}	0.124 ev	2.86
Near infrared	10^{-4}	1	10^4	10^4	3.3×10^{14}	2×10^{-12}	1.24 ev	28.6
Visible red		0.8	8000	1.25×10^4	4.2×10^{14}	2.5×10^{-12}	1.54 ev	35.7
Visible violet		0.4	4000	2.5×10^4	8.3×10^{14}	5×10^{-12}	3.08 ev	71.5
Ultraviolet	10^{-5}	10^{-1}	10^3	10^5	3.3×10^{15}	2×10^{-11}	12.4 ev	286
	10^{-6}	10^{-2}	10^2	10^6	3.3×10^{16}	2×10^{-10}	12.4 ev	2.86×10^3
Soft x-rays	10^{-7}	10^{-3}	10	10^7	3.3×10^{17}	2×10^{-9}	1.24 ev	2.86×10^4
X-rays	10^{-8}	10^{-4}	1	10^8	3.3×10^{18}	2×10^{-8}	12.4 kev	2.86×10^5
γ rays	10^{-9}		10^{-1}	10^9	3.3×10^{19}	2×10^{-7}	124 kev	2.86×10^6
	10^{-10}		10^{-2}	10^{10}	3.3×10^{20}	2×10^{-6}	1.24 Mev	2.86×10^7
	10^{-11}		10^{-3}	10^{11}	3.3×10^{21}	2×10^{-5}	12.4 Mev	2.86×10^8
Cosmic rays	10^{-12}		10^{-4}	10^{12}	3.3×10^{22}	2×10^{-4}	1.24 Mev	2.86×10^9
	10^{-13}		10^{-5}	10^{13}	3.3×10^{23}	2×10^{-3}	1.24 Gev	2.86×10^{10}

[a] The currently accepted values of c and h are given in Appendix Table A1-1 with added significant figures.   kev $= 10^3$ ev, Mev $= 10^6$ ev, Gev $= 19^9$ ev.

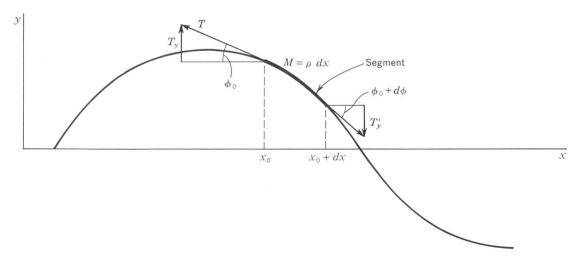

Fig. 16-3   The relation of tension to acceleration for waves in a stretched elastic string.

Since the dimensions of T are dynes, or g cm sec^{-2}, and the dimensions of ρ are g cm^{-1}, T/ρ has the dimensions cm^2 sec^{-2} and may be replaced by a constant u^2, where u has the dimensions of velocity. Thus Eq. (16-7) can be written

$$\frac{\partial^2 y}{\partial x^2} = \frac{1}{u^2}\frac{\partial^2 y}{\partial t^2} \qquad (16\text{-}8)$$

This is the important differential equation for a wave traveling in a medium like the stretched string which has only one dimension (length) available for the passage of waves.

As a limited solution for Eq. (16-8) a sine function may be used. If the zero of the x coordinate axis is located at a point where the amplitude of the wave is zero, then as x increases, the next point where the wave pattern repeats itself will be at the point where $x = \lambda$; but as the argument of the sine function increases, the function repeats itself when the argument equals 2π. In the same way, as time progresses, the wave repeats itself when $t = \tau$, as shown in Fig. 16-4. Thus, the expression for a trial solution of Eq. (16-8) is

$$y = A \sin\left(\frac{2\pi}{\lambda}x - \frac{2\pi}{\tau}t\right) \qquad (16\text{-}9)$$

Since $\tau = 1/\nu$, this can be written

$$y = A \sin\left(\frac{2\pi}{\lambda}x - 2\pi\nu t\right) \qquad (16\text{-}10)$$

Differentiating with respect to x,

$$\frac{\partial y}{\partial x} = A\frac{2\pi}{\lambda}\cos\left(\frac{2\pi}{\lambda}x - 2\pi\nu t\right) \qquad (16\text{-}11)$$

and

$$\frac{\partial^2 y}{\partial x^2} = -A\frac{4\pi^2}{\lambda^2}\sin\left(\frac{2\pi}{\lambda}x - 2\pi\nu t\right) \qquad (16\text{-}12)$$

Differentiating with respect to t,

$$\frac{\partial y}{\partial t} = -A2\pi\nu\cos\left(\frac{2\pi}{\lambda}x - 2\pi\nu t\right) \qquad (16\text{-}13)$$

$$\frac{\partial^2 y}{\partial t^2} = -A4\pi^2\nu^2\sin\left(\frac{2\pi}{\lambda} - 2\pi\nu t\right) \qquad (16\text{-}14)$$

Combining Eqs. (16-12) and (16-14),

$$\lambda^2\frac{\partial^2 y}{\partial x^2} = \frac{1}{\nu^2}\frac{\partial^2 y}{\partial t^2} \qquad (16\text{-}15)$$

Since $\nu\lambda = u$,

$$\frac{\partial^2 y}{\partial x^2} = \frac{1}{u^2}\frac{\partial^2 y}{\partial t^2} \qquad (16\text{-}16)$$

and a comparison with Eq. (16-8) shows that u is the real velocity of wave travel.

This solution of the equation is general enough to illustrate quantization in vibration. More general solutions will be found in a number of the references.

Quanta

In the early part of the nineteenth century, experiments on the interference of light beams were carried out by the English physician and physicist Thomas Young which established on a firm experimental basis the wave theory of light postulated earlier by the Dutch physicist Christian Huygens in 1678. In the latter part of the nineteenth century the work of the

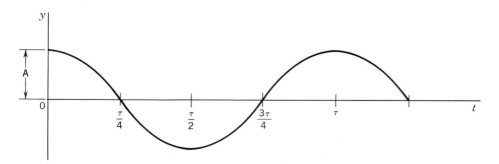

Fig. 16-4   Graph of y as a function of t at $x = \lambda$ in Fig. 16-2.

English theoretical physicist James Clark Maxwell provided this theory with a detailed mathematical structure, in which there were relatively few inconsistencies.

One of the remaining unsolved problems was concerned with the prediction of the intensity of light in different parts of the spectrum when the light came from a source which had no characteristic "coloring" but emitted light with intensities in the different ranges governed solely by the temperature of the source. For example, if the radiation inside a hollow body of uniform temperature is observed through a small hole in the wall of the body, the distribution of the intensity of radiation in different parts of the spectrum depends only on the temperature of the body and is independent of the material of which the body is made and of the size and shape of the interior. Thus, two hollow enclosures at the same temperature exchange energy by exchanging radiation and are in thermal equilibrium with each other. This type of radiation is called *blackbody radiation*. Figure 16-5 shows the distribution of intensity as a function of wavelength in radiation from *blackbodies* at 1000, 1500, and 2000°K. The classical wave theory of light did not yield any equation capable of explaining these graphs.

At the beginning of the twentieth century, Max Planck proposed a bold new idea which led to a satisfactory explanation of these relations. He suggested that light, instead of being emitted in continuous waves, was emitted in small fixed amounts called *quanta* of energy. This relationship is set forth in the expression

$$\epsilon = h\nu \qquad (16\text{-}17)$$

where ϵ is the energy in the discrete bundle of light and ν is the frequency; the constant of proportionality relating these two quantities is denoted by h, today called *Planck's constant*. If the energy is measured in ergs and the frequency in sec^{-1}, Planck's constant has the value 6.6256×10^{-27} erg sec. Planck proposed that the intensity of radiation is related to the wavelength and temperature by

$$I_\lambda = \frac{2\pi c^2 h}{\lambda^5(e^{hc/kT\lambda} - 1)} = \frac{2\pi c^2 h}{\lambda^5(e^{h\nu/kT} - 1)} \qquad (16\text{-}18)$$

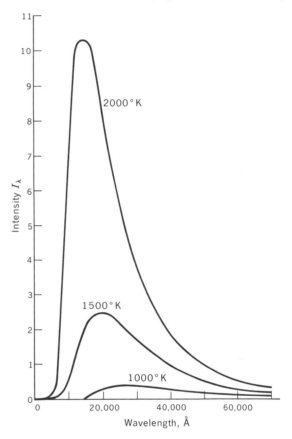

Fig. 16-5 Emission of radiation from a blackbody at different temperatures. The area under the curve between specified wavelengths, divided by 10^4, gives the energy in calories per second radiated from 1 cm^2 of a blackbody in the given range of wavelengths. (Plotted from data in "International Critical Tables," McGraw-Hill Book Company, New York, 1926–1930.)

where I_λ = intensity of radiation
c = velocity of light
k = Boltzmann's constant = gas constant R divided by Avogadro's number
T = temperature, °K

Planck's hypothesis was so bold that even the agreement of Eq. (16-18) with experimental results was hardly enough to convince most scientists that light had a particle aspect as well as a wave aspect. But the correctness of Planck's new view was soon supported by another experimental observation on the emission of electrons from the surfaces of a

Fig. 16-6   The photoelectric effect.

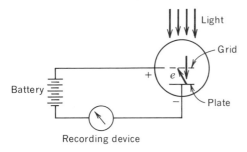

Recording device

number of different metals such as Li, Na, and K, on which light in the blue, violet, or ultraviolet region impinged. The type of apparatus used in an experiment of this sort is shown in Fig. 16-6. The metallic plate is placed in a vacuum and has a negative charge with respect to the grid. Light of variable frequency impinges on the plate. At comparatively low frequencies a recording device shows that there is no current flowing and therefore that no electrons are passing from the plate to the grid. As the frequency of the incident light is increased, a value of the frequency is reached at which a small current of electrons from the grid to the plate is registered. Apparently, electrons jump through the vacuum from the plate to the grid.

In 1905 this effect was explained by Einstein on the basis of Planck's quantum hypothesis. Einstein suggested that there was no mechanism by which energy could slowly accumulate in an atom and finally reach an amount sufficient to expel an electron; instead, he argued, the energy must be received in a single impact of a quantum light. He proposed the equation

$$\tfrac{1}{2}mv_{\max}^2 = h\nu - \phi \qquad (16\text{-}19)$$

where $v_{\max}$ = maximum velocity at which electrons are ejected from surface

h = Planck's constant
ν = frequency of light falling on surface
ϕ = work function

The work function is the amount of work which must be performed to remove the electron from the surface. The American physicist R. A. Millikan made a number of careful measurements of $v_{\max}$, completely confirming Einstein's equation. When $h\nu$ was less

than ϕ, no electrons were ejected. In other words, there is a threshold value ϕ of energy necessary to expel the electron. When the frequency is high enough for $h\nu$ to become greater than ϕ, electrons are expelled. The name *photon* has been adopted for the bundle of light; the energy of the photon is equal to $h\nu$.

During this same period, Einstein showed that the concept of quantization of energy also explained the way in which values of heat capacity decrease as temperature falls toward 0°K. His theory was described in Chap. 6 and will be discussed in further detail in relation to molecular spectra (Chap. 20).

Table 16-2 lists values of λ, ν, σ, and $h\nu = E$ for various parts of the electromagnetic spectrum.

Thus, the idea gained general acceptance that light has a wave aspect, as shown in phenomena like interference, and *also* has a particle aspect, as shown in the photoelectric effect.

De Broglie waves

From considerations of symmetry, which play such an important role in so many areas of science such as crystallography and the theory of so-called fundamental particles, it seems obvious that if light waves have a particle aspect, then particles like electrons and protons might have a wave aspect. But scientists were slow to appreciate the force of such logic, and almost 20 years elapsed between the Planck-Einstein particle theory for light waves and the wave theory for fundamental particles first proposed by the French physicist Louis de Broglie in 1923. By this time there was considerable evidence to support Einstein's theory of relativity, in which the equivalence between mass and energy is given by the famous equation

$$E = mc^2 \qquad (16\text{-}20)$$

where E = energy of particle
m = mass of particle
c = velocity of light

If this equation is combined with the relationship between energy and frequency expressed in Planck's equation,

$$E = h\nu \qquad (16\text{-}21)$$

then the relation between mass, velocity squared, and frequency is

$$mc^2 = h\nu \tag{16-22}$$

Since $\nu = \sigma c$ and $p = mc$, Eq. (16-22) can be put in the form

$$p = h\sigma \tag{16-23}$$

which is de Broglie's equation; it closely resembles Planck's fundamental equation of quantization, Eq. (16-21).

Planck's equation (16-21) relates the frequency ν of a continuous wave train of light through h to the energy E of a discrete quantum of light, the photon particle. De Broglie's equation relates the momentum p of a particle through h to the wave number σ of a wave train. The problem at this point is to discover the nature of this new variety of wave defined by de Broglie's equation (16-23).

De Broglie's idea that a particle might have a wave aspect was just as revolutionary as Planck's suggestion that a wave might have a particle aspect. Since one of the most convincing proofs of the wave nature of light was the reflection of light from a surface with a series of grooves that produce diffraction, the two American physicists C. J. Davisson and L. H. Germer immediately thought of the possibility of reflecting electrons from the surface of a crystal, where the ordered arrangement of the atoms in the crystal produced something that might act like the series of grooves on the diffraction grating. When the experiment was performed, it was found that electrons are reflected from the crystal surface exactly as if they were waves and did *not* bounce off the surface like elastic particles. This was the first experimental confirmation of the de Broglie hypothesis.

If particles have a wave aspect, there should be phenomena in which quantized aspects of these waves could be detected. In order to see the nature of such phenomena more clearly it is helpful to examine the quantization of familiar waves like those in an elastic stretched string.

Example 16-1

The wavelength of the line in the solar spectrum called Fraunhofer's B is 6867 Å. Calculate the frequency and the wave number associated with this line.

The frequency is given by

$$\nu = \frac{c}{\lambda} = \frac{2.998 \times 10^{10} \text{ cm sec}^{-1}}{6867 \times 10^{-8} \text{ cm}}$$

$$= 4.366 \times 10^{14} \text{ sec}^{-1}$$

The wave number is the reciprocal of the wavelength

$$\sigma = \frac{1}{\lambda} = \frac{1}{6867 \times 10^{-8}} = 14,562 \text{ cm}^{-1}$$

Example 16-2

Calculate the energy in ergs and in calories of one photon associated with the wavelength of $6,867 \times 10^{-8}$ cm.

The energy in ergs is given by

$$\epsilon = h\nu = 6.6256 \times 10^{-27} \text{ erg sec} \times 4.366 \times 10^{14} \text{ sec}^{-1} = 2.893 \times 10^{-12} \text{ erg}$$

Ergs can be converted into calories by the expression

$$2.893 \times 10^{-12} \text{ erg} \times 10^{-7} \text{ joule erg}^{-1} \times 0.2390 \text{ cal joule}^{-1} = 6.914 \times 10^{-20} \text{ cal}$$

Example 16-3

Calculate the energy equivalent of the mass in an atom of hydrogen.

This is given by

$$E = mc^2 = 1.008(6.02 \times 10^{23})^{-1} \times (2.998 \times 10^{10} \text{ cm sec}^{-1})^2$$

$$= 1.67 \times 10^{-24} \times 8.99 \times 10^{20} \text{ erg}$$

$$= 1.50 \times 10^{-3} \text{ erg}$$

16-2   QUANTIZED VIBRATION

In physics and chemistry waves are often found under conditions where the motion is restricted at the boundaries of the vibrating system. Under these conditions the wave motion can take place only at certain discrete values of the frequency and is said to be *quantized*.

One-dimensional vibrator

As an example of vibration in one dimension, consider a string of finite length s stretched between two fixed points. Such a string is regarded as a one-dimensional vibrator because the single dimension of length primarily determines the nature of the vibration. Coordinates are defined as shown in Fig. 16-7. Because the ends of the string are fixed, there can be no vertical motion there, a condition which is expressed mathematically by the equations

$$y = 0 \qquad \text{at } x = 0 \qquad\qquad (16\text{-}24a)$$
$$y = 0 \qquad \text{at } x = s \qquad\qquad (16\text{-}24b)$$

The effect of these boundary conditions is to restrict the possible patterns of vibratory motion to certain discrete forms. The equations describing these forms are obtained from the solutions of the differential equation of motion, Eq. (16-16), which satisfy the boundary conditions expressed in Eqs. (16-24a) and (16-24b). A particular solution useful for this purpose is the one in which the time variable t and the distance variable x appear in separate terms

$$y = A_m \sin\left(\frac{2\pi}{\tau}t\right)\sin\left(\frac{2\pi}{\lambda}x\right) \qquad (16\text{-}25)$$

where A_m represents the maximum amplitude of vibration when both t and x are arbitrarily varied.

The period τ is defined as the length of time during which a point on the string rises and falls through one complete cycle of vibration, as shown in Fig. 16-8. As t increases indefinitely from zero to positive values, the value of $\sin[(2\pi/\tau)t]$ starts at zero and oscillates between the maximum value of $+1$ and the minimum value of -1 as follows:

$$\sin 2\pi\nu t = \begin{cases} 0 & \text{when } t = 0 \\ +1 & \text{when } t = \frac{1}{4}\tau \\ 0 & \text{when } t = \frac{1}{2}\tau \\ -1 & \text{when } t = \frac{3}{4}\tau \\ 0 & \text{when } t = \tau \end{cases} \qquad (16\text{-}26)$$

These relations are shown graphically in Fig. 16-8. Thus, in general, $y = 0$ when $t = (n/2)\tau$, where $n = 0$, 1, 2, 3, 4, ...

Consider now the shape of the string at the fixed point in time when $t = \frac{1}{4}\tau$, so that $\sin[(2\pi/\tau)t] = 1$. Equation (16-25) now takes the form

$$y = A_m \sin\left(\frac{2\pi}{\lambda}x\right) \qquad (16\text{-}27)$$

When $x = 0$, then $y = 0$, and the first boundary condition [Eq. (16-24a)] is satisfied. In order to have the second boundary condition [Eq. (16-24b)] satisfied, however, the value of λ must be one which will fulfill the relation

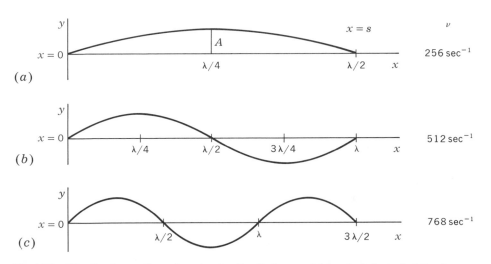

Fig. 16-7 Vibration in one dimension, showing the displacement (at a single instant of time) as a function of x for three different frequencies. ν_0 is assumed to be 256 sec^{-1}.

Fig. 16-8   The motion of one point on a vibrating stretched string as a function of time.

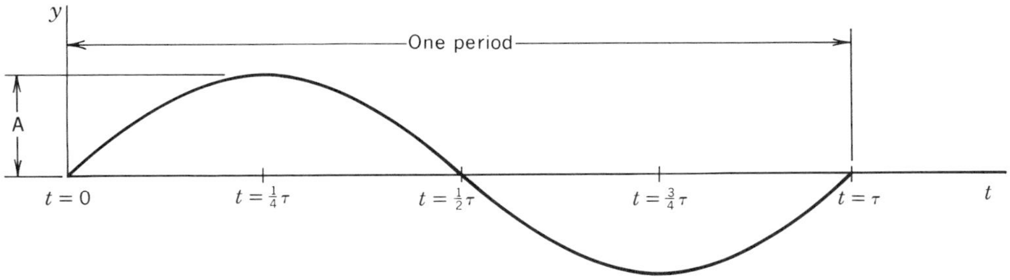

$$\frac{\lambda}{2} = \frac{s}{n} \qquad n = 1, 2, 3, 4, \ldots \qquad (16\text{-}28)$$

since this puts the function at $x = s$ in the form

$$y = A_m \sin\left(\frac{2\pi n}{2s} s\right) = A_m \sin n\pi \qquad (16\text{-}29)$$

which is equal to zero when $n = 1, 2, 3, 4, \ldots$.

In other words, in order to vibrate in a manner which satisfies the boundary conditions, the half wavelength $\lambda/2$ must be equal to an integral fraction of the length of the string s/n. Otherwise, at the end of the string ($x = s$), y would have values different from zero; this would mean that the string was vibrating up and down at the end, which is impossible since the string is fixed at the end. Thus, the condition that the ends are fixed quantizes the values of the possible wavelengths

$$\lambda = \frac{2}{1}s, \frac{2}{2}s, \frac{2}{3}s, \frac{2}{4}s, \ldots, \frac{2}{n}s$$

$$\tau = \tau_0, \frac{\tau_0}{2}, \frac{\tau_0}{3}, \frac{\tau_0}{4}, \ldots \qquad (16\text{-}30)$$

When the string vibrates with one of the allowed wavelengths and with constant energy, i.e., no friction, it is said to be in a *steady state* of vibration. While there is no apparent travel of a wave in the x or $-x$ direction, such a steady state is the equivalent of a wave train traveling along the string and being reflected back upon itself at each end. When the $\lambda = \lambda_0/n$, $n = 1, 2, 3, 4, \ldots$, then the waves traveling in opposite directions exactly cancel each other out at points regularly spaced along the string, separated from each other by a distance of $\lambda/2$. At these points the string is motionless. Halfway between each pair

of points, the waves reinforce each other, and the string vibrates with maximum amplitude. The points where the string is motionless are called *nodal points* or *nodes*. The pattern of motion between two nodes is called a *loop*.

Since the frequency ν is equal to the velocity of travel of the wave u divided by the wavelength λ,

$$\nu = \frac{u}{\lambda} \qquad (16\text{-}31)$$

the frequency, like the wavelength, can have only a *quantized* set of values. The longest possible wavelength is $2s$, as shown in Fig. 16-9a. With this wavelength the string vibrates like a violin string plucked at the center. The pattern varies from a single arch upward (convex) to a single trough downward (concave). Thus, when $\lambda = 2s$, the string vibrates with a single loop. The value of the frequency in this pattern of vibration is called the *fundamental frequency* ν_0. Thus

$$\nu_0 = \frac{u}{\lambda_0} = \frac{u}{2s} \qquad (16\text{-}32)$$

When $\lambda = s$, then, as shown in Fig. 16-9b, the string vibrates with two loops and $\nu = 2u/\lambda_0 = u/s = 2\nu_0$; when, as in Fig. 16-9c, the string vibrates with three loops, $\nu = 3u/\lambda_0 = 3u/2s = 3\nu_0$. Thus, the general formula for all possible values of the frequency is

$$\nu = n\nu_0 \qquad \text{where } n = 1, 2, 3, 4, \ldots \qquad (16\text{-}33)$$

Under the above conditions when the string vibrates in a steady state, the pattern of vibration is called a *mode* of vibration.

Fig. 16-9 The first three modes of vibration of a stretched string.

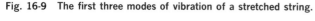

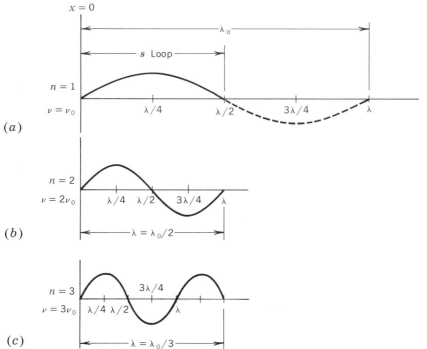

To summarize, the string of *infinite* length shown in Fig. 16-1 can have values of v ranging from zero to infinity; the string of *finite* length can have only the quantized set of values of v given by Eq. (16-33), where each possible value is an integral multiple of the fundamental frequency, $v_0 = u/2s$.

The value of n associated with each pattern of vibration is called the *quantum number* of that pattern; it is equal to the number of *loops*. In acoustics, the sound emitted by the string vibrating with $v = v_0$ and $n = 1$ is called the fundamental tone, and the sounds emitted when the string vibrates with $n > 1$ are called the overtones.

Two-dimensional vibrator

Just as a string provides an example of a one-dimensional vibrator with the single significant dimension of length, a drumhead provides an example of a two-dimensional vibrator with the significant dimensions of length and breadth.

For the simplest approach to an understanding of

the mathematics involved, a drum with a square head is a good example. The drumhead is assumed to be an elastic membrane of uniform thickness and density, stretched under uniform tension and with fixed edges in the form of a square. Such a membrane is shown in Fig. 16-10. The detailed analysis leading to the formulas for the steady-state modes of vibration can be found in some of the references. Only the patterns of vibration and their frequencies will be considered here.

The fundamental mode of vibration for a simple, uniform, square membrane is one where the whole membrane rises and falls as a unit. The shape of the membrane shown in Fig. 16-10 is the form it assumes when it moves below the plane defined by the square boundary. A cross section of the membrane through the center, parallel to the x axis, is shown at the right of the figure. The cross section parallel to the y axis has a similar form. Thus, as the membrane vibrates above and below the plane defined by the boundary, it exhibits one loop in cross section parallel to the x axis and one loop in cross section parallel to the y axis. In order to designate the mode of motion a multiple quantum number is assigned consisting of two integers n_1 and n_2. Thus, for the fundamental mode of motion, $n_1, n_2 = 1,1$, denoting *one* loop in the cross section parallel to the x axis and *one* loop in the cross section parallel to the y axis.

Just as the fundamental mode of vibration with the quantum numbers 1,1 resembles the fundamental mode of vibration of the string with the single quantum number 1, the drumhead vibrates in a mode where there are two loops in the cross section parallel to the x axis and two loops in the cross section parallel to the y axis. This mode of motion is shown in Fig. 16-11a. In this mode, $v = 2v_0$, and the quantum numbers are 2,2. Figure 16-11b shows the pattern of the mode of vibration corresponding to that of the string with three loops. In this mode, $v = 3v_0$, and the quantum numbers are 3,3. Thus, for these and all the higher overtones of the string, there are analogous overtones for the square membrane.

Degenerate modes

Because the square membrane has the two dimensions of length and breadth, there are possibilities for other modes of motion which have no counterparts in the vibration of the one-dimensional string. For example, the square membrane can vibrate with two loops in the plane parallel to the x axis and one loop in the plane parallel to the y axis, as shown in Fig. 16-12a; or it can vibrate with one loop parallel to the x axis and two loops parallel to the y axis, as shown in Fig. 16-12b. Note that these modes of motion are less symmetrical than the three modes where the first quantum number is equal to the second quantum number; in any of these three modes (1,1; 2,2; 3,3) the drumhead can be rotated through 90° and the figure obtained is identical with the original one. By way of contrast, rotating mode 2,1 through 90° changes it into mode 1,2. Both these modes have the same frequency. In view of this relationship, they are called *degenerate* modes. Two such modes can be combined simultaneously to produce patterns of vibration like those shown in Fig. 16-12c and d.

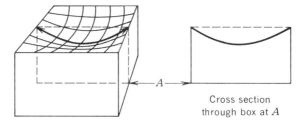

Fig. 16-10   **The fundamental mode of vibration of a square membrane.** $v = v_0$; $n_1, n_2 = $ **1, 1.**

Cross section through box at A

The general formula relating the values of the frequency to the values of the quantum numbers is

$$\frac{v}{v_0} = \frac{\sqrt{n_1^2 + n_2^2}}{\sqrt{1^2 + 1^2}} \qquad (16\text{-}34)$$

A drawing convention has been adopted to show modes of motion in two dimensions. Those parts of the membrane which at any moment may be above the plane determined by the rim are marked $+$, while those parts of the membrane which at the same moment are below this plane are marked $-$. Six of the modes of motion of the square membrane are shown in this manner in Fig. 16-13.

The circular membrane

In progressing from vibration in one dimension (the string) to vibration in two dimensions, the possibility now arises for boundaries of different shapes. While the vibrating membrane with a square boundary bears the closest resemblance in the frequency of its modes to the stretched string, the membrane with the circular edge shows different patterns of motion which are helpful later in understanding the patterns of vibration of the de Broglie waves in the atom. Because of this relationship, a different notation is adopted to designate the modes, some of which are shown in Fig. 16-14.

The fundamental mode is designated by $1s$; the letter s is selected because of the association of this mode with the s lines in atomic spectra, where s signifies *sharp*. In the fundamental $1s$ mode the whole membrane rises and falls as a single unit. There is another mode of motion with the same circular symmetry which is designated $2s$ and which

Fig. 16-11   Overtone modes of a square membrane.

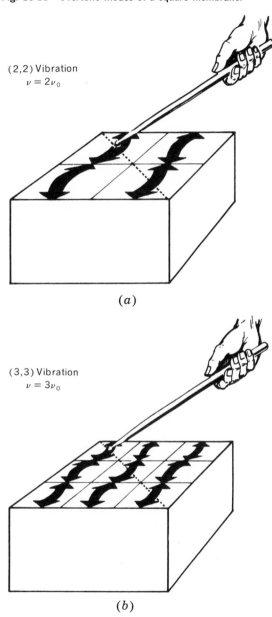

(2,2) Vibration
$\nu = 2\nu_0$

(a)

(3,3) Vibration
$\nu = 3\nu_0$

(b)

corresponds to the overtone of the stretched string vibrating at twice the frequency of the fundamental vibration; however, in the circular drum, this mode vibrates somewhat faster than twice the fundamental vibration. In this mode, (Fig. 16-14) a circular inner

portion of the membrane falls below the median plane while the outer portion rises. In the mode of the square membrane, where $\nu = 2\nu_0$, there are two nodal lines along which the membrane does not move; one of these bisects the membrane vertically and the other horizontally; in the $2s$ mode of motion, the nodal line is circular.

Contrasting with these modes of motion having circular symmetry are the modes designated by the symbol p, which are degenerate modes. In the first set of degenerate modes for the square membrane one nodal line is horizontal and the other is vertical. In the first set of degenerate modes for the circular membrane there is the same relationship. Where the nodal line is normal to the x axis, the mode is designated by the symbol $2p_x$, and where the nodal line is normal to the y axis, it is designated by the symbol $2p_y$. The relations of the values of the frequency to the fundamental frequency are given by somewhat complex formulas which can be found in the texts listed in the References.

Three-dimensional vibrator

The type of three-dimensional vibrator most closely resembling the stretched string is the elastic cube. The modes of motion have three quantum numbers, and the frequency is related to the fundamental frequency by a formula which is the obvious extension of Eq. (16-34) and which can be written

$$\frac{\nu}{\nu_0} = \frac{\sqrt{n_1{}^2 + n_2{}^2 + n_2{}^3}}{\sqrt{1^2 + 1^2 + 1^2}} \qquad (16\text{-}35)$$

Formulas extended in the same pattern apply to the analogs of the cube in higher dimensions.

From the point of view of atomic quantum theory, the most interesting vibrator in three dimensions is the elastic sphere. The problem of boundary conditions in a vibrating sphere is somewhat more complex than in the stretched string and membrane. In the string, the ends are fixed. In the membrane, the rim is fixed. In the sphere, one three-dimensional analogy consists of the vibrations of a gas in a spherical cavity. The details of the boundary conditions are not important; the most useful aspects are the patterns of the modes, which are illustrated in Fig. 16-15.

In depicting the pattern of motion, the convention

Fig. 16-12 (*a*) and (*b*) Degenerate modes; (*c*) and (*d*) other degenerate modes.

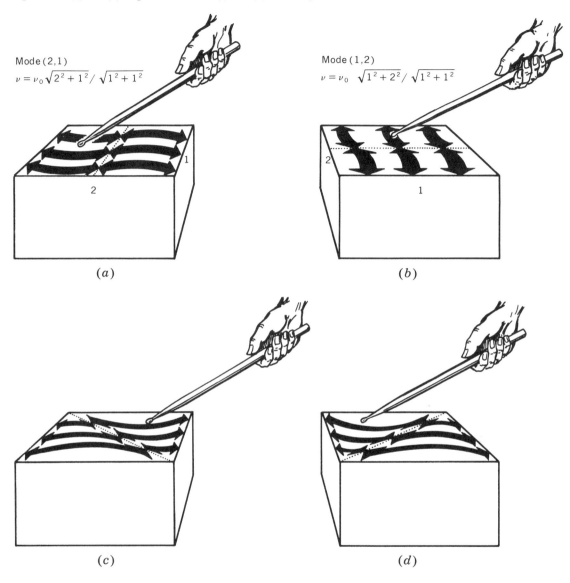

Mode (2,1)

$$\nu = \nu_0 \sqrt{2^2 + 1^2} / \sqrt{1^2 + 1^2}$$

Mode (1,2)

$$\nu = \nu_0 \sqrt{1^2 + 2^2} / \sqrt{1^2 + 1^2}$$

(*a*)

(*b*)

(*c*)

(*d*)

is followed that when one portion of the material is expanding while another contracts, the former is marked with a plus sign and the latter with a minus sign. The nature of the nodes in three-dimensional vibration is a logical extension of the nodes in one and two dimensions. In the one-dimensional vibration, the nodes are points; in the two-dimensional vibration, the nodes are lines; in the three-dimensional vibra-

tion, the nodes are surfaces. Because of the difficulties of showing expansion and compression in three dimensions, the nodes are illustrated with drawings in cross section in Fig. 16-15. The nodes are designated by a notation similar to that used for the modes of the circular membrane and related to the designation of spectral lines.

The fundamental vibration is the 1*s* vibration, where

Fig. 16-13 Fundamental and five overtone modes of the square membrane.

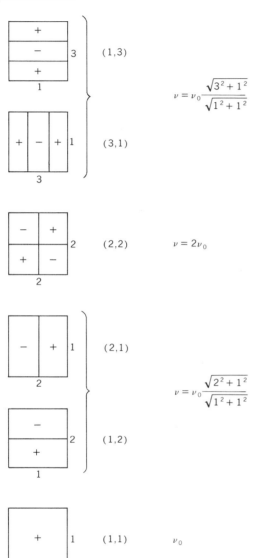

$$\nu = \nu_0 \frac{\sqrt{3^2 + 1^2}}{\sqrt{1^2 + 1^2}}$$

(1,3)

(3,1)

$$\nu = 2\nu_0$$

(2,2)

$$\nu = \nu_0 \frac{\sqrt{2^2 + 1^2}}{\sqrt{1^2 + 1^2}}$$

(2,1)

(1,2)

$$\nu_0$$

(1,1)

the elastic material may be thought of as expanding and contracting in a single *phase* with spherical symmetry. The term *phase* is used here in a different sense than in thermodynamics; e.g., when the square membrane vibrates in mode 1,1 all parts of the membrane either rise together or fall together; they are said to be in the same phase. On the other hand, when the membrane vibrates in mode 2,1 the left half of the membrane rises when the right falls or vice versa. The two parts of the membrane are said to be in different phases or to be moving in opposite phase. Similarly, in the vibrating sphere, all parts marked with a plus sign are in the same phase with each other but in opposite phase with respect to the parts marked with a minus sign.

Ascending the scale of frequency, the next mode with spherical symmetry is designated as the 2*s* mode; as shown in Fig. 16-15, there is a spherical nodal surface halfway between the center and the exterior; with respect to this surface, the inner portion and the outer portion are in opposite phase. The frequency is equal to twice the fundamental frequency. As the number of spherical nodal surfaces increases, the frequencies of the mode correspond precisely to the frequencies of the overtones of the stretched string.

Again, ascending the frequency scale from the fundamental mode, the first degenerate mode is designated as 2*p*. In this mode, there is a nodal surface which is a plane. The motions on either side of this plane are in opposite phases. The plane is shown in cross section in Fig. 16-15. The frequency of this mode is $1.43\nu_0$. Just as the nodal *line* in the 2*p* mode for the square membrane could be oriented normal to the *x* axis or normal to the *y* axis, in this case the nodal *plane* may be oriented normal to the *x* axis, normal to the *y* axis, or normal to the *z* axis. Thus, there are three possible independent modes of vibration all having the frequency $1.43\nu_0$. The nodal plane can be introduced into the 2*s* mode giving the 3*p* mode as shown in Fig. 16-15. Similarly, the introduction of a nodal plane into the 3*s* mode gives the 4*p* mode. In each of these there are three independent modes of motion corresponding to the normality to the three coordinate axes, *x*, *y*, and *z*. One mode is said to be independent of another mode when there is no component of nodal lines or planes in the first which is parallel to the direction of nodal lines or planes in the second.

When two nodal planes are introduced at right angles to each other, modes are obtained as shown in Fig. 16-15.

To summarize, the *number* in the designation of one of the spherical modes corresponds to the number of nodal surfaces if the exterior nodal surface is counted in the sum. Thus, in the $1s$ mode there is only the one node, namely, the exterior surface; in the $2s$ mode there is the exterior surface and the interior spherical nodal surface; in the $2p$ mode there is the exterior surface and the planar nodal surface. The *letter* in the designation symbol indicates the symmetry character of the mode; the s modes are spherical; the p modes have a single planar nodal surface, and the d modes have dual planar nodal surfaces. As will be discussed in Chap. 17 on atomic structure, the symmetry character of the modes of motion of the electron waves resembles closely the symmetry character of the modes of the elastic sphere and plays a key role in determining the periodicity in the periodic table of the chemical elements.

Example 16-4

Calculate the ratio of the frequency to the fundamental frequency for the mode of the square elastic membrane in which there are two interior nodal lines so that the vibrator looks like a flag with three horizontal stripes (in Fig. 16-12a).

$$\frac{\nu}{\nu_0} = \frac{\sqrt{3^2 + 1^2}}{\sqrt{1^2 + 1^2}} = \sqrt{\frac{10}{2}} = 2.24$$

Example 16-5

In the modes of the elastic sphere, how many independent modes of the same frequency are associated with the $4p$ mode?

There are three, one normal to each of the three coordinate axes in space: $4p_x$, $4p_y$, $4p_z$.

16-3 THE SCHRÖDINGER EQUATION

The German physicist Erwin Schrödinger was the first to point out the consequences of combining the de Broglie concept of a wavelength associated with momentum, $p = h/\lambda$, with the familiar differential equation for a wave

$$\frac{\partial^2 \psi}{\partial x^2} = \frac{1}{u^2} \frac{\partial^2 \psi}{\partial t^2} \tag{16-36}$$

The symbol ψ is used to denote the displacement of the particle-wave, analogous to the displacement of the stretched string normal to its length when a wave passes along it. This wave equation has only one

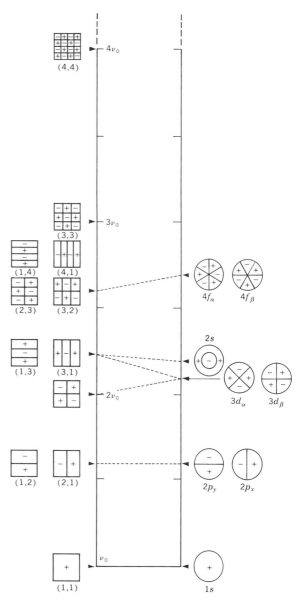

Fig. 16-14   A comparison of the frequencies of the square membrane and the circular membrane. (After W. Kauzmann, "Introduction to Quantum Chemistry," p. 80. Copyright © 1957. Academic Press, Inc., New York. Used by permission.)

Fig. 16-15 The modes of the elastic sphere. (After W. Kauzmann, "Introduction to Quantum Chemistry." Copyright © 1957. Academic Press, Inc., New York. Used by permission.)

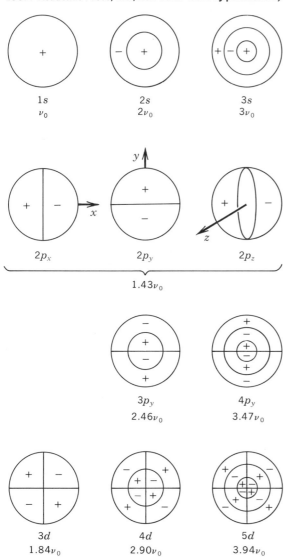

The electron in a box

One of the simplest cases of quantization is found when the motion of an electron is restricted to a straight line between potential barriers that produce total reflection. This is analogous to the case of a perfectly elastic gas molecule moving back and forth along a straight line and reflected at each end of the line by a completely elastic and impenetrable wall. This motion was discussed in Chap. 2 as a means of deriving the relation between the pressure produced by the molecule and its energy.

An electron moving back and forth along a straight line between two impenetrable potential barriers is shown schematically in Fig. 16-16. The length of the path is designated by a. The assumption is made that the amplitude of the vibration of the wave must be zero at each end of the path, as was the case in the vibration of the stretched string ($\psi = 0$, $x = 0$; $\psi = 0$, $x = a$). Denoting the displacement of the wave where $x = 0$ by the symbol ψ_0 and where $x = a$ by the symbol ψ_a, both ψ_0 and ψ_a must be independent of the time since these quantities always have the value zero. A solution of Eq. (16-36) satisfying this condition is one where the variable x and the variable t appear in separate terms which multiply each other

$$\psi = A_m \sin\left(\frac{2\pi}{\lambda}x\right) \sin 2\pi\nu t \tag{16-37}$$

When $x = 0$, $\sin[(2\pi/\lambda)x]$ is identically zero at all times

$$\psi_0 = A_m \sin 0 \sin 2\pi\nu t \tag{16-38}$$

In the same way, at $x = a$, the term $\sin[(2\pi/2a)2a]$ is identically zero since this is equal to $\sin 2\pi$. Thus, ψ_a is given by

$$\psi_a = A_m \sin\left(\frac{2\pi}{2a}2a\right)\sin 2\pi\nu t \tag{16-39}$$

and will be identically zero at all values of t, as shown by

$$\psi_a = A_m \sin 0 \sin 2\pi\nu t \tag{16-40}$$

Differentiating Eq. (16-37) twice with respect to the time gives

$$\frac{\partial^2\psi}{\partial t^2} = -4\pi^2\nu^2\psi \tag{16-41}$$

space variable x and therefore is called *one-dimensional*. Equations with two variables (x, y) and three variables (x, y, z) will be discussed later in connection with atomic structure. Schrödinger showed that these equations explain the *quantization* of particle-waves in various types of potential fields.

If this expression is inserted in Eq. (16-36), it gives

$$\frac{\partial^2 \psi}{\partial x^2} = \frac{1}{u^2}(-4\pi^2 \nu^2 \psi) \tag{16-42}$$

Which can be rearranged to give

$$\frac{\partial^2 \psi}{\partial x^2} + \frac{4\pi^2 \nu^2}{u^2}\psi = 0 \tag{16-43}$$

This procedure is an example of solution by separation of variables.

Assuming that the general wave relationship applies which relates frequency, wavelength, and the velocity of travel of the wave,

$$\frac{\nu}{u} = \frac{1}{\lambda} \tag{16-44}$$

Eq. (16-43) can be put in the form

$$\frac{\partial^2 \psi}{\partial x^2} + \frac{4\pi^2}{\lambda^2}\psi = 0 \tag{16-45}$$

Because of the relation

$$\sigma = \frac{1}{\lambda} \tag{16-46}$$

the de Broglie equation, complementary to Planck's equation $E = h\nu$, can be put in either of the equivalent forms

$$p = h\sigma \tag{16-47}$$

$$p = \frac{h}{\lambda} \tag{16-48}$$

The kinetic energy of a particle E_k is given by

$$E_k = \tfrac{1}{2}mv^2 \tag{16-49}$$

where m is the mass of the particle which is traveling with the velocity v. Since $p = mv$, it follows that

$$p^2 = 2mE_k \tag{16-50}$$

Using the relation from Eq. (16-48) that $\lambda^2 = p^2/h^2$, Eq. (16-45) can be put in the form

$$\frac{\partial^2 \psi}{\partial x^2} + \frac{4\pi^2 p^2}{h^2}\psi = 0 \tag{16-51}$$

Using the value of p^2 given in Eq. (16-50), this can be transformed to

$$\frac{\partial^2 \psi}{\partial x^2} + \frac{4\pi^2(2mE)}{h^2}\psi = 0 \tag{16-52}$$

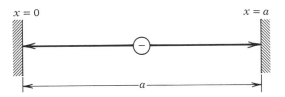

Fig. 16-16   Electron moving back and forth between two impenetrable potential barriers.

For many purposes, it is convenient to denote $h/2\pi$ by the single symbol $\hbar$. Using this notation, Eq. (16-52) can be written

$$\frac{\partial^2 \psi}{\partial x^2} + \frac{2m}{\hbar^2}E\psi = 0 \tag{16-53}$$

The expressions either in the form of Eq. (16-52) or (16-53) are called *Schrödinger's equation in one dimension,* applicable when the potential energy is zero.

Just as the vibration of the string was quantized by the boundary conditions, the possible values of the wavelengths for the de Broglie wave associated with the electron in the box are quantized and given by

$$\lambda = \frac{2a}{n} \qquad n = 1, 2, 3, 4, \ldots \tag{16-54}$$

As a result of Eq. (16-48), values of p are quantized

$$p = \frac{h}{\lambda} = \frac{nh}{2a} \qquad n = 1, 2, 3, 4, \ldots \tag{16-55}$$

The patterns of the wave under these conditions are shown in Fig. 16-17. The nature of the wave will be discussed in the following section.

The wave with the longest possible wavelength is that in which the wavelength is equal to twice the length of the electron path. Denoting this wavelength for the *fundamental* frequency by the symbol λ_0, the relation can be expressed

$$\lambda_0 = 2a \tag{16-56}$$

This defines the lowest value of the momentum p_0 by

$$p_0 = \frac{h}{\lambda_0} = \frac{h}{2a} \tag{16-57}$$

Thus, the quantization of the momentum can be put in the simple form

Fig. 16-17 De Broglie wavelengths for the electron moving back and forth between two impenetrable barriers.

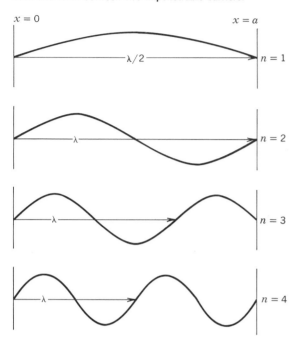

$x = 0$

$x = a$

$n = 1$

$n = 2$

$n = 3$

$n = 4$

$$p = np_0 \qquad n = 1, 2, 3, 4, \ldots \qquad (16\text{-}58)$$

Since the electron has no potential energy, the value of its energy in the nth state is given by

$$E_n = \frac{p_n^2}{2m} = n^2 \frac{p_0^2}{2m} \qquad (16\text{-}59)$$

The value of the energy in the lowest energy level E_0 is given by

$$E_0 = \frac{p_0^2}{2m} = \frac{h^2/4a^2}{2m} = \frac{h^2}{8ma^2} \qquad (16\text{-}60)$$

Thus, the expression showing the quantization of the energy of the electron can be put in the simple form

$$E_n = n^2 E_0 \qquad n = 1, 2, 3, 4, \ldots \qquad (16\text{-}61)$$

The molecule in a box

Schrödinger's equation can be applied in an analogous way to show the quantization of the energy of a molecule moving on a path similar to that described for the electron in the preceding discussion. The value of the energy in the nth energy level can be found by combining Eqs. (16-60) and (16-61) to give

$$E_n = \frac{n^2 h^2}{8ma^2} \qquad (16\text{-}62)$$

The partition function for this simple case is

$$Q = \sum_n e^{\,l - \epsilon_n/kT} = \sum_{n=0}^{\infty} e^{-n^2 h^2/8m a^2 kT} \qquad (16\text{-}63)$$

For a gas molecule in a box where a has a dimension of the order of 1 cm, the energy levels lie so close together that this expression can be integrated

$$Q = \int_0^{\infty} e^{-n^2 h^2/8m a^2 kT} \, dn = \frac{(2\pi mkT)^{1/2}}{h} a \qquad (16\text{-}64)$$

For a gas in a box with a volume V this takes the form

$$Q = \frac{(2\pi mkT)^{3/2}}{h^3} V \qquad (16\text{-}65)$$

This equation, serves as the basis for the Sackur-Tetrode equation for the entropy of a monatomic gas discussed in Chap. 6. Experimental measurements of the heat capacity of monatomic gases from temperatures in the neighborhood of absolute zero to room temperature lead to values for the entropy in agreement with those calculated on the basis of the quantization of energy levels given by Eq. (16-65).

Example 16-6

Calculate the value of E_0 for an electron moving back and forth between potential barriers 10^{-7} cm apart ($m = 9.1 \times 10^{-28}$ g).

$$E_0 = \frac{h^2}{8ma^2} = \frac{(6.63 \times 10^{-27})^2}{8 \times 9.1 \times 10^{-28} \times (10^{-7})^2}$$

$$= 6.04 \times 10^{-13} \text{ erg}$$

Example 16-7

Calculate the value of the difference between the energy of a molecule of He moving back and forth on a path 1 cm long when $n = 2$ and $n = 3$ at $T = 298°$K.

$$E_2 = \frac{2^2 (6.63 \times 10^{-27})^2}{8 \times 4(6.02 \times 10^{23})^{-1} \times 1^2}$$

$$E_3 = \frac{3^2 (6.63 \times 10^{-27})}{8 \times 4(6.02 \times 10^{23})^{-1} \times 1^2}$$

$$E_3 - E_2 = (9 - 4)(8.27 \times 10^{-31})$$

$$= 4.13 \times 10^{-30} \text{ erg}$$

16-4   THE MEANING OF DE BROGLIE WAVES

In the preceding section some of the evidence was reviewed which indicates that any situation affecting the wave associated with a particle affects the properties of the particle. For example, when the electron is confined between two potential barriers so that the wave is reflected from them, only certain values of the wavelength are permitted; this, in turn, quantizes the kinetic energy of the electron. The mathematical approach to the analysis of this quantization implies that the wave exists simultaneously through the entire region between the barriers in a steady state like the wave of vibration of a stretched string. In other words, the wave is *spread out* in space between $x = 0$ and $x = a$. Contrasted with this picture, the particle point of view implies that the electron is something like a particle which is at a *particular* place at any particular time.

There has been much discussion about the best way to reconcile the localization in space of the electron when it behaves as a particle and the delocalization when the electron wave is in a quantized steady state. The opinion seems to be that these are two contrasting aspects of a total pattern of behavior; this is expressed by the phrase that the particle aspect *complements* the wave aspect of behavior. There is also reason to believe that any effort to *picture* the electron as localized and delocalized at the same time is doomed to failure. The two contrasting aspects of the electron have to be accepted as complementary to each other. As long as the mathematical relations between the nature of the wave and the behavior of the particle can be formulated in a way that agrees with experiment, the lack of any picture is not significant.

In the simple situation where the electron moves back and forth between two potential barriers along a line, the equation giving the value of the quantity ψ is relatively simple. ψ may be regarded as analogous to the quantity y, denoting the displacement of the string for the electron in the box where ψ is quantized in *one* dimension

$$\psi = A \sin\left(\frac{2\pi}{\lambda}x\right) \tag{16-66}$$

This equation expresses ψ as a function of x; for this reason, ψ is called a *wave function*.

When an electron is quantized by boundary conditions in *three* dimensions, the wave function may be far more complicated. Also, in many situations, electrons interact, and the wave function may represent quantization affected by the interaction of several electrons. These more complex problems will be considered in the chapters on atomic and molecular structure. In general, the wave function may involve the mathematical quantity $i = \sqrt{-1}$ and is, therefore, in the nomenclature of mathematics, a *complex* quantity.

The simple wave function ψ shown in Eq. (16-66) consists of the product of a constant A, called the amplitude, by a sine function, which gives the variation with x. In the case of an electromagnetic wave like that of visible light, A is thought of as the measure of a quantity like a displacement; A is the value of this displacement when $\sin[(2\pi/\lambda)x] = 1$. Roughly speaking, A^2 is proportional to the energy carried by the light wave.

For the de Broglie wave A has a very different meaning. This again can be expressed in two different ways depending on whether the *wave* or the *particle* aspect of the electron is emphasized. This meaning is generally described from the *wave* point of view by saying that the value of $\int_{x_1}^{x_2} \psi^*\psi\, dx$ is a measure of the part of the charge of the electron in the range over which the integration is carried out. From this point of view the electron is thought of as delocalized, or spread out, over the distance through which the standing wave exists. If the *particle* aspect is emphasized, this same integral is regarded as a measure of the probability of finding the electron in the range x_1 to x_2. In this integral, ψ^* is the complex conjugate of ψ.

For the electron in the box, the probability of finding the electron *somewhere* in the box is unity if the barriers confine it completely to this portion of space. If the following relation holds

$$\int_0^a \psi^*\psi\, dx = 1 \tag{16-67}$$

then ψ is said to be *normalized*. In other words, a value of A in Eq. (16-66) is selected so that Eq. (16-67) is valid.

Is it justifiable to speak of de Broglie waves as

Fig. 16-18 The uncertainty in the momentum of a particle caused by diffraction of the particle-wave.

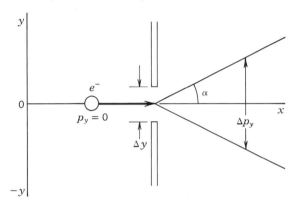

having physical reality? To a large extent this is a question of semantics depending on how physical reality itself is defined. The important fact is that mathematical equations of the form suitable for describing waves provide a suitable mathematical structure for correlating many aspects of the experimental behavior of electrons.

The uncertainty principle

In 1926 the German physicist Werner Heisenberg pointed out that the wave aspect of the behavior of particles introduces a fundamental limit to the precision of the relations between certain of the observed properties of these particles. For example, suppose that the problem is to define with a high degree of precision one of the space coordinates of an electron, say the y coordinate. This can be accomplished by passing the electron through a hole which has a size comparable to that of the electron, as shown schematically in Fig. 16-18. Heisenberg pointed out that if the electron has a wave aspect, one expects the electron wave to be diffracted in passing through the hole, just like a familiar light wave. This diffraction results in a delocalization of the wave with respect to the y axis, since the passage through the hole causes the wave to have a component of motion which spreads it out both in the $+y$ and the $-y$ direction.

Since before the particle approached the hole, the momentum was wholly along the x axis, the momen-

tum is linked to the wavelength by the relation $p = h/\lambda$; the passage through the hole introduces a component of momentum along the y axis. This is expressed quantitatively by stating that the uncertainty in the y component of momentum Δp_y is given by

$$\Delta p_y = 2p \sin \alpha \qquad (16\text{-}68)$$

where α is defined by

$$\alpha \equiv \pm \frac{\lambda}{\Delta y} \qquad (16\text{-}69)$$

p is the momentum before the particle passes through the hole and Δy is the diameter of the hole.

The diameter of the hole fixes the limits of the knowledge of where the particle is on the y axis. It is the uncertainty in the y space coordinate of the particle. Δp_y is the uncertainty in the y component of the momentum. The product of the two uncertainties thus has the value

$$\Delta y \, \Delta p_y = 2p\lambda = 2h \qquad (16\text{-}70)$$

More precisely, according to Heisenberg's general statement of the uncertainty principle,

$$\Delta y \, \Delta p_y \geqslant \frac{h}{4\pi} \qquad (16\text{-}71)$$

Heisenberg pointed out that there are a number of pairs of variables, called *conjugate variables,* related in such a way that the product of the uncertainties in the two variables must be equal to or greater than $h/4\pi$. For example, this is true of the relation between energy and time

$$\Delta E \, \Delta t \geqslant \frac{h}{4\pi} \qquad (16\text{-}72)$$

when considering the behavior of a particle.

Zero-point energy

The uncertainty principle limits the minimum amount of energy a particle can have. For example, for a harmonic oscillator, the earlier quantum theory proposed by Einstein assigned the value of zero energy to the lowest possible energy level; but if the oscillator particle were not moving at all, its position would be completely known and its momentum would be completely known because the momentum would be equal to zero. This would violate the uncertainty principle. Because of the necessity of preserving the relationship expressed in Eq. (16-71), the energy in the lowest energy level must be $\frac{1}{2}h\nu$, where ν is the frequency with which the oscillator vibrates. This leads to the relation for the energy levels of the oscillator

$$E_n = (n + \tfrac{1}{2})h\nu \tag{16-73}$$

instead of the relation originally proposed by Einstein, where $E_0 = 0$ and $E_n = nh\nu$.

The evidence supporting the validity of the uncertainty principle can be found in a number of textbooks on quantum mechanics and leaves no doubt concerning the truth expressed in this relationship. There is still considerable debate about the philosophical meaning of the principle, and the assertion is sometimes made that there is no physical uncertainty in the relation between the position of the particle and its momentum; but any attempt to measure this relation disturbs the system to an extent which leaves the uncertainty expressed by Heisenberg's equation. Quite apart from the philosophical aspects, the uncertainty principle is important because it emphasizes the *wave* aspect of behavior. The discussion in the following chapters will stress the necessity of regarding atoms, molecules, and macromolecules as complexes of waves rather than as structures of *particle* electrons orbiting about particle protons. The wave aspect emphasizes the *wholeness* of the molecule and the holistic features of behavior which arise from the interaction of the waves. The suggestion has even been made that electrons do not exist as individual entities in atoms and molecules, being combined together so intimately that the behavior of the system can never be correlated completely with individual wave functions but can be totally correlated only with a single unified (though admittedly complex) wave. The bearing of this question on physicochemical problems in biochemistry will be discussed in more detail in the later chapters.

Example 16-8

An electron has a velocity of 6.0×10^9 cm sec^{-1}. Calculate the de Broglie wavelength.

The wavelength is given by

$$\lambda = \frac{h}{mv}$$

$$= \frac{6.62 \times 10^{-27} \text{ erg sec}}{9.11 \times 10^{-28} \text{ g} \times 6 \times 10^9 \text{ cm sec}^{-1}}$$

$$= 1.21 \times 10^{-8} \text{ cm}$$

Example 16-9

If the electron in Example 16-8 is confined in a one-dimensional box, how far apart must the walls of the box be when five loops of the de Broglie wave span the distance from one wall to the other?

A loop is just half as long as the wavelength, or 0.605×10^{-8} cm. If there are five loops between the walls of the box, the distance between the walls must be five times the length of the loop, or 3.03×10^{-8} cm.

Problems

1. In the solar spectrum the wavelength of the sodium line D_1 is 5896 Å. Calculate the value of the frequency of this line when the light passes through a vacuum.

2. The wavelength of the hydrogen F line in the solar spectrum is 486.1×10^{-7} cm. How many waves lie within the range of 1 cm?

3. Calculate the energy in a single photon with a wavelength corresponding to the hydrogen C line in the spectrum 6563 Å.

4. A certain group of hydrogen atoms in DNA absorb light in the infrared, where the wave number is 3450 cm^{-1}. The frequency of this light corresponds to the frequency of vibration of these hydrogen atoms in DNA when the C—H bonds are directly stretched; calculate the frequency.

5. Calculate the energy absorbed in calories if 6.02×10^{23} atoms of hydrogen are set in vibration when each atom absorbs one photon of light with a wave number 3450 cm^{-1}.

6. The compound adenine is one of the constituents of DNA. The ring atoms in this compound can be made to vibrate through the absorption of infrared light with a wave number of 950 cm^{-1}. Calculate the increase in the energy in calories of a mole of adenine molecules when 6.02×10^{23} photons of light with this wave number are absorbed.

7. The sugar component of DNA has its strongest infrared absorption band at 3300 cm^{-1}. Calculate the wavelength in centimeters, microns, and angstroms of this absorption band.

8. Calculate the energy in ergs in a single photon that can be absorbed in the absorption band of ribose in the infrared at 3300 cm^{-1}.

9. If 1 g mole of ribose absorbs one photon per molecule in the infrared band at 3300 cm^{-1}, calculate the total amount of energy absorbed in electron volts.

10. DNA has a strong absorption band in the infrared at 1670 cm^{-1}, due to the vibration of molecules connected by double bonds. Calculate the amount by which this first energy level lies above the ground state. Using the Boltzmann formula, calculate the percentage of molecules lying in this excited energy level at body temperature, 37°C.

11. Another absorption band in the infrared spectrum of DNA lying at 700 cm^{-1} is due to the bending of certain parts of the DNA chain. With the help of the Boltzmann formula calculate the percentage of molecules executing this motion in the first vibrational state.

12. An alpha particle expelled from radium has a velocity of 1.51×10^9 cm sec^{-1}; the mass of the particle is 6.6×10^{-24} g. Calculate the de Broglie wavelength.

13. If a rifle bullet has a mass of 2 g and travels with a velocity of 3×10^4 cm sec^{-1}, what is the wavelength of the de Broglie wave?

14. If a helium molecule has a de Broglie wavelength of 1 Å, what is the velocity of the molecule? If this velocity lies at the rms value of the velocity, what is the equivalent temperature of the molecule?

15. An oxygen molecule in the lungs at 37°C has the rms average velocity for this type of molecule at this temperature. Calculate the de Broglie wavelength of the molecule.

16. An elastic square membrane vibrates with a fundamental frequency of 256 sec^{-1}. Calculate the frequency of vibration for the mode with the quantum numbers 2,3 and draw a diagram showing the nodal lines.

17. How many independent degenerate modes are associated with types of vibration of the circular membrane or sphere having the following quantum numbers: 2,3; 3p; 4d; 3,5; 3,3?

18. Calculate the frequency of the vibration in a mode of the elastic cube with the quantum number 2, 3, 4 if $v_0 = 1000$ sec^{-1}.

19. An atom of oxygen is expelled from a molecule of hemoglobin in a way that fixes its position parallel to the principal plane of the hemoglobin with an uncertainty Δy of 1.5 Å. What is the uncertainty in the momentum Δp_y along the y coordinate according to the Heisenberg indeterminacy principle?

20. If an electron passes through a crystal defect slit 2×10^{-8} cm in width Δy, what is the uncertainty in the y component of its momentum?

REFERENCES

General

Heisenberg, W.: "The Physical Principles of the Quantum Theory," trans. by C. Eckart and F. C. Hoyt, The University of Chicago Press, Chicago, 1930.

Ruark, A. E., and H. C. Urey: "Atoms, Molecules and Quanta," McGraw-Hill Book Company, New York, 1930.

Heitler, W.: "Elementary Wave Mechanics," Oxford University Press, London, 1945.

Feynman, R. P.: "Theory of Fundamental Processes," W. A. Benjamin, Inc., New York, 1962.

Gamow, G.: "Thirty Years That Shook Physics," Doubleday & Company, Inc., Garden City, N.Y., 1966.

Ludwig, G.: "Wave Mechanics: Selected Readings," Pergamon Press, New York, 1968.

Wave mathematics

Rayleigh, Lord: "The Theory of Sound," Dover Publications, Inc., New York, 1945.

Coulson, C. A.: "Waves," Interscience Publishers, Inc., New York, 1947.

Morse, P. M.: "Vibration and Sound," 2d ed., McGraw-Hill Book Company, New York, 1948.

Brillouin, L.: "Wave Propagation and Group Velocity," Academic Press, Inc., New York, 1960.

Screaton, G. R.: "Dispersion Relations," Interscience Publishers, Inc., New York, 1961.

Quantum theory

Mott, N. F., and I. N. Sneddon: "Wave Mechanics," Oxford University Press, London, 1948.

Kramers, H. A.: "Quantum Mechanics," North-Holland Publishing Company, Amsterdam, 1957.

Kauzmann, W.: "Introduction to Quantum Chemistry," pp. 51–99, Academic Press, Inc., New York, 1957.

Messiah, A.: "Quantum Mechanics," North-Holland Publishing Company, Amsterdam, 1961.

Merzbacher, E.: "Quantum Mechanics," John Wiley & Sons, Inc., New York, 1961.

Hameka, H. F.: "Introduction to Quantum Theory," Harper & Row, Publishers, Inc., New York, 1967.

17-1 THE BASIC NATURE OF THE ATOM

The first significant knowledge of the inner structure of the atom came as the result of a series of remarkable discoveries made during the last 5 years of the nineteenth century. In 1896, the French physicist H. Becquerel observed that uranium minerals emit invisible radiation which penetrates objects opaque to familiar visible light. Within a short time, the French chemists Pierre and Marie Curie extracted from the minerals two new chemical elements, polonium and radium, that gave off this new kind of radiation with an intensity far surpassing uranium. The British physicists Ernest Rutherford

atomic structure

and F. Soddy then found that new radiation was the result of atomic disintegration. The phenomenon was called *radioactivity*.

Further investigation showed that the radiation consists of three distinct types. One part is made up of particles of atomic size ejected at high speed; these are called *alpha particles;* each particle is made up of two protons and two neutrons bound together with nuclear forces and has two units of positive electric charge. Thus, the alpha particle is identical with the nucleus of a helium atom that has been stripped of its outer electrons. A second part of the radiation consists of electrons also ejected at high speed; these are called *beta particles.* The third part is electromagnetic radiation, the components of which are called *gamma rays;* the wavelengths are less than 1 Å unit and lie in or beyond the hard x-ray region of the spectrum.

Rutherford immediately recognized the possibility of using alpha particles to probe the interior of the atom. He directed a beam of alpha particles at a piece of gold foil and found that while the majority of the particles passed through the foil in straight lines, a significant number were deflected through large angles or even reflected back in the direction from which they had come. He suggested that an alpha particle would be deflected only when it collided with another particle of comparable or greater mass. Since almost all the alpha particles passed through the gold foil without deflection, he concluded that the mass of each gold atom must be concentrated at a very small point in the center of the atom.

Starting with this evidence, Rutherford and his young associate, the Danish physicist Niels Bohr, suggested a model of the atom in which the mass is concentrated at the center in a region 10^{-12} to 10^{-13} cm in diameter, called the *nucleus* of the atom. The nucleus also contains a positive charge consisting of a number of units equal to the number of electrons in the atom. These electrons circulate around the nucleus like planets in orbit about the sun and through their motions give the atom the apparent diameter of 10^{-7} to 10^{-8} cm observed in atomic collisions. These diameters are also of the correct order of magnitude to explain the volume occupied by atoms when they are packed together in a crystal lattice.

The total energy E of an electron moving in orbit about a single proton, as in a hydrogen atom, is given by

$$E = -\frac{e^2}{r} + \frac{p_a{}^2}{2mr^2} \qquad (17\text{-}1)$$

where e = charge on electron

r = distance of electrons from nucleus

p_a = angular momentum

m = mass of electron

The first term represents the potential energy of the electron due to the attraction between the electron and the proton; the second term is the kinetic energy.

According to classical electrodynamics, the electron should radiate continuously while moving in an orbit because of the acceleration associated with the deviation from a linear path. Since radiation is the cause of a continuous loss in energy, the electron should move more and more slowly, circulating in orbits of smaller and smaller radius, and finally come to rest in the proximity of the nucleus. This should result in a decrease in the radius of the atom. In order to account for the fact that contrary to this line of argument, atoms maintain an effectively constant diameter of 10^{-7} to 10^{-8} cm over indefinite periods of time, Bohr suggested that for certain values of the angular momentum radiation does not occur. These are the stable or steady-state orbits. In symbolic form Bohr's first postulate is

$$mvr \equiv p_a = \frac{nh}{2\pi} \qquad n = 1, 2, 3, \ldots \qquad (17\text{-}2)$$

Thus, Bohr introduced quantization of the orbits as an arbitrary postulate. When Eq. (17-1) is combined with Eq. (17-2), an expression is obtained for the radii of circular orbits

$$r_n = \frac{n^2 h^2}{4\pi^2 m e^2} \qquad (17\text{-}3)$$

The energy of such an orbit is given by

$$E_n = -\frac{2\pi^2 m e^4}{n^2 h^2} \qquad (17\text{-}4)$$

In Fig. 17-1 the pattern of these orbits is shown schematically together with corresponding energy levels.

Bohr also suggested that when an electron moves from a higher (n_2) to a lower (n_1) energy level, light is emitted by the atom and when the movement is from a lower to a higher energy level, light is absorbed. The relation between the light frequency and the energy change is

$$E_{n_2} - E_{n_1} = h\nu \qquad (17\text{-}5)$$

in which ν is the frequency of the light emitted. Using the relation $\nu\lambda = c$, the velocity of light in a vacuum, Eq. (17-5) can be put in the form

$$E_{n_2} - E_{n_1} = \frac{hc}{\lambda} \qquad (17\text{-}6)$$

or the alternative form

$$\frac{1}{\lambda} = \frac{E_{n_2} - E_{n_1}}{hc} \qquad (17\text{-}7)$$

According to Eq. (17-4), the energy change in going from orbit 1 to orbit 2 is

$$E_{n_2} - E_{n_1} = -\frac{2\pi^2 m e^4}{h^2}\left(\frac{1}{n_2{}^2} - \frac{1}{n_1{}^2}\right) \qquad (17\text{-}8)$$

Combining this with Eq. (17-7) gives

$$\frac{1}{\lambda} = \left[\frac{2\pi^2 m e^4}{h^3 c}\right]\left(\frac{1}{n_1{}^2} - \frac{1}{n_2{}^2}\right) \qquad (17\text{-}9)$$

Almost 30 years earlier, the German spectroscopist J. J. Balmer had observed sharp lines in the light emitted by hydrogen at high temperatures for which the wavelengths were given by the empirical formula:

Fig. 17-1 Bohr orbits for the hydrogen atom and corresponding energy levels showing the transitions which produce the Balmer lines.

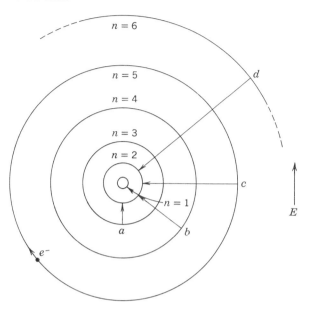

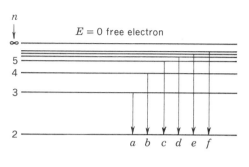

$$\frac{1}{\lambda} = R_0 \left(\frac{1}{2^2} - \frac{1}{n^2} \right) \qquad n = 2, 3, 4, 5 \ldots \qquad (17\text{-}10)$$

A comparison of the observed wavelengths and the values calculated from Eq. (17-9) are given in Table 17-1.

As observations of the spectrum of hydrogen were extended both into the ultraviolet and the infrared, evidence was accumulated to show that formula (17-10) can be written in a form which correlates the wavelengths of lines in a variety of series of lines

$$\frac{1}{\lambda} = R_0 \left(\frac{1}{n_1^2} - \frac{1}{n_2^2} \right) \qquad n = 1, 2, 3, 4 \ldots \qquad (17\text{-}11)$$

The value of R_0 agrees almost precisely with the value of the term in brackets in Eq. (17-9).

The concept of orbits was extended by the German theoretical physicist Arnold Sommerfeld, who suggested that elliptical orbits were possible with the eccentricities also quantized. Thus, Bohr's basic idea served as a foundation for the correlation of many varieties of spectra; but the theory remained on an empirical basis since no satisfactory explanation of the orbits was found.

TABLE 17-1 The wavelengths of the Balmer lines in the spectrum of hydrogen

	λ, A	
n	Observed[a]	Calculated[b]
3	6564.6	6564.8
4	4862.7	4862.8
5	4341.7	4341.8
6	4102.9	4103.0

[a] The values are corrected to correspond to wavelengths in a vacuum.
[b] From Eq. (17-9) based on $m = 9.1033 \times 10^{-28}$ g.

17-2 ATOMIC QUANTUM MECHANICS

The first step toward finding a logical basis for the quantization of the energy of an electron in the atom was taken in 1923 by the French theoretical physicist Louis de Broglie, as described in Chap. 16; he suggested that there might be a wave associated with the electron with the wavelength equal to Planck's constant divided by the momentum. Schrödinger then developed the differential equation for such a wave and showed that the solutions of this equation led to quantization which in the simpler orbits agreed with Bohr's earlier empirical equation. At the same time, Heisenberg formulated a matrix mechanics which, although different in mathematical form, was equivalent to Schrödinger's wave mechanics and led to the same results. Since the concepts in the Schrödinger equation are somewhat easier to visualize in terms of mechanical analogies like the vibrating string and sphere, the discussion in this chapter will follow this approach.

When an electron moves in the electrostatic field produced by the charge on the nucleus of an atom, the electron has both kinetic energy E_k and potential energy U. This introduces an additional term into the equation derived in the preceding chapter. The total energy E is given by

$$E = E_k + U = \tfrac{1}{2} mv^2 + U = \frac{p^2}{2m} + U \qquad (17\text{-}12)$$

Thus, the momentum p can be written as an expression which contains the difference between the total energy E and the potential energy U

$$p = [2m(E - U)]^{1/2} \qquad (17\text{-}13)$$

The expression for the wavelength λ then becomes

$$\lambda = h[2m(E - U)]^{-1/2} \qquad (17\text{-}14)$$

The general wave equation in one dimension is

$$\frac{\partial^2 y}{\partial x^2} = \frac{1}{u^2} \frac{\partial^2 y}{\partial t^2} \qquad (17\text{-}15a)$$

where u is the velocity with which the wave travels. Letting $\phi(x,t) = \psi(x) \sin 2\pi \nu t$, the differential equation for the space variation of y is

$$\frac{d^2 \psi}{dx^2} + \frac{4\pi^2 \nu^2}{u^2} \psi = \frac{d^2 \psi}{dx^2} + 4\pi^2 \lambda^2 \psi = 0 \qquad (17\text{-}15b)$$

since $\nu/u = \lambda$.

When the expression for λ in Eq. (17-14) is substituted in Eq. (17-15b), we obtain the Schrödinger equation in one dimension

$$\frac{d^2 \psi}{dx^2} + \frac{2m}{\hbar^2} (E - U) \psi = 0 \qquad (17\text{-}16)$$

where $\hbar = h/2\pi$. To write equations of this sort in three dimensions, it is convenient to define an operator ∇^2 as follows:

$$\nabla^2 \equiv \frac{\partial^2}{\partial x^2} + \frac{\partial^2}{\partial y^2} + \frac{\partial^2}{\partial z^2} \qquad (17\text{-}17)$$

Using this notation, the Schrödinger equation in three dimensions is

$$\nabla^2 \psi + \frac{2m}{\hbar^2} (E - U)\psi = 0 \qquad (17\text{-}18)$$

The *Hamiltonian operator H* is defined as

$$H \equiv -\frac{\hbar^2}{2m} \nabla^2 + U \qquad (17\text{-}19)$$

Using this notation, Eq. (17-18) can be written

$$H\psi = E\psi \qquad (17\text{-}20)$$

The form of this equation reveals interesting similarities to the problem of finding the characteristic frequencies of vibration in molecules, discussed in Chap. 20.

In deriving solutions for Eq. (17-18) in situations where the boundary conditions may or may not affect the potential energy, the nature of the physical system imposes certain restrictions on the mathematical nature of ψ. If $\psi^*\psi \, dx$ is interpreted as the probability of finding the electron the range dx, then ψ must be single-valued since at any point the probability of finding the electron can have only one value; also, the value of ψ must always be finite since the probability always has finite values. There are also limitations on discontinuity.

In solving the equation for the electron in a box in the previous chapter, potential barriers essentially infinite in height were assumed in order to confine the electron completely within the box. If the potential barriers are not infinite in height, a solution shows that there is a probability of finding the electron beyond the barrier even though the electron may not have sufficient energy to pass over the barrier. This situation is completely unlike any encountered in classical mechanics, where an object like a ball rolling toward a hill must have kinetic energy equal to or greater than the increase in potential energy in climbing the hill if it is to pass over the hill to the other side. The passage of an electron "through" a potential-energy barrier which classically it is unable to surmount is sometimes called the *tunnel effect*.

The hydrogen atom

An electron in an atom moving under the influence of the positive charge on the nucleus has patterns of quantized waves which differ from those of the electron in the box because the boundary conditions are different. In the first place, the problem is not one-dimensional like movement in the box but three-dimensional. The force field about the nucleus has a spherical symmetry. Instead of infinitely high potential-energy barriers at each end of the path, the potential energy in the field of the nucleus is given by

$$U = -\frac{e^2}{r} \tag{17-21}$$

where r is the distance of the electron from the nucleus. Under these conditions, the Schrödinger equation takes the form

$$\nabla^2 \psi + \frac{2m}{\hbar^2}\left(E + \frac{e^2}{r}\right)\psi = 0 \tag{17-22}$$

Because of the spherical symmetry, it is often convenient to use spherical coordinates, in which the equation is

$$r^{-2}\frac{\partial}{\partial r}\left(r^2 \frac{\partial \psi}{\partial r}\right)$$
$$+ (r^2 \sin^2 \theta)^{-1}\left[\frac{\partial^2 \psi}{\partial \phi^2} + \frac{\partial}{\partial \theta}\left(\sin\theta \frac{\partial \psi}{\partial \theta}\right)\right]$$
$$+ \frac{2m}{\hbar^2}\left(E + \frac{e^2}{r}\right)\psi = 0 \tag{17-23}$$

The forms of ψ which satisfy these equations under the boundary conditions are called *eigenfunctions*. These solutions are characterized by quantum numbers in much the same way that the solutions of the wave equation for the vibrating string, membrane, and sphere were characterized by quantum numbers. It is frequently helpful to use wave equations in which there are three terms, each containing only a single variable; mathematically, this is much like the form of the wave equation in Chap. 16, where the time variable and the space variable appeared in different terms. This separation of variables can be denoted by

$$\psi(r,\theta,\phi) = R(r)\Theta(\theta)\Phi(\phi) \tag{17-24}$$

The general symbols for the quantum numbers are n, l, and m.

The details of the solutions will not be discussed here; the more important solutions can be found in textbooks of quantum mechanics (see References); the types of functions together with the quantum numbers associated with them are listed in Table 17-2, and the mathematical expressions for a few of the simpler functions are given in Table 17-3. Attention will be focused largely on the symmetrical properties of the wave functions which are eigenfunctions for the problem under discussion.

The simplest of these eigenfunctions are those applicable to the hydrogen atom, where a single electron moves in the field of the singly charged proton which constitutes the nucleus. The patterns of these eigenfunctions are shown schematically in Fig. 17-2, where they are also compared with analogous patterns of vibration in the string, the circular membrane, and the sphere.

For the stretched string vibrating in quantized motion the amplitude of vibration at any point on the string is given by an equation of the form

$$A_x = A_m \sin\frac{2\pi}{\lambda_n}x \qquad \lambda_n = \frac{2a}{n} \qquad n = 1, 2, 3 \ldots \tag{17-25}$$

The pattern of vibration is illustrated in Fig. 17-3. In this equation, A_x is the amplitude with which the string vibrates over the course of time at the point x. As shown in the figure, the two points of maximum amplitude are at $x = \lambda/4$ and at $x = 3\lambda/4$. At any other point x the amplitude A_x is smaller and becomes zero at $x = 0$, $x = \lambda_1 2$, and $x = \lambda$. For three-dimensional vibration it is not possible to show the amplitude directly, as for the stretched string; instead, the amplitude is denoted by the density of the dots which indicate vibration. The nodes are surfaces. Since the pattern is displayed in cross section only, the linear portions of these surfaces at the cross section are shown as dashed lines.

In Fig. 17-4 stylized representations of the wave

TABLE 17-2   Types of wave functions with three loops

Spin i	Loops n	Symmetry l	Orientation m	Type
$+\frac{1}{2}$	3	0	0	$3s$
$-\frac{1}{2}$	3	0	0	$3s$
$+\frac{1}{2}$	3	1	$+1$	$3p_x$
$-\frac{1}{2}$	3	1	$+1$	$3p_x$
$+\frac{1}{2}$	3	1	0	$3p_z$
$-\frac{1}{2}$	3	1	0	$3p_z$
$+\frac{1}{2}$	3	1	-1	$3p_y$
$-\frac{1}{2}$	3	1	-1	$3p_y$
$+\frac{1}{2}$	3	2	$+2$	$3d_{z^2}$
$-\frac{1}{2}$	3	2	$+2$	$3d_{z^2}$
$+\frac{1}{2}$	3	2	$+1$	$3d_{xz}$
$-\frac{1}{2}$	3	2	$+1$	$3d_{xz}$
$+\frac{1}{2}$	3	2	0	$3d_{yz}$
$-\frac{1}{2}$	3	2	0	$3d_{yz}$
$+\frac{1}{2}$	3	2	-1	$3d_{x^2-y^2}$
$-\frac{1}{2}$	3	2	-1	$3d_{x^2-y^2}$
$+\frac{1}{2}$	3	2	-2	$3d_{xy}$
$-\frac{1}{2}$	3	2	-2	$3d_{xy}$

18 types

TABLE 17-3   Wave functions for atoms with one electron[a]

$$\psi(1s) = \frac{\alpha^{3/2}}{\sqrt{\pi}} e^{-\alpha r}$$

$$\psi(2s) = \frac{\alpha^{3/2}}{4\sqrt{2\pi}} (2 - \alpha) e^{-\alpha r}$$

$$\psi(2p_x) = \frac{\alpha^{5/2}}{4\sqrt{2\pi}} e^{-\alpha r/2} \sin\theta \cos\phi$$

$$\psi(2p_y) = \frac{\alpha^{5/2}}{4\sqrt{2\pi}} e^{-\alpha r/2} \sin\theta \sin\phi$$

$$\psi(2p_z) = \frac{\alpha^{5/2}}{4\sqrt{2\pi}} e^{-\alpha r/2} \cos\theta$$

[a] z = number of unit charges on nucleus

$$a_0 = \frac{h^2}{4\pi^2 mc^2} = 0.529 \times 10^{-8} \text{ cm}$$

$$\alpha = \frac{z}{a_0}$$

functions are shown in three dimensions. These pictures are useful in showing the symmetry of the wave functions. The surfaces in the drawings show the contour of the hypothetical surface at which the amplitude of the wave function has a certain arbitrarily selected value. The function itself has a finite value in the region outside of the surface, all the way to $r = \infty$.

In discussing the wave functions, the first quantum number is denoted by the symbol n and is called the *principal quantum number*. It is closely related to the quantum number denoted by n in the Bohr theory of the hydrogen atom. The number of nodes in the wave function is equal to $n - 1$, where n does not include the hypothetical node which lies at $R = \infty$. These nodes can appear either in the radial part of the wave function $R(r)$ or in the azimuthal part of the function $\Theta(\theta)$.

The second quantum number l is called the *azimuthal quantum number*. It is equal to the number of nodes in $\Theta(\theta)$, which is the number of nodal surfaces passing through the origin including the case where a nodal surface degenerates into a nodal line. Thus the permitted values of l are the integers (inclusive) between zero and $n - 1$. If $l = 0$, all the nodes are in $R(r)$ and the wave function is spherically symmetrical about the central nucleus. The angular momentum of the electron p_θ is quantized, and the values are given by

$$p_\theta = \frac{h}{2\pi} \sqrt{l(l + 1)} \qquad l = 0, 1, \ldots, n - 1 \quad (17\text{-}26)$$

Spectroscopic notation, which was developed long before the advent of the quantum theory, is generally used to denote the states associated with the different values of l. s denotes $l = 0$; p denotes $l = 1$; d denotes $l = 2$; and f denotes $l = 3$. Thus, the quantum state where $n = 3$ and $l = 2$ is assigned the symbol $3d$.

The third quantum number m is called the *magnetic quantum number*. It is associated with the direction of the vector denoting the angular momentum. If the atom is placed in a magnetic field, only certain angles are permitted between the direction of the momentum vector and the field axis. For example, if the momentum vector is denoted by p_θ, and the direction of the

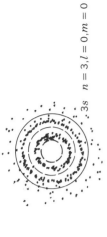

Vibrating string

$n = 3$

$n = 2$

$n = 1$

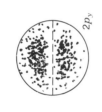
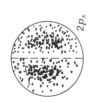
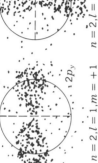
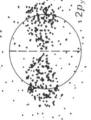

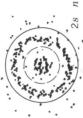

3s

$2p_x$

$2p_y$

2s

1s

Vibrating circular membrane

$3s \quad n = 3, l = 0, m = 0$

$2p_z$

$n = 2, l = 1, m = -1$

$2p_x$

$n = 2, l = 1, m = 0$

$2p_y$

$n = 2, l = 1, m = +1$

$2s \quad n = 2, l = 0, m = 0$

$1s \quad n = 1, l = 0, m = 0$

Atomic wave functions

Fig. 17-2 Mechanical vibrators and wave functions. Dashed lines are nodal lines in two-dimensional vibrations and are cross sections of nodal surfaces in three-dimensional vibrations. Maximum density of dots indicates regions of maximum amplitude.

Fig. 17-3   Time-exposure pattern of a stretched elastic string $n = 2$.

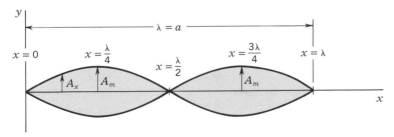

magnetic field is along the x axis, the x component of the momentum vector $p_{\theta,x}$ will be given by

$$p_{\theta,x} = m \frac{h}{2\pi} \qquad m = l, \ldots, 0, \ldots, -l \qquad (17\text{-}27)$$

m may take integral values ranging from $+l$ through 0 to $-l$.

The fourth quantum number s is called the *spin quantum number*. The concept of *electron spin* was proposed by two Dutch physicists, G. E. Uhlenbeck and S. Goudsmit, in 1924 to explain certain aspects of the fine structure of spectra. There are only two possible spin states, one having $s = +\frac{1}{2}$ and the other $s = -\frac{1}{2}$. In view of the wave nature of the electron, the picture of a spinning electron is more an analogy than an accurate portrayal of physical reality. This aspect of electron energy is discussed in Chap. 18.

The wave function in which the electron has the lowest value of the energy is the $1s$ type, shown in cross section in Fig. 17-5. Suppose that the radius selected lies along the x axis; then, as r increases, the figure shows the amplitude of the wave function on this x axis. The zero point in the figure lies at the center of the nucleus; as r goes to negative values, the corresponding point moves out along the $-x$ axis.

Note that the point of maximum amplitude for the wave function is at the value $r = 0$, at the nucleus itself. Then, in moving out from the nucleus in any direction, the amplitude of the wave function decreases, becoming zero at $r = \infty$.

It is interesting to compare this behavior with the corresponding function for the vibrating string, where $n = 1$. Here also the point of maximum amplitude

is at the center of the string when it is vibrating in this mode. There is no nodal point (if one excludes the ends of the string). For the vibrating circular membrane, the point of maximum amplitude is again at the center of the membrane; this is indicated in Fig. 17-2 where the density of the dots is roughly proportional to the amplitude through which the membrane vibrates at any given point. In the same way, for the $1s$ atomic wave function, the maximum density of the dots is at the center, as shown in the cross-sectional diagram in Fig. 17-2.

Figure 17-6 shows the amplitude of vibration for a $2s$ type of wave function for a hydrogen atom. This also has spherical symmetry and can be compared with the pattern for the vibrating string (Fig. 17-2). For the string, there are now two points of maximum amplitude and a node at the center. In the analogous vibration for the circular membrane, however, there are two points of maximum amplitude along the radius. One of these points is at the center, as shown by the maximum density of the dots at the center; the other is three-quarters of the way out from the center to the edge of the membrane. There is one nodal line shown as a dashed circle lying between these two regions of maximum amplitude of vibration. Thus, the wave function of a circular membrane is closer in form to the $2s$ wave function for the hydrogen atom than the $n = 2$ wave function for the vibrating string.

The $2s$ wave function for the hydrogen atom is shown in Fig. 17-2 in cross section, where the maximum density of the dots corresponds to the maximum amplitude of the wave function. Here again there is one region of maximum amplitude at the center and a spherical zone (shown in cross section) lying out at a short distance from the nucleus. This is indicated in

Orbital:

Orbital:

$s\,(l=0)$

$s\,(l=0)$

$p_x\,(l=1)$

$p_x\,(l=1)$

$p_y\,(l=1)$

$p_y\,(l=1)$

$p_z\,(l=1)$

$p_z\,(l=1)$

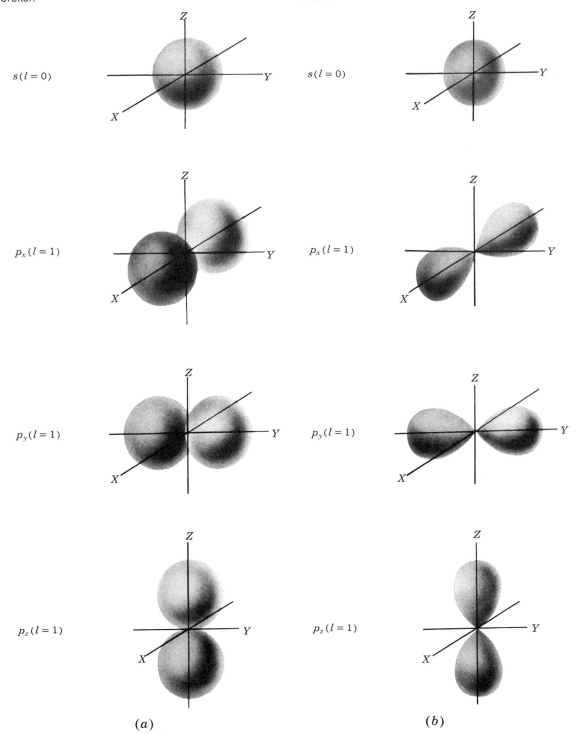

(a)

(b)

Fig. 17-4 (a) Three-dimensional contour surfaces for the angular part of the hydrogen wave function $\theta\phi$; (b) three-dimensional contour surfaces for the angular part of the hydrogen wave function $\theta\phi^2$. The value of the charge density is constant on the surface.

Fig. 17-5   Radial amplitude of 1s type ψ wave in a hydrogen atom (one loop, spherical symmetry). $n = 1, l = 0.$

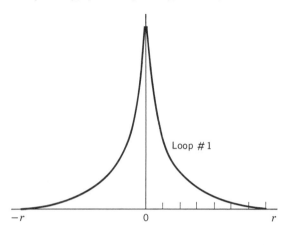

Loop #1

$-r$ 0 r

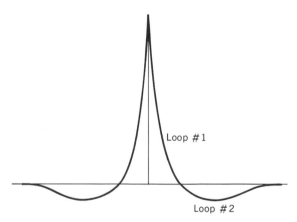

Loop #1

Loop #2

Fig. 17-6   Radial amplitude of a 2s type wave (two loops, spherical symmetry). $n = 2, l = 0.$

Fig. 17-6 by the fact that the curve rises to a maximum height above the horizontal axis at the center and falls to a maximum depth below the horizontal axis about halfway out from the center to the edge of the drawing.   For the membrane, this phase difference corresponds to the fact that when the center of the membrane has risen above the plane defined by the circular rim, the part of the membrane three-quarters of the way toward the rim has fallen below this plane.   The position of the node is shown by the point where the graph in Fig. 17-6 crosses the hori-

zontal axis.   This corresponds to the distance of the spherical nodal surface from the center of the atom.

In Fig. 17-7 the same plot of the amplitude is shown for a 3s type of wave function for a hydrogen atom.   This is shown in cross section in Fig. 17-2 at the upper right-hand corner.   Here again the density of the dots is roughly proportional to the amplitude of the wave function.   As can be seen, there are three zones of maximum amplitude, one lying at the center, one lying about one-third of the way out and the other lying two-thirds of the way out from the center. These correspond to the three zones of maximum amplitude in the vibrating circular membrane with a 3s wave function.   When the center of the membrane is rising, the circular zone of maximum amplitude closest to the center corresponds to the membrane below the plane defined by the circular rim and the third zone of maximum amplitude corresponds to a position of the membrane above the plane.   There are two nodal circles in this mode in the circular membrane and two nodal spherical surfaces in this mode in the wave function of the atom.   In all three of these s modes, $l = 0$, and the symmetry is spherical.

When $n = 2$ and $l = 1$, the symmetry is no longer spherical.   As can be seen in Fig. 17-2, there is now a nodal line.   There are two positions for this line which correspond to independent degenerate vibrations, the $2p_x$ mode and the $2p_y$ mode.   In the first of these, the nodal line is normal to the x axis, and in the second it is normal to the y axis.

For the wave function of the electron in the hydrogen atom, there are three modes of independent vibration.   The assignment of numbers to the quantum number m is entirely arbitrary; there are three values permitted, namely, $+1$, 0, and -1; in Fig. 17-2, $m = +1$ has been assigned to the $2p_x$ vibration, $m = 0$ has been assigned to the $2p_y$ vibration, and $m = -1$ has been assigned arbitrarily to the $2p_z$ vibration.   The important point is that there are three types of vibration each having a nodal plane and differing from each other with respect to the axis to which this nodal plane is normal.   This orientation is also shown in Fig. 17-4.

The space pattern of the wave function has an important bearing on its interpretation.   As pointed out previously, the maximum amplitude for the 1s type of wave function lies at the center of the atomic nucleus. Classically, it is impossible for the electron to move through the nucleus. Quantum-mechanically speaking, one can say that, of all the places in space where one might find the electron, the maximum probability is actually in the nucleus itself. This, however, has unreasonable implications from the particle point of view. One cannot hope to construct a model of the atom in classical terms locating the electron at a certain point moving in a certain direction and at the same time have this model consistent with the principles of quantum theory. As mentioned previously, the degree of reality to be assigned to the electron wave is largely a question of semantics. The wave is "real" largely in the sense that the mathematical formulation of the wave provides a consistent and logical mathematical structure from which the properties of the atom can be derived to a very considerable extent in a consistent and logical way.

In Fig. 17-8 the probability-density factor $(R_{n,1})^2$ is plotted as a function of the distance r between the electron and the nucleus. Wherever the amplitude of the de Broglie wave becomes zero at a node, the probability likewise becomes zero; wherever the amplitude of the de Broglie wave has a value different from zero, whether positive or negative, the probability takes on a positive value since it is effectively the square of the amplitude of the de Broglie wave.

Electric spin

This fourth number indicates the value of an angular momentum which is associated with an intrinsic spin motion of the electron. The nature of this spin will be discussed in more detail in Chap. 18 dealing with atomic spectra. The spin quantum number is designated by the symbol s and can have only two values, $+\frac{1}{2}$ and $-\frac{1}{2}$ corresponding to values of the spin of $\frac{1}{2}\frac{h}{2\pi}$ and $\frac{-1}{2}\frac{h}{2\pi}$.

Example 17-1

A violin string is stretched so that it has a tension resulting in a fundamental frequency of 256 sec⁻¹. Calculate the frequencies for the

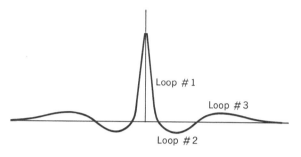

Fig. 17-7 Radial amplitude of a $3s$ type wave in a hydrogen atom (three loops, spherical shape). $n = 3, l = 0$.

Loop #1

Loop #3

Loop #2

patterns of vibration with quantum numbers 2, 3, 4, and 5.

The vibrations are given by the following formulas:

Quantum no.	ν
2	$2 \times 256 = 512$
3	$3 \times 256 = 768$
4	$4 \times 256 = 1024$
5	$5 \times 256 = 1280$

Example 17-2

The membrane of a square drum is tuned so that it sounds the fundamental tone at 256 sec⁻¹. The frequency of an overtone is given by

$$\nu_x = \frac{256}{\sqrt{2}} \sqrt{n_1^2 + n_2^2}$$

where n_1 is the first quantum number and n_2 is the second quantum number. Calculate the frequency for the vibration with the quantum numbers 2, 2.

This frequency is calculated as follows:

$$\nu_{2,2} = \frac{256}{\sqrt{2}} \sqrt{2^2 + 2^2} = \frac{256}{\sqrt{2}} \sqrt{4 + 4}$$

$$\nu_{2,2} = \frac{256}{\sqrt{2}} \sqrt{8} = 256 \times 2 = 512$$

Fig. 17-8 The probability-density factor $(R_{n,1})^2$ plotted (dashed lines) as a function of the electron-nuclear distance r (r is given in units $a_1 = 0.53$ Å, the radius of the first Bohr circular orbit). Density distribution curves $D = 4\pi r^2 (R_{n,1})^2$ are shown as solid lines. (From H. E. White, "Introduction to Atomic Spectra," p. 68, McGraw-Hill Book Company, New York, 1934.)

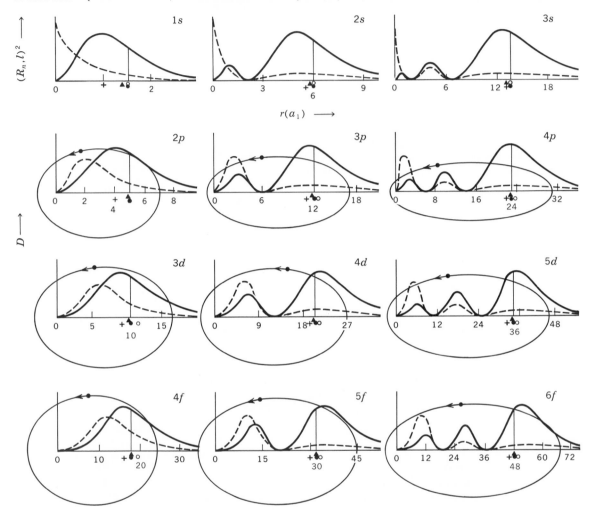

Note that the type of vibration with the quantum numbers 2, 2 has exactly the same frequency as the overtone of the string with the quantum number 2, namely, 512, the frequency of high C, denoted by the symbol c'.

Example 17-3

Calculate the value of the amplitude A_x in a stretched 45-cm string at a distance 15 cm from the left end when the string is vibrating in the mode for which $n = 2$ and $A_m = 1$ cm.

$$A_x = A_m \sin \frac{2\pi}{\lambda n} x = 1 \sin 2\pi \frac{15}{45} = 0.5$$

since for $n = 2$, $\lambda = 2a/n = a$.

Example 17-4

Calculate the value of p_θ for the $3d$ wave function of the hydrogen atom.

For the $3d$ wave function $l = 2$.

$$p_\theta = \frac{h}{2\pi} \sqrt{l(l + 1)}$$

$$= \frac{6.626 \times 10^{-27}}{2 \times 3.1416} \sqrt{2(2 + 1)}$$

$$= 2.58 \times 10^{-27} \text{ erg sec}$$

17-3   THE PERIODIC TABLE OF CHEMICAL ELEMENTS

The hydrogen atom with its set of wave functions serves as the prototype for all the more complex atoms of the other elements in the periodic table. Recall that the forms of the wave functions of hydrogen are determined solely by mathematical and physical principles. The mathematical principles apply to all types of waveforms, ranging from the vibrating stretched string through the two-dimensional elastic membrane and the three-dimensional sphere to the "elastic" electron. The boundary conditions vary, but the basic differential equation is the same. Consequently, it is to be expected that these same mathematical principles will determine the wave functions of the electrons all the way from hydrogen through uranium and the transuranic elements for atoms considered as single units and not linked to other atoms.

We turn from the simple hydrogen atom to examine the wave patterns in the type of atom next in sequence in the order of ascending complexity, the helium atom, with two positive charges on the nucleus and two electrons.

The Pauli exclusion principle

The first question to be answered is whether both electrons have the same identical $1s$ wave function when the helium atom is in its lowest energy state. Shortly after de Broglie made his suggestion that a wave was associated with the electron, the Austrian physicist W. Pauli in 1924 suggested that in an atom no two electrons can have identical wave patterns. This is called the *Pauli exclusion principle*. Specifically this principle states that when two or more electrons are present in the same atom, none of them can have all four quantum numbers n, l, m, and s identically the same.

The underlying mathematical structure of the Pauli exclusion principle can be seen by examining the form of the combined wave function for two electrons in the same atom. As noted previously, when two electrons are in the same atom, it is impossible to regard the system as made up of two waves each vibrating with complete independence. If, from the particle point of view, the electrons are regarded as circulating each in its own orbit around the nucleus, it is obvious that the repulsion of the negative charge on one electron for the negative charge on the other electron will influence the pattern of the orbits. This logic suggests that the waves interact, and the general pattern of mathematics for waves leads to the conclusion that the wave differential equation has a solution must satisfy the boundary conditions when the wave represents *jointly* the probability pattern for both electrons.

One of the basic conclusions drawn from considerations of the joint behavior of fundamental particles is that such particles are indistinguishable from each other. In deducing statistical results the conclusion is always reached that electrons are indistinguishable. To see more clearly the significance of indistinguishability, contrast a statistical experiment with electrons and one with colored balls. Suppose that there are two boxes which can be distinguished by their positions in space, box A on the left and box B on the right. Suppose that someone is blindfolded and tosses a red ball and a green ball at the boxes until one ball is in each box. If the red ball is in the left box and the green ball is in the right box, this is clearly a different case from the situation where the green ball is in the left box and the red ball is in the right box. On the other hand, if the balls are not colored and have no distinguishing marks, there is no way of telling the two cases apart; the system with the first distribution is indistinguishable from the system with the second and must be regarded as in the same state no matter which ball is in which box. Electrons are not colored; they have no distinguishing marks by which they can be identified; this may seem to be a trivial concept, but it has profound consequences.

Just as the wave function for a vibrating stretched

string can be written as the product of a term dependent on the space coordinate x and the time coordinate t, the total wave function for one electron can be written as the product of a term ϕ containing the space coordinates (x, y, and z) and a term σ containing the spin coordinate s. In symbolic form

$$\psi = \phi(x,y,z)\sigma(s) \tag{17-28}$$

the Pauli principle states that a wave function for a system of electrons must be *antisymmetric* for the exchange of the spacial and the spin coordinates for any pair of electrons; i.e., the function changes sign if the electrons are transposed. Such an antisymmetric function might be

$$\psi = {}^1\psi_{n_1,l_1,m_1,s_1} {}^2\psi_{n_2,l_2,m_2,s_2} - {}^2\psi_{n_1,l_1,m_1,s_1} {}^1\psi_{n_2,l_2,m_2,s_2} \tag{17-29}$$

In this expression, n_1,l_1,m_1,s_1 are the quantum numbers of the first electron and n_2,l_2,m_2,s_2 are those of the second. Suppose that for a given point in space at a given time the insertion of these numbers results in the value for ψ_I

$$\psi_\text{I} = ab - cd \tag{17-30}$$

If the electrons are exchanged so that their quantum numbers are exchanged in Eq. (17-29), the value of ψ_II will be

$$\psi_\text{II} = cd - ab = -\psi_\text{I} \tag{17-31}$$

This shows that the function ψ is an antisymmetric function with regard to the exchange of the electrons. This function mathematically is the determinant based on the terms a, b, c, and d. If the wave function is written as the determinant of an algebraic matrix of such a set of terms, the exchange of two columns changes the sign of the determinant. If the quantum numbers are identical, the value of the determinant becomes identically zero. Thus, if the joint wave function made up of the individual wave functions is antisymmetric, none of the sets of quantum numbers for the individual electrons can be identical.

Periodic structure

As a result of the Pauli principle combined with the sets of possible forms for bounded Schrödinger waves, an explanation is provided for the periodicity in the table of chemical elements. As the charge on the atomic nucleus increases unit by unit in passing from hydrogen to the heavier atoms, each added electron takes on the waveform satisfying two conditions. First, in accordance with the Pauli exclusion principle, the waveform must not already be present among any of the other electrons; second, if the atom is to be in the lowest possible energy level, the waveform taken on by the added electron must be that with the lowest energy among the set of all waveforms not already occupied.

Two methods of waveform shorthand have been developed for denoting the waveforms in any particular atom. In one of these, the quantum numbers for each added electron are denoted by the spectroscopic notation written for each electron in a linear sequence; in the other, a triangular matrix of numbers is written where the position of the number indicates the type of waveform. The two types of notation are illustrated for the uranium atom in Fig. 17-9. In Fig. 17-10 the spectroscopic notation and the matrix notation are compared for the first 21 chemical elements. In Fig. 17-11 the matrix notation is shown for the entire periodic table. The wave

```
      1s 2s 2p 3s 3p 3d 4s 4p 4d 4f 5s 5p 5d 5f 6s 6p 6d 7s
U:  2  2  6  2  6 10  2  6 10 14  2  6 10  3  2  6  1  2
```

(a)

$$
\begin{array}{c|ccccccc}
 & U & & & & & & \\
f & & & & 14\ 3 & & & \\
d & & & 10\ 10\ 10\ 1 & & & \\
p & & 6\ 6\ 6\ 6\ 6 & & & \\
s & 2\ 2\ 2\ 2\ 2\ 2\ 2 & & & \\
\hline
n = & 1\ 2\ 3\ 4\ 5\ 6\ 7 & & &
\end{array}
$$

(b)

Fig. 17-9 Representations of the electronic structure of uranium. (a) Spectroscopic notation; (b) triangle matrix notation.

Fig. 17-10 The electronic structure of the first 21 elements of the periodic table in (a) conventional linear and (b) triangle matrix notation.

functions applicable to the outer electrons of the first 10 elements in the periodic table are given in Table 17-4 (assuming no electron interaction).

From the particle point of view, one can say that as the charge on the nucleus increases, the electrons with lower energy are pulled closer and closer to the nucleus. The electrons are said to group themselves in shells. The shell lying nearest the nucleus consisting of electrons in the $1s$ state is called the K shell. The electrons with the next highest values of the energy are those in the $2s$ and $2p$ states, and the shell in which they are grouped is called the L shell. The shell containing the electrons with $n = 3$ is the M shell, with $n = 4$ is the N shell, etc.

The energy levels for the different states are shown in Fig. 17-12. In progressing further up the periodic table, it is found that as a result of the interaction of the electrons the $4s$ shell lies below the $3d$ shell. Thus, in the elements potassium and calcium the last two electrons occupy $4s$ states and only after these shells have been filled do electrons take on the $3d$ patterns. This explains the presence of the transition elements scandium through nickel. A similar situation in the relation of the energy levels explains the series of lanthanide elements.

The structure of a helium atom can now be explained in terms of these principles. Since s can have the value of either $+\frac{1}{2}$ or $-\frac{1}{2}$, the increase in nuclear charge by one unit in passing from H to He causes the addition of one electron to the outer shell. The lowest value of the energy will be achieved if this electron is in the $1s$ state but with the spin of the sign opposite to that of the original electron. Under the Pauli exclusion principle, there are only two $1s$ wave functions possible; these are now both represented by electrons, and the K shell is closed, for which $n = 1$.

When an additional unit charge is added to the nucleus, raising the charge to three units, the resultant *lithium* atom has three electrons. Since the first two fill the $1s$ shell, the next electron has a $2s$ wave function. As more unit charges are added, the $2s$ wave function of opposite spin is filled, and then the six $2p$

(a)

Element	Configuration	(b)
H	$1s(1)$	1
He	$1s(2)$	2
Li	$1s(2)\,2s(1)$	2 1
Be	$1s(2)\,2s(2)$	2 2
B	$1s(2)\,2s(2)\,2p(1)$	1 / 2 2
C	$1s(2)\,2s(2)\,2p(2)$	2 / 2 2
N	$1s(2)\,2s(2)\,2p(3)$	3 / 2 2
O	$1s(2)\,2s(2)\,2p(4)$	4 / 2 2
F	$1s(2)\,2s(2)\,2p(5)$	5 / 2 2
Ne	$1s(2)\,2s(2)\,2p(6)$	6 / 2 2
Na	$1s(2)\,2s(2)\,2p(6)\,3s(1)$	6 / 2 2 1
Mg	$1s(2)\,2s(2)\,2p(6)\,3s(2)$	6 / 2 2 2
Al	$1s(2)\,2s(2)\,2p(6)\,3s(2)\,3p(1)$	6 1 / 2 2 2
Si	$1s(2)\,2s(2)\,2p(6)\,3s(2)\,3p(2)$	6 2 / 2 2 2
P	$1s(2)\,2s(2)\,2p(6)\,3s(2)\,3p(3)$	6 3 / 2 2 2
S	$1s(2)\,2s(2)\,2p(6)\,3s(2)\,3p(4)$	6 4 / 2 2 2
Cl	$1s(2)\,2s(2)\,2p(6)\,3s(2)\,3p(5)$	6 5 / 2 2 2
A	$1s(2)\,2s(2)\,2p(6)\,3s(2)\,3p(6)$	6 6 / 2 2 2
K	$1s(2)\,2s(2)\,2p(6)\,3s(2)\,3p(6)\,4s(1)$	6 6 / 2 2 2 1
Ca	$1s(2)\,2s(2)\,2p(6)\,3s(2)\,3p(6)\,4s(2)$	6 6 / 2 2 2 2
Sc	$1s(2)\,2s(2)\,2p(6)\,3s(2)\,3p(6)\,4s(2)\,3d(1)$	1 / 6 6 6 / 2 2 2 2

Fig. 17-11 Electronic structures of the chemical elements in triangle matrix notation.

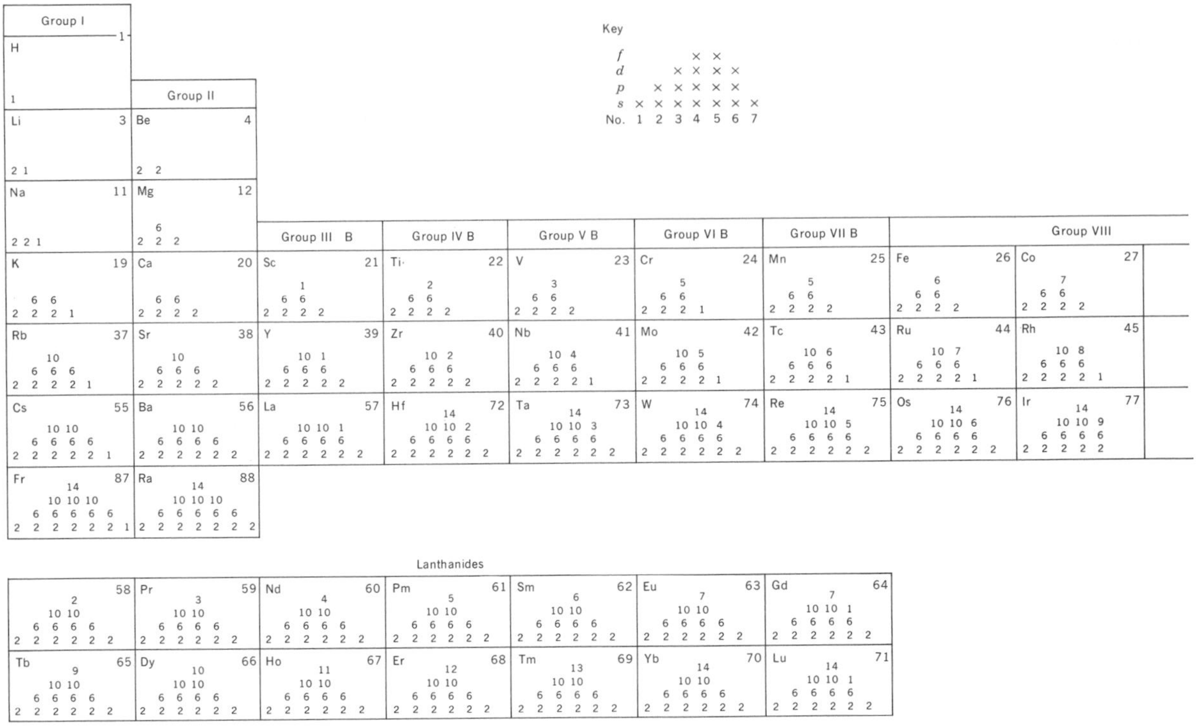

Key

f			× ×			
d		× ×	× ×			
p	×	× ×	× ×			
s	× ×	× ×	× ×	×		
No.	1 2	3 4	5 6	7		

Group I

H — 1
1

Li 3
2 1

Na 11
2 2 1

K 19
6 6
2 2 2 1

Rb 37
10
6 6 6
2 2 2 2 1

Cs 55
10 10
6 6 6 6
2 2 2 2 2 1

Fr 87
14
10 10 10
6 6 6 6
2 2 2 2 2 1

Group II

Be 4
2 2

Mg 12
6
2 2 2

Ca 20
6 6
2 2 2 2

Sr 38
10
6 6 6
2 2 2 2 2

Ba 56
10 10
6 6 6 6
2 2 2 2 2 2

Ra 88
14
10 10 10
6 6 6 6
2 2 2 2 2 2

Group III B

Sc 21
1
6 6
2 2 2 2

Y 39
10 1
6 6 6
2 2 2 2

La 57
10 10 1
6 6 6 6
2 2 2 2 2

Group IV B

Ti 22
2
6 6
2 2 2 2

Zr 40
10 2
6 6 6
2 2 2 2

Hf 72
14
10 10 2
6 6 6 6
2 2 2 2 2

Group V B

V 23
3
6 6
2 2 2 2

Nb 41
10 4
6 6 6
2 2 2 2 1

Ta 73
14
10 10 3
6 6 6 6
2 2 2 2 2

Group VI B

Cr 24
5
6 6
2 2 2 1

Mo 42
10 5
6 6 6
2 2 2 2 1

W 74
14
10 10 4
6 6 6 6
2 2 2 2 2

Group VII B

Mn 25
5
6 6
2 2 2 2

Tc 43
10 6
6 6 6
2 2 2 2 1

Re 75
14
10 10 5
6 6 6 6
2 2 2 2 2

Group VIII

Fe 26
6
6 6
2 2 2 2

Co 27
7
6 6
2 2 2 2

Ru 44
10 7
6 6 6
2 2 2 2 1

Rh 45
10 8
6 6 6
2 2 2 2 1

Os 76
14
10 10 6
6 6 6 6
2 2 2 2 2

Ir 77
14
10 10 9
6 6 6 6
2 2 2 2 2

Lanthanides

58
2
10 10
6 6 6 6
2 2 2 2 2

Pr 59
3
10 10
6 6 6 6
2 2 2 2 2

Nd 60
4
10 10
6 6 6 6
2 2 2 2 2

Pm 61
5
10 10
6 6 6 6
2 2 2 2 2

Sm 62
6
10 10
6 6 6 6
2 2 2 2 2

Eu 63
7
10 10
6 6 6 6
2 2 2 2 2

Gd 64
7
10 10 1
6 6 6 6
2 2 2 2 2

Tb 65
9
10 10
6 6 6 6
2 2 2 2 2

Dy 66
10
10 10
6 6 6 6
2 2 2 2 2

Ho 67
11
10 10
6 6 6 6
2 2 2 2 2

Er 68
12
10 10
6 6 6 6
2 2 2 2 2

Tm 69
13
10 10
6 6 6 6
2 2 2 2 2

Yb 70
14
10 10
6 6 6 6
2 2 2 2 2

Lu 71
14
10 10 1
6 6 6 6
2 2 2 2 2

TABLE 17-4 Outer electron quantum numbers and wave function types.[a]

Spin i	Loops n	Symmetry l	Orientation m	Wave function type	Atom
$+\frac{1}{2}$	1	0	0	$\psi_{1s} = \dfrac{1}{\sqrt{\pi}}\left(\dfrac{z}{a_0}\right)^{3/2} e^{-\sigma}$	H
$-\frac{1}{2}$	1	0	0		He
$+\frac{1}{2}$	2	0	0	$\psi_{2s} = \dfrac{1}{4\sqrt{2\pi}}\left(\dfrac{z}{a_0}\right)^{3/2}(2-\sigma)e^{-\sigma/2}$	Li
$-\frac{1}{2}$	2	0	0		Be
$+\frac{1}{2}$	2	1	$+1$	$\psi_{2p_x} = \dfrac{1}{4\sqrt{2\pi}}\left(\dfrac{z}{a_0}\right)^{3/2}\sigma e^{-\sigma/2}\sin\theta\cos\phi$	B
$-\frac{1}{2}$	2	1	$+1$		C
$+\frac{1}{2}$	2	1	0	$\psi_{2p_z} = \dfrac{1}{4\sqrt{2\pi}}\left(\dfrac{z}{a_0}\right)^{3/2}\sigma e^{-\sigma/2}\cos\theta$	N
$-\frac{1}{2}$	2	1	0		O
$+\frac{1}{2}$	2	1	-1	$\psi_{2p_y} = \dfrac{1}{4\sqrt{2\pi}}\left(\dfrac{z}{a_0}\right)^{3/2}\sigma e^{-\sigma/2}\sin\theta\sin\phi$	F
$-\frac{1}{2}$	2	1	-1		Ne

[a] $\sigma = (z/a_0)r$ r is the distance out from the nucleus, and $a_0 = h^2/4\pi^2\mu e^2$ is the Bohr radius.

Fig. 17-11 (cont'd)   Electronic structures of the chemical elements in triangle matrix notation.

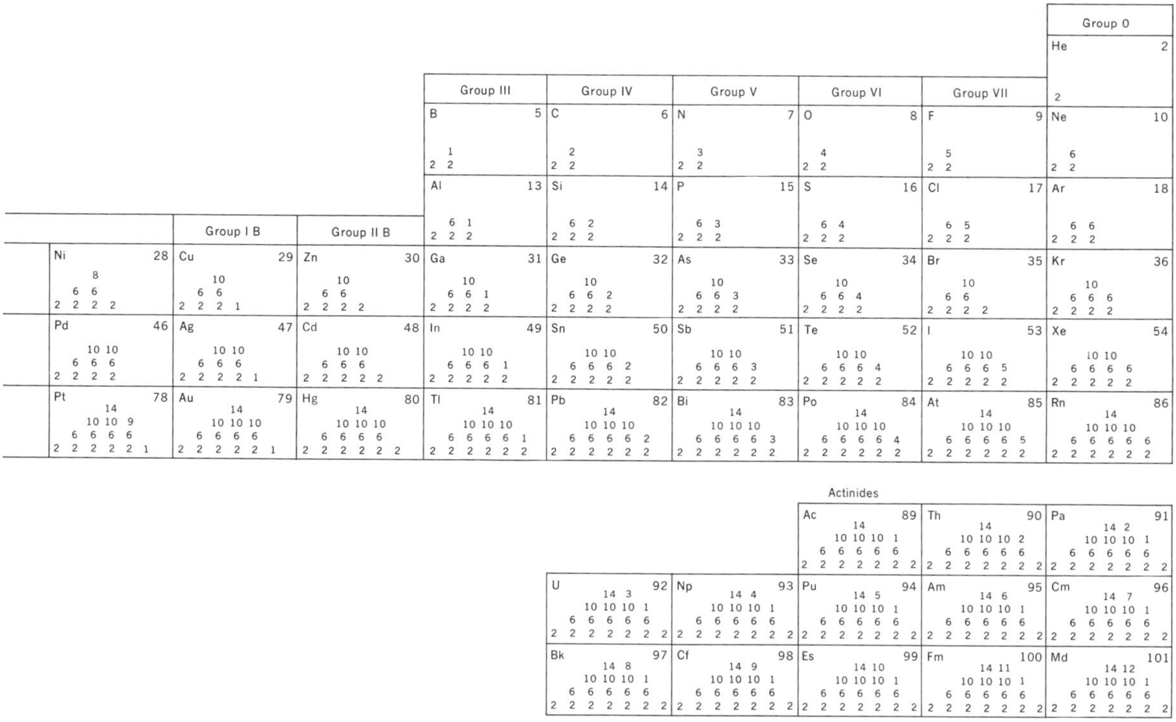

Group 0
He   2
2

Group III	Group IV	Group V	Group VI	Group VII
B 5	C 6	N 7	O 8	F 9	Ne 10

B 5: 1 / 2 2
C 6: 2 / 2 2
N 7: 3 / 2 2
O 8: 4 / 2 2
F 9: 5 / 2 2
Ne 10: 6 / 2 2

Al 13: 6 1 / 2 2 2
Si 14: 6 2 / 2 2 2
P 15: 6 3 / 2 2 2
S 16: 6 4 / 2 2 2
Cl 17: 6 5 / 2 2 2
Ar 18: 6 6 / 2 2 2

Group I B   Group II B

Ni 28: 8 / 6 6 / 2 2 2 2
Cu 29: 10 / 6 6 / 2 2 2 1
Zn 30: 10 / 6 6 / 2 2 2 2
Ga 31: 10 / 6 6 1 / 2 2 2 2
Ge 32: 10 / 6 6 2 / 2 2 2 2
As 33: 10 / 6 6 3 / 2 2 2 2
Se 34: 10 / 6 6 4 / 2 2 2 2
Br 35: 10 / 6 6 / 2 2 2 2
Kr 36: 10 / 6 6 6 / 2 2 2 2

Pd 46: 10 10 / 6 6 / 2 2 2 2
Ag 47: 10 10 / 6 6 / 2 2 2 2 1
Cd 48: 10 10 / 6 6 / 2 2 2 2 2
In 49: 10 10 / 6 6 1 / 2 2 2 2
Sn 50: 10 10 / 6 6 2 / 2 2 2 2
Sb 51: 10 10 / 6 6 3 / 2 2 2 2
Te 52: 10 10 / 6 6 4 / 2 2 2 2
I 53: 10 10 / 6 6 5 / 2 2 2 2
Xe 54: 10 10 / 6 6 6 / 2 2 2 2

Pt 78: 14 / 10 10 9 / 6 6 6 6 / 2 2 2 2 2 1
Au 79: 14 / 10 10 10 / 6 6 6 6 / 2 2 2 2 2 1
Hg 80: 14 / 10 10 10 / 6 6 6 6 / 2 2 2 2 2 2
Tl 81: 14 / 10 10 10 / 6 6 6 6 1 / 2 2 2 2 2 2
Pb 82: 14 / 10 10 10 / 6 6 6 6 2 / 2 2 2 2 2 2
Bi 83: 14 / 10 10 10 / 6 6 6 6 3 / 2 2 2 2 2 2
Po 84: 14 / 10 10 10 / 6 6 6 6 4 / 2 2 2 2 2 2
At 85: 14 / 10 10 10 / 6 6 6 6 5 / 2 2 2 2 2 2
Rn 86: 14 / 10 10 10 / 6 6 6 6 6 / 2 2 2 2 2 2

Actinides

Ac 89: 14 / 10 10 10 1 / 6 6 6 6 / 2 2 2 2 2 2
Th 90: 14 / 10 10 10 2 / 6 6 6 6 / 2 2 2 2 2 2
Pa 91: 14 2 / 10 10 10 1 / 6 6 6 6 / 2 2 2 2 2 2

U 92: 14 3 / 10 10 10 1 / 6 6 6 6 / 2 2 2 2 2 2
Np 93: 14 4 / 10 10 10 1 / 6 6 6 6 / 2 2 2 2 2 2
Pu 94: 14 5 / 10 10 10 1 / 6 6 6 6 / 2 2 2 2 2 2
Am 95: 14 6 / 10 10 10 1 / 6 6 6 6 / 2 2 2 2 2 2
Cm 96: 14 7 / 10 10 10 1 / 6 6 6 6 / 2 2 2 2 2 2

Bk 97: 14 8 / 10 10 10 1 / 6 6 6 6 / 2 2 2 2 2 2
Cf 98: 14 9 / 10 10 10 1 / 6 6 6 6 / 2 2 2 2 2 2
Es 99: 14 10 / 10 10 10 1 / 6 6 6 6 / 2 2 2 2 2 2
Fm 100: 14 11 / 10 10 10 1 / 6 6 6 6 / 2 2 2 2 2 2
Md 101: 14 12 / 10 10 10 1 / 6 6 6 6 / 2 2 2 2 2 2

wave functions are occupied, filling the L shell with $n = 2$. Just as He with the K shell filled is an inert atom, the element Ne with the L shell filled is also relatively inert.

Because the state of an element with a completed shell has a value of the energy so much lower than the state with several electrons in an incomplete shell, there is the thermodynamic tendency for the elements in groups I and II to lose the outer electrons and form ions either in the crystal lattice or in solution. In the same way, the elements like the halogens which need only one electron to complete the shell have the thermodynamic tendency to take on this electron to form negatively charged ions. Thus the periodicity in the table of elements is explained, and the number of elements in each period follows from the number of wave patterns.

For the same thermodynamic reason, two hydrogen atoms tend to combine to form the hydrogen molecule, which has a lower value of the energy. These areas of chemical action will be discussed in Chaps. 19 and 20.

Example 17-5

Draw the triangle electron-structure diagrams for the two kinds of ions in a crystal of calcium chloride.

These diagrams are drawn by removing the two electrons from the $3s$ shell of calcium and adding one electron to the $3p$ shell of chlorine.

Ca^{++}
  6 6
2 2 2

Cl^{-}
  6 6
2 2 2

Fig. 17-12 Approximate energy levels for hydrogen-type wave functions showing filling order.

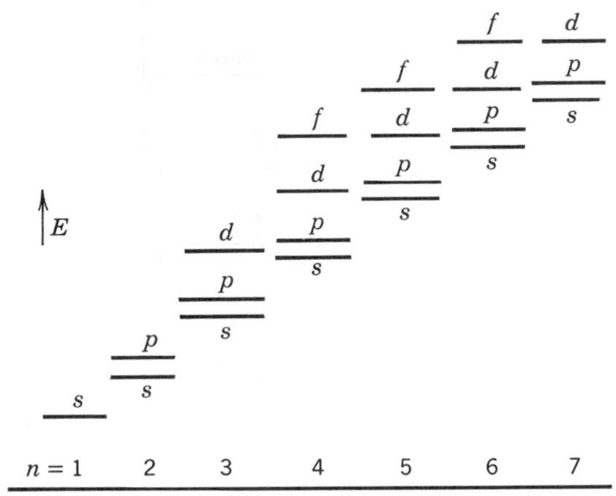

Example 17-6

Draw the triangle electron-structure diagrams for Al^{3+} and Fe^{3+}.

$$\begin{array}{|l}
Al^{3+} \\
\\
\quad 6 \\
2 \quad 2 \\
\hline
\end{array}
\qquad
\begin{array}{|l}
Fe^{3+} \\
\qquad 5 \\
\quad 6 \quad 6 \\
2 \quad 2 \quad 2 \\
\hline
\end{array}$$

Example 17-7

Calculate the value of $\psi(1s)$ in a hydrogen atom at $r = 1.0 \times 10^{-8}$ cm.

Using the formula from Table 17-3,

$$\alpha = 1.890 \times 10^8 \text{ cm}^{-2}$$

$$\alpha^{3/2} = 2.598 \times 10^{12}$$

$$\log \psi(1s) = \log \alpha^{3/2} - \tfrac{1}{2} \log \pi - \frac{\alpha r}{2.303}$$

$$= 12.4147 - 0.2486 - 0.8207$$
$$\times 10^8 \times 10^{-8} = 11.345$$

$$\psi = 2.214 \times 10^{11} \qquad Ans.$$

Example 17-8

Calculate the value of $\psi^2(2s)$ at $r = 1.058$ Å.

$$\psi(2s) = \frac{1}{4}\left(\frac{1}{2\pi}\right)^{1/2}\left(\frac{z}{a_0}\right)^{3/2}\left(2 - \frac{zr}{a_0}\right)e^{-zr/a_0}$$

$$= \frac{1}{4}\left(\frac{1}{2\pi}\right)^{1/2}\left(\frac{z}{a_0}\right)^{3/2}$$

$$\left(2 - \frac{1.058 \times 10^{-8}}{0.529 \times 10^{-8}}\right)e^{-zr/a_0}$$

$$= \frac{1}{4}\left(\frac{1}{2\pi}\right)^{1/2}\left(\frac{z}{a_0}\right)^{3/2}(2-2)e^{-zr/a_0} = 0$$

$$\psi^2(2s) = 0$$

The radius of the nodal surface is 1.058 Å.

Problems

1. A violin string with its fundamental note tuned to vibrate at 256 sec^{-1} is made to sound three tones the frequencies of which are 1792, 2048, and 2304 sec^{-1}. Calculate the quantum numbers for each of these frequencies.

2. A violin string is stretched so that it sounds its fundamental note with a frequency of 256 sec^{-2}. An electronic oscillator is placed next to it and sounds notes having frequencies at 2816, 2999, and 3328 sec^{-1}. Which of these tones will set the string in a steady-state oscillation with one of its overtones? Calculate the number of loops in each of the overtones which appears.

3. Calculate the frequency of the vibration in the overtone pattern for a square drum which makes it appear in the pattern of the flag of France (three vertical stripes) when the fundamental frequency of the drum is 256 sec^{-1}.

4. Calculate the frequency of vibration in the square drum having the quantum numbers 3, 2 when the fundamental frequency is 256 sec^{-1}.

5. The formula for the vibration of an elastic cube is merely an extension of the pythagorean formula for the vibration of a square drum, namely,

$$\nu_{n_1,n_2,n_3} = \frac{256}{\sqrt{3}}\sqrt{n_1^2 + n_2^2 + n_3^2}$$

when

$$\nu_{1,1,1} = \nu_{\text{fund}} = 256 \text{ sec}^{-1}$$

Calculate the frequency when the cube vibrates with a pattern having the quantum numbers 2, 2, 2.

6. Calculate the frequency of vibration when the cube in Prob. 5 executes the pattern of motion with the quantum numbers 1, 2, 3.

7. In a space of four dimensions, the analog of the cube is the *tesseract*. The formula for the frequency of tesseract vibration is

$$\nu_{n_1,n_2,n_3,n_4} = \frac{\nu_{\text{fund}}}{\sqrt{4}} \sqrt{n^2 + n_2^2 + n_3^2 + n_4^2}$$

Calculate the frequency of vibration of a tesseract with the quantum numbers 2, 2, 2, 2 if the fundamental vibration (1, 1, 1, 1) has a frequency of 256 sec^{-1}.

8. If a tesseract has the fundamental frequency of 256 sec^{-1}, calculate the frequency of vibration with the quantum numbers 1, 2, 3, 4.

9. The element hafnium can exist as the ion Hf$^+$. Write the triangle diagram for this ion.

10. Write the triangle diagram for Mg^{3+}, which occurs in the gaseous state.

11. Write the triangle diagram for the ion Ba^{++}.

12. Write the triangle diagram for the compound CsI, where both atoms are in the form of ions.

13. Write the triangle diagram for BaBr$_2$, where all the atoms are in the form of ions.

14. What ion is represented by the following triangle diagram?

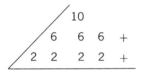

15. What ion is represented by the following triangle diagram?

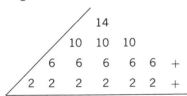

16. Calculate the value of $\psi(1s)$ in a helium ion, He$^+$, at $r = 0$.

17. What is the ratio of the value of $\psi(1s)$ at $r = a_0$ to the value of $\psi(1s)$ at $r = 2a_0$?

18. Calculate the value of r in Li^{++} at the nodal surface.

19. What is the value of $\psi(2p_z)$ at a point on the z axis where $r = 2a_0$ in a hydrogen atom?

20. What is the value of $\psi(2p_x)$ at a point on the x axis where $r = 2a_0$ in a helium ion (He$^+$)?

REFERENCES

Pitzer, K. S.: "Quantum Chemistry," Prentice-Hall, Inc., Englewood Cliffs, N.J., 1953.

Heitler, W.: "Elementary Wave Mechanics," Oxford University Press, New York, 1956.

Fano, U., and L. Fano: "Basic Physics of Atoms and Molecules," John Wiley & Sons, Inc., New York, 1959.

Daudel, R., R. Lefebvre, and C. Moser: "Quantum Chemistry," Interscience Publishers, Inc., New York, 1959.

Slater, J. C.: "Quantum Theory of Atomic Structure," 2 vols., McGraw-Hill Book Company, New York, 1960.

Cumper, C. W. N.: "Wave Mechanics for Chemists," Academic Press, Inc., New York, 1966.

Löwdin, P.-O. (ed.): "Quantum Theory of Atoms, Molecules and the Solid State," Academic Press, Inc., New York, 1966.

18-1 THE NATURE OF ATOMIC SPECTRA

The term *atomic spectra* is used to denote the wavelengths (or frequencies) of the light emitted or absorbed when an electron changes from one energy level to another among the set of energy levels in a *free* atom. The wavelengths of light emitted or absorbed in changes between *molecular* energy levels will be discussed in Chap. 20.

In 1704, Sir Isaac Newton wrote: "Do not all fixed Bodies when heated beyond a certain degree, emit Light and shine; and is not this Emission formed by the vibrating motion of their parts?" Newton was led to the study of spectroscopy through problems

Chapter Eighteen
atomic spectra

encountered in his experimental work in the construction of telescopes. He observed that light of different frequencies was reflected by varying amounts causing the phenomenon of chromatic aberration. In 1666, he purchased a glass prism and observed that a beam of visible white light from a slit was spread out by the prism into what he called a *spectrum*, as a result of the variation in the refractive index of the glass with the variation in color of light. This effect had been observed for many years, but Newton gave the first correct interpretation. The existence of *invisible* rays with wavelengths shorter than the violet was demonstrated by a German physicist, J. W. Ritter, and an English one, W. H. Wollaston, at the beginning of the nineteenth century. At about the same time, Sir William Herschel, an English astronomer, proved the existence of rays with wavelengths longer than the visible red by their heat effect on a thermometer bulb; to these two regions of the spectrum the terms *ultraviolet* and *infrared* were applied.

The first precise quantitative results in spectroscopy were obtained by J. Fraunhofer, a Bavarian optician and physicist, a few years later by viewing the slit with a theodolite telescope placed behind the prism, permitting an accurate measurement of the angles. He observed over 700 dark lines in the spectrum of sunlight, which provided the first precise standards for measuring the dispersion of optical glass. In 1848 a French physicist, L. Foucault, found that a sodium flame emitted a sharp yellow line (now called the D line) and that this flame would also absorb light of the same wavelength from a sodium arc. This relationship was generalized by the German physicist G. R. Kirchhoff in 1859 in the law which states:

The relation between the power of emission and the power of absorption for rays of the same wavelength is constant for all bodies at the same temperature.

He concluded that the Fraunhofer lines are caused by the absorption of certain wavelengths by the sun's atmosphere; knowing from laboratory observation the elements emitting these lines, he could verify the presence of these elements in the sun.

It was in 1885 that Balmer found the relationship between the frequencies of spectral lines emitted by atomic hydrogen which provided the basis for Niels Bohr's theory of atomic structure. Bohr was also helped by the suggestion of two other physicists who had been trying to discover the mechanism by which

spectral lines were produced. In 1907 the Irish physicist Arthur Conway suggested that a single atom produces a single spectral line at a time when a single electron in an "abnormal state" can cause vibrations of a specific frequency. The following year the Swiss physicist Walter Ritz pointed out that the observed frequencies are the differences between spectral terms taken in pairs. It was just at this time that Rutherford proposed a model of an atom with electrons in orbit about a nucleus and the British physicist and mathematician John Nicholson at the University of Cambridge suggested that a line is emitted or absorbed when an electron moves suddenly between two states which correspond to the term values of Ritz. On this foundation, Bohr was able to construct his theory of the structure of the hydrogen atom, extend this model to more complicated atoms, and deduce the first quantitative laws relating atomic spectra and atomic structure.

A prism spectrograph is diagrammed in Fig. 18-1. Examples of the spectral lines emitted by magnesium atoms are shown in Fig. 18-2.

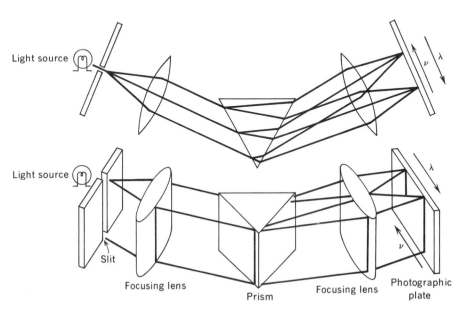

Fig. 18-1 A prism spectrograph (schematic). (After G. M. Barrow, "Physical Chemistry," 2d ed., McGraw-Hill Book Company, New York, 1961.)

Fig. 18-2   The atomic spectrum of magnesium. [From P. D. Foote, W. F. Meggers, and O. C. Mohler, *Phil. Mag.,* **42:1002 (1921); 43:639 (1922).]** (*Note: only the principal lines are shown.*)

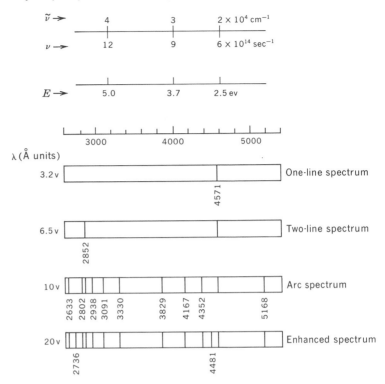

18-2   THE ATOMIC SPECTRA OF HYDROGEN

In Fig. 18-3*a* portions of the Bohr orbits are shown for the states 1*s*, 2*s*, 3*s*, and 4*s*. In Fig. 18-3*b* the cross sections of the wave function diagrams are shown; the density of the dots is proportional to the square of the part of the wave function which gives the amplitude of the wave as a function of the radius. Recall that in the 1*s* state the density is highest at the center of the atom and continually decreases in moving out along the radius toward infinity. With the 2*s* wave function, the square of the amplitude decreases to zero at a short distance from the center at the spherical nodal surface. Proceeding further along the radius, the square of the amplitude increases, going through a maximum and then falling off continually as the value of *r* increases toward infinity. The 3*s* wave function has two spherical nodal

surfaces, and the 4*s* has three. In Fig. 18-3*c* the values of the energy levels for the wave functions from $n = 1$ to $n = 7$ are shown. The energy is calculated as the energy difference between the value for the wave function and the value at $n = \infty$. The value at $n = \infty$ is set equal to zero so that the other energy levels have negative energy values. The values are given in terms of calories per mole and in electron volts (1 ev = 2.3060×10^4 cal mole^{-1}); since $E = h\nu = hc\sigma$, one can derive the value of the wave number σ in cm^{-1} from the relation $\sigma = E/hc$.

Figure 18-4 shows the process by which a hydrogen atom with an electron in the 2*s* state changes to the 1*s* state and emits a photon. On the energy-level diagram in Fig. 18-3*c* this corresponds to the line extending from $n = 2$ to $n = 1$. This photon has the wave number corresponding to $\sigma_2 - \sigma_1 = -27,400 - (-109,700) = 82,300$ cm^{-1}. The value of the wave-

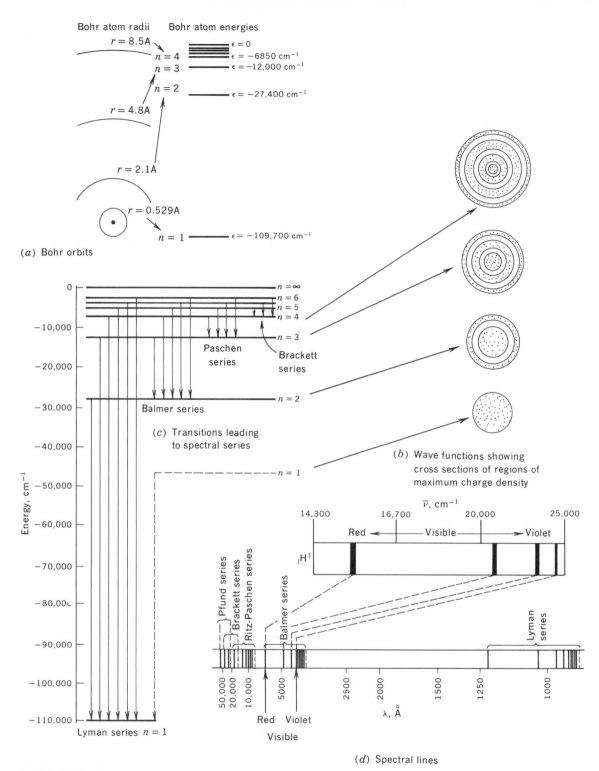

Fig. 18-3 The atomic spectrum of hydrogen for transitions between Bohr orbits. (After G. M. Barrow, "Physical Chemistry," 2d ed., McGraw-Hill Book Company, New York, 1963.)

length is the reciprocal of the wave number and is 1210 Å. This is the first member of the Lyman series, the series emitted in the ultraviolet by the hydrogen atom when an electron falls to the $n = 1$ level from the $n = 2$, 3, 4, 5, 6, and 7 energy levels, respectively. If, instead of falling into the $n = 1$, the electron falls only to the $n = 2$ level, the Balmer series of lines is emitted in the visible, as shown in Fig. 18-3c. Three series of lines have been observed in the infrared and named for their dis-coverers, the Paschen series, the Brackett series, and the Pfund series (not shown).

If a photon is absorbed by a hydrogen atom in the $1s$ state and the atom is raised to the $2s$ state, the process is like that pictured schematically in Fig. 18-5. If the absorbed photon has a value of the wave num-ber greater than 110,000 cm^{-1}, corresponding to energy greater than the change from $n = 1$ to $n = \infty$, the process is similar to that drawn in Fig. 18-6, where the electron is torn completely away, leaving a bare proton.

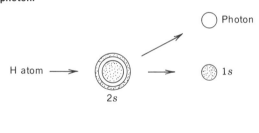

Fig. 18-4   A hydrogen atom changes from $2s$ to $1s$ and emits a photon.

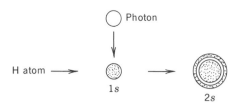

Fig. 18-5   A photon is absorbed by a hydrogen atom in state $1s$ which is raised to state $2s$.

Example 18-1

From the data in Fig. 18-3, calculate the wave-length of the line emitted when an electron falls from $n = 4$ to $n = 3$ in the hydrogen atom.

$$\sigma = \sigma_4 - \sigma_3 = -6850 - (-12,200)$$
$$= 5350 \text{ cm}^{-1}$$
$$\lambda = \frac{1}{\sigma} = \frac{1}{5350} = 1.869 \times 10^{-4} \text{ cm}$$
$$= 1.869 \ \mu = 18,690 \text{ Å}$$

Example 18-2

Calculate in kilocalories per mole the amount of energy needed to raise an electron in the hydrogen atom from the $2s$ to the $4s$ energy level.

$$\sigma = \sigma_4 - \sigma_2 = -6850 - (-27,400)$$
$$= 20,550 \text{ cm}^{-1}$$
$$E_a = hc\sigma = 6.626 \times 10^{-27} \times 2.998 \times 10^{10}$$
$$\times 2.055 \times 10^{4}$$
$$= 4.082 \times 10^{-12} \text{ erg atom}^{-1}$$
$$E_m = E_a N_A \times 10^{-7} \text{ joule erg}^{-1}$$
$$\times 0.239 \text{ cal joule}^{-1}$$
$$= 4.082 \times 10^{-12} \times 6.02 \times 10^{23} \times 10^{-7}$$
$$\times 0.239 = 58.7 \text{ kcal mole}^{-1}$$

18-3   THE ATOMIC SPECTRA OF MULTIELECTRON ATOMS

The calculation of wave functions for atoms with more than one electron is far more complicated than for the hydrogen atom. The interaction of electrons in-troduces a number of factors into the determination of energy levels which must be taken into account before the spectra of these atoms can be interpreted properly. Even in the case of helium, with only two electrons, the electronic interaction leads to compli-cations which make analytical solutions of the wave equations impossible. Among the approximation methods which have been tried, two of the more successful are the *perturbation method* and the *vari-ation method*.

The perturbation method

In order to express the mathematical relations in a concise way, the hamiltonian operator $H°$ is desig-nated by

Fig. 18-6 A photon is absorbed by a hydrogen atom in state
1s, the electron is ejected, and a bare nucleus (proton) remains.

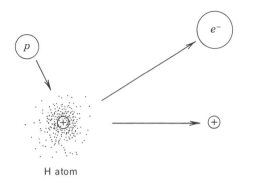

H atom

$$H^\circ = -\frac{\hbar}{2m}\nabla^2 + V \tag{18-1}$$

The generalized wave equation for the hydrogen atom
can then be put in the form

$$H^\circ\psi^\circ = \epsilon^\circ\psi^\circ \tag{18-2}$$

as explained in the previous chapter, where a number of the solutions were listed. For another atom
not very different from the hydrogen atom, the
operator can be written as a series

$$H = H^\circ + \lambda H' + \lambda^2 H'' + \cdots \tag{18-3}$$

where the perturbation constant λ approaches zero
as the new system becomes more and more like the
hydrogen atom. For example, the new atom may
have some additional terms in the potential-energy
expression.

In a similar way, perturbation corrections may be
applied to ψ and ϵ

$$\psi_k = \psi_k^\circ + \lambda\psi' + \lambda^2\psi'' + \cdots \tag{18-4}$$

$$\epsilon_k = \epsilon_k^\circ + \lambda\epsilon' + \lambda^2\epsilon'' + \cdots \tag{18-5}$$

where the values ψ_k° and ϵ_k° apply to the kth state of
the unperturbed system. Through this approach val-
ues of the energy levels can be found. Details of the
calculation are given in several of the texts cited in
the References. Values of the total energy in the
lowest perturbed level calculated by this method are
given in Table 18-1 for four atoms with two electrons.
The percentage disagreement between the observed

value and the value calculated by the perturbation
method diminishes with increasing nuclear charge.

The variation method

The last column of Table 18-1 gives values for the
energy levels calculated by the variation method.
Suppose that it is possible to construct a function as
a linear sum of the other functions ψ_n° which describe
simpler systems related to the system under consideration. For example, one can try to represent a
helium atom as the linear combination of two deuterium atoms. The general expression for this combination is

$$\psi = \sum_n a_n\psi_n^\circ \tag{18-6}$$

(The details of the calculation are found in several
texts listed in the References.) As can be seen from
Table 18-1, the variation method leads to values of
the total energy which are considerably closer to the
values observed for several atoms with two electrons.
The lower the energy, the better are the values calculated by this method.

The self-consistent-field model

In the perturbation method, it is necessary to add a
potential-energy term e^2/r_{12}, where the term in the
denominator is the distance between the two electrons
as they circulate. In the variation method, it is
necessary to modify the term Z/r in the wave equation. These considerations led the British physicist
D. Hartree to treat multielectron atoms as systems in
which a single electron moves independently in a
central field created by the other electrons and the
nucleus. This is called the *self-consistent-field* (SCF)
model of the atom. The assumption is first made
that the charge distribution of the nucleus and all the
electrons save one may be treated as spherically symmetrical, and the motion of the remaining electron in
this field is calculated. When this calculation has
been carried out for each electron in the system, the
patterns of motion so obtained are used as the basis
for a new charge distribution; through repetitions of
this process a final charge distribution is found which
remains unchanged by a further repetition of the
process and is therefore called a self-consistent field.

This approach is particularly suitable because the contributions of electrons in closed subshells lead to spherically symmetric fields. A closed subshell contains all the electrons having a given value of l with all the permitted values of m represented. For example, when $l = 1$, there are six such electrons, since there are electrons with positive and negative spin for each of the permitted values of m: $m = +1$, $m = 0, m = -1$. The subshell for $l = 1$ together with that for $l = 0$, containing two electrons, makes up the total L shell containing all the electrons for which $n = 2$. In the K shell, inside the L shell, there are only two electrons since $n = 1$. In the M shell, just outside the L shell, there are 18 electrons since $n = 3$.

Shell energy levels

As the nuclear charge increases, the attraction between the electron and the nucleus increases correspondingly; consequently, the energy required to remove an electron from the K shell nearest the nucleus becomes so great that it can be provided only by a photon where λ is so small (ν is so high) that the corresponding line for this photon lies in the x-ray region of the spectrum. This energy is called the *ionization energy* for the K shell. When an external electron falls into this level, a photon is emitted with energy corresponding to a frequency in the x-ray region of the spectrum. Thus, measurements of the frequency of x-ray spectra lead to values of the energy E for the levels in the K shell through the basic relationship

$$E = h\nu \tag{18-7}$$

Removal of an electron from the $2s$ level requires considerably less energy; this is called the *ionization potential* of the L_I shell; the energy necessary to remove a $2p$ electron is the ionization potential of L_{II}, etc. The ionization potentials for the L shell are less than those for the K shell because the electron in the K shell *screen* the L electrons from being attracted by the total charge on the nucleus. Using the variation method, the American theoretical physicist C. Zener was able to calculate a set of screening constants to express this effect. This calculation was extended by another American physicist, J. C. Slater. Table 18-2 lists the screening constants calculated by

TABLE 18-1   Total energy of some two-electron systems[a]

Species	Observed	Total energy, ev Calculated Perturbation method	Total energy, ev Calculated Variation method
H	13.6		
He	79.0	74.9	77.5
Li+	198.0	193.6	196.1
Be++	370.5	367	367
B3+	599.5	594	596

[a] From E. D. Kaufman, "Advanced Concepts in Physical Chemistry," p. 57. Copyright 1966. McGraw-Hill Book Company. Used by permission. The observed values are calculated from the results listed in G. Herzberg, "Atomic Spectra and Atomic Structure," p. 200, Dover Publications, Inc., New York, 1944.

Slater. It expresses the fraction of screening effect exerted by an electron of the type listed at the head of each column in shielding an electron of the type listed in the column at the left.

Electron spin

Shortly after Bohr proposed his theory of atomic structure, attempts were made to interpret the thousands of previous observations of spectral lines in terms of Bohr orbits. As the precision of spectroscopic measurements increased, many spectral lines which had been originally thought to have a single wavelength were found to consist of two or more lines very close together. In particular, there were many instances of two lines existing for a given set of quantum numbers (n, l, m). As stated earlier in Chap. 17, Uhlenbeck and Goudsmit, suggested that there might be an intrinsic angular momentum associated with each electron. According to this proposal, the electron has two components of angular momentum, one associated with the orbit of the electron, and the other with its intrinsic motion consisting of a spin. Solutions of the equation of motion show that the components of spin angular momentum are re-

TABLE 18-2 Slater screening constants[a]

Shielded electron	Shielding electron				
	1s	2s	2p	3s	3p
1s	0.35	0	0	0	0
2s	0.85	0.35	0.35	0	0
2p	0.85	0.35	0.35	0	0
3s	1.00	0.85	0.85	0.35	0.35
3p	1.00	0.85	0.85	0.35	0.35
4s	1.00	1.00	1.00	0.85	0.85

[a] From J. C. Slater, "Quantum Theory of Matter," p. 476. Copyright 1951. McGraw-Hill Book Company. Used by permission.

stricted to values of $\frac{1}{2}\hbar$ and $-\frac{1}{2}\hbar$ along any axis, so that $s = +\frac{1}{2}$ or $s = -\frac{1}{2}$.

As soon as the solutions for the Schrödinger equation were found that were applicable to the hydrogen atom, it was apparent that the wave equation for a spinning electron is

$$\phi = \psi(x,y,z)\sigma(s) \tag{18-8}$$

where ψ is a function of the space coordinates x, y, and z and σ depends only on the spin quantum number.

As soon as progress was made in defining wave functions for multielectron atoms, more and more evidence accumulated that in nature only states represented by antisymmetric wave functions exist. From this, Pauli deduced the *exclusion principle*, discussed in the previous chapter. Thus, the study of atomic spectra led to the discovery of the exclusion principle, which in turn explains so beautifully the structure of the periodic table of chemical elements.

Angular momenta

In the spectrum of atomic sodium, there is a strong line in the yellow region of the visible which on close examination can be seen to have two intensity maxima at 5890 and 5896 Å. This splitting is due to the interaction of the angular momentum of electron spin with the orbital angular momentum and shows that electrons are not independent of each other to the

degree implied by the SCF method. They interact with each other and also with externally applied magnetic or electric fields.

The new quantum mechanics based on the extension of the Schrödinger equation leads to the conclusion that the orbital angular momentum I_1 for a single electron is quantized according to the expression

$$I_1 = \hbar\sqrt{l(l+1)} \tag{18-9}$$

The resultant total angular momentum I for an atom with several electrons is given in terms of a total quantum number L

$$I = \hbar\sqrt{L(L+1)} \tag{18-10}$$

The angular momenta of individual electrons add together vectorially. This process is indicated schematically in Fig. 18-7.

In a way analogous to a single-electron system, the values of L for a two-electron system range from $l_1 + l_2$ to the absolute value $|l_1 - l_2|$. Since l_1 and l_2 have values which are either positive integrals or zero the values of L are integral or zero. An example of the values of L for a system containing a p electron and a d electron are shown in Table 18-3.

By convention, the term symbol for a multielectron atom is S, when $L = 0$, just as the hydrogen atom is said to be in the s state when $l = 0$. When $L = 1$, the term symbol is P; when $L = 2$, the term symbol is D, etc. In the S level, each electron contributes either $+\frac{1}{2}$ or $-\frac{1}{2}$ to the total spin quantum number; the smallest possible value of S depends on whether the total number of electrons N is even or odd. The total spin angular momentum I_s is given by

$$I_s = \hbar\sqrt{S(S+1)} \tag{18-11}$$

The splitting or multiplicity M_s of a line is given by

$$M_s = 2S + 1 \tag{18-12}$$

When, in a case like that of Na, $S = \frac{1}{2}$, Eq. (18-12) becomes

$$M_s = 2S + 1 = 2(\tfrac{1}{2}) + 1 = 2 \tag{18-13}$$

and the line is a doublet.

Example 18-3

Calculate the total electron angular momentum in an atom in the $2P$ state.

Fig. 18-7   The addition of the angular momenta l_1 and l_2 of two electrons (1 and 2) to form the several possible states of L.   (After H. E. White, "Introduction to Atomic Spectra," McGraw-Hill Book Company, New York, 1934.)

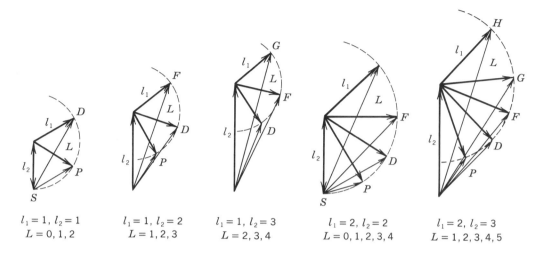

| $l_1 = 1, l_2 = 1$ | $l_1 = 1, l_2 = 2$ | $l_1 = 1, l_2 = 3$ | $l_1 = 2, l_2 = 2$ | $l_1 = 2, l_2 = 3$ |
| $L = 0, 1, 2$ | $L = 1, 2, 3$ | $L = 2, 3, 4$ | $L = 0, 1, 2, 3, 4$ | $L = 1, 2, 3, 4, 5$ |

In the P state, $L = 1$.   From Eq. (18-10)

$$I = \hbar\sqrt{L(L + 1)}$$
$$= 1.055 \times 10^{-27} \text{ erg sec } \sqrt{1(1 + 1)}$$
$$= 1.055 \times 10^{-27} \sqrt{2} \text{ erg sec}$$
$$= 1.491 \times 10^{-27} \text{ erg sec}$$

Example 18-4

Calculate the multiplicity of a state where $S = \frac{3}{2}$.

From Eq. (18-13), $M_s = 2S + 1 = 2(\frac{3}{2}) + 1 = 4$.

TABLE 18-3   Values of the total angular momentum L in terms of the individual angular momenta l_1 and l_2 for a two-electron system

General case	When $l_1 = 1$ and $l_2 = 2$
$l_1 + l_2$	$1 + 2 = 3$
$l_1 + l_2 - 1$	$1 + 2 - 1 = 2$
$l_1 + l_2 - 2$	$1 + 2 - 2 = 1$
. . . .
$|l_1 - l_2|$	$1 - 2 = 1$

Total angular momentum

In 1925 two American astrophysicists, H. N. Russell and F. A. Saunders, proposed a vector relationship to explain the way in which the angular momenta of electrons interact.   This principle is called Russell-Saunders coupling.   Just as the individual spin vectors s_i combine to form a resultant spin S and the individual orbital angular momenta l_i couple to form a resultant total orbital angular momentum L, these two quantities couple magnetically to form the total angular momentum J, which is called the *inner quantum number*.   The quantized value of the total angular momentum I_{tot} is given by

$$I_{\text{tot}} = \hbar\sqrt{J(J + 1)} \tag{18-14}$$

The values of J run through the integers from $L + S$ to the absolute value of $L - S$, thus:

$$J = L + S, L + S - 1, L + S - 2, \ldots, |L - S| \tag{18-15}$$

The designations of the energy levels (term values) are given by symbols of the form $^M L_J$, where M is the multiplicity $2S + 1$, J is the inner quantum number, and the value of L is denoted by letters instead of numbers; for $L = 0, 1, 2, 3, \ldots$, the letter designations are $S, P, D, F, \ldots$. The use of S in this connection must not be confused with the other use of this letter to denote the total spin momentum. Thus, the symbol $^2S_{1/2}$ denotes a state with the total angular momentum $S = \frac{1}{2}$, where $M = 2S + 1 = 2$; since the center symbol is S, $L = 0$; and according to Eq. (18-15), $J = L + S = 0 + \frac{1}{2} = \frac{1}{2}$. The upper left superscript is normally read *singlet* when the value is unity, *doublet* when the value is 2, *triplet* when the value is 3, etc. When L is greater than S, the number of possible values of J is equal to the multiplicity M.

Transition rules

An electron can change from one state to another only if the states are related in such a way that the value of the transition-moment integral T_M is not zero

$$T_M = \int \psi_{\text{init}}^* T \psi_{\text{final}} \neq 0 \qquad (18\text{-}16)$$

where ψ_{init}^* = complex conjugate of wave function for initial state

ψ_{final} = wave function for final state

T = transition-moment operator

This relation leads to the selection rules which say that for a transition to take place the change in J must be either zero or ± 1 (excluding the case where initially $J = 0$ and in the final state also $J = 0$); the change in L must be ± 1; and the change in S must be zero. As the value of the nuclear charge increases, weak lines are found for transitions where there is a change in the value of S. Figure 18-8 shows a number of the permitted transitions; in Fig. 18-9 the changes between energy levels are indicated, together with the pattern of spectral lines which results.

A relation pointed out by the German physicist F. Hund, now known as Hund's rule, states that when there are terms resulting from equivalent electrons, those with the highest multiplicity have the lowest values of energy and of these the lowest is that with the largest value of L. Two possibilities for the distribution of the $2p$ electrons in nitrogen are shown in Table 18-4, together with the actual distribution found in accordance with Hund's rule. In Table 18-5 the values of S, L, and J are shown together with the corresponding term value.

Lists of the term values are given in Tables 18-6

to 18-8 for three atoms of the greatest importance in biochemical molecules. Diagrams showing the origin of the line spectra of these atoms are given in Fig. 18-10.

When atoms are placed in a relatively strong magnetic field, the projection of the total angular momentum in the field direction must be an integral multiple of $h/2\pi$ and there is a splitting of the lines, discovered by the Dutch physicist P. Zeeman in 1896 and known as the *Zeeman effect*. In 1913, the German physicist J. Stark found that every line of the Balmer series of hydrogen when excited in a strong electric field of at least 10^5 volts cm^{-1} is split into a number of components. This is called the *Stark effect*.

Example 18-5

Calculate the energy in kilocalories per mole of a photon emitted by a Li atom for the change 4^2P to 2^2S.

From Fig. 18-10, the wavelength of the photon emitted in this change is 2741 Å.

$$E_a = h\nu = \frac{hc}{\lambda} \text{ erg atom}^{-1}$$

$$= \frac{6.626 \times 10^{-27} \times 2.998 \times 10^{10}}{2741 \times 10^{-8}}$$

$$= 7.247 \times 10^{-12} \text{ erg atom}^{-1}$$

$$E = E_a N_A = 7.247 \times 10^{-12} \times 6.02 \times 10^{23}$$
$$= 4.36 \times 10^{+12} \text{ erg mole}^{-1}$$

$$E_{kc} = 4.36 \times 10^{+12} \text{ erg mole}^{-1}$$
$$\times 2.39 \times 10^{-11} \text{ kcal erg}^{-1}$$
$$= 104.2 \text{ kcal mole}^{-1}$$

Example 18-6

A certain change in the atomic structure of a neutral oxygen atom requires 51.2 kcal mole^{-1}. From Fig. 18-10, identify the spectral line.

$$\lambda = \frac{hc}{E} 6.02 \times 10^{23} \text{ atoms mole}^{-1} \times 2.39$$

$$\times 10^{-11} \text{ kcal erg}^{-1}$$

$$= \frac{6.626 \times 10^{-27} \times 2.998 \times 10^{10}}{51.2}$$

$$\times 6.02 \times 10^{23} \times 2.39 \times 10^{-11}$$

$$= 5577$$

This is the auroral line $^1S_0 \rightarrow {}^1D_2$.

Fig. 18-8   Transitions allowed by selection rules. (After E. D. Kaufman, "Advanced Concepts in Physical Chemistry," McGraw-Hill Book Company, New York, 1966.)

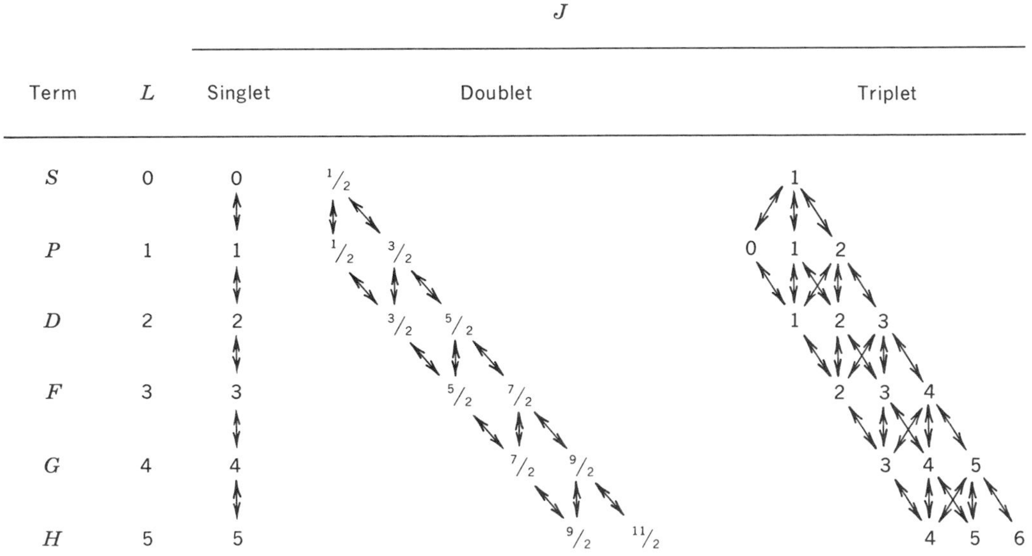

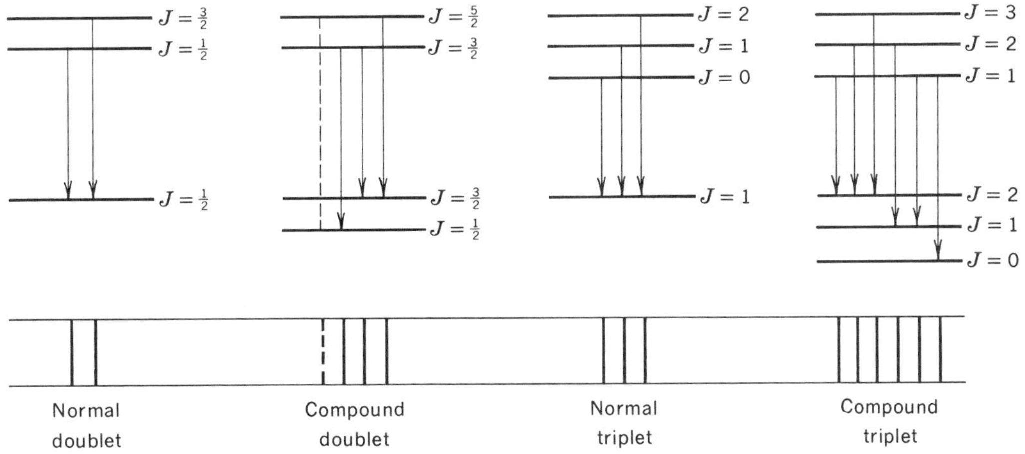

Fig. 18-9   The relations between transitions and the associated spectra. (After E. D. Kaufman, "Advanced Concepts in Physical Chemistry," McGraw-Hill Book Company, New York, 1966.)

18-4   CHEMICAL APPLICATIONS OF ATOMIC SPECTRA

The spectrograph was first developed into a practical laboratory instrument in the middle of the nineteenth century and was used almost immediately in the dis-covery of many new chemical elements. Among these were cesium and rubidium, identified by two Germans, R. W. Bunsen and G. R. Kirchhoff; the ele-ments helium, gallium, indium, and thallium were also discovered and identified by means of their atomic spectra during the next few decades.

TABLE 18-4 An application of Hund's rule to the electron structure of nitrogen showing two possible distributions

Structure not observed		Structure observed	
Level	Spin	Level	Spin
$1s$	$+\frac{1}{2}$	$1s$	$+\frac{1}{2}$
$1s$	$-\frac{1}{2}$	$1s$	$-\frac{1}{2}$
$2s$	$+\frac{1}{2}$	$2s$	$+\frac{1}{2}$
$2s$	$-\frac{1}{2}$	$2s$	$-\frac{1}{2}$
$2p_x$	$+\frac{1}{2}$	$2p_x$	$+\frac{1}{2}$
$2p_x$	$-\frac{1}{2}$	$2p_y$	$+\frac{1}{2}$
$2p_y$	$+\frac{1}{2}$	$2p_z$	$+\frac{1}{2}$

TABLE 18-5 Term values for carbon

$l_1 = 1, l_2 = 1, s_1 = \pm\frac{1}{2}, s_2 = \pm\frac{1}{2}$

$L = 2, 1, 0; S = 1, 0$

	L	S	J
$^1S_0$	0	0	0
$^1P_1$	1	0	1
$^1D_2$	2	0	2
$^3S_1$	0	1	1
$^3P_0$	1	1	0
$^3P_1$	1	1	1
$^3P_2$	1	1	2
$^3D_1$	2	1	1
$^3D_2$	2	1	2
$^3D_3$	2	1	3

The French philosopher Auguste Comte once remarked that "there are some things of which the human race must remain forever in ignorance, for example, the chemical constitution of the heavenly bodies." But by studying the atomic spectra in the light from the stars astronomers have been able to compile an extensive body of knowledge about the chemical elements present in the stars and their energy states. Sixty-six different chemical elements have been found spectroscopically in the sun, many of them in higher energy levels not ordinarily attainable in laboratories here on earth.

In biology and medicine, studies of atomic spectra have made possible the determination of minute traces of metallic elements in biochemical substances, cells, and tissues. This has been of special importance in the investigation of toxicological problems, e.g., determining the accumulation of lead and copper in the blood and tissues of persons exposed to these elements in their work. Spectrographic studies have been useful in archeology; the unusual color of gold in certain ancient Egyptian art objects has been found to be due to impurities. The ability to detect such small amounts of metals from atomic spectra also has been useful in connection with police investigations.

A knowledge of the energy values associated with different energy levels is particularly helpful in the study of activated states in chemical kinetics. As an example, D. J. LeRoy and E. W. R. Steacie (see References) investigated the reaction of excited mercury with ethylene to produce acetylene and hydrogen according to the reaction

$$Hg^*(^3P_1) + C_2H_4 \rightarrow C_2H_4^*(triplet) + Hg(^1S_0) \quad (18\text{-}17)$$

The investigators proposed the mechanism

$$Hg^*(^3P_1) + C_2H_4 \rightarrow C_2H_4^*(triplet) + Hg(^1S_0) \quad (18\text{-}18)$$

$$C_2H_4^* + C_2H_4 \rightarrow 2C_2H_4 \quad (18\text{-}19)$$

$$C_2H_4^*(triplet) \rightarrow C_2H_2^*(triplet) + H_2(^1\Sigma_g{}^+) \quad (18\text{-}20)$$

As will be discussed in more detail in the chapter on molecular spectra, the hydrogen molecule in the above equation has a molecular-orbital angular momentum resultant of zero and a spin-momentum resultant of zero. Previous knowledge of the atomic spectra of mercury permits a calculation of the resultant energy when the state of the mercury changes, as shown in Eq. (18-18),

$$Hg^*(^3P_1) \rightarrow Hg(^1S_0) \qquad \Delta E = -112.2 \text{ kcal mole}^{-1} \quad (18\text{-}21)$$

Previous measurements also give data for the equation

$$C_2H_4 \rightarrow C_2H_2 + H_2 \qquad \Delta E = +85 \text{ kcal mole}^{-1} \quad (18\text{-}22)$$

TABLE 18-6   Spectral terms in neutral carbon[a,b] atoms[c]

$1s^22s^22p^2$; $^3P_0$; first ionization potential $= 11.217$ volts

Configuration	Symbol	J	Term value, cm^{-1}	$\Delta\nu$
$2s^22p^2$	^3P	0	90,878.3	
		1	90,863.5	14.8
		2	90,836	27.5
$2s^22p^2$	^1D	2	80,686	
$2s^22p^2$	^1S	0	69,231	
$2s^22p3s$	$^3P^0$	0	30,547.0	
		1	30,527.0	20.0
		2	30,486.9	40.1
$2s^22p3s$	$^1P^0$	1	28,898	
$2s2p^3$	$^3D^0$	2	26,792	
$2s^22p3p$	^1D	2	22,780	
$2s^22p3p$	^3D	1	21,190.2	
		2	21,169.0	21.2
		3	21,135.3	33.7
$2s^22p3p$	^3S	1	20,139	
$2s2p^3$	$^3P^0$	1, 2	15,626	

[a] S. B. Ingram, *Phys. Rev.,* **34**:421 (1929).
[b] F. Paschen and G. Kruger, *Ann. Physik,* **7**:1 (1930).
[c] From R. Bacher and S. Goudsmit, "Atomic Energy States." Copyright 1932. McGraw-Hill Book Company. Used by permission. Notation: Under configuration, $2s^22p^3$ indicates 2 electrons in the $2s$ state and 3 electrons in the $2p$ state; $^3D^0$ indicates a total angular momentum with $L = 3$, a multiplicity of 3, and an odd state. When the sum of all the l's is odd, this is indicated by a small zero as an upper right superscript. $\Delta\nu$ gives the difference between related term values in cm^{-1}.

where all the molecules are in the ground state. Assuming that the energy from the excited mercury molecule is transferred to the ethylene molecule, it is possible to calculate an upper limit for the excitation energy of the acetylene triplet, which turns out to be 27.2 kcal mole^{-1}. Thus, a knowledge of atomic spectra contributes to the knowledge of molecular spectra.

X-ray spectra

When electrons impinge on a metal target after falling through a potential of even a few kilovolts, their energy is frequently high enough to cause the ejection of electrons from some of the energy levels in the atoms making up the metal target. The slowing down of the original beam of electrons by the impact on the target produces a background radiation called *bremsstrahlung* or braking radiation. After electrons are ejected from the bombarded metal atoms, other electrons in these atoms drop down into the energy levels which have been left vacant. As a result of this process, the atom emits radiation in the x-ray region of the spectrum. The name x-ray was given to this radiation because at the time of its discovery the atomic mechanism for its production was not understood.

As discussed earlier in this chapter, the energy of the photon emitted depends on the difference of the energy values of the levels between which the electron makes the transition. These term values, in turn, depend upon the electron structure of the

Fig. 18-10 Energy levels in neutral atoms and wavelengths of observed lines resulting from transitions between them. (From H. E. White, "Introduction to Atomic Spectra," McGraw-Hill Book Company, New York, 1934.)

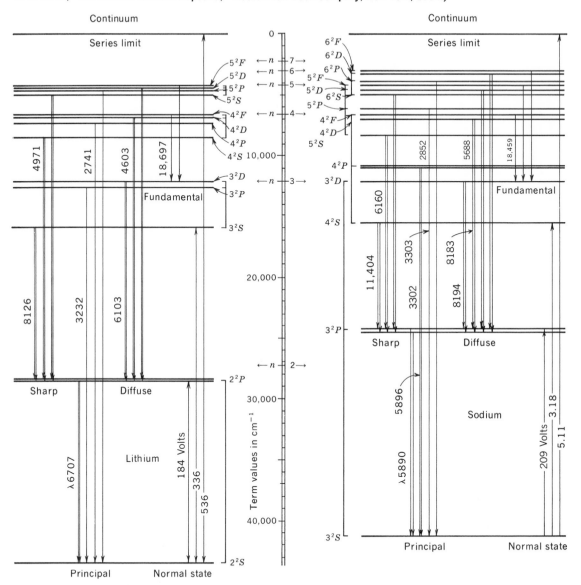

Fig. 18-10   (Cont.)

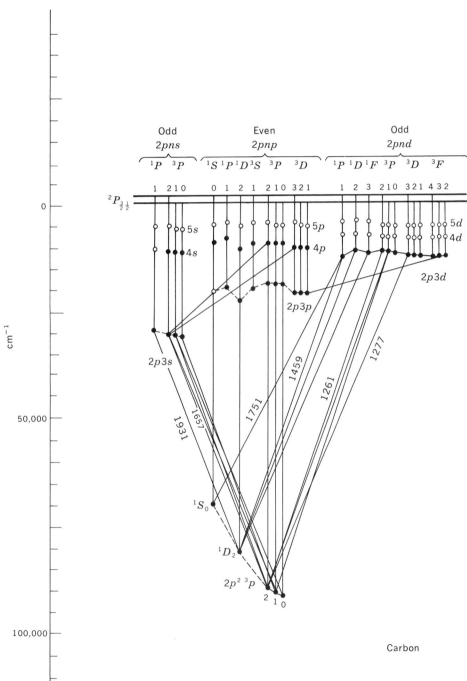

Carbon

Fig. 18-10 **(Cont.)**

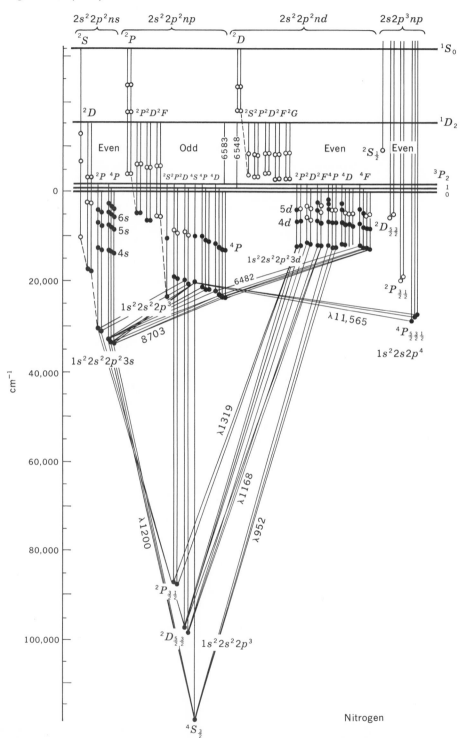

Nitrogen

Fig. 18-10   (Cont.)

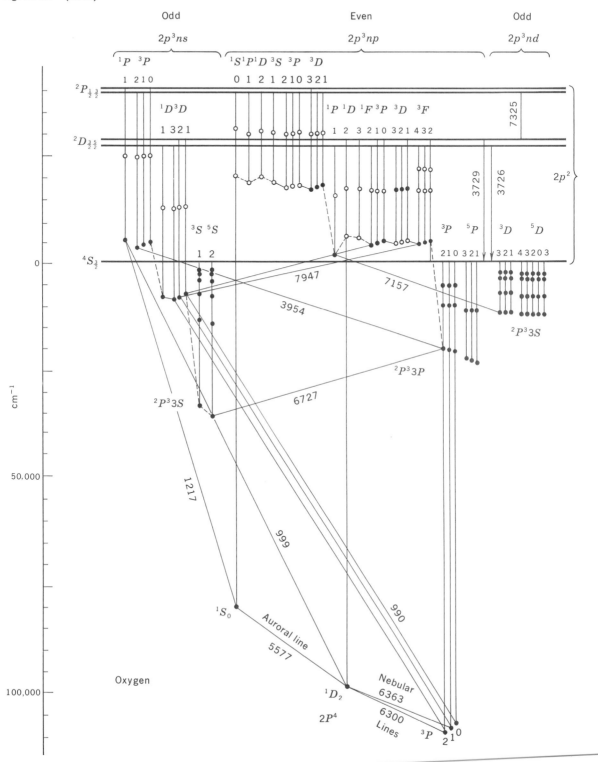

Oxygen

TABLE 18-7 Spectral terms in neutral nitrogen[a,b] atoms[c]

$1s^2 2s^2 2p^3$; $^4S_{1\frac{1}{2}}$; first ionization potential = 14.48 volts

Configuration	Symbol	J	Term value, cm^{-1}	$\Delta\nu$
$2s^2 2p^3$	$^4S^0$	$1\frac{1}{2}$	117,345	
$2p^3$	$^2D^0$	$1\frac{1}{2}, 2\frac{1}{2}$	98,143	
$2p^3$	$^2P^0$	$\frac{1}{2}, 1\frac{1}{2}$	88,537	
$2p^2(^3P)3s$	^4P	$\frac{1}{2}$	34,059.5	
		$1\frac{1}{2}$	34,025.7	33.8
		$2\frac{1}{2}$	33,979.0	46.7
$(^3P)3s$	^2P	$\frac{1}{2}$	31,239.1	
		$1\frac{1}{2}$	31,156.0	83.1
$2s2p^4$	^4P	$2\frac{1}{2}$	29,235.5	
		$1\frac{1}{2}$	29,191.6	−43.9
		$\frac{1}{2}$	29,172.0	−19.6
$2s^2 2p^2(^3P)3p$	$^2S^0$	$\frac{1}{2}$	23,794.7	
$(^3P)3p$	$^4D^0$	$\frac{1}{2}$	22,572.8	
		$1\frac{1}{2}$	22,550.2	22.6
		$2\frac{1}{2}$	22,512.9	37.3
		$3\frac{1}{2}$	22,461.9	51.9

[a] S. B. Ingram, *Phys. Rev.*, **34**:421 (1929).
[b] J. J. Hopfield, *Phys. Rev.*, **36**:789 (1930).
[c] From R. Bacher and S. Goudsmit, "Atomic Energy States." Copyright 1932.
McGraw-Hill Book Company. Used by permission. See Table 18-6 for notation.

TABLE 18-8 Spectral terms in neutral oxygen[a] atoms[b]

$1s^2 2s^2 2p^4$; $^3P_2$; first ionization potential = 13.550 volts

Configuration	Symbol	J	Term value, cm^{-1}	$\Delta\nu$
$2s^2 2p^4$	^3P	2	109,837.3	
		1	109,679.17	−158.13
		0	109,610.52	−68.65
$2p^4$	^1D	2	109,610.52	
$2p^4$	^1S	0	93,969.5	
$2p^3(^4S)3s$	$^5S^0$	2	76,044.5	
$2p^3(^4S)3s$	$^3S^0$	1	36,069.0	
$2p^3(^4S)3s$	^5P	1	33,043.3	
		2	23,211.9	
		3	23,209.2	2.7
$2p^3(^4S)3p$	^3P	0, 1	23,205.8	3.4
		2	21,207.7	
$2p^3(^4S)4s$	$^5S^0$	2	21,207.2	
$2p^3(^4S)4s$	$^3S^0$	1	14,358.5	
			13,612.5	

[a] J. J. Hopfield, *Phys. Rev.*, **37**:160 (1931).
[b] From R. Bacher and S. Goudsmit, "Atomic Energy States." Copyright 1932.
McGraw-Hill Book Company. Used by permission. See Table 18-6 for notation.

TABLE 18-9   Potential difference in kilovolts across the x-ray tube required to excite spectral lines characteristic of various elements[a]

Element	K	L	M	N
92U	115	21.7	5.54	1.44
90Th	109	20.5	5.17	1.33
83Bi	90.1	16.4	4.01	0.96
82Pb	87.6	15.8	3.85	0.89
81Tl	85.2	15.3	3.71	0.86
80Hg	82.9	14.8	3.57	0.82
79Au	80.5	14.4	3.43	0.79
78Pt	78.1	13.9	3.30	0.71
77Ir	76.0	13.4	3.17	0.67
76Os	73.8	13.0	3.05	0.64
74W	69.3	12.1	2.81	0.59
73Ta	67.4	11.7	2.71	0.57
72Hf	65.4	11.3	2.60	0.54
71Lu	63.4	10.9	2.50	0.51
70Yb	61.4	10.5	2.41	0.50
69Tm	59.5	10.1	2.31	0.47
68Er	57.5	9.73	2.22	0.45
67Ho	55.8	9.38	2.13	0.43
66Dy	53.8	9.03	2.04	0.42
65Tb	52.0	8.70	1.96	0.40
64Gd	50.3	8.37	1.88	0.38
63Eu	48.6	8.04	1.80	0.36
62Sm	46.8	7.73	1.72	0.35
60Nd	43.6	7.12	1.58	0.32
59Pr	41.9	6.83	1.51	0.30
58Ce	40.3	6.54	1.43	0.29
57La	38.7	6.26	1.36	0.27
56Ba	37.4	5.99	1.29	0.25
55Cs	35.9	5.71	1.21	0.23
53I	33.2	5.18	1.08	0.19
52Te	31.8	4.93	1.01	0.17
51Sb	30.4	4.69	0.94	0.15
50Sn	29.1	4.49	0.88	0.13
49In	27.9	4.28	0.83	0.12
48Cd	26.7	4.07	0.77	0.11
47Ag	25.5	3.79	0.72	0.10

TABLE 18-9   (Cont.)

Element	K	L	M	N
46Pd	24.4	3.64	0.67	0.08
45Rh	23.2	3.43	0.62	0.07
44Ru	22.1	3.24	0.59	0.06
42Mo	20.0	2.87	0.51	0.06
41Nb	19.0	2.68	0.48	0.05
40Zr	18.0	2.51	0.43	0.05
39Yt	17.0	2.36		
38Sr	16.1	2.19		
37Rb	15.2	2.05		
35Br	13.5	1.77		
34Se	12.7	1.64		
33As	11.9	1.52		
32Ge	11.1	1.41		
31Ga	10.4	1.31		
30Zn	9.65	1.20		
29Cu	8.86			
28Ni	8.29			
27Co	7.71			
26Fe	7.10			
25Mn	6.54			
24Cr	5.98			
23Va	5.45			
22Ti	4.95			
21Sc	4.49			
20Ca	4.03			
19K	3.59			
17Cl	2.82			
16S	2.46			
15P	2.14			
14Si	1.83			
13Al	1.55			
12Mg	1.30			
11Na	1.07			

[a] M. Siegbahn, "Spektroskopie der Röntgenstrahlen," Julius Springer, Berlin, 1924.

atoms in the metal so that the principal lines in the x-ray spectrum of a metal are characteristic of that metal. In Fig. 18-11 the characteristic x-ray spectrum of tungsten is superimposed on the "white" spectrum of the bremsstrahlung. For each metal, specific lines are found associated with the shells of energy levels; these lines are labeled K, L, M, and N according to the shell into which the electron falls. The potential difference in kilovolts required to excite the spectral lines is given for a number of the elements in Table 18-9. The use of beams of x-rays in the elucidation of crystal structure will be discussed in Chap. 22.

One of the most important applications of x-rays is in the identification of chemical elements. The British physicist H. G. J. Moseley made the first systematic study of the K lines of the x-ray spectra of a number of the chemical elements between Ca and

Fig. 18-11 Characteristic x-ray spectrum of tungsten super-imposed on "white" spectrum. (After W. P. Davey, "A Study of Crystal Structure and Its Applications," McGraw-Hill Book Company, New York, 1934.)

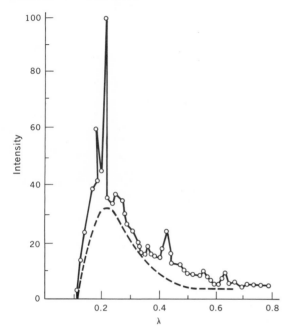

Zn; some of his results are plotted in Fig. 18-12. When the square root of the frequency of one of these lines divided by the spectroscopic constant R (the Rydberg constant) is plotted against the charge on the atomic nucleus, a graph is obtained which is almost a straight line (Fig. 18-13). Because of this linear relationship, the knowledge of the frequency of the x-ray line in an unknown element immediately provides a measure of the charge on the nucleus and thus identifies the place of the element in the periodic table.

Example 18-7

Calculate the minimum energy in kilocalories per mole required to excite the K line in the x-ray spectrum of Cu.

From Table 18-9 the minimum potential through which the electron must fall to excite this line is 8.86 kv.

$$E = 8.86 \times 10^3 \text{ ev} \times 23.06 \text{ kcal mole}^{-1} \text{ ev}^{-1}$$
$$= 2.04 \times 10^5 \text{ kcal mole}^{-1}$$

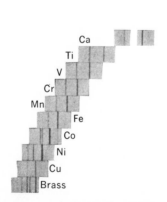

Fig. 18-12 The K radiation spectra of several elements. [From H. G. J. Moseley, *Phil. Mag.*, 26:1024 (1913); 27:703 (1914).]

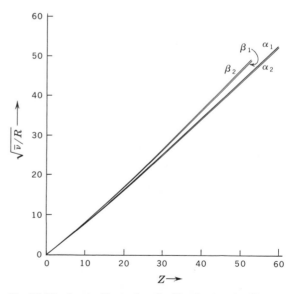

Fig. 18-13 Graphs illustrating the Moseley law for K series x-ray lines. α_1, α_2, β_1, and β_2 denote the transitions which produce the lines. (After H. E. White, "Introduction to Atomic Spectra," p. 302, McGraw-Hill Book Company, New York, 1934.)

Example 18-8

What x-ray line would provide a photon with exactly the right amount of energy to raise an atom to the excited state corresponding to 2306 kcal mole^{-1}?

$\mathcal{E} = E(23.06$ kcal mole^{-1} ev$^{-1})^{-1} = 100$ ev
$= 0.10$ kv

According to the data in Table 18-9, this corresponds to the x-ray line in the N series for Ag.

Problems

1. Calculate the frequency and the wavelength for the line in the Balmer series where the transition is from the $n = 4$ to the $n = 2$ shell. What is the approximate color of this line?
2. Calculate the wave number of the line in the Brackett series for which $n_1 = 5$.
3. As shown in Fig. 18-2, there is a line in the arc spectrum of magnesium with a wavelength of 5168 Å. Calculate the wave number of this line and the amount of energy in a photon corresponding to this line where energy is expressed in electron volts.
4. Calculate the wavelength of the light emitted when an electron falls from an $n = 6$ energy level to the energy level where $n = 2$ in the hydrogen atom, as shown in Fig. 18-7. What is the approximate color of the light?
5. The hydrogen atoms in DNA absorb infrared radiation quite strongly in the region between 2900 and 3200 cm^{-1}. Identify a line in the hydrogen atom spectral series (Table 18-1) which would be absorbed directly by the hydrogen atoms in DNA in this region of the infrared spectrum.
6. The sugar component of DNA, ribose, has a strong absorption band centering around the wave number 1080 cm^{-1}. This motion involves a bending of a ring of atoms constituting the sugar. Identify at least one line in the emission spectrum of hydrogen (Fig. 18-7 and Table 18-1) which would be absorbed by ribose at this deep absorption band, assuming absorption if the line is within 20 cm^{-1} of the band.
7. What is the approximate color of the light emitted when a sodium atom changes from the 3^2P state to the 3^2S state (see Fig. 18-10)?
8. Calculate the approximate frequency of the line when the lithium atom changes from the 4^2D state to the 2^2P state (see Fig. 18-10).

9. Calculate the energy difference in calories per mole between 1 g mole of neutral carbon atoms in the $2S^2 2p3p$ state and the $2S2p^3$ state (see Table 18-6).
10. Calculate the energy difference in electron volts between a neutral nitrogen atom in the $^4D^0$ ($J = \frac{1}{2}$) state and the $^2D^0$ state.
11. Calculate the energy difference in electron volts between a neutral oxygen atom in the $^3S^0$ state and the $^3P(0,1)$ state. (*Note:* Data in Table 18-8.)
12. Using the data shown in Table 18-9, calculate the approximate frequency of the K x-ray line for calcium.
13. Calculate the wavelength of the K line for zirconium using the data in Table 18-9.
14. Calculate the wavelengths for all four (K, L, M, and N) of the x-ray lines for Sn.
15. In the single-line spectrum of magnesium (Fig. 18-1) the wavelength of the line is 4571 Å. If a photon corresponding to this line were absorbed by a carbon atom in the lowest ground state ($2s^2 2p^3$), the energy would be almost sufficient to raise the carbon atom to one of its excited states. What is that state?
16. From the data in Table 18-3, calculate the wavelength in angstrom units of light where a photon has energy just sufficient to ionize a neutral carbon atom.
17. Calculate the ionization potential for carbon in electron volts.
18. Calculate the wave number of the line in the Lyman series for which $n_1 = 2$.
19. What line in the Lyman series will have photons with energy sufficient to ionize a neutral carbon atom?
20. Using the values shown in Fig. 18-10, calculate the wavelength of the light emitted when an electron in lithium drops from the level 5^2P to the level 2^2S.
21. From the data in Fig. 18-13, estimate the value of $\bar{\nu}$ for the K line for element 34 and also calculate the wavelength for this line.

REFERENCES

Russell, H. N., and F. A. Saunders: The Coupling of the Momenta of Electrons, *Astrophys. J.,* **61**:38 (1925).

Pauling, L., and S. Goudsmit: "Structure of Line Spectra," McGraw-Hill Book Company, New York, 1930.

Bacher, R., and S. Goudsmit: "Atomic Energy States," McGraw-Hill Book Company, New York, 1932.

White, H. E.: "Introduction to Atomic Spectra," McGraw-Hill Book Company, New York, 1934.

Herzberg, G.: "Atomic Spectra and Atomic Structure," trans. by J. W. T. Spinks, Prentice-Hall, Inc., New York, 1937.

LeRoy, D. J., and E. W. R. Steacie: The Reaction of Excited Mercury Atoms with Ethylene, *J. Chem. Phys.*, **9:**829 (1941).

Condon, E. U., and G. H. Shortley: "The Theory of Atomic Spectra," Cambridge University Press, New York, 1951.

Laidler, K. J.: "The Chemical Kinetics of Excited States," Oxford University Press, Fair Lawn, N.J., 1955.

Sandorfy, C.: "Electronic Spectra and Quantum Chemistry," Prentice-Hall, Inc., Englewood Cliffs, N.J., 1964.

Kaufman, E. D.: "Advanced Concepts in Physical Chemistry," pp. 50–91, McGraw-Hill Book Company, New York, 1966.

Hindmarsh, W. R.: "Atomic Spectra, with Selections from Early Papers in Spectroscopy," Pergamon Press, New York, 1967.

The *molecule* represents one of the most basic concepts in chemical science, and it may be helpful to recall precisely what this word means. A *molecule* is a group of atoms so tightly bound together that the atoms stay together as a unit during changes of physical state. For example, the organic chemical compound *methane* consists of a carbon atom sur-rounded by four hydrogen atoms symmetrically placed at the points of a regular tetrahedron. These hydrogen atoms are so tightly bound to the central carbon atom that they remain with it when a crystal of methane melts and becomes a liquid at $-184°C$ and when the liquid boils at 1 atm and turns to vapor at $-161.5°C$.

Chapter Nineteen

molecular structure

The significance of the molecule as a unit in the kinetic behavior of matter was shown by the relations among the properties of ideal gases. As first made clear by John Dalton in 1808, at a fixed volume and temperature, the pressure of every kind of ideal gas depends only on the number of molecules present:

$$PV = nRT \tag{19-1}$$

The law for the ideal gas in *mole* form also can be put in *molecular* form since the number of molecules N present in volume V is equal to the number of moles n multiplied by Avogadro's number N_A; a gas constant per molecule r can be defined as equal to R/N_A; Eq. (19-1) can then be written

$$PV = NrT \tag{19-2}$$

There are analogous relations for solutes in ideal solutions. For example, the osmotic pressure is given by the relation which closely resembles the ideal-gas law

$$\pi V = nRT \tag{19-3a}$$

and also can be written

$$\pi V = NrT \tag{19-3b}$$

As another example, the concept of *mole fraction* implies the stability of molecules particularly in the observations of colligative properties which lead to Raoult's law

$$P = P^\circ X = P^\circ \frac{n_1}{n_{\text{tot}}} = P^\circ \frac{N_1}{N_{\text{tot}}} \qquad (19\text{-}4)$$

As the significance of the concept of molecule became so firmly established in these and other ways, the nature of the forces holding a group of atoms together in the form of a molecule became a more and more important question. As early as 1812, the Swedish chemist J. J. Berzelius suggested that all chemical combinations were caused by electrostatic attractions. This rather qualitative idea was refined and made more quantitative as the concept of a unit of electricity, the electron, emerged and was experimentally established; and the belief gained ground that electrons were responsible for chemical binding. When in 1913 the work of Moseley, discussed in the previous chapter, made possible the calculation of the number of electric charges on the nucleus of each chemical element, the number of electrons associated with each kind of atom in the neutral state could also be found and this made possible explanations of molecular binding in terms of electrons exchanged or shared between atoms.

In 1916, W. Kossel, a German physicist, suggested the formation of stable ions by the tendency of atoms to gain or lose electrons until a rare-gas configuration was formed. This explained the association of NaCl as a molecule in the gaseous state and paved the way to understanding the association of positive and negative ions in the solid state as crystals.

In the same year, the American physical chemist G. N. Lewis suggested that the binding in nonpolar compounds was the result of the sharing of pairs of electrons between atoms that again resulted in the formation of stable rare-gas configurations, as in Kossel's theory. And there was recognized also the possibility of the presence of both types of bindings in one and the same bond, especially after the development of the wave theory of the electron eliminated the necessity of thinking of the electron wholly in terms of the concept of a localized particle.

19-1 FUNDAMENTAL BOND STRUCTURE

As a basis for discussing the nature of the forces that tie atoms together in the form of molecules, the patterns of force in the two extreme types of bonds, ionic and covalent, serve as a helpful basis for the analysis of structure of more complex types of binding.

The ionic bond

The simplest case of ionic binding is found when an electron is transferred completely from one atom to another. For example, suppose that the $3s$ electron in a neutral sodium atom is transferred to a chlorine atom (Fig. 19-1), resulting in the formation of Na^+ and Cl^-, where the outer shell on each ion has the 2,6 rare-gas configuration. There is now a coulombic attraction between the positively and negatively charged ions. The potential energy E_a due to this attraction is given by

$$E_a = -\frac{e^2}{r} \qquad (19\text{-}5)$$

where e is the charge on the electron and r is the distance between the nuclei of the two atoms. The graph of E_a is shown as the solid line in the lower half of Fig. 19-2. The state has zero potential energy when the sodium ion and the chlorine ion are so far apart that there is no significant force of attraction operating between them. As they come closer and closer together, work is done by the force of attraction so that the potential energy is *decreased*.

When the two nuclei approach within 4 Å of each other, the shells of negatively charged electrons surrounding each atom begin to repel each other. As the atoms get nearer together, this *increases* the potential energy E_r due to repulsion according to the expression

$$E_r = be^{-ar} \qquad (19\text{-}6)$$

In this equation a is a constant that is almost the same for all singly charged pairs of ions, b is a constant determined empirically, r is the distance of the separation of the nuclei, and e is the basis of the system of natural logarithms and is not to be confused with the same symbol in the previous formula, which represents the charge on the electron. The value of E_r is plotted as a solid line in the upper part of the diagram. The total potential energy E which is the result of the action of both types of force is given by

$$E = E_a + E_r = be^{-ar} - \frac{e^2}{r} \qquad (19\text{-}7)$$

Fig. 19-1 The formation of a sodium chloride molecule by electron transfer.

$$1s^2 2s^2 2p^6 3s^1 + 1s^2 2s^2 2p^6 3s^2 3p^5 \rightarrow 1s^2 2s^2 2p^{6+} + 1s^2 2s^2 2p^6 3s^2 3p^{6-}$$

(a)

(b)

(c)

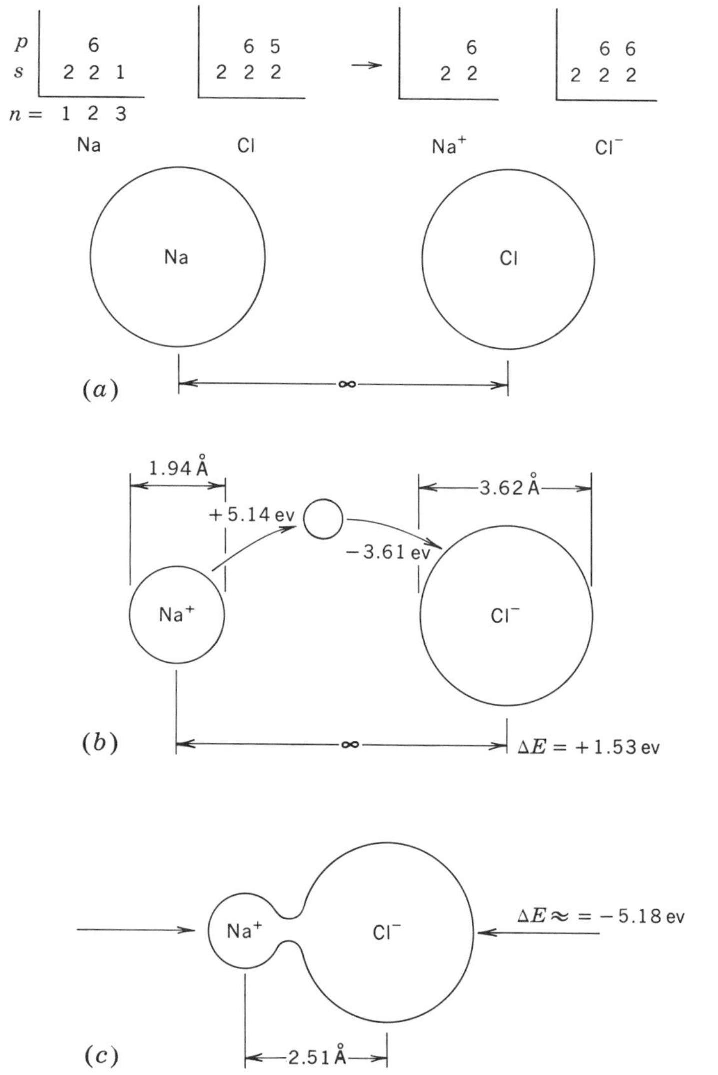

As may be seen in the upper part of Fig. 19-2, the term be^{-ar} falls to zero when r is greater than 3 Å. When r is less than 3 Å, the value of the total potential energy is shown as a dotted line. The total potential energy is at a minimum E_{min} when $r = r_{min} = 2.5$ Å. If there were no thermal energy and the system went to the state of lowest potential energy like a ball rolling to the bottom of a concave dish, the ions would remain separated by the distance r_m. In other words, the effect of coupling the force of attraction (varying inversely as the first power of r) with the force of repulsion (varying exponentially with $-a$) is to pro-

Fig. 19-2 Experimentally observed and theoretically calculated potential energy of Na$^+$ and Cl$^-$ ions as a function of the distance of separation. (After G. M. Barrow, "Physical Chemistry," 2d ed., McGraw-Hill Book Company, New York, 1966.)

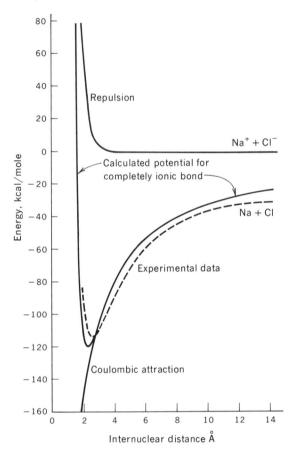

for all the electrons in such a pair of ions produces a situation where other forces also are at work. These effects will be discussed more in detail later.

Example 19-1

The frequency of vibration of the gaseous NaCl molecule in the ground state is 1.1×10^{13} sec^{-1}. Calculate the change in potential energy when the distance of separation of the two atoms is increased from the equilibrium distance (minimum potential energy) to a distance 0.5 Å greater, assuming that the force varies with distance according to the relation for a simple harmonic oscillator. From the graph in Fig. 19-2, estimate the difference between this value and (a) the value given by Eq. (19-7) and (b) the value given by the curve based on experimental data.

The relation between frequency and elastic force constant is

$$k = 4\pi^2 \nu^2 \mu$$
$$= 4 \times 9.87 \times 1.2 \times 10^{26} \times 2.31 \times 10^{-23}$$
$$= 1.1 \times 10^5 \text{ dynes cm}^{-1}$$

The potential energy per molecule at $x = 2$ Å is

$$E_p = \tfrac{1}{2} kx^2 = \tfrac{1}{2}(1.1) \times 10^5 \times 0.25 \times 10^{-16}$$
$$= 1.4 \times 10^{-12} \text{ erg molecule}^{-1}$$

The potential energy per mole in kilocalories is

$$E_p = 1.4 \times 10^{-12} \times 6.02$$
$$\qquad \times 10^{23}(2.39 \times 10^{-11} \text{ kcal erg}^{-1})$$
$$= 20 \text{ kcal mole}^{-1}$$

This is in agreement with the rise in potential energy shown by the two graphs for an increase in nuclear separation of 0.5 Å.

Covalent bonding

In contrast to ionic bonding, which can be explained in terms of classical electrodynamics, covalent bonding is due to a type of electron interaction that can be understood only in terms of quantum mechanics. Bonding of the pure ionic type is approximated as the result of the transfer of an electron from one atom to another as in the formation of Na$^+$Cl$^-$ from a sodium atom or a chlorine atom; bonding of the pure covalent

duce a force between the two ions much like that of a spiral spring. If the ions are pushed closer together than 2.5 Å, it is like the compression of the spring, which tends to force them apart. If they are pulled further apart than 2.5 Å, the force is like that produced when the spiral spring is stretched and tends to pull the two ions back together. When the two ions are separated by a distance greater than 30 Å, the force of attraction becomes negligible and they may be regarded as completely dissociated.

In actuality, the pattern of forces is never quite as simple as this. The overlapping of the wave functions type is approximated as the result of the sharing of a pair of electrons as in the formation of the Cl—Cl molecule from two atoms of chlorine, as shown in Fig. 19-3. By sharing a pair of electrons it is possible for each chlorine atom to have the equivalent of eight electrons in the outer shell with the resultant stable rare-gas configuration, which has a minimum of potential energy.

In quantum mechanics a simple covalent bond is associated with a wave function which is a solution for the Schrödinger equation applied to a system consisting of the kernels of the two atoms and the two electrons which are shared. The simplest case of covalent bonding is found in the hydrogen molecule, H_2, which is frequently symbolized as H:H to indicate the shared pair of electrons. In a more complex type of covalent bonding like that found in the chlorine molecule, Cl:Cl, the exact solution of the Schrödinger equation would have to take into account the action of all the electrons surrounding each of the atoms; and the resultant complexity is so great that an exact solution is not possible for Cl:Cl with presently available methods of mathematical analysis. However, relatively exact solutions have been obtained for the simpler case of H:H and will be discussed in detail in a following section.

Qualitatively the solution indicates that, as shown in Fig. 19-4, a shift in the distribution of electric charge takes place as two H atoms approach each other under conditions which permit the formation of a covalent bond. A portion of the charge cloud concentrates in the region between the two nuclei. The net result is a force between the two nuclei which varies with distance much like the force in a coiled spiral spring. The variation of force as a function of internuclear distance is plotted in Fig. 19-5, and the potential energy is shown as a function of distance in Fig. 19-6.

When the atoms are separated by about 0.75 Å, the potential energy is at a minimum; this situation resembles the mechanical model made of two balls joined by a spiral spring where the spring is neither stretched nor compressed and therefore exerts no force. If the two nuclei are moved together closer than 0.75 Å, the potential energy increases and there is a force of repulsion similar to that exerted by the

Fig. 19-3 The sharing of a pair of electrons in the formation of Cl:Cl from 2Cl.

$$\text{Before sharing} \begin{array}{|cc} 6\ 5 & 5\ 6 \\ 2\ 2\ 2 & 2\ 2\ 2 \end{array}$$

Cl   Cl

$$\text{After sharing} \begin{array}{|cc} 6(6\ :\ 6)6 \\ 2\ 2\ 2\ \ \ 2\ 2\ 2 \end{array}$$

Cl : Cl

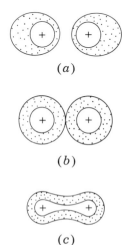

(a)

(b)

(c)

Fig. 19-4 Redistribution of charge as two hydrogen atoms approach (a) and form (b) an H_2 molecule (c).

spring when it is compressed. If the two nuclei are pulled apart farther than 0.75 Å, the potential energy is also increased and a force of attraction tends to pull the two nuclei back to the equilibrium position. However, if the two nuclei are pulled still farther apart, the force of attraction goes through a maximum and then decreases effectively to zero as the potential energy rises and levels off, at a distance of about 3 Å, corresponding to the complete dissociation of the molecule into atoms.

A real hydrogen molecule is always vibrating, the two nuclei moving toward and away from each other in repetitive cycles. In the state of lowest vibrational energy (the zero energy level) the potential energy E_p of the system at the extremes of amplitude has a

Fig. 19-5   Qualitative nature of force exerted by bond between the atoms in H:H (not to scale).

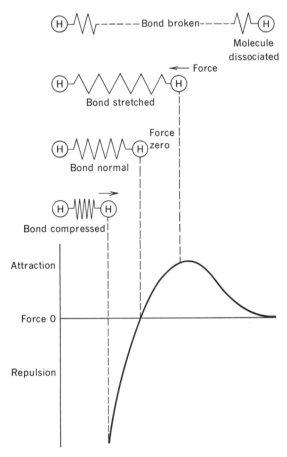

the average charge distribution shown in Fig. 19-3. Therefore, it is possible to derive solutions for the wave functions for stationary nuclei and apply them to the real molecule without introducing serious error.

19-2   APPROXIMATE WAVE FUNCTIONS

There are two principal methods for deriving molecular wave functions, which, though only approximate, are still close enough to the exact solution of the Schrödinger equation to provide valuable insight into the nature of the wave function and the character of the bonding. In the first approach called the *valence-bond method,* the two atoms to be bonded are first considered separately; and then the changes in the wave functions are calculated as the two atoms approach each other and the charge clouds of the two bonding electrons interact. In the second approach, called the *molecular-orbital method,* a wave function is devised for an electron moving in the field of the two nuclei of the atoms to be bonded. In this section the valence-bond method is presented; in Sec. 19-3 the molecular-orbital approach is discussed, and in Sec. 19-4, it is applied to a number of different molecules.

As the start for deriving approximate molecular wave functions, consider the Schrödinger equation (17-18) for one electron with the wave function ψ in three dimensions

$$\nabla^2\psi + \frac{2m}{\hbar^2}(E - U)\psi = 0 \qquad (19\text{-}8)$$

As in the previous chapter, the hamiltonian function H is defined by

$$H \equiv -\frac{\hbar^2}{2m}\nabla^2 + U \qquad (19\text{-}9)$$

so that Eq. (19-8) can be written

$$H\psi = E\psi \qquad (19\text{-}10)$$

The nature of the molecule is too complex to permit an exact calculation of the precise forms of ψ which satisfy this equation. As the first step in deriving an approximate form of ψ, a trial form of ψ is devised; multiplying each side of Eq. (19-10) on the left by this trial ψ gives

value of $\frac{1}{2}h\nu$, where ν is the frequency of the mechanical vibration. This is the zero-point energy of the system. According to Heisenberg's uncertainty principle, there can never be a state where the product of the uncertainties in position and momentum is less than h, Planck's constant. Therefore, the system can never be motionless; it must have a total energy of at least $\frac{1}{2}h\nu$.

The German theoretical physicist M. Born and the American theoretical physicist J. R. Oppenheimer pointed out that this frequency, however, is comparatively slow compared with the frequency of the orbiting of the electrons regarded as particles circulating about the two nuclei in a way that produces

$$\psi H \psi = \psi E \psi \tag{19-11}$$

Since H is an operator, $\psi H \psi$ in general is not equal to $\psi^2 H$; but since E is a number, $\psi E \psi = \psi^2 E$. Thus Eq. (19-11) can be put in the form

$$E = \frac{\psi H \psi}{\psi^2} \tag{19-12}$$

The average value of E is found by integrating over the entire volume τ in which the charge cloud is found

$$E = \frac{\int \psi H \psi \, d\tau}{\int \psi^2 \, d\tau} \tag{19-13}$$

The hydrogen molecule ion

A simple system to which this method can be applied is provided by the molecule H—H⁺, which is formed by removing one electron from the hydrogen molecule H:H. If the charge cloud of the single remaining electron were in the state of lowest potential energy and confined entirely to the nucleus on the left, the wave function would be of the type $1s$, which is spherical with only one loop; this wave function is designated by the symbol $^{1s}\psi_A$, where the subscript indicates the nucleus about which the charge cloud is centered as shown in Fig. 19-7(a). For this case, the expression from Table 17-3 is

$$^{1s}\psi_A = \frac{1}{\sqrt{\pi}} \left(\frac{1}{a_0} \right)^{3/2} e^{-r_A/a_0} \tag{19-14a}$$

If the electron charge cloud is similarly confined about, and influenced only by, the nucleus on the right B, it takes the form shown in cross section in Fig. 19-7b, and the equation is

$$^{1s}\psi_B = \frac{1}{\sqrt{\pi}} \left(\frac{1}{a_0} \right)^{3/2} e^{-r_B/a_0} \tag{19-14b}$$

If the charge cloud is influenced by both nuclei, the form of ψ will be a kind of average of the two independent ψ's such as

$$\psi = \frac{1}{\sqrt{2}} \left(^{1s}\psi_A + {}^{1s}\psi_B \right) \tag{19-15}$$

where the term $1/\sqrt{2}$ correctly normalizes the value of ψ so that $\int \psi^2 \, d\tau = 1$. When this value of ψ is inserted in Eq. (19-13), the value of E shown as the solid line in Fig. 19-8 is found; the dotted line gives the values of E derived from spectra.

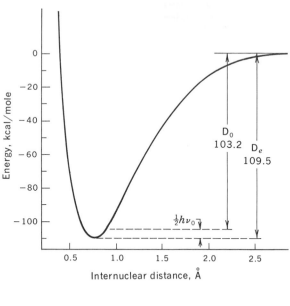

Fig. 19-6 The potential energy of a hydrogen molecule as a function of internuclear distance.

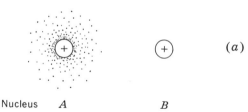

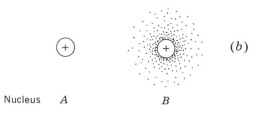

Fig. 19-7 The hydrogen molecule ion with the single electron confined (a) to the single nucleus on the left A, and (b) to the single nucleus on the right B.

The hydrogen molecule

The system consisting of two proton nuclei and two electrons is, of course, the hydrogen molecule. For this case a slightly more extended notation is necessary to denote the different quantities involved. The

Fig. 19-8 The potential energy of the hydrogen molecule ion as a function of the internuclear distance: (a) derived from spectra; (b) calculated from Eq. (19-15). (After G. M. Barrow, "Physical Chemistry," 2d ed., McGraw-Hill Book Company, New York, 1966.)

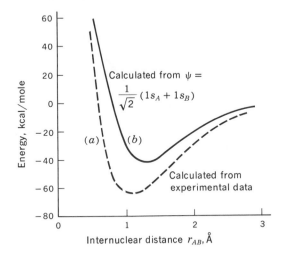

Fig. 19-9 The designation of coordinates in the hydrogen molecule.

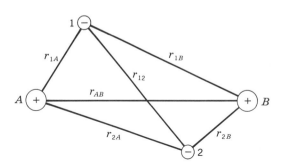

electrons are designated by the numbers 1 and 2 and the nuclei by the letters A and B. The subscripts on τ indicate the two particles which are separated by the distance denoted by r, as shown in Fig. 19-9.

The wave function for the charge cloud of electron 1 around nucleus A is

$$^{1s}\psi_{1A} = \frac{1}{\sqrt{\pi}} \left(\frac{1}{a_0}\right)^{3/2} e^{-r_{1A}/a_0} \tag{19-16}$$

The other three wave functions relating the electrons to the nuclei are precisely similar in form, differing only in subscripts.

If one electron is associated only with one nucleus and the other electron with the other nucleus, the appropriate trial function is

$$\psi = {}^{1s}\psi_{1A} \times {}^{1s}\psi_{2B} \tag{19-17}$$

The potential energy calculated from this function is compared with the experimentally observed energy in Fig. 19-10.

Electron spins

When two electrons are present, it is necessary to take into account the spin properties of the electrons in order to obtain a proper wave function. The spin functions are denoted by α and β with a subscript to indicate the electron with which the spin function is associated. α_1 indicates that electron 1 has a spin of $+\frac{1}{2}$; β_1 that electron 1 has a spin of $-\frac{1}{2}$; α_2, that electron 2 has a spin of $+\frac{1}{2}$; and β_2 that electron 2 has a spin of $-\frac{1}{2}$. The total spin function is the product of the two spin functions. Thus, there are four possible total spin functions: $\alpha_1\alpha_2$, $\alpha_1\beta_2$, $\beta_1\alpha_2$, and $\beta_1\beta_2$; of these, the second and the third are called *antiparallel* because the spins have opposite directions.

The Heitler-London method

About a year after Schrödinger proposed his famous equation, two German physicists W. Heitler and F. London applied the equation to the hydrogen molecule in a way that demonstrated clearly that the covalent-bond force originates from the steady-state solution for two electrons shared by two nuclei. They started with Eq. (19-17) and pointed out that the true wave function must express the relation that the two electrons are indistinguishable. Put another way, the two electrons are now embodied in a common wave pattern in which it is impossible to say that any particular part is due to one or to the other. Not only are the electrons delocalized; they are also de-individualized. Heitler and London concluded that there are two joint wave functions which satisfy this condition

$$\psi_s = \psi_{1A}\psi_{2B} + \psi_{2A}\psi_{1B} \qquad (19\text{-}18)$$

$$\psi_a = \psi_{1A}\psi_{2B} - \psi_{2A}\psi_{1B} \qquad (19\text{-}19)$$

where it is understood that all the ψ's are of the $1s$ type. The type of wave function defined by Eq. (19-18) has the subscript s to denote that it is symmetric, i.e., an interchange of coordinates between the two terms on the right of the equal sign does not change the sign of ψ. The wave function defined by Eq. (19-19) has the subscript a denoting antisymmetric because the interchange of subscripts does change the sign

$$\psi_{1A}\psi_{2B} - \psi_{2A}\psi_{1B} = -(\psi_{2A}\psi_{1B} - \psi_{1A}\psi_{2B}) \qquad (19\text{-}20)$$

A similar situation is found with respect to the spin functions. There are three symmetric combined functions $\alpha_1\alpha_2$, $\beta_1\beta_2$, and $\alpha_1\beta_2 + \alpha_2\beta_1$; there is only one antisymmetric function, $\alpha_1\beta_2 - \alpha_2\beta_1$.

According to the Pauli exclusion principle, every allowable wave function for a system of two or more electrons must be antisymmetric for the simultaneous interchange of the position and spin coordinates of any pair of electrons. Thus, the total wave function must be either antisymmetric in the orbitals or in the spins but not in both. There are four total wave functions satisfying this condition, as shown in Table 19-1. Because the molecular state is made up of two atomic states (nucleus A and nucleus B) each of which has s-type wave functions, the total angular momentum is zero and the term is denoted by Σ, the capital Greek letter corresponding to s. The upper left subscript signifies the multiplicity ($2s + 1$) of the wave functions corresponding to this term, where s is the total spin.

Figure 19-11 shows a plot of the potential energy for ψ_s and ψ_a. The symmetric wave function shows a deep potential minimum which results in the bonding force between the nuclei; the antisymmetric functions have no such minimum; the force is always one of repulsion, and there is no bonding. Since the one bonding state and the three antibonding states are all equally probable, there is only 1 chance in 4 that a molecule will be formed when two hydrogen atoms approach each other. The potential-energy curve for the bonding state based on evidence from spectra is shown as a dotted line.

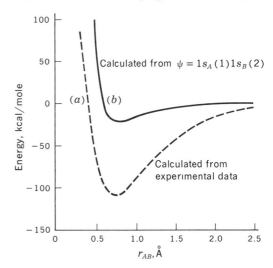

Fig. 19-10   The potential energy for H:H for the simple wave function $\psi_{1A}\,\psi_{2B}$: (a) as derived from spectra; (b) calculated from Eq. (19-17). (After G. M. Barrow, "Physical Chemistry," 2d ed., McGraw-Hill Book Company, New York, 1966.)

TABLE 19-1   Antisymmetric wave functions

Total spin	Spin function	Orbital function	Term
0 (singlet)	$\alpha_1\beta_2 - \alpha_2\beta_1$	ψ_s	$^1\Sigma$
1 (triplet)	$\alpha_1\alpha_2$	ψ_a	$^3\Sigma$
	$\beta_1\beta_2$		
	$\alpha_1\beta_2 + \alpha_2\beta_1$		

Example 19-2

If the hydrogen molecule vibrated as a simple harmonic oscillator, it would have a frequency of 1.32×10^{14} sec^{-1}. Calculate the energy in the fourth energy level and compare with the observed energy as shown in Fig. 19-6 at 1.3 Å, the separation for $n = 4$.

$$E_4 = 4h\nu = 4 \times 6.63 \times 10^{-27} \times 1.3 \times 10^{14}$$
$$= 34 \times 10^{-13} \text{ erg molecule}^{-1}$$
$$E_4 = 34 \times 10^{-13} \text{ erg molecule}^{-1} \times 2.39$$
$$\times 10^{-11} \text{ kcal erg}^{-1} \times 6.02$$
$$\times 10^3 \text{ molecule mole}^{-1}$$
$$= 49 \text{ kcal mole}^{-1}$$

**Fig. 19-11   The potential energy for the hydrogen molecule:
(*a*) from spectra; (*b*) for the Heitler-London bonding orbital $^1\Sigma$;
(*c*) for the antibonding orbital $^3\Sigma$.**

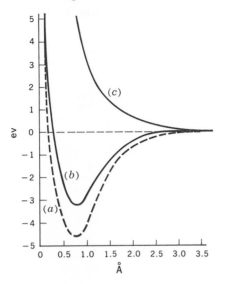

On the diagram in Fig. 19-6, $E_4 = 45$ kcal mole^{-1}.

19-3   THE MOLECULAR-ORBITAL METHOD

As contrasted with the valence-bond method, which starts with two atoms, each treated independently, with its electron or electrons, the molecular-orbital method (MO) starts with two nuclei and considers what the nature of the wave function will be if an electron is influenced by the electrostatic fields of these two nuclei *jointly*. Just as a wave function for a single electron in a single individual atom is called an atomic orbital, the wave function of an electron in the joint field of nuclei in a molecule is called a *molecular orbital*.

X—X orbitals

The simplest molecular orbitals are found when the electron is shared between two identical nuclei (X—X) in the formation of the molecule. The simplest example of this is the hydrogen molecule. The approximately correct form of the orbitals can be found by considering the system under conditions where the two nuclei are separated by various distances. If the separation is large, the system degenerates into the case of two hydrogen atoms, each with its own nucleus surrounded by its own electron, the situation used as the start for the valence-bond approach. If the distance of separation becomes vanishingly small, the system degenerates into that of the atom where a single nucleus with a double charge is surrounded by two electrons. If the electrons have opposite spins (as they must in the same orbital in the helium atom), then at an intermediate distance of separation, the system has a minimum of potential energy and there is a binding force which makes the system into a stable molecule.

The molecular orbital for two nuclei 10 Å apart would be, as a first approximation, the sum of the atomic orbitals for each nucleus

$$\psi^{(1)} = {}^{1s}\psi_a + {}^{1s}\psi_b \tag{19-21}$$

where the superscript (1) indicates the first approximation and where the subscripts a and b refer to the nuclei.

The molecular orbital ψ_{MO} for the system with two electrons under the influence of two nuclei about 1 Å apart can be taken as the product of two ψ's of the form in Eq. (19-21)

$$\psi_{MO}{}^{(1)} = ({}^{1s}\psi_{a1} + {}^{1s}\psi_{b1})({}^{1s}\psi_{a2} + {}^{1s}\psi_{b2}) \tag{19-22}$$

where the subscripts 1 and 2 refer to the electrons. If ψ_{MO} is substituted in the equation

$$E = \frac{\int \psi^* H \psi \, d\tau}{\int \psi^* \psi \, d\tau} \tag{19-23}$$

and the integration carried out over the space coordinate τ, the distance of separation for minimum potential energy E_{min} is found to be 0.850 Å. If the potential energy is set equal to zero when the separation is sufficiently large to prevent atomic interaction, then E_{min} has the value 61.8 kcal. In Fig. 19-6, E_{min}, determined experimentally from spectra, has the value 109.5 kcal and $r_0 = 0.732$ Å. While quantitatively the disagreement is substantial, qualitatively this MO does have a form that yields a binding force and a stable molecular configuration.

In order to make the calculated orbital conform more closely to the experimentally observed state, a *scale factor Z* can be introduced into the exponent

$$\psi^{(2)} = e^{-Zr_a} + e^{-Zr_b} \tag{19-24}$$

The minimum value for E is found when $Z = 1.197$ and $r_0 = 0.732$ Å, giving a value of $E_{min} = 80$ kcal mole^{-1}. The addition of further approximations can bring the calculated values into almost exact agreement with the experimentally observed values.

There are various ways of interpreting the physical meaning of the final form of the MO. One can say that one electron screens the other from the influence of the nucleus. The distance of separation of the nuclei must also be taken into account. The American theoretical physicists H. M. James and A. S. Coolidge devised an MO expressed by 13 terms which yielded a value differing by only 0.7 percent from the observed value.

If the wave function is antisymmetric, there is a force of repulsion between the two atoms, for all distances of separation of the nuclei, as shown by the curve marked $^3\Sigma$ in Fig. 19-11.

The same difference between the kinds of force is found in the case of one electron held between two nuclei, as illustrated in the hydrogen molecule ion. In Fig. 19-12 the cross-sectional map of charge distribution is shown for the two types of functions.

When the two atoms in the X—X molecule become more complex, the pattern of charge distribution changes quantitatively but keeps close to the same qualitative form, as shown in Fig. 19-13.

Frequently it is helpful to consider the symmetry properties of molecular orbitals and to use a graphic notation for indicating these properties. The nodes in a wave function constitute an important factor in determining the symmetry. As an example, consider the $1s$ functions of two independent hydrogen atoms and the wave functions formed when these are, first, added and, second, subtracted. In Fig. 19-14 these two ways of combining the wave functions are shown graphically. The $1s$ function of a hydrogen atom is indicated by a circle with a plus sign. The circle denotes the spherical symmetry (circular in the planar cross section); the plus sign indicates that a plot of the value of ψ at every point in the cross-sectional

plane will be positive; there is no node. If the two wave functions are added together (*i*), there are still no nodes, but the shape has now assumed a kind of dumbbell form. With respect to the axis between the two nuclei, however, rotation does not alter the pattern; there is circular symmetry just as there was in the $1s$ atomic orbital. Consequently, this orbital is denoted by the Greek letter σ, to indicate its relation to s. If the wave functions are subtracted (*ii*), a nodal plane appears at the midpoint between the two nuclei. At this point the two wave functions have the same value since the value is a function only of the distance from the nucleus. Therefore, $\psi = 0$ in the plane at the midpoint of the line joining the two nuclei and perpendicular to it; and this plane is a region of no charge. As a result, the charge on one side of the plane repels the charge on the other; the total charge is spread throughout a larger volume of space than when distributed around each nucleus according to the pattern based on the $1s$ atomic orbital; consequently, the two parts of the molecule repel each other; there is no boundary force; and the orbital is called an *antibonding orbital,* denoted by σ^*, where the asterisk indicates the antibonding quality. As mentioned in an earlier section, the interchange of the electrons reverses the sign of the wave function, a characteristic property of antisymmetric functions. The change in potential energy upon combining two s orbitals to form a σ or a σ^* molecular orbital is shown in Fig. 19-15.

It is necessary to have two kinds of molecular orbitals formed when atomic orbitals are combined. Each atomic orbital can accommodate two electrons, one with plus and the other with minus spin. Therefore, in two atomic orbitals there is room for four electrons. When these are combined to form molecular orbitals, the number of accommodated electrons must not decrease. By forming both a σ orbital and a σ^* orbital, each with room for two electrons, the number of accommodated electrons is unchanged.

X—Y molecules

When the atoms constituting a diatomic molecule are no longer of the same kind, a dissymmetry is present which affects shape of the charge distribution. As discussed in the previous section on the ionic bond,

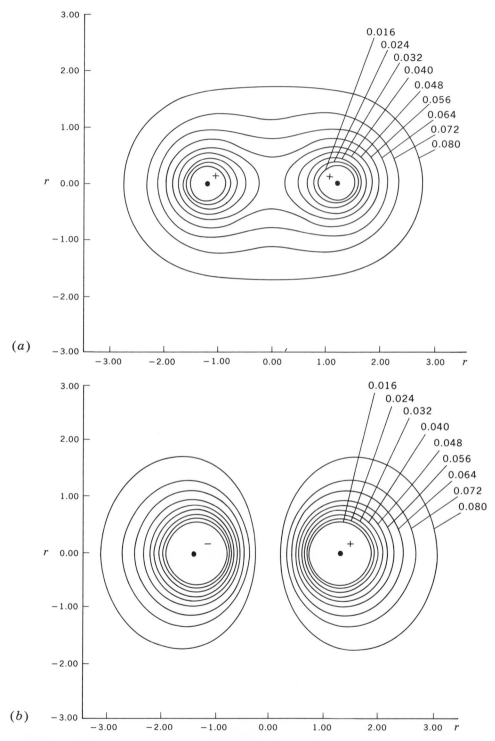

Fig. 19-12 Charge-density maps of H_2^+: (a) bonding orbital; (b) antibonding orbital. The nuclei, indicated by dots, are 2.5 au apart. The contours represent selected values of the charge density also in atomic units. [After D. R. Bates, K. L. Ledsham, and A. L. Stewart, *Phil. Trans.*, 246:215 (1953).]

Fig. 19-13   Calculated charge cloud density maps.   The innermost contour corresponds to 1 acdu, except for H_2 where it is 0.25 acdu; each contour outward is one-half of the nearest inner neighbor value.   [After A. C. Wahl, *Science*, **131**:966 (1966).]

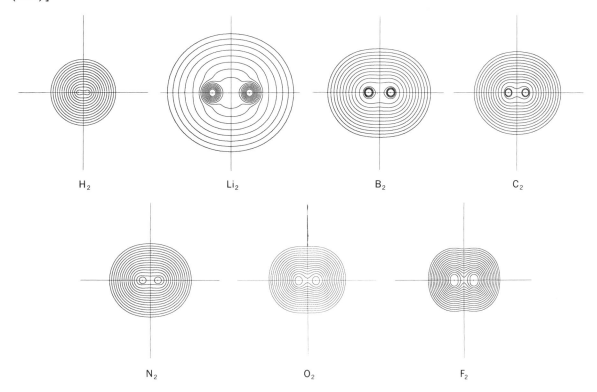

the extreme result of dissymmetry would be the complete transfer of an electron from an atom like sodium to an atom like chlorine when the two approach each other closely.   If such a transfer were complete, a large dielectric moment would result, the value of which can be calculated.   In the case of LiF, the calculated moment (in terms of dielectric *displacement* units D) is 7.51 D; but the observed moment is 6.28 D. This indicates that the charge has not been completely transferred and that there is an orbital representing some sharing of charge.

In order to see more clearly the nature of this sharing, it is helpful to plot a contour map showing the *difference* between the charge density in the partially covalent orbital and the charge density for two atomic orbitals of the type found in single atoms. Such a plot for H:H is shown in Fig. 19-16.   The positive values (solid lines in the center) indicate the

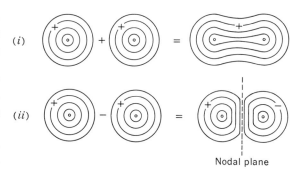

Fig. 19-14   A schematic representation of the effect of (*i*) adding and (*ii*) subtracting atomic orbitals to form the molecular orbitals shown on the right side.

areas where the covalent-bonding interaction has increased the density of charge.   A similar plot is shown in Fig. 19-17 for the difference between the LiF molecule and Li and F atoms and in Fig. 19-18 for the dif-

Fig. 19-15 The change in energy resulting from the formation of σ and σ^* orbitals in H_2. (After M. Orchin and H. H. Jaffé, "The Importance of Antibonding Orbitals." Copyright © 1967 by Milton Orchin and Hans H. Jaffé. Published by Houghton Mifflin Company, Boston, 1967. Used by permission.)

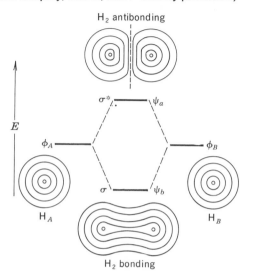

ference between the LiF molecule and Li^+ and F^- ions. The charge-distribution map for LiF is shown in Fig. 19-19.

Table 19-2 lists the distance r between the nuclei, the frequency of vibration ω expressed in wave numbers (cm^{-1}), and the energy of dissociation H_d for a number of diatomic molecules.

Example 19-3

From Fig. 19-19 calculate the value of the quotient which is the distance from the F nucleus to the point of minimum charge density lying on the axis between the nuclei divided by the distance from the Li nucleus to the same midpoint.

The point of minimum density of charge lies between the contour for 0.09 au (atomic units) surrounding the F nucleus and the contour for 0.13 au surrounding the Li nucleus; the ratio of the two distances is 1.9.

Example 19-4

Using the graphs of wave-function components in Appendix 2, calculate in relative wave-function units: (a) the value of $\psi_{MO} = {}^{1s}\psi_A + {}^{1s}\psi_B$ for the hydrogen molecule at a point midway between nucleus A and nucleus B; (b) the value at nucleus A.

The distance from nucleus A to nucleus B is 0.7414 Å.

$$\frac{0.7414 \text{ Å}}{0.5292 \text{ bohr Å}^{-1}} = 1.400 \text{ bohrs}$$

At the midpoint, $r = 0.700$ bohr.

(a) The value of the exponential component $\psi_{MO}{}^e$ of ψ_{MO} at 0.700 bohr is given by

$$\psi_{MO(1)}{}^e = {}^{1s}\psi_A{}^e + {}^{1s}\psi_B{}^e = 2^{1s}\psi^e = 2 \times 0.5 = 1$$

$$\psi_{MO} = \frac{\alpha^{3/2}}{\sqrt{\pi}}\psi_{MO(1)}{}^e = \frac{1}{\sqrt{\pi}}$$

$$= 0.564 \text{ rwfu}$$

(b) The value of ψ_{MO} at nucleus A is given by

$$\psi_{MO(2)}{}^e = {}^{1s}\psi_A{}^e + {}^{1s}\psi_B{}^e = 1 + 0.25 = 1.25$$

$$\psi_{MO} = \frac{1.25}{\sqrt{\pi}} = 0.705 \text{ rwfu}$$

TABLE 19-2 Distance between the two atomic nuclei in diatomic molecules, the frequency of vibration (ground state), and heat of dissociation[a]

Molecule	r, Å	ω, cm^{-1}	H_d, ev
Cl_2	1.989	564.9	2.481
CO	1.1284	2168.2	9.144
H_2	0.7414	4405.3	4.4776
HBr	1.414	2649.7	3.60
HCl	1.2747	2988.95	4.431
HF	0.9166	4141.3	6.4
HI	1.604	2309.5	2.75
I_2	2.667	214.36	1.5422
KH	2.244	983.3	1.9
Li_2	2.6723	351.3	1.14
N_2	1.095	2359.6	7.384
O_2	1.2076	1580.4	5.082
P_2	1.890	780.43	5.033

[a] From "Molecular Spectra and Molecular Structure," 2d ed., by G. Herzberg. Copyright © 1950 by Reinhold Publishing Corporation, by permission of Van Nostrand Reinhold Company.

19-4   DIRECTED VALENCE

In discussing the nature of the force produced by the covalent bond, it is helpful to keep in mind the analogy between the bond and a spiral spring. In both there is an equilibrium length at which no force is exerted; stretching produces a force tending to make the sys- tem retract to the equilibrium length. In diatomic molecules these are the only forces exerted by the bond; but in polyatomic molecules, other forces come into play. For example, the H_2O molecule has the shape shown in Fig. 19-20, where the angle between the bonds is 104.5°. What happens if the two hydro- gen atoms are pushed toward each other, making the

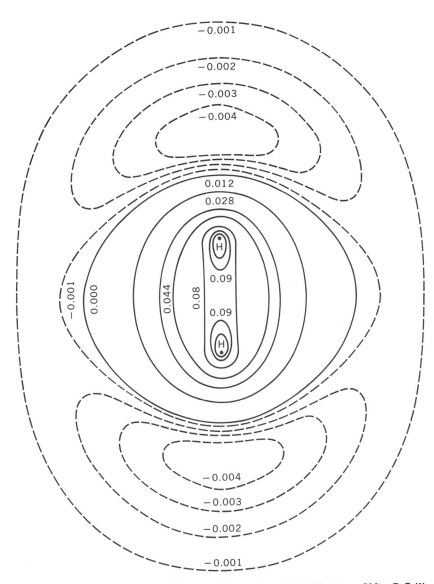

Fig. 19-16   The charge-density difference between H_2^+ and two H atoms.   [After R. F. W. Bader and W. H. Henneker, *J. Am. Chem. Soc.*, **87:3063** (1965).   Used by permission.]

Fig. 19-17   The charge-density difference between the LiF molecule and the Li and F atoms.   [After R. F. W. Bader and W. H. Henneker, *J. Am. Chem. Soc.*, 87:3063 (1965).   Used by permission.]

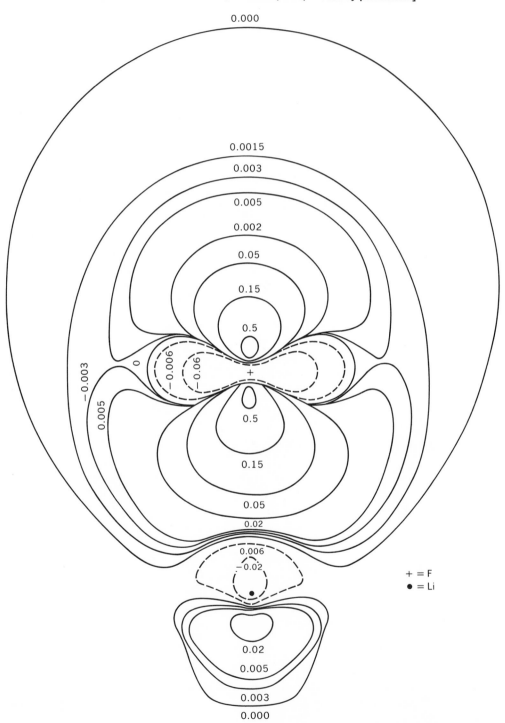

angle smaller?   The evidence is that a force is brought into play tending to restore the angle to its equilibrium value.   If the two hydrogen atoms are moved apart from the equilibrium position, an opposite force is produced tending to make the angle contract to the original equilibrium position.   In other words, as first pointed out in 1925 by Paul Ehrenfest, Professor of Theoretical Physics at the University of Leiden in Holland, the chemical bond resembles a spiral spring not only in its resistance to extension or compression but also in its resistance to bending.   This leads to the concept of a covalent bond *directed* in space in such a way that the force of a bond resembles the force of a directed spiral spring located along a line in space much more than it resembles the force field of an electric charge spread out in space and not

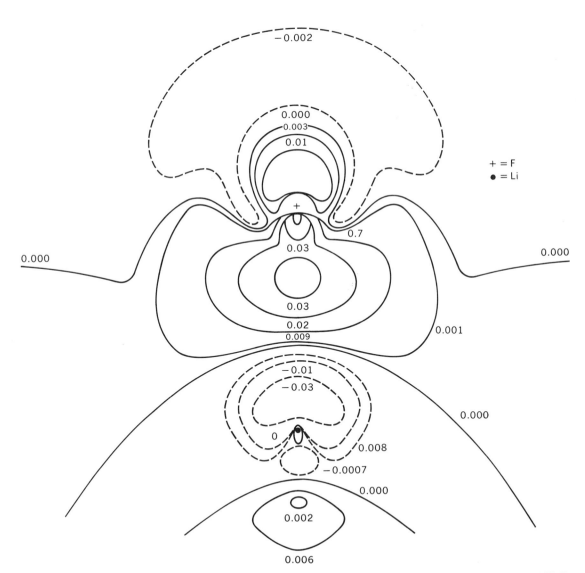

Fig. 19-18   The charge-density difference between the LiF molecule and Li$^+$ and F$^-$ ions.   [After R. F. W. Bader and W. H. Henneker, *J. Am. Chem. Soc.*, 87:3063 (1965).   Used by permission.]

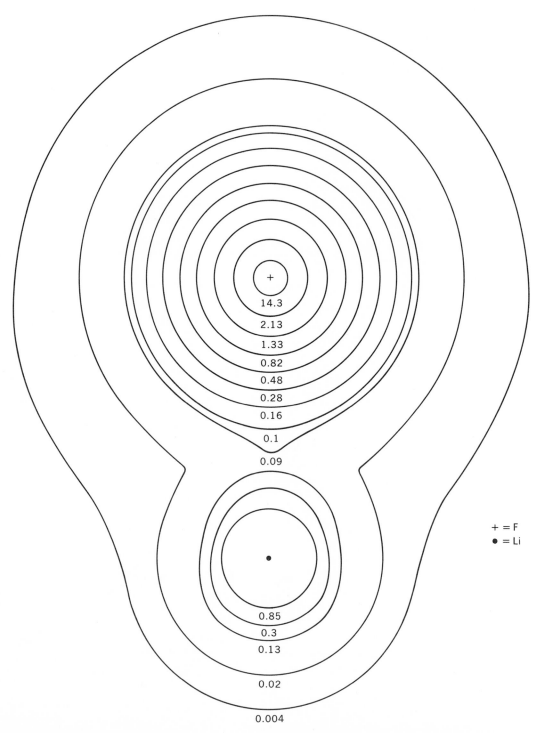

14.3

2.13

1.33

0.82

0.48

0.28

0.16

0.1

0.09

+ = F

● = Li

0.85

0.3

0.13

0.02

0.004

**Fig. 19-19   The charge distribution in a LiF molecule planar cross section.   Values are in atomic units ($e/a_0{}^3$).
[After R. F. W. Bader and W. H. Henneker, _J. Am. Chem. Soc._, 87:3063 (1965).   Used by permission.]**

normally thought of as concentrated in any particular direction.

In order to see the relation between the structure of the H$_2$O molecule and the shape of the atomic orbitals in the oxygen atom, recall the electronic structure of oxygen given by the sequence $1s^2 2s^2 2p^4$. For the moment we can neglect the $1s^2$ and $2s^2$ electrons, which, roughly speaking, make up the inner part of the oxygen atom, and consider only the outer shell. Here there are the four $2p$ electrons. According to Hund's rule, two of these will be in one of the orbitals, say p_x, one each will be in the p_y and p_z orbitals.

If two hydrogen atoms, each with a single $1s$ electron, approach these two unfilled oxygen orbitals, there is a tendency for the hydrogen atoms to move in to the distance from the oxygen nucleus that results in the minimum potential energy for the system. At this distance, the *overlap* of the hydrogen $1s$ orbitals with the oxygen $2p$ orbitals is of just the right amount to produce the minimum in the potential energy of the system which results in a stable molecule. However, this is a somewhat oversimplified picture. Actually the total orbital is not a simple sum of the patterns of the atomic orbitals but is formed by a process called *hybridization* which changes the character of the charge distribution.

Hybridization

Consider next the situation when four hydrogen atoms each with a $1s$ electron approach a carbon atom, as in the formation of methane, CH$_4$. The electronic structure of an unbound carbon atom in the ground state is $1s^2 2s^2 2p^2$. According to Hund's rule, the two p electrons will be in two p orbitals which are differently oriented in space; they will not be in the same orbital with opposite spins. This suggests that these two electrons are *free* to be shared with other atoms in the formation of two covalent bonds. But in methane, carbon forms four covalent bonds instead of two. Why?

The answer to this question is found when the form of the total wave function is derived in which the conditions of Schrödinger's equation are satisfied for a molecule like CH$_4$. Remember that in such a molecule there must be a total pattern of electron

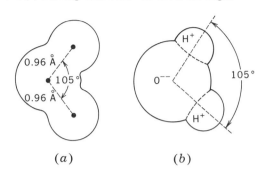

Fig. 19-20 A molecule of H$_2$O in cross section (a) and schematic (b) showing internuclear distances and angle.

wave vibration present simultaneously for all the s and all the p electrons so that a steady state exists. Just as the steady states are found for the s patterns and for the p patterns in individual atoms, similar steady states are found for molecules; but in the molecule new patterns appear which are a mixture of an s and a p pattern, the mixing, or *hybridization,* being brought about by the new boundary conditions due to the arrangement of the several nuclei in the molecule.

When wave functions are simultaneously combined, new patterns of charge distribution emerge. For example, if two wave functions are added, the resultant at any given point in space is the sum of the amplitudes of each function at that point

$$\psi_{MO} = \psi_1 + \psi_2 \qquad (19\text{-}25)$$

The pattern of charge distribution for this new ψ_{MO} is determined by the value of ψ_{MO}^2 at each point in space

$$\psi_{MO}^2 = (\psi_1 + \psi_2)^2 \qquad (19\text{-}26)$$

But it is essential to keep in mind that

$$(\psi_1 + \psi_2)^2 \neq \psi_1^2 + \psi_2^2 \qquad (19\text{-}27)$$

The charge distribution is given not by adding the distributions of the individual wave functions, ψ_1^2 and ψ_2^2, but by taking the square of the sum, $(\psi_1 + \psi_2)^2$.

As a specific example, suppose that ψ_1 is a $2s$ type of wave for which the charge cloud is shown in Fig. 19-21a; suppose that ψ_2 is a $2p$ type with the charge cloud shown in Fig. 19-21b. If the *two* charge clouds were added, the result ($\psi_1^2 + \psi_2^2$) would be a cloud

Fig. 19-21 (*i*) The values of ψ and (*ii*) charge density in cross section, for (*a*) a $2s$ wave function, (*b*) a $2p$ wave function, and (*c*) a ($\frac{1}{2}$) ($2s + 2p$) wave function.

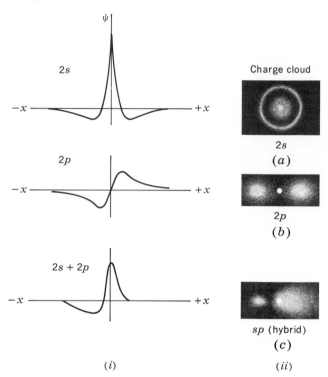

symmetrical about a plane passing vertically through the nucleus. Instead, to get the new charge cloud pattern, one must take $(\psi_1 + \psi_2)^2$; if this is the pattern for a *single* electron, with a wave function that is a combination of ψ_1 and ψ_2, then the combined charge cloud is $\frac{1}{2}(\psi_1 + \psi_2)^2$ and the pattern is that shown in Fig. 19-21*c*.

For a carbon atom there are three principal types of hybridization:

1. Each of the four electrons with two loops ($2s$ or $2p$) has an orbital where the wave derives one-quarter of its character from the s-type orbital and three-quarters from the p-type orbital. This is denoted as sp^3 hybridization. This provides four single bonds each at an angle of 109° to the others, as in methane, CH_4.

2. Three of the electrons are in a $\frac{1}{3}s$, $\frac{2}{3}p$ orbital,

called sp^2 hybridization, and one electron remains with a pure p type of orbital. This provides a double bond and two single bonds as in ethylene, $H_2C=CH_2$; each bond lies in a plane with the other two at an angle of 120° to each of the others.

3. Two of the electrons are in an orbital which is $\frac{1}{2}s$, $\frac{1}{2}p$; this is called sp hybridization. The other two electrons are in pure p orbitals at right angles to each other, as in acetylene, $H—C\equiv C—H$. All the atoms lie along a single line.

Pi bonds

When p atomic orbitals are combined to form a molecular orbital, the result is called a π orbital, with an asterisk if the orbital is antibonding. Unlike an s orbital, which has no node, a p orbital has a nodal plane normal to the line which runs through the center of each of the pear-shaped regions of charge density.

If, as shown in Fig. 19-22, two p orbitals are added, there is still only one node in the π orbital so formed. This is a bonding orbital. If, as shown in Fig. 19-23, one p orbital is subtracted from another, a new nodal plane appears and the π^* orbital so obtained is antibonding.

An important symmetry property of molecular orbitals is the invariance or lack of invariance in the sign of orbital amplitude in the transformation process called *inversion*. In this process a point component of the orbital (shown as a small circle in Fig. 19-22) is moved along a line passing through the center point

of the orbital (the path shown as a dotted line in the figure) and carried to an equal distance on the other side of the center. As can be seen by inspection of the figure, the inversion of all the points in the orbital changes the sign of the upper half from plus to minus, and of the lower half from minus to plus. Thus, the orbital is said to be antisymmetric with respect to inversion, a property frequently denoted by the German term *ungerade,* meaning uneven; a subscript u is sometimes added to the orbital symbol to denote this property, for example π_u. If on the other hand, inversion does not change the sign of the orbital, then it is called symmetrical with respect to inversion, or *gerade,* and the subscript g is used.

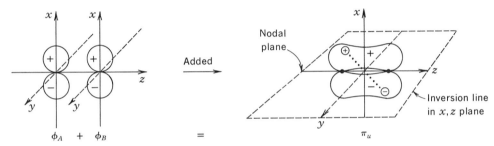

Fig. 19-22   The addition of two atomic $2p$ orbitals to form one molecular π orbital, each having a nodal plane through the center normal to the x axis.   Note that this molecular orbital is antisymmetric with respect to inversion, i.e., *ungerade,* signified by π_u.

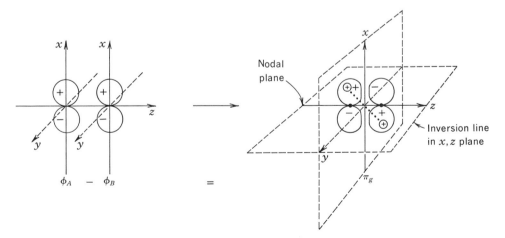

Fig. 19-23   The subtraction of one $2p$ atomic orbital from a similar $2p$ atomic orbital (each having a nodal plane normal to the x axis) to give a molecular orbital with a nodal plane normal to the x axis and a nodal plane normal to the z axis.   Note that this molecular orbital is symmetric with respect to inversion, i.e., *gerade,* signified by π_g.

Fig. 19-24 The orbital structure of ethylene, $H_2C{=}CH_2$. Note that the plane containing the four hydrogen atoms is normal to the x axis. (From C. R. Noller, "The Chemistry of Organic Compounds," W. B. Saunders Company, Philadelphia, 1952.)

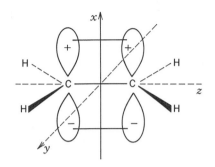

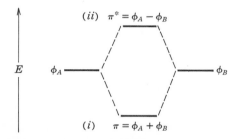

Fig. 19-25 The change in energy produced by (i) adding a $2p$ orbital to another $2p$ orbital and (ii) subtracting a $2p$ orbital from another $2p$ orbital.

The π-orbital bonding results in the formation of a double bond when two CH_2 radicals are combined to form ethylene. The bonding is indicated schematically in Fig. 19-24. The shift in energy accompanying bond formation is shown in Fig. 19-25.

The formation of the nitrogen molecule provides an interesting example of π bonding. The electronic structure of the nitrogen atom is $1s^2 2s^2 2p^3$. The atomic orbitals on each N atom combine to form essentially two molecular orbitals in N_2. If there were no hybridization, the changes in energy would be represented as shown in Fig. 19-26. However, hybridizing does take place, as indicated in Fig. 19-27. The result is the energy diagram shown in Fig. 19-28. This is discussed in detail by Orchin and Jaffé (see References).

Bond radii

Although the complexities in the structure of nearly all molecules are so great that they render impossible any exact calculation of molecular properties from molecular-orbital theory, the semiempirical and approximate approach has yielded many deep insights into the character of molecular structure. One of the most successful results of this approach has been the correlation and prediction of bond radii. The basis for the calculation of bond radii with the help of molecular-orbital theory was established largely as the result of the work of the theoretical physicist and chemist R. S. Mullikan, who received the Nobel Prize for Chemistry in 1966. Extensive research in this field was also carried out by Nobel Laureate Linus Pauling, who summarized the most important conclusions in his book on the chemical bond (see References). One of the most recent summaries has been prepared by A. C. Hurley and included in a book published as a tribute to Mullikan in 1964. Some recent values for bond radii are listed in Table 19-3.

Unbonded electron pairs

When a molecule has in the outer shell one or more pairs of electrons which are not participating in normal bonding, these pairs can exert forces on other atoms to produce the kind of semibonding found in water and called hydrogen bonding. This was discussed earlier in Chap. 13, and can be described at this point in somewhat greater detail with the help of molecular-orbital theory.

In Fig. 19-29 are shown the hybrid orbitals in H_2O and in NH_3 that make possible the attraction of external protons to the molecule by means of hydrogen bonding. This contributes to the formation of the hydronium ion, H_3O^+, and the ammonium ion, NH_4^+.

Hydrogen bonding plays an especially important role in biochemistry. In aqueous solution an amino acid like glycine quickly assumes the ionic form according to the reaction

$$NH_2CH_2COOH \rightarrow NH_3^+CH_2COO^-$$

In the ladderlike macromolecule DNA the cross-links, or rungs, consist of two parts held together at the center by the sharing of hydrogen atoms by π orbitals.

This will be described in detail in Chap. 24, which deals with biochemical macromolecules.

Example 19-5

From Table 19-3 and values of bond angles given in the text, calculate the distance between the H nuclei in the acetylene molecule.

The acetylene molecule is linear, and the bond radii are

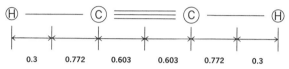

Hence, the distance between the hydrogen nuclei is the sum of these values, which is 3.35 Å.

Example 19-6

Calculate in wave-function units the amplitude of ψ_{MO} formed from the sum of $^{1s}\psi$ and $^{2p}\psi$ at

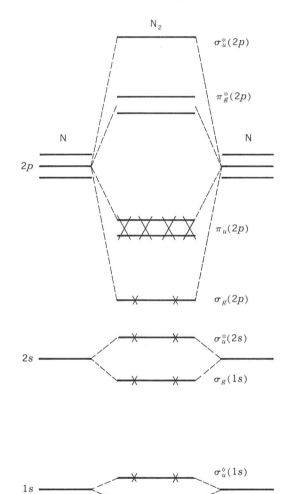

Fig. 19-26 The energy changes in the formation of a nitrogen molecule from two nitrogen atoms of the orbitals are combined but not hybridized. (From H. H. Jaffé and M. Orchin, "Theory and Applications of Ultraviolet Spectroscopy," John Wiley & Sons, Inc., New York, 1962.)

a point halfway between the hydrogen nucleus and the carbon nucleus forming a normal CH bond.

The bond radius of H— is 0.30 Å and of C— is 0.772 Å; the bond length is 1.07 Å = 2.02 bohrs.

Using the graphs in Appendix 2, the ampli-

TABLE 19-3	Radial length of covalent bonds[a]		
Bond	**Å**	**Bond**	**Å**
(H—	0.30)	F—	0.64
C—	0.772	Cl—	0.99
C=	0.667	Cl=	0.89
C≡	0.603	Br—	1.14
Si—	1.17	Br=	1.04
Si=	1.07	I—	1.33
Si≡	1.00	I=	1.23
N—	0.70	Ge—	1.22
(N=	0.60)	Ge=	1.12
P—	1.10	Sn—	1.40
P=	1.00	Sn=	1.30
P≡	0.93	As—	1.21
O—	0.66	As=	1.11
(O=	0.56)	Sb—	1.41
S—	1.04	Sb=	1.31
S=	0.94	Te—	1.37
S≡	0.87	Te=	1.27

[a] From L. Pauling, "The Nature of the Chemical Bond," 3d ed., Cornell University Press, Ithaca, N.Y., 1960, except for values in parentheses, which were estimated by the author.

Fig. 19-27 Hybridization and combination processes. Unhybridized orbitals are shown in the left column. The result of adding or subtracting these orbitals is shown in the right column.

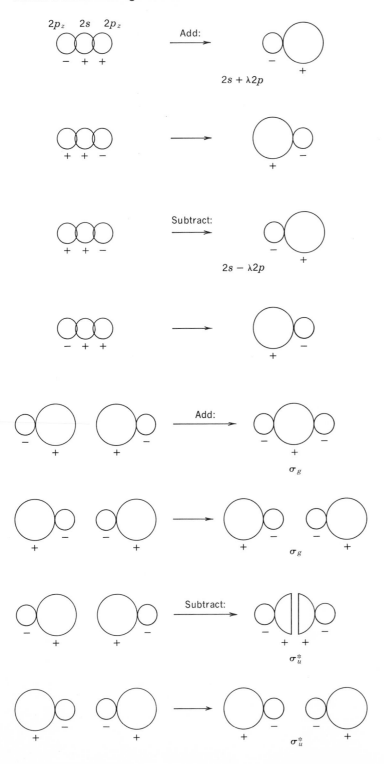

tude of the waves in wave-function units can be estimated at the midpoint, 1.01 bohrs from each nucleus. For $^{1s}\psi$ the value is 0.37 wfu; for $2p$ the exponential factor is $-r_b/2 = 0.61$. Since the hydrogen nucleus lies on the p-bond axis, $\cos \theta = 1$. The preexponential factor is 0.10. The value of the amplitude for $^{1s}\psi$ is $0.61 \times 0.10 = 0.06$; $^{1s}\psi + ^{2p}\psi = 0.37 + 0.06 = 0.43$ wfu.

19-5 DELOCALIZED ORBITALS

The first step in delocalizing the electron was taken when de Broglie proposed that the action of an elec-

tron was influenced by a wave associated with the electron. True, this delocalization was relative. Even for the classical electron as envisaged in the first Rutherford-Bohr model of the atom, the field due to the charge was thought of as spread out, mathematically speaking, to infinity; but in the case of a free electron, this charge field had a spherical symmetry so that one could define a point at the center of the spherical field and regard that point as representing the location of the electron; the coordinates of the point could be evaluated, and the motion of the point expressed by algebraic equations, with the coordinates as variables.

On the other hand, when it became necessary to write wave functions as solutions of Schrödinger's

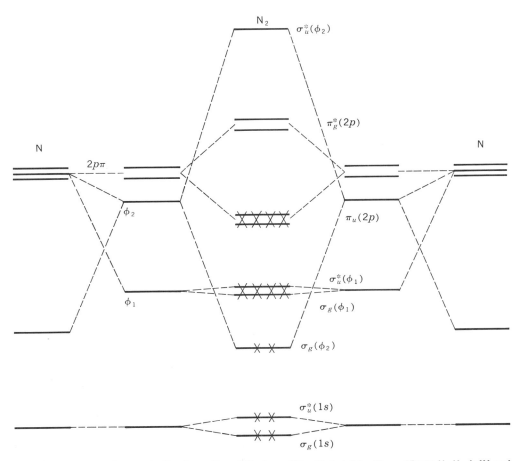

Fig. 19-28 Energy changes in the formation of N_2 from 2N with hybridization. (From H. H. Jaffé and M. Orchin, "Theory and Applications of Ultraviolet Spectroscopy," John Wiley & Sons, Inc., New York, 1962.)

Fig. 19-29   Partial representation of orbitals: (*a*) H_2O; (*b*) NH_3.   The points where hydrogen bonding can occur are indicated by arrows.   The negative charge cloud of the hybridized π orbital creates a zone which attracts the positive charge in a neighboring hydrogen atom.

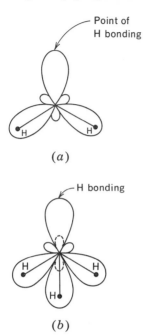

Point of H bonding

(*a*)

H bonding

(*b*)

equation in order to describe the action of the electron, the concept of a point charge lost something of its localized character.   Particularly when Heisenberg's uncertainty principle made plain the limit on any attempt to define simultaneously the exact momentum and the exact location, considerable doubt arose about the significance of ascribing any kind of point location to the electron in the steady state associated with an atomic orbital.   As many theoretical physicists expressed it, the point-localized electron was "smeared out" through space.

When atomic orbitals were combined to form *molecular* orbitals, this smearing-out process was extended to include the concept of an orbital embracing not one but two atomic nuclei; and so, as a further step, it was logical to raise the question whether an electron could be smeared out still more to embrace many atomic nuclei.

Benzene orbitals

One of the first suggestions of the possibility of a polynuclear orbital was made in connection with the problem of benzene.   This important molecule consists of a planar hexagonal ring of six carbon atoms to each of which is attached a hydrogen atom, also lying in the plane of the ring.   In the nineteenth century, even before the exact geometrical form of benzene was known, chemists recognized the peculiar chemical nature of this compound.   Studies of many aspects of the chemistry of carbon compounds indicated that in the vast majority of molecules carbon had a valence of 4 and hydrogen had a valence of 1. The problem was how to reconcile these rules with the formula of benzene, C_6H_6.

The German chemist F. A. Kekule proposed in 1865 that the formula of benzene should show the molecule as being derived from the theoretical straight-chain double-bonded hydrocarbon called hexatriene

$$H\underset{H}{\overset{}{\diagdown}}C{=}C{-}\overset{H}{\underset{}{C}}{=}\overset{H}{\underset{}{C}}{-}\overset{H}{\underset{}{C}}{=}C\overset{H}{\underset{H}{\diagup}}$$

By removing a hydrogen atom from each end and then linking the ends together, one might expect to form one or the other of the equivalent ring compounds corresponding to cyclohexatriene

(*a*)   (*b*)

But the chemical properties of benzene, especially with respect to the addition of molecules like halogens and halogen acids, differed strikingly from hexatriene and suggested that there were no ethylenic double bonds in benzene.   Because of this, many attempts were made to explain the chemical properties of benzene in terms of a modified bond structure.

The first quantitatively successful approach came largely as a result of the work of Pauling and the American physicist J. C. Slater who developed what has come to be known as the *resonance* theory of double bonds. From the diagrams (a) and (b) above, it is clear that, given the same hexagonal ring of carbon atoms, bonds can be distributed either in pattern (a) or in pattern (b); both patterns should have identical physical and chemical properties. In such a situation the bonds were sometimes thought of as alternating very rapidly between the (a) distribution and the (b) distribution, a phenomenon to which the term *resonance* was applied. Coulson (see References) points out that physically there is no actual resonance or alternation between these two forms and therefore it would be more appropriate to use a term like *mesomerism,* which denotes a state intermediate between different forms. In other words, when conventional bond structures for a molecule can be written in two or more different ways for a molecule, the pattern of molecular orbitals corresponds to an intermediate type of structure. For benzene, in addition to the two alternating double-bond structures there are also a number of other arrangements of bonds which contribute slightly to the mesomerism such as the three Dewar formulas

For benzene, the bonds of lowest energy are the single bonds formed by the overlap of hybrid orbitals joining the carbon atoms together in the regular hexagon and bonding the hydrogen atoms to the carbon atoms, as shown in Fig. 19-30a. This leaves a p_z orbital for each carbon oriented with the pear-shaped charge clouds above and below the plane of the ring and with the major axis of each p_z orbital normal to the plane, as shown in Fig. 19-30b. These six p_z atomic orbitals can merge to form six molecular orbitals; in the unexcited molecule, two electrons occupy each of the three orbitals that have the lowest energy. This results in two ring-shaped charge clouds, one above and one below the plane of the ring, formed by the merging of the individual p_z atomic charge clouds, as shown in Fig. 19-30c. The schematic representation of the charge clouds of each of the three wave functions of lowest energy are shown in Fig. 19-31. In Fig. 19-31a, all six atomic orbitals merge in a nodal plane; in Fig. 19-31b there is a nodal plane normal to

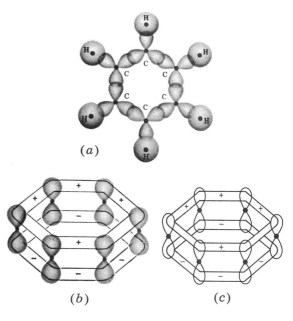

Fig. 19-30 Bonding in benzene: (a) sp^2 hybrid orbitals; (b) p_z atomic orbitals; (c) bonding, with carbon atoms shown as black dots and original p_z atomic orbitals shown as light lines, with signs showing the phase of the wave. (After J. E. Spice, "Chemical Binding and Structure," Pergamon Press, New York, 1964.)

(a)

(b)   (c)

the x axis; and in Fig. 19-31c there is a nodal plane normal to the y axis.

A helpful analogy can be drawn between an electron in an orbital in benzene and the electron in a box, which was discussed in Chap. 16. When an electron is confined in a one-dimensional "box" between two potential barriers, the orbital for the ground state consists of a single loop, as shown in Fig. 19-32a; the next highest state has two loops (Fig. 19-32b); and the state above that has three loops (Fig. 19-32c). The longer the box, the longer the wavelength in each state; a longer wavelength corresponds to a lower energy. When an electron in benzene is confined to a single atomic orbital, this corresponds to enclosure in a small box; when the atomic orbitals merge to form the molecular orbital 6 times longer, the frequency is 6 times less and the energy drops accordingly. In terms of this analogy, one can see why benzene should have a lower energy than cyclohexatriene.

Fig. 19-31 Bonding orbitals in benzene: (a) ψ_1 with no nodal plane; (b) ψ_2 with nodal plane normal to x axis; (c) ψ_3 with nodal plane normal to y axis. (After C. W. Cumper, "Wave Mechanics for Chemists." Copyright © 1966. Academic Press, Inc., New York. Used by permission.)

(a)

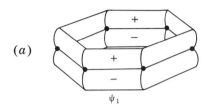

ψ_1

(b)

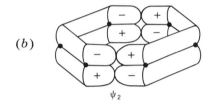

ψ_2

(c)

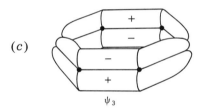

ψ_3

Measurements of the heat liberated upon hydrogenation provide proof of the lower energy actually present in benzene. As shown in Fig. 19-33a, the hydrogenation of cyclohexene lowers the value of H by 28.6 kcal. In cyclohexadiene (Fig. 19-33b) there are two double bonds; if each bond had the same energy as the one bond in cyclohexene, H should be lowered by $28.6 \times 2 = 57.2$ kcal; experimentally $H = 55.4$ kcal. For benzene, if there were really three ethylenic double bonds present, the lowering of H should be $28.6 \times 3 = 85.8$ kcal; the observed lowering is only 49.8 kcal; consequently the mesomerism produced by the merging of the π orbitals lowers the energy of benzene by $85.8 - 49.8 = 36$ kcal.

In addition to this striking deviation, there is other evidence to show the mesomerism in benzene. If there were really three ethylenic double bonds present, the bond lengths should alternate around the ring, the single bonds having a length of 1.54 Å and the double bonds having a length of 1.33 Å according to the data in Table 19-3. Instead, evidence from x-ray measurements shows that all bonds have the same length, 1.39 Å. The molecular spectra of benzene also confirms the mesomerism, as discussed in Chap. 20. The molecular orbitals extending around the ring make it possible for electrons to "flow" around the ring; this accounts for magnetic anomalies which have been observed experimentally. These currents provide a means for extending chemical influence over distances in the molecule far greater than would be expected on the basis of normal orbital interaction; this extended chemical influence is of great importance in many biochemical molecules, as discussed in Chap. 24 on macromolecules.

Conjugated systems

Molecules of the type in which the mutual influence of mesomeric structures delocalizes electrons are classified as *conjugated systems*. Benzene is just one example, but perhaps the most intensively studied one, of this molecular class. Another important example is *butadiene,* where the presence of two adjacent double bonds results in abnormal chemical behavior, recognized by organic chemists long before the development of the molecular-orbital theory. The formula for butadiene is shown in Fig. 19-34, together with representations of the atomic and molecular orbitals.

The four p_z atomic orbitals merge to produce a kind of elongated box into which the four electrons associated with these orbitals can be placed. The first two electrons have the de Broglie waveform ψ_1, in which there is only one loop in the box, the molecular orbital of lowest energy, spread out over all four carbon atoms. The other two electrons have the ψ_2 wave function with two loops. The other two orbitals, with three loops and four loops, are also shown. As a result of this charge distribution, a molecule like bromine combines with butadiene much more according to reaction (a) than to reaction (b):

(a) $\quad H_2C{=}CH{-}HC{=}CH_2 + Br_2 \rightarrow$
$\qquad\qquad\qquad H_2BrC{-}CH{=}CH{-}CBrH_2$

(b) $H_2C\!=\!CH\!-\!HC\!=\!CH_2 + Br_2 \rightarrow$
$$H_2BrC\!-\!CHBr\!-\!CH\!=\!CH_2$$

In order to show the energy changes produced by delocalization of electrons, a convention has been devised in which the energy is expressed in two terms denoted by the Greek symbols α and β, which are functions of terms in the formulas of the variation method. In Sec. 19-2, the variation method was described as a means of obtaining the approximate form of molecular orbitals. The energy ϵ of the approximate orbital falls to lower values and approaches the value E_0 of the actual physical orbital as a minimum as the molecular orbital approximates more closely the form which is the true solution of the Schrödinger equation. This is the wave pattern which represents the true steady state of vibration under the conditions of potential energy found in the real molecule. As shown in Sec. 19-2, the relation between ϵ and E_0 is

$$\epsilon = \frac{\int \psi H \psi \, d\tau}{\int \psi^2 \, d\tau} \geqslant E_0 \qquad (19\text{-}28)$$

where the equality holds when the form of ψ is the true steady-state waveform.

In applying the variation method to the butadiene molecule, the total molecular orbital ψ is taken as the product of two molecular orbitals Φ_σ, Φ_π. Φ_σ is assumed to be the product of molecular orbitals, each of which embraces only two nuclei, corresponding to single bonds joining two adjacent atoms. Φ_π is taken as the product of molecular orbitals which embrace all four carbon atoms and are linear combinations of the p_z atomic orbitals of these atoms. There are four of these molecular orbitals ψ_1, ψ_2, ψ_3, ψ_4, as shown diagrammatically in Fig. 19-34. Written as linear combinations of the atomic orbitals ϕ_1, ϕ_2, ϕ_3, ϕ_4, they are

$$\psi_1 = C_{11}\phi_1 + C_{12}\phi_2 + C_{13}\phi_3 + C_{14}\phi_4$$
$$\psi_2 = C_{21}\phi_1 + C_{22}\phi_2 + C_{23}\phi_3 + C_{24}\phi_4 \qquad (19\text{-}29)$$
$$\psi_3 = C_{31}\phi_1 + C_{32}\phi_2 + C_{33}\phi_3 + C_{34}\phi_4$$
$$\psi_4 = C_{41}\phi_1 + C_{42}\phi_2 + C_{43}\phi_3 + C_{44}\phi_4$$

where the subscripts on the ϕ's refer to the different carbon atoms.

In minimizing the value of ϵ, with the help of Eq.

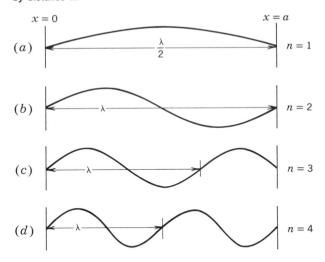

Fig. 19-32   One-dimensional de Broglie waves associated with an electron in a box between two potential barriers separated by distance a.

(19-28), four terms H_{11}, H_{22}, H_{33}, H_{44} are found, which are called *coulomb integrals* and represent closely the energy of the electron in each of the four p_z atomic orbitals. It is assumed that they are all equal, and the value of one of them is denoted by α.

There are also found 12 other terms which represent the reduction in energy due to the interaction of the electrons in the smearing out process; these are H_{12}, H_{13}, H_{14}, H_{21}, H_{23}, H_{31}, H_{32}, H_{34}, H_{41}, H_{42}, H_{43}. Expressing these by a general symbol H_{rs}, and assuming that all H_{rs} for *bonded* atoms interacting are equal, this value is denoted by the symbol β. With respect to the energy of an electron so far removed from the molecule that there is no interaction with it (the usual zero reference point on the potential-energy scale), both α and β are negative quantities. However, β may lie either below or above α. If the wave pattern to which β refers is a bonding pattern, β lies below α; if it is an antibonding pattern, β lies above α.

It is now possible to express the energy of an electron in a given molecular orbital such as ψ_1 as a function of α and β

$$\epsilon_1 = \alpha + m_1\beta \qquad \epsilon_2 = \alpha + m_2\beta$$
$$\epsilon_3 = \alpha + m_3\beta \qquad \epsilon_4 = \alpha + m_x\beta \qquad (19\text{-}30)$$

Fig. 19-33 A comparison of the experimental heat of hydrogenation and the heat calculated on the basis of the presence of ethylenic double bonds for cyclohexadiene and benzene. (After "Introduction to Molecular-Orbital Theory," by A. Liberles. Copyright © 1966 by Holt, Rinehart and Winston, Inc., New York. Reprinted by permission of Holt, Rinehart and Winston, Inc.)

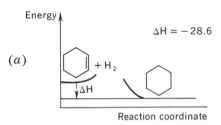

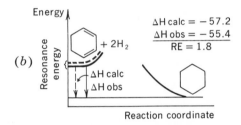

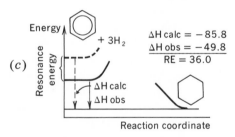

where the values of the coefficients m_1, m_2, m_3, and m_4 depend on the type of total interaction associated with each of the orbitals. The values of ϵ for butadiene are shown in Fig. 19-35 for the four wave functions illustrated schematically in Fig. 19-36. The values of ψ_1 and ψ_2 lie below that of α; these are bonding orbitals, and each contains two electrons in the ground state of the molecule. The values of ψ_3 and ψ_4 lie above; these are antibonding orbitals and contain electrons only when the molecule is activated.

Just as the removal of a hydrogen atom from each end of hexatriene makes possible the formation of benzene, the removal of a hydrogen atom from each end of butadiene makes possible the formation of cyclobutadiene

$$\begin{array}{c}
H \\ \diagdown \\ C{=}CH{-}CH{=}C \\ \diagup \\ H
\end{array}
\begin{array}{c}
H \\ \diagup \\[6pt] \diagdown \\ H
\end{array}
\rightarrow
\begin{array}{c}
HC{-}CH \\ \| \quad \| \\ HC{-}CH
\end{array} + H_2$$

This four-carbon ring, however, does not show a reduction in energy at all comparable to that which results in the aromaticity of benzene. The distortion of the angles between the carbon atoms is much greater in the formation of this smaller ring, and this reduces the interaction of the p_z electrons. The character of the ψ_z functions in cyclobutadiene is illustrated in Fig. 19-36.

The reduction in energy due to the interaction of the p_z electrons is frequently called the *resonance energy*. Values of this quantity for a number of mesomeric molecules are listed in Table 19-4.

Example 19-7

The energy needed to dissociate a normal ethylenic bond is 145 kcal mole^{-1}. According to Table 19-10, the energy necessary to dissociate a C—C bond is 83 kcal mole^{-1}. Calculate the energy necessary to dissociate all the bonds between carbon atoms in benzene (a) if three bonds are normal ethylenic bonds and three are single bonds; (b) if, as suggested by MO theory, each bond is the equivalent of a single bond plus $(1 + 1.05 + 1 \times 0.33)$ times the p_z equivalent of an ethylenic bond.

(a) $3 \times 145 + 3 \times 83 = 684$ kcal mole^{-1}
(b) $6 \times 83 + (1 + 0.5 + 0.33)(145 - 83)$
       $= 498 + 226.92 = 611$ kcal mole^{-1}

According to this rough calculation, the resonance energy is $684 - 611 = 73$ kcal mole^{-1}.

19-6 COORDINATION COMPOUNDS

During the nineteenth century theories of molecular structure were developed largely on the basis of the concept of valence; and both in organic and inorganic chemistry a clear understanding was achieved of the structural form of the vast majority of known mole-

Fig. 19-34 The molecular-orbital theory of butadiene: (*a*) conventional formula; (*b*) hybrid sp^2 orbitals with p_z orbitals omitted; (*c*) p_z atomic orbitals with molecular wave functions; (*d*) charge delocalization associated with merged p_z orbitals. [(*c*) from L. F. Phillips, "Basic Quantum Chemistry," John Wiley & Sons, Inc., New York, 1965. (*b*) and (*d*) from J. E. Spice, "Chemical Binding and Structure," Pergamon Press, New York, 1964.]

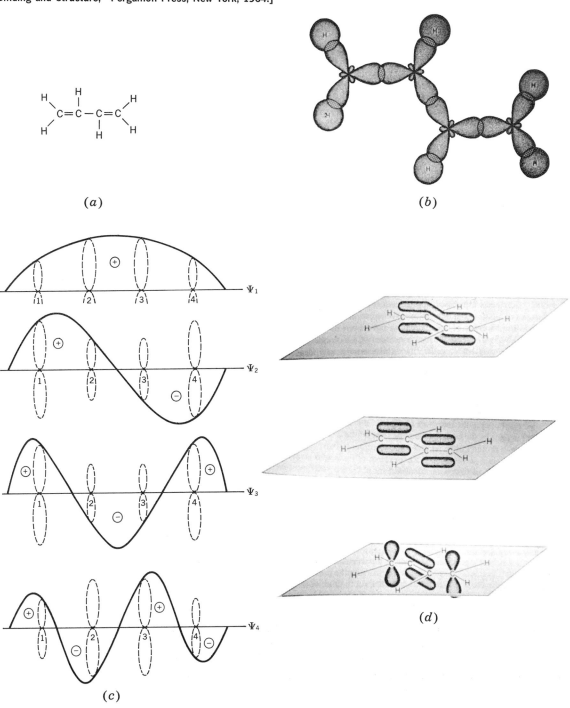

(*a*)

(*b*)

(*c*)

(*d*)

Fig. 19-35   Values of resonance energy for butadiene shown as a function of α and β. (After A. Streitwieser, "Molecular Orbital Theory for Organic Chemists," John Wiley & Sons, Inc., New York, 1961.)

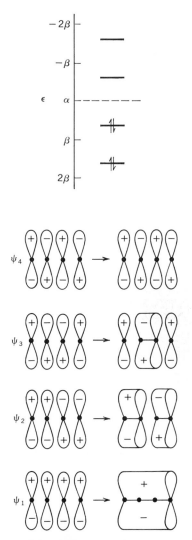

Fig. 19-36   The combination of atomic orbitals (*left*) in butadiene to form molecular orbitals (*right*). (After C. W. Cumper, "Wave Mechanics for Chemists." Copyright © 1966. Academic Press, Inc., New York. Used by permission.)

cules. But as this insight was enlarged, it became increasingly clear that the simple concept of valence was unable to explain many compounds like salt hydrates, double salts, metal amines, and carbonyls. In

1893, the German chemist Alfred Werner suggested the idea of secondary valence to explain these anomalies, which were frequently called *complexes*. In 1915, G. N. Lewis applied the theory of the electron-pair bond to this group of substances with considerable success; but it was not until the development of MO theory, especially as applied to d orbitals, that the basic structure of complexes really was clarified.

d orbitals

In Figure 19-37 the electronic structure of the chemical elements from potassium to krypton are shown as triangle diagrams. Note that after filling the $4s$ orbitals in K and Ca, the next electron added in forming scandium goes into a d orbital. This behavior is explained by referring to the energy values of the different orbitals as shown in Fig. 19-38. After filling the $3s$ and $3p$ orbitals in the formation of argon, the next available level is the $4s$, which lies *lower* than the $3d$. After filling the two places for electrons available in $4s$, then the ten places in $3d$ are filled, and finally the six places in $4p$, yielding the completed shell structure of krypton.

The $3d$ orbital patterns are drawn schematically in Fig. 19-39. These are the patterns of charge distribution in three dimensions produced by the de Broglie waves that are solutions of the Schrödinger equation having *two* nodal planes. Recall that the number of nodal planes is equal to the quantum number l and the number of independent patterns is equal to $2l + 1$. With $l = 0$, there is only one independent pattern of de Broglie wave vibration; this pattern has spherical symmetry; it is an s orbital. With *one* nodal plane ($l = 1$) there are $2 \times 1 + 1 = 3$ independent patterns, the p orbitals, with pear-shaped symmetry. With two nodal planes there are $2 \times 2 + 1 = 5$ independent patterns, the d orbitals with double pear-shaped patterns. Remember that these patterns are all derived from the mathematics of wave vibration in the steady state as expressed by the Schrödinger equation.

The mathematical expressions for the d orbitals are listed in Table 19-5. Like the p orbitals, the d orbitals are the product of an angular function. $\psi(\theta,\phi)$ and a radial function $\psi(r)$. The total function $\psi = \psi(\theta,\phi)\psi(r)$ and has the units of (electron charge

TABLE 19-4   Resonance energies[a]

Substance	Formula	Resonance energy, kcal/mole^{-1}
Benzene		37
Naphthalene		75
Pyridine		43
Pyrrole	HC—CH HC CH N H	31
Furan	HC—CH HC CH O	23
Thiophene	HC—CH HC CH S	31
Carboxylic acids	R—C—OH (=O)	28
Amides	R—C—NH$_2$ (=O)	21
Carbon dioxide	$O=C=O$	36
Phenol	—OH	7[b]
Benzaldehyde	—C—H (=O)	4[b]
Acetophenone	—C—CH$_3$ (=O)	7[b]
Benzophenone	—C— (=O)	10[b]

[a] Data from L. Pauling, "The Nature of the Chemical Bond," 3d ed., Cornell University Press, Ithaca, N.Y., 1960.
[b] Extra resonance energy not including that within the benzene ring.

Fig. 19-37   Matrix triangle diagrams of the electronic structure of the chemical elements from potassium through krypton.

	K 19	Ca 20	Sc 21	Ti 22	V 23	Cr 24	Mn 25	Fe 26	Co 27
d			1	2	3	5	5	6	7
p	6 6	6 6	6 6	6 6	6 6	6 6	6 6	6 6	6 6
s	2 2 2 1	2 2 2 2	2 2 2 2	2 2 2 2	2 2 2 2	2 2 2 1	2 2 2 2	2 2 2 2	2 2 2 2

n 1 2 3 4

	Ni 28	Cu 29	Zn 30	Ga 31	Ge 32	As 33	Se 34	Br 35	Kr 36
d	8	10	10	10	10	10	10	10	10
p	6 6	6 6	6 6	6 6 1	6 6 2	6 6 3	6 6 4	6 6 5	6 6 6
s	2 2 2 2	2 2 2 1	2 2 2 2	2 2 2 2	2 2 2 2	2 2 2 2	2 2 2 2	2 2 2 2	2 2 2 2

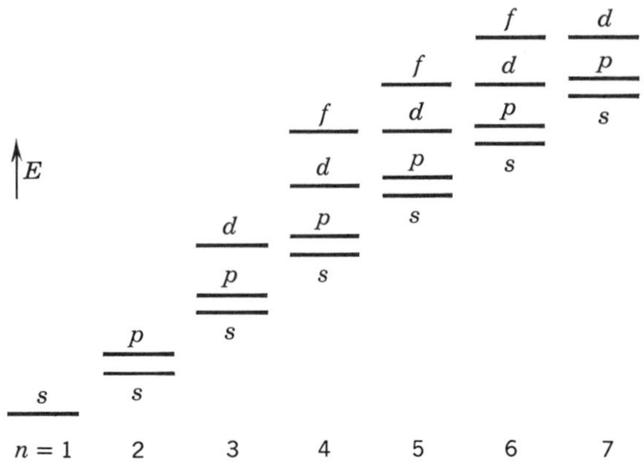

Fig. 19-38   Energy levels associated with orbitals in atomic hydrogen.

p orbitals which can overlap and interact with the d orbitals of the ion, forming secondary valence bonds which result in a complex molecule or ion, like $Fe(CN)_6^{3-}$, which has great chemical stability.

Some of the principal arrangements of ligands about central metallic atoms are shown in Fig. 19-40. When Ni^{++} combines with $4CN^-$, the result is a planar complex molecule; when Ni combines with 4CO, the result is a tetrahedral complex molecule; when Ti^{3+} combines with $6F^-$, the result is an octahedral complex molecule.

When such linking takes place, the d orbitals hybridize. Some of the possible hybridized orbitals are

per cubic bohr)$^{1/2}$. The charge density is ψ^2 and has the units of electron charge per cubic bohr. Since a bohr is equal to 0.529 Å, the charge density gives the fraction of an electron charge that would be found in a cube with an edge equal to 0.529 Å if the charge density within the cube were equal to the charge density at the point for which ψ^2 is calculated.

The relations of the values of θ and ϕ to the coordinate axes are shown in Fig. 19-39b. The variation of ψ with the radial distance r is shown in Fig. 19-39c. A semilog plot for determining the value of the radial part of the function is given in Appendix 2.

Ligands

When an atom like iron is in the ionized state Fe^{3+}, there is a tendency for it to link with molecules like NH_3 or ions like CN^-, which group around it in symmetrical patterns. These surrounding substances are called *ligands*. They are characterized by free

TABLE 19-5 d-orbital wave functions, showing angular part multipled by $\psi(r)$[a]

$$\psi(d_{z^2}) = \left[\left(\frac{5}{16\pi}\right)^{1/2}(3\cos^2\theta - 1)\right]\psi_r$$

$$\psi(d_{xz}) = \left[\left(\frac{15}{4\pi}\right)^{1/2}\sin\theta\cos\theta\cos\phi\right]\psi_r$$

$$\psi(d_{yz}) = \left[\left(\frac{15}{4\pi}\right)^{1/2}\sin\theta\cos\theta\sin\phi\right]\psi_r$$

$$\psi(d_{x^2-y^2}) = \left[\left(\frac{15}{4\pi}\right)^{1/2}\sin^2\theta\cos^2\phi\right]\psi_r$$

$$\psi(d_{xy}) = \left[\left(\frac{15}{4\pi}\right)^{1/2}\sin^2\theta\sin^2\phi\right]\psi_r$$

[a] $\psi_r = \dfrac{1}{9\sqrt{30}}\left(\dfrac{z}{a_0}\right)^{3/2}\sigma^2 e^{-\sigma/2}$

$\sigma = \dfrac{2zr}{na_0}$

$a_0 = \dfrac{h^2}{4\pi^2 me^2}$

Fig. 19-39 (*a*) Boundary diagrams of d orbitals in atomic hydrogen; (*b*) angle variables; (*c*) the radial functions $\psi_d(r)$ in d orbitals and the radial charge density $4\pi^2 r^2 \psi_d{}^2(r)$ in electron charge per cubic bohr.

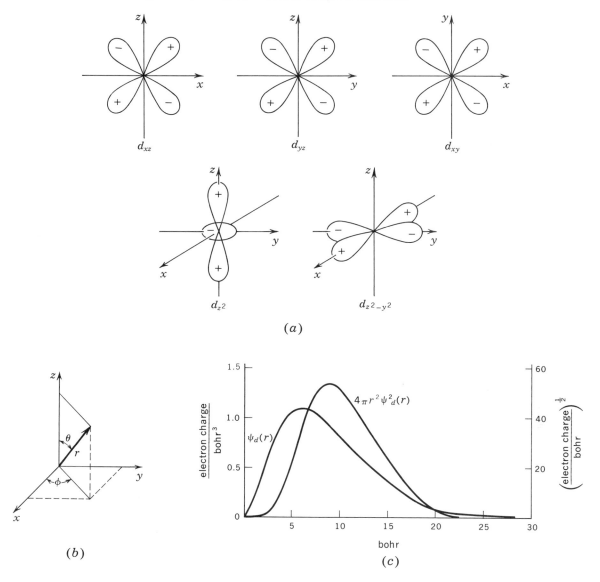

listed in Table 19-6. The coordination number is the number of ligand molecules making up the complex.

d-orbital splitting

When the patterns of charge density rearrange in the formation of a complex, there is a shift in the energy levels of the resulting orbitals, much like that pro-

duced by resonance in benzene and butadiene. Note that the d orbitals as represented in Fig. 19-39 are of two kinds. The main axes of the pear-shaped charge clouds lie *between* the coordinate axes for the d_{xz}, d_{xy}, and d_{yz} orbitals; the main axes of the charge clouds lie *along* the coordinate axes for the d_{z^2} and $d_{z^2-y^2}$ orbitals. The result is that the molecular orbitals fall

Fig. 19-40   The arrangement of ligands in complex molecules.   (After C. J. Ballhausen and H. B. Gray, "Molecular Orbital Theory," W. A. Benjamin, Inc., New York, 1964.)

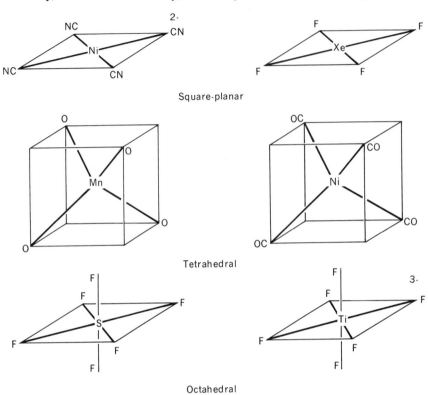

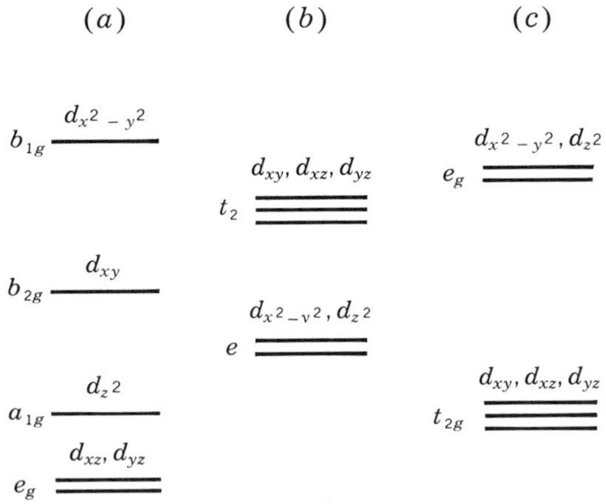

into different energy categories.   If an atomic orbital points *toward* a negatively charged ligand, it has a higher potential energy as an orbital in the complex molecule; if it points *between* two negatively charged ligands, it has a lower potential energy.   This splitting is shown diagrammatically in Fig. 19-41.   As electrons rearrange in the formation of the complex molecule, the lower orbitals are occupied first.   According to Hund's rule, there will be one electron in each level as long as the levels do not differ too much in energy and as long as there are enough levels.   The final arrangement of the electrons depends on the magnitude of the splitting produced by the ligand field.

For an octahedral field produced by six NH_3 molecules, the $d_{x^2-y^2}$ and d_{z^2} orbitals are raised while the d_{xy}, d_{yz}, and d_{xz} orbitals are lowered.   On the other

Fig. 19-41   Energy-level splitting in complex molecules as a function of shape: (*a*) square; (*b*) tetrahedral; (*c*) octahedral.

TABLE 19-6 Orbitals formed by hybridization[a]

Coordination number	Atomic orbitals	Arrangement	Angles between hybrid orbitals
2	dp	Linear	180°
	ds	Angular	
3	dp^2, ds^2	Trigonal plane	120°
	d^2p	Trigonal pyramid	90°
4	d^3s	Tetrahedral	109°28′
	dsp^2, d^2p^2	Tetragonal plane	90°
	d^2sp, dp^3, d^3p	Irregular tetrahedron	
5[b]	dsp^3, d^3sp	Trigonal bipyramid	120°, 90°
	dsp^3, d^2sp^2, d^4s,	Tetragonal pyramid	
	d^4p, d^2p^3		
6	d^2sp^3	Octahedral	90°
	d^4sp, d^5p	Trigonal prism	90°, 60°
	d^3p^3	Trigonal antiprism	
7	d^3sp^3, d^5sp	ZrF_7^{3-}	
	d^4sp^2, d^4p^3, d^5p^2	TaF_7^{--}	
8	d^4sp^3	Dodecahedron	
	d^5p^3, d^5sp^2	Prisms	

[a] After C. W. N. Cumper, "Wave Mechanics for Chemists," Academic Press, Inc., New York, 1966.
[b] dsp^3 hybridization gives a trigonal bipyramid with the d_{z^2} orbital but a tetragonal pyramid with the $d_{x^2-y^2}$ orbital.

hand, in a tetrahedral field the change is exactly opposite.

As an example of octahedral complexing, consider chromium carbonyl, $Cr(CO)_6$. There are a pair of CO molecules on each coordinate axis, one at each side of the central metal atom. Along the x axis the d-orbital lobes overlap with the antibonding π^* lobes of the orbitals of CO, as shown in Fig. 19-42. The secondary valences in the molecular complexes are produced by overlaps of this sort which result in orbitals that envelop the whole molecule in much the same way that the π orbitals envelop the whole ring in benzene.

Example 19-8

Using the graphs in Appendix 2, calculate in relative wave function units (rwfu) the value of ψ for d_{z^2} at the values of $r = 4$-, 6-, and 7-bohr units. Assume that $\theta = 0°$.

At $\theta = 0°$,

$$\psi = \left(\frac{5}{16\pi}\right)^{1/2}(3\cos^2\theta - 1)(0.02a_0{}^{-3/2})\,r^2 e^{-r/3}$$

$$= \left(\frac{5}{16\pi}\right)^{1/2}(3 - 1)(0.02a_0{}^{-3/2})\,r^2 e^{-r/3}$$

$$= (0.631)b\,r^2 e^{-r/3}$$

Since relative values are sought, the constant terms may be dropped.

$$For\ r = 4:\quad \psi = (16)(0.260)$$
$$= 4.16\ \text{rwfu}$$
$$For\ r = 6:\quad \psi = (0.0005687)(36)(0.135)$$
$$= 4.86\ \text{rwfu}$$
$$For\ r = 8:\quad \psi = (0.0005687)(49)(0.0975)$$
$$= 4.77\ \text{rwfu}$$

Fig. 19-42 Overlap along the x axis in $Cr(CO)_6$. (From M. Orchin and H. H. Jaffé, "The Importance of Antibonding Orbitals." Copyright © 1967 by Milton Orchin and Hans H. Jaffé. Published by Houghton Mifflin Company, Boston, 1967. Used by permission.)

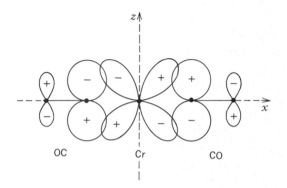

A cd unit expresses charge density in terms of electron charges per cubic bohr. Thus ψ^2 goes through a maximum on the z axis in the neighborhood of 6 bohrs, or 3.175 Å, from the nucleus.

19-7   ELECTRONEGATIVITY

From the preceding discussion of the character of bonds, it is clear that there is a variation all the way from the strongly ionic type like the Na—Cl bond to the pure covalent type like the C—C bond. Obviously, the bond character depends on the nature of the atoms being bonded. When a bond is formed between X and Y to give X—Y, the bond is more ionic in character if X and Y lie at opposite sides of the periodic table of chemical elements; but the bond is more covalent in character if X and Y lie closer together, like C—O, or are identical like C—C. In order to predict in as quantitive a way as possible the character of the bond, efforts have been made to set up a scale on which the tendency of atoms to attract an electron is measured. This is called the scale of *electronegativity*.

The values of the electronegativity are denoted by the symbols x_A and x_B; these values are selected in such a way as to maximize the validity of the equation

$$\Delta'_{A-B} = 30(x_A - x_B)^2 \tag{19-31}$$

where x_A and x_B are the values of the electronegativity for atoms A and B and Δ'_{A-B} is the ionic resonance energy of the A—B bond. This is related to the dissociation energy D_{A-B} of the A—B bond and the dissociation energies D_{A-A} and D_{B-B} of the A—A and B—B bonds by the expression

$$\Delta'_{A-B} = D_{A-B} - (D_{A-A}D_{B-B})^{1/2} \tag{19-32}$$

where the dissociation energies are expressed in kilocalories. The difference between x_A and x_B is close numerically to the dipole moment associated with the A—B bond.

The Amercian chemists N. B. Hannay and C. P. Smyth proposed that the percent of ionic character C_i may be related to electronegativity by the expression

$$C_i = 16 \sqrt{(x_A - x_B)^2} + 3.5 (x_A - x_B)^2 \tag{19-33}$$

When an electron is brought from infinity and introduced into an atomic shell, the potential energy change is called the *electron affinity*.

Values of the electron affinity A are listed in Table 19-7; values of the ionization potential I_A are given in Table 19-8; values of the electronegativity x are shown in Table 19-9; and values of bond energy D_{AB} are included in Tables 19-10.

Example 19-9

Calculate the percent of ionic character C_i of the bond in KI, using the values of the electronegativity in Table 19-9.

$$
\begin{aligned}
C_i &= 16 \sqrt{(x_A - x_B)^2} + 3.5 (x_A - x_B)^2 \\
&= 16 \sqrt{(0.8 - 2.4)^2} + 3.5 (0.8 - 2.4)^2 \\
&= 16(1.6) + 3.5(2.56) = 25.6 + 9.0 \\
&= 34.6\%
\end{aligned}
$$

TABLE 19-7   Electron affinity

Element	E_a, ev
H	−0.7
F	−3.5
Cl	−3.5
Br	−3.4
I	−3.1

TABLE 19-8   Ionization potentials

Element	Atomic radius, Å	$I_{\frac{1}{2}}^{I}$, ev	$I_{\frac{1}{2}}^{II}$, ev	$I_{\frac{1}{2}}^{III}$, ev	$I_{\frac{1}{2}}^{IV}$, ev	Electronic structure		
						d	p	s
Group I:								
Li	1.50	5.4						2 1
Na	1.86	5.1					6	2 2 1
K	2.27	4.3					6 6	2 2 2 1
Rb	2.43	4.2				10	6 6 6	2 2 2 2 1
Cs	2.62	3.9				10 10	6 6 6 6	2 2 2 2 2 1
Third row:								
Na	1.86	5.14	47.29				6	2 2 1
Mg	1.60	7.65	15.03	80.12			6	2 2 2
Al	1.48	5.99	18.82	28.44	120		6 1	2 2 2
Si	1.17	8.15	16.34	33.49	45.13		6 2	2 2 2
P	1.0	10.98	19.65	30.16	51.35		6 3	2 2 2
S	1.06	10.35	23.41	35.05			6 4	2 2 2
Cl	0.97	12.96	23.80	39.91			6 5	2 2 2
Ar		15.76	27.6				6 6	2 2 2
(K)	2.27	4.34	31.81	45.7			6 6	2 2 2 1
Fourth row:								
K	2.27	4.34						
Ca	1.97	6.1						
Sc	1.51	6.7						
Ti	1.45	6.8						
V	1.31	6.7						
Cr	1.27	6.7						
Mn	1.25	7.4						
Fe	1.25	7.8						
Co	1.25	7.8						
Ni	1.24	7.6						
Cu	1.28	7.7						
Zn	1.33	9.4						
Ga	1.33	6.0						
Ge	1.22	8.1						
As	1.25	10.5						
Se	1.16	9.7						
Br	1.13	11.8						

Problems

1. Calculate the wavelength in angstrom units of a photon just sufficient to put a molecule of KH in the first vibrational energy level.

2. Calculate the wavelength in microns of the photon in the infrared region which has the energy equivalent to the energy difference between the zero vibrational level and the first vibrational level for the molecule CO.

3. Calculate the wavelength in microns of the photon in the infrared region which would be absorbed

TABLE 19-9   Electronegativities of atoms

Li	Be	B	C	N	O	F
1.0	1.5	2.0	2.5	3.0	3.5	4.0
Na	Mg	Al	Si	P	S	Cl
0.9	1.2	1.5	1.8	2.1	2.5	3.0
K	Ca	Sc	Ge	As	Se	Br
0.8	1.0	1.3	1.7	2.0	2.4	2.8
Rb	Sr	Y	Sn	Sb	Te	I
0.8	1.0	1.3	1.7	1.8	2.1	2.4
Cs	Ba	H				
0.7	0.9	2.1				

TABLE 19-10   Approximate bond lengths and bond energies

Bond	Bond energy, kcal	Bond length, Å
F—F	37	1.28
Cl—Cl	58	1.99
Br—Br	46	2.28
I—I	36	2.68
H—H	104	0.74
C—C	83	1.54
O—O	33	1.32
N—N	38	1.40
S—S	51	2.08
Si—Si	42	2.34
P—P	51	2.20
C—F	105	1.41
N—Cl	48	1.47
C—Br	66	1.91
I—Cl	50	2.32
C—H	99	1.07
S—Cl	60	2.03
H—O	111	0.96
N—H	93	1.00
C—Cl	78	1.76
Si—I	51	2.50
P—Br	65	2.24

by a molecule of HBr in going from the zero vibrational energy level to the second vibrational energy level.

4. Calculate the wavelength in microns of a photon emitted by a molecule of HF in dropping from the first vibrational energy level to the zero vibrational energy level.

5. Calculate the approximate wavelength in angstroms of a photon emitted by the molecule HF in dropping from the fourth vibrational energy level to the zero vibrational energy level, assuming that the vibrational levels are equally spaced. Will this radiation lie in the infrared, the visible, or the ultraviolet part of the spectrum?

6. Calculate the wavelength in angstroms of a photon having just sufficient energy to dissociate a molecule of HF which is in the zero vibrational energy level. Will this correspond to light in the infrared, visible, or ultraviolet portion of the spectrum?

7. Calculate the wavelength in angstroms of a photon having energy equivalent in amount to that necessary to dissociate a molecule of N_2 which is in the zero vibrational energy level.

8. Calculate the wavelength in angstroms of a photon having just sufficient energy to dissociate a molecule of HI when this molecule is in the second vibrational energy level.

9. Calculate the value of the energy change when independent ions are formed from neutral atoms by the transfer of electrons from Al to the Br atoms.

10. Starting with neutral atoms, calculate the energy change involved when independent K^+ and I^- ions are formed.

11. Draw the triangle diagrams showing the formation of the compound Cs^+I^- from the neutral atoms.

12. Draw the triangle diagrams showing the formation of the compound $Ba^+I_2^-$ from the neutral atoms by electron transfer.

13. Draw the triangle diagrams showing the formation of the compound $SiCl_4$ from the neutral atoms, indicating how electrons are shared to complete the outer shells in all the atoms.

14. Calculate Δ'_{A-B} and C_i for the formation of the compound NaF.

15. Calculate Δ'_{A-B} and C_i for the compound CsBr.

16. Using values of bond energy, calculate the energy change when CCl_4 is formed from graphite and Cl_2. Assume that two C—C bonds are broken for each carbon atom extracted from the graphite.

17. Calculate from bond energies the energy change when one molecule of Cl_2 and one molecule of I_2 exchange atoms and form two molecules of ICl.

18. Using the values of bond energy, calculate the energy change when a molecule of $H_3C—CH_3$ combines with a molecule of H_2 to form two molecules of CH_4.

19. Using the graphs in Appendix 2, (*a*) calculate the value of the sum ψ of ψ_s in H and ψ_s in F at the midpoint between the two nuclei; assume that ψ_s in F has the same variation with r as ψ_s in H; (*b*) calculate ψ^2.

20. Using the graphs in Appendix 2, (*a*) calculate the sum ψ of ψ_s in H and ψ_p in N at the midpoint of the NH bond, assuming that ψ_p for N is the same as ψ_p for H; (*b*) calculate ψ^2.

21. Using the graphs in Appendix 2, (*a*) calculate the sum ψ of ψ_d for Cr and ψ_p for C at the point where the axis of the d lobe of Cr intersects the axis of the p lobe of C in the $Cr(CO)_6$ complex; estimate the radial distance from Fig. 19-42, using the CO distance to establish the scale; (*b*) calculate ψ^2.

REFERENCES

Kauzmann, W.: "Quantum Chemistry," Academic Press, Inc., New York, 1957.

Pauling, L.: "The Nature of the Chemical Bond," 3d ed., Cornell University Press, Ithaca, N.Y., 1960.

Coulson, C. A.: "Valence," Oxford University Press, Fair Lawn, N.J., 1961.

Streitwieser, A.: "Molecular Orbital Theory for Organic Chemists," John Wiley & Sons, Inc., New York, 1961.

Roberts, J. D.: "Notes on Molecular Orbital Calculations," W. A. Benjamin, Inc., New York, 1962.

Simpson, W. T.: "Theories of Electrons in Molecules," Prentice-Hall, Inc., Englewood Cliffs, N.J., 1962.

Parr, R. G.: "The Quantum Theory of Molecular Electronic Structure," W. A. Benjamin, Inc., New York, 1963.

Pullman, B., and A. Pullman: "Quantum Biochemistry," John Wiley & Sons, Inc., New York, 1963.

Ballhausen, C. J., and H. B. Gray: "Molecular Orbital Theory," W. A. Benjamin, Inc., New York, 1964.

Spice, J. E.: "Chemical Binding and Structure," Pergamon Press, New York, 1964.

Löwdin, P. O., and B. Pullman: "Molecular Orbitals in Chemistry, Physics and Biology," Academic Press, Inc., New York, 1964.

Sinanoglu, O. (ed.): "Modern Quantum Chemistry: part I: Orbitals," Academic Press, Inc., New York, 1965.

Phillips, L. F.: "Basic Quantum Chemistry," John Wiley & Sons, Inc., New York, 1965.

Cumper, C. W.: "Wave Mechanics for Chemists," Academic Press, Inc., New York, 1966.

Lagowski, J. J.: "The Chemical Bond," Houghton Mifflin Company, Boston, 1966.

Liberles, A.: "Introduction to Molecular-Orbital Theory," Holt, Rinehart and Winston, Inc., New York, 1966.

Löwdin, P. O.: "Quantum Theory of Atoms, Molecules and the Solid State," Academic Press, Inc., New York, 1966.

Chu, B.: "Molecular Forces," John Wiley & Sons, Inc., New York, 1967.

Orchin, M. O., and H. H. Jaffé: "The Importance of Antibonding Orbitals," Houghton Mifflin Company, Boston, 1967.

Pohl, H. A.: "Quantum Mechanics for Science and Engineering," Prentice-Hall, Inc., Englewood Cliffs, N.J., 1967.

Strauss, H. L.: "Quantum Mechanics: An Introduction," Prentice-Hall, Inc., Englewood Cliffs, N.J., 1968.

Aside from data derived directly from chemical behavior, the most helpful sources of information about molecular structure are the spectra produced when molecules emit or absorb radiation.

20-1 SPECTRA AND STRUCTURE

As the first step in discussing the relation between molecular structures and their spectra, it is helpful to recall a few of the basic facts concerning the nature of radiation. Electromagnetic radiation has two complementary aspects. In refraction, diffraction,

Chapter Twenty

molecular spectra

and reflection it changes direction in a way which suggests that it is composed of trains of waves with the parameters wavelength λ, frequency ν, and velocity c, as shown in Fig. 20-1a. The basic relation between these parameters is

$$\lambda \nu = c \tag{20-1}$$

By way of contrast, in the photoelectric effect and in many aspects of emission and absorption, radiation behaves in a way that suggests a stream of particles, the *photons* postulated by Einstein on the basis of Planck's quantum hypothesis, as shown in Fig. 20-1b. Each photon has an amount of energy E_p related to the frequency ν_p by the expression

$$E_p = h\nu_p \tag{20-2}$$

where h is Planck's constant and the subscript p stands for *photon*. Moreover, in its interaction with matter, the photon behaves as if it consisted of a positive and a negative electric charge oscillating toward and away from each other with the frequency ν_p.

Molecular energy levels

An important example of the interaction of a photon with matter is provided by the absorption of radiation by a single atom like a hydrogen atom, which consists of a single positively charged proton (the hydrogen nucleus) and a negatively charged electron. According to the de Broglie hypothesis, the electron,

Fig. 20-1 The (*a*) wave and (*b*) particle aspects of light.

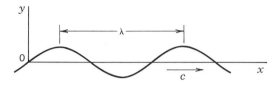

ν = waves per second passing a fixed point
(*a*) c = cm per second at which a wave crest moves

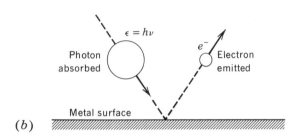

(*b*)

although acting like a particle when it is in the free state, behaves like a three-dimensional wave when it is bound to the nucleus. This wave can assume many different kinds of patterns, each pattern corresponding to a standing steady-state wave defined as a solution of the Schrödinger equation. With each type *i* of wave, there is associated a definite value of the energy E_i. The values of E_i constitute the *energy levels* of the atom. When a photon is absorbed by the atom, raising it from the zero energy level, where the atom has energy E_0, to the first energy level E_1, the relation between the energy E_p of the photon absorbed and the energy gained by the atom ($E_1 - E_0$) is

$$E_p = E_1 - E_0 \qquad (20\text{-}3)$$

In the same way, the electrons in a molecule, for example, HCl, can assume many different patterns of waves, each with an associated value of the energy E_i; but as contrasted with an atom where there is only a single nucleus, the molecule has two or more nuclei; and the patterns of de Broglie waves embody not only different electronic energy values but also additional amounts of energy corresponding to the rotation of the nuclei about their center of mass and

to the vibration of the nuclei toward and away from each other. Each of these patterns *i* has a value of the energy E_i associated with it; all the values of E_i constitute the set of energy levels of the molecule.

As an example, consider the various ways in which a molecule like HCl can acquire values of energy lying above the lowest, or ground, energy level E_0. The molecule can rotate at different speeds; the faster the rotation, the higher the energy; but since the possible speeds of rotation are quantized, there is a discrete set of energy levels for rotations, as shown in Fig. 20-2. As another alternative, the molecule can vibrate, the two nuclei moving toward and away from each other at a constant frequency. This motion is also quantized, and there is a discrete set of energy levels corresponding to vibration, as shown in Fig. 20-3. As still another alternative, the pattern of the de Broglie waves can change, from the ground pattern where the two *s* orbitals of the hydrogen atom and the chlorine atom merge to form a common σ orbital to a different molecular-orbital pattern which has a higher value of energy (Fig. 20-4), i.e., to a higher *electronic* energy level.

Note that the spacing between the vibrational levels in Fig. 20-3 is much larger than that between rotational levels in Fig. 20-2; likewise, the spacing between electronic levels in Fig. 20-4 is much greater than that between the vibrational levels in Fig. 20-3. This variation in the spacing of energy levels is characteristic of nearly all molecules. The three types of levels can be combined into a single diagram (Fig. 20-5), in which each level is labeled with three quantum numbers as subscripts on E. For example, the first subscript can denote the electronic level, the second the vibrational level, and the third the rotational level. The energy in the state having the lowest possible amount of energy, i.e., the *ground* state, is $E_{0,0,0}$. In this state there is the minimum amount of vibration and rotation, and the de Broglie wave has the pattern with minimum energy for a molecular orbital.

Suppose the molecule absorbs a photon having energy E_p and then shifts to the first electronic level but does not increase its vibration or rotation; the new value of the energy will be $E_{1,0,0}$ and $E_p = E_{1,0,0} - E_{0,0,0}$. But if the molecule absorbs a photon with a little higher energy E'_p and also starts to vibrate more rapidly as it shifts to the new electronic level, the new quantum number will be 1,1,0 and $E'_p = E_{1,1,0} - E_{0,0,0}$. If, in addition, the change involves an increased amount of rotation due to the absorption of a still more energetic photon E''_p, then $E''_p = E_{1,1,1} - E_{0,0,0}$. These three changes are shown in Fig. 20-5. The regions of the spectrum in which absorption or emission generally occur when these changes take place are shown in Fig. 20-6.

If the molecule is *originally* in the 1,1,1 state and then changes back to the 0,0,0 state, the energy of the *emitted* photon E''_p will be equal to the energy difference between the two states, $E''_p = E_{1,1,1} - E_{0,0,0}$.

Such changes in state are called *transitions;* not all transitions are possible; the rules governing the possibilities of transition are called *selection rules* and are derived from the probability relations associated with the de Broglie wave patterns for the initial and the final states. Various aspects of these different types of transition will be discussed in the following sections.

Potential functions

Both the rotation and the vibration of a molecule alter the distance of separation of the nuclei. In determining the energy levels associated with these types of motion, it is helpful to know the relation between the distance of separation of a pair of bonded nuclei and the potential energy. As a simple example of this relation, consider the hydrogen molecule. As discussed in the previous chapter, the chemical bond between the two hydrogen atoms behaves much like a spiral spring. When the atoms are separated by a distance r_e, the force of repulsion balances the force of attraction and the potential energy is at a minimum, as shown in Fig. 20-7. When the nuclei move closer together, the force of repulsion becomes greater than the force of attraction and the potential energy is increased. When the nuclei move farther apart, the force of attraction becomes greater than the force of repulsion and the potential energy likewise increases. But when the distance of separation is sufficiently great, the force of attraction becomes vanishingly small and the potential energy approaches a constant value, equivalent to the state corresponding to the

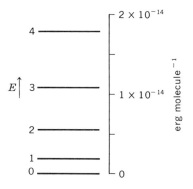

Fig. 20-2   The rotational energy levels for gaseous HCl.

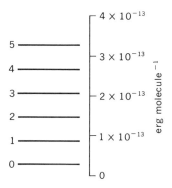

Fig. 20-3   The vibrational energy levels for gaseous HCl treated as an ideal harmonic oscillation.

complete chemical separation of the two nuclei, the breaking of the bond.

The American physicist Philip Morse suggested a simple empirical algebraic expression which expresses the variation of potential energy with distance quite accurately

$$U = D_{eq}(1 - e^{-\beta(r - r_{eq})})^2 \tag{20-4}$$

where U = potential energy
  D_{eq} = potential energy at infinite separation
  r = distance of separation of nuclei
  r_{eq} = distance of separation at equilibrium
  β = empirical constant

According to this equation, $U = 0$ at $r = r_{eq}$ and $U = D_{eq}$ at $r = \infty$; however, it is more usual to define $U = 0$ at $r = \infty$ and $U = -D_{eq}$ at $r = r_{eq}$. With

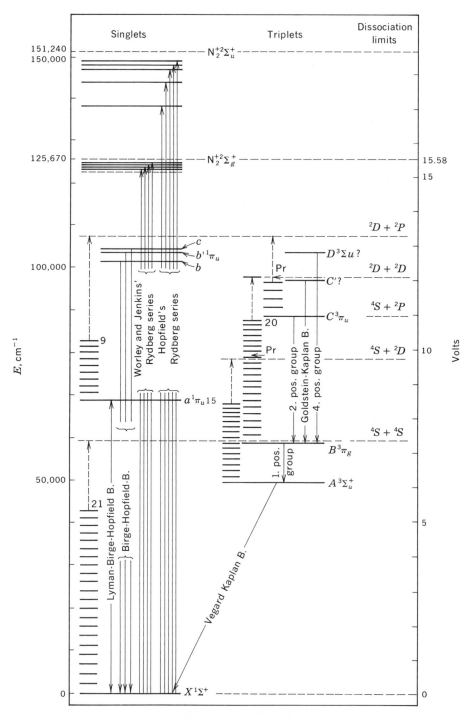

Fig. 20-4 Electronic energy levels in gaseous N_2. (*Note:* a few of the vibrational levels are shown in short lines above some of the electronic levels.) (*After* "Molecular Spectra and Molecular Structure," vol. 1, Spectra of Diatomic Molecules, 2d ed., by G. Herzberg. Copyright © 1950 by Reinhold Publishing Corporation, by permission of Van Nostrand Reinhold Company.)

the help of spectra, it is possible to determine the actual values of the energy levels and the distance of separation of the nuclei associated with each level. This makes possible the calculation of the true potential-energy function, shown as a solid line in Fig. 20-7. The values calculated on the basis of the Morse function are shown as a dotted line. The two curves coincide quite closely over a considerable portion of the range. The potential-energy curve for HCl is shown in Fig. 20-8.

When light is passed through gaseous hydrogen or hydrochloric acid, the molecules actually are moving rapidly in space as the radiation is absorbed or emitted. However, in devising expressions for the potential energy, the center of mass of the molecule can be taken as the zero point of the coordinate system and the motions of the nuclei expressed in terms of such coordinates.

20-2 INFRARED SPECTRA

As shown in Fig. 20-6, infrared spectra are found in the wavelength range bounded on one side by the red limit of visible radiation (about 0.75 μ) and on the other side by the lower limit of the region generally called *microwave* (about 500 μ). The portion of this range between 0.75 and 2.5 μ is called the *near infrared,* and the portion between 2.5 and 500 μ is called the *far infrared.* These boundaries are empirically set, and there is no generally accepted precise definition of their location. The measurements of infrared spectra require different types of instruments, of design appropriate to the region under investigation.

Infrared spectrometers

At room temperature, substances do not emit sufficiently intense infrared radiation to permit the direct observation of their emission spectra with common types of spectrometers. Figure 20-9 shows plots of the intensity of infrared emission in different parts of the spectrum for blackbodies at different temperatures. From these curves it can be seen that in order to get a sufficient intensity of radiation (about 1,000 units) for the measurement of the emission spectra of a substance like benzene, the molecules

must be heated to a temperature so high that the benzene disintegrates rapidly. For this reason, nearly all determinations are made from the measurement of the frequencies absorbed when infrared radiation from an external source is passed through the vapor, through a thin layer of the substance in the liquid or

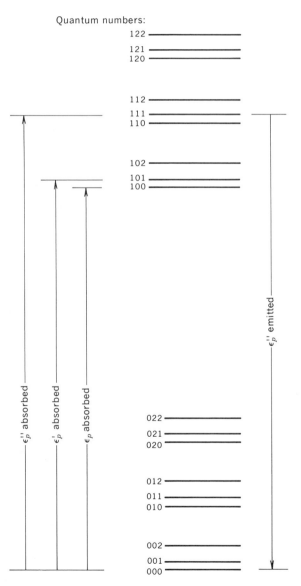

Fig. 20-5 Electronic, vibrational, and rotational energy levels (not to scale).

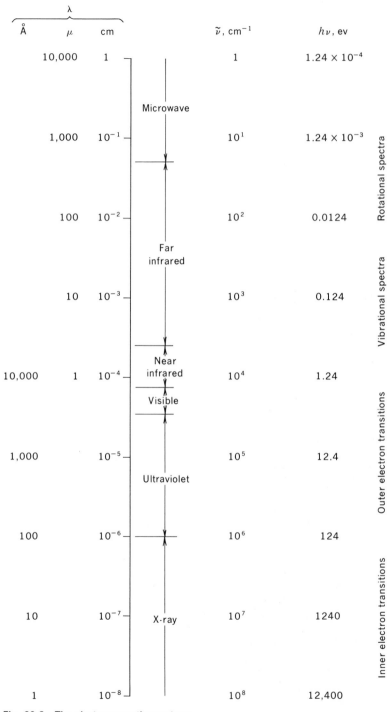

Fig. 20-6 The electromagnetic spectrum.

crystalline state, or in solution in a solvent for which the infrared spectrum is known.

In Fig. 20-10 the optical path in a typical infrared spectrograph is shown. The source of radiation is generally a heated piece of metal or refractory material. The radiation is formed into a beam which is then passed through the sample and refracted or diffracted into a spectrum. On passing through the sample of the substance under investigation, certain frequencies are absorbed in varying amounts, and the amount of absorption is measured as a function of wavelength.

In the near infrared a prism can be used as the means of refraction. As the prism is turned to increase refraction, radiation of decreasing wavelength falls on the detector. The wavelength can be read on a scale calibrated to express wavelength as a function of angle of rotation. In the region just beyond the visible a glass prism can be used; but the absorption of infrared radiation by glass increases rapidly with increasing wavelength so that, for observations in the region much beyond 1-μ wavelength, the prism is generally made of rock salt or some similar substance. For observations at still greater wavelengths, the prism is replaced by a diffraction grating.

The amount of absorption by the sample can be measured by the amount of blackening on the photographic film in the portion of the near infrared lying closest to the visible range. But over the greater portion of the infrared spectrum a detector customarily is used which changes temperature as the amount of radiation falling on it increases. For example, there are many types of thermocouples and bolometers from which the electrical output is a function of temperature. Usually radiation is interrupted at a steady frequency by a rotating chopper, producing a pulsed electric signal which can be amplified electronically. It has been possible to measure changes as small as 10^{-7} erg falling on a detector in a millisecond. In the most precise instruments the beam of radiation from the source is split, half of the beam passing through the sample and falling on one detector and the other half passing directly to another similar detector. The signals from the detectors are opposed so that as the different portions of the spectrum sweep over the detectors, the instrument measures the difference in intensity caused by the passage through the sample. A roll of paper is pulled beneath a pen at a rate geared to the rate of change of wavelength; the position of the pen on the vertical scale measures the amount of absorption. The components of such an arrangement are shown in Fig. 20-11.

Fig. 20-7 The potential-energy curve for H—H (solid line) showing the vibrational energy levels, and the potential-energy curve (dotted line) calculated from the Morse equation. (After "Molecular Spectra and Molecular Structure," vol. 1, Spectra of Diatomic Molecules, 2d ed., by G. Herzberg. Copyright © 1950 by Reinhold Publishing Corporation, by permission of Van Nostrand Reinhold Company.)

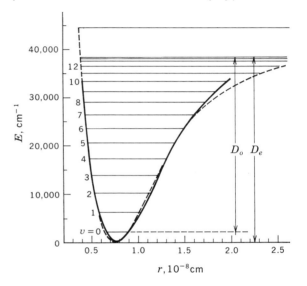

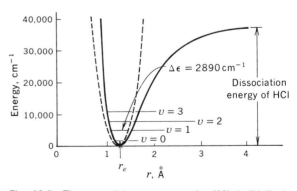

Fig. 20-8 The potential-energy curve for HCl (solid line) showing vibrational energy levels, and the theoretical potential-energy curve (dotted line) for an ideal harmonic oscillation. (After G. M. Barrow, "Physical Chemistry," 2d ed., McGraw-Hill Book Company, New York, 1966.)

Fig. 20-9   The emission of electromagnetic radiation from blackbodies at different temperatures. (From ''Infrared Spectroscopy,'' by Robert T. Conley, Fig. 3.3, p. 31. Copyright © 1966 by Allyn and Bacon, Inc., Boston. Reprinted by permission of the publisher.)

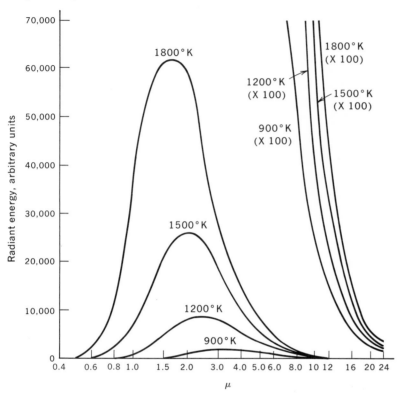

Rotational spectra

The simplest rotational spectra are those produced by the absorption of infrared radiation when diatomic molecules pass from a state of lower to a state of higher rotational energy, as discussed in the previous section. A diatomic molecule is shown schematically in Fig. 20-12. The moment of inertia I is given by

$$I = m_1 r_1^2 + m_2 r_2^2 = \mu r^2 \qquad (20\text{-}5)$$

where m_1 and m_2 are the masses of the atoms, r_1, r_2, and r are the distances shown in the figure, and μ is the reduced mass, equal to $m_1 m_2 (m_1 + m_2)^{-1}$.

A solution of the Schrödinger equation for rotatory motion shows that the energy values of the different rotational states are given by

$$E_J = \frac{h^2 J(J + 1)}{8\pi^2 I} \qquad (20\text{-}6)$$

where J is the rotational quantum number, which can have the values 0, 1, 2, 3, 4, ...

According to classical mechanics the energy E of a rotating body is

$$E = \tfrac{1}{2} I \omega^2 = \frac{M^2}{2I} \qquad (20\text{-}7)$$

where ω is the rotational frequency in radians per second and M is the angular momentum $I\omega$. From Eqs. (20-6) and (20-7),

$$M = \frac{h}{2\pi} \sqrt{J(J + 1)} \cong \frac{h}{2\pi} J \qquad (20\text{-}8)$$

Because of the fundamental equation,

$$E = h\nu = hc\bar{\nu} \qquad (20\text{-}9)$$

it is possible to divide Eq. (20-6) on both sides by hc

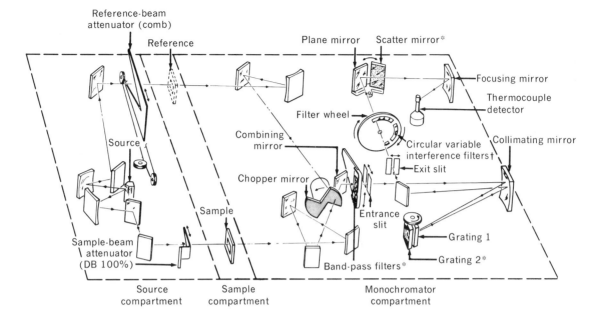

Reference-beam attenuator (comb)

Reference

Plane mirror   Scatter mirror*

Focusing mirror

Thermocouple detector

Filter wheel

Source

Combining mirror

Circular variable interference filters†

Collimating mirror

Chopper mirror

Exit slit

Sample

Entrance slit

Grating 1

Sample-beam attenuator (DB 100%)

Band-pass filters*

Grating 2*

Source compartment

Sample compartment

Monochromator compartment

*IR-20 only
†Manufactured to Beckman specifications by Optical Coating
Laboratory, Incorporated, Santa Rosa, California.

(a)

(b)

Fig. 20-10   (a) **An infrared spectrometer (diagrammatic);** (b) **photograph of the spectrometer.   (Courtesy of Beckman Instruments, Inc.)**

Fig. 20-11 Components of a typical infrared spectrometer. (From Robert T. Conley, "Infrared Spectroscopy," Fig. 3.1, p. 29. Copyright © 1966 by Allyn and Bacon, Inc., Boston. Reprinted by permission of the publisher.)

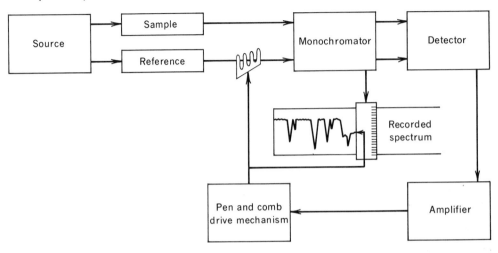

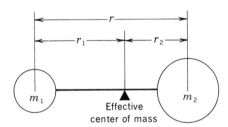

Fig. 20-12 The coordinates of a diatomic molecule.

and express the values of the energy levels in wavenumber $\tilde{\nu}$ units

$$E(\text{cm}^{-1}) = \frac{hJ(J+1)}{8\pi^2 cI} = BJ(J+1) \qquad (20\text{-}10)$$

where $B = h/8\pi^2 cI$. The energy levels can thus be written in terms of units of B, as shown in Fig. 20-13.

If transitions take place only between adjacent energy levels, as shown in the figure, the absorption lines corresponding to these transitions are evenly spaced on a cm^{-1} scale, each lying beyond its neighbor by an amount $2B$. The observed pure rotation spectrum of HCl is shown in Fig. 20-14. The first absorption line, corresponding to $2B$, lies at 20.7 cm^{-1}; the next line, corresponding to $4B$, lies at 41.4 cm^{-1}; the third line, corresponding to $6B$, lies at 62.1 cm^{-1}; and so on. Thus each line is separated from its neighbor by $2B$. These observed values permit an extremely accurate determination of the value of B, and from this it is possible to calculate an accurate value of the moment of inertia of HCl. Since the masses of the nuclei are known accurately, this makes it possible to calculate accurately the distance r by which the two nuclei are separated, i.e., the length of the HCl bond.

Example 20-1

Calculate the length of the ^{1}H^{35}Cl bond from the separation of the lines in the rotational spectrum. *Note:* the superscripts are the mass numbers of the atoms.

From the spectrum the average value of B is found to be 10.4 cm^{-1}. Thus

$$I = \frac{h}{8\pi^2 cB}$$

$$= \frac{6.63 \times 10^{-27}}{8 \times 9.87 \times 2.99 \times 10^{10} \times 10.4}$$

$$= 2.70 \times 10^{-40}$$

$$\mu = \frac{m_1 m_2}{m_1 + m_2} = \frac{1 \times 35}{(1 + 35)6.02 \times 10^{23}}$$

$$= 1.61 \times 10^{-24}$$

$$r = \sqrt{\frac{I}{\mu}} = \sqrt{\frac{2.70 \times 10^{-40}}{1.61 \times 10^{-24}}}$$

$$= \sqrt{1.68 \times 10^{-16}} = 1.30 \times 10^{-8} \text{ cm}$$

Matrix-isolated molecules

If a molecule like HCl is trapped in a hole in a crystal lattice of an inert substance like frozen neon, there is evidence that this molecule can rotate freely, much as if it were free in space. The American physical chemist D. W. Robinson has obtained far-infrared spectra of such matrix-isolated molecules which resemble pure rotation spectra. For example, in the gaseous state, a line for HF has been observed at 41.9 cm^{-1} corresponding to the shift from $J = 0$ to $J = 1$. With HF trapped in a matrix, the following lines have been observed: 39.8 (neon), 44.0 (argon), 45.0 (krypton), 50.5 (xenon). A molecule of H_2O trapped in a matrix shows similar "lines" which actually appear as absorption bands of varying widths due to interaction with the matrix; these bands are shown in Fig. 20-15.

Band intensity

The intensity of absorption due to transitions between rotational energy levels depends on several factors.

First of all, the molecule must have a dipole moment so that there can be interaction with the electromagnetic vibration in the radiation and a transfer of energy by this means from the radiation to the molecule. Thus, molecules like HF and HCl show pure rotational spectra while molecules without dipole moments like H_2 and N_2 do not.

There are also relative differences in the intensities of the different bands in the same molecule due to the different populations in the different rotational levels. The relative population at thermal equilibrium is given by the Boltzmann formula

$$N_i = N_0 g_i e^{-E_i/kT} \qquad (20\text{-}11)$$

Fig. 20-13 Rotational energy levels in gaseous HCl for which $B_v = 10.35$ cm^{-1}.

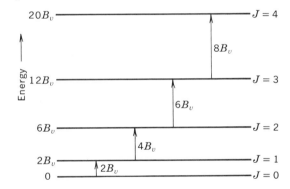

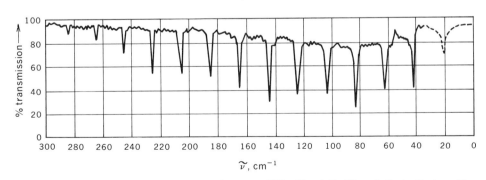

Fig. 20-14 The pure notation spectrum of gaseous HCl. The dotted lines indicate the transition from $J = 0$ to $J = 1$ for which $\bar{\nu}$ was too small to permit observation. (After G. M. Barrow, "Physical Chemistry," 2d ed., McGraw-Hill Book Company, New York, 1966.)

Fig. 20-15   Observed absorption spectra of H$_2$O condensed with a 500-fold molar excess of inert gas onto a window cooled with liquid helium.   [From D. W. Robinson, *J. Chem. Phys.*, 39:3431 (1963).]

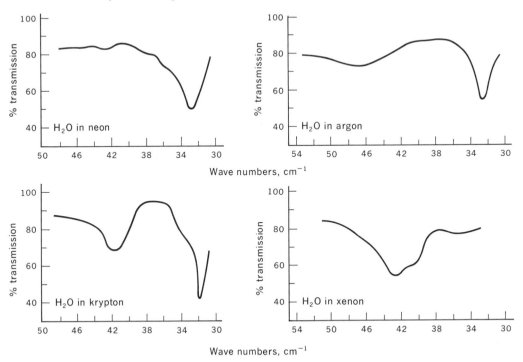

where N_i = number of molecules in ith rotational
  level

$\quad N_0$ = number of molecules in zeroth level

$\quad g_i$ = degeneracy of ith level

$\quad E_i$ = energy of the ith level

$\quad k$ = Boltzmann constant

Thus the band for the transition from the J level to the $J + 1$ level is more intense if J is a relatively small number, because there are more molecules in that level ready to absorb the radiation.

Vibration spectra

In order to understand the nature of the bands produced in the infrared spectrum when a molecule passes from one vibrational level to another, consider the case of a diatomic molecule like HCl which behaves much like an ideal harmonic oscillator in its lower vibrational levels and deviates from this behavior in a relatively simple way at higher levels.

In the harmonic oscillator, the variation of force with distance is given by the equation expressing Hooke's law

$$F = -k_h x \qquad (20\text{-}12)$$

where F = force, dynes

$\quad k_h$ = elastic constant

$\quad x$ = displacement from equilibrium position, cm

Note: the subscript h in k_h denotes Hooke's law and distinguishes this from Boltzmann's constant k. For a diatomic molecule, Hooke's law can be expressed

$$F = -k_h(r - r_{eq}) \qquad (20\text{-}13)$$

where according to the previously used notation, r_{eq} is the distance of separation of the nuclei at *equilibrium* and r is the distance of separation at which force F is exerted. Such an oscillator vibrates with a frequency ν given by

$$\nu = \frac{1}{2\pi}\left(\frac{k_h}{\mu}\right)^{1/2} \qquad (20\text{-}14)$$

where μ is the reduced mass. The frequency can be expressed in wave numbers (cm^{-1}) instead of reciprocal seconds (sec^{-1}) by dividing by the velocity of light c:

$$\bar{\nu}(\text{cm}^{-1}) = \frac{\nu(\text{sec}^{-1})}{c} = \frac{1}{2\pi c}\left(\frac{k_h}{\mu}\right)^{1/2} \qquad (20\text{-}15)$$

For an ideal harmonic oscillator, the solution of the Schrödinger equation shows that the vibrational energy levels expressed in cm^{-1} and denoted by E_v are given by

$$E_v = \omega_{eq}(v + \tfrac{1}{2}) \qquad (20\text{-}16)$$

where v is the vibrational quantum number, which may be equal to 0, 1, 2, 3, 4, . . . , and ω is the frequency in cm^{-1}, where the subscript eq signifies that the quantity has the value corresponding to the equilibrium position r_{eq}.

As contrasted with the rotational energy levels, which become more and more widely separated as the value of J increases (Fig. 20-13), the vibrational levels of an ideal harmonic oscillator are all spaced the same distance apart (Fig. 20-16a). But in the real molecule, the force of attraction between the nuclei falls off as r increases; consequently, the distance between the levels actually becomes less and less as v increases, as shown in Fig. 20-16b. This can be expressed quite satisfactorily by

$$E_v = \omega_{eq}(v + \tfrac{1}{2}) - \omega_{eq}x_{eq}(v + \tfrac{1}{2})^2 \qquad (20\text{-}17)$$

where ω_{eq} is the frequency of vibration approached as a limit as the energy of vibration approaches zero and x_{eq} is an empirical constant for the particular molecule to which the equation is applied. Values of ω_{eq} and $\omega_{eq}x_{eq}$ are given in Table 20-1 for a number of diatomic molecules.

The vibrational levels for polyatomic molecules are far more complicated and will be discussed in Sec. 20-6.

Example 20-2

From the data in Table 20-1, calculate the value of the energy in cm^{-1} of the fifth vibrational level of $^1H^{35}Cl$.

$$\begin{aligned}
E_v &= \omega_{eq}(v + \tfrac{1}{2}) - \omega_{eq}x_{eq}(v + \tfrac{1}{2})^2 \\
&= 2988.95\,(5 + \tfrac{1}{2}) - 51.65\,(5 + \tfrac{1}{2})^2 \\
&= 1.644 \times 10^4 - 0.156 \times 10^4 \\
&= 1.488 \times 10^4 \text{ cm}^{-1} \qquad Ans.
\end{aligned}$$

Vibration-rotation spectra

As pointed out at the beginning of Sec. 20-2, a molecule can undergo a transition in which both the vibration quantum number and the rotation quantum number change. A collection of absorption bands in which such changes take place is known as a vibration-rotation spectrum. In Fig. 20-17a the energy levels are drawn for $v = 0$ and $v = 1$, with the associated levels for $J = 0$, 1, 2, and 3. In Fig. 20-17b the energy of the absorbed photon in cm^{-1} is shown for various transitions from the set of levels associated

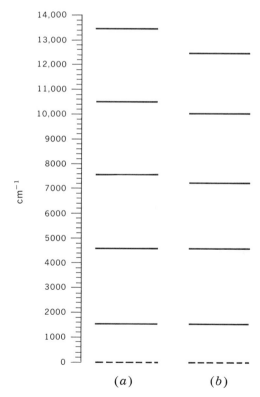

cm^{-1}

14,000
13,000
12,000
11,000
10,000
9000
8000
7000
6000
5000
4000
3000
2000
1000
0

(a) (b)

Fig. 20-16   The vibrational energy levels in gaseous HCl calculated (a) as an ideal harmonic oscillator; (b) according to Eq. (20-16), which takes into account the decrease in the force of attraction with increasing bond length.

TABLE 20-1   Molecular constants[a]

Molecule	ω_{eq}, cm^{-1}	$\omega_{eq}x_{eq}$, cm^{-1}	B_{eq}, cm^{-1}	α_{eq}, cm^{-1}	r_{eq}, Å
Ag^{1}H	1760.0	34.05	6.453	0.203	1.618
^{27}Al^{35}Cl	481.3	1.95	0.242	0.002	2.14
^{27}Al^{1}H	1682.57	29.145	6.3962	0.188	1.6461
^{197}Au^{1}H	2305.01	43.12	7.2401	0.2136	1.5239
Ba^{1}H	1172	16	3.382	0.066	2.232
BaO	671.48	2.20	0.3644	0.0016	1.797
^{79}Br^{81}Br	323.2	1.07	0.08091	0.00027	2.284
^{12}C^{12}C	1641.70	11.71	1.6320	0.01659	1.3121
^{40}Ca^{1}H	1299.	19.5	4.278	0.096	2.002
^{35}Cl^{35}Cl	564.9	4.0	0.2438	0.0017	1.989
^{74}Ge^{16}O	985.7	4.3	0.4704	0.0029	1.651
^{1}H^{1}H	4405.3	125.325	60.872	3.0671	0.7414
^{2}H^{2}H	3118.8	64.15	30.429	1.0492	0.7414
^{1}H^{80}Br	2649.67	45.21	8.471	0.226	1.414
^{1}H^{35}Cl	2988.95	51.65	10.5909	0.3019	1.2747
^{1}H^{19}F	4141.305	90.866	20.967	0.879	0.9166
^{1}H^{127}I	2309.53	39.73	6.551	0.183	1.604
^{127}I^{127}I	214.36	0.593	0.03736	0.00012	2.667
^{39}K^{1}H	983.3	14.40	3.407	0.0673	2.244
^{14}N^{14}N	2351.61	14.445	2.007	0.018	1.095
^{16}O^{16}O	1580.36	12.073	1.4456	0.0158	1.2076
^{32}S^{32}S	725.8	2.85	0.296	0.0016	1.89
^{32}S^{16}O	1123.73	6.116	0.70894	0.00562	1.4935

[a] (After "Molecular Spectra and Molecular Structure," vol. 1, Spectra of Diatomic Molecules, 2d ed., by G. Herzberg. Copyright © 1950 by Reinhold Publishing Corporation, by permission of Van Nostrand Reinhold Company.)

with $v = 0$ to various levels associated with $v = 1$. The selection rule states that the change in J must be $+1$ or -1. Denoting the initial rotational quantum number by J and the final one by J', the lines for which $J' - J = -1$ lie almost evenly spaced at the left (the P branch) in the total band as shown in Fig. 20-18 for HCl. The lines for which $J' - J = +1$ lie almost evenly spaced at the right (the R branch). If it were possible for a transition to take place in which $v = 0$ changes to $v = 1$ with *no change* in J, the absorption band would lie in the center, as shown by the symbol ν_0; however, this line does not appear, since such a change is forbidden. Since the value of the constant B is inversely proportional to the moment of inertia, the amount of separation of the lines depends slightly on the speed of rotation [see Eq. (20-10)]; the greater the rotation, the larger the value

of I and the smaller the value of B; the factor B was assumed to be constant in the earlier discussion. Thus denoting the wave number of the bands by the symbol P where $J' - J = -1$ and by R where $J' - J = +1$, the values of P and R are given by:

$$P = \omega_0 - (B_1 + B_0)J + (B_1 - B_0)J^2 \qquad (20\text{-}18a)$$
$$R = \omega_0 + 2B_1 + (3B_1 - B_0)J + (B_1 - B_0)J^2$$
$$(20\text{-}18b)$$

Example 20-3

The elastic force constant k_h for HF is 9.67×10^5 dynes cm^{-1}. Calculate the frequency of vibration in sec^{-1} and cm^{-1}.

The frequency of vibration is given by

$$\nu = \frac{1}{2\pi}\left(\frac{k_h}{\mu}\right)^{1/2} \qquad \frac{1}{\mu} = \frac{1}{m_1} + \frac{1}{m_2}$$

$$\frac{1}{\mu} = \frac{6.02 \times 10^{23}}{1.008} + \frac{6.02 \times 10^{23}}{19}$$

$$= 6.29 \times 10^{23}$$

$$\nu = (0.1592)(9.67 \times 10^5 \times 6.29 \times 10^{23})^{1/2}$$
$$= (0.1592)(7.80 \times 10^{14})$$
$$= 1.242 \times 10^{14} \text{ sec}^{-1}$$

$$\omega = \frac{\nu}{c} = 4153 \text{ cm}^{-1}$$

Example 20-4

Calculate the energy in cm^{-1} of the photon absorbed when HF goes from the state $v = 0$, $J = 0$ to $v' = 1$, $J' = 1$ using the data in Table 20-1 and neglecting the change in B.

Since $J' - J = +1$, this corresponds to the first member of the R series; Eq. (20-16) can be used to calculate $E_1 - E_0$, the change due to increased amplitude of vibration.

$$(E_0)_v = \omega_e(0 + \tfrac{1}{2}) - \omega_e x_e(0 + \tfrac{1}{2})^2$$
$$= 4141.3 \times 0.5 - 90.87 \times 0.25$$
$$(E_1)_v = 4141.3 \times 1.5 - 90.87 \times 2.25$$
$$(E_1 - E_0)_v = 4141.3 - 90.87 \times 2$$
$$= 3959.6 \text{ cm}^{-1}$$

The increase in energy due to the change from $J = 0$ to $J = 1$ is given by Eq. (20-10)

$$(E_1)_r = B_v 1(1 + 1) \equiv 2B_v$$
$$(E_0)_r = B_v 0(0 + 1) = 0$$
$$(E_1 - E_0)_r \equiv 2B_v$$

$$B_v = \frac{h}{8\pi^2 cI} \qquad I = \mu r^2$$

$$\mu = \frac{m_1 m_2}{m_1 + m_2}$$

$$= \frac{1 \times 19}{1 + 19} \frac{1}{6.02 \times 10^{23}}$$

$$= 1.578 \times 10^{-24}$$

$$I = \mu r^2 = 1.578 \times 10^{-24}$$
$$\times (0.9166 \times 10^{-8})^2$$
$$= 1.326 \times 10^{-40}$$

$$B_v = \frac{6.626 \times 10^{-27}}{8 \times 9.870 \times 2.998}$$
$$\times 10^{10} \times 1.326 \times 10^{-40}$$

$$= 21.11 \text{ cm}^{-1}$$

$$(E_1 - E_0)_r = 42.2$$
$$R = 3961.6 + 42.2 = 4003.8 \text{ cm}^{-1}$$

Ans.

20-3   RAMAN SPECTRA

In 1928 Sir C. V. Raman, a physicist working in India, reported the discovery of spectra in scattered light. When radiation having primarily a single frequency is passed into the side of a tube containing a liquid and the scattered light emitted from the end of the tube is observed with a spectrograph, a number of sharp lines appear in the spectrum, displaced slightly from the line corresponding to the frequency of the incident light. A diagram of a Raman spectrograph is shown in Fig. 20-19 and one of Raman's lines is shown in Fig. 20-20.

Raman scattering mechanism

The phenomenon of light scattering had been studied for many years before Raman's discovery, notably by the British physicists J. Tyndall and Lord Rayleigh. Scattering due to inhomogeneities such as smoke particles is called *Tyndall scattering;* when due to local density fluctuations, it is called *Rayleigh scattering.* For example, if the tube shown in Fig. 20-19 is filled with carbon disulfide, CS_2, and the intense beam of light at 22,938 cm^{-1} from a mercury arc is passed into the side of the tube, the light coming out of the end of the tube at an angle of 90° to the incident light is made up almost completely of radiation with wave number 22,938 cm^{-1} and is due to Rayleigh scattering, as shown in Fig. 20-20. But as Raman observed, there are faint sharp lines on either side of the strong Rayleigh line. The wave-number differences between these Raman lines and the Rayleigh line depend only on the nature of the substance through which the light is scattered and are independent of the frequency of the incident light. The lines for which the wave number has been lowered by the scattering are called *Stokes lines* in honor of the British physicist W. Stokes, who made extensive studies of energy interchange between particles. Denoting the wave number of the incident light by $\bar{\nu}_i$ and of the Raman line by $\bar{\nu}_r$, the relation is

Fig. 20-17   (*a*) Energy levels for $v = 1$ ($J = 0, 1, 2, 3$) and for $v = 2$ ($J = 0, 1, 2, 3$); (*b*) the absorption lines corresponding to various transitions between these levels.   (After N. B. Colthup, L. H. Daly, and S. E. Wiberly, "Introduction to Infrared and Raman Spectroscopy."   Copyright © 1964. Academic Press, Inc., New York.   Used by permission.)

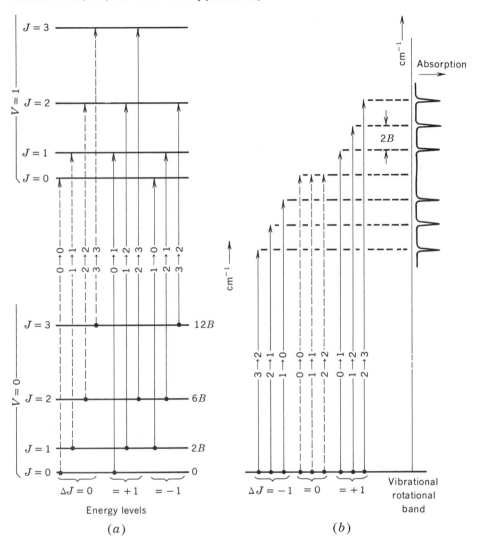

Energy levels

(*a*)

Vibrational rotational band

(*b*)

$$\tilde{\nu}_i - \tilde{\nu}_m = \tilde{\nu}_r \qquad (20\text{-}19)$$

where $\tilde{\nu}_m$ is the quantity dependent only on the nature of the molecule causing the scattering.   If the wave number of the Raman line lies higher than that of the incident line, the emitted light is called an anti-Stokes line and the relation is

$$\tilde{\nu}_i + \tilde{\nu}_m = \tilde{\nu}_r' \qquad (20\text{-}20)$$

where $\tilde{\nu}_r'$ is the wave number of the anti-Stokes line.
  Since $c\tilde{\nu} = \nu$ and $E = h\nu$, the exchange of energy is

Stokes: $hc\tilde{\nu}_i - hc\tilde{\nu}_m = hc\tilde{\nu}_r$

Anti-Stokes: $hc\tilde{\nu}_i + hc\tilde{\nu}_m = hc\tilde{\nu}_r'$
                                                     $(20\text{-}21)$

Thus, in the Stokes lines, the photon leaves the liquid with less energy than it had on entering, and the amount $hc\bar{\nu}_m$ has been left with the molecule. This suggests that the molecule has been transformed to an activated state. In the case illustrated in Fig. 20-20, the numerical relations are

$$hc(22{,}938\text{ cm}^{-1}) - hc(656\text{ cm}^{-1}) = hc(22{,}282\text{ cm}^{-1}) \tag{20-22}$$

The amount of energy left in the molecule corresponds to a wave number in the far infrared that might be associated with a transition to an activated vibrational state. Suppose that the frequency of vibration is $\bar{\nu}_m$; then if E_0 is the energy in the ground vibrational state and E is the energy in the first vibrational state,

$$E_1 - E_0 = h\nu_m = hc\bar{\nu}_m \tag{20-23}$$

For an anti-Stokes line the incident photon with energy $hc\bar{\nu}_i$ picks up energy $h\nu_m$ from a molecule in the activated state and leaves with energy $hc\bar{\nu}_i + hc\bar{\nu}_m$, and the molecule drops from E_1 to E_0.

Following Raman's original discovery, thousands of observations have been made of the Raman spectra of many substances both in the liquid, gaseous, and crystalline states and in solution. Not only can molecules exchange energy with photons and undergo changes in vibrational states, they can also, by the same process, change rotational states.

Transition probabilities

The intensity of a Raman line depends on the relative number of molecules in the state which serves as the initial state in the production of the line. At room temperature, nearly all molecules are in the ground vibrational state; the energy change $E_1 - E_0$ to the first vibrational state is so large compared with RT that the relative number of molecules even in the first vibrational state is very small

$$\frac{N_1}{N_0} = e^{-(E_1-E_0)/RT} \tag{20-24}$$

Consequently, the anti-Stokes line is very much weaker than the Stokes line. However, the value of $E_1 - E_0$ is much smaller for rotational energy levels; the lower levels are well populated; and the anti-Stokes lines appear strongly in the rotational Raman spectra.

The probability of energy exchange also depends on the nature of the vibration or rotation. In order to have Raman interaction with the photon, the polarizability of the molecule must change in the motion under consideration. Polarizability α is defined as the relative efficiency with which a force field F induces a dipole m in a molecule

$$m = \alpha F \tag{20-25}$$

The transition probability between two states 1 and 2 can be denoted by $|p|^{1,2}$ and is given by the quantum-mechanical expression

$$|p|^{1,2} = |F| \int \psi_2^* \alpha \psi_1 \, d\tau \tag{20-26}$$

If an atom has spherical symmetry, the value of α for rotation is zero; for a nonspherical molecule like O_2, α has a sufficiently large value for the rotational Raman lines to be clearly emitted.

As previously discussed, molecules do not show the direct absorption of infrared radiation when the motion does not involve a change in the dipole moment. In many cases, however, this motion does involve a change in polarizability, and Raman lines are observed, even though infrared bands are not. The important role of both Raman and infrared spectra in the study of the structure of polyatomic molecules will be discussed in Sec. 20-6.

During the 1960s the development of the *laser* as a source of monochromatic radiation has vastly improved the possibilities of producing Raman lines of high intensity. In addition, methods have been developed for measuring Raman lines photoelectrically instead of by changes in intensity on a photographic

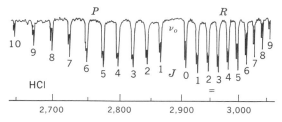

Fig. 20-18 The vibration-rotation bands in gaseous HCl. (After N. B. Colthup, L. H. Daly, and S. E. Wiberly, "Introduction to Infrared and Raman Spectroscopy." Copyright © 1964. Academic Press, Inc., New York. Used by permission.)

Fig. 20-19 Diagram of a Raman spectrometer showing entrance slit S, mirror M, collimator axis A, parabolic mirror C, grating G, lenses L, and photographic recording plate P. The sample tube and arc are drawn on a larger scale than the spectrograph. (After G. R. Harrison, R. C. Lord, and J. R. Loofbourow, "Practical Spectroscopy." Copyright 1948. By permission of Prentice-Hall, Inc., Englewood Cliffs, N.J.)

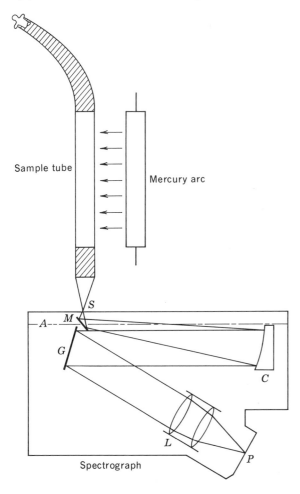

Sample tube

Mercury arc

S

A -- M

G

C

L

P

Spectrograph

plate. This permits the use of much smaller samples for study, a development of great importance for research on substances of biochemical interest.

Example 20-5

Calculate the value of N_1/N_0 for carbon disulfide at 298°K in the vibration with a Raman line for which $\tilde{\nu}_i - \tilde{\nu}_r = 796$ cm^{-1}.

$$E_1 - E_0 = hc\tilde{\nu}_m N_A$$
$$= 6.6256 \times 10^{-27} \times 2.998$$
$$\times 10^{10} \times 796 \times 6.02 \times 10^{23}$$
$$= 9.52 \times 10^{10} \text{ ergs}$$
$$= 9.52 \times 10^3 \text{ joules} = 2.27 \text{ kcal}$$
$$RT = 1.987 \times 298° = 0.592 \text{ kcal}$$

$$\log \frac{N_0}{N_1} = 2.303^{-1} \frac{2.27}{0.592} = 1.66$$

$$\frac{N_0}{N_1} = 45.8 \qquad \frac{N_1}{N_0} = 0.0218 \qquad Ans.$$

Example 20-6

Using the line from the mercury arc at 18,303 cm^{-1}, a Raman line is observed at 17,652 cm^{-1} in carbon disulfide. Calculate (a) the value of $\tilde{\nu}_m$ and (b) the wavelength of the absorption band in the infrared, if this frequency caused the direct absorption of infrared radiation.

(a) $\tilde{\nu}_m = 18,303 - 17,652 = 651$ cm^{-1}

(b) $\lambda = \dfrac{1}{\tilde{\nu}} = 0.00154$ cm $= 15.4 \ \mu$

20-4 ELECTRONIC SPECTRA

Among all possible types of transitions between molecular states, the transition from one electronic configuration to another involves the greatest energy change. For the majority of such transitions involving electrons in the outer shell of the atom, the value of $E_2 - E_1$ lies between 1 and 10 ev, corresponding to 23 to 230 kcal mole^{-1}. A photon with 1 ev of energy has a wavelength of 12,340 Å; a photon with 10 ev has a wavelength of 1234 Å; thus, the majority of such transitions absorb or emit photons with wavelengths in the visible range (about 4000 to 8000 Å) or in the ultraviolet range with wavelengths shorter than 4000 Å.

Spectrometers

For observations of spectra in the visible range, the light is generally refracted by a glass prism or diffracted by a grating. In the ultraviolet, with the exception of the part lying very close to the visible range, glass absorbs too strongly to be used for refraction.

The optical path in a typical prism spectrometer is shown in Fig. 20-21; either a photographic plate or a photoelectric cell can be used to detect and measure the intensity in various parts of the spectrum. The optical path in a grating spectrometer is shown in Fig. 20-22. Photographs of portions of the spectra of P_2 and $(COF)_2$ are shown in Fig. 20-23.

Theory of electronic spectra

As can be seen in Fig. 20-5, each molecular state has at least three quantum numbers. Considering the simplest case of the diatomic molecule, these are the rotational quantum number J, the vibrational quantum number v, and the electronic quantum number usually specified by more complex symbols, indicating the nature of the state. As discussed in the preceding sections, changes between electronic states can occur which involve changes in all three quantum numbers. This results in an ordered structure in the band of absorption or emission spectra associated with the electronic transition. Within this structure, the lines spaced most closely together correspond to the changes between states having different rotational quantum numbers, since the rotational states are separated by the smallest amounts of energy. The features of structure involving the next wider spacings are due to changes in vibrational quantum number as these have the next wider separation on the energy scale (Fig. 20-5). Measurements of these spacings permit a calculation of the differences in energy due to differences in J and v, which can be compared with the values from direct observations of pure rotation spectra in the very far infrared region and of vibration spectra in the infrared with considerably shorter wavelengths.

Franck-Condon principle

Certain features of the electronic spectra associated with the dissociation of the molecule are of special interest for this reason. The German physicist James Franck and the American physicist E. U. Condon suggested that the passage from one electronic state to another is generally so fast that the nuclei of the atoms move only by negligible amounts during the

time of the transition, even though in the process of vibration. Depending on where the nuclei happen to be in the vibration cycle, different effects are produced (Fig. 20-24). In each figure the variation of potential energy with distance is assumed to be the same in the state with lower energy.

In Fig. 20-24a, the variation of potential energy with distance in the electronic state with higher energy has approximately the same shape as that for the ground state, and the curve is merely displaced to a higher level. The horizontal line just above the bottom of the well shows the ground vibrational level for the diatomic molecule. The vertical arrow on the left (i) indicates the change of state if a photon with energy E_a is absorbed when the molecule happens to

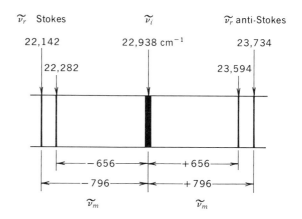

Fig. 20-20 Raman lines observed for two vibrational modes of motion in carbon disulfide, CS_2.

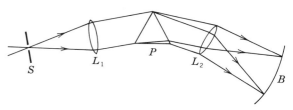

Fig. 20-21 Simple prism spectrograph. S = slit, L_1 = collimator lens, P = prism, L_2 = camera lens, and B = photographic plate. (From "Spectroscopy and Molecular Structure," by G. W. King. Copyright © 1964 by Holt, Rinehart and Winston, Inc., New York. Reprinted by permission of Holt, Rinehart and Winston, Inc.)

Fig. 20-22   Grating spectrograph.   Concave grating mounts. The slit S_1 and plate B_1 are in a typical Paschen-Runge mount on the Rowland circle of the grating G. S_2 and B_2 are in the Eagle mounting, which can be contained in an evacuated tube, indicated by dotted lines, for work in the vacuum region. (From "Spectroscopy and Molecular Structure," by G. W. King.   Copyright © 1964 by Holt, Rinehart and Winston, Inc., New York.   Reprinted by permission of Holt, Rinehart and Winston, Inc.)

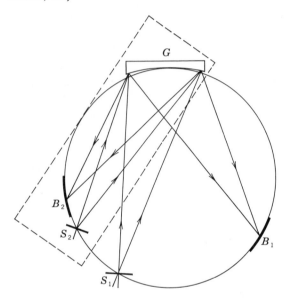

be in the part of the vibration cycle where the nuclei are closest together.   The vertical arrow at the right (*ii*) shows the change which takes place when the photon is absorbed at the part of the cycle where the nuclei are farthest apart.   In either case, after the photon is absorbed and the molecule makes the transition to the state indicated by the upper potential-energy curve, the molecule continues to vibrate in much the same way as it did before the absorption of the photon took place.   There is no dissociation.

In Fig. 20-24*b*, the assumption is made that the de Broglie wave pattern of the molecular electrons in the excited state is such that the upper potential-energy well is shifted to the right, so that the position of minimum potential energy (the bottom of the well) corresponds to a larger separation of the nuclei. Now, if the photon is absorbed when the nuclei are at the distance of *maximum* separation in the vibration cycle (vertical arrow *ii*), the nuclei are separated by a

distance in the new electronic state such that vibration takes place much as in the ground state.   But if the photon is absorbed when the nuclei are at the distance of *minimum* separation (*i*), then in the excited state the nuclei have more potential energy than that required to separate them completely (the horizontal dashed line) when they move apart.   As a result, the molecule dissociates.

In case Fig. 20-24*c*, it is assumed that there is no well in the potential-energy curve for this excited state. Consequently, when the photon is absorbed either at the position of minimum separation (*i*) or maximum separation (*ii*), the molecule in the excited state has enough potential energy to pass immediately to a dissociated state.

When a photon contains enough energy to dissociate a molecule, there is no observable quantization of absorption since the dissociated parts can have practically any values of kinetic energy.   In theory, the fact that these parts are confined in a limited space does impose quantization on the kinetic energy, but (as noted in the discussion of the de Broglie waves associated with a particle in a box) the energy levels associated with this quantization are spaced so closely together that the resultant spectrum is effectively a continuum.   In an observed spectrum, the point at which this continuum begins provides a measure of the dissociation energy for the excited state of the molecule.   In Fig. 20-25, the electronic band spectra of N_2 illustrate this effect.

Example 20-7

From Fig. 20-24*b* estimate the kinetic energy with which the molecular fragments will fly apart if the molecule absorbs a photon when in the position of minimum separation for $v = 1$.

The line representing the $v = 1$ level in the ground state will lie at a height three times that of the level for $v = 0$, since $E = (v + \frac{1}{2})h\nu_e$. At the distance of minimum separation $r = 1.0$ Å.   In the excited state, the potential energy of separation of the molecule will be approximately 4 ev.   The energy of dissociation is 3.3 ev.   Thus the kinetic energy after dissociation will be 0.7 ev.

20-5 MICROWAVE AND RADIOFREQUENCY SPECTRA

If a molecule has a large moment of inertia, the spectral lines associated with changes in rotational energy levels can lie at frequencies even smaller than found in the far-infrared range, so that they are detectable by microwave techniques. Two examples of such spectra are shown in Fig. 20-26.

The theory of spectra associated with angular momenta either of the electrons or of the total molecule is closely related to still another type of spectrum associated with the spinning of the nuclei and also found beyond the far infrared.

Nuclear magnetic resonance spectra

Each nucleus in an atom can spin around its own axis in a manner similar to the spinning of an electron. The spectra associated with spins of this sort are called *nuclear magnetic resonance spectra* (NMR) when observed in a strong magnetic field.

For electrons, the total angular-momentum vector has the magnitude $(h/2\pi)\sqrt{l(l+1)}$ in the quantization of orbital angular momentum; the magnitude of the component parallel to the magnetic field is $(h/2\pi)m$, where $m = -l, \ldots, 0, \ldots, +l$. For the nucleus, the total spin angular-momentum vector I has the magnitude $(h/2\pi)\sqrt{I_n(I_n + 1)}$, where I_n is the nuclear-spin quantum number. The angular momentum in the direction of a magnetic field is quantized, so that it is equal to $(h/2\pi)m_n$, where $m_n = -I_n, \ldots, 0, \ldots, +I_n$; thus, there are $2I_n + 1$ different possible orientations for this vector. The vector and several spinning nuclei of different shapes are shown in Fig. 20-27.

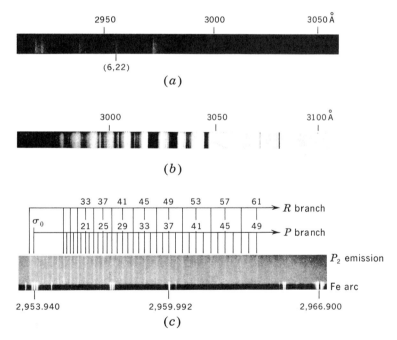

Fig. 20-23 Band spectra: (a) portion of the emission spectrum of the diatomic phosphorus molecule, resulting from a $^1\Sigma_u^+ \longrightarrow {}^1\Sigma_g^+$ electronic transition. The rotational structures of the bands are partially resolved; the band labeled (6,22) in (a). (b) The discrete absorption spectrum of oxalyl fluoride, $(COF)_2$, showing a complex band structure. (c) The (6,22) band in part (a) is shown in higher resolution. (From "Spectroscopy and Molecular Structure," by G. W. King. Copyright © 1964 by Holt, Rinehart and Winston, Inc., New York. Reprinted by permission of Holt, Rinehart and Winston, Inc.)

Fig. 20-24 Potential-energy curves for the ground state (*lower*) and for three types of excited states (*upper*) illustrating the Franck-Condon principle and showing the dissociation energy D.

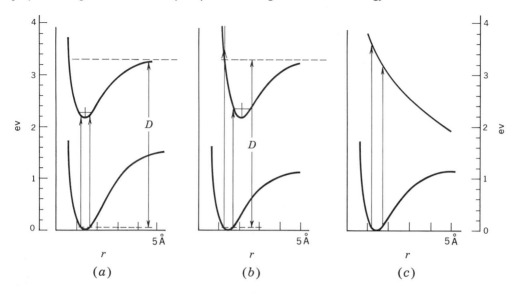

(a)   (b)   (c)

Fig. 20-25 Band progression in the electronic spectrum of N_2. (After G. R. Harrison, R. C. Lord, and J. R. Loofbourow, "Practical Spectroscopy." Copyright 1948. By permission of Prentice-Hall, Inc., Englewood Cliffs, N.J.)

The magnitude of the magnetic moment μ divided by $hI/2\pi$ is called the gyromagnetic ratio γ. Since the maximum value of m is I_n, and since the maximum value of I in the direction of the field is $I_n(h/2\pi)$, the maximum value of μ in this direction is $\gamma I_n(h/2\pi)$.

The simplest nucleus is that of the hydrogen atom, consisting of a single proton; it has a value of μ equal to $\dfrac{h}{2\pi}\dfrac{e}{2\pi Mc}$ according to classical theory; this is called the classical nuclear magneton μ_c and is equal to 5.05×10^{-24} erg gauss^{-1}. The quantization results in a value of the magnetic moment for the proton μ_p which is $2.79277\ \mu_c$.

The values of the energy E_μ for different orientations of the vector in a magnetic field H_0 are given by

$$E_\mu = -\gamma m \frac{h}{2\pi} H_0 \qquad (20\text{-}27)$$

Because of the selection rule $\Delta m = \pm 1$, the values of ΔE_μ are

$$\Delta E_\mu = \frac{h}{2\pi} H_0 \qquad (20\text{-}28)$$

When a quantum of electromagnetic radiation (photon in the microwave region) is absorbed, the energy of the photon is equal to $\Delta E = h\nu$, where ν is the frequency in the microwave region. Thus,

$$\nu = \frac{1}{2\pi} \gamma H_0 \qquad (20\text{-}29)$$

The wavelengths associated with these levels are given in Table 20-2. The wavelengths of a few of these spectra are plotted in Fig. 20-28 to show the relation to the more familiar parts of the total electromagnetic spectrum.

The apparatus for measuring NMR spectra is shown schematically in Fig. 20-29. In the uniform magnetic field, the nuclei in the sample have only certain quantized directions for the magnetic moment. When a radio frequency ν_r, providing the value of $\Delta E = h\nu_r$

corresponding to a shift from one spin level to another, is impressed on the sample, part of the signal is absorbed by resonance and an absorption peak appears in the emitted radiation much like an absorption peak in infrared radiation passing through a sample.

Because the magnetic fields from neighboring nuclei merge with the imposed field, the effective field for a nucleus depends on its location in the molecule. Thus a hydrogen nucleus in an OH group absorbs radiation at a frequency different from that absorbed by a hydrogen nucleus in a CH_2 group. Typical NMR spectra for these groups are shown in Fig. 20-30.

These spectra provide an especially helpful means for studying the phenomenon of hydrogen bonding, which plays such an important role in many biochemical reactions and is crucial in the synthesis of DNA. Since the exchange of protons between molecules is a rate phenomenon dependent on temperature, the spectrum of protons involved in hydrogen bonding changes with temperature, as shown in Fig. 20-30.

Example 20-8

From the shift observed between 27 and 108°C for the OH bond, calculate the ratio of the shift due to the change in hydrogen bonding to the shift due to the difference in effective magnetic field for a proton in OH and a proton in CH_2.

Calling the difference between the center of the OH bond and the center of the CH_2 bond at 27°C arbitrarily 10 units, the shift of the OH bond is roughly 7 units, and so the ratio is 0.7.

20-6   THE SPECTRA OF POLYATOMIC MOLECULES

In passing from a diatomic molecule to one composed of more than two atoms, an additional property of the covalent bond begins to play an important role in vibration; this is the bending elasticity, which is now coupled to the stretching elasticity in the intramolecular feedback which determines the characteristic modes of vibration of a molecule.

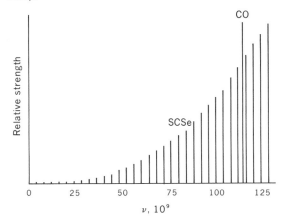

Fig. 20-26   Pure rotation spectra in the microwave region. (After W. Gordy, W. V. Smith, and R. F. Trambarulo, "Microwave Spectroscopy," John Wiley & Sons, Inc., New York, 1953).

(a) Spherical nucleus, not spinning $\mu = 0$

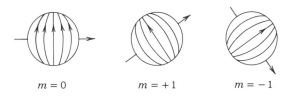

$m = 0$ $m = +1$ $m = -1$

(b) Spherical nucleus spinning $\mu \neq 0$ quadrupole $= 0$

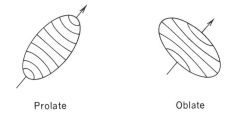

Prolate   Oblate

(c) Ellipsoidal nucleus spinning $\mu \neq 0$ quadrupole $\neq 0$

Fig. 20-27   Spinning nuclei.

Fig. 20-28   Nuclear magnetic resonance frequencies as related to the general electromagnetic spectrum.

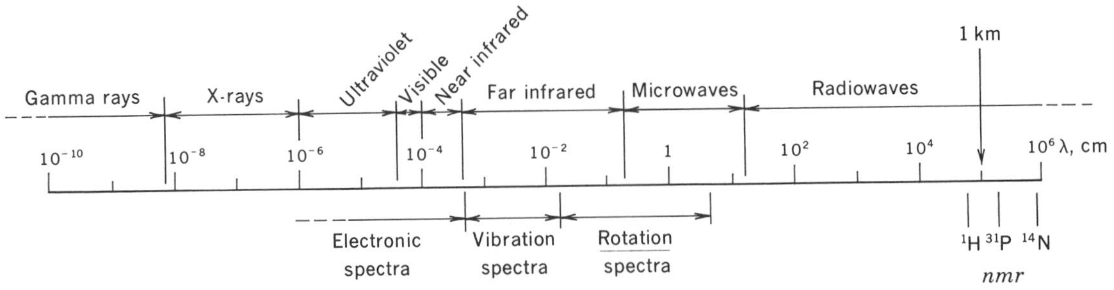

TABLE 20-2   Nuclear magnetic resonance frequencies

Nucleus	λ, m
^1H	704
^{19}F	748
^{31}P	1,739
e^-	1,070
^6Li	4,523
^{14}N	9,743
^{10}B	6,551

Modes of vibration

In the study of the mechanics of molecules treated as assemblies of mass points (the nuclei) coupled together with covalent bonds (the equivalent of spiral springs), the application of Newton's basic laws shows that only certain definite patterns of vibration are possible as steady states. These patterns are called *vibrational modes*. The laws of mechanics together with the geometrical pattern of the bonds lead to a theoretical prediction of these modes.

First of all, to fix the position of any atom, a set of three space coordinates is required (x, y, z). In a molecule containing n atoms, the atoms can be numbered by subscripts; e.g., for CO, where $n = 2$, the carbon atom can be denoted by the subscript 1 and the oxygen atom by the subscript 2. For this molecule there are $3n = 3 \times 2 = 6$ coordinates: $x_1, y_1, z_1, x_2, y_2, z_2$.

But it is not necessary to choose these particular

six coordinates to describe the position of the molecule. For example, to denote the location in space of the molecule, the three coordinates of the center of mass may be used: x_c, y_c, z_c; to describe angles between the axis of the molecule and the coordinate axes, the values of θ and ϕ may be given; finally, to describe the distance apart of the C and O atoms the radius r may be specified. These six coordinates give all the information necessary to describe the configuration of the molecule at any instant in time.

Thus, for a diatomic molecule there are a total of $3n = 3 \times 2$ coordinates. Three refer to the center of mass, two to the angles of position, and one to the vibration. For a polyatomic *linear* molecule there are $3n$ coordinates; three for the position in space, two for the angle, and $3n - 5$ vibrational coordinates.

For a polyatomic *nonlinear* molecule, there are three space coordinates; three angular coordinates are necessary to fix the angles between the axes of the molecule and the coordinate axes; and there are $3n - 6$ coordinates remaining, the coordinates associated with vibration.

The number of coordinates necessary to describe the configuration of the molecule is called the number of *degrees of freedom* and is denoted by f. For a diatomic molecule there are 3 degrees of freedom of movement in space f_{sp}, 2 degrees of freedom of rotation f_{rot}, and 1 degree of freedom of vibration f_{vib}.

Since the set of coordinate axes can always be selected at will, it is convenient in the study of vibrational modes to select a set with the zero point at the center of mass and axes which rotate with the molecule. The effect of rotation, e.g., the production of

centrifugal forces, and the effect of spatial acceleration in general can be neglected.

Triatomic linear molecules

As an example, consider the carbon dioxide molecule
for which the axes are selected as shown in Fig. 20-31a.
According to the laws of mechanics, when vibration
takes place in a steady state, the atoms must move
in such a way that with respect to this special coordinate system (1) there is no motion of the center of
mass which lies in the center of the nucleus of the
carbon atom and (2) there must not be any net
moment of momentum for the entire molecule. The
analysis shows that, in the case of CO_2, there are four
modes of motion satisfying these conditions.

In two of these the bond angles change; i.e., the
bonds, regarded as spiral springs, are *bent*. These
modes are shown in parts i and (i') of the figure.
Since there is no distinction between *up* and *down*
and *in* and *out* in the structure of the molecule, these
two modes of vibration have the same frequency;
they are called *degenerate* modes. As these types of
vibration take place, the dipole moment of the molecule changes; therefore, this mode is active in the infrared spectrum. It is identified with the sharp band
at 667.3 cm^{-1} (14.986 μ).

Another mode of motion is shown in part (ii), in
which the oxygen atoms move alternately out from
and in toward the central carbon atom, which remains
motionless. This alternately stretches and com-
presses the bonds. Because of the symmetry, the
dipole moment does not change with vibration, and
there is no corresponding band in the infrared; on
the other hand, the conditions meet the requirement
for a line in the Raman spectrum, and this is found
at 1388.3 cm^{-1}. Note that stretching the bond involves a much higher elastic constant than bending,
the kind of behavior normally associated with a spiral
spring. Consequently, the frequencies of the stretching modes are almost always much higher than the
frequencies of the bending modes.

The fourth mode of motion, which satisfies the two
conditions imposed by mechanics, is shown in part
(iii). This is an unsymmetrical mode and has a
changing dipole. The corresponding line is found at
2349.3 cm^{-1} in the infrared. Note that in the (iii)

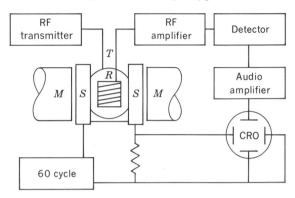

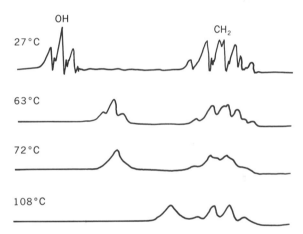

Fig. 20-30 Part of the NMR trace of CH_3CH_2OH at different
temperatures, showing the variation of hydrogen bonding.
(After J. A. Pople, W. G. Schneider, and H. J. Bernstein, "High
Resolution Nuclear Magnetic Resonance," McGraw-Hill Book
Company, New York, 1959.)

mode, the bonds are stretched both by the motion of
the O atom and *also* by the motion of C atom; in the
(ii) mode, only the motion of the O atoms stretches
the bond; consequently, the resultant elastic force is
greater in the (iii) mode, and this unsymmetrical
mode has a higher frequency than the symmetrical
(ii) mode. Thus, there are four vibrational degrees
of freedom: $f_{vib} = 3n - 5 = 4$.

Fig. 20-31 Normal modes of vibration in CO_2. The small circle over C in part (ii) indicates that the atom is not moving.

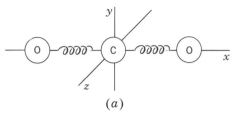

(a)

(i) $\overset{\uparrow}{O}=\underset{\downarrow}{C}=\overset{\uparrow}{O}$ (i') $\overset{\nearrow}{O}=C=\overset{\nearrow}{O}$ 667.3 cm^{-1}

(parallel to y axis)   (parallel to z axis)   (IR)

(ii) $\overset{\leftarrow}{O}=\overset{\circ}{C}=\vec{O}$ 1388.3 cm^{-1}

(Raman)

(iii) $\overset{\leftarrow}{O}=\vec{C}=\overset{\leftarrow}{O}$ 2349.3 cm^{-1}

(IR)

(b)

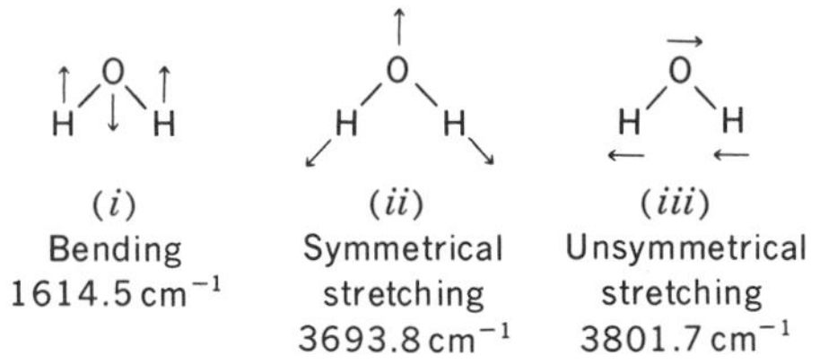

(i)   (ii)   (iii)
Bending   Symmetrical   Unsymmetrical
1614.5 cm^{-1} stretching   stretching
3693.8 cm^{-1} 3801.7 cm^{-1}

Fig. 20-32 The normal modes of vibration in H_2O.

Other examples of linear molecules are HCN, CS_2, and HCCH.

Triatomic nonlinear molecules

An example of this class is H_2O. The modes of motion are shown in Fig. 20-32. The total number of vibrational degrees of freedom are $f_{vib} = 3n - 6 = 3 \times 3 - 6 = 3$.

As in the case of CO_2, the slowest mode (i) is the bending mode; all the modes involve changes in the dipole moment and therefore are represented by bands in the infrared; thus the (i) mode is found at 1595.0 cm^{-1}. The symmetrical stretching mode (ii) is found at 3693.8 cm^{-1}, and the unsymmetrical

stretching mode (iii) at 3801.7 cm^{-1}. Thus the pattern of the ratio of frequencies is much the same as in CO_2; the bending frequency is low; the symmetrical stretching is highest. Because of the nonlinearity, the distinction between the bending and stretching modes is not as precise as in the case of CO_2; there is some bending and some stretching in each of the modes; but (i) has more bending than stretching while (ii) and (iii) have more stretching than bending.

Basic vibration theory

The theory of the vibration of polyatomic molecules is too complex to be treated in an introductory text, but certain aspects of it can be helpful in understanding the behavior of these molecules. The theory is, of course, an extension of the theory of the mechanics of vibration of diatomic molecules discussed in previous sections.

Many of the modes of vibration in polyatomic molecules reduce in practice to cases which can be treated by the simple diatomic theory. As an example, in the symmetrical stretching mode (ii) of CO_2, the carbon atom at the center is motionless. Thus the oxygen atoms behave as if they were attached to an immovable wall, the equivalent of an atom with infinite mass, as shown in Fig. 20-33. For the lower vibrational levels, this system can be regarded as an ideal harmonic oscillator, in which the frequency ν is related to the elastic constant of the bond k_h and to the mass of the oxygen molecule m by the expression

$$\nu = \frac{1}{2\pi}\sqrt{\frac{k_h}{m}} \qquad (20\text{-}30)$$

The observed value of the Raman line corresponding to this frequency permits a direct calculation of the value of k_h, which is

$$k_h = (2\pi\nu)^2 m$$
$$= (2\pi \times 1388.3 \times 2.998 \times 10^{10})^2 \frac{16.000}{6.023 \times 10^{23}}$$
$$= (2.6151 \times 10^{14})^2(2.656 \times 10^{-23})$$
$$= 1.816 \times 10^6 \text{ dynes cm}^{-1} \qquad (20\text{-}31)$$

It is interesting to compare this figure with the value of k_h found for CO, where both the carbon and oxygen atoms move when the molecule vibrates. In this case, the infrared fundamental vibration band is at 2168.2 cm^{-1}, and the frequency is given by the expression

$$\nu = \frac{1}{2\pi}\sqrt{\frac{k_h}{\mu}} \qquad (20\text{-}32)$$

where μ is the reduced mass, given by

$$\frac{1}{\mu} = \frac{1}{m_1} + \frac{1}{m_2} \qquad (20\text{-}33)$$

Inserting the numerical values,

$$\mu = 1.137 \times 10^{-23}$$
$$k_{h_1} = (2\pi\nu)^2\mu$$
$$= (4.084 \times 10^{14})^2 \times 1.137 \times 10^{-23}$$
$$= 1.896 \times 10^6 \text{ dynes cm}^{-1}$$

Whenever one of the masses, say m_2, in Eq. (20-33) is very large with respect to the other m_1, then $\mu \cong m_1$ and an approximate value of the elastic constant can be obtained with the help of Eq. (20-30).

In Fig. 20-34 are plotted values of elastic-force constants calculated by either precise or approximate methods for a number of the bonds. The values for single bonds lie in the range between 2×10^5 and 8×10^5 dynes cm^{-1}. The values for the NH and OH bonds are calculated from the spectra of gaseous NH$_3$ and H$_2$O. The average of the more common single bonds is 5.2×10^5 dynes cm^{-1}; the average of the common double bonds is 10.6×10^5 dynes cm^{-1}; the average of the triple bonds is 16.2×10^5 dynes cm^{-1}. Thus the values of k_h for common single, double, and triple bonds are roughly in the ratio 1:2:3 as might be expected.

Calculation of modes

When this simple theory of vibration is further extended with the help of some of the more advanced theorems of mechanics, it is possible to predict quite accurately the patterns of vibrational modes in rather complex molecules and in some cases to calculate the frequencies of the modes. If the molecule has features of symmetry, the analysis is simplified considerably.

As an example, the benzene, C$_6$H$_6$, molecule has 20 modes of vibration which have been derived on a theoretical basis and confirmed by Raman and infrared spectra. For benzene, $n = 12$; thus, $f_{\rm sp} = 3n - 6 = 36 - 6 = 30$. Of these 30 modes, 10 are doubly degenerate, and 10 are simple modes (Fig. 20-35). Four of the modes involve changes in the dipole moment and are found as absorption bands in the infrared. The positions of these bands are drawn in Fig. 20-35 as dashed lines. The observed bands for these frequencies as determined on an infrared spectrograph are shown in Fig. 20-36. The additional bands in this figure are due to other more complex transitions. Ten of the modes involve changes in the polarizability and are found in the Raman spectrum.

Fig. 20-33   The mechanical system equivalent to the —C=O system in mode (ii) of CO$_2$.

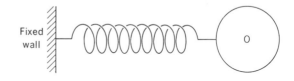

Fixed wall
0

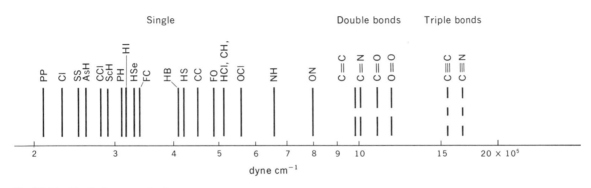

Fig. 20-34   Elastic-force constants.

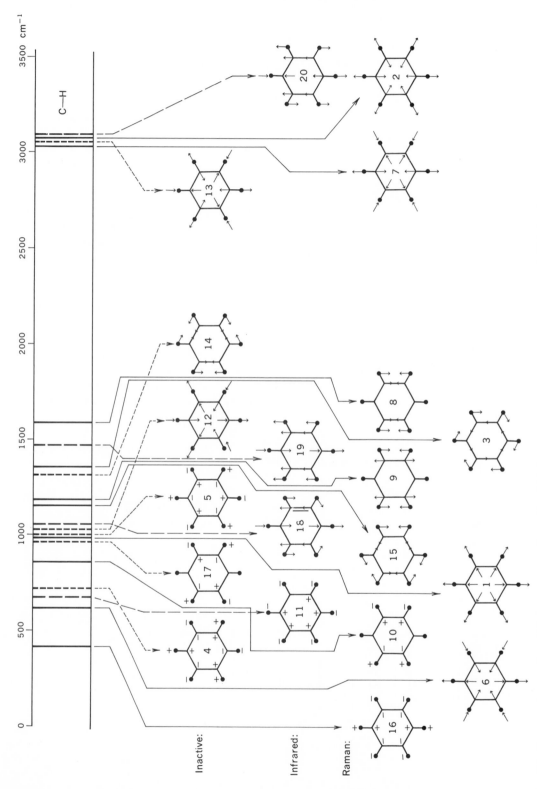

Fig. 20-35 The infrared and Raman spectra of benzene, showing also inactive modes of vibration. The frequencies are numbered according to current usage.

Four are inactive in either the infrared or Raman spectra.

Modes 1 and 2 are highly symmetrical and closely resemble the (ii) mode of CO_2 in this respect. The application of the theorem discussed in the previous section permits a direct calculation of k_h for the C—C and the C—H bonds. The former is found to have the value 7.6, about 1.5 times that of the average single bond value for k_h of 5.2. The theory of bonding in benzene predicts a value of this amount, since the average of six single and six double bonds is the equivalent of 1.5 bonds per pair of carbon atoms. The value of k_h for the C—H bond is 5.05 dynes cm^{-1}, close to the average for the single bonds shown in Fig. 20-34. In a similar way, values of k_h for the bending of the bonds can be calculated, and from the combinations of the values with the values of k_h for stretching the frequencies of the inactive modes can be obtained.

The application of the principle of effective reduced mass shows that many side-chain atoms or groups vibrate with frequencies that change very little from molecule to molecule. For example, since the hydrogen atom is so light, the effective mass for the vibration depends very little on the mass of the molecule to which the hydrogen is coupled. When the motion of hydrogen primarily involves stretching of the bond, the frequencies almost always lie between 2800 and 3600 cm^{-1}. For example, the simple CH stretching for hydrogen attached to the ring lies at 3061.5 cm^{-1} in benzene. As shown in the infrared spectrum of aniline, $C_6H_5NH_2$, in Fig. 20-37, there is a band at 3000 cm^{-1}, which is a C—H stretching frequency, and another band at 3500 cm^{-1}, which is an N—H stretching frequency. This latter band lies higher because the value of k_h for N—H is higher than that for C—H (see Fig. 20-34).

If a nitrogen atom is substituted for one of the carbon atoms in benzene, as in pyridine, C_5NH_5, the position of the C—H stretching bond is not appreciably shifted, as shown in Fig. 20-38. By studying the infrared and Raman lines in a series of related molecules, the shift in bond elasticity can be followed with precision. In Fig. 20-39, the infrared spectrum of adenine is shown; this is the molecule which serves as one of the principal constituents of DNA; it contains an NH group, and the NH absorption frequency can be seen at about 3100 cm^{-1}.

Ribose is another constituent of DNA, containing several OH groups. The characteristic OH absorption is shown at about 3300 cm^{-1} in Fig. 20-40. The infrared spectrum of a species of DNA shows a broad band in this H-stretching region where there are a variety of groups present, as can be seen in Fig. 20-41. The structure of DNA is discussed in detail in Chap. 24.

Over 30,000 different molecules have been studied and the infrared spectra cataloged; this collection provides an extremely helpful way of identifying the molecules and of recognizing features of structure in complex substances.

Spectra and thermodynamic properties

Infrared and Raman spectra provide one of the most helpful aids in the calculation of the thermodynamic properties of molecules. The internal heat capacity C_{int} of a molecule depends almost completely on the frequencies of its modes of vibration. For any one mode j, the relation is

$$C_j = \text{ein}\frac{\theta_j}{T} \tag{20-34}$$

where ein denotes the Einstein heat capacity function defined in Eq. (6-38), and $\theta_j = hc\omega_j/k = 1.4399\ \omega_j$ cm^{-1}.

The total internal heat capacity C_{int} is the sum of the heat capacities of all the modes of vibration. If, as in benzene, certain modes are degenerate, there is a term for each of the members of the degenerate mode. Thus

$$C_{int} = \sum_j \text{ein}\frac{\theta_j}{T} \tag{20-35}$$

A comparison of the experimentally observed heat capacity with that calculated with the help of spectra was shown in Fig. 6-19.

The internal entropy S_{int} is given by

$$S_{int} = \sum_j \left[\frac{Rs_j}{e^{xj} - 1} - R \ln (1 - e^{-xj}) \right]$$
$$x_j = \frac{\theta_j}{T} = \frac{hc\omega_j}{kT} \tag{20-36}$$

The values of ω_i from spectra provide the data for theoretically calculating the internal entropy with con-

Fig. 20-36 The infrared absorption bands of benzene. (Courtesy of H. Kaufmann.)

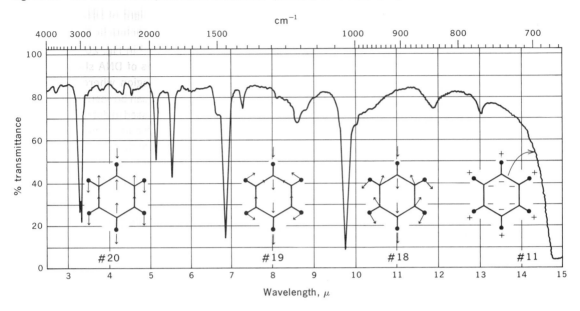

Fig. 20-37 Infrared spectrum of aniline. (Courtesy of H. Kaufmann.)

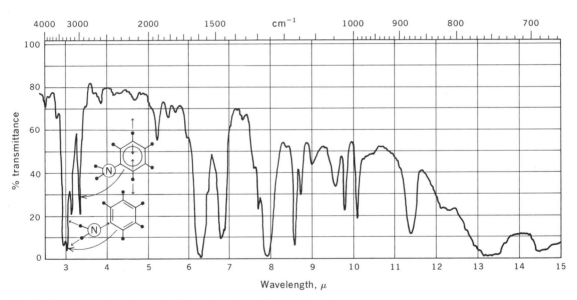

siderable precision. A graph of the value of S as a function of θ/T is given in Fig. 20-42.

For benzene, the theoretical calculation can be compared with the experimental measurements to provide a check on the accuracy of the method; 1 mole of the vapor at 1 atm pressure and at the temperature of the normal boiling point 353.2°K is selected as the reference state. From statistical mechanics the entropy of translation is calculated to be 39.83 eu, and of rotation to be 21.30 eu. Putting the values of ω_j in Eq. (20-34) and calculating S_{int} with the help of Eq. (20-35) gives $S_{int} = 7.01$ eu.

Thus the theoretical and experimental values of S are

$$S_{trans} + S_{rot} + S_{int} = \begin{cases} S_{theor} = 68.14 \text{ eu} \\ S_{exper} = 68.15 \text{ eu} \end{cases} \quad (20\text{-}37)$$

Fig. 20-38 The infrared spectrum of pyridine. (Courtesy of H. Kaufmann.)

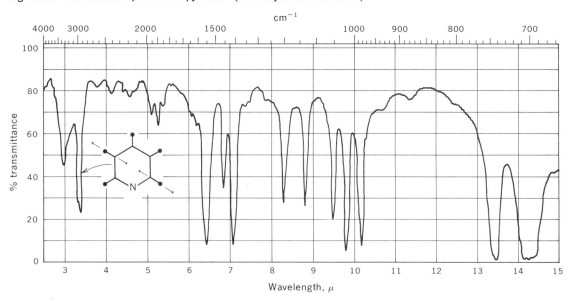

Fig. 20-39 The infrared spectrum of adenine. (Courtesy of H. Kaufmann.)

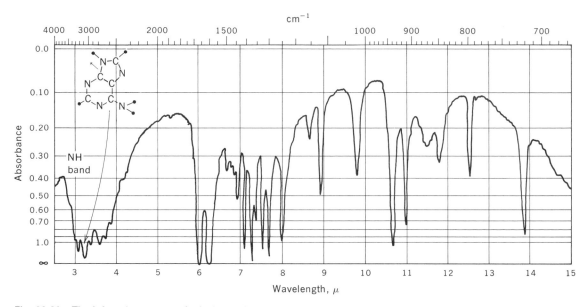

Fig. 20-40   The infrared spectrum of ribose.   (Courtesy of H. Kaufmann.)

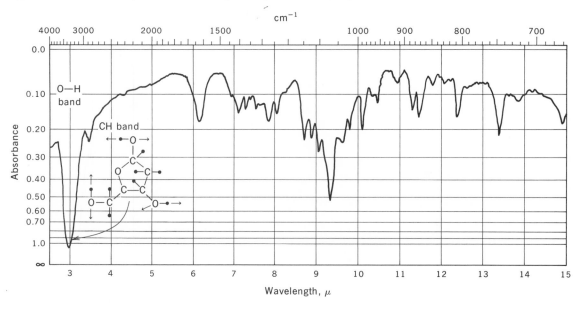

Fig. 20-41   The infrared spectrum of a species of DNA.   (Courtesy of H. Kaufmann.)

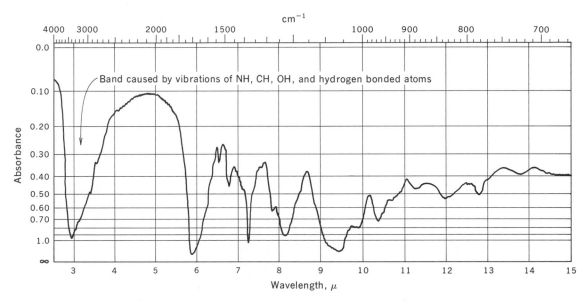

The good agreement shows the accuracy of values of S calculated with the help of spectra.

Experimental determinations of values of the entropy are generally very difficult and time-consuming compared with the observations of spectra.   For this reason the calculation of thermodynamic properties from spectra has provided a wealth of data otherwise unavailable.   The American physical chemist D. R.

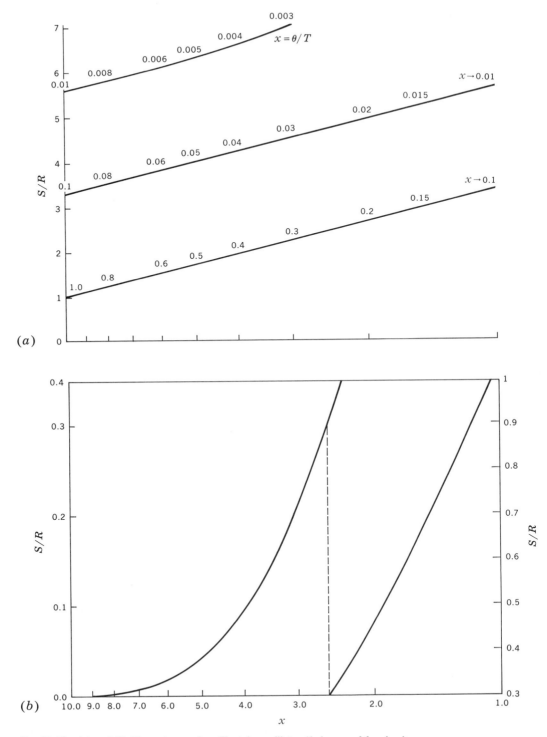

Fig. 20-42 (a) and (b) The entropy of an Einstein oscillator (1 degree of freedom).

Stull and his associates have compiled thermodynamic tables for over 500 substances using the method described above, combined with the best available experimental data. These tables provide an invaluable aid in the application of physical chemistry to a wide variety of problems in science and technology.

Example 20-9

From the data given in Fig. 20-32, calculate the internal heat capacity of H_2O at $1000°K$.

The values of θ_i are obtained by multiplying the values of ω_i by 1.4389 and are: $\theta_i = 2323.1°$, $\theta_{ii} = 5315.0°$, $\theta_{iii} = 5470.3°$. Using the chart in Fig. 6-13, the values of C are $C_i = 1.29$, $C_{ii} = 0.28$, $C_{iii} = 0.28$ cal deg^{-1} mole^{-1}. Thus $C_{int} = \sum_j C_j = 1.85$ cal deg^{-1} mole^{-1}. Note that the bending vibration contributes 75 percent of the internal heat capacity.

Example 20-10

Assuming that the deep absorption band at 730 cm^{-1} in DNA is the same mode as that causing absorption in adenine, pyridine, and benzene in this region (the simultaneous bending of hydrogen bonds on the ring), calculate the contribution of this vibration to the entropy of DNA in *Escherichia coli,* assuming that the *E. coli* DNA molecule contains 6×10^6 binucleotides (see Chap. 24) and that each cross rung has one ring of this type in it.

The wavelength of this band is converted to θ units when multiplied by 1.439, so that $730 \times 1.439 = 1,050 = \theta$ and $x = \theta/T = 1,050/1,000 = 1.05$. According to the chart in Fig. 20-42b, at $x = 1.05$, $S/R = 1.00$. So per oscillator, $S = 1.987$ cal deg^{-1} mole^{-1}. There are 6×10^6 oscillators of this type in one molecule of DNA. Therefore, the entropy per molecule due to these oscillators will be $1.987 \times 6 \times 10^6/6 \times 10^{23} = 2 \times 10^{-17}$ cal deg^{-1} molecule^{-1}.

Vibration and absolute-rate theory

The American physical chemist Henry Eyring and his associates have developed a theory of absolute rates of reaction based on the concept of a transition state in which two reacting molecules are momentarily linked together so that the rate of their rearrangement into the products of the reaction depends on the frequency of vibration of the particular mode which moves the transition complex into dissociation into the products. Assuming that this vibration is classically excited, $\epsilon = kT$; assume also that the probability of dissociation is proportional to the frequency of vibration ν; then there should be a probability factor given by the relation

$$\nu = \frac{\epsilon}{h} = \frac{kT}{h} \tag{20-38}$$

and the rate can be expressed

$$r = \frac{kT}{h} e^{-H\ddagger°/RT} e^{-S*°/R} A** \tag{20-39}$$

where

$$\frac{kT}{h} e^{-S*°/R} A** = A* e^{-S\ddagger°}$$

$A*$ is a dimensionality constant with the numerical value of unity; the choice of rate units determines the value of entropy of activation in the standard state $S\ddagger°$; $A**$ is also a dimensionality constant with the numerical value of unity; and $S*°$ is the value of the entropy of activation in the standard state, with a numerical value dependent on the units selected for r.

Problems

1.   The transition from the $^1\Sigma^+$ state to the $^2\Sigma^+_g$ state in gaseous N_2 requires 15.58 ev. Calculate the energy required in ergs per molecule and kilocalories per mole.

2.   For the transition described in Prob. 1, calculate the frequency in sec^{-1} and the wavelength in angstroms of the light absorbed.

3.   In the molecule $^1H^{79}Br$, the nuclei are separated by a distance of 1.414 Å. Calculate the moment of inertia I.

4.   Using the value of I for $^1H^{79}Br$ obtained in Prob. 3, calculate the separation of the bands in the rotation spectrum, assuming a constant value of I.

5.   For $^1H^{79}Br$ the value of ω_{eq} is 2649.67 cm^{-1} and of $\omega_{eq}x_{eq}$ is 45.21. Calculate the value of the energy in cm^{-1} for the vibration level where $v = 2$.

6. Using the data from Probs. 3 and 5, calculate the value of the elastic-force constant for $^1H^{79}Br$.

7. Calculate in cm^{-1} the change in the energy of a $^1H^{79}Br$ molecule in passing from the state where $v = 0$ and $J = 0$ to $v = 1$ and $J = 1$.

8. The structural formula of adenine is given in Fig. 20-39. Calculate the total number of vibrational degrees of freedom and the number due primarily to the stretching of H—C bonds.

9. The Raman spectrum of methane shows a line corresponding to a vibration at 2914.2 cm^{-1}. In this vibration all four hydrogen atoms move out simultaneously from the central carbon atom, which is motionless. From this information, calculate the stretching constant of the CH bond and compare with the value in Fig. 20-34.

10. The Raman spectrum of CD_4 has a line corresponding to 2084.7 cm^{-1} for the type of vibration described in Prob. 9. Calculate the value of the stretching constant for C—D.

11. The compound ribose (see Fig. 20-40) constitutes the sugar part of DNA. Ribose shows two absorption bands in the infrared spectrum, one at about 3000 cm^{-1} and one at 3400 cm^{-1}. Name the pairs of atoms which have vibrational frequencies with these values.

12. The compound C_6H_5SH has a Raman line corresponding to 188 cm^{-1}. Name the atom primarily responsible for this frequency.

13. In the compound ribose (Fig. 20-40) the OH groups vibrate perpendicular to the ring at a frequency corresponding to 250 cm^{-1}. Calculate the contribution to the heat capacity of 1 mole of ribose at body temperature 37°C due to the perpendicular vibration of the three OH groups.

14. The NH group in adenine absorbs in the infrared at approximately 3100 cm^{-1}. If the mass factor is the mass of the hydrogen atom when this group is treated as a simplified harmonic oscillator, calculate the frequency shift if the hydrogen atoms are replaced by deuterium atoms.

15. In adenine there are two hydrogen atoms attached to carbon atoms in the ring structure. These atoms vibrate perpendicular to the ring at a frequency corresponding approximately to 700 cm^{-1}. Calculate the contribution to the heat capacity of 1 mole of adenine due to these vibrations at body temperature, 37°C.

16. The lowest vibrational frequency in benzene (#16 in Fig. 20-35) is the source of the Raman line at 404 cm^{-1}. A similar vibration is observed in uracil, one of the principal constituents of RNA. Calculate the contribution which this vibration would make to the entropy of 1 mole of uracil at 37°C, assuming that in uracil this is not a degenerate mode.

17. In the species of DNA found in *E. coli*, there are approximately 3×10^6 nucleotides of the single-ring type like thymine, $C_4N_2H_3O_2(CH_3)$, in a single molecule of DNA. Each of these has a nondegenerate frequency of vibration corresponding roughly to 400 cm^{-1}. Calculate the contribution of these vibrations to the entropy of 1 mole of DNA of this type at 37°C.

18. Calculate the increase in the energy of a molecule of thymine when it passes from the state where $v = 0$ to the state where $v = 4$ in the vibration mode corresponding to 400 cm^{-1}, assuming an ideal harmonic oscillator. Using the Boltzmann equation, calculate the number of thymine nucleotides in a molecule of the DNA from *E. coli* which will be in this activated state assuming that the total number of nucleotides of this type is 2×10^6 molecule^{-1}.

19. Using the data from Table 19-10, calculate the wavelength in angstroms of a photon with energy just sufficient to break the NH bond in a molecule like adenine.

20. Using the data in Fig. 20-4, what is the wavelength of a photon with energy just sufficient to raise an N_2 molecule from the ground level to the level for $v = 0$ in the Lyman-Birge-Hopefield band?

REFERENCES

Herzberg, G.: "Infrared and Raman Spectra of Polyatomic Molecules," D. Van Nostrand Company, Inc., New York, 1945.

Harrison, G. R., R. C. Lord, and J. R. Loofbourow: "Practical Spectroscopy," Prentice-Hall, Inc., Englewood Cliffs, N.J., 1948.

Herzberg, G.: "Molecular Spectra and Molecular Structure," vol. 1, Spectra of Diatomic Molecules, 2d ed., Van Nostrand Reinhold Company, 1950.

Wilson, E. B., Jr., J. C. Decius, and P. C. Cross: "Molecular Vibrations," McGraw-Hill Book Company, New York, 1955.

Bellamy, L. J. "The Infrared Spectra of Complex Molecules," John Wiley & Sons, Inc., New York, 1958.

Brügel, W.: "An Introduction to Infrared Spectroscopy," John Wiley & Sons, Inc., New York, 1962.

Allen, H. C., Jr., and P. C. Cross: "Molecular Vibrotors" John Wiley & Sons, Inc., New York, 1963.

King, G. W.: "Spectroscopy and Molecular Structure," Holt, Rinehart and Winston, Inc., New York, 1964.

Colthup, N. B., L. H. Daly, and S. E. Wiberly: "Introduction to Infrared and Raman Spectroscopy," Academic Press, Inc., New York, 1964.

Cross, A. D.: "Introduction to Practical Infrared Spectroscopy," Butterworth & Co. (Publishers), Ltd., London, 1964.

Spice, J. E.: "Chemical Binding and Structure," Pergamon Press, New York, 1964.

Dixon, R. N.: "Spectroscopy and Structure," John Wiley & Sons, Inc., New York, 1965.

Conley, R. T.: "Infrared Spectroscopy," Allyn and Bacon, Inc., Boston, 1966.

Sonnessa, A. J.: "Introduction to Molecular Spectroscopy," Reinhold Publishing Corporation, New York, 1966.

Rao, C. N. R.: "Ultraviolet and Visible Spectroscopy," 2d ed., Plenum Press, New York, 1967.

Hadni, A.: "Essentials of Modern Physics Applied to the Study of the Infrared," Pergamon Press, New York, 1967.

Harrick, N. J.: "Internal Reflection Spectroscopy," John Wiley & Sons, Inc., 1967.

Wollrab, J. E.: "Rotational Spectra and Molecular Structure," Academic Press, Inc., New York, 1967.

Dunford, H. B.: "Elements of Diatomic Molecular Spectra," Addison-Wesley Publishing Company, Reading, Mass., 1968.

In the preceding chapter on molecular spectra, a survey was made of the effects on *light* during passage through aggregates of molecules; in general, the molecules absorb photons of specific frequencies, resulting in regions of decreased intensity in the light spectra. In this chapter, the emphasis is reversed, as a survey is made of effects on aggregates of *molecules* produced by the absorption of light; in general, the molecules absorb photons and are raised to activated chemical states.

This branch of science has been called photochemistry. During the last decade or two, there also has been a growing interest in the effects on matter

Chapter Twenty-One

photochemistry and radiation chemistry

of electromagnetic radiation of frequency in the x-ray and gamma-ray region far beyond the range of visible light. This field is generally called radiation chemistry, although the distinction between this and photochemistry is rather artificial. Actually, effects of radiation on matter are observed over the whole spectrum from the gamma rays even to the radio range. These effects are particularly significant in the study of the physics and chemistry of living matter.

During the last few years, research in photo and radiation chemistry has grown to such an extent that only a very limited coverage is possible in a single chapter of a general text. In 1967 alone, seven books totaling over 2,300 pages were published entirely devoted to this field; there were a number of shorter review articles; and 362 journal publications were listed in just one of these review articles summarizing the year's research in this field. In this chapter the discussion is focused primarily on the role of light in processes related to chemical reactions in living matter.

Fig. 21-1 Excitation of electrons by light: (*a*) completely paired electrons in singlet state S_0; (*b*) singlet state S_n with one activated electron after absorption of photon; (*c*) triplet state T_n after nonradiative change of state.

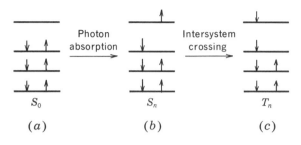

(a)   *(b)*   *(c)*

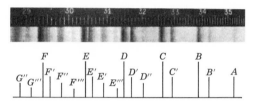

Fig. 21-2 Ultraviolet absorption spectrum of HCHO. [After V. Henri and S. A. Schou, *J. Chim. Phys.*, **25**:665 (1928).]

21-1 LIGHT ABSORPTION

The first significant research in photochemistry was carried out in the eighteenth century by an English physiologist, Stephen Hales, who investigated the nature of photosynthesis but lacked both the theoretical concepts and the experimental techniques necessary to ascertain the exact nature of the light-absorption process. In 1817 the German physicist C. J. T. T. von Grotthus recognized that only the light actually absorbed by a system can cause a chemical reaction to occur in it. The first quantitative correlation of the amount of absorption with the amount of reaction was made by the British chemist J. W. Draper in 1841 in a study of the reaction of hydrogen with chlorine in the presence of light. After an induction period, the rate of the reaction was directly proportional to the intensity of the light.

When light passes through any medium, the amount of light absorbed depends on the distance traversed in the medium. This principle was first stated in 1729 by Pierre Bougeur, a French mathematician and hy-

drographer; it was rediscovered in 1758 by J. H. Lambert, a German physicist, and is usually called Lambert's law. Stated mathematically,

$$\frac{-dI}{I} = a \, dx \qquad (21\text{-}1)$$

where I = intensity of light
x = distance traveled in medium
a = const

a is called the extinction coefficient. If I_0 is the intensity of the light as it enters the medium, the intensity I at distance x in the medium is

$$I = I_0 e^{-ax} \qquad (21\text{-}2)$$

If light is absorbed by a substance in solution, a is proportional to the concentration of the substance in solution and Eq. (21-2) can be written

$$I = I_0 e^{-\epsilon' cx} \qquad (21\text{-}3)$$

where ϵ' is called the *molar extinction coefficient* and c is the concentration in molar units. This is called Beer's law after its discoverer, August Beer, a German physicist.

Quantum principles

The concept of a quantum of energy was applied to photochemical reactions by J. Stark, a German physicist, in 1908 and by Albert Einstein in 1912. This led to the conclusion that each quantum of radiant energy absorbed activates precisely one molecule in the primary step of a photochemical process. The amount of radiant energy E_e is called one *einstein* when

$$E_e = N_A h\nu \qquad (21\text{-}4)$$

The *quantum yield* of a photochemical reaction is the number of the moles of product formed per einstein absorbed. Because of diversion of the energy in different steps of the reaction, the yield is generally less than 1.

21-2 PHOTOEXCITATION

Activation

The term *photochemical absorption* is generally restricted to a process in which the absorbed photon raises the absorbing molecule to a new electronic state which has a greater energy than the state before the absorption. The molecule is then said to be in an *excited* or *activated* state. In nearly all cases the molecule has an even number of electrons, half with spin $+\frac{1}{2}$ and half with spin $-\frac{1}{2}$ so that the electrons are *paired*, as shown in Fig. 21-1. Since there is no net spin angular momentum, this is called a *singlet* state S_0. After the absorption, one of the electrons may be in a higher energy level (Fig. 21-1b) but still with the same spin so that the state is still a singlet S_n. However, the electron can reverse its spin (Fig. 21-1c) giving a state with net angular momentum of 1 unit, where the vector can be oriented with respect to a magnetic field to give any of the three values: $+1$, 0, or -1. This state is called a *triplet* state T_n. The change from Fig. 21-1b to c is called an intersystem crossing. This is a nonradiative process.

The activated triplet state can change to other states of the system by a number of different processes:

1. The molecule can dissociate if the energy of this activated state is equal to or greater than the energy of dissociation.
2. The molecule can change to still another state by a nonradiative process.
3. The energy can be lost to other molecules through collisions.
4. The molecule can exchange energy with another molecule by reacting with it.
5. The molecule can transform the energy of activation back into radiant energy by phosphorescence or fluorescence in one or more steps.

As an example, consider a molecule of formaldehyde, HCHO. Like most molecules with π bondings, this has characteristic absorption bands, as shown in Fig. 21-2. The orbitals of HCHO are portrayed in Fig. 21-3a and the electronic configuration in Fig. 21-3b. The energy levels of the molecular orbitals are drawn in Fig. 21-4. The transition requiring the least energy is between the n level (p_x, p_y) and the π^* level and is denoted by the symbol $n \rightarrow \pi^*$. This causes absorption at 2700 Å. The absorption band

Fig. 21-3 The configuration of formaldehyde, HCHO: (*a*) orbitals; (*b*) electron spins. (After M. Orchin and H. H. Jaffé, "The Importance of Antibonding Orbitals," Houghton Mifflin Company, Boston, 1967.)

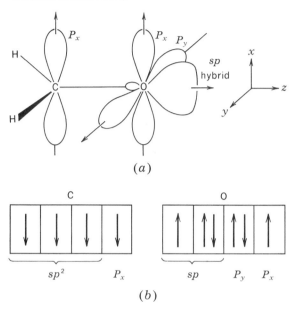

(*a*)

(*b*)

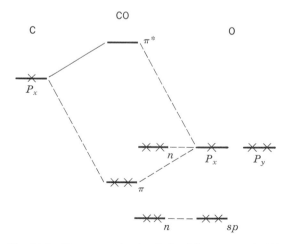

Fig. 21-4 The energy levels of the CO group in HCHO. (After M. Orchin and H. H. Jaffé, "The Importance of Antibonding Orbitals," Houghton Mifflin Company, Boston, 1967.)

Fig. 21-5 The shift of electrons in formaldehyde induced by the absorption of a photon. (After "Mechanistic Organic Photochemistry," by D. C. Neckers. Copyright © 1967 by Reinhold Publishing Corporation, New York, by permission of Van Nostrand Reinhold Company.)

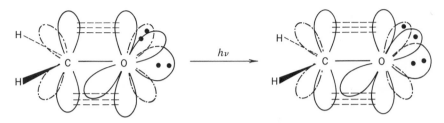

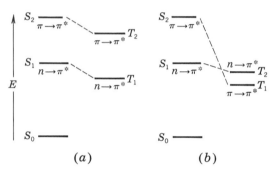

Fig. 21-6 The relative amounts of energy needed to induce transitions in formaldehyde: (a) normal; (b) m solvents where $\pi \rightarrow \pi^*$ is lowered. (After H. H. Jaffé and M. Orchin, "Theory and Applications of Ultraviolet Spectroscopy," John Wiley & Sons, Inc., New York, 1962.)

is weak, because the probability of the transition is low. The π^* orbital is in the xz plane, and the n orbitals are in a plane perpendicular to this, so that there is little overlap. The $n \rightarrow \pi^*$ can be displayed diagrammatically as indicated in Fig. 21-5.

The $n \rightarrow \pi^*$ excitation actually leads to the singlet state S_1 shown in Fig. 21-6; there is also a triplet state T_1 with somewhat lower energy. A nonradiative cross-over can occur between these two states, as depicted in Fig. 21-7.

Because nearly all excited states in molecules have relatively short lives, there is no practical way to determine the shape of the molecule in such a state by x-ray or other diffraction methods. For this reason, the theoretical methods for calculating configurations

are especially important, as C. A. Coulson[1] has pointed out. Table 21-1 presents a few of the results based on MO calculations for three simple molecules. Although such calculations are usually difficult, the German physicist H. Hellmann and the American physicist R. P. Feynman independently have shown that if the true wave function is known for any given positions of the nuclei of a molecule, the actual force exerted on any nucleus has the value which would be computed classically on the assumption that the other nuclei and the total charge cloud interact classically.

Coulson uses the example of CO, as depicted in Fig. 21-8. The top σ cloud attracts the C nucleus more strongly than the O nucleus, but the π cloud pulls the C and O nuclei toward each other. As shown in Table 21-1, a change from σ to π shortens the internuclear distance, but if a π orbital is removed, there should be an increase in bond length.

The study of excited states of proteins with the help of observations of fluorescence leads to helpful knowledge of their properties. This will be discussed in Sec. 21-6. There is considerable evidence that these activated states play an important role in the biochemical reactions taking place within the cell.

Dissociation

As discussed in the previous chapter, a molecule can absorb a photon and change to a state with a new potential-energy curve; the molecule in this final state may have sufficient energy to move directly to dis-

[1] See References, "Proceedings of the Solvay Institute: 13th Conference on Chemistry."

sociation if the potential-energy curve has no minimum; or the final state may be at a point on the compression side of the curve at a level above that of dissociation. In this case, the molecule moves through the potential well and then dissociates. These cases were illustrated in Fig. 20-24.

If there are two activated states with potential-energy curves like those shown in Fig. 21-9, the molecule can make the transition first to the state with the potential well and then shift to the state B, where the curve shows no minimum. This is known as *predissociation*.

Photochemical reactions

When a molecule like Cl_2 is dissociated by the absorption of a photon, the free Cl atoms have great chemical reactivity and can link to many different kinds of molecules. Reactions initiated in this way are members of a general class called *photochemical reactions*, in which photons play the key role.

Chlorine gas absorbs radiant energy over a continuous range in the spectrum below 4785 Å. The maximum absorption is at about 3300 Å, a region where the photons have just sufficient energy to lift the molecule from the $^2\Sigma_g$ state to $^1\Pi_u$ or $^3\Pi_u$ state. A photon at this point in the spectrum has energy corresponding to 86.6 kcal mole^{-1}. The dissociation energy of Cl_2 is only 59.4 kcal mole^{-1}. When the molecule dissociates, the Cl atoms leave with a kinetic energy corresponding to about 27 kcal mole^{-1}. This interpretation of the ultraviolet spectrum of chlorine

by the German physicist James Franck was the first satisfactory analysis of transitions in molecular electronic spectra.

Once these excited chlorine atoms are formed,

Fig. 21-7 Nonradiative intersystem transition from singlet state S_1 to triplet state T_1.

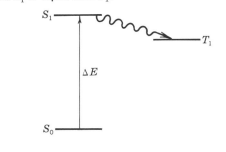

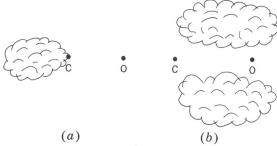

(a)   *(b)*

Fig. 21-8 Charge cloud in CO: *(a)* normal state σ orbital; *(b)* occupied π orbital. (After C. A. Coulson, Electronic Excitation in Molecules, in "Proceedings of the Solvay Institute: 13th Conference on Chemistry," p. 3, John Wiley & Sons, Inc., New York, 1967.)

TABLE 21-1   Change of internuclear distance in transitions to excited states[a]

Molecule	Nature of ground orbital	Bond length, Å		Increase due to excitation, Å
		Ground state	Excited state	
H_2	$1s\sigma$ (bonding)	0.741	1.070	+0.329
CO	3σ (nonbonding around C)	1.128	1.115	−0.013
O_2	$2p\pi$ (antibonding)	1.208	1.123	−0.085

[a] From C. A. Coulson, Electonic Excitation in Molecules, in "Proceedings of the Solvay Institute: 13th Conference on Chemistry," p. 3, John Wiley & Sons, Inc., New York, 1967.

Fig. 21-9 Morse curves associated with predissociation. (After M. Orchin and H. H. Jaffé, "The Importance of Antibonding Orbitals," Houghton Mifflin Company, Boston, 1967.)

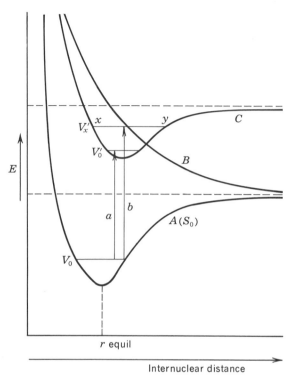

TABLE 21-2   Approximate wave numbers activating radiation for some photochemical reactions

$\bar{\nu} \times 10^{-5}$ cm^{-1}	Reaction
5	$NH_3 \rightarrow \frac{1}{2}N_2 + \frac{3}{2}H_2$
3.5	$HI \rightarrow \frac{1}{2}H_2 + \frac{1}{2}I_2$
3.3	$(CH_3)_2CO \rightarrow CO + C_2H_6$
3.2	$CH_3CHO \rightarrow CO + CH_4$
2.8	$2C_{14}H_{10} \rightarrow C_{28}H_{20}$
2.2	$2NO_2 \rightarrow 2NO + O_2$

they react with a variety of other molecules. For example, the two consecutive reactions take place

$H_2C{=}CH_2 + Cl \rightarrow H_2ClC{-}CH_2 \cdot$
$H_2ClC{-}CH_2 \cdot + Cl \rightarrow H_2ClC{-}CH_2Cl$

Other examples of photochemical reactions are given in Table 21-2.

Chain reactions

There are many instances where a single molecule dissociates and one of the fragments reacts in such a way that a chain of dissociation results. As an example,

$Cl_2 + h\nu \rightarrow 2Cl$
$Cl + H_2 \rightarrow HCl + H$
$H + Cl_2 \rightarrow HCl + Cl$

Such a chain reaction continues until free atoms unite, terminating the chain:

$H + Cl \rightarrow HCl$

Once free atoms with excess energy are produced by the photon, the chain continues until the energy is diverted to other locations where it no longer acts as a dissociating agent. Thus, the termination reaction above does not take place unless the excess energy in the product HCl is passed on almost immediately to another molecule or to the wall of the vessel by collision.

Photolysis

In order to make the products of a photochemical reaction in sufficient quantity to permit accurate observations, it is sometimes necessary to use light of extremely high intensity. Although the production of a sustained intense beam may be impractical, the intensity can often be reached in a single short flash. This method is called *flash photolysis* and was first developed during the 1940s at Cambridge University in England by R. G. W. Norrish and George Porter; they received the Nobel Prize in 1967. Using electronic techniques, an intense flash is produced which causes the photochemical reaction. Then a second flash passes light through the reaction vessel and into a spectrometer, where the spectrum of the products is observed. A series of repeated flashes permits observations of the change of intensity of lines as a function of time as the products decay. A diagram of the apparatus for flash photolysis is shown in Fig. 21-10.

This technique has been applied successfully to

Fig. 21-10   Apparatus developed for observations of flash photolysis by E. J. Bair, Indiana University.   (Courtesy of E. J. Bair.)

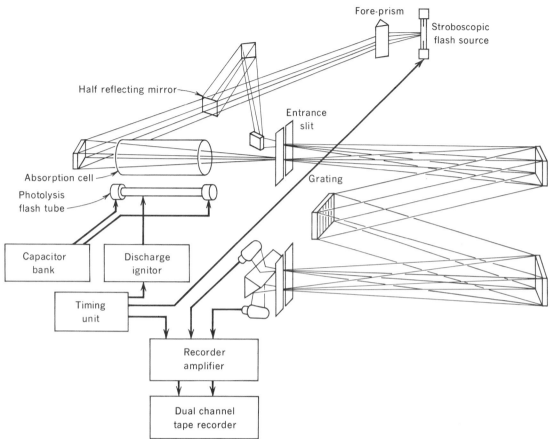

the study of many molecules of biochemical interest. When egg albumin is observed, one of the principal transient products is the phenoxyl radical, C_6H_5O, from tyrosine, $HOC_6H_4CH_2CH(NH_2)(COOH)$, and the spectrum is almost the same as that of the phenoxyl radical obtained directly from phenol, C_6H_5OH. Flash-photolysis measurements have provided four separate rate constants for the hemoglobin-oxygen reaction. A detailed review of this work has been compiled by Ashmore and his associates (see References).

Photoconductivity

When a light beam falls on certain organic crystals, they show electric conductivity. In the dark the energy gap Σ_g for conductivity in aromatic crystals is of the order of 1 to 3 ev. But anthracene, for example, becomes conducting on the surface, where the light is absorbed mainly in the first 1000 Å of penetration. The number of carriers is proportional to the intensity of light, and the quantum yields are proportional to the extinction coefficient of the crystal.

When the photon falls on the crystal, a free electron may be produced, leaving a corresponding positive charge that can be regarded as a hole in the electron pattern of the crystal. The hole and the electron pair are called an *exciton*. They may move through the crystal with comparative freedom, but are attracted to each other by the usual coulombic force. The production of an exciton by a photon is shown in Fig. 21-11.

In anthracene the first exciton band occurs between

Fig. 21-11   The absorption of a photon near the surface of a crystal of anthracene causing the formation of an exciton consisting of a free electron e and a hole h. (a) Initial formation; (b) migration of the pair in the lattice, held loosely together by coulombic force. (After C. Kittel, "Introduction to Solid State Physics," John Wiley & Sons, Inc., New York, 1966.)

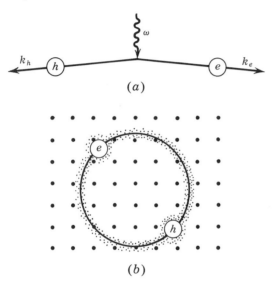

(a)

(b)

4150 and 4550 Å.   The photocurrent induced by this weakly absorbed light is temperature-dependent and has an activation energy of about 0.2 ev.   The energy-level diagram of solid anthracene is shown in Fig. 21-12.   Anthracene crystals also can emit light in the form of *laser* beams, a phenomenon discussed in detail in Chap. 22.

Semiconductivity in aromatic compounds is of special interest because of Szent-Györgyi's suggestions that electron-transfer reactions may play an important role in the behavior of macromolecules like DNA in biological systems.   The aromatic character of the nucleotides might make possible phenomena of this sort which could influence the course of reactions like DNA and RNA synthesis.

Example 21-1

Light of an intensity of 3.2×10^{-6} einstein sec^{-1} falls on the surface of an anthracene crystal.   The absorption coefficient is 3×10^6 cm^{-1} at the wavelength of this light.   Calculate the intensity of the light at a depth of 100 Å below the surface of the crystal.

Putting Eq. (21-2) in logarithmic form,

$$\log \frac{I_0}{I} = \frac{ax}{2.303} = \frac{3 \times 10^6 \times 10^{-6}}{2.303} = 1.3$$

$$\frac{I_0}{I} = 20 \qquad \frac{I}{I_0} = 0.050$$

$$I = 3.2 \times 10^{-6} \times 0.050$$
$$= 1.6 \times 10^{-7} \text{ einstein}$$

Example 21-2

Calculate the activation energy in kilocalories per mole for the production of excitons in anthracene by light between 4150- and 4550-Å wavelength and calculate the percentage increase of the rate constant for this reaction between normal body temperature (37°C) and 42°C.

0.2 ev $\times$ 23.06 kcal mole^{-1} ev^{-1}
$$= 4.61 \text{ kcal mole}^{-1}$$

$$\log \frac{k_{42°}}{k_{37°}} = -(4610)(2.303 \times 1.987)^{-1}$$
$$\times (\tfrac{1}{315} - \tfrac{1}{310}) = 0.052$$

$$\frac{k_{42°}}{k_{37°}} = 1.13$$

The increase is 13 percent.   *Ans.*

21-3   PHOTOSENSITIZATION

When an atom or a molecule is changed to an activated state and then induces a photochemical reaction while itself returning to the original inactive state, the process is called *photosensitization*.

Mechanisms

One of the simplest examples of photosensitization is the dissociation of hydrogen induced by atoms of mercury vapor when illuminated by light at 2536 Å wavelength.   The reaction takes place in two steps:

$$Hg + h\nu \rightarrow Hg^*$$
$$Hg^* + H_2 \rightarrow Hg + 2H$$

The energy of the photon is 112 kcal einstein^{-1}; the energy of dissociation of hydrogen is 102.4 kcal mole^{-1}. Activated mercury atoms also decompose ammonia and many organic compounds. The free hydrogen atoms reduce metallic oxides, nitrous oxide, carbon monoxide, ethylene, and many other substances.

There are many organic reactions which can be induced by photosensitization. When benzophenone, $(C_6H_5)_2CO$ (briefly denoted by ϕ_2CO), and butadiene C_4H_6 are mixed and irradiated at 3500 Å, the following series of changes occur:

$$\phi_2CO(S_0) \xrightarrow{h\nu} \phi_2CO(S_1)$$
$$\phi_2CO(S_1) \rightarrow \phi_2CO(T_1)$$
$$\phi_2CO(T_1) + (C_4H_6)(S_1) \rightarrow \phi_2CO(S_0) + C_4H_6(T_1)$$

In the triplet state, butadiene is able to absorb radiation to which it is completely transparent in the singlet state and thus can undergo photochemical reactions otherwise impossible directly.

Biophotosensitization

Many forms of living matter are viable in intense light but suffer deleterious effects from the light when photosensitized. The first significant investigation of this phenomenon was made by the German biologist O. Rabb in 1900, who found that paramecia suspended in solutions of acridine were viable in the dark but died when exposed to light.

Protein photooxidation

Several of the amino acid constituents of proteins are oxidizable when photosensitized with dyes, and this may be one of the mechanisms which render living cells photosensitive. An example is the reaction with methionine

$$(NH_2)(COOH)CHCH_2CH_2SCH_3 + O_2 + H_2O \xrightarrow[\text{dye}]{h\nu}$$
$$(NH_2)(COOH)CHCH_2CH_2SOCH_3 + H_2O_2$$

The rate of photosensitized oxidation often depends on the amount of ionization of the amino acid and hence on the value of pH, as shown in Fig. 21-13.

Nucleic acids

Both in vivo and in vitro nucleic acids are susceptible to photosensitization. This leads to chromosomal

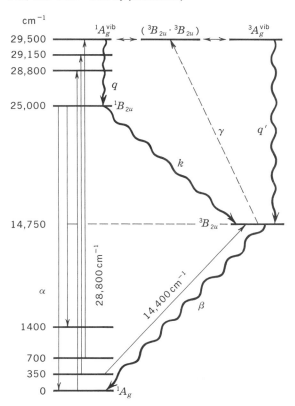

Fig. 21-12 Energy levels in crystalline anthracene. (After W. Siebrand, "Laser Generation and Transport Properties of Triplet Excitations in Anthracene Crystals," in "Modern Quantum Chemistry." Copyright © 1965. Academic Press, Inc., New York. Used by permission.)

changes, altered mitotic activity, and mutagenesis. The guanine nucleotide in DNA oxidizes readily in the presence of a dye and ultraviolet light with a high quantum efficiency (Fig. 21-14). In guanine oxidation a number of different dyes act as photosensitizers, including toluidine blue, rose bengal, and riboflavin, but acridine orange does not photosensitize the guanine in DNA.

Virus photosensitization

The Australian biochemist F. M. Burnett and his associates studied the photosensitization of a number of viruses and found that they could be rendered inactive by this process. They concluded that the protein components necessary for the infective process are

Fig. 21-13   The rate of oxidation of histidine as a function of pH.   [After L. A. Ae. Sluyterman, *Biochem. Biophys. Acta.*, 60:557 (1962).]

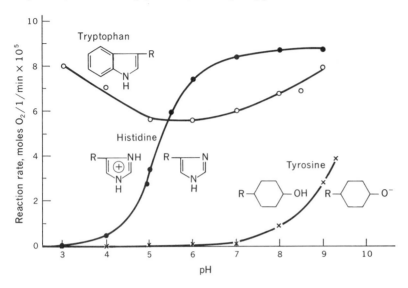

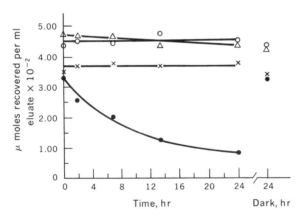

Fig. 21-14   Change in the composition of T-4 DNA at various stages of photooxidation: thymine (triangles); adenine (open circles); x cytosine; guanine (solid circles).   [After Simon and Van Vunakis, *J. Mol. Biol.*, 4:488 (1962).   Copyright © 1962.   Academic Press, Inc., New York.   Used by permission.]

photooxidized.   J. N. Welsh and M. H. Adams (see References) found that mutations and genetic recombinations were induced in the virus by this means.

Immunology studies have confirmed the photosensitivity of the viralphage DNA.   Even after weak irradiation, the phage nucleic acid lost over 80 percent of its ability to react with anti-DNA sera.

Furocoumarins

An extensive study has been made of the photosensitizing action of this class of compounds, of which 8-methoxypsorolan

is an example.   These compounds bind to DNA and to RNA and when irradiated are found to be extremely mutagenic.

Such polycyclic hydrocarbon derivatives are of considerable interest because they induce cancer and this activity has been correlated with their photosensitizing potential.   When ^{3}H-benzpyrene, made radioactive by the tritium ^{3}H content, was bound to DNA and irradiated, a firm binding between the two molecules was induced by light normally absorbed by the benzpyrene.   The adduct of benzpyrene and cytosine was isolated and tentatively identified:

Example 21-3

Does the line in the mercury spectrum at 22,938 cm^{-1}, extensively used in obtaining Raman spectra, correspond to an excited state of Hg with sufficient energy to dissociate hydrogen through photosensitization?

$$\frac{22{,}938 \text{ cm}^{-1}}{349.82 \text{ kcal mole}^{-1}} = 65.57 \text{ kcal mole}^{-1}$$

To dissociate H_2, 102.4 kcal mole^{-1} is required. Mercury atoms in the excited state used to produce Raman spectra lack $102.4 - 65.6 = 36.8$ kcal mole^{-1} to dissociate hydrogen molecules.

21-4 PHOTOSYNTHESIS

Photosynthesis is the name applied to reactions by which light energy is absorbed and converted into chemical energy through the synthesis of complex molecules from simpler molecules, with a resultant *increase* in the free energy of the chemical system. The absorption of the light drives the reaction in a direction which is the reverse of the normal direction in which the reaction would proceed if the only driving force were a decrease in free energy.

Photosynthesis indirectly provides the ultimate driving force for virtually all life processes, which need energy in order to proceed; through a long and complex chain of reactions the energy from photosynthesis is transferred to many species of molecules which distribute the energy to all parts of the living organisms where eventually it is transformed to meet the organism's needs.

The study of photosynthesis began about two centuries ago in 1772, when the English chemist, Joseph Priestley, noted that plant leaves could restore to air the ability to support combustion. Eight years later, Jan Ingenhousz, a Dutch physician and plant physiologist, showed that light was necessary for this process. In 1845, J. R. von Mayer (see Chap. 7) recognized that the process stored energy by synthesizing chemical products. Since that time, many notable contributions have elucidated a number of the steps in this extremely complex chemical chain, some of the most important of which are not completely understood even today.

Photosynthesis mechanisms

The overall result of photosynthesis in plants can be expressed by the equation

$$6CO_2 + 6H_2O \xrightarrow{nh\nu} C_6H_{12}O_6 + 6O_2$$

$$\Delta S = -43.6 \frac{\text{cal}}{\text{deg mole}}$$

$$\Delta G^\circ = +686 \text{ kcal mole}^{-1} \qquad \Delta H = 673 \text{ kcal mole}^{-1}$$

where $C_6H_{12}O_6$ stands for the basic synthetic product of the reaction, which has glucose structure. The combustion reaction, complementary to the above process, can be written

$$C_6H_{12}O_6 + 6O_2 \rightarrow 6CO_2 + 6H_2O$$
$$\Delta S = 43.6 \text{ cal deg}^{-1} \text{ mole}^{-1}$$
$$\Delta G^0 = -686 \text{ kcal mole}^{-1}$$
$$\Delta H = -673 \text{ kcal mole}^{-1}$$

To make the direct photosynthetic process take place, light with a wavelength of 2300 Å or shorter is needed. A photon at this wavelength has 124 kcal einstein^{-1}. The ΔH requirement for the above photosynthesis is 116 kcal mole^{-1} of CO_2. Therefore if one $h\nu$ per CO_2 were needed, the direct photosynthesis might work with light at 2300 Å. A study of evolutionary changes in living matter suggests that in the early stages of our planet's history, light of this wavelength may have penetrated to the earth's surface in appreciable amounts. But today, so much of the ultraviolet radiation is absorbed by the atmosphere that the direct photosynthetic process is impossible. However, by photosensitizing the material in plant cells through molecules called chlorophylls, nature is able to use light of much longer wavelengths in the visible red part of the spectrum. Since the value of $h\nu$ for such light is only 40 kcal mole^{-1}, the photosynthetic process must take place in small steps.

For purposes of discussion, the overall process can be broken down into the following stages:

1. Primary absorption of light by chlorophyll.
2. Transfer of energy to decompose water.
3. Secondary absorption of light by chlorophyll.
4. Reduction of CO_2.
5. Synthesis of carbohydrates.

Chlorophyll

This pigment, which acts as a photosensitizer, is found in two different forms, denoted a and b, both having roughly the same structure. The structural formulas of chlorophyll are shown in Fig. 21-15. There is a fairly rigid ring, containing a magnesium atom at the center and made up of carbon and nitrogen atoms linked by alternating single and double bonds. This is a resonance structure in which electron wave functions extend over a number of atoms, producing the kind of action at a distance which Szent-Györgyi has suggested as an essential ingredi-

ent of the life process. This type of wave function also provides excited-state electronic energy levels, to which molecules can be promoted by photons with the energy of only 40 kcal mole^{-1} found in the visible red.

The ring and its smaller attached groups is called *chlorophyllin;* the long side chain is called *phytol.* There is evidence that several hundred of the molecules are combined into a cooperative unit in the plant cell. Possibly, as a result of this close interconnection, the energy of an absorbed photon can be dissipated in an extremely short time and the chlorophyll can return to the original unactivated state in less than 10^{-8} sec. The absorption spectrum of chlorophyll a is shown in Fig. 21-16.

Energy transfer

When the energy of the absorbed photon is not dissipated but used to effect photosynthesis, a chain of reactions is set in motion, as illustrated schematically in Fig. 21-17. The first role (a) of the radiant energy is to be absorbed by chlorophyll, providing energy which passes (I) to two molecules of H_2O and (II) to a molecular complex E. The water molecules interact to produce three kinds of products: (1) Protons go to

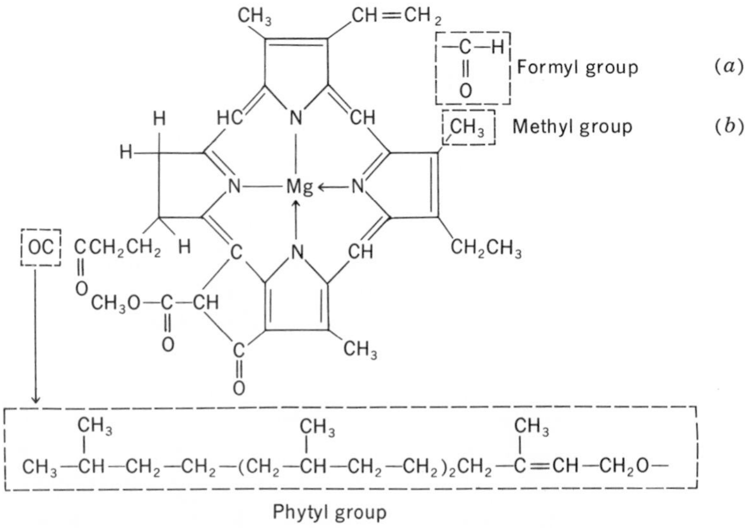

Fig. 21-15 The structure of the chlorophyll molecule, showing side groups for the (a) and the (b) species.

adenosine diphosphate (ADP) and bring about linkage with another phosphate radical to produce adenosine triphosphate (ATP). This reaction was discussed in Chap. 14, and the structural formula of ATP is shown in Fig. 14-16. The ATP serves as an energy source for many other biochemical reactions in the plant. (2) Molecules of O_2 are formed and pass out of the system as free oxygen gas. (3) Electrons pass to a molecular complex E and change it to a reduced form. This complex then receives additional energy from the chlorophyll(II) and transfers the electrons to another complex A, which finally passes them to a third complex P after the latter is activated by radiant energy (b), with the photosensitizing action of chlorophyll again assisting. The P complex passes the electrons on to an X^{3+} complex, which is reduced and then provides the energy for the CO_2 reduction process that changes the latter into carbohydrates. A suggested arrangement of these foci of action is shown in Fig. 21-18.

Dark reactions

Studies by the American chemist S. Ruben in 1940 indicated that the reactions reducing the CO_2 to carbohydrates continue for several minutes after light has been shut off. The American chemists M. Calvin and A. A. Benson used ^{14}C-labeled CO_2, added after the light was extinguished, and found that the ^{14}C was also reduced. They showed that the entire path from CO_2 to carbohydrate takes place through *dark* reactions guided by enzymes. This carbon reduction cycle is shown schematically in Fig. 21-19.

Photopotentials

The changes in the redox potentials during various stages of photosynthesis provide a great deal of information with regard to the driving force. This is shown in the form of a scale diagram in Fig. 21-20. The first stage of radiation (Fig. 21-20) raises the free energy from the value of $\mathcal{E}'_0$ at $+0.8$ to the higher free energy corresponding roughly to $\mathcal{E}'_0 = 0$. The free energy then falls as ATP is synthesized. In the second stage (Fig. 21-20) the absorbed radiant energy then activates the system to a value of $\mathcal{E}'_0$ of about -0.6 volt, and the free energy falls once more in the reduction of CO_2 to carbohydrates.

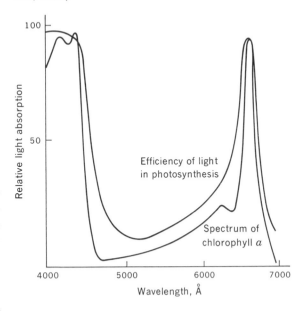

Fig. 21-16 Absorption spectrum of chlorophyll *a*. (After A. L. Lehninger, "Bioenergetics," W. A. Benjamin, Inc., New York, 1965.)

Example 21-4

If the glucose molecule were synthesized in a single step from $6CO_2$ and $6H_2O$ by a single photon, in what part of the spectrum would the photon have to be in order to correspond in its energy to the free energy required for this reaction?

686 kcal mole^{-1} $\times$ 350 cm^{-1} kcal^{-1} mole
$$= 240{,}100 \text{ cm}^{-1}$$

$$\frac{1}{240{,}100} \text{ cm}^{-1} = 4.164 \times 10^{-6} \text{ cm} = 416.4 \text{ Å}$$

Ans.

21-5 VISION

Basic principles

In order to provide a human being with the sense of sight, there must be a mechanism which absorbs radiant energy and transforms it with high selectivity into nerve impulses. The American biologist George Wald, of Harvard University, through a series of in-

Fig. 21-17 The preliminary reactions in chlorophyll photosynthesis. The heavy arrows indicate energy transfer. E, A, P, and X^{+++} are electron-transfer compounds. (From "How Light Interacts with Living Matter" by Sterling B. Hendricks. Copyright © September, 1968 by Scientific American, Inc. All rights reserved.)

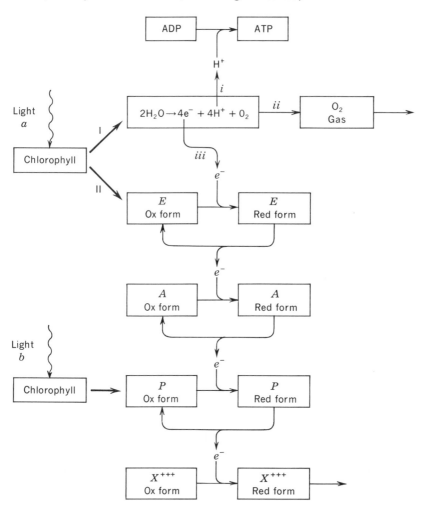

vestigations begun in 1935, found that the absorbing molecule is vitamin A aldehyde (11-*cis*-retinal) associated with a protein called *opsin*. This *chromophore* molecule is contained in the specialized cells on the retina of the eye and has a structure of conjugated bonds which provide relatively low-energy excited electronic states. When light energy is absorbed, there appears to be a change from a cis to a trans configuration, triggering the nerve impulse.

Photoreceptor cells

The cells which absorb the light for human vision are of two kinds. In the eye there are about 100 million rod-shaped cells and 5 million cone-shaped cells. There is a synapse, or junction, between the cell and a nerve fiber leading to the brain. Within the cell, the chromophores are attached to a membrane in which a chemical change is induced by the photo-activated chromophore, initiating the signal in the nerve fiber.

The outer segment of a rod or cone is composed of a sac of double-membrane disks, enclosed in a plasma membrane, as shown in Fig. 21-21. The molecular arrangement in an outer disk is sketched in Fig. 21-22. The visual-pigment chromophores lie perpendicular to the long axis of the rod, in the same plane as the disks.

Chromophores

The two structural formulas of the chromophore molecule are shown in Fig. 21-23. The chromophore is a straight-chain molecule with alternating double bonds. This straight-chain configuration contrasts with the ring structure of chlorophyll, but the conjugation in both provides the low-energy activated electronic energy levels that are essential for the absorption of light in the visible spectrum. The inactivated chromophore has the cis configuration and is relatively rigid. The transition to the activated state appears to relax the rigidity, producing a shift to the trans configuration. By lowering the temperature of the pigment the reaction rates of the stages following photoabsorption can be reduced so that the inter-

mediate states can be identified. These are shown schematically in Fig. 21-24. The two positions of the chromophore in the lipid pocket are drawn in Fig. 21-25.

The phenomenon of human color vision is possible because of the three different types of opsin in the cones. The iodopsin pigments in the cones have maximum absorption at three different locations in the spectrum, as shown in Fig. 21-26.

Example 21-5

Calculate the energy in kilocalories per einstein of the photons which correspond to maximum absorption of the iodopsin pigments in the human eye.

$Yellow:$ 5700×10^{-8} cm $= 17,543$ cm^{-1}

$$\frac{17,543 \text{ cm}^{-1}}{349.8 \text{ cm}^{-1} \text{ kcal}^{-1} \text{ einstein}^{-1}}$$
$$= 50.2 \text{ kcal einstein}^{-1}$$

$Green:$ 5350×10^{-8} cm

53.4 kcal einstein^{-1}

$Blue:$ 4450×10^{-8} cm

64.2 kcal einstein^{-1}

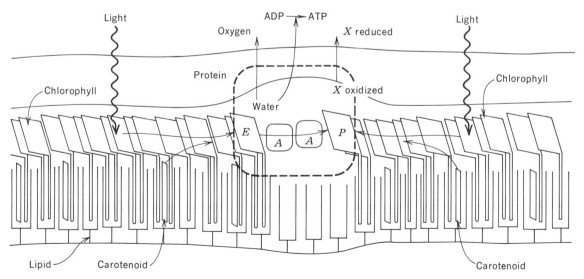

Fig. 21-18 Schematic representation of the active site of a cell for photosynthesis in chlorophyll. (From "How Light Interacts with Living Matter" by Sterling B. Hendricks. Copyright © September, 1968 by Scientific American, Inc. All rights reserved.)

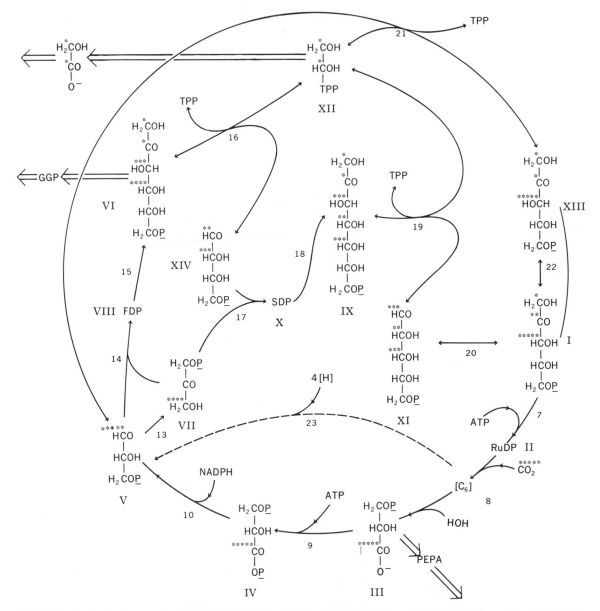

Fig. 21-19 The Calvin carbon reduction cycle of photosynthesis. Open arrows indicate the initiation of biosynthetic paths. Asterisks show relative degree of labeling after a few seconds of photosynthesis. (After J. A. Bassham, *Surv. Prog. Chem.*, 1966:2–51. Copyright © 1966. Academic Press, Inc., New York. Used by permission.)

21-6   FLUORESCENCE

After a molecule has absorbed a photon and is raised to an activated state, it may emit part or all of this extra energy as radiation. If the emission takes place because of a transition to a state of like multiplicity, the process is called *fluorescence*. Usually the emission follows less than 10^{-8} sec after the absorption. If spin inversion is involved, so that the multiplicity changes, the process takes longer and is called *phosphorescence*.

The name fluorescence was adopted because fluorite, CaF_2, exhibits this phenomenon in a striking way. Many organic compounds also show fluorescence, including chlorophyll and dyes such as eosin and fluorescein; a number of inorganic vapors also fluoresce, like sodium, mercury, and iodine. Phosphorescence is more likely to occur in solids, such as the sulfides of the alkaline earths.

Theory

G. B. Beccari, an Italian physician, reported in 1757 that he had observed his hand glowing in a dark room shortly after exposure to intense sunlight. During the nineteenth and early twentieth century extensive studies were made of this afterglow, especially when induced by ultraviolet light. With the advent of modern atomic theory, fluorescence has become a valuable tool for investigating the nature of excited states.

As an example, the fluorescent spectra of tryptophan under different conditions are shown in Fig. 21-27. The peak can be shifted considerably by altering the nature of the solvent and the amount of ionization.

Some of the possible steps in producing fluorescence can be summarized in equations. The primary absorption may be made by a molecule D in a singlet state ^1D:

$$^1D + h\nu \rightarrow {}^1D^*$$

This may directly emit by fluorescence

$$^1D^* \rightarrow h\nu + {}^1D^F$$

Or it may change to a triplet state

$$^1D^* \rightarrow {}^3D^*$$

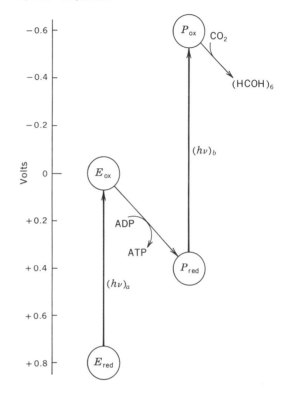

Fig. 21-20   Skeletal path of energy in photosynthesis indicating changes in redox potentials. Reprinted from G. Govindjie, G. Papageorgiou, and E. Rabinowitch, chap. 12, in G. G. Guilbault (ed.), "Fluorescence," by courtesy of Marcel Dekker, Inc., New York, 1967.

The triplet state may phosphoresce

$$^3D^* \rightarrow h\nu^P + {}^1D$$

Either the singlet or triplet state can transmit the energy by radiationless exchange to an acceptor molecule A:

$$^1D^* + {}^1A \rightarrow {}^1D + {}^1A^*$$
$$^3D^* + {}^3A \rightarrow {}^3D + {}^3A^*$$

The acceptor molecule may emit

$$^3A^* \rightarrow {}^1A + h\nu^P$$

The fluorescence spectra of proteins have yielded valuable information about their structure. Extensive studies have been made of the fluorescence due to the

Fig. 21-21   Ocular photoreceptor cell (schematic). [From A. I. Cohen, *Biol. Rev.,* Cambridge Philosophical Society, 38(1963), 427.]

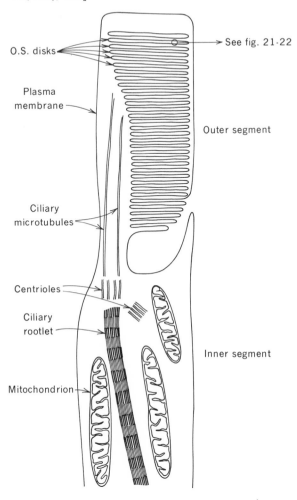

O.S. disks

Plasma membrane

See fig. 21-22

Outer segment

Ciliary microtubules

Centrioles

Ciliary rootlet

Inner segment

Mitochondrion

molecules can be triggered into an almost simultaneous joint transition back to the inactive state, with the resulting emission of a beam of intense coherent radiation.   Such a device is now called a *laser,* an acronym for *l*ight *a*mplification by *s*timulated *e*mission of *r*adiation.

The four steps associated with such a device are shown in Fig. 21-28.   In a the molecule is inactive; in b the molecule is activated by a photon; in c the molecule returns spontaneously to its initial state, emitting radiation; and in d the transition to the ground state is triggered by an incoming beam, which is then amplified by the emitted radiation.   For effective laser operation the amount of emission from c must be small compared with d.

In a typical solid laser like a ruby crystal, the active atoms are those of chromium, which are held in a lattice of aluminum oxide.   A flash of intense light raises a large percentage of the chromium atoms to the excited state.   Then radiation can trigger the return transition, resulting in the emission of light which is coherent both in space and in time because the stimulating photon on entering the atom causes it to emit a photon which is precisely "in step" with it. In this way, monochromatic beams of great intensity can be produced which travel great distances without spreading.   There are many important applications

presence of tryptophan, and the evidence indicates that this ring differs widely in its physicochemical state depending on the particular protein in which it is present.   In Table 21-3 some of the peaks of fluorescence for different proteins are listed.

Laser techniques

When a large percentage of an ensemble of molecules can be brought into an excited state, a way frequently can be found by which large numbers of these excited

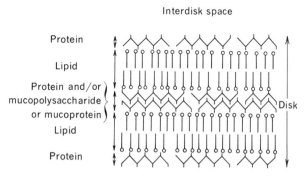

Interdisk space

Protein

Lipid

Protein and/or mucopolysaccharide or mucoprotein

Lipid

Protein

Disk

Interdisk space

Fig. 21-22   The molecular configuration of a disk in an ocular photoreceptor cell (see Fig. 21-21).   [After C. D. R. Bridges, "Biochemistry of Visual Processes" (Visual Pigments in Solution) in "Photobiology," Elsevier Publishing Company, Amsterdam, 1967.]

possible for such beams both in science and technology. They are valuable in studying Raman scattering and are certainly going to be an asset in communication.

Since the laser principle applies over the entire range of the spectrum, it has been used extensively in the generation of microwaves. The device is then called a *maser*, *m*icrowave *a*mplification by *s*timulated *e*mission of *r*adiation. In Fig. 21-29, some of the wavelengths of various laser and maser sources are shown.

Example 21-6

Calculate the energy in kilocalories per einstein for the shortest and the longest wavelengths shown in Fig. 21-29.

For the shortest wavelength

$$\log \lambda = 3.371$$
$$\lambda = 2350 \text{ Å} \qquad \omega = 42{,}500 \text{ cm}^{-1}$$
$$E = \frac{42{,}500 \text{ cm}^{-1}}{349.82 \text{ kcal einstein}^{-1} \text{ cm}^{-1}}$$
$$= 122 \text{ kcal einstein}^{-1} \qquad Ans.$$

Fig. 21-23   The two configurations of 11-*cis*-retinal, as transformed from left to right by the absorption of a photon.

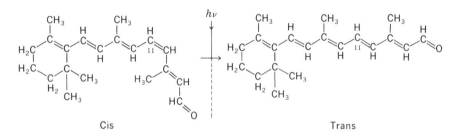

Fig. 21-24   The vision cycle of retinal opsin. The temperatures at which the intermediate compounds were traced are shown below the designations, together with absorption spectra peak wavelengths. Note 1 NM = 10 A = 10^{-7} cm. (From "How Light Interacts with Living Matter" by Sterling B. Hendricks. Copyright © September, 1968 by Scientific American, Inc. All rights reserved.)

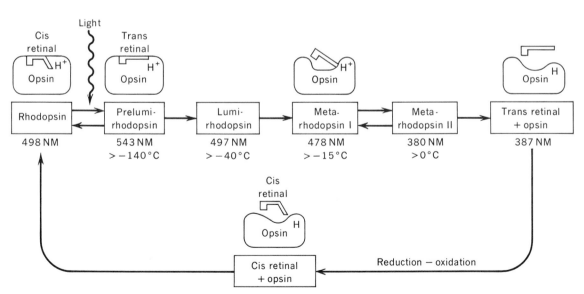

TABLE 21-3   Wavelengths of protein fluorescence spectra[a]

Protein	Wavelength, Å	
	In aqueous solution	In 8 M urea
Chymotrypsin	3340	3500
Chymotrypsinogen	3310	3500
Edestin (1 M NaCl)	3280	3500
Fibrinogen	3370	3500
Fumarase	3350	
Hemoglobin globin	3350	3480
Pepsin	3420	3500
Trypsin	3320	3500
Trypsinogen	3320	3500
Urease	3300	

[a] From F. W. J. Teale, *Biochem. J.*, **76**:18 (1960).

Fig. 21-25 Positions of retinal in the opsin cavity. 11-*cis* (solid line); 9-*cis* (dotted line); trans (dashed line). (From R. Hubbard, *Nat. Phys. Lab. Symp.* No. 8, p. 153, H. M. Scientific Office, London, 1958.)

For the longest wavelength

$\lambda = 10^6$ Å

$\omega = 100$ cm^{-1}

$E = \frac{100}{350} = 0.285$ kcal einstein^{-1} *Ans.*

21-7   HIGH-ENERGY RADIOLYSIS

Almost immediately after x-rays were discovered at the close of the nineteenth century and means for producing this extremely shortwave radiation were devised, scientists were aware that x-rays provided one of the most powerful tools for causing atomic and molecular transitions to highly activated states. During the twentieth century, scientists have constructed other increasingly powerful devices for producing high-energy particles which cause high-energy states in matter.   Table 21-4 lists the energy attainable by means of some recent instruments.

In familiar photochemistry one photon usually activates only one acceptor atom or molecule at the most.   In high-energy radiolysis the entering particle may have so much energy that tens or hundreds of neighboring atoms are activated.

Some of the secondary processes attending the absorption of a high-energy particle p^* by an atom or molecule A can be indicated as follows:

A $\xrightarrow{p^*}$ A*

A* + B → A + B*

A* → radicals

A* → A*′ + B*

B* + C → B + C*

Other possibilities are

$$A \xrightarrow{p^*} A^+ + e^{*-}$$
$$A^+ + e^{*-} \rightarrow A^*$$
$$e^{*-} + B \rightarrow B^{*-}$$
$$e^{*-} + B \rightarrow B^* + e^-$$

When H_2O is irradiated with high-energy photons, there is almost always a reaction

$$H_2O \xrightarrow{h\nu} H_2O^+ + e^-$$
$$H_2O^+ + H_2O \rightarrow H_3O^+ + OH$$

Frequently

$$2OH \rightarrow HOOH$$

When high-energy electrons are passed into water, the reaction

$$H_2O(l) + e^-(aq) \rightarrow H(aq) + OH^-(aq)$$

frequently occurs. The thermodynamic properties of some of these states are listed in Table 21-5.

For some time there has been awareness of the fact that both high-energy photons and high-energy particles can have profound and usually destructive effects on key biochemical molecules like DNA. Studies of such effects are important not only as a way of assessing damage done to genes from radiation exposure following the detonation of nuclear bombs, but also because this provides a valuable probe for studying the structure of DNA itself. For example, the American scientists L. S. Myers, Jr., M. L. Holles, and

L. M. Theard studied the transient species produced by the reaction of OH free radicals with DNA with the help of pulse radiolysis. Pulses between 2×10^{-9} and 5×10^{-7} sec of 10-Mev electrons were used. The rate constant for the formation of the DNA transient was found to be 6×10^8 mole^{-1} sec^{-1} when calculated on the basis of nucleotide molecular weight

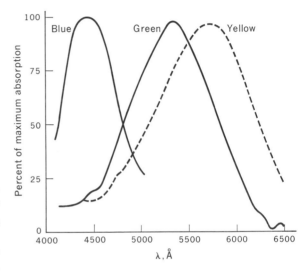

Fig. 21-26 Relative absorption of visible light by the three types of cones in the human eye. (From "How Light Interacts with Living Matter" by Sterling B. Hendricks. Copyright © September, 1968 by Scientific American, Inc. All rights reserved.)

TABLE 21-4   Sources of excitation for radiation chemistry

Particle	Source	Maximum energy	
		kcal mole^{-1}	Mev
Photon	X-ray	1.5×10^7	0.6
Electron	Linear accelerator	2×10^8	10
	Betatron	4×10^8	20
Proton	Linear accelerator	2×10^8	10
	Cyclotron	8×10^8	40
Neutron	Cyclotron	7×10^8	35
Deuteron	Cyclotron	4×10^8	20
Alpha particle	Cyclotron	2×10^8	10

Fig. 21-27 The fluorescence spectra of tryptophan at room temperature: (1) in neutral aqueous solution; (2) in a mixture of 4% NaOH and 0.5% HCHO; (3) in 4% NaOH solution. [After E. A. Chernitskii and S. V. Konev, *Zh. Prik. Spektroskopii*, 2:261 (1965).]

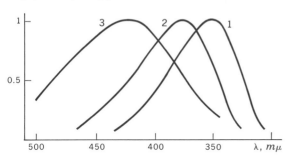

wavelengths was used to establish the rates of decay. For thymine at pH 12.4 and below, the decay rates at 3250 and 5500 Å differed, suggesting that more than one species of transient was present.

Example 21-7

Calculate the weight of the product formed and the energy absorbed when one 10^{-8}-sec pulse of 10-Mev electrons produces a DNA transient at the rate of 2×10^{-9} mole sec^{-1} (mol wt = 350), assuming 100 percent efficiency.

$$2 \times 10^{-9} \text{ mole sec}^{-1} \times 10^{-8} \text{ sec}$$
$$\times 350 \text{ g mole}^{-1} = 7 \times 10^{-16} \text{ g}$$

$$10 \text{ Mev} \times 10^6 \frac{\text{ev}}{\text{Mev}} \times 23.06 \text{ kcal mole}^{-1}$$
$$\times 2 \times 10^{-9} \text{ mole sec}^{-1} \times 10^{-8} \text{ sec}$$
$$= 4.6 \times 10^{-9} \text{ cal}$$

Problems

1. Calculate the energy increase in a system which in 7.2×10^{-3} sec absorbs the radiation in a beam with an intensity of 4.987×10^{-5} einstein sec^{-1} in the region around 2536 Å.

2. A leaf absorbs energy from a lamp in the red part of the spectrum which has an intensity of 1.532 watts in the region with an average wavelength of 7576 Å. In what interval of time will the leaf absorb 1 cal?

3. HCHO absorbs radiation at 30,090 cm^{-1}. If 0.0621 mole is activated as the result of an absorption

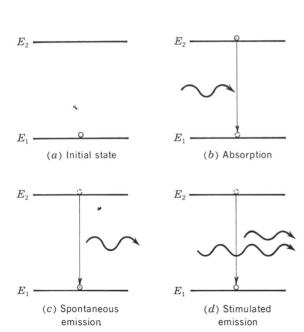

(a) Initial state

(b) Absorption

(c) Spontaneous emission

(d) Stimulated emission

Fig. 21-28 Four stages of laser action. [After A. F. Haught, *Ann. Rev. Phys. Chem.*, 19:343 (1968).]

of 350, using about 100 rad pulse^{-1} of radiation. The points of attack on thymine, 5-methylcytosine, and thymidine depend on pH.

The first-order rate constants were also determined for the decay of the transients. Typical transient-decay curves are shown in Fig. 21-30. Light of varying

TABLE 21-5 Thermodynamic properties of $H_2O(l)$ and principal aqueous products in radiation chemistry[a]

Species	ΔG_f°	ΔH_f°	$\bar{S}_0$	V
$H_2O(l)$	−56.69	−68.32	16.72	18.0
$H(aq)$	+53.03	51.09	9.08	2.5
$OH^-(aq)$	−141.41	−153.79	−1.35	−3.8
$H^+(aq)$	103.81	98.82	−1.17	−1.5
$e^-(aq)$	−37.54	−36.61[b]	3.13[b]	99.3[b]

[a] From R. M. Noyes, Radiation Chemistry, chap. 4, in *Advan. Chem. Ser.*, **81** (1968).
[b] Predicted value.

Fig. 21-29   Laser and maser radiation sources available in 1968.   (From "Laser Light" by Arthur L. Schawlow.   Copyright © September, 1968 by Scientific American, Inc.   All rights reserved.)

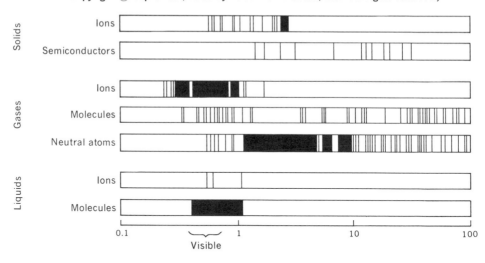

Wavelength in microns

of 1.562 einstein of light, what is the quantum efficiency?   How much energy has been used in activating the molecules?

4.   Anthracene crystals absorb ultraviolet light at 3390 Å.   Calculate the height of the activated energy level above the ground state in kilocalories per mole and in cm^{-1}, corresponding to this absorption.

5.   An anthracene crystal absorbs 1 einstein of radiation at 3390 Å, and then the activated molecules drop to an energy level at 14,750 cm^{-1} by a nonradiative transmission; finally, all activated molecules drop from there to the ground level, emitting radiation in a laser beam of 10^{-9} sec duration.   Calculate the wavelength of the beam in angstroms and the intensity of the beam in calories per second.

6.   If the laser beam described in Prob. 5 is used to produce the Raman spectrum of benzene, what will be the lowest wavelength found in the emitted Raman light?

7.   Using the data in Fig. 21-16, estimate the quantum efficiency of the primary reaction if the laser beam described in Prob. 5 is used for photosynthesis by chlorophyll a molecules.

8.   Assume that in Fig. 21-20, the radiation absorption a corresponds to the change in the redox potential of two electrons.   Calculate the change in free energy.

9.   Calculate the energy per einstein in the light at the wavelength corresponding to maximum absorption

in chlorophyll and compare with the answer in Prob. 8.

10.   Calculate the energy in kilocalories per mole and in ergs per photon at which the yellow-, green-, and blue-sensitive cones of the human eye have maximum absorption.

11.   In Fig. 21-9 assume that the height of the minimum in the curve corresponding to predissociation is 22,516 cm^{-1} above the ground level.   Estimate the dissociation energy for a molecule in the ground state and in the predissociated state if no transfer to state B occurs.

12.   Assuming that the line structure in the ground level of anthracene (Fig. 21-12) is due to variations in the vibrational quantum number, calculate the wavelength of the line corresponding to this vibration and the frequency of the vibration.

13.   Calculate the wavelength of the Raman line corresponding to the frequency of the vibration in Prob. 12 if the Raman spectrum is excited by the laser beam with highest frequency available from ions in liquids as shown in Fig. 21-29.

14.   From the decay curve c in Fig. 21-30, estimate the half-life of the transient molecules in microseconds and calculate a value of the reaction rate constant.

15.   If the transient activated species are present immediately following the incident radiation at a concentration of 8.649×10^{-9} mole liter^{-1}, what will be the concentration after 10^{-5} sec?

Fig. 21-30 Oscillograph tracings (millivolts against time) showing the decay of transients in 5×10^{-4} M (saturated N_2O) solutions of uracil. Pulse width 5.00×10^{-9} sec. [After L. S. Meyers, Jr., Radiation Chemistry, I, *Advan. Chem. Ser.,* 3(24) (1968).]

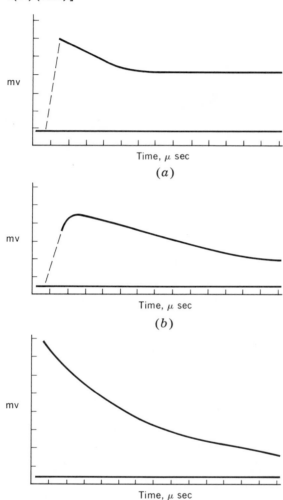

Time, μ sec

(*a*)

Time, μ sec

(*b*)

Time, μ sec

(*c*)

REFERENCES

Welsh, J. N., and M. H. Adams: *J. Bacteriol.,* **68**:122 (1954).

Claus, W. D.: "Radiation Biology and Medicine," Addison-Wesley Publishing Company, Inc., Reading, Mass., 1958.

Saltiel, J.: The Mechanisms of Some Photochemical Reactions of Organic Molecules, *Surv. Prog. Chem.,* **1964**:240–328.

Sinanoglu, O. (ed.): "Modern Quantum Chemistry, pt. III: Action of Light and Organic Crystals," pp. 93–310, Academic Press, Inc., New York, 1965.

Vernon, L. P., and G. R. Seely (eds.): "The Chlorophylls," pp. 523–639, Academic Press, Inc., New York, 1966.

Bassham, J. A.: Photosynthesis, *Surv. Prog. Chem.,* **1966**:2–51.

Ausloos, P.: Radiation and Photochemistry, *Ann. Rev. Phys. Chem.,* **1966**:205–236.

Spikes, J. D., and R. Straight: Sensitized Photochemical Processes in Biological Systems, *Ann. Rev. Phys. Chem.,* **1967**:409–436.

Neckers, D. C.: "Mechanistic Organic Photochemistry," Reinhold Publishing Corporation, New York, 1967.

Florkin, M., and E. H. Stotz (eds.): "Photobiology, Ionizing Radiations," Elsevier Publishing Company, Amsterdam, 1967.

Pizzarello, D. J., and R. L. Witcofski: "Basic Radiation Biology," Lea & Febiger, Philadelphia, 1967.

Ashmore, P. G., F. S. Dainton, and T. M. Sugden: "Photochemistry and Reaction Kinetics," Cambridge University Press, London, 1967.

Reactivity of the Photo-excited Organic Molecule, in "Proceedings of the Solvay Institute: 13th Conference on Chemistry," John Wiley & Sons, Inc., New York, 1967.

Guilbault, G. G. (ed.): "Fluorescence," Marcel Dekker, Inc., New York, 1967.

Konev, S. V.: "Fluorescence and Phosphorescence of Proteins and Nucleic Acids," Plenum Press, New York, 1967.

Haught, A. F.: Lasers and Their Applications to Physical Chemistry, *Ann. Rev. Phys. Chem.,* **1968**:343–370.

Hart, E. J. (ed.): Radiation Chemistry, I, *Advan. Chem. Ser.,* 81, 1968.

Hart, E. J. (ed.): Radiation Chemistry, II, *Advan. Chem. Ser.,* 82, 1968.

Chapter Twenty-Two

solid states

When a portion of matter can retain its shape without being confined in a vessel, it is said to be in the *solid state.* If the atoms or molecules are regularly arranged, it is called *crystalline;* if they are not, it is called *amorphous* or *glassy.*

In the crystalline state, atoms are arranged with the highest degree of order to be found in any of the states of matter. In some crystals the pattern of order is based on the arrangement of individual atoms not linked as molecules; this is true in crystals of many metals like iron, where the atoms stack together very much like marbles arranged regularly in a box. In other crystals the atoms exchange electrons and are present in the form of ions; e.g., in the sodium chloride lattice each sodium atom has lost an electron, leaving it as a positive ion, and each chlorine has gained an electron, leaving it as a negative ion; and the sodium ions and chlorine ions are packed together alternately making a checker-board pattern of positive and negative ions. In still other crystals the particles which are packed together are molecules. This is true of most organic substances. For example, a crystal of benzene consists of benzene molecules regularly stacked together in a pattern much like a box packed with dinner plates, except that not all the benzene molecules are fitted in at exactly the same angles; there is a periodic variation in the angles of the molecules in passing through the crystal.

When the atoms or ions are essentially spherical in character, the packing in the crystal is quite similar to the packing of marbles. In Fig. 22-1 are shown two different ways of packing spherically: cubic close packing and hexagonal close packing. A glance at the cubic close packing shows that the group of balls in the picture forms an array with square sides and a square top and bottom; hence it resembles a cube. In Fig. 22-1*b* the top and the bottom resemble hexagons. The atoms of the rare gases (He, Ne, Ar, etc.) are spherical in shape and pack together in the crystalline state in this manner.

Another interesting example of crystalline packing is found in the mineral *diamond.* As in the crystals of the rare gases, the particles packed together in diamond are individual atoms; but unlike the rare gases, where the atoms in the lattice are held together only by the van der Waals forces, the carbon

Fig. 22-1 Close packing: (*a*) face-centered cubic with unit cell; (*b*) hexagonal with unit cell.

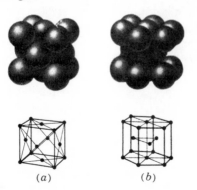

(*a*)   (*b*)

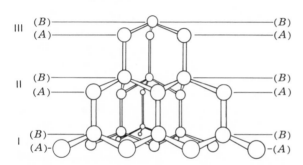

Fig. 22-2 A portion of a diamond crystal showing the carbon atoms stacked in three major planes (I, II, III). Each major plane consists of two subplanes (*A*) and (*B*) with plane (*B*) slightly above plane (*A*).

atoms in diamond are bound to each other by covalent bonds making tetrahedral angles around each atom. This is the reason why diamond has such a high melting point and low vapor pressure and is so hard. Figure 22-2 is a drawing of part of a diamond crystal lattice. In Fig. 22-3 is shown the arrangement of the sodium and chlorine atoms in a crystal of sodium chloride.

22-1 CRYSTAL THEORY

Such regular arrays, in which particles can be packed, are called *crystal lattices*. A significant character-istic of these lattices are the planes in which there is

an especially high density of molecules. This can be seen clearly in the representation of the diamond lattice in Fig. 22-2. In the plane *A* at the bottom of the figure are found the four carbon atoms at the very front of the model. If the plane is extended back in space, similar lines of atoms are found. Recall that in a diamond of the familiar size worn in rings there may be 10^{20} carbon atoms. Thus, extending a plane throughout such a crystal, one finds trillions of atoms in the plane. In this figure, there is another plane *B* slightly above the bottom plane *A* which contains the same density of atoms. Then, moving upward there is a vertical distance of about 1 Å where a plane ex-tended back through the crystal, parallel to *A* or *B*, would contain no atoms at all. At a distance of 1.54 Å above the I plane there are then two more planes II*A* and II*B* lying close together and also containing trillions of carbon atoms. Planes like these play a key role in the optical phenomena by which the nature of crystals has been determined.

Crystal nomenclature

In order to express the relationships which are the basis of the theory of crystal lattices precisely, a number of terms and symbols have been defined, linking these relationships to several branches of mathematics such as vector algebra and group theory. It is helpful to understand a few of the primary con-cepts of these fields.

A *vector* is a quantity which has both a numerical value and a direction. In the two-dimensional coordi-nate system shown in Fig. 22-4, the distance from point *a* to point *b* is a vector and is denoted by a bold-face letter **r**. In order to specify the exact nature of the arrow vectors shown in the figure, both the direc-tion and length of the vector and the angle between the vector and one of the coordinate axes must be specified. The application of the vector **r** to a particle at point *a* moves it to point *b*. The particle also can be moved from *a* to *b* in two steps, first by applying $\mathbf{r}_x$ and then by applying $\mathbf{r}_y$. An equation which is the sum of two or more vectors implies the application of one vector first, followed by the appli-cation of the others

$$\mathbf{r} = \mathbf{r}_x + \mathbf{r}_y \qquad\qquad (22\text{-}1)$$

A basic ideal crystal is defined as a body composed of atoms arranged in a lattice with certain regularities. First, the crystal is made up of repeated units called unit cells. Second, if a certain vector **r** is applied to cell i, carrying cell i to the position of cell j, and the vector **-r'** is applied to cell j, the group of atoms in cell i is interchanged with the group of atoms in cell j and the crystal lattice looks exactly the same after the exchange as it did before. The vector **r** is defined as

$$\mathbf{r} = n_a\mathbf{a} + n_b\mathbf{b} + n_c\mathbf{c} \tag{22-2}$$

where n_a, n_b, and n_c are arbitrary integers and **a**, **b**, **c** are the crystal axes called the primitive axes; the parallelepiped defined by these axes is called the *primitive* cell and is a type of unit cell.

If the lattice extends to infinity in all directions, the application of vector **T** does not change the lattice when **T** is defined

$$\mathbf{T} \equiv n_a\mathbf{a} + n_b\mathbf{b} + n_c\mathbf{c} \tag{22-3}$$

T is called a *translation vector* since it merely moves the lattice without rotating it.

There also may be rotations about certain axes which will leave the lattice unchanged. For example, in a crystal composed of N_2 molecules, the rotation of each molecule through 180° about its own principal axis merely interchanges the N atoms and leaves the lattice unchanged. The rotation of a cubic lattice about the x, y, or z axes by 90, 180, or 270° leaves it unchanged if **a**, **b**, and **c** lie along x, y, and z axes, normally the case; of course, rotating by 360° leaves any lattice unchanged.

There are lattices with axes about which rotations of 120 and 60° leave the lattice unchanged. The rotations resulting in identity are limited to the values as shown in Table 22-1, where names and numerical designations are given. There are also three other types of transformation which may result in identity; these are reflection, inversion, and rotation followed by inversion. There are 14 types of lattice, defined by the types of transformations which leave them unchanged. These are called the *Bravais lattices* and are shown schematically in Fig. 22-5.

A notation, called a *Miller index,* has been devised to designate planes in a lattice. The intercepts of the

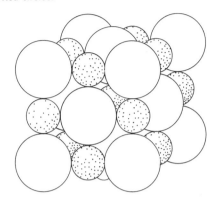

Fig. 22-3 The lattice of crystalline sodium chloride. Na^+ is represented by the empty circles, and Cl^- is represented by the dotted circles.

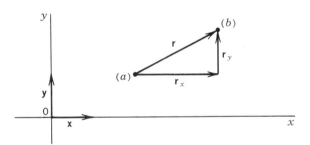

Fig. 22-4 A vector r lying in a two-dimensional space with coordinate axes x and y with unit vectors x and y. The drawing also illustrates $\mathbf{r} = \mathbf{r}_x + \mathbf{r}_y$.

plane on the three basic axes are first found. Then the reciprocals of the values of n_i are taken, and the smallest set of integers having the same ratios are the index, as shown in Fig. 22-6. For example, a plane which only intersects the x axis at the distance from the origin defined by **a** has $n_a = 1$, $n_b = 0$, $n_c = 0$. The Miller index is therefore (100). The plane which intersects the x axis at the point defined by **a**, the y axis at the point defined by **b**, and the z axis at the point defined by **c** has $n_a = 1$, $n_b = 1$, and $n_c = 1$ and thus has a Miller index of (111). A plane which intersects the x axis at the point defined by 2**a**, the y axis at **b**, and the z axis at **c** has the values $n_a = \frac{1}{2}$, $n_b = 1$, and $n_c = 1$; and so the Miller index is (122).

TABLE 22-1   Permissible rotations of lattices and their designations

Angle, deg	Numerical designation	Name
360	1	Monad
180	2	Diad
120	3	Triad
90	4	Tetrad
60	6	Hexad

Symmetry groups

The property of symmetry plays an important role in many areas of physical chemistry. The natural logarithm of the symmetry number of a molecule multiplied by R gives the amount per mole by which the entropy in the liquid or gaseous state is reduced because of symmetry. The melting point of a crystal is raised by the symmetry of the molecules of which it is composed. Symmetry plays a significant part in determining the nature of wave functions also. In order to understand the nature and properties of crystals, it is essential to know the basic principles of symmetry.

From the mathematical point of view, symmetry is closely related to the concept of an algebraic *group*. An algebraic group is defined as a set of distinct elements which satisfy four conditions. Before stating them, it may be helpful to offer an example. Consider the set of transformations which carry a triangular molecule like cyclopropane, C_3H_6, from one configuration to another. The structure of cyclopropane is shown in Fig. 22-7. Since each of the CH_2 clusters is identical in shape with the others, differing only in position, they can be denoted by circles labeled a, b, and c, as a kind of geometrical shorthand.

Suppose, first of all, that the molecule lies with the carbon atoms in the plane of the paper, and that cluster a lies at the top apex with b and c at the bottom as shown. (The term *cluster* will be used instead of the more familiar term *group*, in order to avoid confusion with the mathematical group.) This special configuration is designated as i. The i configuration can be changed by rotation or inversion of

the molecule into five other configurations each identical physically and chemically with the i configuration and differing from the others only in the CH_2 cluster positions. The six configurations are shown in the figure.

The transformation which carries the i configuration into *itself* is called the *identity* transformation and is labeled I. This could be a rotation around the axis perpendicular to the plane of the paper passing through the center of the molecule, rotating either clockwise or counterclockwise by $n \times 360°$, where n is an integer: 1, 2, 3, 4, ... The transformation which carries i into ii is labeled II. This could be a similar rotation clockwise by 120° or counterclockwise by 240°. The rotation which carries i into iii is labeled III. This could be a similar rotation clockwise by 240° or counterclockwise by 120°. The transformation which carries i into iv is labeled IV. This could be a rotation about the axis passing through a and the center of the molecule by 180° either clockwise or counterclockwise. The transformation V, which carries i into v, is a similar rotation about an axis passing through b and the center; the transformation VI, which carries i into vi, is a similar rotation about an axis passing through c and the center.

The set of these six transformations constitute the representation of a group. The group is a set of distinct elements (in this case, represented by the six transformations) satisfying the following conditions:

1.   A composition rule designated as the *product* must connect the elements. (In this case it is the application of one transformation *followed* by another transformation written I × II or simply I II.) The final result of any product of elements (in this case the sequential applications of transformations such as I × II) must equal some element of the group (the application of some single transformation).

2.   The associative law holds for all elements in the set: $(XY)Z = X(YZ)$.

3.   The set must contain an identity element E such that its application leaves the configuration unchanged. (This is transformation I.)

4.   For each element in the set X, there must be

Fig. 22-5 The 14 Bravais space lattices. (After C. Kittel, "Elementary Solid State Physics," John Wiley & Sons, Inc., New York, 1962.)

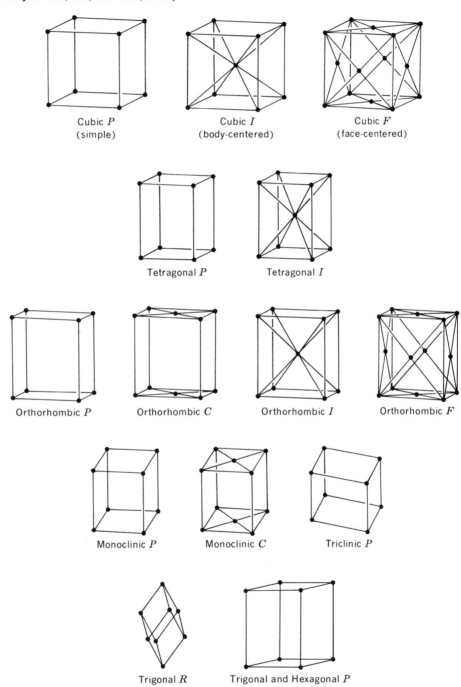

Cubic P
(simple)

Cubic I
(body-centered)

Cubic F
(face-centered)

Tetragonal P

Tetragonal I

Orthorhombic P

Orthorhombic C

Orthorhombic I

Orthorhombic F

Monoclinic P

Monoclinic C

Triclinic P

Trigonal R

Trigonal and Hexagonal P

an inverse X^{-1} such that $XX^{-1} = E$. (Note that the application of II followed by the application of III leaves the molecule in its original position so that II $\times$ III $= E$ and III $=$ II^{-1}.)

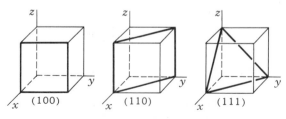

Fig. 22-6 Miller indices to denote planes.

In order to show the nature of each transformation more clearly, a notation can be used which denotes directly the change to which element corresponds. The identity element can be denoted by $\bigcirc$, meaning no change at all. Element II is denoted by a triangle $\triangle$, indicating that all three of the clusters move, and that rotation is clockwise by 120°. Element III is denoted by a blackened triangle $\blacktriangle$ indicating that rotation is counterclockwise by 120°. Element IV is denoted by a triangle with a heavy line at the base $\underline{\triangle}$ indicating that clusters b and c are transposed. Element V similarly is $\triangle$ and element VI is $\triangle$.

The results of the product of any two elements can be listed in a group table, as shown in Table 22-2. If

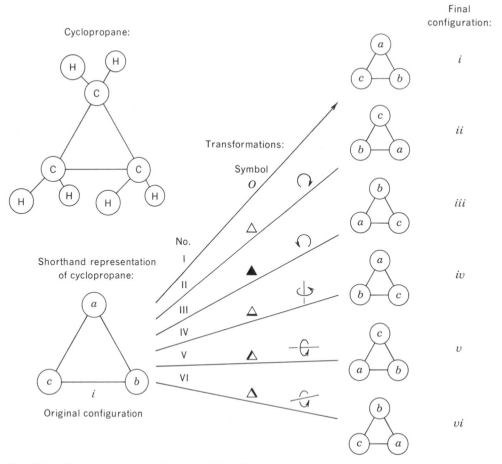

Fig. 22-7 The space-symmetry transformations of cyclopropane.

a transformation in the extreme left-hand column, for example, △, is applied first, followed by the application of a transformation appearing in the top row, for example, ▲, the result of the sequence is shown at the intersection of the row containing the first (△) and the column containing the second (▲). In this case, the result is (◯). It is customary to arrange the order of the elements so that the (◯) results are in a diagonal line running from the upper left corner to the lower right corner. Such a multiplication table conveys at a glance many of the important features of the structure of the group. Since the genetic code is isomorphous with a group, this is an additional reason for acquiring a little familiarity with these group relations which many regard as the most basic part of all mathematics.

Permutation groups

The set of relations which constitute a group comprise a mathematical entity which is so basic that it goes beyond any of the individual glimpses we get of it as it appears in various examples. Thus, the set of transformations of the positions of the CH_2 clusters in cyclopropane is not *the group* itself but only a *representation* of this particular group, which is sometimes designated by the symbol D_3. An equally valid representation is the set of all possible permutations of three symbols, say, a, b, and c. These three symbols can be transformed from the original alphabetical order in six different ways which correspond to the rearrangements of the CH_2 clusters in cyclopropane. These permutation transformations are listed in Table 22-3. In this notation each symbol is replaced by the symbol following it in the sequence. Thus in permutation notation, △ is (acb) since in clockwise rotation by 120°, a is replaced by c, c is replaced by b, and b is replaced by a. △ in full permutation notation is $(bc)(a)$, since b is replaced by c, c is replaced by b, but a remains unchanged. Frequently a shorter notation is used which omits the letters which remain unchanged; thus, △ = (bc).

Groups can also be represented by algebraic matrices. Such a representation of D_3 is listed in the column at the far right. The matrices combine under the laws of matrix algebra in a pattern similar to the permutations.

TABLE 22-2   Multiplication table of the D_3 group

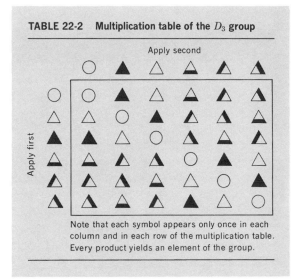

Note that each symbol appears only once in each column and in each row of the multiplication table. Every product yields an element of the group.

TABLE 22-3   Representations of the group D_3

Space symmetry	Permutation		Matrix
	Full	Brief	
◯	$(a)(b)(c)$	0	100 010 001
△	(acb)	(acb)	010 001 100
▲	(abc)	(abc)	001 100 010
△	$(a)(bc)$	(bc)	100 001 010
△	$(ac)(b)$	(ac)	001 010 100
△	$(c)(ab)$	(ab)	010 100 001

TABLE 22-4   Representations of the permutation group for $n = 4$ (S_4)

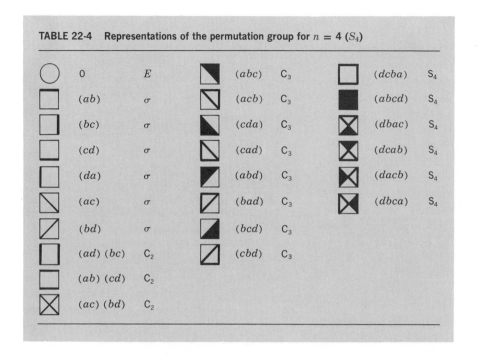

	0	E		(abc)	C_3		$(dcba)$	S_4
	(ab)	σ		(acb)	C_3		$(abcd)$	S_4
	(bc)	σ		(cda)	C_3		$(dbac)$	S_4
	(cd)	σ		(cad)	C_3		$(dcab)$	S_4
	(da)	σ		(abd)	C_3		$(dacb)$	S_4
	(ac)	σ		(bad)	C_3		$(dbca)$	S_4
	(bd)	σ		(bcd)	C_3			
	$(ad)(bc)$	C_2		(cbd)	C_3			
	$(ab)(cd)$	C_2						
	$(ac)(bd)$	C_2						

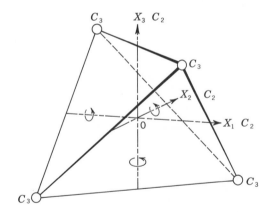

Fig. 22-8   Rotational axes of symmetry in a tetrahedron.

A set of all permutations of n symbols is called a full permutation group of degree n. It contains $n!$ elements where $n! = 1 \times 2 \times 3 \times \cdots \times n$. Thus group D_3 is a full permutation group, having the number of elements equal to $3! = 1 \times 2 \times 3 = 6$.

The permutation group of four symbols is an important group, playing a basic role in space symmetry, mathematics, and genetics. It can be represented as the rotations and inversions of the CH_2 clusters in cyclobutane or as the rotations and reflections of a tetrahedron. Using the space notation, the elements are shown in Table 22-4. The tetrahedral axes of rotation are shown in Fig. 22-8. The elements of the group corresponding to transformations of the five CH_2 clusters in planar cyclopentane are shown in Fig. 22-9; this is the full permutation group with $n = 5$.

Subgroups

In each of the complete permutation groups, certain elements can be selected from the group which by themselves fulfill the group conditions and therefore constitute a group. Such a group is called a subgroup of the larger group from which the elements are selected.

As an example, the elements $\bigcirc$, $\triangle$, and $\blacktriangle$ from the full permutation group are a subgroup, as can be verified from Table 22-2. Each of the pairs $(\bigcirc, \triangle)$, $(\bigcirc, \triangle)$ and $(\bigcirc, \triangle)$ are also subgroups isomorphic with the permutation group with $n = 2$.

The permutation group with $n = 4$ contains a number of subgroups, one of which is especially important. The elements are listed in Table 22-5. This group is an example of a *cyclic group,* so called because it can be generated by taking successive powers of one of its elements ($\sqrt{-1}$). For ($\sqrt{-1}$)^4 the cycle is completed, and the elements repeat themselves beyond this power.

Crystallographic groups

By the application of the group principles to the symmetry of the crystal lattices, it is found that there are 11 crystallographic rotational point groups. They are derived by considering all the axes of rotation about which rotations less than 360° can be made which yield superpositions of the lattice. These groups, their symbols, and elements of symmetry are listed in Table 22-6. If all the elements of each of these groups are multiplied by the element corresponding to an inversion and added to the original elements, 11 more crystallographic groups are obtained. These groups also contain 10 subgroups which do not contain inversion explicitly. Adding these 10 to the 11 original rotation groups and to the 11 rotation-inversion groups gives a total of 32 groups, which are called the *crystallographic conventional point groups* (Table 22-7).

Fig. 22-9 The 120 elements of the complete permutation group with $n = 5$, corresponding to planar cyclopentane.

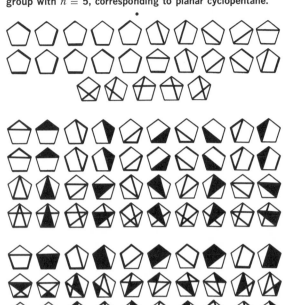

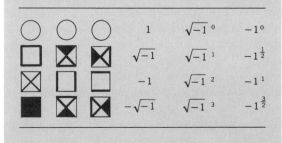

Example 22-1

Give the elements in four different types of subgroups found in the complete permutation group where $n = 4$.

Four of the types of subgroups are

Example 22-2

List the elements in the cyclic permutation group generated by the powers of ($abcde$).

They are ($abcde$)^2 = ($acebd$); ($abcde$)^3 = ($adbec$); ($abcde$)^4 = ($aedcb$); ($abcde$)^5 = ($abcde$) = 0.

TABLE 22-5 The subgroup of the full permutation group with $n = 4$, representable by powers of 1 and -1

○	○	○	1	$\sqrt{-1}\,^0$	-1^0
□	⊠	⊠	$\sqrt{-1}$	$\sqrt{-1}\,^1$	$-1^{\frac{1}{2}}$
⊠	□	□	-1	$\sqrt{-1}\,^2$	-1^1
■	⊠	⊠	$-\sqrt{-1}$	$\sqrt{-1}\,^3$	$-1^{\frac{3}{2}}$

TABLE 22-6   Rotational groups

	Class		
	Rotation	Symbol	Elements
1	360°	C_1	E
2	180	C_2	$E\ C_2$
3	120	C_3	$E\ C_3$
4	90	C_4	$E\ C_4\ C_2\ C_4$
5	60	C_6	$E\ C_6\ 2C_3\ C_2\ C_6$
6	(180)	D_2	$E\ C_2\ C_2'\ C_2''$
7	(120, 180)	D_3	$E\ C_3\ 3C_2\ C_3$
8	(90, 180)	D_4	$E\ C_4\ C_2\ 2C_2'\ 2C_2''\ C_4$
9	(60, 120, 180)	D_6	$E\ C_6\ 2C_3\ C_2\ 3C_2'\ 3C_2''$
10	(120, 180)	T	$E\ 8C_3\ 3C_2$
11	(90, 120, 180)	O	$E\ 8C_3\ 3C_2\ 6C_2\ 6C_4$

Compare with Table 22-4 and Fig. 22-8.

22-2   EXPERIMENTAL CRYSTALLOGRAPHY

During the last quarter of the nineteenth century, the American physicist Henry A. Rowland constructed a ruling engine that scratched parallel lines on the highly polished surface of a metal plate to form a pattern called an optical diffraction grating. When the distance between the lines was of the order of magnitude of the wavelength of light, a beam of light reflected from the plate spread out to form a spectrum in a manner similar to the rainbow formed by a prism.

In 1912, when the German physicists P. P. Ewald and A. Sommerfield were studying the passage of light waves through crystals, their results suggested to the German theoretical physicist Max von Laue that x-rays might be diffracted by a crystal lattice in the same way that light was diffracted by a grating. The diffraction grating acts as it does because the scratches are separated by a distance comparable to the wavelength of visible light. The planes of atoms in a crystal are also separated by a distance comparable to the wavelength of x-rays. When von Laue's hypothesis was tested by passing x-rays through a crystal, the x-rays were diffracted in a regular way, showing that the crystal was acting like a grating.

The Bragg method

As the result of von Laue's suggestion, a number of ways of diffracting x-rays from crystals were studied. One of the simplest was developed by the English physicist Sir William Bragg and his son Lawrence. They used a monochromatic beam of x-rays having a single wavelength, in order to simplify the action, and reflected it from the surface of a crystal at different angles as shown in Figs. 22-10 and 22-11. In order to see the nature of the diffraction mechanism, consider the two parallel paths of x-ray waves marked 1 and 2 in Fig. 22-10. Beam 1 strikes the surface of the crystal at the point marked X, the angle of incidence θ_i is equal to the angle of reflection θ_r, so that the beam leaves the crystal as shown in the drawing. Beam 2 penetrates into the crystal and is reflected from the layer immediately *beneath* the surface. If we follow the common wavefront in these two beams, we see that at the moment when beam 1 is being reflected from the surface of the crystal, beam 2 has not yet reached the layer beneath the surface. The result is that the waves in beam 2 lag behind the waves in beam 1 since they follow a path longer by the distance ABC. The Braggs pointed out that if this distance corresponds to the wavelength of the x-radiation, then the two waves will be in phase as they start back up on the paths at the angle of reflection. If the waves are in phase, then they reinforce each other. This is the situation shown in Fig. 22-10.

If, on the other hand, as shown in Fig. 22-11, the beam of x-rays strikes the crystal at a different angle so that the extra length ABC of the path for beam 2 is equal not to 1 but to $1\frac{1}{2}$ wavelengths, then when beam 2 reaches point C, it is exactly *out* of phase, with beam 1. When the waves are out of phase, they

TABLE 22-7   Point groups[a]

| | Class | | |
	Rotation	Symbol	Elements of symmetry
Triclinic	360°	C_1	E
		C_i	$E\ i$
Monoclinic	180°	C_s	$E\ \sigma_h$
		C_2	$E\ C_2$
		C_{2h}	$E\ C_2\ i\ \sigma_h$
Orthorhombic		C_{2v}	$E\ C_2\ \sigma_v'\ \sigma_v''$
		D_2	$E\ C_2\ C_2'\ C_2''$
		D_{2h}	$E\ C_2\ C_2'\ C_2''\ i\ \sigma_h\ \sigma_v'\ \sigma_v''$
Tetragonal	90°	C_4	$E\ 2C_4\ C_2$
		S_4	$E\ 2S_4\ C_2$
		C_{4h}	$E\ 2C_4\ C_2\ i\ 2S_4\ \sigma_h$
		C_{4v}	$E\ 2C_4\ C_2\ 2\sigma_v'\ 2\sigma_v''$
		D_{2d}	$E\ C_2\ C_2'\ C_2''\ \sigma_v'\ 2S_4\ \sigma_v''$
		D_4	$E\ 2C_4\ C_2\ 2C_2'\ 2C_2''$
		D_{4h}	$E\ 2C_4\ C_2\ 2C_2'\ 2C_2''\ i\ 2S_4\ \sigma_h\ 2\sigma_v'\ 2\sigma$
Rhombohedral	120°	C_3	$E\ 2C_3$
		C_{3i}	$E\ 2C_3\ i\ 2S_6$
		C_{3v}	$E\ 2C_3\ 3\sigma_v$
		D_3	$E\ 2C_3\ 3C_2$
		D_{3d}	$E\ 2C_3\ 3C_2\ i\ 2S_6\ 3\sigma_v$
Hexagonal	60°	C_{3h}	$E\ 2C_3\ \sigma_h\ 2S_3$
		C_6	$E\ 2C_6\ 2C_3\ C_2$
		C_{6h}	$E\ 2C_6\ 2C_3\ C_2\ i\ 2S_3\ 2S_6\ \sigma_h$
		D_{3h}	$E\ 2C_3\ 3C_2\ \sigma_h\ 2S_3\ 3\sigma_v$
		C_{6v}	$E\ 2C_6\ 2C_3\ C_2\ 3\sigma_v'\ 3\sigma_v''$
		D_6	$E\ 2C_6\ 2C_3\ C_2\ 3C_2'\ 3C_2''$
		D_{6h}	$E\ 2C_6\ 2C_3\ C_2\ 3C_2'\ 3C_2''\ i\ 2S_3\ 2S_6\ \sigma_h\ 3\sigma_v'\ 3\sigma_v$
Cubic	(120°, 180°)	T	$E\ 8C_3\ 3C_2$
		T_h	$E\ 8C_3\ 3C_2\ i\ 8S_6\ 3\sigma$
		T_d	$E\ 8C_3\ 3C_2\ 6\sigma\ 6S_4$
		O	$E\ 8C_3\ 3C_2\ 6C_2\ 6C_4$
		O_h	$E\ 8C_3\ 3C_2\ 6C_2\ 6C_4\ i\ 8S_6\ 3\sigma\ 6\sigma\ 6S_4$

[a] After S. Bhagavantam, "Crystal Symmetry and Physical Properties," Academic Press, Inc., New York, 1966.

Fig. 22-10   The reflection of waves from parallel planes in a crystal lattice where $\lambda = 2d \sin \theta$ giving reinforcement after reflection.

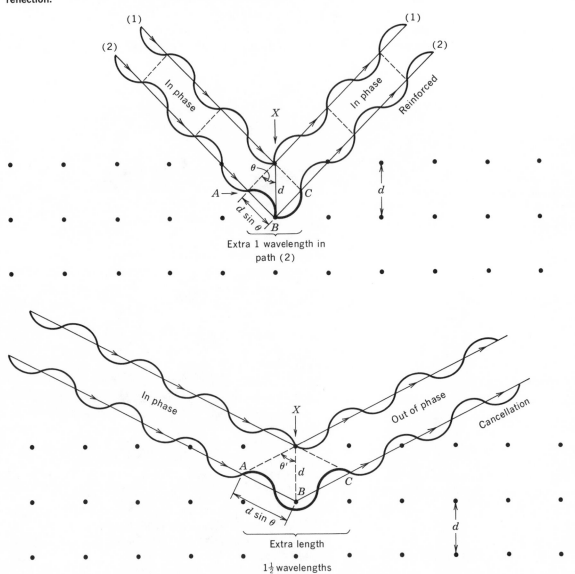

Fig. 22-11   The reflection of waves from parallel planes in a crystal lattice where $(3/2)\,\lambda = 2d \sin \theta'$ giving cancellation after reflection.

cancel each other. For this reason, when a photographic plate is used to measure the intensity of the waves reflected at different angles from the crystal surface, there is a spot on the plate showing maxi-

mum intensity at the point corresponding to the angle at which the extra path ABC is equal to a full wavelength. If the angle is such that the extra length of path ABC is equal to $1\frac{1}{2}$ wavelengths, there is a minimum intensity or even no intensity at all of the x-ray beam. Since the extra path is equal to $2d \sin \theta$, where d is the distance between the two planes in the crystal, the condition for the reinforcement of the beams to produce intense x-rays on the photographic plate is

$$\lambda = 2d \sin \theta \qquad (22\text{-}4)$$

Of course, the waves will also reinforce if the distance ABC is 2 full wavelengths, 3 full wavelengths, and so on. Thus, the general equation giving the angle at which reinforcement will occur is:

$$n\lambda = 2d \sin \theta \qquad n = 1, 2, 3, \ldots \qquad (22\text{-}5)$$

Thus, the determination of the angles and a knowledge of the wavelength of the x-rays permits a measurement of the distance between the planes in the crystal.

In any crystal lattice there are generally a number of different planes which contain atoms at a relatively high density. For example, in the cubic lattice there are the (111) and the (110) planes and the three equivalent planes designated as (100), (010), and (001). A study of crystal structure by the Bragg method usually includes measurements of the intensity of reflection from a number of these planes. Results from two such studies are shown in Fig. 22-12.

Powder method

Since the discovery of the diffraction of x-rays by crystals, many different methods have been developed for using this phenomenon to obtain information about the nature of crystal structures. One of the most widely used is the *powder method*, a procedure which is simple technically but gives the least amount of information. It is valuable for the identification of crystals and determination of lattice constants and particle sizes.

In this method the specimen of the material to be studied is usually prepared as a powder of tiny crystals compressed in the form of a cylinder; the material also can be placed in thin-walled glass capillaries, extruded in the form of a cylindrical rod, or made to adhere to the surface of a glass fiber. The specimen is accurately centered on a rotating axis and turned during the period of exposure to the

Fig. 22-12 The intensities of different orders of Bragg diffraction from (100), (110), and (111) planes. (After W. P. Davey, ''A Study of Crystal Structure and Its Applications,'' McGraw-Hill Book Company, New York, 1934.)

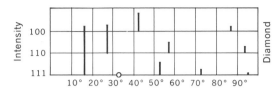

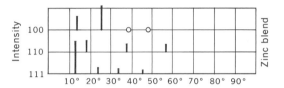

x-rays. Since the specimen is made up of a large number of small crystals oriented in various ways, the beam of x-rays is diffracted only when each crystal arrives at just the right angle with respect to the x-rays. A series of lines obtained by this method is shown in Fig. 22-13.

Rotating crystal method

Another widely used method for crystal-structure determinations uses a single crystal, which is slowly rotated in an x-ray beam so that during the exposure a succession of planes passes through the angle necessary for the Bragg reflection. A pattern obtained by this method is shown in Fig. 22-14. Sometimes the film is moved back and forth with a period synchronized with the rotation of the crystal in order to simplify the pattern obtained on the photographic film.

Fourier synthesis

In the discussion of the nature of a crystal lattice, an expression was developed for expressing the positions of the atoms or molecules in terms of a vector equation. In an idealized lattice these positions can be regarded as density points, where ρ is a density vector given in terms of the basic lattice vectors **a**, **b**, and **c** as

Fig. 22-13   X-ray powder photograph of α-quartz.   (After
J. E. Spice, "Chemical Binding and Structure."   Copyright
© 1964.   The Macmillan Company, New York.   Used by
permission.)

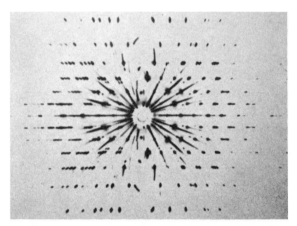

Fig. 22-14   X-ray rotation photograph of tri-ortho-thymotide.
(After J. E. Spice, "Chemical Binding and Structure."   Copyright © 1964.   The Macmillan Company, New York.   Used
by permission.)

$$\rho = m\mathbf{a} + n\mathbf{b} + p\mathbf{c} \tag{22-6}$$

where m, n, and p are integers.

An electromagnetic wave passing through a lattice
can be represented by a vector $\mathbf{k}$ which is perpendicular to the wavefront and has a length proportional to
the amplitude of the wave.   When the wave is
scattered from the electrons at the points where ρ
has the high values given by Eq. (22-6), the new wave
may be represented by a vector $\mathbf{k}'$.   The original wave
in a cubic lattice is depicted in Fig. 22-15 and the
scattered wave in Fig. 22-16.

For a strong diffracted beam, von Laue showed that
the following relations must hold simultaneously

$$\mathbf{a} \cdot (\mathbf{k}' - \mathbf{k}) = 2\pi q \qquad \mathbf{b} \cdot (\mathbf{k}' - \mathbf{k}) = 2\pi r$$
$$\mathbf{c} \cdot (\mathbf{k}' - \mathbf{k}) = 2\pi s \tag{22-7}$$

where q, r, and s are integers and the heavy dot denotes an inner vector product (see Appendix 4).

The problem is to take the data on the positions of
the diffracted beams and from there find the points
at which ρ has the maximum values, i.e., to find the
positions of the atoms and molecules in the lattice.
One of the most effective ways to accomplish this
employs the concept of a *reciprocal lattice*.   In this
new lattice, the basic vectors $\mathbf{A}$, $\mathbf{B}$, and $\mathbf{C}$ are in the
same direction as the original vectors $\mathbf{a}$, $\mathbf{b}$, and $\mathbf{c}$ but
the length of each is equal to 2π divided by the length
of the original complementary vector.   In vector
notation

$$\mathbf{A} \cdot \mathbf{a} = 2\pi \qquad \mathbf{B} \cdot \mathbf{b} = 2\pi \qquad \mathbf{C} \cdot \mathbf{c} = 2\pi \tag{22-8}$$

Using this lattice, a quantity $\mathbf{G}$ complementary to ρ
is defined

$$\mathbf{G} = h\mathbf{A} + k\mathbf{B} + l\mathbf{C} \tag{22-9}$$

where h, k, and l are integers.

Then the von Laue conditions for wave diffraction
can be written simply

$$\mathbf{k}' - \mathbf{k} = \mathbf{G} \tag{22-10}$$

This approach makes it possible to construct electron-density maps showing the location of atoms and
molecules in the crystal lattice.   Such a map of a
crystal of hexamethylbenzene is shown in Fig. 22-17.

Neutron diffraction

Still another method of crystal-structure determination
has been developed in which neutrons instead of
x-ray photons are used as the wave components to be
diffracted.

The de Broglie equation for the wavelength λ of a
particle is

$$\lambda = \frac{h}{p} \tag{22-11}$$

where h is Planck's constant and p is the momentum
equal to mv, or $\sqrt{2m\epsilon}$.   Thus,

$$\epsilon = \frac{h^2}{2m\lambda^2} \tag{22-11a}$$

and

$$\lambda \text{ (Å)} = \sqrt{\frac{0.28}{\epsilon \text{ (ev)}}} \qquad (22\text{-}11b)$$

Thus, if $\epsilon = 0.08$ ev, $\lambda = 1$ Å. This is of the order of magnitude of lattice spacings. Since neutrons have a magnetic moment, they interact with the magnetic moments in the lattice. Neutrons also interact with the atomic nuclei. The combination yields a valuable tool for investigating crystal structures.

Example 22-3

A monochromatic beam of x-rays enters a cubic crystal at an angle of 16.5° from the normal to the surface and produces a reflected beam of maximum intensity corresponding to $n = 1$. The surface corresponds to the (111) plane. The wavelength of the x-ray beam is 1.25 Å. What is the distance of spacing between the (111) planes?

The Bragg equation is

$$\lambda = 2 \, d \sin \theta$$

so that

$$d = \frac{\lambda}{2 \sin \theta} = \frac{1.25}{2 \times 0.2840} = 2.20 \text{ Å} \qquad Ans.$$

Example 22-4

In a certain cubic lattice the base vectors each have a length of 2.32 Å. Calculate the length of each base vector in the reciprocal lattice and the length of G when $h = k = l = 1$.

$$A = B = C = \frac{2\pi}{a} = 2.708 \text{ Å}^{-1}$$

$$G = 2.708 \, (\cos 45°)^{-1} \, (\cos 35°15')^{-1}$$

$$= 4.69 \text{ Å}^{-1} \qquad Ans.$$

22-3 CRYSTAL STRUCTURES

Crystal structures can be classified first of all with respect to the types of binding holding the substituent particles together. As discussed at the beginning of this chapter, in a crystal of neon the lattice is made up of individual atoms which are held together by van der Waals forces. This is called an atomic

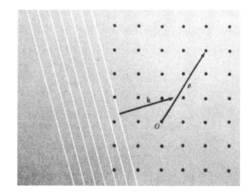

Fig. 22-15 A wavefront represented by the vector k as it passes through a crystal lattice represented by the vector ρ. (After C. Kittel, "Introduction to Solid State Physics," 3d ed., John Wiley & Sons, Inc., New York, 1966.)

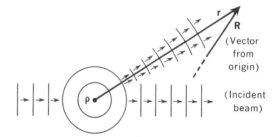

Fig. 22-16 The atom at ρ will scatter some of the radiation from the incident beam. The amplitude of the scattered radiation as seen at a point distant R from ρ will be proportional to $(F_0 \exp ik \cdot \rho) \, (r^{-1} \exp ik \cdot R)$, where the first parenthesis contains the amplitude and phase factor of the incident beam, and the second parenthesis describes the spatial variation of the radiation scattered from a point atom at ρ.

crystal. In a crystal of benzene the lattice is made up of molecules, also bound to each other by van der Waals forces. This is called a molecular crystal; other examples range all the way from crystals of hydrogen, H_2, to crystals of extremely large molecules like proteins. The largest molecular unit found in crystalline form occurs in crystals like diamond; the lattice is composed of individual carbon atoms. In a sense, a single crystal of diamond is really one giant molecule.

Contrasted with all these types, a crystal of copper consists of individual atoms bound together primarily

Fig. 22-17 (*a*) **Electron-density map of nine hexamethylbenzene molecules in** *C* **axis projection.** (From J. E. Spice, "Chemical Binding and Structure." Copyright © 1964. The Macmillan Company, New York. Used by permission.) (*b*) **Diagram of hexamethylbenzene molecules in the crystal.** (From Charles Bunn, "Crystals," Academic Press, Inc., New York, 1964.)

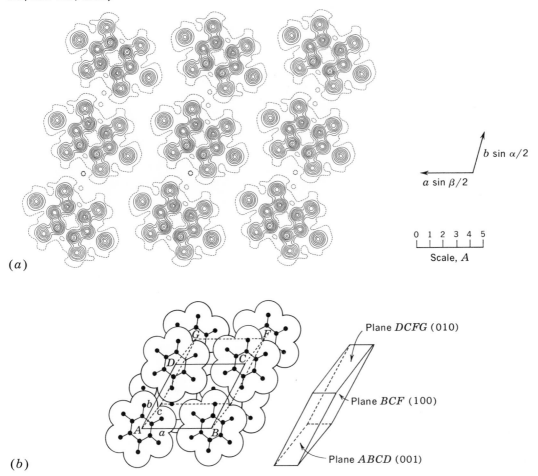

(*a*)

(*b*)

by the forces resulting from the de Broglie waves of the conduction electrons. Like the electrons which produce aromatic bonding in the benzene ring, which are shared by all six carbon atoms, these conduction electrons have wave functions which have significant amplitudes throughout volumes containing hundreds of atoms.

In a crystal of an inorganic salt like sodium chloride, the particles are ions, which are held together largely by the forces of coulombic attraction.

Another characteristic in which crystals differ is the pattern of the covalent-bond linkage. For example, there may be long chains of atoms or atom clusters bound together in a single thread by covalent bonds, the threads being held together by covalent forces. Such a structure is found in the crystals of long straight-chain hydrocarbons. In one sense, such a single chain is a kind of one-dimensional crystal itself. Similarly, the long chain of atomic clusters found in a protein may be regarded as a single long-chain crystal, and a DNA molecule is a kind of twin-spiral long-chain crystal.   These more complex cases will be discussed in more detail in Chap. 24, dealing with macromolecules.

Inorganic crystal structures

Examples of structure consisting of long covalent-bonded chains are shown in Fig. 22-18.   The units making up the chains may be written as SiS_2, SeO_2, and $PdCl_2$.   The chains may be denoted by symbols such as $(SiS_2)_n$, $(SeO_2)_n$, and $(PdCl_2)_n$.

Graphite provides an example of a two-dimensional sheet of atoms held together internally by covalent bonding, with the sheets held to each other by van der Waals forces.   Mica is another example of this type of structure, and for this reason can be sliced into very thin sheets.   These structures are shown in Fig. 22-19.

SiO_2 clusters can also be held together in three-dimensional structures that also constitute a giant ion, as shown in Fig. 22-20; this is a cross section of the mineral talc.

Ionic lattices can be made up of individual ions as in NaCl (Fig. 22-21c) or of clusters of atoms bound covalently (CO_3^-) into an ion interspersed with other ions of the opposite sign, as in $CaCO_3$ (Fig. 22-22).   In Fig. 22-23, a few other examples of inorganic crystal structures are shown.   Metallic lattices will be discussed in a later section of this chapter.

Intermediate between inorganic and organic molecules are the organometallic compounds, many of which play a most significant role in biochemistry, e.g., chlorophyll.   In the solution of its structure, methods of structure determination using x-rays played an important part.   The structure of another organometallic compound recently studied is shown in Fig. 22-24.   An example of a coordination compound called a *clathrate* is shown in Fig. 22-25 in which nickel cyanide is linked to benzene.

An interesting example of structure determination

Chain ions $(SiO_3)_n^{2n-}$, as in pyroxenes

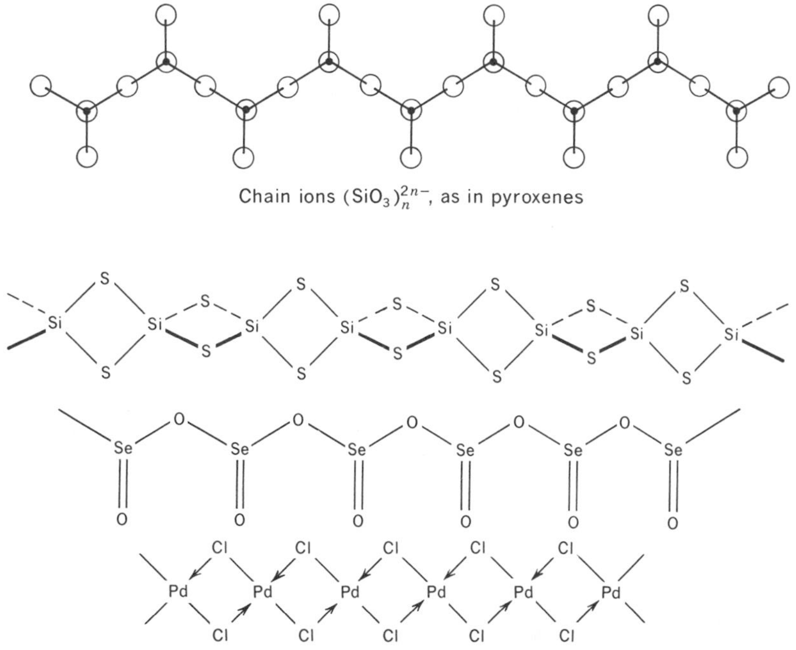

Fig. 22-18   Examples of long covalent-bonded straight-chain molecules.   (After J. E. Spice, "Chemical Binding and Structure."   Copyright © 1964.   The Macmillan Company, New York.   Used by permission.)

Fig. 22-19 Covalent-bonded sheets of atoms. (After J. E. Spice, "Chemical Binding and Structure." Copyright © 1964. The Macmillan Company, New York. Used by permission.)

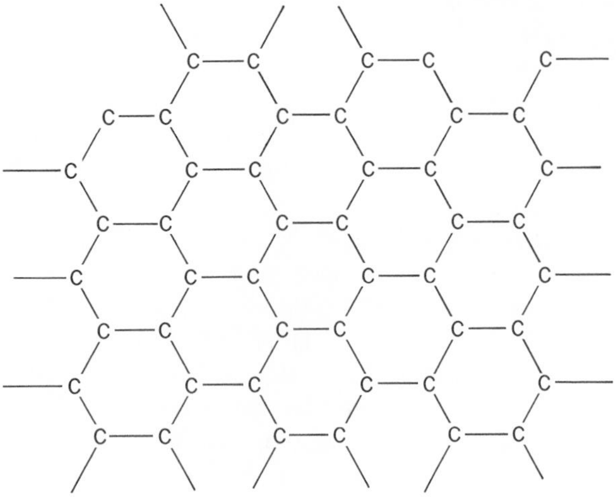

Arrangement of carbon atoms in each sheet of graphite

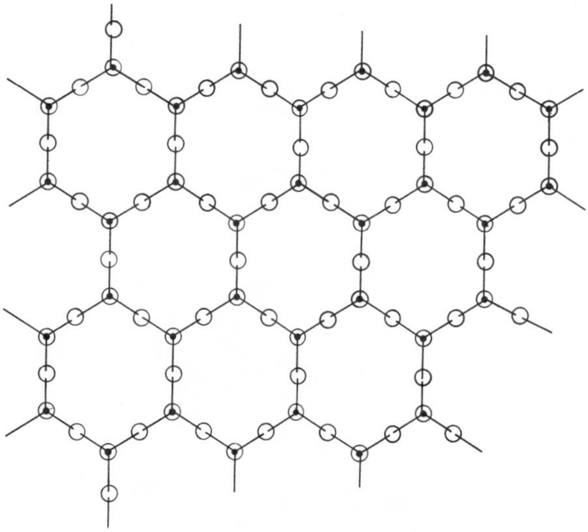

Sheet ions $(Si_2O_5)_n^{2n-}$, as in micas

is found in the studies on nickel phthalocyanine. The unit cell has two molecules, each with one atom of nickel. All reflections from $(0k0)$ with k odd and from $(h0l)$ with h odd are absent, showing that the

space group has a twofold screw axis and a glide plane (Fig. 22-26). Each nickel atom must be at the crystallographic center of symmetry. This information coupled with the knowledge of the chemistry of the compound shows that the structural formula is that drawn in Fig. 22-27.

Organic structures

The applications of the principles of crystallographic analysis have been helpful in elucidating the structures of a host of organic compounds and their crystal lattices. A few examples, including some from the recent literature, are given in Fig. 22-28. The cage structure in the adamantine-type molecule is especially interesting. The geometry is closely related to the symmetry, and the symmetry affects the entropy, both the thermodynamic entropy and the entropy of activation, which can play the dominant role in determining the rate of a reaction. Geometric structure rather than strictly chemical factors plays the key role in many of the primary aspects of genetic action. These areas of the interplay of structure, thermodynamics, and kinetics will be discussed at greater length in several of the remaining chapters.

Example 22-5

Name three types of axes of rotation and the number of each in the simple cubic lattice.

There are three axes, each collinear with one of the base vectors of the lattice $(3C_4)$; there are six lying in the plane defined by two of the base vectors, at 45° to these and at 90° to the third $(6C_2)$; there are four at 45° to each of the three base vectors $(4C_3)$.

Example 22-6

Write the elements of the group isomorphic with the transformations produced by the rotation of a single type of C_4.

Select one point in the axis as the origin of the coordinate framework. Then designate the six nearest neighbors as a, b, c, d, e, and f. Rotating about the axis passing through e and f, the elements of the group will be 0, $(abcd)$, $(ac)(bd)$, $(adcb)$.

22-4   CRYSTAL DEFECTS

Because the crystalline state is the most highly ordered state of matter, it is subject also to the most subtle deviations from order. These *crystal defects* play an important role not only in the growth of crystals but in many other aspects of crystal behavior.

First of all, defects cause a change in the free energy of the crystal, usually resulting in an increase of the free energy per mole. When molecules make the transformation spontaneously from some other state, such as a liquid or gas, to form a perfect

crystal lattice, the reaction proceeds in this direction because the lattice has a lower free energy than the gas or liquid. However, a perfect lattice must be infinite in extent. If there is a surface, as there must be if the crystal is not infinite, the energy of molecules at the surface will be different from the energy of molecules in the interior because the binding of the molecules is different from that in the interior. This is illustrated in Fig. 22-29 for a simple cubic lattice. Suppose, for the sake of simplicity, that in the crystal there is one atom at each point of a cubic lattice and that each atom is bound to each

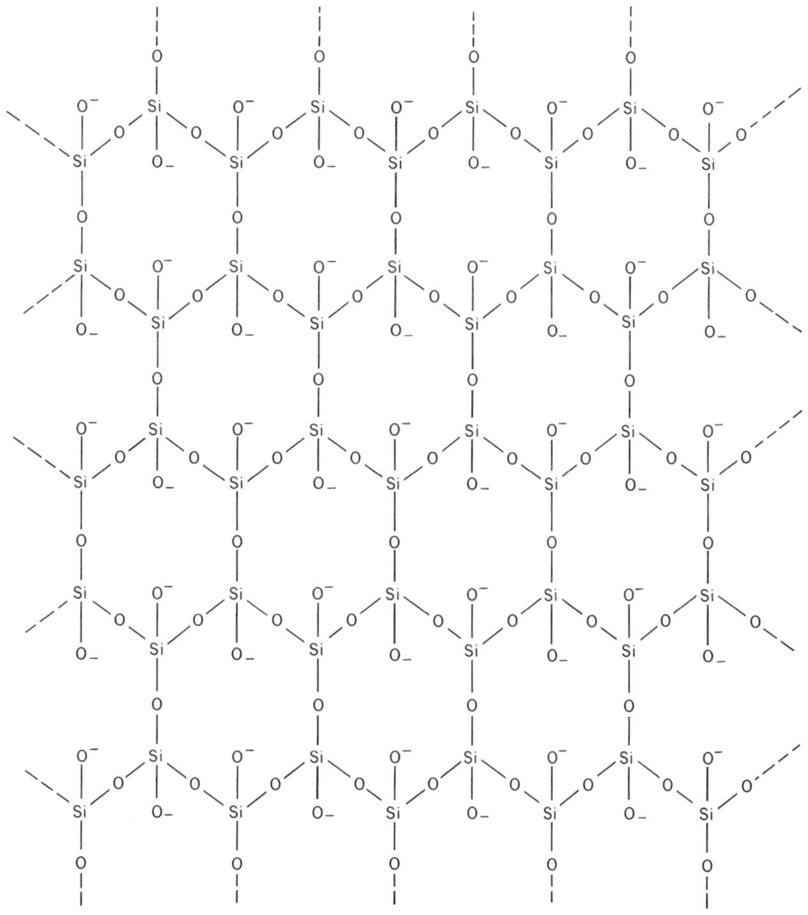

Fig. 22-20   Part of a giant two-dimensional ion sheet in talc. Mg^{++} ions lie above and below the sheet. (After A. K. Galwey, "Chemistry of Solids," Chapman & Hall, Ltd., London, 1967.)

Fig. 22-21   Three types of crystals: (*a*) atomic; (*b*) molecular; (*c*) ionic.   (After C. Bunn, "Crystals," Academic Press, Inc., New York, 1964.)

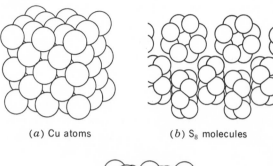

(*a*) Cu atoms   (*b*) S_8 molecules

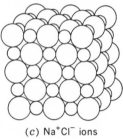

(*c*) Na^+Cl^- ions

of its six nearest neighbors by a single bond.   Then an atom in the interior is held by six bonds, but an atom on a plane surface is held by only five.   An atom protruding above the surface is held by one bond only.   If the bonds are formed with a loss of energy, the free energy at the surface is higher as shown in Fig. 22-30.   The normal melting point might be defined as that temperature at which the free energy of the liquid is equal to the free energy of the same material in the interior of the crystal.   But at this temperature, the atoms bound by only five bonds or less have a larger free energy than the liquid molecules and consequently escape into the liquid.   Thus, if the surface free energy is significant, the nature of the true equilibrium between crystal and liquid or crystal and vapor is somewhat complicated.   From this point of view the surface itself is a crystal defect.

Point defects

In forming a crystal from a liquid or gas at a finite growth rate, there is always a chance that as the crystal grows, some site on the lattice may fail to acquire an atom and the crystal is left with a hole in

it.   There is a great deal of experimental evidence that such holes exist.   In fact, an atom from a site adjacent to the hole can shift to fill the hole, thus leaving another hole at the point where it was originally.   Thus, in effect, the hole migrates.   Such a hole is a point defect.   The importance of these point defects will be discussed further in connection with the abnormal behavior of metal crystals.   These holes are shown in Fig. 22-31*a*.

One of the types of point defect in a lattice is that caused by the absence of an electron, producing an electron hole which acts like a positively charged particle.   These defects are especially important in metallic lattices.   The generation of electric conductivity by electron holes in a silicon crystal is shown in Fig. 22-31*b*.

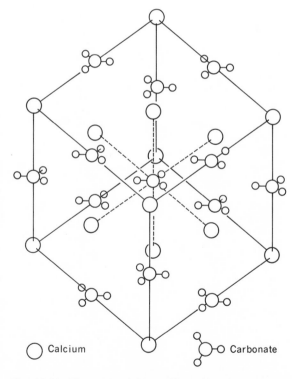

○ Calcium   Carbonate

Fig. 22-22   The calcite lattice.   The carbonate groups in each horizontal plane are oriented in the same manner, whereas the carbonates in the adjacent layers are rotated by $\pi/3$ rad.   (After P. B. Dorain, "Symmetry in Inorganic Chemistry," Addison-Wesley Publishing Company, Inc., Reading, Mass., 1965.)

Fig. 22-23   Examples of inorganic crystal lattices. [(*a*), (*b*), and (*c*) after P. B. Dorain, "Symmetry in Inorganic Chemistry," Addison-Wesley Publishing Company, Inc., Reading, Mass., 1965. (*d*) after C. D. Garner and S. C. Wallwork, *J. Chem. Soc.*, 1966A:1496.]

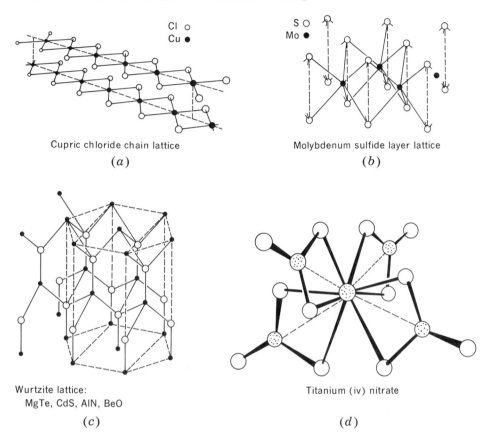

Cl o
Cu ●

Cupric chloride chain lattice
(*a*)

S O
Mo ●

Molybdenum sulfide layer lattice
(*b*)

Wurtzite lattice:
MgTe, CdS, AlN, BeO
(*c*)

Titanium (iv) nitrate
(*d*)

An opposite kind of imperfection is that caused by the intrusion of a foreign atom or molecule into a lattice. Frequently atoms of an impurity can intrude into the spaces between the atoms in the regular crystal lattice without greatly distorting the lattice. This produces an unstable state where the free energy in general is higher than in the pure state, even though the increase in entropy due to the foreign atom tends to lower the free energy. Impurities of this type are shown in Fig. 22-32.

If the foreign atoms are sufficiently similar to the atoms occupying the sites on the lattice, the intruders can displace some of the lattice atoms from their sites. Both phosphorus and aluminum atoms can displace silicon atoms in a silicon crystal, as shown in Fig. 22-33.

Dislocations

In many cases, the geometric nature of the interatomic or intermolecular forces is such that a perfect lattice is the state of minimum free energy at temperatures below the melting point. There are also many cases, however, in which the regular geometric order, corresponding to the lattice, distorts the geometry of the interatomic forces and introduces a strain force into a lattice. After the lattice has extended through growth to a sufficient distance, this strain may cause the displacement of a whole plane of atoms from its

Fig. 22-24 The structure of organometallic compounds as determined by crystallographic analyses. (*a*) *Bis* (tricobalt enneacarbonyl) acetone. (From G. Allegra, E. M. Peronaci, and R. Ercoli, *Chem. Comm.*, 1966:549; (*b*) a cyclic ligand (perfluorocyclopentadiene) bridging a Co (CO)$_3$ and Co (CO)$_4$ fragment in perfluorocyclopentadienedicobalt heptacarbonyl. From P. B. Hitchcock and R. Mason, *Chem. Comm.*, 1966:503.)

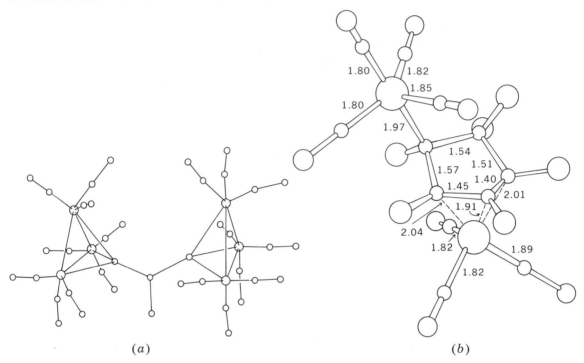

(a) (b)

normal position. Such an irregularity is called a *dislocation.*

There are many types of dislocations, some of which are illustrated in Fig. 22-34. The surface irregularities produced by these dislocations play an important role in crystal growth. As previously pointed out, a single atom protruding from the surface is bound to the surface on only one face and may have a free energy enough above that of the neighboring liquid so that it is almost instantly dissolved in the liquid and cannot serve as a focus for further growth. By way of contrast, when a step irregularity exists on the surface, an atom fitting into the angle of the step is bound to the crystal on two faces and has a correspondingly lower free energy. This may be sufficiently low to keep the atom in place; thus the step is extended and the crystal continues to grow.

Frequently a plane of atoms may terminate at some point in the interior of a crystal, leaving a gap beyond which the neighboring planes come in contact and continue the lattice. This not only introduces strain but leaves space where impurities are easily absorbed.

In the growth of a crystal, an irregularity on the surface may start growth with the lattice oriented in an entirely different direction. Zones of regular lattice structure surrounded by other zones of regular lattices differently oriented are called grains. Figure 22-35 shows two aspects of grain imperfections.

Because many atoms along a dislocation surface are not bound as tightly to their neighbors as in the main body of the crystal, dislocations are zones of weakness and reduce the tensile strength of the crystal considerably. On the other hand, dislocations are zones of enhanced chemical activity. When the surface of a crystal is etched, the particles in the dislocation zone may be removed more rapidly than those from the main body of the crystal, leaving a surface where the dislocation can be seen plainly, as shown in Fig. 22-36. In Fig. 22-37, the surface of a paraffin crystal is portrayed where growth started from a single dislocation and produced a striking pattern of surface steps.

In many ways active sites on the surface of enzymes may resemble dislocations, and for this reason the theory of dislocations has a bearing on our understanding of the nature of enzyme action and the general nature of biogrowth.

Fig. 22-26   The combination of glide plane and twofold screw axis used to determine the structure of nickel phthalocyanine (Fig. 22-27). ac is reflection plane; glide along a axis. b is screw axis. Operation of these two symmetry elements on particle A necessitates there being particles at B, C, D; that is, four general positions per unit cell. Centers of symmetry are in ab planes. (After J. E. Spice, "Chemical Binding and Structure." Copyright © 1964. The Macmillan Company, New York. Used by permission.)

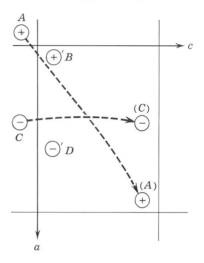

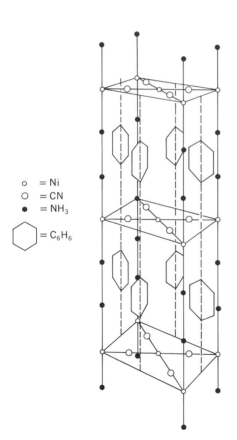

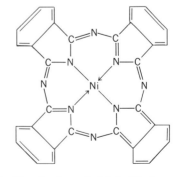

Fig. 22-25   Ammonia-nickel-cyanide-benzene clathrate. (After J. E. Spice, "Chemical Binding and Structure." Copyright © 1964. The Macmillan Company, New York. Used by permission.)

Fig. 22-27   The structure of nickel phthalocyanine. (After J. E. Spice, "Chemical Binding and Structure." Copyright © 1964. The Macmillan Company, New York. Used by permission.)

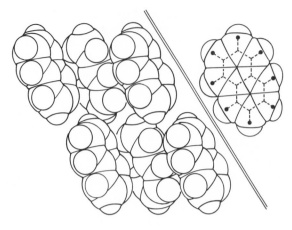

(a) Naphthalene

(b) Formycin B with pyrazolo [4,3-d] pyrimidine base and ribofuranose. Top ring planar, bottom ring puckered.

(c) Antibiotic blasticidine S

(d) Tetradonic acid

(e) Diacetylanhydrotetradotaxin

Fig. 22-28 Original x-ray structure determinations. [(a) after C. Bunn, "Crystals," Academic Press, Inc., New York, 1964; (b) after G. Koyama, K. Maeda, H. Umezawa, and Y. Iitaka, *Tetrahedron Letters*, 1966:597; (c) after S. Chuma, Y. Nawata, and Y. Saito, *Bull. Chem. Soc. Japan*, 39:1091 (1966); (d) after C. Tamura, O. Amakasu, Y. Sasada, and K. Tsuda, *Acta. Cryst.*, 20:219, 226 (1966).]

Example 22-7

The free energy of crystalline benzene can be expressed as a linear function of the temperature for a range of 20° below the freezing point at 278.5°K. Over the same range the free energy of the liquid can be expressed by an analogous equation:

Crystal: $G_c = 15{,}773 - 30.89T$ cal mole^{-1}
Liquid: $G_l = 18{,}121 - 39.32T$ cal mole^{-1}

If the free energy of a molecule attached to the surface of a benzene crystal is the same as that of a molecule in the interior of the crystal plus 391 cal mole^{-1}, at what temperature will such a molecule be in equilibrium with the neighboring liquid state?

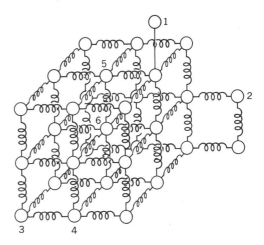

Fig. 22-29   Stylized binding forces in a cubic lattice showing atoms held by 1, 2, 3, 4, 5, and 6 units of force.

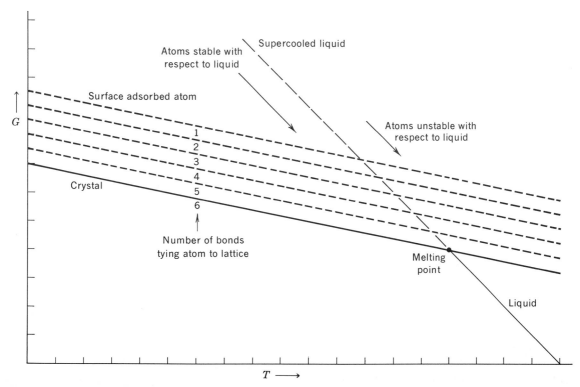

Fig. 22-30   The free energy as a function of temperature for atoms tied to the crystal lattice by different numbers of bonds.

Fig. 22-31 (*a*) Holes caused by missing atoms; (*b*) irregularities in electron placement producing positively charged holes in a crystal lattice. (After N. B. Hannay, "Solid State Chemistry." ⓒ 1967. By permission of Prentice-Hall, Inc., Englewood Cliffs, N.J.)

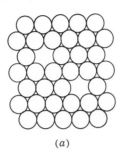

(*a*)

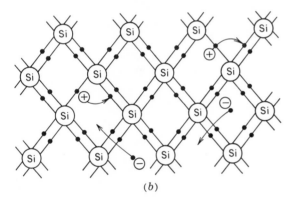

(*b*)

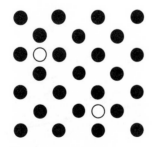

Fig. 22-32 Impurities in a crystal lattice, located in interstices.

For the surface molecule the free energy can be assumed to be roughly the free energy of a molecule in the interior, plus 391 cal mole^{-1} so that

$$G_s = 391 + 15,773 - 30.89T$$
$$= 16,164 - 30.89T$$

At the temperature of equilibrium between the surface molecule and the liquid

$$G_s = G_l$$
$$16,164 - 30.89T = 18,121 - 39.32T$$
$$8.43T = 1957$$
$$T = \frac{1957}{8.43} = 232°K \qquad Ans.$$

The conclusion is that a simple molecule of benzene attached to the surface of the crystal by only one-sixth of the normal force of attraction holding a molecule in the interior of the crystal would be dissolved in the supercooled liquid at any temperature above 232°K.

22-5   SOLID-STATE ENERGY

Unlike the gaseous state, in which molecules are on the average relatively far apart, liquids and crystals are made of molecules which are, in effect, touching one another practically all the time. Any one molecule (or atom or ion) is bound to a number of its neighbors—tightly to the nearest neighbors, more loosely to the more distant neighbors. But because the binding force involves a number of particles, the free energy can no longer be considered as the property of a single molecule, atom, or ion but must be treated as a joint property.

Fig. 22-33 Impurities in crystal lattice located in lattice sites. (After N. B. Hannay, "Solid State Chemistry." ⓒ 1967. By permission of Prentice-Hall, Inc., Englewood Cliffs, N.J.)

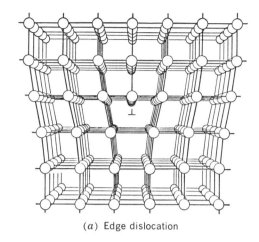

(*a*) Edge dislocation

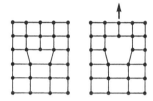

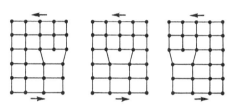

(*b*) Production and annihilation of vacancies by edge dislocation climb

(*c*) Production and annihilation of vacancies by edge dislocation slip

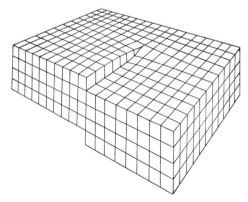

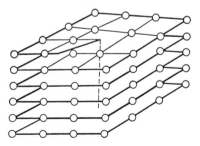

(*d*) Screw dislocation surface view

(*e*) Screw dislocation showing axis

Fig. 22-34   Types of lattice dislocation.   [(*a*) after J. Weertman and J. R. Weertman, "Elementary Dislocation Theory."   Copyright © 1965.   The Macmillan Company, New York.   Used by permission; (*b*) to (*d*) after N. B. Hannay, "Solid State Chemistry."   © 1967.   By permission of Prentice-Hall, Inc., Englewood Cliffs, N.J.]

Fig. 22-35 Granular dislocations. [(a) to (c) after A. R. Ubbelhode, "Melting and Crystal Structure," Oxford University Press, New York, 1965; (d) and (e) from N. B. Hannay, "Solid State Chemistry." © 1967. By permission of Prentice-Hall, Inc., Englewood Cliffs, N.J.]

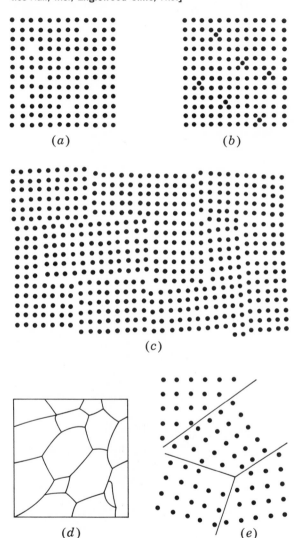

(a)   (b)

(c)

(d)   (e)

Lattice energy

The calculation of a joint property is nearly always more difficult than the calculation of the property of a single atom, ion, or molecule. One of the more successful, as well as one of the most interesting, is the calculation of the lattice energy when a crystal is composed of ions.

In this case the energy E may be regarded as due to the algebraic sum of coulombic force due to the charges and a force due directly to the interaction of the de Broglie waves. This leads to an approximation given by

$$E = -\frac{Aq^2}{r} + \frac{B}{r^n} \tag{22-12}$$

In this equation B and n are constants characteristic of a given type of structure. A, the Madelung constant, also depends on structure. For a sodium chloride lattice, where each atom has 6 nearest neighbors of opposite sign and 12 of the same sign at a greater distance,

$$A = 6 - \frac{12}{\sqrt{2}} + \frac{8}{\sqrt{3}} - \frac{6}{\sqrt{4}} + \cdots = 1.747 \cdots \tag{22-12a}$$

Since the lattice at equilibrium corresponds to a minimum at $r = a$ in the energy curve,

$$\frac{dE}{dr} = N\left(\frac{Aq^2}{a^2} - \frac{nB}{a^{n+1}}\right) = 0 \tag{22-13}$$

and

$$B = \frac{Aq^2a^{n-1}}{n} \tag{22-14}$$

Equation (12-12) thus takes the form for $2N$ ions

$$E = \frac{-NAq^2(1 - 1/n)}{a} \tag{22-15}$$

If the compressibility K of a crystal is measured, n can be expressed as

$$n = 1 + \frac{9ca^4}{Kq^2A} \tag{22-16}$$

where c is a geometrical factor for a given lattice. n varies from 6 to 12 for the alkali halides for which values of K have been measured at low temperatures.

Two Germans, the physicist M. Born and physical chemist F. Haber, devised a theoretical cycle of measurements

$$Na(s) \rightarrow Na(g) - E_1$$

sublimation energy of Na

$$Na(g) \rightarrow Na^+ + e^- - E_2$$

ionization energy of Na

$$\tfrac{1}{2}Cl_2 \rightarrow Cl - \tfrac{1}{2}E_3$$

dissociation energy of Cl_2

$$Cl + e^- \rightarrow Cl^- + E_4$$

electron affinity of Cl

$$Na^+ + Cl^- \rightarrow NaCl + E_l$$

lattice energy of NaCl

The sum of these equations is

$$Na(s) + \tfrac{1}{2}Cl_2 \rightarrow NaCl - E_1 - E_2$$
$$- \tfrac{1}{2}E_3 + E_4 + E_l = E_{final}$$

where E_{final} is the total energy change in forming NaCl from its elements. This is calculatable from thermodynamic data. The value of E_l so obtained checks well with the theoretical value of E from Eq. (22-15).

Bond theory

When atoms with loosely bound electrons in the external shell (metallic elements) are grouped together in a crystal lattice, they form a solid which exhibits the property of electrical conductivity. This conductivity is due to a resonance of the electron waves which is produced by the periodic character of the lattice. This is the same kind of resonance that is found when six carbon atoms are linked into the hexagonal ring with three double bonds that is the core of benzene and its derivatives.

In order to see the basis for resonance in a lattice it is helpful to review briefly some of the wave patterns produced when electrons are shared between atoms.

Recall, first of all, the basic wave patterns of the electron in the shell of a single hydrogen atom. Three of these waveforms are shown in Fig. 22-38, both as wave functions and as density diagrams. When two hydrogen atoms approach each other, the electron wave on one atom begins to envelop the nucleus of the other; finally, when the atoms are as close as 1 Å, the two electron waves merge and, if the spins are opposite, form the new bonding pattern with a combined energy which is lower than that of two independent waves considered separately. The two $1s$ energy levels, one on each atom when the atoms

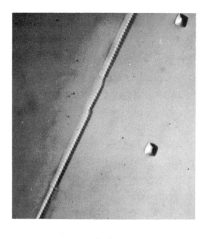

(a)

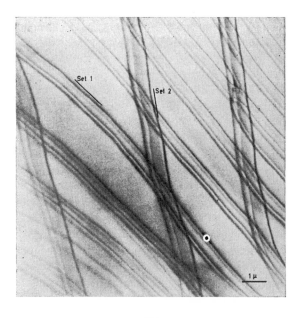

(b)

Fig. 22-36   Direct observations of imperfections in crystals: (a) dislocation etch pits on the (100) face of a germanium crystal observed by F. L. Vogel, Jr., *Acta Metallurgica,* 3:245 (1955); (b) dislocation ribbons in talc observed by S. Amelinckx and P. Delavingnette and reported in "Dislocations in Layer Structures" ("Direct Observations of Imperfections in Crystals," J. B. Newkirk and J. H. Wernick, eds.), John Wiley & Sons, Inc., New York, 1962.)

Fig. 22-37 A single crystal of $n\text{-}C_{36}H_{74}$ exhibiting the pattern of growth from dislocation at center. Electron micrograph by I. M. Dawson and V. Vand, *Proc. Roy. Soc.* (London), A206:555 (1951).

were separated, now shift from the same value for each to two new energy levels for the joint wave, one level being lower than the level for the single $1s$ wave and the other being higher. In Fig. 22-38e to h, the same type of wave modification is shown in the formation of Li_2.

If it were possible to arrange a number of hydrogen atoms in a row, as shown in Fig. 22-39, then as each atom is added to the row, the number of different energy levels would increase.

In Fig. 22-40, there is a drawing of the lattice arrangement in a crystal of metallic lithium. Like hydrogen, lithium has only one electron in the outer shell. Consider the lithium atom in the center of the cube shown in the drawing. It has eight closest neighbors, four above it and four below. There is no reason why it should share its electron with any one of its neighbors more exclusively than with any other. Consequently, a communal wave is formed. When this sharing is extended all the way through the lattice in a pattern quite similar to that shown for the sharing of electrons between the row of hydrogen atoms, there is a similar shift in the energy levels of the electrons and the original single levels spread out into bands, as shown in Fig. 22-41. The band formation is a function of the distance of separation of the

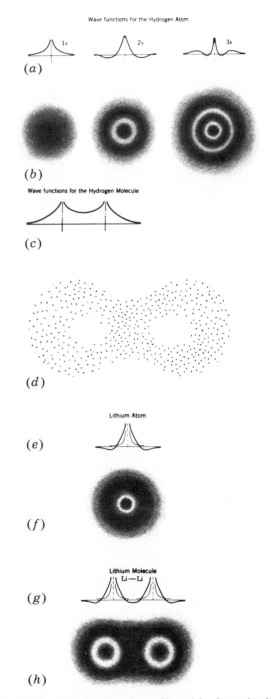

Fig. 22-38 Wave functions in profile and in charge-density diagrams. (After D. H. Andrews and R. J. Kokes, "Fundamental Chemistry," 2d ed., John Wiley & Sons, Inc., New York, 1965.)

atoms, the bands getting wider as the atoms approach each other. Finally the bands spread to the point where, for example, the $2s$ bond overlaps the $2p$ bond.

When the atoms are widely separated, the Pauli exclusion principle allows only one pair of electrons in any level of each individual atom. When the atoms pack together in a crystal lattice, the exclusion principle still allows only one pair of electrons in any one of the band levels. As suggested in Fig. 22-41, the shift, due to the closer packing of the atoms, moves the majority of the levels up to much higher positions on the energy scale. This has a drastic effect on the properties of the electrons.

According to the classical theory of electric conduction, electrons are thought of as moving quite freely through the lattice of a metallic crystal, much like gas molecules. If this were really the case, one might expect that a mole of electrons would have a heat capacity of $\frac{3}{2} R$ equal to the molar heat capacity of a monatomic gas. But at ordinary temperatures, no such term is found experimentally in the observed heat capacity of a metal. From the quantum-mechanical point of view, it is clear that the Pauli exclusion principle forces the electrons into the band energy levels much higher than the level of the average energy in a monatomic gas at the same temperature as that of the crystal. The probability of an electron's absorbing energy at room temperature and moving into a still higher level becomes negligibly small. Consequently, the electrons show no appreciable heat capacity at room temperature and move into higher levels in significant amounts only at temperatures of the order of 1000°C. In fact, electrons would not become truly gaslike until a temperature of roughly 50,000°K was attained, a temperature far beyond the vaporization point of a metal.

Brillouin zones

The band structure which determines the nature of the conductivity is itself determined by the nature of the reflections of the electron waves. The first clear description of these relations was given by the French theoretical physicist Leon Brillouin in 1934.

In a one-dimensional lattice, the electron wave can be represented by a scalar (nondirectional) quantity k which is the equivalent of the vector **k** defined in con-

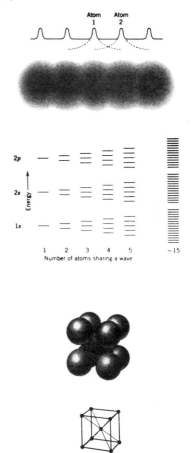

Fig. 22-39 Resonance interaction in a row of hydrogen atoms. (After D. H. Andrews and R. J. Kokes, "Fundamental Chemistry," 2d ed., John Wiley & Sons, Inc., New York, 1965.)

Fig. 22-40 Lithium atoms in a body-centered cubic lattice showing eight nearest neighbors. (After D. H. Andrews and R. J. Kokes, "Fundamental Chemistry," 2d ed., John Wiley & Sons, Inc., New York, 1965.)

nection with Eq. (22-7) applied in three dimensions. k is 2π times the wave number and is therefore proportional to the energy.

In the quantization of the permissible energy levels for an electron in a box, the resonance conditions were such that only a wave with a wavelength λ equal to $2d$ or an integral fraction of $2d$ could exist as a steady state in a one-dimensional box of length d.

This was expressed by the relation $\lambda = 2d/n$, where $n = 1, 2, 3, \ldots$. Waves of other wavelengths had the amplitude reduced to zero in the course of multiple reflection from the ends of the box.

Fig. 22-41 Increasing multiplicity of energy levels forming band structure as the separation of atoms in a crystal lattice becomes smaller: (*a*) width of band as a function of distance of separation; (*b*) wave-function profile with atoms widely separated; (*c*) wave-function profile with resonance. (After D. H. Andrews and R. J. Kokes, "Fundamental Chemistry," 2d ed., John Wiley & Sons, Inc., New York, 1965.)

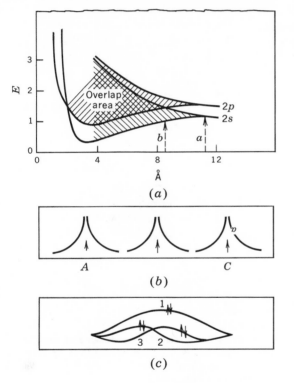

(*a*)

(*b*)

(*c*)

In much the same way, waves with $\lambda = 2\,a/n$ are forbidden in a periodic lattice where the points are separated by the distance a. This leads to the relation

$$k = \frac{n\pi}{a} \tag{22-17}$$

where $n = \ldots, -3, -2, -1, 0, 1, 2, 3, \ldots$. If n is positive, the wave travels left to right if negative, right to left. These forbidden wave numbers form points of discontinuity in the energy parameter as shown in Fig. 22-42. The regions where the waves are permitted are called *Brillouin zones*. The zones for a two-dimensional lattice are shown in Fig. 22-43 and for a three-dimensional lattice in Fig. 22-44.

The consideration of the electron waves in various types of crystals leads to the conclusion that there are types of energy-band distribution which determine the electrical properties of the material. These are shown in Fig. 22-45. If the electrons fill one or two bonds completely, a slight increase in energy due to the imposition of a potential will not move any electrons into a higher zone because so much energy is required to cross the forbidden gap. In this case (Fig. 22-45*a*) the material is an insulator. If, as in Fig. 22-45*b*, one of the bands is only half filled, only a minute increase in energy from an external potential is required to push electrons into higher levels and the material is a metallic conductor. If there are only a few electrons in a band (Fig. 22-45*c*), or if the electrons fill a band almost full (Fig. 22-45*d*), only a few electrons can move into higher energy levels and the material is a semiconductor.

Since so many of the key molecules in living cells have configurations with alternating double bonds,

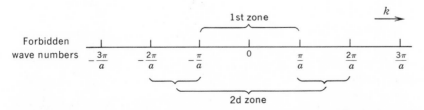

Fig. 22-42 A plot of the wave number (multiplied by 2π) of the de Broglie waves in a one-dimensional lattice, showing the values of the wave numbers which are forbidden because of interaction with the crystal lattice and the Brillouin zones lying in between.

the energy levels in these molecules have configurations that resemble the one-dimensional Brillouin zones. No extensive progress has been made in interpreting the role of the zones in the functioning of biochemical reactions, but it is clear that this may be an important part of biological quantum mechanics.

In a sense, molecules like DNA and RNA may be regarded as being made up in part of semiconductors. Since semiconductors in transistors and similar devices are now playing such a key role in so many industrial applications, there is some hope that our rapidly advancing knowledge of the solid state may contribute to our understanding of basic biochemistry.

Example 22-8

If a long straight-chain molecule is the equivalent of a one-dimensional crystal lattice with a spacing a of 1.56 Å, calculate the first six values of k which will be forbidden to electron waves in this molecule according to Brillouin zone theory.

According to Eq. (22-17), the forbidden values of k are $n\pi/a$. Since $a = 1.56 \times 10^{-8}$ cm, the forbidden values of k are given by

$$k = \frac{n\,3.1416}{1.56 \times 10^{-8}}$$

or

$$2.01,\ 4.03,\ 6.04,\ 8.06,\ 10.07,$$
$$\text{and } 12.08 \times 10^8\ cm^{-1} \qquad Ans.$$

22-6   DISORDERED CRYSTALS

In the discussion of crystal defects the evidence was reviewed which shows that in all crystals of finite size there are departures from lattice regularity that constitute crystal defects. These include the crystal surfaces, screw dislocations, edge dislocations, holes, and impurities. There are a number of states of matter in which these defects in themselves are sufficiently regular to make possible the definition of these states as quasi-crystalline forms of matter. Thus, when a single species of impurity is present in considerable amounts, the state may be called a *solid*

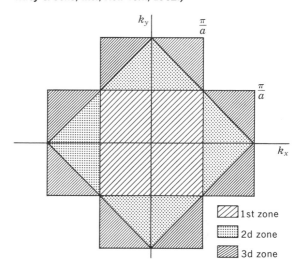

Fig. 22-43   Brillouin zones in a two-dimensional lattice. (After C. Kittel, "Elementary Solid State Physics," John Wiley & Sons, Inc., New York, 1962.)

1st zone
2d zone
3d zone

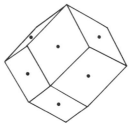

(a) First zone, body-centered-cubic lattice

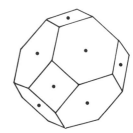

(b) First zone, face-centered-cubic lattice

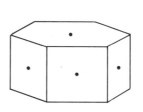

(c) First zone, hexagonal-close-packed lattice

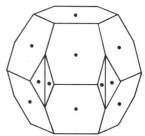

(d) Second zone, hexagonal-close-packed lattice

Fig. 22-44   Brillouin zones in three dimensions.

Fig. 22-45   A schematic representation of typical Brillouin zone arrangements which result in (a) insulator, (b) metal, and (c) semiconductors. (After C. Kittel, "Elementary Solid State Physics," John Wiley & Sons, Inc., New York, 1962.)

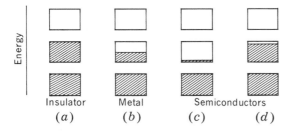

Insulator   Metal   Semiconductors
(a)   (b)   (c)   (d)

solution. Again, a species of impurity may be present in solution in a liquid but may have a molecular shape like a long rod which tends to orient the solute molecules with respect to each other; this state is frequently called a *liquid crystal.* Again, molecules in the lattice may become disoriented with respect to each other to the extent that almost no crystalline order remains; this state is called a *glass.* Thus, while the term *disordered crystals* may seem self-contradictory, there are a number of intermediate states of matter having simultaneously both order and disorder to a degree that justifies this special classification.

Glasses

Some liquids can be cooled below the freezing-point temperature in such a way that crystallization does not take place. Usually, the liquid must be cooled with some rapidity, so that the factors tending to bring about crystallization cannot function. When such a *supercooled liquid* is cooled far enough below the freezing temperature, it develops rigidity and may retain its shape independent of the shape of its container. This supercooled state can persist without crystallization for weeks, years, or centuries. Under these circumstances the material is said to constitute a *glass;* it is in the *glassy* state. A diagrammatic representation of the crystalline and glassy states of an R_2O_3 oxide is shown in Fig. 22-46.

In normal freezing the molecules must pass from the relatively disordered state in the liquid to the almost perfectly ordered state in the crystal. If a

liquid is cooled below the freezing point, and if there are no crystals present and in contact with the liquid, then before crystallization can begin several molecules of the liquid must orient themselves properly with respect to each other to form the *nucleus* of the incipient crystal lattice. As pointed out in the discussion of crystal defects in connection with the nature of crystal surfaces, the free energy of just two molecules oriented with respect to one another as in the lattice is apt to be higher than the free energy of the same molecules disoriented in the liquid state. It is only when a number of molecules assemble in an oriented three-dimensional lattice that the free energy of the cluster drops below that of the liquid. From the thermodynamic point of view, the free energy G is made up of two factors

$$G_l = H_l - TS_l \tag{22-18}$$
$$G_c = H_c - TS_c \tag{22-19}$$

and the free-energy change on freezing is

$$\Delta G_{freez} = G_c - G_l \tag{22-20}$$

Since $S_l > S_c$ the entropy factor ($TS_c - TS_l$) favors the change from crystal to liquid, i.e., to a state of higher disorder On the other hand, $H_l > H_c$, and the enthalpy factor ($H_c - H_l$) favors the change from liquid to crystal, i.e., to a state of lower enthalpy. At the melting temperature, these factors are equal

$$H_c - H_l = TS_c - TS_l \tag{22-21}$$

But before the full three-dimensional crystal can form, two molecules must first get together properly oriented, then three, then four, and so on. Because the average value of G per molecule is higher for two molecules crystal-oriented than for two molecules randomly oriented, as in the liquid, the net tendency is to prevent the formation of the original group or crystal-oriented molecules, the nucleus about which the crystal can grow. This makes possible the formation of the thermodynamically unstable supercooled state.

As temperature is lowered, a point is finally reached at which the free energy favors the formation of a nucleus. In Example 22-7, this temperature for benzene was roughly 75° below the melting point. But at this reduced temperature the movements of the molecules necessary to effect the crystal orientation become so sluggish that even if a nucleus is formed, the crystal growth rate may be negligibly slow. Thus, the molecules may remain indefinitely in the glassy state.

Observations on supercooled liquids have disclosed a number of factors that influence their transformation into crystals. First of all, the presence of small particles of extraneous matter can cause crystal nuclei to form. Scratching the surface of the container can also induce crystallization.

The shape of the molecule also influences the probability of nucleus formation. Symmetrical molecules can rotate more easily in the liquid state and orient more readily to the positions necessary to form a nucleus. Long rod-shaped molecules orient less easily.

Studies of the growth rate of single crystals have thrown considerable light on the factors governing the transformation from the supercooled liquid to the crystalline state. A single crystal of a substance can be placed in a thin layer of liquid and quickly cooled to a selected temperature below the freezing point. Then the growth rate can be observed with a microscope. When this growth rate is plotted as a function of temperature, a curve is obtained like that shown in Fig. 22-47 based on observations of salol crystals growing in thin-walled capillaries.

The theory of the variation of growth rate with temperature is of interest not only in physical chemistry but in biochemistry and biology also because biological growth is in essence the transfer of molecules from a less-ordered state resembling a liquid to a more-ordered state resembling a crystal. When the surface of the less-ordered medium is in contact with the surface of the more-ordered medium, molecules move from the former to the latter and vice versa. The rate of movement in each direction can be expressed by the usual Arrhenius equations

$$k_f = k_0 e^{-E_f^{\ddagger}/RT} e^{S_f^{\ddagger}/R} \tag{22-22a}$$

$$k_b = k_0 e^{-E_b^{\ddagger}/RT} e^{S_b^{\ddagger}/R} \tag{22-22b}$$

In these equations k_f is the rate for movement from liquid to crystal (forward) and k_b for movement from crystal to liquid (backward). Since $k_f = k_b$ at the melting-point temperature, the units of k can be

selected so that $k_0 = 1$ at the melting-point temperature. Since there is no concentration factor, there is no distinction between rate and rate constant.

The rate at which the crystal grows is $k_f - k_b$. At the temperature where this rate reaches a maximum T_1 the derivatives with respect to T and with respect

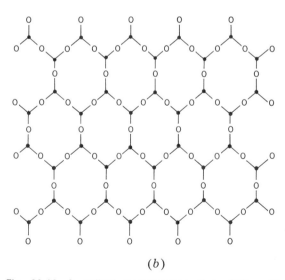

(a)

(b)

Fig. 22-46 A stylized representation of an R_2O_3 oxide arranged. (a) In the form of a glass; (b) in a regular crystal lattice. (After A. K. Galwey, "Chemistry of Solids," Chapman & Hall, Ltd., London, 1967.)

Fig. 22-47 (*a*) The observed rate of crystal growth for salol, in thin-walled glass capillary tubes (solid line); (*b*) values calculated from the Arrhenius equations (triangle with circle); (*c*) Arrhenius values corrected for crystal surface smoothing in region between maximum rate and melting point (square with circle). [Data from K. Neumann and G. Micus, *Z. Phys. Chem.*, **2**:25 (1954).]

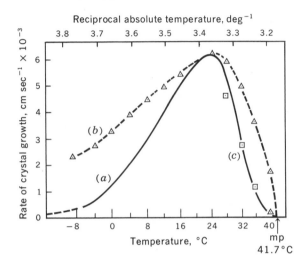

$$E_b^\ddagger = \frac{\Delta H_{\text{fus}}}{1 - W} \tag{22-26b}$$

$$S_f^\ddagger = \frac{E_f^\ddagger}{T_0} \tag{22-26c}$$

$$S_b^\ddagger = \frac{E_b^\ddagger}{T_0} \tag{22-26d}$$

Using these equations, the values of the energy of activation and entropy of activation have been calculated for the crystallization of salol from the supercooled liquid. The graphs of the rates of the forward and the back reactions are plotted as functions of $1/T$ in Fig. 22-48. The theoretical values of the rate of crystallization based on these values of $E_f^\ddagger$ and $E_b^\ddagger$ are shown in Fig. 22-47.

According to the simple Arrhenius theory, the curve for the rate of crystallization as a function of temperature should intersect the temperature axis at a finite angle. The experimental values, however, show a tangential approach. If the measured rate of crystallization depends on the number of irregularities on the crystal surface, the deviation of the experimental curve from the theoretical may be due to the decrease in number of these irregularities as the temperature of the crystal approaches the melting point and the protruding molecules dissolve in the supercooled liquid phase.

The various types of surface molecules are shown in Fig. 22-49, and the free-energy relations are illustrated in Fig. 22-50. The lowering of the rate by the smoothing of the crystal surface can be expressed approximately by

$$k_{\text{obs}} = k_{\text{Arr}} \frac{T_0 - T}{T_0 - T_1} \tag{22-27}$$

to T^{-1} are both equal to zero. This result leads to the expression

$$
\begin{aligned}
0 &= \frac{d(k_f - b_f)}{dT^{-1}} \\
&= \frac{E_f^\ddagger}{R} e^{-E_f^\ddagger/RT_1} e^{S_f^\ddagger/R} + \frac{E_b^\ddagger}{R} e^{-E_b^\ddagger/RT_1} e^{S_b^\ddagger/R}
\end{aligned} \tag{22-23}
$$

Also, $S_f^\ddagger = E_f^\ddagger/T_0$ and $S_b^\ddagger = E_f^\ddagger/T_0$, where T_0 is the melting-point temperature. Therefore Eq. (22-23) reduces to

$$\log \frac{E_f^\ddagger}{E_b^\ddagger} = \frac{-\Delta H_0}{2.303R} \left(\frac{1}{T_1} - \frac{1}{T_0} \right) \tag{22-24}$$

where ΔH_0 is the heat of fusion. If this quantity and T_1 and T_0 are all known, this equation makes possible the calculation of $E_f^\ddagger$, $E_b^\ddagger$, $S_f^\ddagger$, and $S_b^\ddagger$. Denoting $E_f^\ddagger/E_b^\ddagger$ by W and solving Eq. (22-24) for W gives

$$\Delta H_{\text{fus}} = E_b^\ddagger - E_f^\ddagger = \frac{E_f}{W} - E_f \tag{22-25}$$

$$E_f^\ddagger = \Delta H_{\text{fus}} \frac{W}{1 - W} \tag{22-26a}$$

where k_{obs} = observed rate of growth

k_{Arr} = rate calculated from Arrhenius equation

T_1 = temperature at which maximum growth rate is observed

T_0 = melting temperature

When the effect of surface smoothing is taken into account, the calculated curve agrees quite well with the experimental observation, as shown in Fig. 22-47. In the range below the maximum-growth-rate temperature, the calculated curve lies considerably above the experimentally observed growth rate. Since the simple Arrhenius equation does not take into account the variation of $E^{\ddagger}$ with T, some deviation of the calculated curve from the experimental is to be expected. However, the simple Arrhenius theory does predict the qualitative nature of the variation of growth rate with temperature.

Since $E_b^{\ddagger}$ is considerably larger than $E_f^{\ddagger}$, the rate of the back reaction becomes negligibly small with respect to the rate of the forward reaction at temperatures considerably below the melting-point temperature. Figure 22-51 is a graph showing the decrease in $\log k$ as a function of the temperature in reciprocal degrees below the melting point for various values of $E^{\ddagger}$.

In a glass which is stable over relatively long periods of time, values of k_f must be very low. For example, many kinds of glass remain stable for years or even centuries. Also, when no crystals are present, many organic compounds in the supercooled liquid state remain for years without crystallizing, showing that k has a value far less than 10^{-7} cm sec^{-1} even at temperatures only 20° below the melting-point temperature. Yet at the same temperature when crystals *are* present, the growth rate is of the order of centimeters per second. There seems to be no way in which the presence of the seed crystal could decrease the value of $E^{\ddagger}$ by an amount sufficient to account for this change in growth rate. The conclusion is that the presence of the crystals raises the activation entropy by an amount sufficient to alter the growth rate by a factor of 10^8. In other words, the presence of a crystal surface, on which growth can take place, raises the probability of growing by this large factor.

In view of the relation between entropy and information, one can say that the supercooled liquid fails to change to the crystal state because of lack of information; the liquid does not "know how" to crystallize. When a seed crystal is introduced, it imparts the needed information, raising $S^{\ddagger}$ to the value which permits rapid growth. In a similar way, macromolecules like DNA and RNA control biological growth by controlling the values of $S^{\ddagger}$ for the various reactions. These relations will be discussed in further detail in Chaps. 25 and 26.

Liquid crystals

As noted briefly in Chap. 4, there are many examples of substances in a state which is intermediate between the liquid and the crystalline states. In 1888 the

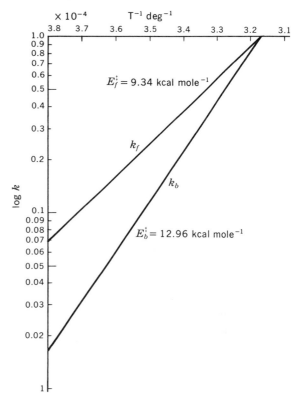

Fig. 22-48 Values of $\log k_f$ and $\log k_b$ for the transfer of molecules between liquid and crystal for salol based on the Arrhenius equations.

$E_f^{\ddagger} = 9.34$ kcal mole^{-1}

k_f

k_b

$E_b^{\ddagger} = 12.96$ kcal mole^{-1}

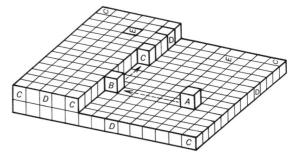

Fig. 22-49 Types of binding for molecules causing irregularities in a crystal surface.

Fig. 22-50 The free energy of various types of molecules on the crystal surface, showing stability with respect to the liquid state.

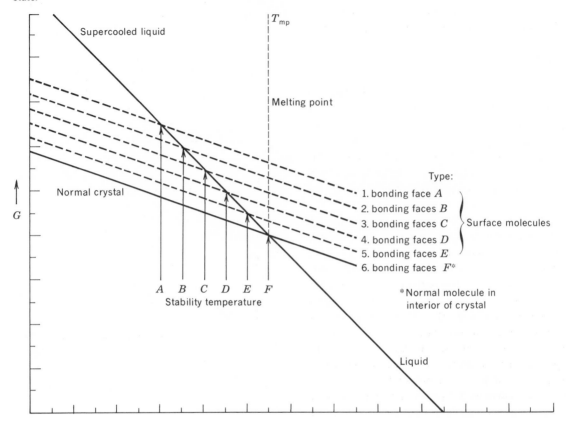

German chemist F. Reinitzer prepared a number of esters of cholesterol which exhibited a peculiar kind of double melting. At 145.5°C, the crystals melted sharply, but the melt was opaque; at 178.5°, the opacity disappeared sharply, giving a true isotropic liquid. This opaque form is an example of the *paracrystalline* state, less ordered than a true crystal but more ordered than a true liquid. Substances in this state are frequently called *liquid crystals,* and the different phases corresponding to this state of matter are called *mesophases*. In Fig. 22-52, the various kinds of structures are drawn schematically to indicate the nature of these phases.

The long-chain fatty acid salts in aqueous solution form various paracrystalline phases. Detergent action is associated with this type of order. Many other compounds with long thin cylindrical shapes form liquid crystals; examples are ethyl *p*-anisalaminocinnamates

$$CH_3O \langle \rangle -CH{=}N \langle \rangle -CH{=}CH-COOC_2H_5$$

with three paracrystalline phases between 83 and 139°C and *p*-azoxyanisole

$$CH_3O \langle \rangle -\overset{\overset{O}{\uparrow}}{N}{=}N- \langle \rangle -OCH_3$$

(mp 84°, liquid crystal mp 150°C).

The paracrystalline state is of great importance in

biochemistry because so many of the phases of living matter such as cytoplasm are paracrystalline. In view of the key role played by entropy of activation in biochemistry, it is not surprising to find these various phases where differences in order and the differences in entropy associated with them are guiding influences in biological action.

Order-disorder transformations

Other examples of crystals with partial disorder are found when the transformation from one crystalline phase to another does not take place like a normal melting at a single point on the temperature scale but is extended over a temperature range quite a few degrees in width. Figure 25-53 is a graph of the heat capacity of methane over the range 13 to 28°K. Instead of a heat capacity increasing at a relatively constant rate up to the melting point and a heat of transition absorbed effectively over a range of less than a degree in width, there is found a heat capacity showing a rapid increase over a range of 8° below the temperature of maximum C_p. The transition appears to be due to the onset of rotation of the molecules of CH_4 in the crystal lattice. The van der Waals forces tying the molecules together are so weak that when libration reaches a certain intensity, more and more molecules start to rotate almost freely while still remaining at sites in the crystal lattice. Since the number of molecules rotating varies with temperature, there is a gradual transition from one state to the other over the range of 8°.

Another example is the transition in the alloys of copper and zinc which are called *beta-brasses*. These are found when the atomic concentration of zinc lies between 46 and 49 percent, essentially equal numbers of Zn and Cu atoms. Below 200°C, these atoms are practically all at regular sites in the crystal lattice, one regular set of sites having Cu and the remaining regular set having Zn. This arrangement is called a superlattice. As temperature rises above 200°C, some Zn and Cu atoms swap places. While the lattice sites still form an ordered pattern, the Cu and Zn are randomly distributed over these sites, introducing a special kind of disorder into the lattice. Such a change is called an *order-disorder transition* or *transformation*. This transformation introduces an extra term

into the heat capacity much like the rotation term in the heat capacity of methane. The graph is shown in Fig. 22-54.

From the thermodynamic point of view, these transitions differ from the transition from one lattice to another at a distinct temperature like the change from rhombic to monoclinic sulfur because of the possibility of the intermediate states between the two forms. For methane, it is not necessary that all molecules be librating on the one hand or all spinning on the other; the percentage spinning can vary all the way from zero to a hundred depending on the temperature. For brass, it is not necessary that all molecules be ordered or all randomly arranged on the sites. In fact, at each temperature throughout the transition range there is an intermediate state of a distinct kind which has the lowest free energy at that temperature and therefore is the equilibrium state at that temperature.

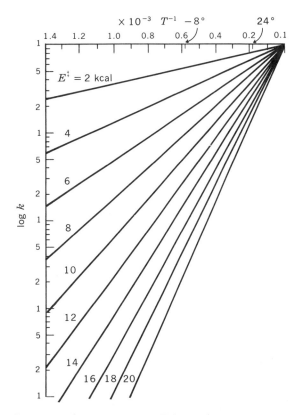

Fig. 22-51 k as a function of T^{-1} and $E^{\ddagger}$.

Fig. 22-52 The paracrystalline states showing (a) ideal crystal; (b) smectic paracrystal; (c) nematic paracrystal; (d) liquids.

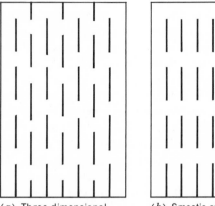

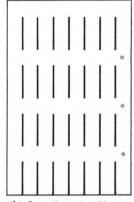

(a) Three-dimensional order in a crystal

(b) Smectic state with glide planes at *

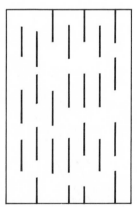

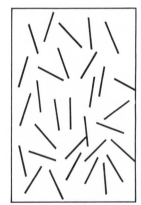

(c) Nematic state with order in one dimension only

(d) Isotropic liquid with virtually complete disorder

As the transition is made, the free energy (G = H − Ts) is determined by the difference between the enthalpy H and the temperature-entropy product Ts. Since disordering increases the entropy, there will be just enough disorder at each temperature to keep G at a minimum.

When each molecule or atom in the lattice can be in one of two states, e.g., on the right site or the wrong site for a superlattice, the relation between the entropy per mole due to disorder S_d and the number of molecules or atoms contributing to the disorder can be expressed by an equation of the form

$$S_d = R \ln (1 + d) \tag{22-28}$$

where d is the fraction of molecules in the disordered state and ranges from zero to one.

Since the paracrystalline state is found in cytoplasm and similar biological media, there are many possibilities of partial disorder dependent on temperature in a way resembling these order-disorder transformations.   While the data available are not sufficient to reveal the details of such transformations, the theory of order-disorder transformations is almost certain to be a helpful guide in investigating biophase equilibria.

Example 22-9

The temperature T_1 of maximum rate of crystallization for benzophenone, $(C_6H_5)_2CO$, is 29.50°C.   The melting point T_0 is 47.92°C. The heat of fusion ΔH_0 is 4282 cal deg^{-1} mole^{-1}. Calculate the values of $E_f^{\ddagger}$ and $E_b^{\ddagger}$ on the basis of the simple Arrhenius equation.

Using Eq. (22-25),

$$\log W = -\frac{\Delta H_0}{2.303R}\left(\frac{1}{T_1} - \frac{1}{T_0}\right)$$

the values are

$$\log W = -\frac{4282}{2.303 \times 1.987}$$
$$\times \left(\frac{1}{302.65} - \frac{1}{321.07}\right) = -0.1774$$

$$W = 0.665 \qquad 1 - W = 0.335$$

$$E_f^{\ddagger} = \frac{\Delta H_0 W}{1 - W} = 8500 \text{ cal mole}^{-1}$$

$$E_b^{\ddagger} = \frac{\Delta H_0}{1 - W} = 12{,}782 \text{ cal mole}^{-1}$$

Example 22-10

The variation of the degree of disorder d can be approximated roughly by

$$d = 1 - \left(\frac{T}{T_c}\right)^3$$

Fig. 22-53 The heat capacity of methane over the range where crystal I changes to crystal II through internal rotation. [Data from K. Clusius, *Z. Physik. Chem.*, **133**:41 (1929).]

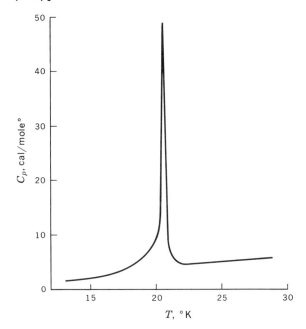

For β brass $T_c = 622°K$, calculate the entropy due to disorder at 500°K.

$$d = 1 - \left(\frac{500°}{622°}\right)^3 = 0.48$$

$$S_d = R \ln (1 + d) = 2.303R \log 1.48$$
$$= 0.779 \text{ cal deg}^{-1} \text{ mole}^{-1}$$

Problems

1. The distance d between two crystal planes in diamond is 1.54 Å. Calculate the wavelength required for a beam of x-rays which will be diffracted by these planes with $n = 1$ and $\theta = 30°$.

2. If in a crystal there are crystallographic planes spaced 4.75 Å apart and a beam of x-rays with a wavelength of 2 Å is reflected from the planes, calculate the values of θ at which there will be maximum reinforcement of the reflected rays when $n = 1, 2, 3$, and 4.

3. If a beam of x-rays strikes a crystal and is reflected with maximum reinforcement at an angle of 62°, calculate the spacing of the planes causing the reflection if $n = 1$ and $\lambda = 2.65$ Å.

4. A beam of x-rays is reflected from the planes of a crystal which are spaced 1.7 Å apart when the angle of reflection is 55°. Calculate the wavelength of the x-rays if $n = 3$.

5. Make a sketch of the cubic crystal lattice and show the orientation of the planes corresponding to (001), (011), (101), and (210). Write down the crystallographic numbers designating the two planes which are at right angles to the (110) plane.

6. Draw a sketch of the face-centered-cubic lattice and draw in the plane passing through the atom in the center of the face that is facing you and the atom that is in the center of the face on the right-hand side of the cube. What is the numerical designation of this plane?

7. The hydrocarbon propane, C_3H_8, shows a heat of transition in the crystalline state at $-165°C$. The heat of transition per mole is 8.506 cal g^{-1}. Calculate the entropy of transition.

8. Calculate the molar entropy of the phase transition in methane, for which the change in enthalpy is 18.08 cal mole^{-1} and the transition temperature may be approximated as 20.5°K.

9. The energy of activation for the forward reaction in the crystallization of diphenyl ether from the supercooled liquid is 9.48 kcal mole^{-1}. Calculate the

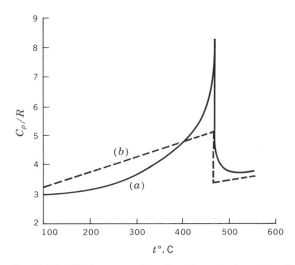

Fig. 22-54 The heat capacity of beta-brass in the region of the order-disorder transformation: (*a*) observed; (*b*) calculated. (After A. H. Wilson, "Thermodynamics and Statistical Mechanics," Cambridge University Press, London, 1957.)

entropy of activation if at the melting point (26.9°C) the rate is unity.

10. The temperature of maximum growth rate for diphenylamine, $(C_6H_5)_2NH$, is 40.6°C; the melting point is 53.3°C; the heat of fusion is 4.19 kcal mole^{-1}. Calculate $E_f^‡$ for crystallization.

11. Calculate $S_f^‡$ and $S_b^‡$ for the forward and backward crystallization processes in Prob. 10 at the melting point.

12. The hydrocarbon 2,3-dimethylbutane has a crystalline transition point at -137.09°C. The heat of transition is 18.010 cal g^{-1}. Calculate the entropy change for the transition.

13. Triphenylmethane, $(C_6H_5)_3CH$, melts at 92.9°C. Use units for the forward and backward rate of reaction such that $k_f = k_b = 1$ at the temperature of the melting point. $E_f^‡ = 11.95$ kcal mole^{-1} and $E_b^‡ = 16.98$ kcal mole^{-1}. Make a plot of R in k_f and R in k_b against $1/T$, covering the range from $1/T = 0$ to about 50° below the melting point. From the interception of the lines with the ordinate at $1/T = 0$, estimate the values of $S_f^‡$ and $S_b^‡$.

14. By measuring the distance between the two straight-line graphs in the plot for Prob. 13 at every 5° below the melting point, make a plot of the theoretical rate of crystallization as a function of temperature. Estimate the temperature at which a maximum rate of crystallization occurs.

15. The compound p-tolylbenzophenone, p-CH$_3$ $(C_6H_4)CO(C_6H_5)$, has a melting point of 56.1°C and shows a maximum rate of crystallization at 31.8°C. The heat of fusion is 4.7 kcal mole^{-1}. Make a calculation of $E^‡$ and $S^‡$ for both the forward and the backward reactions.

16. Apply Eq. (22-27) to the data obtained in Prob. 15 and plot the curve for rate of growth when the active sites on the crystal surface are partially dissolved.

17. List the rotations which transform cyclobutane into itself.

18. Write the group elements corresponding to the rotational transformation of cyclobutane in permutation notation.

19. Using permutation notation, list the subgroups in the cyclobutane rotation group.

20. Using permutation notation, write the elements of the group corresponding to rotation of benzene about an axis through the center and perpendicular to the plane of the ring.

21. Using the graphical notation, list the elements of the group associated with rotations of the tetragonal P lattice (see Fig. 22-5).

22. The forbidden value of the wave "vector" for $n = 2$ in a one-dimensional crystal lattice is 3.90×10^8 cm^{-1}, what is the spacing of the lattice points?

REFERENCES

Smoluchowski, R., J. E. Mayer, and W. A. Weyl (eds.): "Phase Transformations in Solids," John Wiley & Sons, Inc., New York, 1951.

Nyburg, S. C.: "X-ray Analysis of Organic Structures," Academic Press, Inc., New York, 1961.

Kittel, C.: "Elementary Solid State Physics," John Wiley & Sons, Inc., New York, 1962.

Gray, G. W.: "Molecular Structure and the Properties of Liquid Crystals," Academic Press, Inc., New York, 1962.

Newkirk, J. B., and J. H. Wernick (eds.): "Direct Observations of Imperfections in Crystals," John Wiley & Sons, Inc., New York, 1962.

Fox, D., M. M. Labes, and A. Weissberger (eds.): "Physics and Chemistry of the Organic Solid State, I," John Wiley & Sons, Inc., New York, 1963.

Bunn, C.: "Crystals," Academic Press, Inc., New York, 1964.

Spice, J. E.: "Chemical Binding and Structure," The Macmillan Company, New York, 1964.

Fine, M. E.: "Phase Transformations in Condensed Systems," The Macmillan Company, New York, 1964.

Knox, R. S., and A. Gould: "Symmetry in the Solid State," W. A. Benjamin, Inc., New York, 1964.

Dorain, P. B.: "Symmetry in Inorganic Chemistry," Addison-Wesley Publishing Company, Inc., Reading, Mass., 1965.

Ubbelohde, A. R.: "Melting and Crystal Structure," Oxford University Press, New York, 1965.

Weertman, J., and J. R. Weertman: "Elementary Dislocation Theory," The Macmillan Company, New York, 1965.

Brout, R. H.: "Phase Transitions," W. A. Benjamin, Inc., New York, 1965.

Kittel, C.: "Introduction to Solid State Physics," John Wiley & Sons, Inc., New York, 1966.

Bhagavantam, S.: "Crystal Symmetry and Physical Properties," Academic Press, Inc., New York, 1966.

Strickland-Constable, R. F.: "Kinetics and Mechanism of Crystallization," Academic Press, Inc., New York, 1967.

Galwey, A. K.: "Chemistry of Solids," Chapman & Hall, Ltd., London, 1967.

Hannay, N. B.: "Solid State Chemistry," Prentice-Hall, Inc., Englewood Cliffs, N.J., 1967.

Bondi, A.: "Molecular Crystals, Liquids and Glasses," John Wiley & Sons, Inc., New York, 1968.

23-1 SURFACE FORCES

In considering the nature of the crystallization process, we discussed the fact that at the surface of the crystal there are some molecules which have higher free energy because they are bound to the crystal by a smaller field of force than molecules in the interior of the crystal, which are completely surrounded by neighboring molecules. There is a similar situation at the interface between the liquid and its vapor. In the cross-sectional drawing in Fig. 23-1 this is indicated by arrows showing a molecule at the surface bound to its immediate neighbors. If we followed the convention used in discussing the

Chapter Twenty-Three

surface chemistry

cubical molecules on the crystal, we would say that the surface molecule has neighbors on five of its six faces and thus has one face where it does not experience attraction, the face exposed to the vapor.

Drops

In order to see the consequences of this special surface-force pattern, consider a spherical globule of mercury lying on a metallic surface, as shown in Fig. 23-2a. The surface of this globule has the area $4\pi r^2$, where r is the radius of the sphere; this is the minimum possible surface to surround this volume of matter. Now suppose that the globule of mercury is flattened out somewhat by pushing down on it from above (Fig. 23-2b). In its new oblate shape, it has a larger surface area. As a consequence, some atoms have been moved from the interior of the mercury, where there were neighbors on all sides, to the surface, where there is one side of the atom without a neighbor, the side exposed to the gas. Work must have been done to pull a neighboring atom away and expose this surface. Therefore, in the oblate shape with the larger surface the mercury globule has a greater free energy. This is precisely analogous to the increased free energy of the atom on the surface of the crystal as shown in Fig. 22-49.

The coefficient which expresses the increase in free energy for each unit of surface formed is called the *surface tension*. This surface tension can be felt

Fig. 23-1 Cross-sectional representation of molecules (the small cubes) at the surface of a liquid. A force of attraction is exerted horizontally and downward but not upward.

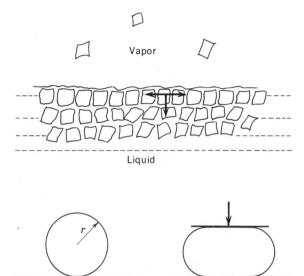

Vapor

Liquid

Fig. 23-2 A drop of mercury lying on a flat surface: (a) with no vertical pressure; (b) slightly flattened by pressure from above.

if a finger is placed on top of a drop of mercury to push downward. When the finger is lifted, the mercury springs back from the oblate to the spherical shape because the surface tension pulls the molecule into the form which has the minimum surface and, therefore, the minimum free energy.

The relation between the increase in free energy dG and the increase in surface area dA can be expressed

$$dG = \gamma \, dA \qquad (23\text{-}1)$$

A small Greek gamma γ is used to denote the surface tension. Thus

$$\gamma = \frac{dG}{dA} \qquad (23\text{-}2)$$

$$\frac{\text{ergs}}{\text{cm}^{-2}} \quad \frac{\text{ergs}}{\text{cm}^2}$$

Force is related to energy by

$$F \, dx = dE_p \qquad (23\text{-}3)$$

dynes cm   ergs

where F = force

dx = distance through which force acts

dE_p = increase in potential energy due to action of force

The dimensional relations are

$$\text{Force} = \text{dynes} = \text{ergs cm}^{-1}$$
$$\gamma = \text{ergs cm}^{-2} = \text{dynes cm}^{-1} \qquad (23\text{-}4)$$

Thus, surface tension γ has the dimensions of dynes per centimeter.

Bubbles

A soap bubble behaves much like the globule of mercury. A soap bubble has a roughly spherical shape if it is not too large. One can press gently on the top of the bubble and deform it from the spherical shape into the shape of the oblate spheroid; and in so doing, one can feel the force required to produce the deformation resulting in the larger surface.

For the bubble, the force that resists the surface tension arises from the pressure P_1 of the gas inside the bubble which is in excess over that of the atmosphere pressing on the outside of the bubble. Consider a bubble like that shown in Fig. 23-3. We can regard the thickness of the bubble as so small that the area of the outer surface exposed to the air is the same as the area of the inner surface exposed to the slightly higher pressure of the air inside the bubble. On both these surfaces there will be molecules of H_2O which are roughly only five-sixths in contact with the neighboring molecules. Thus, the total area A producing the surface tension is related to the radius of the bubble by the expression

$$A = 4\pi r^2 \quad + \quad 4\pi r^2 \qquad (23\text{-}5)$$

outside area   inside area

When the bubble is slightly flattened, producing a small increase in the area, this becomes

$$dA = 16\pi r \, dr \qquad (23\text{-}6)$$

obtained by differentiating Eq. (23-5) and combining terms. Thus, the increase in free energy dG is given by

$$dG = \gamma \, dA = 16\gamma\pi r \, dr \qquad (23\text{-}7)$$

This increase in free energy must be balanced by the increase in free energy dG_1 of the gas inside the bubble, which will be $P_1\,dV = P_1(4\pi r^2)\,dr$. Equating these two expressions for the free-energy increase, we get

$$P_{\text{ext}(2)} = \frac{4\gamma}{r} \qquad (23\text{-}8)$$

For a drop like the drop of mercury, where there is only one interface, the internal pressure is only half that which we find for the soap bubble, with an interface both inside and outside; so that for one interface

$$P_{\text{ext}(1)} = \frac{2\gamma}{r} \qquad (23\text{-}9)$$

When a caisson worker is taken out of the high-pressure chamber under a river too suddenly, bubbles of air form in his bloodstream. Such a bubble has only one interface since the blood is on the outside and the air is on the inside. Thus, Eq. (23-9) is applicable.

Values of the surface tension for a number of substances are given in Table 23-1. In general, the surface tension depends both on the material forming the surface and on the kinds of molecules in the vapor or gas above the surface. As might be expected, it is also a function of the temperature. The surface tension of water is changed markedly when various substances are dissolved in it.

As the size of a drop of water decreases, the surface decreases as the square of the radius while the volume decreases as the cube of the radius. Consequently, the smaller the droplet, the larger the proportion of surface molecules to interior molecules; thus the free energy is higher, and, consequently, the vapor pressure is higher. This relationship was first deduced by Kelvin and is given by

$$\log \frac{P}{P_0} = \frac{2M\gamma}{2.303RT\rho r} \qquad (23\text{-}10)$$

where P = vapor pressure when droplet has radius r
   P_0 = vapor pressure of liquid in bulk
   γ = surface tension
   R = gas constant
   T = temperature, °K
   ρ = density
   M = molecular weight

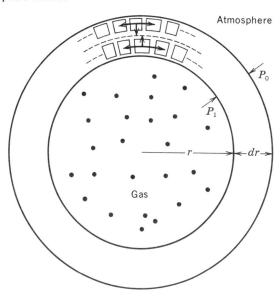

Fig. 23-3 Cross-sectional representation of molecules (the small cubes) at the inner and outer surfaces of a soap bubble, showing the direction of the forces of attraction and the forces due to the pressure of the gas within and the atmosphere without.

TABLE 23-1   Surface-tension coefficent γ

A	against	B	γ, dynes cm^{-1}	t, °C
Water	Air		74.22	10
Water	Air		72.75	20
Water	Air		71.18	30
Water	Air		69.56	40
Water	Air		67.91	50
Water	Benzene		35.00	20
Water	CCl_4		45.00	20
Water	$(C_2H_5)_2O$		10.7	20
Water	$n\text{-}C_6H_{14}$		31.1	20
Water	$n\text{-}C_8H_{13}OH$		8.5	20
Benzene	Air		28.9	20
CCl_4	Air		26.8	20
C_2H_5OH	Air		22.3	20
CH_3OH	Air		22.6	20
C_6H_5OH	Air		40.9	20
$C_6H_5CH_3$	Air		28.4	20
CH_3COOH	Air		27.6	20
NH_3	$NH_3(g)$		23.0	11.1

Fig. 23-4   Water in a capillary tube.   The circle at the left shows the radius of curvature r_c of the meniscus.

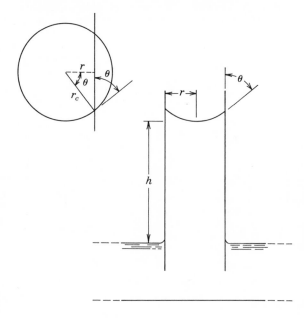

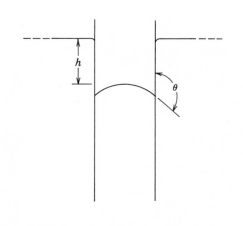

Fig. 23-5   Mercury in a capillary tube.

Capillary action

When a liquid is placed in a tube of relatively small diameter, the forces of surface tension cause the surface of the fluid to curve (Fig. 23-4); this surface is called the meniscus.   There is a balance between the forces of interaction among the liquid molecules

themselves and between the liquid and the walls of the tube; and when, as shown in Fig. 23-4, the angle θ is less than 90°, the liquid is said to wet the surface of the tube and the meniscus is concave.   If, on the other hand, the angle θ is greater than 90°, as shown in Fig. 23-5, the liquid does not wet the tube and the meniscus is convex.

When the liquid wets the tube and a concave meniscus is formed, the forces of attraction between the liquid and the tube are greater than the forces of attraction between the molecules of the liquid themselves and the liquid is pulled up into the tube.   It rises to a height where the force of gravity pulling down on the portion of the liquid above the normal surface is just sufficient to balance the upward pull due to the force of wetting.   For very narrow tubes, the column height is great enough to be easily measurable.

In order to simplify the mathematical treatment, let us regard the surface of the concave meniscus as having the same curvature as the portion of a sphere, with radius of curvature r_c as shown in Fig. 23-4. Thus the radius of the capillary tube r and the radius of curvature are related to the angle of incidence θ of the meniscus

$$\cos \theta = \frac{r}{r_c} \tag{23-11}$$

The upward force per unit area on the meniscus due to surface tension must be balanced by a downward force to produce equilibrium.   For a soap bubble with the same radius of curvature this downward force would be due to the pressure of the gas and is given by Eq. (23-9):

$$P_1 = \frac{2\gamma}{r_c} = \frac{2\gamma \cos \theta}{r} \tag{23-12}$$

The weight of the cylindrical column of the liquid is

$$W = \pi r^2 gh\rho \tag{23-13}$$

where the weight of the air is neglected.   The force per unit area is therefore

$$F_a = gh\rho \tag{23-14}$$

This is, of course, the force which must be balanced against the effective force per unit area upward due to the surface tension, and is given in Eq. (23-12). Thus

$$2\frac{\gamma}{r}\cos\theta = gh\rho \qquad (23\text{-}15)$$

giving the value of the surface tension in terms of the capillary rise as

$$\gamma = \tfrac{1}{2}gh\rho\,\frac{r}{\cos\theta} \qquad (23\text{-}16)$$

In many instances the wetting forces are so strong that effectively $\theta = 0$ and Eq. (23-16) becomes

$$\gamma = \tfrac{1}{2}gh\rho r \qquad (23\text{-}17)$$

Thus, the measurement of capillary rise is a useful way of determining surface tension.

Example 23-1

Calculate the excess free energy produced when a drop of mercury 1 mm in diameter is changed from a sphere to an oblate spheroid where the surface area is doubled ($t = 20°C$). $\gamma = 520$ dynes cm^{-1}

The area of a sphere 1 mm in diameter is $4\,\pi(\tfrac{1}{2})^2 = 3.14 \times 10^{-2}$ cm^2. The increase in free energy is $\Delta G^s = \gamma\,\Delta A = 520$ dynes cm$^{-1} \times 3.14 \times 10^{-2}$ cm$^2 = 16.3$ ergs.   *Ans.*

Example 23-2

Water rises in a capillary tube with a radius of 0.100 cm when the lower end is immersed in a dish of water at 30°C. The density of water at this temperature is 0.996. Assuming the contact angle with the glass wall to be zero, calculate the height of the meniscus.

Using Eq. (23-17),

$$h = \frac{2\gamma}{g\rho r} = \frac{2 \times 71.18}{981 \times 0.996 \times 0.100}$$
$$= 1.46 \text{ cm} \qquad Ans.$$

23-2   SURFACE ADSORPTION

If instead of the surface of pure water we consider the surface of an aqueous solution, certain new problems arise. Will the surface action affect the concen-

tration of the solute at the surface? The answer is to be found in the thermodynamic relations which involve surface action.

Adsorption thermodynamics

We shall designate the thermodynamic properties of the layer of solution at the surface by superscript s, and consider the surface layer to consist of those molecules close enough to the surface area A^s to be acted upon fully by the surface forces. The free energy G^s of the surface layer is made up of the free energy due to the surface tension γA^s and the free energy of the solute $\mu_1 n_1{}^s$ and the solvent $\mu_2 n_2{}^s$. Thus

$$G^s = \gamma A^s + \mu_1 n_1{}^s + \mu_2 n_2{}^s \qquad (23\text{-}18)$$

A change in the free energy dG^s can be brought about (1) by varying the surface area, which will cause the change $\gamma\,dA^s$; (2) by varying the amount of solute, which will cause the change $\mu_1\,dn_1{}^s$; or (3) by varying the amount of solvent, which will cause the change $\mu_2\,dn_2{}^s$. Thus

$$dG^s = \gamma\,dA^s + \mu_1\,dn_1{}^s + \mu_2\,dn_2{}^s \qquad (23\text{-}19)$$

But the total differential of G^s is

$$dG^s = \gamma\,dA^s + A^s\,d\gamma + \mu_1\,dn_1{}^s$$
$$+ n_1{}^s\,d\mu_1 + \mu_2\,dn_2{}^s + n_s\,d\mu_2 \quad (23\text{-}20)$$

Therefore

$$A^s\,d\gamma = -(n_1{}^s\,d\mu_1 + n_2{}^s\,d\mu_2) \qquad (23\text{-}21)$$

If the concentration of solute is small, a change in concentration will not affect the free energy of the solvent, so that $d\mu_2 = 0$. Dividing Eq. (23-21) by A^s expresses the change in surface tension as a function of the moles per square centimeter of solute, a kind of surface concentration ($\Gamma_2 = n_2/A^s$). Thus

$$d\gamma = -\Gamma_2\,d\mu_2 \qquad (23\text{-}22)$$

The free energy of the solute can be expressed in terms of the activity (a_2) as

$$d\mu_2 = -RT\,d\ln a_2 \qquad (23\text{-}23)$$

so that

$$d\gamma = -RT\Gamma_2\,d\ln a_2 \qquad (23\text{-}24)$$

For an ideal solution $a_2 = X_2$, the mole fraction, and

Fig. 23-6   The action of oil on a water surface. (After D. J. Shaw, "Introduction to Colloid and Surface Chemistry," Butterworth & Co. (Publishers), Ltd., London, 1966.)

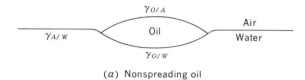

(a) Nonspreading oil

(b) Saturated solution of n-hexanol in water

$$\Gamma_2 = -\frac{1}{RT}\frac{d\gamma}{\partial \ln X_2} \qquad (23\text{-}25)$$

In a dilute solute X_2 is proportional to the concentration c_2 so that

$$\Gamma_2 = -\frac{1}{RT}\frac{d\gamma}{d \ln c_2} \qquad (23\text{-}26)$$

This expression, called the Gibbs adsorption isotherm, was first derived by Gibbs in 1872; it shows that when an increase in bulk concentration ($d \ln c_2$ positive) lowers the surface tension ($d\gamma$ negative), there will be an increase in concentration at the surface (Γ_2 positive). Values of Γ have been measured experimentally by the American physical chemist J. W. McBain, who was able to scoop off thin layers of the surface of a solution and prove the increase in the solute there.

Surface films

One of the most striking examples of the adsorption of solute molecules by a surface is found when the solute molecules have a group of atoms at one end with a tendency to form hydrogen bonds. Such a molecule is stearic acid, $CH_3(CH_2)_{18}COOH$, which is relatively insoluble in water. However, when stearic acid is added to water, the molecules are strongly adsorbed at the surface. The COOH group forms hydrogen bonds with the H_2O molecules and so is held just inside the surface; this group is called *hydrophilic*. The long hydrocarbon tail then sticks up above the water; this part of the molecule is called *hydrophobic*. The reason for the insolubility of the hydrocarbon tail can be seen by recalling the deviation from ideal-solution behavior when a hydrocarbon is shaken with water, as described in Chap. 11. When water and a typical hydrocarbon like octane are mixed, two layers of solution form. There is one layer almost purely hydrocarbon with a few water molecules wandering around in it and another layer which is almost pure water with a few hydrocarbon molecules wandering around in it. The reason for this behavior is, of course, that the water molecules are so strongly attracted to each other by the hydrogen bonding that they force out of the solution any intruding hydrocarbon molecules; and only a few water molecules can escape the attraction to their own kind and go off into the hydrocarbon. Figure 23-6 shows the way in which a drop of substance tends to spread when the molecules are hydrophilic on one end and hydrophobic on the other.

The American physical chemist Irving Langmuir in 1916 developed a device by which the two-dimensional pressure due to this spreading tendency could be measured (Fig. 23-7). A ribbon is moved across the surface of such a solution to push the stearic acid molecules closer and closer together until finally all the molecules are touching like a stack of pencils with the erasers down in water and the main length of the pencils in an upright position. By mounting the ribbon in such a way that precise measurements could be made of its tension, Langmuir found the tension to increase markedly when all the molecules were pushed so closely together that they were touching. If a measurement is made to show the precise number of molecules on the surface, the area of surface just sufficient to hold this monolayer of molecules supplies the information necessary to derive the value of the cross-sectional area of the molecules when held in this upright position. In the case of stearic acid, it turns out that the cross-sectional layer is 20.5 Å^2 molecule^{-1}. Knowing the molecular volume of stearic acid, 556 Å^3, one can then calculate the length of the stearic acid molecule, $556/20.5 = 27.1$ Å.

In the crystalline state, stearic acid molecules are also stacked together like pencils with all the erasers in one plane and all the points in another. By x-ray diffraction studies like those discussed in the previous chapter, it is possible to measure the distance between these two planes and to estimate the length of the molecule. The value obtained from x-ray diffraction studies agrees well with the value obtained by the Langmuir monolayer method.

Solid surface adsorption

As discussed in Chap. 22, the surfaces of solids are the seat of forces that tend to attract other molecules to them. Even the molecules of a gas may be attracted so strongly that they can be retained on the surface as an adsorbed layer.

One of the simplest ways to investigate adsorption of this type is the determination of the weight of a known amount of solid, usually in the form of fine particles with a large total amount of surfaces, when the particles come to equilibrium with a gas or vapor at different pressures. A device for making measurements of this sort is shown in Fig. 23-8. From the pressure and temperature data and the molecular weight of the gas, the volume can be calculated corresponding to the amount of gas adsorbed at different pressures. A graph of this sort is called an adsorption isotherm. A number of different types of such graphs are shown in Fig. 23-9.

The American physical chemist Stephen Brunauer has classified these as follows:

I.   Langmuir type. The first molecules are held very strongly, by forces of the order of magnitude of the forces of chemical bonds. This kind of adsorption is therefore called *chemisorption*. When all the sites on the surface exhibiting such forces are covered, no further adsorption takes place. The surface probably is covered only by a monolayer of adsorbed molecules.

II.   These are *sigmoid isotherms*. They represent the strong adsorption of a monolayer (up to point *B*), followed by the weaker adsorption of superimposed layers.

III. There is no strong adsorption or specific monolayer. The gas is absorbed in pores, defects, or as solid solution.

IV.   There is monolayer adsorption; at higher pressures, the isotherms level off near the vapor pressure of material adsorbed in the pores by capillary condensation.

V.   There is no monolayer adsorption, but the pore absorption is the same as in IV.

Fig. 23-7 Langmuir trough for measuring surface adsorption on liquids. (*a*) Schematic cross section. (After D. J. Shaw, "Introduction to Colloid and Surface Chemistry," Butterworth & Co. (Publishers), Ltd., London, 1966.) (*b*) Photograph showing how the horizontal force on the movable barrier can be measured as a function of the distance from the float; a sharp break occurs when the molecules are pushed together in the stacking position. (Courtesy of C. Smart, © Unilever Research Laboratory, Port Sunlight, Birkenhead, Cheshire, England.)

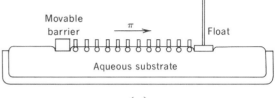

(*a*)

(*b*)

Fig. 23-8   McBain-Bakr sorption balance. (After D. J. Shaw, "Introduction to Colloid and Surface Chemistry," Butterworth & Co. (Publishers), Ltd., London, 1966.)

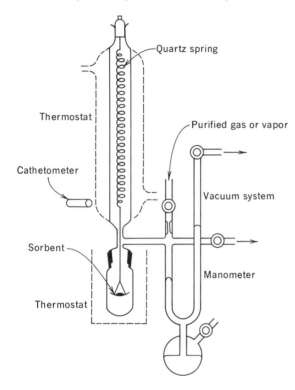

When the gas forms a liquid in a pore which wets the surface, the vapor pressure is reduced by the reduction in free energy due to the surface tension. The vapor pressure P is given by the Kelvin equation

$$RT \ln \frac{P}{P_0} = -\frac{2\gamma \mathrm{v} \cos \theta}{r} \tag{23-27}$$

where P_0 = vapor pressure of bulk liquid
θ = contact angle
r = radius of pore
v = molar volume of liquid

Irving Langmuir suggested that when chemisorption is involved, the final equilibrium depends on balancing the rate of adsorption and the rate of escape

$$P - \left(1 - \frac{v}{v_m}\right)e^{-E_f^{\ddagger}/RT} = k\frac{v}{v_m}e^{-E_b^{\ddagger}/RT} \tag{23-28}$$

where k is a proportionality constant, v is the equilibrium volume of the gas adsorbed per gram adsorbent at pressure P, v_m is the volume of gas required to cover 1 g of adsorbent with a monolayer, and $\Delta H = E_b^{\ddagger} - E_f^{\ddagger}$. Thus

$$P = ke^{\Delta H/RT}\frac{v/v_m}{1 - v/v_m} \tag{23-29}$$

Letting $a = (ke^{\Delta H/RT})^{-1}$, where a is independent of the amount of adsorption but varies with temperature,

$$aP = \frac{v/v_m}{1 - v/v_m} \tag{23-30}$$

$$v = \frac{v_m aP}{1 + aP} \tag{23-31}$$

$$\frac{P}{v} = \frac{P}{v_m} + \frac{1}{av_m} \tag{23-32}$$

At low pressures this reduces to

$$v = v_m aP \tag{23-33}$$

For a number of observed isotherms, this equation fits very well.

Brunauer, in collaboration with another American physical chemist P. H. Emmett and theoretical physicist E. J. Teller, proposed an extension of the Langmuir isotherm in a form called the BET equation. The purpose of this is to explain the nature of type II isotherms. The BET equation is generally written

$$\frac{P}{v(P_0 - P)} = \frac{1}{v_m c} + \frac{c-1}{v_m c}\frac{P}{P_0} \tag{23-34}$$

where P_0 = saturation vapor pressure
v_m = monolayer capacity
$c \sim e^{-(\Delta H_l - \Delta H_1)/RT}$
ΔH_l = enthalpy of evaporation of liquid
ΔH_1 = enthalpy of evaporation of monolayer

Plots of the BET isotherms are shown in Fig. 23-10. The equation fits the experimental data well at pressures between $0.05P_0$ and $0.35P_0$. Generally the monolayer saturation is reached within this range.

Electron-diffraction studies with low-voltage electron beams (LEED = low-energy electron diffraction) have revealed interesting information on the nature of the adsorbed layers. Electron-diffraction photographs taken by P. J. Estrup and J. Anderson of the

Bartol Research Foundation of the Franklin Institute in Philadelphia are shown in Figs. 23-11 and 23-12.

LEED and HEED (high-energy electron diffraction) studies of the interaction of oxygen with single crystal surfaces of copper have also been made and these suggest that oxygen is first adsorbed randomly in the troughs of the (110) surface at 3×10^{-8} torr (1 torr is a pressure of 1 mm Hg). At pressures of 2×10^{-5} torr for 1 min, the photographs indicate that an ordered structure has formed with a spacing of twice the substrate in the (110) direction and the same spacing in the (001) direction. Figure 23-13 shows electron micrographs of the oxide formed.

Since biological growth frequently takes place at surfaces through attachment much like chemisorption, information from studies like those described above may be helpful in evolving a more detailed understanding of the nature of growth.

Example 23-3

Calculate the vapor pressure of a spherical particle of water with a radius r of 10^{-6} cm at 30°C.

According to the equation relating P and r,

$\log P/P_0 = 2 \ M\gamma/2.303RT\rho r = 2 \times 18$ g mole$^{-1} \times 71.18$ dynes cm$^{-1}/2.303 \times 8.3143 \times 10^7$ ergs deg^{-1} mole$^{-1} \times 303° \times 1$ g cm$^{-3} \times 10^{-6}$ cm $= 0.0443$.

$$\frac{P}{P_0} = 1.108$$

$$P = 1.108 \times 31.8 \ \text{mm} = 35.2 \ \text{mm} \qquad Ans.$$

Example 23-4

Calculate the adsorption of 1-aminobutyric acid at the surface of an aqueous solution where $m = 0.1$ at 25°C if $d\gamma/dm = 0.4$ dyne cm^{-1} mole^{-1} cm^3 and $d\gamma/(d \ln m) = (m \ d\gamma)/dm = 4 \times 10^{-2}$ dyne cm^{-1}.

Using Eq. (23-26),

$$\Gamma_2 = -\frac{1}{RT} \frac{d\gamma}{d \ln m}$$

$$= \frac{4 \times 10^{-2} \ \text{dyne cm}^{-1}}{8.31 \times 10^7 \ \text{dyne cm deg}^{-1} \ \text{mole}^{-1} \ 293°}$$

$$\Gamma_2 = \frac{4 \times 10^{-2}}{2.43 \times 10^{10}} = 1.6 \times 10^{-12} \ \text{mole cm}^{-2}$$

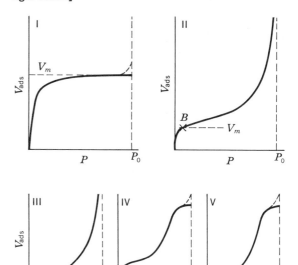

Fig. 23-9   Types of adsorption isotherms. [After S. Brunauer, L. S. Deming, W. E. Deming, and E. J. Teller, *J. Am. Chem. Soc.*, 62:1723 (1940). Copyright 1940 by the American Chemical Society.   Reprinted by permission of the copyright owner.]

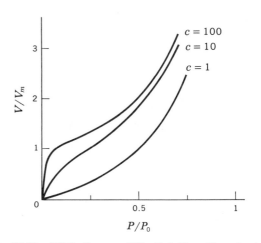

Fig. 23-10   BET isotherms. (After D. J. Shaw, "Introduction to Colloid and Surface Chemistry," Butterworth & Co. (Publishers), Ltd., London, 1966.)

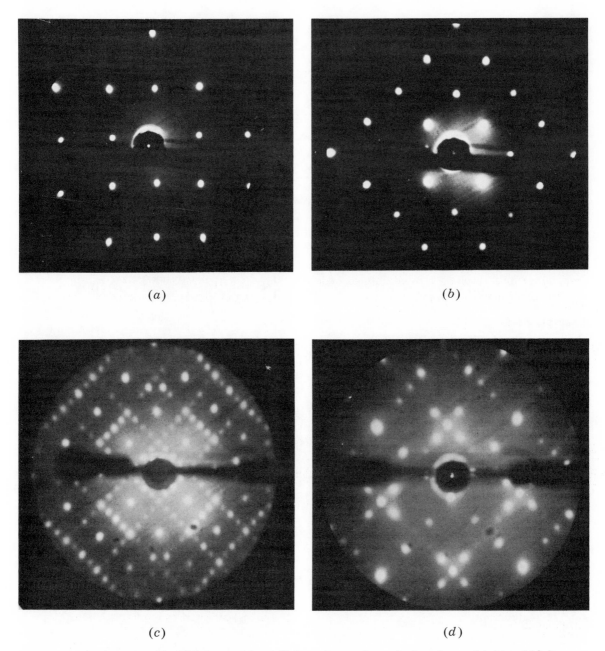

(a)

(b)

(c)

(d)

Fig. 23-11   Patterns produced by LEED beams: (a) and (b) from clean surfaces of a tungsten crystal; (c) and (d) from surfaces on which oxygen is chemisorbed. (From P. J. Estrup and J. Anderson, "Characterization of Chemisorption by LEED," in "The Structure of Surfaces," H. M. Davis and K. R. Lawless, eds., North-Holland Publishing Company, Amsterdam, 1967.)

23-3   COLLOIDS

When particles of matter have a diameter lying roughly in the range between 100 and 10,000 Å, they exhibit special kinds of behavior which depend strongly on the nature of the particle surface, and their properties are so characteristic that they can be regarded as a special state of matter called the *colloidal state*, after the French word for glue.   A colloidal particle is thus quite a few times bigger than an ordinary molecule, with a diameter between 1 and 100 Å, and yet is small enough so that the ratio of surface to volume is millions of times greater than for a grain of sand, with a diameter of a few tenths of a millimeter.   This is why the behavior of a colloidal particle is generally determined far more by its surface properties than by its bulk properties.

One of the most characteristic properties of colloidal particles is the fact that they remain suspended in a medium and do not settle out.   Droplets of water of colloidal size remain suspended in air, producing the atomospheric condition called *fog*.   Colloidal droplets consisting largely of fat and protein remain suspended in water in the liquid called *milk*, produced by mammals.   Colloidal particles of gold suspended in supercooled silica are found in the material called *ruby glass*.

As pointed out at the beginning of the chapter on solutions, the solute in a true solution consists of particles of true molecular size.   For this reason, a mixture of colloidal particles suspended in a medium is not called a solution but a *suspension* or *dispersion*. The particles themselves are called the *disperse* phase or *discontinuous* phase, since in order to pass through any large amount of particle matter it is necessary to cross the surface boundaries and proceed across a large number of particles.   On the other hand, by moving between the particles, one can pass through an indefinite distance in the medium in which the particles are suspended.   This is therefore called the *continuous phase*.   Examples of colloidal systems are listed in Table 23-2.

Sols, gels, and emulsions

When the continuous phase of a colloid is essentially a fluid, the colloid is called a *sol*.   If the fluid is water,

Fig. 23-12   (*a*) LEED pattern produced by room temperature adsorption of oxygen on a c(2 × 2)-Th surface; beam voltage 119 volts; (*b*) LEED pattern resulting from chlorine adsorption on W(100); beam voltage 119 volts.   (Electron diffraction photographs by P. J. Estrup and C. J. Anderson, "Characterization of Chemisorption by LEED," in "The Structure of Surfaces," H. M. Davis and K. R. Lawless, eds., North-Holland Publishing Company, Amsterdam, 1967.)

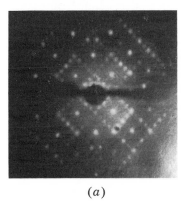

(*a*)

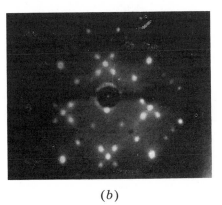

(*b*)

the colloid may be called a *hydrosol*.   If the continuous phase is air, the colloid may be called an *aerosol*. Fog is an example of an aerosol.   There are many examples of colloids where the continuous phase is so viscous or rigid that it may be more properly regarded as a jelly or a solid rather than as a free-flowing liquid.   Under these conditions, the colloid is called a *gel*.

There are many examples of pairs of liquids which are so mutually insoluble that they normally form a two-phase system when mixed together.   If these

Fig. 23-13 (*a*) Optical micrograph of oxide nuclei on the $C_u(110)$ surface, 135X; (*b*) electron micrograph of the surface of the oxide nucleus shown in (*a*), 4,700X. (From Gary W. Simmons, Don F. Mitchell, and Kenneth R. Lawless in "The Structure of Surfaces," H. M. Davis and K. R. Lawless, eds., North-Holland Publishing Company, Amsterdam, 1967.)

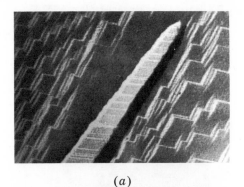

(*a*)

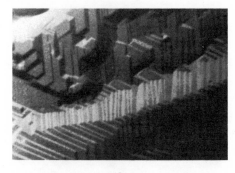

(*b*)

liquids are whipped up with sufficient agitation, a colloidal system called an *emulsion* may form. Sometimes the discontinuous phase may be made up of component *A* with just small amounts of *B* dissolved in it while the continuous phase is made up of *B* with small amounts of *A* dissolved in it. Again, the situation may be completely reversed with the continuous phase consisting primarily of component *A* and the discontinuous phase of component *B*. There are many examples of such pairs of components where a slight variation of conditions will cause an *inversion* of phase (Fig. 23-14).

There are also pairs of liquids which can be whipped up with sufficient agitation to form a colloidal suspension *temporarily,* but after standing a sufficient length of time, the discontinuous particles come in contact with each other and merge to form larger particles which eventually merge to revert to a two-phase system (Figs. 23-15 and 23-16). In many cases, the instability of such a system can be changed to stability by adding some kind of protective coating called a *stabilizer.* If a substance is added which tends to collect at the interface between the discontinuous phase and the continuous phase to form a kind of coating on the surface of the colloidal particle, this makes it more difficult for the surface of one particle to touch the surface of another and thus hinders coalescence; thus, the discontinuous particle may remain stable for a period of time many orders of magnitude longer than if not protected; this is illus-

---

TABLE 23-2   Classification of colloids

Continuous phase	Discontinuous phase	Class	Examples
Gas	Liquid	Aerosol	Fog
	Solid	Aerosol	Smoke
Liquid	Gas	Foam	Beer head
	Liquid	Emulsion	Milk
	Solid	Sol	Gold sol
Solid	Gas	Gas-solid suspension	Pumice
	Liquid	Gel	Silica gel
	Solid	Solid-solid suspension	Ruby glass

trated in Fig. 23-17 together with a photograph of a stabilized emulsion.

There is a diagnostic test for meningitis and syphilis based on the protective action of spinal fluid. If the spinal fluid is added to portions of a specially pre-pared gold hydrosol and a second portion of dilute salt solution is also added, it is found that the range of concentration through which the spinal fluid protects the gold sol characterizes the type of infec-tion in the patient from whom the spinal fluid was taken.

Micelles

When ions are adsorbed at the interface between the discontinuous phase and the continuous phase, a protective action frequently takes place that may be ascribed to the charges on the ions. If there is a residual positive charge on the face of a colloidal particle, this particle will repel any similarly charged particles. Under these conditions a great deal more energy is required to push the two particles together into a position where they can merge and form a larger particle. There is evidence that in colloidal solutions of gold, which can be formed only when ions are present, the chloride ions group around the gold atoms and confer a negative charge which repels other gold colloidal particles similarly charged. Such a charged particle is sometimes called a *colloidal micelle;* the structure of a micelle is shown in Fig. 23-18.

When the protective action is the result of electric charges, one frequently finds that the addition of another electrolyte can neutralize the charge and precipitate the colloid.

All these phenomena play an important part in bio-chemistry. Almost all the body fluids contain aggre-gates of molecules where the diameters are such that the aggregates can be regarded as colloidal particles. The relation of the size of these particles to the porosity of various membranes in the body plays a significant role in promoting or limiting the transfer of molecules between different parts of the body.

Example 23-5

Calculate the number of molecules in an oil drop of dodecane, $C_{12}H_{26}$, which is spherical in

Fig. 23-14   Phase inversion. (*i*) The two types of emulsions: (*a*) oil in water and (*b*) water in oil; (*ii*) the steps in chang-ing from type (*a*) to type (*b*). Portions of the material are labeled to identify their positions before and after the inver-sion. (After A. W. Adamson, "Physical Chemistry of Sur-faces," John Wiley & Sons, Inc., New York, 1967.)

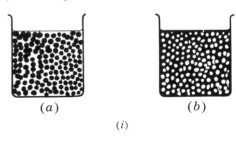

(*a*)   (*b*)

(*i*)

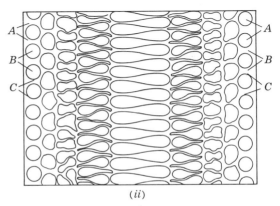

(*ii*)

shape and has a density of 0.75 and a radius of 1.78×10^{-5} cm.

The molecular weight is 170.3. The vol-ume is

$$V = \tfrac{4}{3}\pi r^3 = 2.36 \times 10^{-14} \text{ cm}^3$$

The weight is

$$m = 2.36 \times 10^{-14} \text{ cm}^3 \times 0.75 \text{ cm}^{-3}$$
$$= 1.77 \times 10^{-14} \text{ g}$$

The number of moles is

$$\frac{1.77 \times 10^{-14}}{170.3} = 1.04 \times 10^{-16} \text{ mole}$$

The number of molecules is

1.04×10^{-16} mole $\times 6.02$
$\times 10^{23}$ molecules mole^{-1}
$= 6.26 \times 10^7$ molecules

Fig. 23-15 Ways in which an unstable dispersion can change to nonstable states. (After J. A. Kitchener and P. R. Mussellwhite, "Theory of Stability of Emulsions" in "Emulsion Science," P. Sherman, ed., Academic Press, Inc., New York, 1968.)

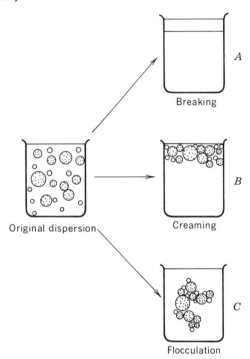

Breaking

A

Original dispersion

Creaming

B

Flocculation

C

Example 23-6

How many molecules of cetyl alcohol will be adsorbed on the surface of the drop in Example 23-5 if the cross section of one molecule is 2.58×10^{-15} cm²?

The area is $4\pi r^2 = 4\pi(1.78 \times 10^{-5})^2 = 3.98 \times 10^{-9}$ cm². The number of adsorbed molecules is 3.98×10^{-9} cm²/2.58×10^{-15} cm² $= 1.54 \times 10^6$ molecules.   *Ans.*

23-4 DIFFUSION

In the basic concept of equilibrium, molecules are thought of as being exchanged across a boundary between the two states under consideration; in a given interval of time, a certain number of molecules pass from left to right across the boundary and an equal number pass from right to left. The general term for molecular movement of this sort is *diffusion;* and this type of action plays a most significant role in biochemical processes taking place in vivo and in many of the techniques used to identify and study biochemical compounds in vitro.

Membranes

One of the most important examples of diffusion is found in the passage of molecules through biological membranes, especially those found both as the outer coverings and in the interior of biological cells. The isolation of bacterial membranes has provided an easy way of obtaining material for study since bacteria lack mitochondrial, microsomal, and nuclear membranes, though sometimes they do have an interior membrane system called the *mesosome* or *chondroid.*

In Fig. 23-19 a human red blood cell membrane is shown in profile, and Fig. 23-20 is a schematic drawing of the cross section. The membrane molecules, called lipids, consist of hydrocarbon chains, which are hydrophobic, terminating in ionizable groups, which are hydrophilic. The structural formulas of some

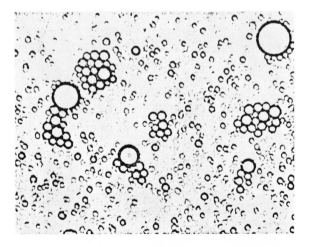

Fig. 23-16 A hypercontrast reproduction of a photograph of flocculation of paraffin oil in water, stabilized by a nonionic surfactant. (After J. A. Kitchener and P. R. Mussellwhite, "Theory of Stability of Emulsions" in "Emulsion Science," P. Sherman, ed., Academic Press, Inc., New York, 1968.)

Fig. 23-17 Stabilization: (a) an oil droplet stabilized by the adsorption of oriented surface molecules; (b) an emulsion particle stabilized by the adsorption of smaller particles; (c) a hypercontrast reproduction of a photograph of oil droplets stabilized by a coating of mineral particles. (a) and (b) after J. A. Kitchener and P. R. Mussellwhite, "Theory of Stability of Emulsions" in "Emulsion Science," P. Sherman, ed., Academic Press, Inc., New York, 1968. [(c) After A. W. Adamson, "Physical Chemistry of Surfaces," 2d ed., John Wiley & Sons, Inc., New York, 1967.]

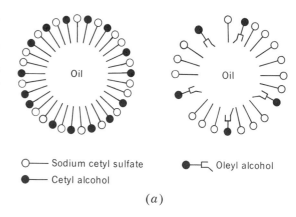

○— Sodium cetyl sulfate   ●⌐ Oleyl alcohol
●— Cetyl alcohol

(a)

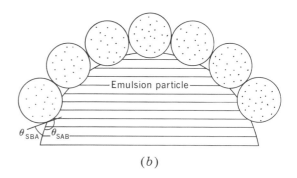

(b)

lipids are shown in Fig. 23-21. The stability of such a structure can be regarded both from the thermodynamic and the kinetic standpoint. The membrane is stable thermodynamically because the enthalpy part of the free energy has a much lower value when the hydrocarbon chains are stacked side by side than when each chain individually is dissolved in the primarily aqueous cell plasma. To dissolve in the cell fluid, the hydrocarbon must break a large number of the hydrogen bonds between the water molecules in the plasma, and this raises the enthalpy to a relatively high value. Kinetically speaking, the energy of activation is high for the process by which a lipid is removed from the membrane and dissolved in the plasma. Hence, the stability of the membrane. On the other hand, the *surface* of the membrane is hydrophilic and can be in contact with the plasma without raising the free energy appreciably.

It has been postulated that there is a tendency of lipids to form micelles (Fig. 23-22) of various shapes that aid in lipid transport through the plasma in growth processes of membranes such as those found on nerves. In order to synthesize a membrane, lipids must first be synthesized themselves and then transported and placed in their proper sites by processes which are made possible by appropriate values of the entropy of activation.

Gibbs-Donnan equilibrium

For a membrane that is the outer covering of a cell there may be quite a difference in composition both qualitatively and quantitatively between the fluid on one side and on the other. As a rule, the permeability of the membrane to the different components of the fluids differs also.

(c)

Fig. 23-18 Examples of micelles. (After A. W. Adamson, "Physical Chemistry of Surfaces," 2d ed., John Wiley & Sons, Inc., New York, 1967.)

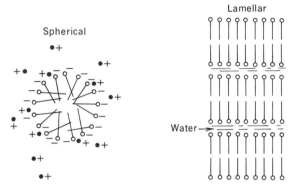

Spherical

Lamellar

Water →

There is probably a variety of mechanisms by which molecules or ions can pass through membranes. Small ions may slip through the interstices between the hydrocarbon chains, although one would expect a fairly high enthalpy of activation for such a process. A membrane is really a two-dimensional crystal. Like all other crystals, a membrane may have areas of dislocation which might permit easier passage of ions. There may even be features of structure, like micro holes which can be traversed by the smaller ions. At any rate, experimental evidence shows that some ions can pass through readily, while to others the membrane is almost impermeable.

This variation in permeability results in a displacement of the concentration values at equilibrium. Suppose that before equilibrium is established, there are ions distributed on the two sides of the membrane as shown in Fig. 23-23. The membrane is permeable

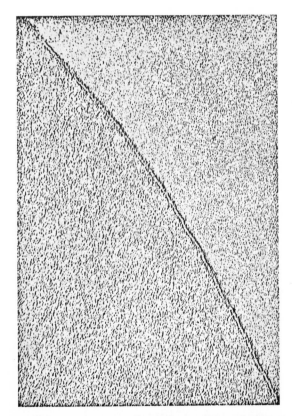

Fig. 23-19   The cross section of the membrane of a human red blood cell, shown in a hypercontrast reproduction from a microphotograph. (From W. D. Stein, "The Movement of Molecules across Cell Membranes," Academic Press, Inc., New York, 1967.)

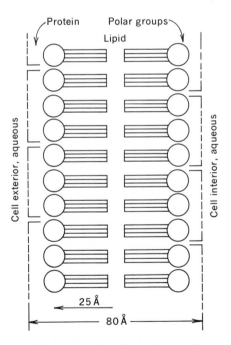

Protein   Polar groups

Lipid

Cell exterior, aqueous

Cell interior, aqueous

25 Å

80 Å

Fig. 23-20   The cross section of a membrane. (From a drawing in "The Movement of Molecules across Cell Membranes" by W. D. Stein, Academic Press, Inc., New York, 1967, based on the concept by J. F. Danielli in "The Permeability of Natural Membranes," by H. Dawson and J. F. Danielli, Cambridge University Press, London, 1943.)

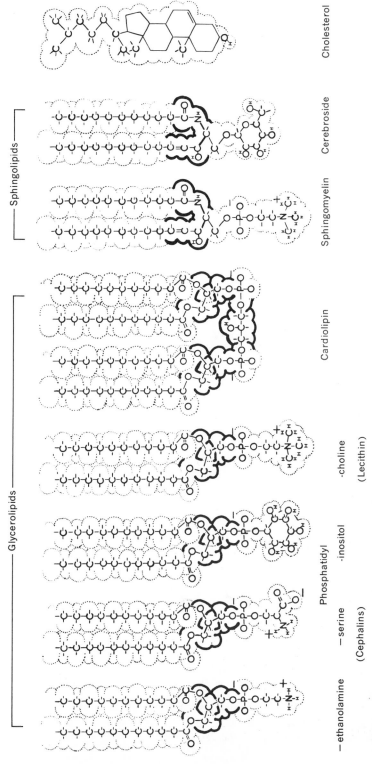

Fig. 23-21 The structure of lipid molecules. (After J. B. Finean, "Chemical Ultrastructure in Living Tissue," Charles C Thomas, Publisher, Springfield, Ill., 1961.)

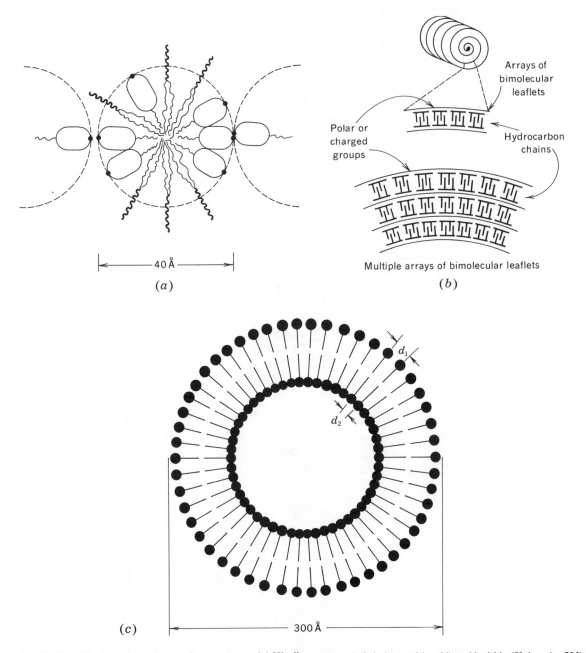

Fig. 23-22 Micelle and membrane ultrastructure. (*a*) Micelle composed of cholesterol (ovoid) and lecithin (*Y* shape). [After J. A. Lucy and A. M. Glauert, *J. Mol. Biol.*, 8:727–748 (1964). Academic Press, Inc., New York, 1964.] (*b*) Micelle spiral ultrastructure. (After D. E. Green and S. Fleischer, "Role of Lipids in Mitochondria Function" in "Metabolism and Physiological Significance of Lipids," R. M. C. Dawson and D. N. Rhodes, eds., John Wiley & Sons, Inc., New York, 1964.) (*c*) Bimolecular structure of lipids bounding a spherical vesicle. (After J. D. Robertson, "A Review of Synaptic Membranes" in "Cellular Membranes in Development," M. Locke, ed., Academic Press, Inc., New York, 1964.)

Fig. 23-23   The change in concentration occurring when a Gibbs-Donnan equilibrium is established between rows on two sides of a semipermeable membrane.

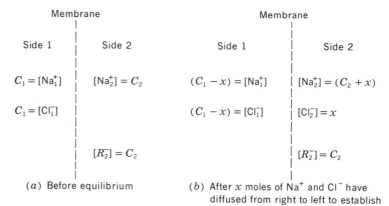

Membrane

Side 1 | Side 2

$C_1 = [Na_1^+]$ | $[Na_2^+] = C_2$

$C_1 = [Cl_1^-]$ |

| $[R_2^-] = C_2$

(a) Before equilibrium

Membrane

Side 1 | Side 2

$(C_1 - x) = [Na_1^+]$ | $[Na_2^+] = (C_2 + x)$

$(C_1 - x) = [Cl_1^-]$ | $[Cl_2^-] = x$

| $[R_2^-] = C_2$

(b) After x moles of Na^+ and Cl^- have diffused from right to left to establish equilibrium

to Na^+ and Cl^- but not to R^-. At the beginning Na^+ and Cl^- are on the left and Na^+ and R^- on the right. In order to establish equilibrium, an amount x of Cl^- must diffuse from left to right since the product of concentrations of the diffusible ions on either side must be equal at equilibrium

$$[Na_1^+][Cl_1^-] = [Na_2^+][Cl_2^-] \tag{23-35}$$

In order to maintain electrical neutrality on both sides of the membrane, an equal amount of Na^+ must diffuse from left to right also. Thus, calling the original concentrations of ions on the left c_1 and on the right c_2, the product of the concentrations at equilibrium is

$$(c_1 - x)(c_1 - x) = (c_2 + x)x \tag{23-36}$$

Solving for x gives

$$x = \frac{c_1^2}{c_2 + 2c_1} \tag{23-37}$$

Rearranging Eq. (23-36), the ion ratios may be written in the following form and set equal to r:

$$\frac{[Na_1^+]}{[Na_2^+]} = \frac{c_1 - x}{c_2 + x} = \frac{x}{c_2 - x} = \frac{[Cl_2^-]}{[Cl_1^-]} = r \tag{23-38}$$

Combining Eqs. (23-37) and (23-38) gives

$$r = \frac{c_1}{c_1 + c_2} \tag{23-39}$$

For a single ion species equilibrium is attained when the free energy per mole (chemical potential $= \mu$) for that species on one side of the membrane is the same as that on the other

$$\mu_1^+ = \mu_2^+ \tag{23-40}$$

Equation (23-40) shows that the concentrations, and therefore the activities, are different on the two sides of the membrane. Activity and chemical potential are related by the expression

$$\mu^+ = \mu_0 + T(R \ln a^+) \tag{23-41}$$

Comparing this with the more general equation for free energy,

$$G = H - Ts \tag{23-42}$$

we see that μ_0 is, in effect, the difference between H and Ts at unity activity, and $TR \ln a^+$ is really $T \Delta s$, where Δs is the change in entropy due to the change in a from unit value. In other words, a change in concentration changes the value of the entropy in a solution, just as the change in concentration of a gas changes its entropy.

The changes in electrochemical potential for the two ions are given by

$$\mathcal{E}^+ = \frac{1}{n^+ \mathcal{F}}(\mu_1^+ - \mu_2^+) = \frac{RT}{n^+ \mathcal{F}} \ln \frac{a_1^+}{a_2^+} \tag{23-43}$$

Fig. 23-24 Separation of free amino acids in the leaves of *Fragaria* **by combined electrophoresis and chromatography on cellulose MN-300 layers. Electrophoresis in 0.7 percent formic acid buffer, 400 volts, 15 min; chromatography, with butanol–acetic acid–water (4:1:2). aspa, aspartic acid; glua, glutamic acid; asp, asparagine; glua, glutamine; am, amino-butyric acid. [After N. Nybom,** *Physiol. Plantarum,* **17:434 (1964), © 1964 by Munksgaard Publishers, Ltd., Copenhagen, Denmark. Used by permission.]**

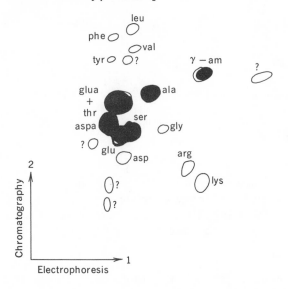

$$\mathcal{E}^- = \frac{1}{n^-\mathfrak{F}}(\mu_1^- - \mu_2^-) = \frac{RT}{n^-\mathfrak{F}}\ln\frac{a_1^-}{a_2^-} \qquad (23\text{-}44)$$

An example of a Gibbs-Donnan equilibrium is found when a protein hydrochloride is separated from hydrochloric acid by a membrane. If both are initially at the same pH, hydrogen ions pass through the membrane. Another example is found on an erythrocyte which contains more protein than plasma does. If its membrane were permeable to all small ions, it would swell and burst; actually, it is almost impermeable to Na^+ and K^+, so that a Gibbs-Donnan equilibrium is established which results in osmotic equilibrium between red cells and plasma.

Chromatography

In 1902, the Russian botanist M. S. Tswett discovered that a solution of chlorophyll extracted from leaves and poured over a column packed with fine particles

of an adsorbent separates into colored bands which move at different rates down the column. This is the result of selective adsorption and re-solution. The rates for this process differ sufficiently for compounds, even closely resembling each other chemically, to make possible separation and identification of the different types of molecules. This method is called *chromatography*. The rate of migration down the column is related to the enthalpy and entropy of activation for adsorption on the surface and deadsorption from the surface. The full report on Tswett's work was made in 1910.

Between 1898 and 1903, the American chemist and geologist D. T. Day carried out a series of similar experiments with petroleum fractions and reported the results in 1900 and 1903. Methods based on these experiments and those of Tswett were refined, especially through the work of A. Tiselius, Swedish chemist and Nobel Laureate. Between 1941 and 1944, the English chemists A. J. P. Martin and R. L. M. Synge developed methods using two liquid phases and paper. Today chromatography has been extended to include thin-layer, ion-exchange, and gas chromatography and electrophoresis.

In moving down a column and passing alternately from the solution to the adsorbed phase and back, a molecule spends only part of its time t_a moving with the solution stream. The rest of the time t_d it is adsorbed on the stationary surface. The zone, characteristic of this molecule, moves at speed Rv_s, where v_s is the velocity of the solvent progress and $R = t_a/(t_a + t_d)$.

The distribution coefficient K giving the ratio between material in the solvent and on the surface is, in effect, an equilibrium constant. Its variation with temperature is given by the usual equation

$$K = e^{\Delta S_a/R}\,e^{-\Delta H_a/RT} \qquad (23\text{-}45)$$

where ΔS_a and ΔH_a are the entropy change and the enthalpy change on adsorption. ΔH_a is normally negative and ranges from -10 to about 1 kcal mole^{-1}. In extreme cases, K is halved by an increase in temperature of only 20°C. A two-dimensional chromatogram by a combination of migration by diffusion and transport in an electrostatic field (electrophoresis) is shown in Fig. 23-24.

In gas chromatography, the stationary phase may be a solid adsorbent (gas-solid chromatography) or a liquid (gas-liquid chromatography). The gas carries the molecules through the column, the molecules being alternately adsorbed and deadsorbed. The gas is analyzed as it leaves the column by methods which identify the different solutes after they pass in sequential blocks out of the column, having been separated by the different rates of adsorption and deadsorption. A typical analysis diagram is shown in Fig. 23-25.

Chromatography constitutes one of the most important applications of surface chemistry.

Example 23-7

A Gibbs-Donnan equilibrium is established across a membrane which is permeable to Na^+ and Cl^- but not to $CH_3CH_2COO^-$. At the beginning the concentration on the left for Na^+ and Cl^- is 0.00100 M and on the right for Na^+ and $CH_3CH_2COO^-$ is 0.00040 M. Calculate the concentrations of Na^+ and Cl^- at equilibrium.

The amount of Na^+ and Cl^- transported across the membrane to establish equilibrium will be

$$x = \frac{c_1^2}{c_2 + 2c_1} = \frac{10^{-6}}{4 \times 10^{-4} + 2 \times 10^{-3}}$$

$$= \frac{10^{-6}}{2.4 \times 10^{-3}} = 0.000417 \ M$$

$[Na_2^+] = c_2 + x = 0.000817 \ M$
$[Cl_2^-] = x = 0.000417 \ M$
$[Cl_1^-] = [Na_1^+] = 0.000583 \ M$ *Ans.*

Example 23-8

Calculate the electrochemical potential across the membrane due to Na^+ in Example 23-7 at 310°C.

$$\mathcal{E}^+ = \frac{2.303RT}{n^+\mathcal{F}} \frac{\log 0.000583}{\log 0.000817}$$

$$= -\frac{4.5757 \times 310}{1 \times 2.306 \times 10^4} \log 1.401$$

$$= \frac{207.5}{2.306 \times 10^4} = 0.009 \ \text{volt}$$ *Ans.*

Problems

1. A small spherical droplet of water rests on the surface of a block of paraffin in air at 20°C. By pushing down with a paraffin-coated spatula, the droplet is deformed and flattened so that the surface area is increased by 50 percent. Assuming that the surface tension between water and paraffin is the same as between water and air, calculate the increase

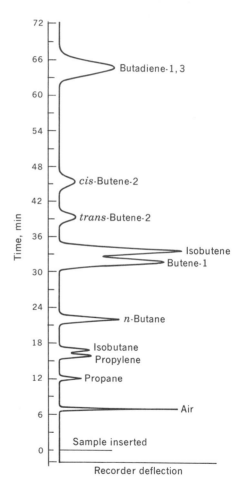

Fig. 23-25 Typical chromatogram of a cracked C_4 gas mixture. Sample size, 0.5 ml at 158 mm Hg; column temperature, 20°C; stationary phase, dimethyl sulfolane; carrier gas, hydrogen 30 ml min^{-1}. (After D. H. Andrews and R. J. Kokes, "Fundamental Chemistry," 2d ed., John Wiley & Sons, Inc., New York, 1965.)

in the free energy of the droplet if the original diameter of the droplet in spherical form was 1 mm.

2. An air bubble is trapped in a cylinder filled with benzene at 20°C. The original diameter of the air bubble is 0.5 mm, and it is spherical. External pressure is brought to bear on the benzene so that the diameter of the bubble decreases by 10 percent. Calculate the change in the free surface energy of the bubble.

3. A scuba diver goes down to a depth of 50 ft in a freshwater lake. He expels a bubble of air at this depth which is 3 mm in diameter and spherical. The bubble rises to the surface of the lake retaining its shape. When it has risen to the point where the diameter has tripled, what is the increase in the surface free energy of the bubble? The temperature of the lake is 20°C.

4. In the condensation of moisture in the air at high altitudes, a spherical droplet of liquid water is formed with a diameter 10^{-6} cm at an altitude where the temperature is 10°C. Calculate the vapor pressure of the water assuming that it is spherical.

5. A simulated biological cell is formed by using a membrane which will pass potassium and chlorine ions but not molecules of butyric acid. Inside the cell there is a concentration of 0.015 M of the potassium salt of the acid fully ionized and outside the cell there is a concentration of an equal value of KCl. When equilibrium is established, what is the ratio of the concentration of Cl^- inside the cell to that outside?

6. When equilibrium is established across a Donnan membrane, it is found that the ratio of the external to the internal concentration of potassium ion is 1.3. If the initial concentration of the external ion was 0.02 M, calculate both the internal and external concentrations after equilibrium is established.

7. A Donnan equilibrium is established across a membrane of a biological cell where initially inside the cell there is a concentration of 0.01 M of $NaOOCCH_2NH_2$ (fully ionized) and outside the cell a concentration of 0.02 NaCl. Calculate the value of $(Cl^-)_{ext}/(Cl^-)_{int}$ at equilibrium.

8. Calculate the vapor pressure of a drop of water 1.75×10^{-6} cm in diameter at 30°C.

9. How small must the radius of a drop of water be at 0°C to increase the vapor pressure by 10 percent?

10. A long-chain hydrocarbon with an acid group at the end is selectively adsorbed at the surface of water. If $d\gamma/(d \ln m) = 1.32$ dynes cm^{-1} at 25°C, calculate the surface excess concentration of this compound.

11. A long-chain organic alcohol is found to have a surface excess concentration of 2.53×10^{-12} mole cm^{-2} at 32°C. Find the value of $d\gamma/dm$ for $m = 0.100$ mole liter^{-1}.

12. The volume of gas necessary to saturate the surface of an adsorber is 1.33 cm^3. Using Eq. (23-30), plot the value of P as a function of v as v varies from zero to v_m when $a = 0.4$ cm^{-1} Hg.

13. Using Eq. (23-33), calculate the value of a for the type I isotherm in Fig. 23-9, assuming $P_0 = 1.52$ atm.

14. Calculate the height to which a column of water will rise against air in a capillary of diameter 0.0065 cm at a temperature of 10°C.

15. If liquid hexane is substituted for the air in Prob. 14, what will be the height of the column at 20°C? Assume that for hexane $\rho = 0.66$ and $\theta = 0$.

16. If water is observed to rise by 0.157 cm in a capillary at 30°C, calculate the radius of the capillary.

17. Calculate the number of molecules of CS_2 in a drop with a radius of 1.589×10^{-4} cm at 20°C where the density is 1.26 g cm^{-3}.

18. How many molecules of cetyl alcohol will be absorbed in a monolayer on a drop of CS_2 with a radius of 1.589×10^{-4} cm if the cross-sectional area per molecule is 2.58×10^{-15} cm^2?

19. Calculate the percentage change in the value of the distribution coefficient for a chromatographic absorption where ΔH_a is -9562 cal mole^{-1} and temperature varies from 15 to 30°C.

20. If the value of the distribution constant K decreases by 75.5 percent in the range 20 to 40°C, calculate the value of the change in enthalpy on adsorption.

REFERENCES

Vold, M. J., and R. D. Vold: "Colloid Chemistry," Reinhold Publishing Corporation, New York, 1964.

Bray, H. G., and K. White: "Kinetics and Thermodynamics in Biochemistry," Academic Press, Inc., New York, 1966.

Shaw, D. J.: "Introduction to Colloid and Surface Chemistry," Butterworth & Co. (Publishers), Ltd., London, 1966.

Stein, W. D.: "The Movement of Molecules across Cell Membranes," Academic Press, Inc., New York, 1967.

Gregg, S. J., and K. S. W. Sing: "Adsorption Surface Area and Porosity," Academic Press, Inc., New York, 1967.

Davis, H. M., and K. R. Lawless (eds.): "The Structure of Surfaces," North-Holland Publishing Company, Amsterdam, 1967.

Adamson, A. W.: "Physical Chemistry of Surfaces," 2d ed., John Wiley & Sons, Inc., New York, 1967.

Heftmann, E. (ed.): "Chromatography," 2d ed., Reinhold Publishing Corporation, New York, 1967.

Sherman, P. (ed.): "Emulsion Science," Academic Press, Inc., New York, 1967.

In ascending the scale of complexity all the way from a single electron, through the atom, through the molecule, through colloidal particles, through individual biological cells, to the organized collection of trillions of these cells which constitute the human body, we pass through a continuous chain of increasing sizes and increasing subtleties of order. At the level of simplest order, the orbitals of electrons merge and interact to produce an atom like carbon. When carbon atoms are bonded together and joined with nitrogen, oxygen, and hydrogen atoms to form a molecule like thymine (Fig. 24-1), again the principal

Chapter Twenty-Four

macromolecules

characteristic of the molecule lies in the pattern of interaction of the atoms. However, a group of atoms containing about the same mass as the thymine molecule might form an aggregate of the same molecular weight with nine carbon atoms bonded together in the structure found in a diamond crystal (Fig. 24-2), characterized by a simple repetitive formula.

Thus, there are two divergent paths to be followed in constructing atomic aggregates of greater and greater complexity. On the one hand, it is possible to build a network of covalent bonds like that found in the diamond crystal, extending this network farther and farther until the whole aggregate contains 10^{20} or so atoms. Or, on the other hand, also using covalent bonds and a few different kinds of atoms like carbon, hydrogen, nitrogen, oxygen, phosphorus, and sulfur, it is possible to construct networks where instead of the simple repetitive pattern found in a crystal lattice like diamond or sodium chloride, one has the increasingly subtle networks of interwoven patterns of order that characterize the various components of living organisms.

There are good reasons for regarding both a unit of hemoglobin (Fig. 24-3) and a fragment of the diamond lattice containing approximately the same number of atoms as true molecules. The carbon atoms in the fragment of diamond are

Fig. 24-1 The structural formula of thymine.

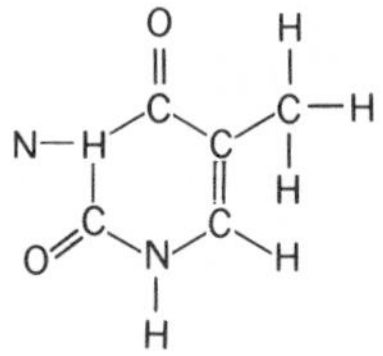

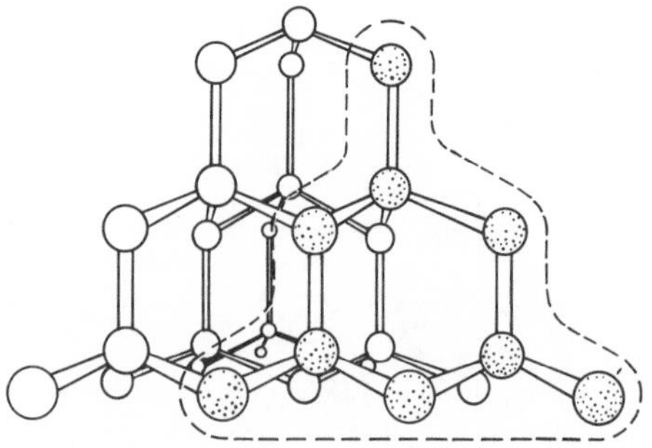

Fig. 24-2 A fragment of a diamond lattice consisting of nine carbon atoms (dotted circle), having approximately the same molecular weight as thymine.

joined together with covalent bonds in much the same way that carbon atoms are joined in hydrocarbons, and so one can argue that diamond-lattice fragments of various sizes are entitled to be called molecules. The unit of hemoglobin is likewise called a molecule since its atoms are linked together in such a stable way that they move around as a stable aggregate. Both the fragment of the diamond lattice and the hemoglobin molecule have characteristic structures. However, the chemical and physical behavior of the fragment of the diamond lattice is not affected very much by the removal of 10 carbon atoms out of the thousand which it might contain. The similar removal of a key group of 10 atoms from the hemoglobin molecule would almost certainly change both its physical and chemical behavior quite drastically.

Recognizing that the term *molecule* can be applied with considerable justification both to aggregates of atoms joined in relatively simple networks like the fragment of the diamond lattice and to the aggregate of atoms joined in the pattern of the hemoglobin unit, chemists are now using the term *macromolecule* to

designate both types of aggregates when, roughly speaking, the diameter of the aggregate lies somewhere between 100 and 10,000 Å. In the preceding chapter, we examined a number of the aspects of physical behavior of particles in this region of size. In this chapter we shall explore in greater detail the inner structure of such aggregates which, because of characteristic physical and chemical behavior, are especially entitled to be called macromolecules.

In the chapter dealing with crystals and crystallization, we surveyed aggregates where the nature of the inner structure resulted in the formation of solids with the regular pattern of faces usually called *crystals*. In this chapter we want first to take a look at the solids with an inner network type of structure characterized by a pattern of chemical bonds where the forces hold the structure together but do not have the same tendency to produce patterns of faces on the exterior. Especially during the last 30 or 40 years, chemical methods have been found for linking atoms together in networks of chains which produce solids of varying degrees of rigidity but which differ from familiar crystals particularly in their malleability and the ease with which they can be molded into different shapes. Such substances are now called *polymers*. Two interesting examples are *polyethylene* and *silicone* which are synthesized in great quantities industrially. There are also naturally occurring substances where the interior structure consists of networks of bonds. Examples of these are polyisoprene, the basic form of natural rubber, and the polysaccharides, of which the constituent of wood, *cellulose,* is an example.

The other major class of macromolecules is made up of substances with far more complex microstructure. Hemoglobin has been mentioned as an example of this type of macromolecule; it has a molecular weight of approximately 65,000. There is a specially important class of biological macromolecules called proteins which are chains or networks woven together principally from about 20 kinds of unit molecules, the amino acids.

Structural classification

While distinctions can be drawn in terms of the molecules serving as the unit building blocks, there is still considerable similarity in structure between the macromolecules called proteins and other somewhat larger molecules which have somewhat different kinds of building blocks. For example, there are the enzymes, which play a leading role in the processes of life by controlling the synthesis of proteins. *Catalase*, with a molecular weight of about 250,000, and *urease*, with a molecular weight of about 170,000, are members of this class. Finally, there are the molecules which represent the greatest complexity to be found in all of biochemistry. These are the molecules associated with the genetic process, of which the most complex is the structure called deoxyribose nucleic acid (DNA). It is sometimes difficult to specify precisely the significant molecular weight for such a molecule. As an example, the rough molecular weight of 1.2×10^8 has been assigned to DNA found in the T-2 virus.

Since our survey of physical chemistry in this book is focused primarily on applications in biochemistry, we shall consider only briefly some examples of covalent polymers and then devote the major portion of this chapter to an examination of the macromolecules of biochemical interest and how their synthesis and chemical reactions are governed by the principles of physical chemistry set forth in the earlier chapters of this book.

Example 24-1

The density of diamond is 3.51 g cm^{-3}. Calculate the length of the edge of a cube of diamond having a molecular weight of 65,000.

Since carbon has an atomic weight of 12 and diamond is pure carbon, a "molecule" of diamond with a molecular weight of 65,000 must contain $65,000/12 = 5,417$ atoms of carbon. Such a molecule weighs $65,000/(6.02 \times 10^{23}) = 1.08 \times 10^{-19}$ g. The volume of this molecule is $(1.08 \times 10^{-19})/3.51 = 3.08 \times 10^{-20}$ cm^3. If it is cubical in shape, the edge of the cube will be $(3.08 \times 10^{-20})^{-3} = 3.13 \times 10^{-7}$ cm.

Example 24-2

Assuming that the density of hemoglobin is 1.5 g cm^{-3} and that the molecular weight is 65,000, calculate the diameter of the molecule if it is spherical.

From the previous example, the weight of a single molecule of hemoglobin will be 1.08×10^{-19} g. The volume will be $(1.08 \times 10^{-19})/1.5 = (7.2 \times 10^{-20})$ cm^3. Since $v = \frac{4}{3}\pi r^3$, $r = \sqrt[3]{\frac{3}{4}v/\pi} = 1.3 \times 10^{-7}$ cm. The diameter is 2.6×10^{-7} cm.

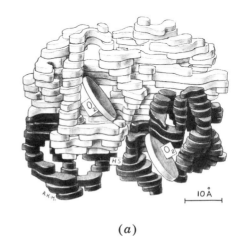

(a)

(b)

Fig. 24-3 Parts of a hemoglobin molecule (MW 65,000). (a) β-chain. [From M. F. Perutz, *Science*, **140**:863 (1963).] (b) Heme, the core of the hemoglobin molecule.

24-2   COVALENT POLYMERS

Polyethylene

One of the simplest covalent polymers consists of a large number of methylene, $-CH_2-$, groups strung together either in long straight chains or branched chains. The ends of the chains may consist of CH_3 groups or of CH_2 with a free electron, denoted by $\dot{C}H_2$. Chains of this sort grow by the addition of ethylene

$$CH_3-(CH_2)_n-\dot{C}H_2 + H_2C\!=\!CH_2 \rightarrow$$
$$CH_3(CH_2)_{n+2}-\dot{C}H_2 \quad (24\text{-}1)$$

Such a process is called *addition polymerization.* When ethylene is condensed to such a product at high temperatures and pressures, the chains tend to be branched. At lower temperatures and pressures in the presence of surface catalysts, essentially linear unbranched products can be obtained.

The length of the chain is determined by the probability of a reaction such as

$$CH_3-(CH_2)_n-\dot{C}H_2 + \dot{C}H_2-CH_3 \rightarrow$$
$$CH_3(CH_2)_{n+2}-CH_3 \quad (24\text{-}2)$$

The rate-constant expression for Eq. (24-1) is

$$R \ln k_1 = \frac{-E_1^{\ddagger}}{T} + S_1^{\ddagger} \quad (24\text{-}3)$$

The rate-constant expression for Eq. (24-2) is

$$R \ln k_2 = \frac{-E_2^{\ddagger}}{T} + S_2^{\ddagger} \quad (24\text{-}4)$$

Suppose that values of $E_1^{\ddagger}$ and $E_2^{\ddagger}$ are almost the same. If conditions are such that $S_1^{\ddagger}$ is very much greater than $S_2^{\ddagger}$, chains of considerable length will form before being terminated by Eq. (24-2). Thus, the catalyst affects $S^{\ddagger}$ rather than $E^{\ddagger}$ as shown in Fig. 24-4.

As might be expected, linear polyethylene is much more crystalline than branched polyethylene, containing far more regions with small spherulitic crystals, of the type shown in Fig. 24-5. The ease of rotation about single bonds in the noncrystalline regions gives polyethylene a considerable degree of flexibility.

Silicone

Another industrially important linear polymer is formed by silicon atoms which have two methyl groups attached to them and are linked together by oxygen atoms so that the compound has the formula $[(CH_3)_2SiO]_n$. The structure of silicone is shown in Fig. 24-6.

Rubber

Another linear polymer is found in natural rubber and is called *polyisoprene.* It consists of chains with the formula $[-CH\!=\!C(CH_3)-CH_2-CH_2-]_n$. When polyisoprene is treated with sulfur monochloride, S_2Cl_2, cross-links are formed

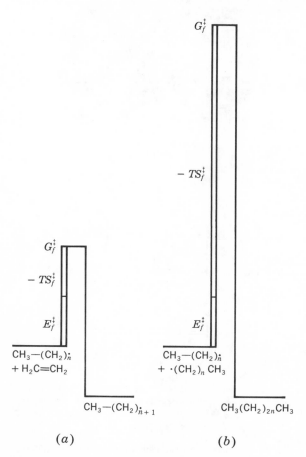

(a)   (b)

Fig. 24-4   A comparison of the free-energy barriers for Eqs. (24-1) and (24-2).

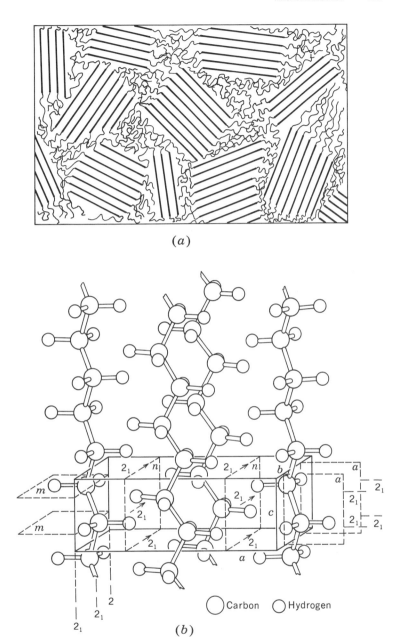

Fig. 24-5 (a) The morphology of semicrystalline polyethylene. (Reprinted from Paul J. Flory, "Principles of Polymer Chemistry." Copyright 1953 by Cornell University Press, Ithaca, N.Y. Used by permission of Cornell University Press.) (b) The structure of the crystalline part of polyethylene. (After S. C. Nyburg, "X-ray Analysis of Organic Structures," Academic Press, Inc., New York, 1961.)

$$\underset{\text{polyisoprene}}{\cdots-CH=\overset{\overset{\textstyle CH_3}{|}}{C}-CH_2-CH_2-CH} \rightarrow$$

$$-CH-\overset{\overset{\textstyle CH_3}{|}}{C}-CH_2-CH_2-\cdots + S_2Cl_2 \rightarrow$$

$$\underset{\text{commercial rubber}}{\begin{array}{c}\cdots-CH-\overset{\overset{\textstyle CH_3}{|}}{\underset{\underset{\underset{}{\underset{}{}}}{|}}{C}}-CH_2-CH_2-\cdots \\ \underset{|}{S}\quad \underset{|}{Cl} \\ \underset{|}{S}\quad \underset{|}{Cl} \\ \cdots-CH-\underset{|}{C}-CH_2-CH_2-\cdots \\ CH_3 \end{array}} \quad (24\text{-}5)$$

By varying the conditions of the reaction, the number of cross-links can be controlled, and this in turn affects the resistance of the product to abrasion, important when it is used in automobile tires.

Cellulose

Another important natural polymer is *cellulose*, constituting an important part of the stalks of plants and the trunks of trees. This is formed by the condensation of glucoside rings, $C_6H_9O_4$, with the structure shown in Figs. 24-7 and 24-8.

Example 24-3

In the formation of linear polyethylene, the catalytic action of a special surface directly influences the rate of production of the polymer. Assuming that the probability of the reaction is proportional to the catalytic surface area, thereby having a logarithmic effect on $S^{\ddagger}$, calculate the increase in $S^{\ddagger}$ when the area of the surface is increased hundredfold, and the effect on k.

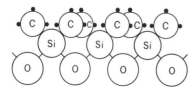

Fig. 24-6   A portion of the chain in silicone $[(CH_3)_2SiO]_n$.

Denoting the probability of the reaction before increasing the surface by Y_1 and after increasing it by Y_2,

$$\Delta S^{\ddagger} = R \ln \frac{Y_2}{Y_1} = 2.303R \log 100$$

$$= 4.5757 \times 2$$

$$= 9.15 \text{ cal deg}^{-1} \text{ mole}^{-1}$$

$$R \ln k = \frac{-E^{\ddagger}}{R} + S^{\ddagger}$$

$$R \ln \frac{k_2}{k_1} = S_2^{\ddagger} - S_1^{\ddagger} = \Delta S^{\ddagger}$$

$$= 9.15 \text{ cal deg}^{-1} \text{ mole}^{-1}$$

$$\log \frac{k_2}{k_1} = \frac{9.15}{2.303R} = 2$$

$$\frac{k_2}{k_1} = 100$$

This illustrates the direct connection between probability and rate constant.

Example 24-4

Calculate the change in $\Delta G^{\ddagger}$ in the reaction in the above example. $T = 300°C$.
    Since $\Delta E^{\ddagger} = 0$,

$$\Delta G^{\ddagger} = \Delta E^{\ddagger} - T \Delta S^{\ddagger}$$
$$= -(300 + 273)(9.15)$$
$$= -5.2 \text{ cal mole}^{-1}$$

24-3   PROTEINS

Proteins are a class of biochemical compounds which are essentially long chains of amino acids linked together by peptide bonds. They constitute about 15 percent of the total mass of biological cells. The cell membranes are about one-half protein. The biological catalysts called enzymes are made up of proteins. Proteins are, therefore, one of the most important kinds of macromolecules in biochemistry.

Amino acids

An amino acid is a compound with the general structure $R-CH(NH_2)COOH$; in a few cases, the structure has minor complications such as another link to the nitrogen atom in place of one of the hydrogen atoms or the presence of two —CH(NH$_2$)COOH groups coupled to R. The group R takes on a variety of forms, sometimes containing acidic groups, sometimes basic groups, sometimes only neutral groups.

The 20 most important amino acids are listed in Table 24-1. The common abbreviations are given after the name. The right-hand column shows the abbreviated structure formulas. The letter Y stands for the group —CH(NH$_2$)COOH which has a kind of Y-shaped configuration and is common to practically all the amino acids. The carbon spine of the group is indicated without showing the hydrogen atoms present at the appropriate places to saturate the valences. A comparison of the abbreviated formulas with the complete formulas in the second column will make clear the meaning of the notation, which is designed to reveal more vividly the essential structure of these compounds. It is important to remember that the amino acids can be classified by the acidity and basicity of the groups and by the presence of oxygen, nitrogen, and sulfur atoms in the radicals.

The peptide bond

Because of the presence of both COOH and NH$_2$ groups, the amino acids have an amphoteric character. Generally speaking, above pH 2 the COOH becomes ionized

$$-COOH \longrightarrow -COO^- + H^+$$

Below pH 9, the NH$_2$ group accepts a proton

$$-NH_2 + H^+ \longrightarrow -NH_3^+$$

At pH 7 (neutrality) the amino acid is largely in the doubly ionized state

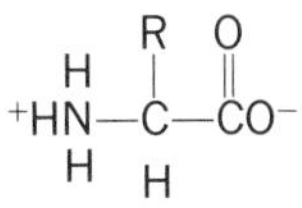

Two amino acids can be coupled together by the reaction

Fig. 24-7   Portion of the chain in cellulose.

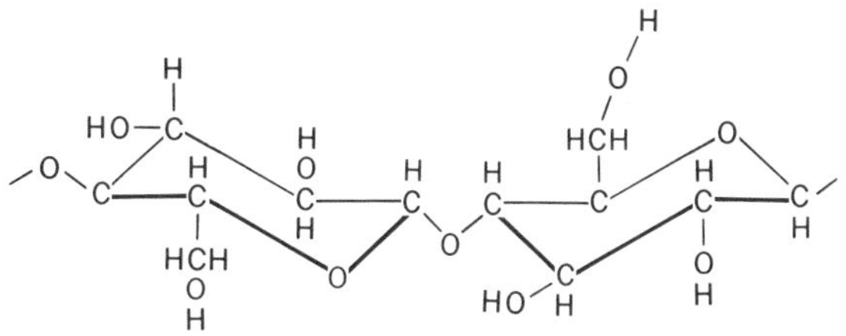

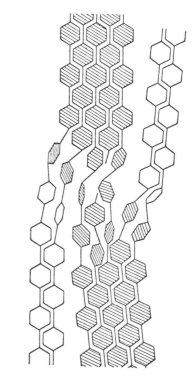

Fig. 24-8   Chains in cellulose. (After H. Mark, "The Investigations of High Polymers with X-rays" in "The Chemistry of Large Molecules," R. E. Burk and O. Grummitt, eds., Interscience Publishers, Inc., New York, 1943.)

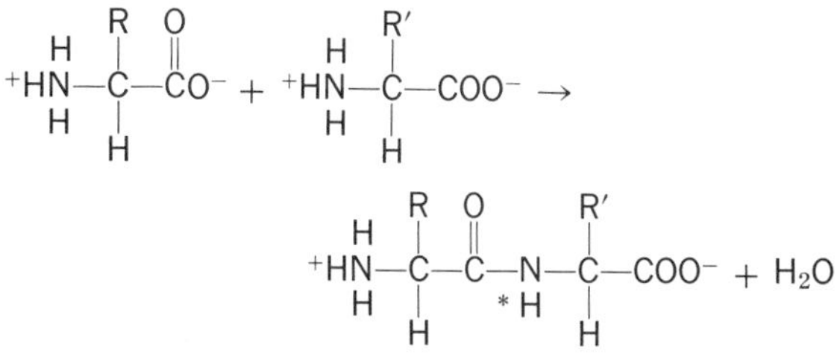

TABLE 24-1   Principal amino acid constituents of proteins

Name		Formula	Short formula
Aliphatic:			
Glycine	Gly	H—CH(NH$_2$)COOH	HY
Alanine	Ala	CH$_3$—CH(NH$_2$)COOH	CY
Valine	Val	(CH$_3$)$_2$CH—CH(NH)$_2$COOH	$\overset{C}{\underset{C}{}}$CY
Leucine	Leu	(CH$_3$)$_2$CHCH$_2$—CH(NH$_2$)COOH	$\overset{C}{\underset{C}{}}$CCY
Isoleucine	Ile	(CH$_3$CH$_2$)(CH$_3$)CH—CH(NH$_2$)COOH	$\overset{CC}{\underset{C}{}}$CCY
Aromatic:			
Phenylalanine	Phe	⬡CH$_2$—CH(NH$_2$)COOH	⬡CY
Tyrosine	Tyr	HO⬡CH$_2$—CH(NH$_2$)COOH	HO⬡CY
Tryptophan	Trp	(indole)CCH$_2$—CH(NH$_2$)COOH	(indole)CY
Hydroxy:			
Serine	Ser	HOCH$_2$—CH(NH$_2$)COOH	HOCY
Threonine	Thr	(CH$_3$)(HO)CH$_2$—CH(NH$_2$)COOH	$\overset{HO}{\underset{C}{}}$CY
Acidic:			
Aspartic acid	Asp	HOOCCH$_2$—CH(NH$_2$)COOH	$\overset{O}{\underset{O}{}}$CCY
Glutamic acid	Glu	HOOC—CH$_2$CH$_2$—CH(NH$_2$)COOH	$\overset{O}{\underset{O}{}}$CCCY
Amide:			
Asparagine	Asn	H$_2$N(CO)CH$_2$—CH(NH$_2$)COOH	$\overset{N}{\underset{O}{}}$CCY
Glutamine	Gln	H$_2$N(CO)CH$_2$CH$_2$—CH(NH$_2$)COOH	$\overset{N}{\underset{O}{}}$CCCY
Basic:			
Lysine	Lys	H$_2$NCH$_2$CH$_2$CH$_2$CH$_2$—CH(NH$_2$)COOH	NCCCCY
Arginine	Arg	H$_2$NC(NH)NHCH$_2$CH$_2$CH$_2$—CH(NH$_2$)COOH	$\overset{N}{\underset{N}{}}$CNCCCY
Histidine	His	HC══C—CH$_2$—CH(NH$_2$)COOH (imidazole)	CCCY NN C
Sulfuryl:			
Methionine	Met	CH$_3$SCH$_2$CH$_2$—CH(NH$_2$)COOH	CSCCY
Cysteine	Cys	HSCH$_2$—CH(NH$_2$)COOH	SCY
Secondary:			
Proline	Pro	(pyrrolidine ring) H$_2$C—N(H)—CHCOOH, H$_2$C—CH$_2$	$\overset{C}{\underset{CC}{}}$ Y

The linkage denoted by the asterisk is called the peptide bond. In its formation according to the above equation, the free energy of the system increases by about 5 kcal mole^{-1}; that is, $\Delta G = +5$ kcal mole^{-1}. Therefore, the thermodynamic tendency of a protein is to dissociate in aqueous solution. Proteins are relatively stable because the free energy of activation for the dissociation $G^{\ddagger}$ is as high as 30 to 50 kcal mole^{-1}.

As an example, a mole of alanine may react with a mole of glycine (both in aqueous solution at standard concentrations) to form the peptide bond and yield H_2O

$$\text{Ala}(aq) + \text{Gly}(aq) \rightarrow \text{Ala—Gly}(aq) + H_2O(l)$$
$$\underset{1\ M}{} \qquad \underset{1\ M}{} \qquad \underset{1\ M}{}$$

$$\Delta G° = 4.13 \text{ kcal mole}^{-1}$$

If the concentrations are varied,

$$0.1\ M \quad 0.1\ M \quad 1.25 \times 10^{-5}\ M$$
$$\Delta G = 0$$
$$1\ M \qquad 1\ M \qquad 0.1\ M$$
$$\Delta G = 2.7 \text{ kcal mole}^{-1}$$

If the reaction takes place and the thermodynamic properties are listed for solid amino acids, then for the linking of DL-Leu and Gly at 25°C, the values are

$$\text{DL-Leu}(s) + \text{Gly}(s)$$
$$= \text{Leu—Gly}(s) + H_2O(l)$$
$$H_f°: \quad -154.16 - 126.66$$
$$= -207.1 \quad -68.32$$
$$\Delta H° = 5.4 \text{ kcal mole}^{-1}$$
$$S°: \quad 49.5 \qquad 26.1$$
$$= \quad 67.1 \qquad 16.72$$
$$\Delta S° = 8.3 \text{ kcal deg}^{-1} \text{ mole}^{-1}$$
$$\Delta G° = 2.9 \text{ kcal mole}^{-1}$$

Primary structure

The sequence in which the amino acids occur in the protein chain is called the *primary* structure of the protein. This sequence was first determined in 1953 for the relatively small protein which constitutes beef insulin and which contains 51 amino acids arranged in two chains called the A fraction and the B fraction (Fig. 24-9). At the left, the chains terminate in NH_2 groups; at the right in COOH groups. The two chains are linked together with two —S—S— bridges at the points where cystine replaces two molecules of cysteine.

$$H_2C\text{—}CH(NH_2)COOH$$
$$|$$
$$S$$
$$|$$
$$S$$
$$|$$
$$H_2C\text{—}CH(NH_2)COOH$$

cystine

Note also that in the A fraction there is another molecule of cystine which forms an —S—S— bridge between two parts of the same chain. This is possible because the chain is coiled rather than stretched straight, as shown in the figure.

One of the largest proteins for which the primary structure has been determined is chymotrypsinogen, which contains 246 amino acids. The sequence is shown in Fig. 24-10.

Where amino acids like Tyr, Ser, and Thr occur, OH groups protrude from the protein chain. With Asp and Glu, there are COOH groups, and with Trp, Asn, Gln, Lys, and Arg there are free NH or NH_2 groups on the surface of the chain. These different types can accept or donate protons depending on the nature of the environment. At many places they play an important role in buffering the solution with which they are in contact. In fact, the protein could not be dissolved in aqueous solution at all without the hydrogen bonds which are formed between these groups and the surrounding H_2O molecules. These groups make the protein mildly hydrophilic instead of strongly hydrophobic, as it would be without them.

Secondary structure

The sequence of the amino acids in a protein molecule is essentially a one-dimensional type of structure. Once that is determined, the question remains of how this strand is arranged in space. Strands may lie side by side like long-chain hydrocarbons in a crystal, or they may be coiled around each other like strands in a rope. This aspect of arrangement is called the *secondary structure.*

The forces holding the amino acids in a strand are the covalent forces in the peptide bonds. Typical dissociation energies of covalent bonds are listed in

Fig. 24-9 The primary structure of beef insulin. (After F. Sanger, "Structure of Insulin" in "Currents in Biochemical Research," D. E. Green, ed., Interscience Publishers, Inc., New York, 1956.)

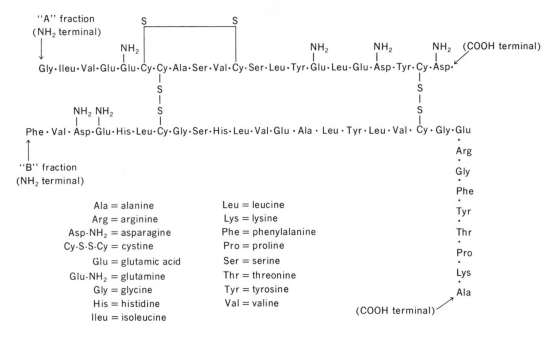

Table 24-2. As contrasted with these relatively strong bonds, the forces binding one strand to another or parts of a strand to itself are usually relatively weak. Except for occasional ----S----S---- bonds, the interstrand attraction consists largely of hydrogen bonds and van der Waals forces with some electrostatic attraction between oppositely charged portions of the chain. As a rough figure, 10 kcal mole^{-1} is required to break hydrogen bonds (not covalent H----H, C----H, or N----H, of course, but ----H$\cdots$H----). Only about 1 kcal mole^{-1} is needed to separate a typical pair of atoms held together by van der Waals forces. If, however, the rupture of the ----H$\cdots$H---- bond takes place in water with the subsequent formation of hydrogen bonds to water molecules, then only about 1.5 kcal mole^{-1} is needed.

Data from x-ray diffraction measurements on peptide chains have been the principal source of information on the patterns in which these chains are packed together. In Fig. 24-11 the dimensions and angles of a peptide bond are shown, and in Fig. 24-12 the fully extended chain is drawn. The simplest amino acid glycine forms a polymer consisting of sheets of chains, as illustrated in Fig. 24-13. *Silk fibroin* has a somewhat similar structure made of 44 percent glycine and 40 percent alanine plus serine, as shown in Fig. 24-14.

As contrasted with these structures, keratin shows a coiled structure of helical form (Fig. 24-15). These helices can be coiled about each other to give the ropelike structure found in the myosin of muscle and the epidermin of skin.

The protein helix may be either right-handed or left-handed; that is, a movement along the chain toward the observer is an advance upward, either when the movement is toward the right or toward the left. There is some evidence that the right-hand spiral is the more stable. The spiral shown in Fig. 24-15a is right-handed. The six-strand spiral shown in Fig. 24-15b is right-handed, and the three-strand spiral is left-handed. There is evidence that *collagen* consists of three strands, each made of a protein helix which is left-handed, the three strands being coiled together in a right-hand spiral.

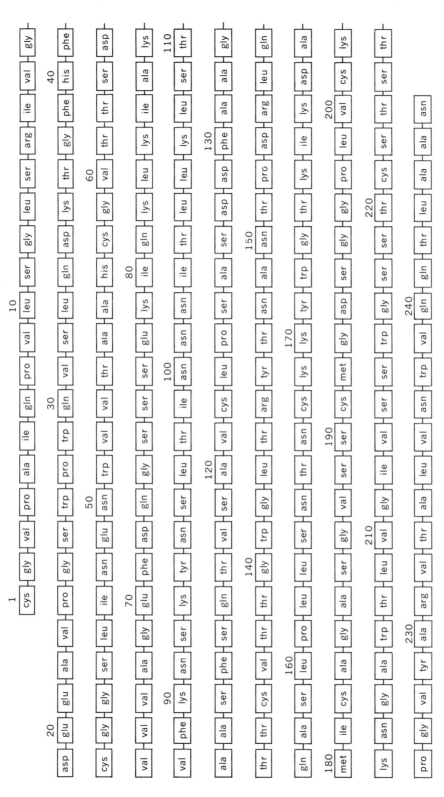

Fig. 24-10 The primary structure of chymotrypsinogen. [After B. S. Hartley, *Nature*, 201:1284 (1964).]

Fig. 24-11   The dimensions of the polypeptide bond. [After R. B. Corey and L. Pauling, *Rend. 1st Lombardo Sci.*, **89**:10 (1955).]

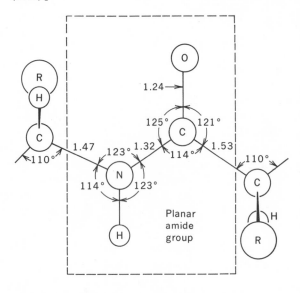

TABLE 24-2   Energy of dissociation for covalent bonds in biochemical compounds[a]

Bond	Energy, kcal mole^{-1}
C=O	173
C=C	145
C—C	80–85
C—H	98
C—O	79
C—N	66
O—H	109
N—H	92

[a] Data from K. S. Pitzer, *J. Am. Chem. Soc.*, **70**:2140 (1948).

Tertiary structure

Frequently the ropes are knotted or supercoiled into patterns of still larger characteristic dimensions, in what is called the *tertiary structure*. Such a pattern is shown in Fig. 24-16, which depicts the myoglobin macromolecule. These units are sometimes packed together in superpatterns referred to as *quaternary structure*.

To summarize, the 20 amino acids provide units ideal for the construction of a variety of patterns which can function in the many different ways necessary for the total life pattern of an organism. The amino acids have enough polar groups to be soluble in the aqueous cell and body media. They can be coupled together with peptide bonds, for which $G^{\ddagger}$ is attainable at temperatures and under conditions available and suitable. Once coupled, the $G^{\ddagger}$ for breaking the bonds is sufficiently high to make the peptide chains stable. The side groups provide a variety of means for holding the chains together in secondary and tertiary structures, and, equally important, these side groups are seats of chemical activity performing a host of buffering and catalytic actions that serve as essential ingredients of the life process.

Example 24-5

If the value of $H^{\ddagger}$ of a peptide bond is assumed to be about 70,000 cal, how much beyond body temperature must the temperature be increased to increase the rate of breaking tenfold?

The ratio of increase in breaking is related to the temperature by the usual Arrhenius equation

$$R \ln \frac{k_2}{k_1} = -H^{\ddagger}\left(\frac{1}{T_2} - \frac{1}{T_1}\right)$$

Putting in numerical values and solving for T_2,

$$2.303 \times 1.987 \log 10$$
$$= -70,000\left(\frac{1}{T_2} - \frac{1}{310}\right)$$

$$\frac{1}{310} - \frac{1}{T_2} = \frac{4.5757}{7.0 \times 10^4} = 6.5 \times 10^{-5}$$

Fig. 24-12 Dimensions of the fully extended polypeptide chain. [After L. Pauling and R. B. Corey, *Adv. Protein Chem.*, 12:133 (1957).]

$$\frac{1}{T_2} = 0.003226 - 0.000065 = 0.003161$$

$$T_2 = 316 \qquad t_2 = 43.2$$

or 5° above body temperature. Thus a high fever would cause a drastic increase in the rate of disruption of protein on the basis of these assumptions.

Example 24-6

Contrast the rate of breakup of secondary structure for the case where (*a*) the chain is held together with 1 hydrogen bond and (*b*) where it is held together with 100 hydrogen bonds if the temperature is increased 10°C above body temperature.

(*a*) $\quad R \ln \dfrac{k_2}{k_1} = -H^{\ddagger}\left(\dfrac{1}{T_2} - \dfrac{1}{T_1}\right)$

$\qquad \log \dfrac{k_2}{k_1} = \dfrac{1,500}{4.5757}\left(\dfrac{1}{310} - \dfrac{1}{320}\right)$

$\qquad\qquad = 0.0330$

$\qquad \dfrac{k_2}{k_1} = 1.08$

(*b*) $\quad \log \dfrac{k_2}{k_1} = \dfrac{150,000}{4.5757}\left(\dfrac{1}{310} - \dfrac{1}{320}\right)$

$\qquad\qquad = 3.30$

$\qquad \dfrac{k_2}{k_1} = 2,000 \qquad Ans.$

Since changes in structure seldom require the breaking of more than a few hydrogen bonds simultaneously, the lower variation of rate with temperature is nearer the truth.

24-4   ENZYMES

A protein can be constructed in such a way that when coiled, its exposed surface with its array of chemically

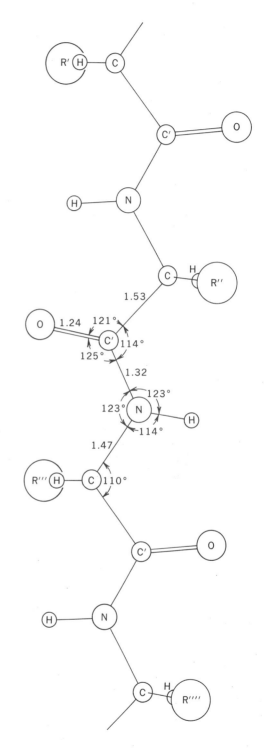

7.23 Å

1.53

1.24   121°

125°

114°

1.32

123°

123°

114°

1.47

110°

Fig. 24-13 The sheet structure of polyglycine. (After G. Orgel, "Biophysical Science," J. L. Oncley, ed., John Wiley & Sons, Inc., New York, 1959.)

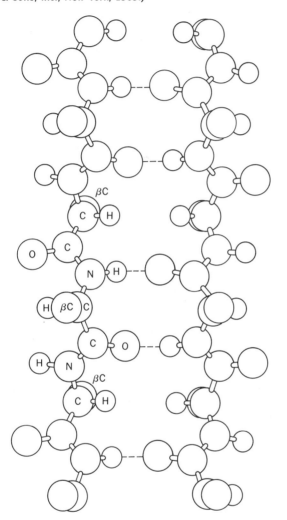

active groups acts as a catalyst. As noted in the previous section, proteins can be arranged in space in many ways. Not only does the secondary and tertiary structure permit the formation of surfaces of different geometric shapes, but the primary structure, the sequence of amino acids, makes possible the arrangement on the surface of a variety of chemically active groups in a wide range of geometric patterns. As a result, there are found in nature many enzymes

which, compared with inorganic catalysts, have a highly *specific* form of catalytic activity.

The material on which the enzyme exerts catalytic action is called the *substrate*. The enzyme *urease* has unique specificity, acting only on the one substrate *urea* and activating this substance in only a single chemical reaction, the production of ammonia and carbon dioxide. It is this specificity which permits enzymes to guide the raw material of cytoplasm into the networks of thousands of reactions which maintain the total bioprocess in its dynamic pattern of energy flow and synthesize the thousands of varieties of molecules and macromolecules needed for growth and replacement.

Enzyme kinetics

The relation between the rate of an enzyme reaction and the concentration of the substrate is frequently found to be represented by the type of curve shown in Fig. 24-17. When the concentration is low, the variation is given by a first-order equation, but as the concentration of the substrate increases, the rate levels off to a constant corresponding to zero order. This suggests that there are a limited number of active sites on the surface of the enzyme. At low concentrations, the rate depends on how often the substrate molecules hit these active sites, absorb, and react. If there are many active sites and few molecules, the rate depends linearly on the concentration of molecules. If the concentration of molecules is so high that the sites become saturated, the rate depends only on the length of time required for the reaction to take place on the surface and is independent of the concentration of the substrate.

Suppose that the overall reaction is the decomposition of molecule AB into A and B

$$AB \longrightarrow A + B$$

This can be thought of as taking place in two steps; first, the molecule AB is absorbed by the enzyme Z forming the complex ABZ at the active site:

Step 1: $AB + Z \underset{^1k_b}{\overset{^1k_f}{\rightleftarrows}} ABZ$

Then the molecule decomposes, and the parts leave the enzyme:

Step 2: $ABZ \underset{^2k_b}{\overset{^2k_f}{\rightleftarrows}} A + B + Z$

The rate ^1r_f for the first step is

$$^1r_f = {}^1k_f[AB][Z] \tag{24-6}$$

where [Z] is the effective "concentration" of active sites.

If the concentration [AB] is small and the second step takes place with relative rapidity, then [Z] is independent of [AB] and may be regarded as a constant and incorporated into k_1, yielding a first-order expression

$$^1r'_f = {}^1k_f[AB] \tag{24-7}$$

The Arrhenius equation for ^1k_f is

$$^1k_f = {}^1k_{f_0}e^{-H_1\ddagger/RT}e^{S_1\ddagger/R} \tag{24-8}$$

In this expression $^1k_{f_0}$ is determined merely by the choice of units in which the rate is measured. $H_1^\ddagger$ is the activation enthalpy of reaction and depends primarily on the amount of energy to move AB from the solution and onto the site. This is probably of the order of the activation energy for the crystallization of an organic compound—relatively small, perhaps 10 kcal mole^{-1}. $S^\ddagger$ depends logarithmically on the number of active sites

$$S^\ddagger = S_0^\ddagger + R \ln [Z] \tag{24-9}$$

where the numerical value of $S_0^\ddagger$ is determined by the units in which [Z] is expressed. This is exactly similar to the dependence of the entropy of a gas on volume

$$S = S_0 + R \ln V \tag{24-10}$$

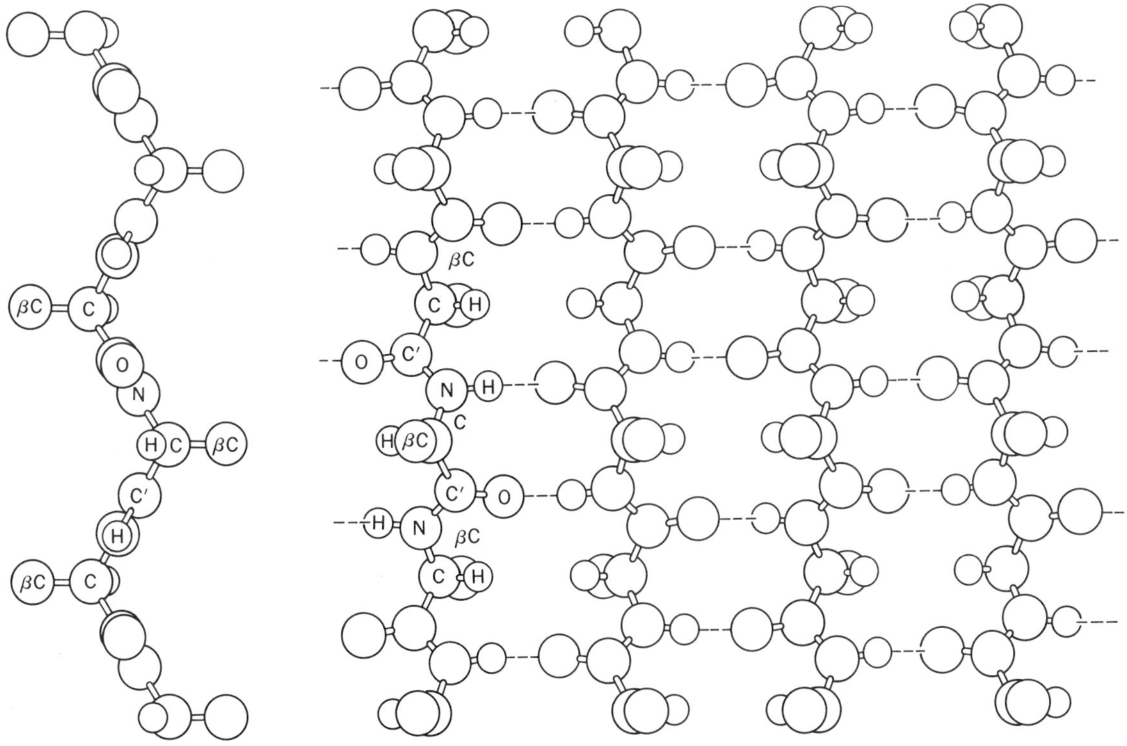

Fig. 24-14 The pleated sheet structure in silk fibroin, 44% glycine, 40% alanine and serine. [After L. Pauling and R. B. Corey, *Proc. Nat. Acad. Sci.*, **37**:729 (1951).]

Fig. 24-15   The secondary structure of proteins.   (*a*) The α-helix in polypeptide chains.   The dashed lines indicate hydrogen bonding.   [After L. Pauling, R. B. Corey, and H. R. Branson, *Proc. Nat. Acad. Sci.,* **37**:205 (1951).] (*b*) Coiling of α-helices into a rope pattern.   (After L. Pauling, "Les Proteines: Rapports et Discussions," R. Stoops, ed.   Copyright 1953.   Institut International de Chimie Solvay, Brussels, Belgium.   Used by permission.)

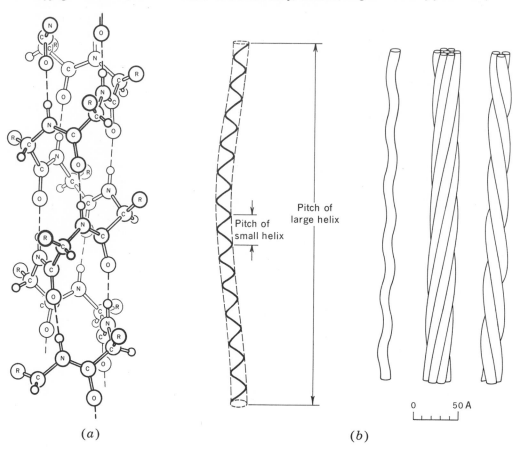

Pitch of large helix

Pitch of small helix

0   50 A

(*a*)   (*b*)

Just as the volume is a measure of the probability of finding the gas molecules in a given domain, [Z] is a measure of the probability of the collision of AB with an active site.   Double the number of sites, and the probability of adsorption doubles.

The rate at which step 2 takes place depends on the surface concentration of AB, that is, on [ABZ], so that

$$^2r_f = {}^2k_f = {}^2k_f[\text{ABZ}] \tag{24-11}$$

This is also a first-order expression; but if the sites are saturated, [ABZ] is a constant and the expression becomes zero order.

If [Z_0] denotes the total number of sites, whether free or bound with an adsorbed AB molecule, then

$$[Z_0] = [Z] + [\text{ABZ}] \tag{24-12}$$

This is called the enzyme conservation equation.

The equilibrium constant for the first reaction ^1K is given by the expression:

$$^1K = \frac{^1k_f}{^1k_b} = \frac{[\text{ABZ}]}{[\text{AB}][\text{Z}]} \tag{24-13}$$

Substituting from Eq. (24-12),

$$[\text{ABZ}] = \frac{^1k_f}{^1k_b}([Z_0] - [\text{ABZ}])[\text{AB}] \tag{24-14}$$

and

$$[ABZ] = \frac{[Z_0][AB]}{{}^1k_b/{}^1k_f + [AB]} \qquad (24\text{-}15)$$

The constant ${}^1k_b/{}^1k_f = K_m$ is the equilibrium for the dissociation

$$ABZ = AB + Z$$

This is called K_m after L. Michaelis and M. L. Menten, the German biochemists who first proposed this approach to enzyme kinetics in 1913.

Using this notation, Eq. (24-15) can be written

$$[ABZ] = \frac{[Z_0][AB]}{K_m + [AB]} \qquad (24\text{-}16)$$

When the rate is at the maximum r_m shown in Fig. 24-17, all the sites are occupied and $[Z_0] = [ABZ]$. Thus

$$r_m = {}^2k_f[Z_0] \qquad (24\text{-}17)$$

Using Eq. (24-16),

$$^2r_f = {}^2k_f[ABZ] = \frac{{}^2k_f[Z_0][AB]}{K_m + [AB]} \qquad (24\text{-}18)$$

Combining Eqs. (24-17) and (24-18), we get

$$^2r_f = \frac{r_m[AB]}{K_m + [AB]} \qquad (24\text{-}19)$$

In the enzyme literature this is generally written

$$v_f = \frac{V_{\max}[AB]}{K_m + [AB]} \qquad (24\text{-}20)$$

where the term *velocity of the reaction* is used so that $v_f = {}^2r_f$ and $V_{\max} = r_m$. Equation (24-20) is called the Michaelis-Menten equation. It reduces to a first-order equation when [AB] is small

$$v_f = \frac{V_{\max}[AB]}{K_m} \qquad (24\text{-}21)$$

since $[AB] \ll K_m$ and can be neglected in the denominator. When $[AB] \gg K_m$, K_m can be eliminated and the zero-order expression is

$$v_f = V_{\max} \qquad (24\text{-}22)$$

Enzyme mechanisms

One of the most important problems in enzyme kinetics revolves around the interpretation of the enzyme entropy of activation. Is the specificity of enzyme action due to a high entropy of activation for the reaction specific to the enzyme and substrate and a low entropy of activation for all other reactions; or is it a question of the reverse kind of difference with regard to the enthalpy of activation?

(a)

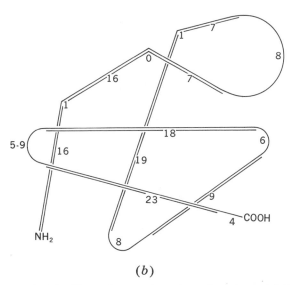

(b)

Fig. 24-16 The tertiary structure of proteins: (a) a model of the myoglobin molecule based on the 6-Å Fourier synthesis; the heme group is the dark disk near the top. [After J. C. Kendrew, *Science*, 139:1261 (1963).] (b) The distribution pattern of amino acids in sperm-whale myoglobin. (After Green, "Comprehensive Biochemistry," vol. 7, "Proteins," M. Florkin and E. H. Stotz, eds., Elsevier Publishing Company, Amsterdam, 1963.)

Fig. 24-17 The variation of rate with concentration in an enzyme catalyzed reaction. (*a*) Plot of data for β-methyl aspartase. [From V. R. Williams and J. Selbin, *J. Biol. Chem.*, 239:1936 (1964).] (*b*) Enzyme with sites almost empty, low concentration of substrate [*AB*]. First-order rate. (*c*) Enzyme with sites almost completely occupied, high concentration of substrate. Zero-order rate.

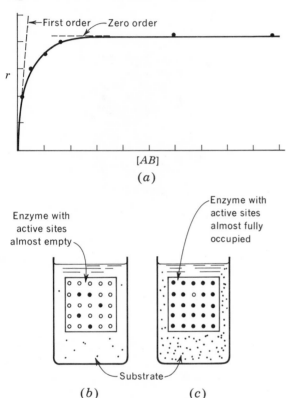

would be required to force a key molecule into a lock which it did not exactly fit. However, the rates for reactions where the substrate does not fit the site are so unobservably small that unreasonably large values of $H^{\ddagger}$ would be required to account for the almost infinitesimally small values of the rates. Unfortunately, in such cases, it is not possible to measure the change of rate with temperature and thus make an experimental determination of $H^{\ddagger}$. It seems more reasonable to think that a lack of fit reduces the probability of reaction to such small values that the consequent lowering of $S^{\ddagger}$ is the cause of the reduction of rate.

This point of view is supported by the phenomenon of catalyst poisons. There are many instances where the introduction of a small amount of substance reduces catalytic activity drastically or stops it altogether. If, as is postulated, the poison molecules are selectively adsorbed on the active sites and prevent the substrate molecules from reaching those sites, the number of active sites is reduced. This affects the rate in just the same way as a reduction in total catalyst surface and is clearly a reduction in the probability of the reaction, reducing $S^{\ddagger}$ according to the relation expressed in Eq. (24-9). This concept is in line with the observed form of the relation of the variation of the rate of the enzyme action with [Z] as expressed in the Michaelis-Menten equation, which assumes the possibility of saturating the active sites and thus implicitly introduces the $S^{\ddagger}$ which is related logarithmically to [Z], as illustrated above.

These relations are shown graphically in a plot of $R \ln k$ against reciprocal temperature in Fig. 24-19. The intercept of the graph with the $R \ln k$ ordinate at $1/T = 0$ is the value of $S^{\ddagger}$. Diminishing the number of sites either by diminishing the amount of the enzyme or by poisoning the enzyme drops the value of $S^{\ddagger}$ according to Eq. (24-9).

In order to explain this specificity in terms of mechanism, the theory has been proposed that the enzyme is like a lock with sites that have a certain specific concave shape, as shown in Fig. 24-18; in order to enter the lock, the substrate molecule (the key) must have the precise complementary convex shape of the enzyme site. There is a rather general acceptance of the idea that at least certain aspects of the shape of the substrate molecule must conform to the geometry of the enzyme surface, but it is not clear whether lack of conformation prevents the reaction from taking place at an appreciable rate by raising $H^{\ddagger}$ or by lowering $S^{\ddagger}$.

There is every reason to believe that a larger $H^{\ddagger}$

Example 24-7

A certain enzyme has a value of $^{1}H^{\ddagger}$ equal to 12.6 kcal mole^{-1} and a value of $^{1}S^{\ddagger}$ equal to 31.3 cal deg^{-1} mole^{-1} when $^{1}k_{f}$ is expressed in moles min^{-1} (mole liter^{-1})$^{-1}$. Calculate the rate at 38 and at 48°C.

The equation for the rate is

$$^1k_f = {}^1k_{f_0}e^{-H^{\ddagger}/RT}e^{S^{\ddagger}/R}$$

Let $^1k_{f_0} = 1$, the rate constant at the temperature where $H^{\ddagger} - TS^{\ddagger} = G^{\ddagger} = 0$. Then, at 38°C

$$R \ln {}^1k_f = \frac{-H^{\ddagger}}{T} + S^{\ddagger}$$

$$\log {}^1k_f = -\frac{H^{\ddagger}}{2.303R}\frac{1}{T} + \frac{S^{\ddagger}}{2.303R}$$

$$= -\frac{12{,}600}{2.303 \times 1.987 \times 310}$$

$$+ \frac{31.3}{2.303 \times 1.987}$$

$$= -8.882 + 6.840$$

$$\log 1/{}^1k_f = 2.042$$
$$1/{}^1k_f = 110$$
$$^1k_f = 9.09 \times 10^{-2} \text{ mole min}^{-1}$$
$$\text{(moles liter}^{-1})^{-1}$$

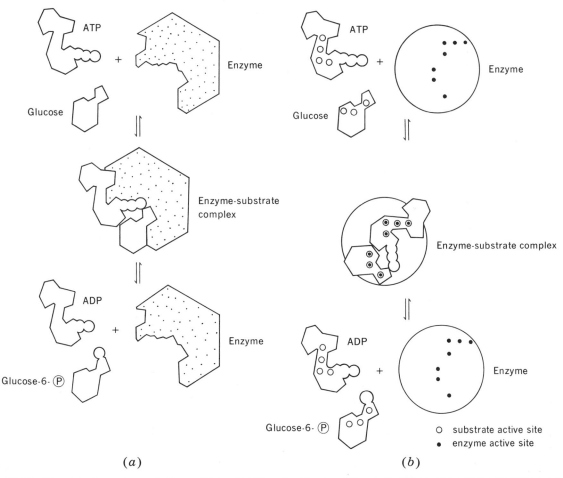

Fig. 24-18 Theories of enzyme mechanism. (a) Lock and key hypothesis. (After J. D. Watson, "The Molecular Biology of the Gene," W. A. Benjamin, Inc., New York, 1965.) (b) Active sites of attraction hypothesis; molecules are adsorbed and held long enough for the reaction to take place only if the pattern of sites on the enzyme matches the pattern on the substrate.

Fig. 24-19   The effect of an enzyme on $S^{\ddagger}$: (*a*) the rate is relatively high when an enzyme brings together the reacting substrate molecules, thereby increasing the probability of the reaction and raising $S^{\ddagger}$ to $S_e^{\ddagger}$; (*b*) if there is no enzyme the probability of the reaction is enormously reduced, lowering $S^{\ddagger}$ to $S_e^{\ddagger} - \Delta S = S_o^{\ddagger}$.

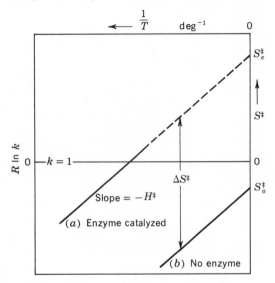

(*a*) Enzyme catalyzed

(*b*) No enzyme

At 48°C

$$\log {}^1k_f = -8.604 + 6.840$$

$$\log \frac{1}{{}^1k_f} = 1.764$$

$${}^1k_f = 1.72 \times 10^{-2} \text{ mole min}^{-1} \text{ (moles liter}^{-1})^{-1}$$

Example 24-8

Calculate the percentage by which the number of sites on the enzyme must be blocked to reduce the rate at 48°C to the rate at 38°C with an unpoisoned enzyme in Example 24-7.

Since the change in temperature raises $\log {}^1k_f$ by 0.278, the value of $S^{\ddagger}/2.303R$ must be reduced by this amount to keep $\log {}^1k_f$ the same; thus $\Delta S^{\ddagger} = -0.278 \times 2.303R$.

$$\Delta S^{\ddagger} = R \ln \left[\frac{Z_p}{Z_0} \right]$$

where Z_p is the concentration of sites when the enzyme is poisoned.

$$\log \frac{Z_0}{Z_p} = 0.278/4.58$$

$$\frac{Z_0}{Z_p} = 1.15$$

$$\frac{Z_p}{Z_0} = 0.87$$

Thus the number of sites must be reduced by $100(1 - 0.87) = 13$ percent in order to keep the rate constant when the temperature is raised by 10°C.

24-5   DNA AND RNA

During the first half of the twentieth century, evidence accumulated from the study of genetics to indicate that in the biological cell the genetic information was concentrated in a single long threadlike macromolecule called *deoxyribonucleic acid* (DNA). The chemical analysis of this molecule showed that it was composed of sugarlike parts, deoxyribose, a number of purine rings, a number of pyrimidine rings, and phosphate radicals. The key problem was the elucidation of the way in which these parts fitted together in a pattern which could explain logically how this molecule guides cell replication and differentiation.

DNA structure

Since there are many kinds of organisms, there are many kinds of DNA, but all forms of DNA have certain elements of structure in common. Electron micrographs show these molecules to be threadlike (Fig. 24-20), which suggests that they may consist of units strung together in long chains much like the proteins. In the terms used to describe proteins, the primary structure consists of a sequence of chemical groups. This concept is in line with the evidence from biology, which supports the picture of a linear sequence of genes.

The electron micrographs also indicate that such a thread is not folded in multiple segments to form a sheetlike fabric, as in the case of some of the proteins, but is a single linearly extended chain. The structure of this chain was revealed by an x-ray diffraction photograph (Fig. 24-21) which suggested that the DNA thread has a spiral structure. On the basis of this evidence (obtained by the English chemists R. E. Franklin and R. Gosling in 1953) and the evidence for uniformity of base composition (provided by the investigations of the American biochemist E. Chargaff), the American biochemist J. D. Watson and the English biologist F. H. C. Crick at the University of Cambridge proposed that the structure of DNA was a double

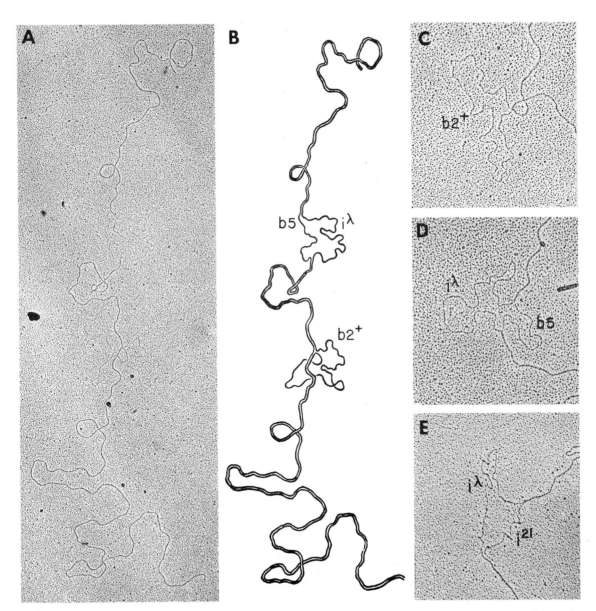

Fig. 24-20 Electron micrographs of samples of DNA with an outline drawing. [From B. C. Westmoreland, W. Szybalski, and H. Ris, *Science*, 163:1343–1348 (1969).]

Fig. 24-21 The key x-ray diffraction pattern which revealed the helical structure of DNA observed by R. E. Franklin and R. Gosling, *Nature,* **171**:740 (1953). (From J. D. Watson, "The Molecular Biology of the Gene," W. A. Benjamin, Inc., New York, 1965.)

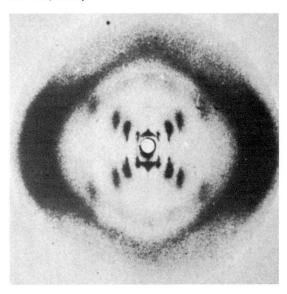

helix, like a twisted length of ribbon candy or a ladder spiraled about its long axis (Fig. 24-22). The structural formula of a section of the ladder is shown in Fig. 24-23. The sides of the ladder are alternate phosphate and deoxyribose groups. Each rung is made of a pyrimidine ring (single hexagon) linked by hydrogen bonds to a purine ring (hexagon-pentagon). There are only two types of rungs; one is made of thymine (one ring) joined to adenine (two rings); the other is made of cytosine (one ring) joined to guanine (two rings). The structural formulas of these ring compounds are shown in Fig. 24-24.

In order to see the structure more clearly, it is frequently helpful to think of the DNA spiral as untwisted and laid flat on a plane (Fig. 24-25). As can be seen, the rungs can be placed either with the single ring on the left and the double ring on the right or vice versa. This provides the possibility of four different choices for the structure of each rung of the ladder. This structure can be designated by the capital letter of the group which appears on the left side of the rung when the molecule is drawn planar as in Fig. 24-23.

DNA replication

When a biological cell is to be replicated, it is necessary to provide the new *daughter* cell with an exact copy of the DNA molecule in the original *mother* cell. The structure proposed by Watson and Crick provides a mechanism for this replication. Assume that just before replication the DNA is in the form of the double helix. Then, in preparation for replication the helix begins to unwind, and the hydrogen bonds between single and double rings begin to break in sequence starting at the bottom. This is a kind of "unzipping" between the left- and right-hand portions of the molecule, as shown in Fig. 24-26. As fast as the hydrogen bonds are broken and the active sites are exposed, free rings, with deoxyribose and phosphate groups attached, move from the cytoplasm and link to these sites. Because a thymine molecule has the possibility of forming two hydrogen bonds, it will link *only* to an adenine molecule and vice versa. Because a cytosine molecule has the possibility of forming three hydrogen bonds, it will link only to a guanine molecule. In addition, the geometry is such that a single ring can link only to a double ring and a double ring can link only to a single ring. Thus, at the completion of the unzipping and relinking, there are now two DNA molecules where only one existed before; the molecule has undergone precise replication, and a new DNA molecule is ready to take its place in the new daughter cell. The genetic *message* has been transmitted from the old molecule to the new in the four-letter language which is denoted as the GACT language.

RNA synthesis

In addition to providing a mechanism for DNA replication, the Watson-Crick model also offers an explanation of how DNA guides the synthesis of the proper kinds of proteins necessary for the building of a new cell. This is accomplished by the synthesis, first, of some molecules called RNA (ribonucleic acid), which are formed from a single strand of DNA unzipped apart from its complementary half and acting as a template. The RNA molecules themselves then act as templates to guide the protein synthesis. The structural formula of a portion of an RNA chain is shown in Fig. 24-27.

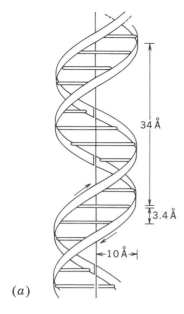

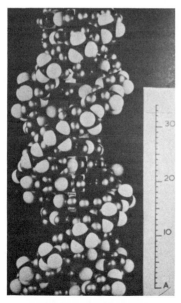

Fig. 24-22 DNA as a helical ladder with rungs: (a) model proposed by J. D. Watson and F. H. C. Crick, *Proc. Roy. Soc.*, **A223:80** (1954); (b) stereomolecular model. (After N. H. F. Wilkins as shown in J. D. Watson, "The Molecular Biology of the Gene," W. A. Benjamin, Inc., New York, 1965.)

Fig. 24-23 A portion of the DNA ladder, untwisted and laid on a plane. (After D. H. Andrews and R. J. Kokes, "Fundamental Chemistry," 2d ed., John Wiley & Sons, Inc., New York, 1965.)

In RNA the single-ring compound *thymine* is replaced by another single-ring compound called *uracil,* which differs from thymine in lacking the methyl group. Thus, when a portion of a DNA chain serves as a template to synthesize RNA, the following substitutions take place:

G ⟶ C
A ⟶ U
C ⟶ G
T ⟶ A

The message thus has been translated from the GACT language to the CUGA language.

Three kinds of RNA have been identified in cells. The species which carries the message is called messenger ribonucleic acid (*m*RNA). This species of

Fig. 24-24 The components of DNA: (*a*) structural formulas; (*b*) mnemonic pattern showing the classification of the nucleotides and their pairing in DNA.

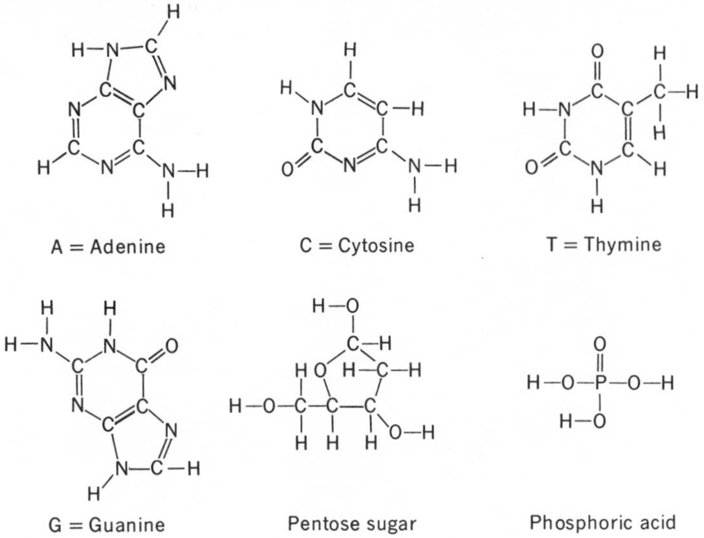

A = Adenine   C = Cytosine   T = Thymine

G = Guanine   Pentose sugar   Phosphoric acid

(*a*)

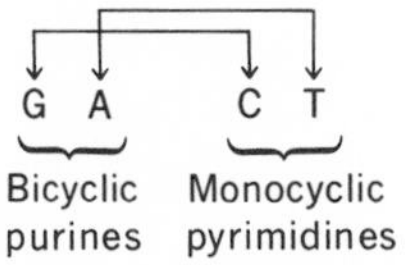

G A   C T

Bicyclic   Monocyclic
purines   pyrimidines

(*b*)

molecule is single-stranded, as compared with the double-stranded DNA, and the members of this species vary greatly in length. The evidence shows that in order to convey the message to stitch a particular amino acid to the end of a growing protein chain, three cross-links of the DNA chain are required. These might be CTA, for example. Translated into the complementary group on *m*RNA, this three-letter "word" becomes GAU. Such a set is known as a *codon*. This particular codon gives instructions to add aspartic acid to the chain of the protein which is being synthesized. If the assumption is made that each *m*RNA molecule carries the message directing the synthesis of a single protein molecule, the factor governing the length of the chain can be deduced.

The shortest protein has about 100 amino acids. This requires 100 codons or 100 × 3 letters; and so an *m*RNA molecule must contain at least 300 nucleotides. In the bacteria *E. coli,* the average length of *m*RNA is 900 to 1,500 nucleotides, corresponding to the 300 to 500 amino acids found in the average protein present in this cell.

After its synthesis on the DNA molecule, the *m*RNA travels to a portion of the cell called a *ribosome*. A schematic drawing of a ribosome is shown in Fig. 24-28. This structure serves as a focal point through which the *m*RNA is fed and where it performs its function in synthesizing the proteins.

According to Watson, there are about 15,000 ribosomes in an *E. coli* cell, each ribosome having a molecular weight just under 3×10^6. The ribosomes constitute about one-fourth of the mass of the cell. They contain a second species of RNA designated as *r*RNA. The function of this kind of RNA has not been discovered.

Protein synthesis

The individual amino acid molecules are brought to the ribosome protein factory by still another species

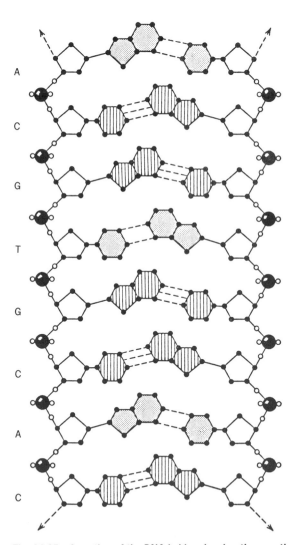

Fig. 24-25 A portion of the DNA ladder showing the genetic message in terms of the initial letter of the left portion of the nucleotide rung, a message spelled in the GACT language.

Fig. 24-26 A portion of the DNA double helix in the process of unzipping and linking with new nucleotides, a proposed mechanism of replication. (After J. D. Watson, "The Molecular Biology of the Gene," W. A. Benjamin, Inc., New York, 1965.)

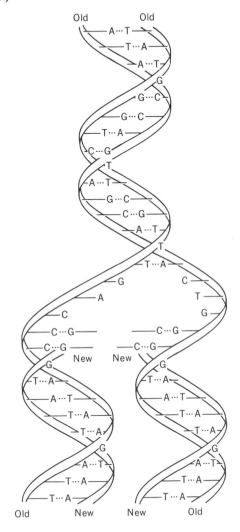

of RNA, called transfer ribonucleic acid (*t*RNA) because of its role in transferring information; the designation *soluble* ribonucleic acid (*s*RNA) is sometimes used also. A schematic drawing of a *t*RNA molecule is shown in Fig. 24-29. All *t*RNA molecules have a molecular weight of about 2.5×10^4. They are characterized by three nucleotides on one end which are arranged to fit on one and only one codon

Fig. 24-27   A portion of the *m*RNA single strand, composed of the units CUGA, complementary to the GACT units of DNA.

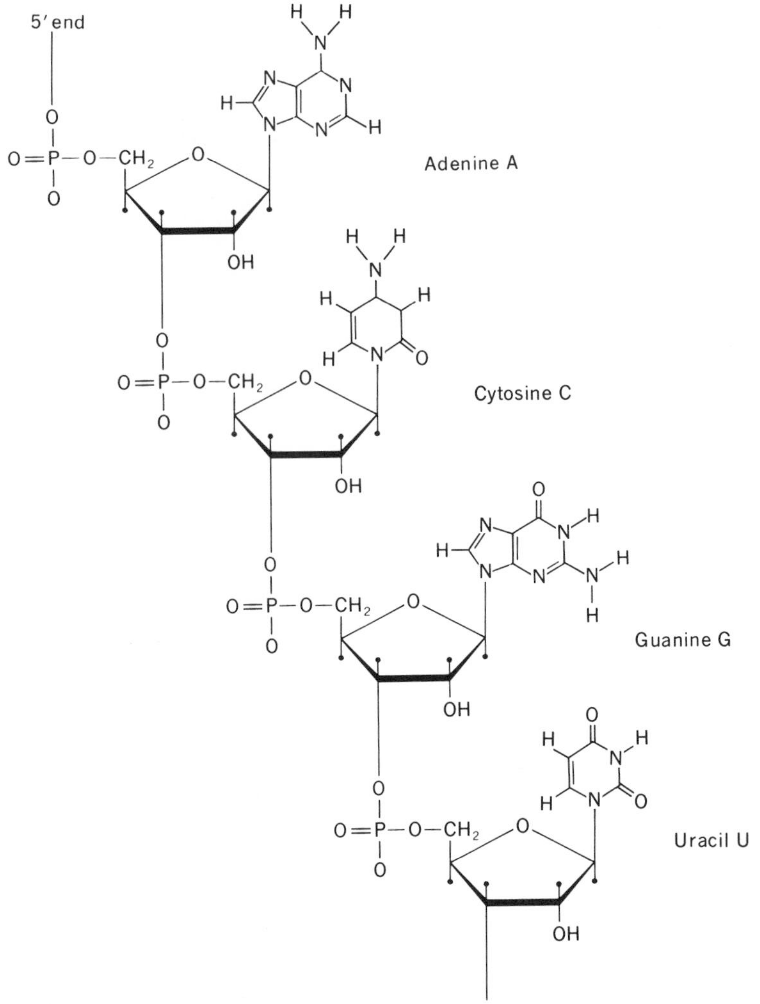

on the *m*RNA molecule; this might be one like the GAU referred to above. On the other end of the *t*RNA molecule there is again a surface which is highly specific so that one and only one amino acid molecule can attach there. The entire code relating codons to amino acids has been determined and will be discussed in Chap. 25; a schematic picture of this process is shown in Fig. 24-30.

As the *m*RNA slides through the ribosome, exactly the proper sequence of *t*RNA molecules matches their

codon ends to the *m*RNA template, thus aligning the amino acids on their upper ends so that the amino acids are joined by peptide bonds to form the protein chain. When the *m*RNA has completed its task, it slides out of the ribosome which is available to start a new synthesis.

Watson estimates that under optimal conditions, an interval of only 10 sec is required to synthesize a protein with a molecular weight of 40,000. Since the average molecular weight of the amino acids occurring in proteins is 141, this means that about 280 amino acids have been linked together in 10 sec in the synthesis of the protein. Since there are 15,000 ribosomes operating, there are 4.2×10^5, or almost $\frac{1}{2}$ million, peptide linkages formed per second in a single microscopic bacterial cell when it is synthesizing proteins at the maximum rate.

Energy and entropy considerations

Both the replication of the cell and the synthesis of proteins raise some interesting physicochemical questions. In order to construct the vast network of linkages and to maintain the intricate pattern of energy flow and molecular interchange found even in a single cell, a number of conditions must be fulfilled.

First of all, there is the question of strength of binding. The DNA molecule must be able to resist disruptive influences that might break it and destroy part or all of the message which it carries. For this reason, covalent binding with relatively high energy of dissociation is required along the two sides of the ladder.

In contrast to this tensile strength, the rungs of the ladder must be able to split in two to provide for the replication process; but the two halves must each have stable binding internally and also have a certain amount of rigidity so that the part capable of hydrogen bonding will link only with its proper complementary counterpart. G must match with C but not with A or T. In the replication process A must match with T but not with C or G or with itself; and in the formation of mRNA, A must match with U. This stability and rigidity is attained by the use of the one- and two-membered rings with resonance structure. These rings with π bonding are far more rigid than saturated hydrocarbon rings like cyclohexane. By introducing a single ring in two of the nucleotides (T and C) and double rings in the other two (A and G), the geometry is made to favor a linking of a single with a double ring. Two single rings would not reach far enough out toward each other from a rung, two double rings would require a greater length than is found between the sides of the ladder.

Apparently the number of places on which hydrogen bonds will form is also a factor. T with its two hydrogen-bond sites will link to A with two hydrogen-

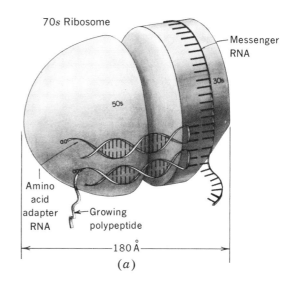

(a)

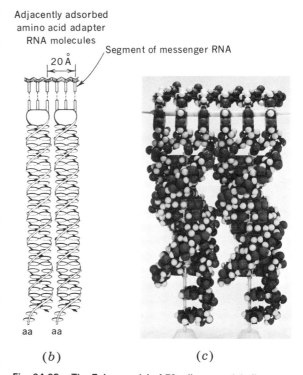

(b)   (c)

Fig. 24-28   The Zubay model of 70s ribosome: (a) ribosome with mRNA on the right and two strands of tRNA, lower left, in process of forming a polypeptide; (b) two segments of mRNA with two tRNA molecules attached; (c) the same molecules shown in model form. [From G. Zubay, *Science*, 140: 1092 (1963). Used by permission.]

Fig. 24-29 A model of _s_RNA. (After J. D. Watson, "The Molecular Biology of the Gene," W. A. Benjamin, Inc., New York, 1965.)

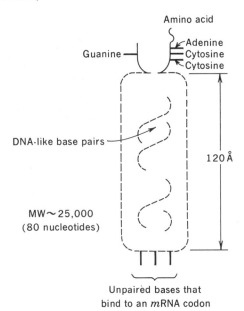

Amino acid

Guanine

Adenine
Cytosine
Cytosine

DNA-like base pairs

120 Å

MW ~ 25,000
(80 nucleotides)

Unpaired bases that
bind to an _m_RNA codon

bond sites but not to G with three hydrogen-bond sites. The details of the stereochemistry have yet to be worked out, but the outlines are clear.

From the kinetic point of view, the hydrogen bonds have a sufficiently weak enthalpy of activation for linking and breaking to be easily achieved at temperatures in the neighborhood of 37°C. As noted earlier, the energy necessary to break a hydrogen bond is about 10 kcal mole^{-1} as compared with 50 to over 100 kcal mole^{-1} to break covalent bonds. Presumably the value of $H^{\ddagger}$ for forming these bonds is of the same order of magnitude. The fact that water does not have too high a viscosity at room temperature shows that single or double hydrogen bonds are easily broken and re-formed. On the other hand, the simultaneous breaking of 100 hydrogen bonds would have a correspondingly higher value of $H^{\ddagger}$. Thus the two halves of DNA will not readily fall apart.

The final criterion of the speed of any of these processes is, of course, the value of $G^{\ddagger}$, which is equal to $H^{\ddagger} - TS^{\ddagger}$. As discussed previously in connection

with the synthesis of proteins and the peptide bond, the influence of geometry may affect both $H^{\ddagger}$ and $S^{\ddagger}$. If two parts which are brought together do not have the proper key-in-lock steric fit, the value of $H^{\ddagger}$ required to force them together may be so high that the reaction under these circumstances has a negligibly small rate. By way of contrast, both the relative numbers of proper sites and the degree of molecular orientation required for the reaction affect the probability of the reaction's taking place and therefore affect $S^{\ddagger}$.

The question of the multiplicity of bonds holding a substrate to an enzyme also affects $S^{\ddagger}$ and may be extremely important in processes involving RNA and DNA. For example, the specificity of the _t_RNA molecule with the codon GAU for aspartic acid may depend on the geometric arrangement of sites with forces of attraction on the end of the molecule opposite the codon. If there are three sites so arranged geometrically that they fit corresponding sites on the aspartic acid and that three bonds can operate simultaneously, then the chance is high that the aspartic acid is absorbed and held. If they do not fit and only one bond can be formed, the probability is much smaller that the molecule will be held on the site long enough for the amino acid to be linked on the end of the growing protein chain.

The action in establishing juxtaposition also affects $S^{\ddagger}$. The chance is almost infinitesimally small that even 1,000 amino acids would ever assemble into a specific sequence by random collision without a template. For such a random reaction $S^{\ddagger}$ is correspondingly small, though $H^{\ddagger}$ is not abnormally high. It is the $S^{\ddagger}$ factor that prevents the synthesis of unwanted DNA or RNA. On the other hand, the presence of the template raises the value of $S^{\ddagger}$ by many powers of 10 for the desired reactions which duplicate the DNA and provide the correct _m_RNA to synthesize the needed proteins. Both the DNA and the _m_RNA are carriers of $S^{\ddagger}$ and thereby carriers of _information_. It is the transmisson of information which directs the orderly progress of the life process. The relation between information and entropy is discussed in Chap. 25.

From the thermodynamic point of view there must be a decline in the value of G for the system if the reactions are to go forward spontaneously. In just the formation of one peptide bond in the synthesis of protein, about 5.5 kcal mole^{-1} is required. This energy is provided by the high-energy bonds in adenosine triphosphate (ATP) which is present in the cytoplasm. About three ATP bonds are used for the formation of each peptide bond.

The process of synthesizing a protein also represents an increase in order, or a decrease in the entropy of the system. For this reason, additional free energy must be provided, since a decrease in S means a less negative value for the term $-T\,\Delta S$, which appears as one of the components of $\Delta G = \Delta H - T\,\Delta S$. If ΔG is to be negative, as thermodynamics requires for a spontaneous reaction, the increase in ΔG caused by the decrease in ΔS must be more than compensated for by some factor which

causes a decrease in ΔG. The production of order must be paid for in terms of energy.

Example 24-9

If a protein is formed from 100 amino acids arranged in a specific sequence, instead of a random sequence of any of the 20 varieties of amino acids, what is the decrease in entropy?

In a random sequence there is a choice of any one of the 20 kinds of amino acids to fill a particular site on the chain. If there are 100 sites, the entropy due to disorder is

$$\Delta S_d = R \ln 20^{100}$$
$$= 1.987 \times 2.303 \times 100 \times \log 20$$
$$= 595 \text{ cal deg}^{-1} \text{ mole}^{-1}$$

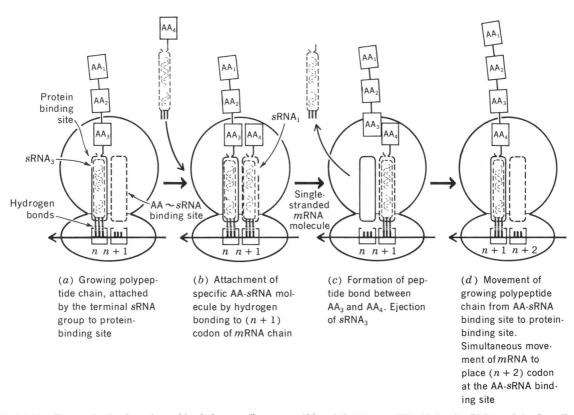

(a) Growing polypeptide chain, attached by the terminal sRNA group to protein-binding site

(b) Attachment of specific AA-sRNA molecule by hydrogen bonding to ($n + 1$) codon of mRNA chain

(c) Formation of peptide bond between AA$_3$ and AA$_4$. Ejection of sRNA$_3$

(d) Movement of growing polypeptide chain from AA-sRNA binding site to protein-binding site. Simultaneous movement of mRNA to place ($n + 2$) codon at the AA-sRNA binding site

Fig. 24-30 The synthesis of a polypeptide chain on a ribosome. (After J. D. Watson, "The Molecular Biology of the Gene," W. A. Benjamin, Inc., New York, 1965.)

Example 24-10

Calculate the decrease in enthalpy necessary to counteract the decrease in entropy caused by the ordering in the above problem at 37°C.

$$\Delta H_d = T \Delta S_d = 310 \times 595 = 184 \text{ kcal mole}^{-1}$$
$$Ans.$$

Problems

1. Calculate the molecular weight of the phosphate-deoxyribose-guanine group which constitutes a nucleotide.

2. Calculate to within 5 percent the average weight of the four nucleotides which are composed of a phosphate and a deoxyribose group coupled respectively to thymine, cytosine, guanine, and adenine.

3. According to Watson, the molecular weight of the DNA in *E. coli* is 2.5×10^9. Calculate the weight in tons of 1 g mole of this variety of DNA.

4. How many cross rungs are there on the ladder of the DNA of *E. coli*?

5. The average molecular weight of the 20 amino acids which commonly occur in proteins is 141. The average molecular weight of the protein in *E. coli* is 40,000. How many amino acid molecules are present on the average in one of these protein molecules?

6. Calculate the entropy difference per gram mole between one of the molecules of protein in *E. coli* with a specific sequence and a molecule with a similar number of amino acids selected at random from the 20 common types, assuming that the number of molecules of amino acid in the protein is the same as the figure obtained in Prob. 5.

7. Calculate the decrease in enthalpy per mole necessary to compensate at 37°C for the lack of entropy of disorder in the protein molecule from *E. coli* considered in Prob. 6.

8. Regarding each rung in the DNA ladder as a unit, calculate the difference in entropy per mole between the specific DNA in *E. coli* and DNA with the same number of nucleotides arranged at random.

9. Calculate the increase in G per mole at 37°C corresponding to the decrease in S due to order in the DNA in *E. coli*.

10. Calculate the increase in energy of a protein with a molecular weight of 1.32×10^5 due to the peptide bonds, using the average molecular weight of the 20 most common amino acids.

11. An *E. coli* cell contains approximately 10^3 molecules of *m*RNA with an average molecular weight of 10^6. Calculate the number of nucleotides in the cell due to the presence of the *m*RNA.

12. A cell of *E. coli* contains approximately 4×10^5 molecules of *t*RNA. The average molecular weight of such a molecule is 25,000. Calculate the ratio of the weight of *m*RNA to that of *t*RNA in the cell.

13. Assuming that the rate of production of a protein with molecular weight of 40,000 is one molecule in every 10 sec at 37°C in the ribosomes of *E. coli* and that $H^{\ddagger} = 8.5$ kcal mole^{-1}, calculate the value of $S^{\ddagger}$ for this reaction if rate is measured in molecules per second.

14. Calculate the rate of production of the protein in Prob. 13 at 40°C.

15. If the protein in Prob. 13 were to be formed by random collisions and $S^{\ddagger}$ were decreased by the amount calculated in Prob. 6, what would the rate of production be at 37°C?

16. If a DNA molecule is replicated at the rate of 2.7 hr^{-1} at 37°C and the rate of 1.2 hr^{-1} at 25°C, calculate the value of $H^{\ddagger}$.

17. Calculate the value of $S^{\ddagger}$ for the process in Prob. 16.

18. Calculate the rate if $S^{\ddagger}$ in Prob. 17 is reduced to the value for a completely random process, assuming that the DNA is similar to that found in *E. coli*.

19. Using the data from the previous problems, calculate the time in years needed to produce one molecule of DNA (*E. coli*) by a random process.

20. If the process described in Prob. 16 could be carried out at 100°C, what would be the length of time needed to produce one molecule?

REFERENCES

Nyburg, S. C.: "X-ray Analysis of Organic Structures," Academic Press, Inc., New York, 1961.

Burstone, M. S.: "Enzyme Histochemistry," Academic Press, Inc., New York, 1962.

Tanford, C.: "Physical Chemistry of Macromolecules," John Wiley & Sons, Inc., New York, 1962.

O'Driscoll, K. F.: "The Nature and Chemistry of High Polymers," Reinhold Publishing Corporation, New York, 1964.

Mandelkern, L.: "Crystallization of Polymers," McGraw-Hill Book Company, New York, 1964.

Walter, C.: "Steady State Applications in Enzyme Kinetics," The Ronald Press Company, New York, 1965.

Lehninger, A. L.: "Bioenergetics," W. A. Benjamin, Inc., New York, 1965.

Watson, J. D.: "The Molecular Biology of the Gene," W. A. Benjamin, Inc., New York, 1965.

Harrow, B., and A. Mazur: "Textbook of Biochemistry," 9th ed., chaps. 2–6, W. B. Saunders Company, Philadelphia, 1966.

De Garilhe, M. P.: "Enzymes in Nucleic Acid Research," Holden-Day, Inc., Publisher, San Francisco, 1967.

Howland, J. L.: "Introduction to Cell Physiology," The Macmillan Company, New York, 1968.

Rich, A., and N. Davidson: "Structural Chemistry and Molecular Biology," W. H. Freeman and Company, San Francisco, 1968.

In the previous chapter, we discussed how DNA is made up of a sequence of four kinds of nucleotides which spell out a genetic message in the GACT language. This message exerts a chemical influence by altering the values of $S^{\ddagger}$ for a network of reactions which synthesize proteins. The resultant chemical action is the tangible expression of the relationship between entropy and information.

25-1 ENTROPY, INFORMATION, AND ORDER

In DNA the genetic information is embodied in the *order* in which the nucleotides are arranged. This order is also closely related to the entropy of the DNA molecule.

Chapter Twenty-Five

entropy and information

Some years ago Schrödinger called attention to the problem of defining mathematically the degree of order in a molecule like DNA. He pointed out that DNA is, in a sense, a one-dimensional crystal. True, the DNA does not have its one-dimensional sequence in any simple periodic order, related to a small mathematical group as in the case of crystals like NaCl and diamond. But although the order is not simple, it is still undeniably *order* even though superficially it may be classed as *aperiodic*.

Entropy and order

In their classic investigation of the heat capacity of crystalline carbon monoxide, CO, described in Chap. 9, Clayton and Giauque pointed out the effect on the entropy due to the increasing lack of order caused by disorientation of the individual CO molecules. In order to see more clearly the relation of this type of disorder to the aperiodicity in DNA, the American chemist D. H. Andrews and the American biologist Manley L. Boss have compared DNA with a stylized crystal of CO where a single completely ordered row of molecules may be represented as:

$$. . . \text{C C C C C C C C C C} . . .$$
$$\text{O O O O O O O O O O} \qquad (i)$$

Such a single row is in effect a single linear periodic crystal. There is no entropy due to disorder.

By way of contrast, it is possible to have a single row of molecules where the orientation is random

$$\cdots \begin{matrix} C\;C\;O\;C\;O\;O\;O\;C\;O\;C \\ O\;O\;C\;O\;C\;C\;C\;O\;C\;O \end{matrix} \cdots \qquad (ii)$$

Clayton and Giauque established the relation between this randomness (disorder) and the entropy by considering the number of possible states in which such a row can be arranged. Any individual CO group may be oriented in one of two ways, either with the carbon on top or with the oxygen on top. The number of ways of orienting two molecules is $2 \times 2 = 2^2 = 4$

C	C		C	O		O	C		O	O
O	O		O	C		C	O		C	C
(a)			(b)			(c)			(d)	

In general, the number of ways of orienting n molecules is 2^n. The entropy S_r due to randomness in orientation is, therefore, defined as

$$S_r \equiv k \ln 2^n \qquad (25\text{-}1)$$

If the row of CO contains a mole of molecules, then $n = N_A$, where N_A denotes Avogadro's number and

$$s = k \ln 2^{N_A} = kN_A \ln 2 = R \ln 2 \qquad (25\text{-}2)$$

In the perfect crystal, where all the molecules are arranged in the same way, the number of randomly oriented molecules n is equal to zero. The entropy of disorder in the ideal crystalline or *completely ordered* state S_{r_0} is thus:

$$S_{r_0} = k \ln 2^0 = k \ln 1 = 0 \qquad (25\text{-}3)$$

The difference in entropy between the disordered state and the ordered state ΔS_r for 1 mole is thus

$$\Delta S_r = S_r - S_{r_0} = k \ln \frac{2^{N_A}}{2^0} = kN_A \ln 2 = R \ln 2 \qquad (25\text{-}4)$$

In order to make the relation between the linear "crystal" of DNA and the hypothetical linear crystal of CO as clear as possible, consider a molecule of DNA made up entirely of two types of rungs, one with thymine-adenine (denoted by TA) and one with adenine-thymine (denoted by AT). A perfectly ordered sequence can be denoted as:

$$\cdots \begin{matrix} A\;A\;A\;A\;A\;A\;A\;A\;A\;A \\ T\;T\;T\;T\;T\;T\;T\;T\;T\;T \end{matrix} \cdots \qquad (iii)$$

This is strictly comparable with the order sequence of CO molecules in (i).

A random sequence of DNA rungs of this sort can be denoted as

$$\cdots \begin{matrix} A\;A\;T\;A\;T\;T\;T\;A\;T\;A \\ T\;T\;A\;T\;A\;A\;A\;T\;A\;T \end{matrix} \cdots \qquad (iv)$$

This is strictly comparable with the order sequence of CO molecules in (ii).

Clayton and Giauque concluded that in 1 mole of CO cooled close to $0°K$ under conditions which resulted in the type of order expressed in (ii), there was a residual entropy of disorder per mole of $R \ln 2$; and the comparison of the experimentally measured difference in entropy between the crystals at $1°K$ and the same material in the gaseous state at $298°K$ appeared to confirm this conclusion. Therefore, does a hypothetical DNA molecule containing N_A nucleotides arranged in disorder as shown in (iv) also have a residual entropy of $R \ln 2$?

As Schrödinger pointed out, there is a fundamental theoretical difference between the order of the molecules in Clayton and Giauque's CO crystal and the aperiodic order in DNA. The CO molecules are brought into random order by the operation of thermal actions which are assumed to represent true theoretical randomness. The *apparently* random order in a species of DNA has been achieved by the processes of evolution, small variations of order introduced over billions of years, but with the sequence transmitted from organism to organism through billions of precise replicative processes. Clayton and Giauque's CO aperiodic order does not contain significant information; Schrödinger's DNA aperiodic order does.

Information and order

The fundamental knowledge of the relation between information and order was developed largely as the result of studies during and just after World War II by a group of scientists interested in a better understanding of ways of transmitting information, including secret intelligence codes, instructions to antiaircraft batteries, and a variety of other applications.

Basically these involve the comparison of the order achieved by the operation of pure chance with the order conveying significant information.

One of the most informative studies of the operation of chance or probability in physical science was carried out by Maxwell during his investigations in the latter part of the nineteenth century on the thermal properties of gases. He suggested the concept of a supernatural being, now generally called *Maxwell's demon,* who could perceive individual molecules and influence their behavior.

In order to see the relation between probability, order, and information for a linear crystal of CO molecules, let us suppose that such a crystal is being formed on a plane surface under the observing eye of such a demon. The demon does not try to influence the position of the molecules but merely observes them as they come in from the gaseous state with purely random orientations. Thus the first CO may fall with the carbon end upward or the oxygen end upward; the second molecule is oriented purely by chance, and so on. Let us consider the case of 10 molecules ($n = 10$) arranged in a line similar to the pattern in (ii).

In the language of information theory, before any of the molecules have landed on the surface (the state designated by subscript zero), there are p_0 possibilities of order. $p_0 = 2^{10}$, or in general $p_0 = m^n$, where $m = 2$, the number of possible orientations of a single molecule. After the molecules have fallen in line (the state designated by subscript 1), there is only one *realized* possibility so that $p_1 = 1$. Before the molecules fell, the demon had no information I_0 about which out of the 2^{10} possibilities might actually be fulfilled ($I_0 = 0$); after the molecules have fallen, the demon knows which sequence has been fulfilled ($I_1 > 0$). Summarizing in this notation, which was suggested by Brillouin,

Initial state: p_0 possibilities $I_0 = 0$
Final state: p_1 possibilities $I_1 > 0$

The relation connecting information and probability is defined as

$$I_1 - I_0 \equiv -K \ln \frac{p_1}{p_0} \qquad (25\text{-}5a)$$

In information theory K is customarily set equal to $1/(\ln 2)$ so that the units of I are the binary units called *bits*. In the case under consideration,

$$I_1 - I_0 = -\frac{1}{\ln 2} \ln \frac{1}{2^{10}}$$

$$= -\frac{1}{\ln 2} 10 \, (-\ln 2) = 10 \text{ bits} \qquad (25\text{-}5b)$$

After the 10 molecules have fallen, the demon has 10 bits of information. Since a majority of the computing machines dealing with information have components of the binary type with only the two positions "off" or "on," binary units are a particularly useful way of measuring information.

Suppose that instead of observing the crystallizing of CO, the demon watches the formation of a molecule of DNA from adenine-thymine nucleotides. The information in hand after 10 nucleotides have formed is exactly the same as for CO. Since the AT nucleotide can fit in place either as AT or TA, $m = 2$ and $n = 10$ and $I_1 - I_0 = 10$ bits.

Unit conversion

In order to compare entropy and information, it is necessary to express both quantities in the same units. Since our interest here is focused on the physicochemical aspects of biochemistry, the use of entropy units (calories per degree per mole) will be helpful. Information in bits may be converted into information in cal deg^{-1} mole^{-1} by multiplying with the factor k/K, where k is Boltzmann's constant (1.38054×10^{-16} erg deg^{-1} molecule^{-1}), or by multiplying by R/K, where R is derived from k by the expression

$$R = 1.38054 \times 10^{-16} \text{ erg deg}^{-1} \text{ molecule}^{-1}$$
$$\times 10^{-7} \text{ joule erg}^{-1} \times 0.23901 \text{ cal joule}^{-1}$$
$$\times 6.02252 \times 10^{23} \text{ molecule mole}^{-1}$$
$$= 1.9872 \text{ cal deg}^{-1} \text{ mole}^{-1}$$

Thus

$$\frac{k}{K} = \frac{1.38054 \times 10^{-16}}{(\ln 2)^{-1}}$$

$$= \frac{1.38054 \times 10^{-16}}{0.69315^{-1}}$$

$$= \frac{1.38054 \times 10^{-16}}{1.44269}$$

$$= 9.56921 \times 10^{-17} \text{ erg deg}^{-1} \text{ molecule}^{-1} \text{ bit}^{-1}$$

and

$$\frac{R}{K} = \frac{1.9872}{1.4427} = 1.3774 \text{ cal deg}^{-1} \text{ mole}^{-1} \text{ bit}^{-1}$$

If Boltzmann's constant is denoted by k_c and is expressed in calories per degree per molecule, then $k_c/K = (1.38054 \times 10^{-16} \text{ erg deg}^{-1} \text{ molecule}^{-1} \times 10^{-7} \text{ joule erg}^{-1} \times 0.23901 \text{ cal joule}^{-1})/1.44269 = 2.2871 \times 10^{-24} \text{ cal deg}^{-1} \text{ molecule}^{-1} \text{ bit}^{-1}$.

Basic interrelations

If information is converted into calories per degree per molecule and entropy is also expressed in the same units, the relation between these two quantities in various species of DNA becomes relatively simple. Where $m = 2$, as in CO or in a DNA composed only of AT and TA,

$$I_1 - I_0 = -k_c \ln 2^n \tag{25-6}$$

$$S_1 - S_0 = k_c \ln 2^n \tag{25-7}$$

Thus the basic relation is clear:

$$\Delta I_{\text{ord}} = (I_1 - I_0) = -(S_1 - S_0) = -\Delta S_{\text{ord}} \tag{25-8}$$

The term on the left is the *increment* of information when for n components a specific order is known as compared with a state of no information about order; the term on the right is the decrement in entropy when n components are ordered as compared with being in a random sequence. This relation suggested to Brillouin the term *negentropy* as an alternative for the term *information*. The *change* in information is the negative of the change in entropy in the passage between two states. These states are degrees of knowledge in the case of information; but in the case of entropy, the relation raises basic questions about the definition of order. As Schrödinger pointed out, the justification for accepting the aperiodic order in a DNA molecule as order in the thermodynamic sense is far from being obvious.

Example 25-1

In the species of DNA found in *E. coli* there are 4.6×10^6 binucleotides (rungs on the ladder).

Calculate the amount of information in bits due to the order in a molecule of DNA of this length if all the rungs are either AT or TA.

$$\Delta I_{\text{ord}} = K \ln 2^n = K \ln 2^{4.5 \times 10^6}$$
$$= 4.5 \times 10^6 \text{ bits} \qquad Ans.$$

Example 25-2

Calculate the information content in the hypothetical species of DNA described in the previous problem both in calories per degree per molecule and in calories per degree per mole.

$$\Delta I_c = \frac{k_c}{K} n = 2.287 \times 10^{-24} \times 4.5 \times 10^6$$

$$= 1.03 \times 10^{-17} \text{ cal deg}^{-1} \text{ molecule}^{-1}$$
$$\Delta I_c = \Delta I_c \times 6.02 \times 10^{23} \text{ molecules mole}^{-1}$$
$$= 6.2 \times 10^6 \text{ cal deg}^{-1} \text{ mole}^{-1} \qquad Ans.$$

25-2   THE INFORMATION CONTENT OF DNA

The primary function of DNA in the genetic process appears to be the storage and transmission of information which guides the synthesis of proteins. Proteins are composed essentially of 20 different kinds of amino acids; on the DNA ladder, there are only four different kinds of rungs, G, A, C, T. This means that more than one rung is required to carry the complete code for any one position on the protein chain. One rung could be coded to specify only four different kinds of protein constituents; even two rungs jointly could specify only 4 × 4, or 16, different kinds of constituents. Consequently, it is clear that there must be at least three rungs to specify unambiguously the kind of amino acid to be added at a particular point in a growing protein chain. This three-rung triplet appears to be the mechanism by which information is stored and transmitted in DNA.

Codons

The triplet, composed of three adjacent rungs in DNA and specifying a single kind of amino acid, is called a codon. Since there are four kinds of rungs, G, A, C, T, a triplet codon can specify 4 × 4 × 4, or 64, different kinds of substituents. Since there are only 20 kinds of amino acids, a code consisting of 64 codons provides more symbols than are needed. Such a code with superfluous symbols is called *redundant*.

The information from the DNA codons is transferred to similar codons in RNA. Because each nucleotide on DNA is matched by a complementary nucleotide on RNA, the new codons on RNA are complementary to the DNA codons, thus: G → C, A → U, C → G, T → A; and the message in the GACT language on DNA becomes a message in the CUGA language on RNA, since each word, or codon, on RNA is made up of three letters each selected from one of the four code letters C, U, G, or A. In discussions of the genetic code, it is customary to use the CUGA codons rather than the GACT codons, although either set is unambiguous and the two sets are in one-to-one correspondence.

The genetic code group

Andrews and Boss have found that the 64 codons constitute a representation of a mathematical group. The codons, the corresponding amino acids, and the group symbols are listed in Table 25-1.

The element of a group designates a transformation from one state to another. Regarded as a permutation, a group element designates the exchange of one or more original symbols for one or more different symbols in a permutation operation. Regarded as a rotation or reflection, as in crystal-lattice theory, the element of a group designates a transformation of state when the operation is performed.

The group element made up of a codon may be regarded as specifying the transformation which occurs when a codon arbitrarily selected as the reference state is transformed to the codon specified by the group element. In Table 25-1, the codon UUU is selected as the reference state. Thus the group element UUU is the identity transformation, symbolized by I. Then, the codon UUC symbolizes the replacement of the third U by C in forming RNA. The codon UCC symbolizes the replacement of both the second and third U by C, and so on.

The group represented by the genetic code is a subgroup of the full permutation duodecic group on 12 letters. This subgroup can be represented by three rotary display disks placed side by side like the

Fig. 25-1 A genetic codon presented by the selective rotation of each of three disks, each inscribed with the letters C, U, G, and A. (*Note:* A is at the back, opposite U, and does not show. The letters are 90° apart.)

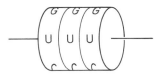

disks in a three-digit counter. As shown in Fig. 25-1, each disk contains the four symbols C, U, G, A. In the 12-letter permutation symbol, the first four letters (a, b, c, d) correspond to the C, U, G, and A on the first disk; (e, f, g, h) corresponds to the symbols C, U, G, A on the second disk; (i, j, k, l) correspond to the symbols C, U, G, A on the third disk. The first, fifth, and ninth symbols are the symbols which are read as the codon. For example, the code symbol UUC indicates the replacement of the third U in the reference state UUU by C. In the reference state the counter reads UUU; the wheel must be rotated forward by 90° to effect this transformation; so U becomes C, G becomes U, A becomes G, and C becomes A. In the usual permutation notation this element is indicated by the symbol (l, k, j, i) since i designates C, j designates U, k designates G, and l designates A on the third wheel.

The code-group table has several structural features of interest. To bring these out more clearly the table has been rearranged in Table 25-2. First of all, the amino acids with the more neutral side chains average 4.4 codons each. By way of contrast, the two amino acids with acidic side chains have only 2 codons each; and two of the amino acids with basic side chains have only 2 codons each, although arg, usually classified as basic, has 6. If the distribution of codons is intended to provide roughly equal probability for the reactions by whch a *t*RNA finds its matching amino acid, this distribution suggests that in the cytoplasm, acidic or basic side chains are easier to recognize than neutral ones.

Another interesting feature is that the redundancy occurs almost completely in the third letter of the codon. For example, UCX always means serine, irrespective of the letter in the X position.

TABLE 25-1 Correlation between symmetric group 12, the genetic code, and the amino acids

Permutation group symbol	Codon	Amino acid	No. of codons	Chemical character
I $(i\,l\,k\,j)$	UUU UUC	Phe	2	 aromatic
$(i\,j)(k\,l)$ $(i\,j\,k\,l)$	UUA UUG	Leu	6	 neutral
$(h\,g\,f\,e)$ $(h\,g\,f\,e)(l\,k\,j\,i)$ $(h\,g\,f\,e)(i\,j)(k\,l)$ $(h\,g\,f\,e)(i\,j\,k\,l)$	UCU UCC UCA UCG	Ser	6	$HO-CH_2-Y$ hydroxyl (neutral)
$(e\,f)(g\,h)$ $(e\,f)(g\,h)(l\,k\,j\,i)$	UAU UAC	Tyr	2	 aromatic
$(e\,f)(g\,h)(i\,j)(k\,l)$	UAA	N2(ocher)	3	Punctuation
$(e\,f)(g\,h)(i\,j\,k\,l)$	UAG	N1(amber)	3	Punctuation
$(e\,f\,g\,h)$ $(e\,f\,g\,h)(l\,k\,j\,i)$	UGU UGC	Cys	2	YCH_2SSCH_2Y sulfur
$(e\,f\,g\,h)(i\,j)(k\,l)$	UGA	N3	3	Punctuation
$(e\,f\,g\,h)(i\,j\,k\,l)$	UGG	Trp	1	
$(d\,c\,b\,a)(l\,k\,j\,i)$ $(d\,c\,b\,a)$ $(d\,c\,b\,a)(i\,j)(k\,l)$ $(d\,c\,b\,a)(i\,j\,k\,l)$	CUU CUC CUA CUG	Leu	6	 neutral
$(d\,c\,b\,a)(h\,g\,f\,e)$ $(d\,c\,b\,a)(h\,g\,f\,e)(l\,k\,j\,i)$ $(d\,c\,b\,a)(h\,g\,f\,e)(i\,j)(k\,l)$ $(d\,c\,b\,a)(h\,g\,f\,e)(i\,j\,k\,l)$	CCU CCC CCA CCG	Pro	4	 secondary
$(d\,c\,b\,a)(e\,f)(g\,h)$ $(d\,c\,b\,a)(e\,f)(g\,h)(l\,k\,j\,i)$	CAU CAC	His	2	 basic
$(d\,c\,b\,a)(e\,f)(g\,h)(i\,j)(k\,l)$ $(d\,c\,b\,a)(e\,f)(g\,h)(i\,j\,k\,l)$	CAA CAG	Gln	2	 amide

Permutation group symbol	Codon	Amino acid	No. of codons	Chemical character
$(d\,c\,b\,a)(e\,f\,g\,h)$ $(d\,c\,b\,a)(e\,f\,g\,h)(l\,k\,j\,i)$ $(d\,c\,b\,a)(e\,f\,g\,h)(i\,j)(k\,l)$ $(d\,c\,b\,a)(e\,f\,g\,h)(i\,j\,k\,l)$	CGU CGC CGA CGG	Arg	6	$NH_2-\overset{\overset{H}{\mid}}{\underset{\underset{H}{\overset{\|}{N}}}{C}}-N-CH_2-CH_2-CH_2-Y$ basic
$(a\,b)(c\,d)$ $(a\,b)(c\,d)(l\,k\,j\,i)$ $(a\,b)(c\,d)(i\,j)(k\,l)$	AUU AUC AUA	Ile	3	CH_3 $\quad\quad\quad CHY$ CH_3-CH_2 neutral
$(a\,b)(c\,d)(i\,j\,k\,l)$	AUG	Met	1	$CH_3-S-CH_2-CH_2-Y$ sulfur
$(a\,b)(c\,d)(h\,g\,f\,e)(l\,k\,j\,i)$ $(a\,b)(c\,d)(h\,g\,f\,e)$ $(a\,b)(c\,d)(h\,g\,f\,e)(i\,j)(k\,l)$ $(a\,b)(c\,d)(h\,g\,f\,e)(i\,j\,k\,l)$	ACU ACC ACA ACG	Thr	4	CH_3-CH-Y $\quad\quad\;\; O$ $\quad\quad\;\; H$ hydroxyl (neutral)
$(a\,b)(c\,d)(e\,f)(g\,h)(l\,k\,j\,i)$ $(a\,b)(c\,d)(e\,f)(g\,h)$	AAU AAC	Asn	2	$\overset{\overset{O}{\|}}{H_2N-C-CH_2-Y}$ basic
$(a\,b)(c\,d)(e\,f)(g\,h)(i\,j)(k\,l)$ $(a\,b)(c\,d)(e\,f)(g\,h)(i\,j\,k\,l)$	AAA AAG	Lys	2	$H_2N-CH_2-CH_2-CH_2-CH_2-Y$ basic
$(a\,b)(c\,d)(e\,f\,g\,h)(l\,k\,j\,i)$ $(a\,b)(c\,d)(e\,f\,g\,h)$	AGU AGC	Ser	6	$HO-CH_2-Y$ hydroxyl
$(a\,b)(c\,d)(e\,f\,g\,h)(i\,j)(k\,l)$ $(a\,b)(c\,d)(e\,f\,g\,h)(i\,j\,k\,l)$	AGA AGG	Arg	6	$\overset{}{\underset{H_2}{N}}-\overset{\overset{}{\underset{\|}{N}}}{C}-\overset{H}{N}-CH_2-CH_2-CH_2-Y$ $\quad\quad\;\; H$ basic
$(a\,b\,c\,d)$ $(a\,b\,c\,d)(l\,k\,j\,i)$ $(a\,b\,c\,d)(i\,j)(k\,l)$ $(a\,b\,c\,d)(i\,j\,k\,l)$	GUU GUC GUA GUG	Val	4	CH_3 $\quad\quad CH-Y$ CH_3 neutral
$(a\,b\,c\,d)(h\,g\,f\,e)$ $(a\,d\,c\,d)(h\,g\,f\,e)(l\,k\,j\,i)$ $(a\,b\,c\,d)(h\,g\,f\,e)(i\,j)(k\,l)$ $(a\,b\,c\,d)(h\,g\,f\,e)(i\,j\,k\,l)$	GCU GCC GCA GCG	Ala	4	CH_3-Y neutral
$(a\,b\,c\,d)(e\,f)(g\,h)$ $(a\,b\,c\,d)(e\,f)(g\,h)(l\,k\,j\,i)$	GAU GAC	Asp	2	$HOOC-CH_2-Y$ acidic
$(a\,b\,c\,d)(e\,f)(g\,h)(i\,j)(k\,l)$ $(a\,b\,c\,d)(e\,f)(g\,h)(i\,j\,k\,l)$	GAA GAG	Glu	2	$HOOC-CH_2-CH_2-Y$ acidic
$(a\,b\,c\,d)(e\,f\,g\,h)$ $(a\,b\,c\,d)(e\,f\,g\,h)(l\,k\,j\,i)$ $(a\,b\,c\,d)(e\,f\,g\,h)(i\,j)(k\,l)$ $(a\,b\,c\,d)(e\,f\,g\,h)(i\,j\,k\,l)$	GGU GGC GGA GGG	Gly	4	$H-Y$ neutral

TABLE 25-2   Codons arranged by the chemical characteristics of the side chains on the corresponding amino acid

No. of codons	Amino acid	Codons
Neutral:		
6	Ser	UCU, UCC, UCA, UCG, AGU, AGC
6	Leu	UUA, UUG, CUU, CUC, CUA, CUG
4	Val	GUU, GUC, GUA, GUG
4	Ala	GCU, GCC, GCA, GCG
4	Gly	GGU, GGC, GGA, GGG
4	Thr	ACU, ACC, ACA, ACG
3	Ile	AUU, AUC, AUA
4.4 av		
Acidic:		
2	Asp	GAU, GAC
2	Glu	GAA, GAG
2 av		
Basic:		
6	Arg	CGU, CGC, CGA, CGG, AGA, AGG
2	Lys	AAA, AAG
2	His	CAU, CAC
3.3 av		
Amides:		
2	Asn	AAU, AAC
2	Gln	CAA, CAG
2 av		
Aromatic:		
2	Phe	UUU, UUC
2	Tyr	UAU, UAC
1	Trp	UGG
1.7 av		
Sulfur:		
2	Cys	UGU, UGC
1	Met	AUG
1.5 av		
Secondary:		
4	Pro	CCC, CCU, CCA, CCG
Punctuation:		UAA, UAG, UGA

Redundancy

The term *redundancy* refers to the use of several symbols or words to denote a meaning for which one would suffice. Because the genetic code employs 64 symbols (the codons) for only 23 meanings (the amino acids plus three punctuation marks), it is redundant. Because of the redundancy, the information content of DNA is not as large as it would be if the 64 codons all had different meanings.

The relation between information and codons can be seen by considering several species of DNA with different meanings. Suppose that there is a kind of DNA composed only of two different kinds of rungs (A, T) and that each rung conveys a separate message. As derived earlier in this chapter, the relation between information content and the number of rungs is

$$\Delta I \text{(bits)} = K \ln 2^n = n \qquad (25\text{-}9)$$

$$\Delta I \text{(cal deg}^{-1}) = k_c \ln 2^n = \frac{k_c}{K} n \qquad (25\text{-}10)$$

where $K = (\ln 2)^{-1}$ and $k_c = 3.300 \times 10^{-24}$ cal deg^{-1} mole^{-1}.

If, instead of two kinds of rungs, there are four (G, A, C, T), then

$$\Delta I \text{(cal deg}^{-1}) = k_c \ln 4^n$$

$$= k_c \ln 2^{2n} = \frac{k_c}{K} 2n \qquad (25\text{-}11)$$

If instead of being conveyed in n rungs, the information is conveyed in terms of $n/3$ codons, of 23 different kinds

$$\Delta I \text{(bits)} = K \ln 23^{n/3} = 1.51n \qquad (25\text{-}12)$$

$$\Delta I \text{(cal deg}^{-1}) = k_c \ln 23^{n/3} \qquad (25\text{-}13)$$

$$= \frac{k_c n}{3} \ln 23$$

$$= 1.0452 k_c n = 3.4488 \times 10^{-24} n$$

The calculation of bioinformation

If the molecular weight of a species of DNA is known, an estimate of the number of rungs n can be made by dividing the molecular weight M by the average molecular weight M_r of the two kinds of double rungs (AT, GC). M_r has the value 610.0. Thus, for *E. coli*, $M = 2.8 \times 10^9$ so that $n = M/M_r = 2.8 \times 10^9/ 610.9 = 4.58 \times 10^6$. Then applying Eq. (25-12) to one *molecule* of DNA in *E. coli*,

$$\Delta I = 2.39 \times 10^{-24} n = 1.09 \times 10^{-17} \text{ cal deg}^{-1} \text{mole}^{-1}$$

For 1 mole of *E. coli* DNA, the information content is

$$\Delta I_M = 1.09 \times 10^{-17} \times 6.02 \times 10^{23}$$
$$= 6.59 \times 10^6 \text{ cal deg}^{-1} \text{ mole}^{-1}$$

Since $\Delta S_M = -\Delta I_M$, the value of ΔS_M is -6.59×10^6 cal deg^{-1} mole^{-1}. This is the decrease in entropy in the system when 1 mole of DNA with 1.52×10^6 codons changes from the state where the codons are arranged in random sequence to the state where in the entire mole all codons are in the sequence corresponding to *E. coli*.

It is interesting to compare the average amount of information contained in the sequence of codons in a DNA molecule with the average amount of information contained in familiar pages of printed language. On a typical page of English, the average word (including the blank space following it) contains 5.5 symbols. The content of information, allowing for redundancy in the language, is 2.14 bits per symbol. The average page of a typical encyclopedia contains about 5×10^3 bits of information. The DNA molecule in *E. coli* contains 4.50×10^6 rungs. Thus

$$\Delta I \text{(bits)} = K \ln 23^{n/3} = n \frac{K}{3} \ln 23$$

$$= n \frac{3.1355}{3 \times 0.69314} = 1.50n$$
$$= 1.50 \times 4.58 \times 10^6$$
$$= 6.9 \times 10^6 \text{ bits} \qquad (25\text{-}14)$$

The equivalent number of pages of English M_p is:

$$\frac{\Delta I \text{(bits in } E.\ coli \text{ DNA)}}{\Delta I \text{(bits per page of English)}} = \frac{6.9 \times 10^6}{5 \times 10^3} = 10^3 \qquad (25\text{-}15)$$

Thus the DNA molecule in *E. coli* contains about as much information as a thousand pages of a typical encyclopedia.

Example 25-3

The average DNA molecule in a human cell is estimated to have about 5×10^8 binucleotides (rungs). Estimate the information content in the nucleotide sequence.

Using Eq. (25-12),

$$\Delta I \text{(bits)} = 1.51 \times 5 \times 10^8 = 8 \times 10^8 \text{ bits}$$
$$Ans.$$

Example 25-4

Calculate the number of pages n_p of a typical encyclopedia which would contain information that is the equivalent of the information in a typical molecule of human DNA.

$$n_p = \frac{8 \times 10^8 \text{ bits in DNA}}{5 \times 10^3 \text{ bits per page}}$$
$$= 1.6 \times 10^5 \text{ pages} \quad Ans.$$

25-3 THE ENTROPY OF DNA

From the biochemical point of view there is far more interest in knowing how the information *governs* other chemical processes than in knowing how *much* information is present. Because information is so closely related to entropy, and because entropy affects both thermodynamic free energy G and the free energy of activation $G^\ddagger$, this information-entropy relationship provides a bridge between the concepts of information theory and the concepts of biokinetics.

Thermodynamic bioentropy

In order to establish quantitative relationships between the information content and the bioentropy of DNA, it is necessary, first of all, to define precisely what is meant in speaking of DNA.

From the genetic standpoint the term DNA is used as a kind of collective noun to embrace many species of molecules all of which have certain common elements of structure such as the phosphate-ribose ladder sides and the binucleotide ladder rungs. In order to develop quantitative thermodynamic or kinetic equations it is necessary to specify the kind of DNA in question and the amount of DNA to which quantities such as G and $G^\ddagger$ refer. In speaking of *E. coli* DNA, the assumption is made that in *E. coli* there is a molecule with DNA structure having a precise number n of binucleotide rungs consisting of AT, TA, CG, or GC, arranged in a defined sequence. Almost all thermodynamic and kinetic calculations deal with enthalpy, entropy, and free energy per mole. If *mole* refers to a mole of *E. coli* DNA, then s means, for example, the entropy of a set of 6.02×10^{23} molecules, each having the exact length and sequence of nucleotides characteristic of *E. coli*. Since the molecular weight of *E. coli* DNA is 2.5×10^9, this set of molecules weighs over a million kilograms.

In theory, the entropy s of this mole of *E. coli* DNA could be determined by having these molecules (or an appropriate portion of them, suitable for a calorimeter of reasonable size) assembled as a perfect or essentially perfect crystal. By measuring the heat capacity C_p from about 1 to 298°K, the entropy at the latter temperature (25°C) could be determined by the relation:

$$s = s_0 + \int_0^{298} C_p \, d \ln T \qquad (25\text{-}16)$$

It is, of course, assumed that the heat capacity between 0 and 1°K is so small that the entropy increases between 0 and 1°K can be neglected. It is also assumed that this DNA is a perfect crystal so that s_0, the entropy at 0°K, is zero.

The comparison of this perfect crystal, state (*i*), and a crystal of DNA with another sequence, state (*ii*), is informative. Suppose that all the AT and TA in *E. coli* are disassembled and put together again in the form of new molecules in a sequence at the beginning of each new molecule and all oriented to the AT position; similarly, the CG and GC groups are all put in the latter half of the new molecules, and oriented CG, state (*ii*). The original state (*i*) can be represented as something like

$$\cdots \begin{array}{c} \text{A T G A C T G A C C} \\ \text{T A C T G A C T G G} \end{array} \cdots \qquad (i)$$

After the reorientation and resequencing, the molecule, state (*ii*), looks like

$$\cdots \begin{array}{c} \text{A A A A A} \\ \text{T T T T T} \end{array} \cdots \begin{array}{c} \text{C C C C C} \\ \text{G G G G G} \end{array} \cdots \qquad (ii)$$

Schrödinger asked the crucial question: Does (*i*) have any greater entropy than (*ii*)? In other words, is the order in (*i*) just as perfect as the order in (*ii*)? There seems no reason to think that (*i*) actually is any more random than (*ii*), although admittedly it does look that way.

State (*ii*) is comparable to Clayton and Giauque's ideal CO crystal in which all CO molecules are oriented in a single way. If state (*i*) is regarded as *significantly* oriented and not random, it *also* is comparable to the perfect crystal of CO with ordered alignment. Then what state of DNA is similar to Clayton and Giauque's *randomly* oriented CO crystals? If the question is asked with regard to the nucleotide sequence, the answer is that mole of DNA, with each molecule having the same *number* of AT and GC groups as in *E. coli,* might be assembled with the *sequence* in each molecule randomly determined.   Under these circumstances, there would probably be a different sequence in almost every molecule.   Each molecule would then represent a different species of DNA and there would be an entropy of mixing at $0°K$, somewhat like the entropy of randomness in Clayton and Giauque's randomly oriented CO.

Thus, if the reference is to a single species of DNA in a perfect crystal, the entropy term $\Delta S_{\text{ord}} = -\Delta I$ will not appear in the entropy determined by heat-capacity measurements of the type made by Clayton and Giauque.

The third-law paradox

If one were to take Brillouin's definition of information as negentropy too broadly, application of the third law of thermodynamics to DNA might lead to a paradox.   A perfect crystal of DNA cooled to $0°K$ has no entropy according to the third law.   If information is negentropy, or the negative of entropy, and the entropy is zero, then the information content is zero; and this is obviously false.   The sequence of nucleotides certainly is present at $0°K$, and the information is there just as much as at room temperature.

This paradox emphasizes the need to keep in mind that the equations relating information content and entropy apply to *changes* of state only and not to absolute values of information or entropy.   Clayton and Giauque's crystal of perfectly oriented CO contains a message and therefore information at $0°K$ even though the entropy is zero.   Using a notation similar to that previously employed to denote the sequence in DNA, the CO crystal with all C atoms on top of the line can be written:

$$\text{Molecules:} \quad \cdots \begin{array}{c} \text{C C C C C C C} \\ \text{O O O O O O O} \end{array} \cdots \qquad (iii)$$

$$\text{Message:} \quad \cdots \text{C C C C C C C} \cdots \qquad (iv)$$

A comparable message in the AT form of DNA is

$$\text{Molecules:} \quad \cdots \begin{array}{c} \text{A A A A A A A} \\ \text{T T T T T T T} \end{array} \cdots \qquad (v)$$

$$\text{Message:} \quad \cdots \text{A A A A A A A} \cdots \qquad (vi)$$

By way of contrast, the line of molecules in randomly oriented CO or AT might be:

$$\text{Molecules:} \quad \cdots \begin{array}{c} \text{C C O C O C C O} \\ \text{O O C O C O O C} \end{array} \cdots \qquad (vii)$$

$$\text{Message:} \quad \cdots \text{C C O C O C C C} \cdots \qquad (viii)$$

or

$$\text{Molecules:} \quad \cdots \begin{array}{c} \text{A A T A T A A T} \\ \text{T T A T A T T A} \end{array} \cdots \qquad (ix)$$

$$\text{Message:} \quad \cdots \text{A A T A T A A T} \cdots \qquad (x)$$

Both (iv) and (vi) as well as $(viii)$ and (x) are specific messages selected out of the set of 2^n possible messages which could be formed from n molecules of this binary type.   Although the order appears simpler in (iv) and (vi) than in $(viii)$ and (x), from neither the informational nor the thermodynamic point of view is there *significant* simplicity; the greater complexity of order in (ix) and (x) is significant only if this order is established by processes in which temperature action makes the order random.

In Clayton and Giauque's CO, when randomness due to thermal motion comes into play, the random order has measurable entropy.   Suppose that Clayton and Giauque warmed a perfect crystal of CO to the gaseous state at $298°K$ and found a value ΔS_p. If a randomly oriented crystal of CO were warmed in the same way and the entropy increase were ΔS_r, then $\Delta S_p - \Delta S_r = S_r$, the positive entropy due to randomness at $0°K$.   Similarly, if a crystal of perfectly ordered AT were warmed from $0°K$ to a temperature T_g, where the DNA hypothetically disintegrated into AT binucleotides in the gaseous state, there might be an entropy increase ΔS_p; if another mole with the same numbers and kinds of binucleotides were formed so that the nucleotides were in random sequences, and if this random kind of DNA were warmed from effectively $0°K$ to the same temperature, the measured increase in entropy would be ΔS_r.   As in the case of CO, $\Delta S_p - \Delta S_r = S_r$ which is the entropy corresponding to the randomness of this kind of DNA at $0°K$, contrasted with the entropy of a perfect crystal formed from a single species; in the latter case, the entropy at $0°K$ is $S_0 = 0$.

Allotropic nature of DNA

The above considerations indicate the degree to which definitions of state are important in establishing absolute values either of entropy or of information. In this connection, it is important to keep in mind that almost all substances are in thermodynamically nonequilibrium states when referred to the equilibrium state that would be established if sufficient time were allowed. Clayton and Giauque's random CO would certainly revert to an ideal ordered crystal if left long enough at 1°K. The time required might be x years, where $x = 10$ raised to the power of 10^{10}; but in theory, if other reactions did not take place, equilibrium eventually would be established resulting in almost perfect order.

In many ways even a single species of DNA resembles Clayton and Giauque's random CO, as the comparison of the sequences expressed in message form in ($viii$) and (x) makes plain. Assuming that the only reaction possible is a rearrangement of sequence, a species of DNA like that in *E. coli* probably would revert to a sequence like that shown in (ii) if left long enough, for it is really a kind of unstable form like the allotropic monoclinic sulfur. From that point of view, the sequence in *E. coli* has an entropy that does differ from the entropy of the simply ordered sequence (ii). Of course, the final equilibrium state will be the state with minimum possible free energy under the conditions of restriction on the kinds of reactions permitted during the period over which equilibrium is attained. For nearly all purposes in thermodynamics, the most useful basis for the calculation of absolute entropy is one where the nonequilibrium character of many molecular species is neglected. Thus *n*-heptane might rearrange to form another isomer if left long enough at 1°K; but there is no ambiguity in assuming that the entropy of *each* isomer is zero at 0°K when in the form of a perfect crystal. Different species of DNA, all formed from the same number of AT and GC nucleotides, are really isomers just like the varieties of heptane. Only when the relation between entropy and information must be considered do the distinctions of greater or lesser *simple* order enter into the definitions of entropy and information content.

25-4   INFORMATION AND $S^{\ddagger}$

In the discussion of crystallization from supercooled melts, the role of $G^{\ddagger}$ in determining the rate of crystallization was examined. In general, a high enthalpy of activation $H^{\ddagger}$ raises a kinetic barrier which decreases the rate of a reaction; a low value of the probability of the reaction produces a low value of $S^{\ddagger}$, which increases $G^{\ddagger}$ because of the negative sign in the relation $G^{\ddagger} = H^{\ddagger} - TS^{\ddagger}$, and thus also raises a kinetic barrier. In a supercooled melt, the absence of a seed crystal lowers $S^{\ddagger}$ and raises $G^{\ddagger}$ to the point where in many cases no crystallization takes place.

Genetic $S^{\ddagger}$ barriers

The role of DNA in the synthesis of RNA appears to be much the same as that of a seed crystal, raising $S^{\ddagger}$ and lowering $G^{\ddagger}$ barriers. As an example, suppose that a molecule of *m*RNA is only partly synthesized; it is on the way toward the form which will ultimately direct the synthesis of chymotrypsinogen, for which the primary structure is drawn in Fig. 24-10. The first five codons in this molecule might be

$$\underline{UG}\underline{UGG}\underline{UGU}\underline{UCC}\underline{UGCU} \qquad (i)$$

Cys   Gly   Val   Pro   Ala

The next codon needed is the one that will place isoleucine (Ile) in the protein chain. This can be either AUU, AUC, or AUA. If the fragment of *m*RNA shown in (i) were to wander around in the cytoplasm in a random manner colliding with free nucleotides, there would be only 1 chance in 4 that A would attach as the next nucleotide in its chain; again, there is only 1 chance in 4 that A would be followed by U; there would be 3 chances out of 4 that U would be followed by U, C, or A to form one of the three codons required to code for *isoleucine* as listed above. Thus the probability of getting the correct codon at this point is $(4 \times 4 \times \frac{4}{3})^{-1} = 21.3^{-1} = 0.0469$. However, if the *m*RNA is being formed on the face of the DNA, where the complementary codon for AUU is TAA, then the probability becomes almost unity that the next codon will be AUU.

There are 246 codons required to specify the primary structure of the chymotrypsinogen chain. The probability varies for the random formation of these codons, but it will be of the order of $(\frac{1}{20})^n$ where n is the number of codons. For $n = 246$, the probability will be $(\frac{1}{20})^{246} = 10^{-320}$. The chance that the protein chain will form randomly has the same value.

If, on the other hand, the mRNA is formed on the DNA surface, the probability of its formation may be taken as essentially unity. The difference between the entropy of activation $S_d^{\ddagger}$ for formation on DNA and the entropy of activation $S_r^{\ddagger}$ for random formation is related to the respective probabilities p_d and p_r by the equation:

$$\Delta S^{\ddagger} = S_d^{\ddagger} - S_r^{\ddagger} = R \ln \frac{p_d}{p_r}$$

$$= 2.303 \times 1.987 \log \frac{1}{10^{-320}}$$

$$= 4.5757 \times 320$$
$$1464 \text{ cal deg}^{-1} \text{ mole}^{-1} \qquad (25\text{-}17)$$

The height of the free-energy barrier $\Delta G^{\ddagger}$ for the random reaction as compared with the DNA guided reaction at normal body temperature is

$$\Delta G^{\ddagger} = G_r^{\ddagger} - G_d^{\ddagger} = TS_d^{\ddagger} - TS_r^{\ddagger} = T \Delta S^{\ddagger}$$
$$= 310 \times 1464 = 454 \text{ kcal mole}^{-1} \qquad (25\text{-}18)$$

The reaction-rate constant ratio for the two reactions is given by

$$\log \frac{k_d}{k_r} = \frac{\Delta G^{\ddagger}}{2.303RT} = 43$$

$$\frac{k_d}{k_r} = 10^{43} \qquad (25\text{-}19)$$

Of course the value of k_d/k_r may be even higher because of the effect of concentration in the cytoplasm on the entropy. At any rate, from these figures it is clear that the role of the DNA and of the mRNA is connected with changes in $S^{\ddagger}$ which guide the reactions of chain synthesis. The DNA raises the value of $S^{\ddagger}$ for the reactions which will form the complementary mRNA and thus lowers the $G^{\ddagger}$ barriers so that the appropriate reactions take place. The mRNA in turn performs the same role in the formation of the protein.

The transmission of bioinformation

In discussing the failure of a supercooled melt to crystallize in the absence of a seed crystal, the statement was made that the melt did not "know how" to crystallize. The seed crystal supplied the information that allowed the reaction to proceed at a significant rate. The role of $S^{\ddagger}$ in directing biosynthesis, coupled with the relation between $\Delta S^{\ddagger}$ and ΔI, shows that the same description is applicable to the formation of protein chains under the direction of DNA carried out through mRNA and the reactions with tRNA. Without DNA, the amino acid units do not know how to unite to form the protein. Changing the values of $S^{\ddagger}$ is equivalent to imparting information.

The same kind of transmission of information takes place when a DNA chain is replicated. The organized sequence of nucleotides brings about changes in the values of $S^{\ddagger}$ for the reactions by which the replications are carried out. These changes lower the free-energy barriers in a catalytic manner for each required reaction at the appropriate time in the sequence of replication and leave the free-energy barriers intact for the unwanted reactions. In this way the information is transmitted, and as the cells continue to replicate, the information content increases, spelled out in the four-letter GACT language with its three-letter words, the codons.

Concurrently, the synthesis of mRNA molecules converts the information to the CUGA language, also in three-letter words. Through the coupling with tRNA in the ribosomes, the information is then translated to the amino acid language (AMAC) having 20 letters, and words (the proteins) vary in length from a few dozen to a few hundred letters in length.

All these changes represent increasing organization in the pattern of the atoms which are affected. And there certainly is dissemination of information as the organism grows. Thus there appears to be an increase in the total amount of information contained in the atoms which were in a more random state before being organized into the new cells. Because of the relation between change in information content and change in entropy content, this suggests that the growth process lowers the entropy of disorder in the material which it absorbs into the organism. However, as the discussion of the third law paradox showed, there is a need to define reference states with care before evaluating the absolute entropy of an organism. In spite of these difficulties, the conclusion is clear that information content will prove to be a valuable parameter in the search to interpret the patterns of change in life processes.

In conclusion, we should recall that in order for any sequence reactions to proceed, there must be a net decrease in free energy; that is, $\Delta G < 0$. Since $\Delta H - T \Delta S$, it is clear that if ΔG is to be negative, any decrease in ΔS due to increase in information content must be accompanied by a decrease in ΔH large enough to ensure that $|\Delta H| > T|\Delta S|$, where $|\Delta H|$ and $|\Delta S|$ are the absolute values (disregarding sign) of the changes. Growth, in order to be growth, must result in a net change of the pattern of arrangement of the atoms involved, and therefore it cannot be an equilibrium process. The relation between thermodynamic quantities and kinetic parameters such as $H^{\ddagger}$ and $S^{\ddagger}$ in nonequilibrium processes is important for this reason.

Example 25-5

Calculate the amount of information contained in the next to the last amino acid of the beef insulin chain (Fig. 24-9).

This amino acid is Lys. The codons for Lys are AAA and AAG. The probability of a random selection of A for the first letter is $1/4$; the same value applies for the second letter; the chance of the random selection of A or G for the third letter is $1/2$. Therefore, the probability of finding this codon in this place is

$$p_r = \left(\frac{1}{4} \frac{1}{4} \frac{1}{2} \right) = \frac{1}{32} = \frac{1}{2^5}$$

The information in this codon, therefore, is 5 bits.   *Ans.*

Example 25-6

Calculate the decrease in the height of the free-energy barrier for the formation of the last codon in beef insulin mRNA as compared with random formation at $37°$K.

$$S_d^{\ddagger} - S_r^{\ddagger} = R \ln \frac{p_d}{p_r}$$
$$= R \ln 16 = 4.5757 \times 1.2041$$
$$= 5.51$$
$$G_r^{\ddagger} - G_d^{\ddagger} = T(S_d^{\ddagger} - S_r^{\ddagger}) = 1.71 \text{ kcal mole}^{-1}$$
    Ans.

Problems

1. Translate the first four amino acids in the A part of beef insulin at the NH_2 terminal into the codons in the CUGA code and then into the GACT code.

2. Translate the following sequence of codons in the GACT code into the CUGA code and then into the amino acid sequence: AGAACAACCTGATTA.

3. Calculate the information content in bits of the codons found in Prob. 1, allowing for redundancy.

4. Calculate the information content in bits of the codons specified in Prob. 2, allowing for redundancy.

5. Calculate the number of lines of typical English-language symbols equivalent in information content to the codons found in Prob. 1.

6. Calculate the number of words of typical English-language symbols equivalent in information content to the codons found in Prob. 2.

7. Calculate the difference in $S^{\ddagger}$ in calories per degree per mole for the codons in Prob. 1 formed randomly and formed with DNA.

8. Calculate the difference in the height of the $G^{\ddagger}$ barrier for the synthesis of the part of beef insulin described in Prob. 1 at $37°$C if carried out randomly as compared with synthesis guided by DNA.

9. Calculate the difference in the height of the $G^{\ddagger}$ barrier for DNA guided and for random syntheses at $37°$C of the part of the protein chain specified by the codons in Prob. 2.

10. By what amount would ΔH have to change to offset the effect on ΔG due to a change in ΔS caused by the increase in information content of 152 bits per molecule at $37°$C?

11. According to Watson, the production of a chain of molecular weight of 40,000 requires only 10 sec in a ribosome in *E. coli* at approximately $37°$C. Using the figure for the average molecular weight of an amino acid, calculate the bits per second of information produced in this process, neglecting redundancy.

12. Using Watson's figure for the rate of production of a peptide chain, calculate the average number of amino acid units added per second to the chain if the process took place with a reduction in $S^{\ddagger}$ of 25 cal deg^{-1} mole^{-1}.

13. Using Watson's figure for the rate of production of a protein chain, calculate the length of time required to produce one molecule of chymotrypsinogen by a random process at $37°$C, neglecting the redundancy effect and volume effect.

14. Assuming that the rate of replication of DNA in *E. coli* corresponds to 300 binucleotides (rungs) per second, a rate comparable to Watson's figure for amino acid units per second added to a protein chain, calculate the length of time necessary to replicate a molecule of DNA of the sort found in *E. coli.*

15.   Assuming the conditions stated in Prob. 14, what is the length of time required to replicate a complete strand of human DNA?

16.   Neglecting redundancy and concentration effects, what would be the length of time necessary to synthesize a complete strand of human DNA by the random process.

17.   If the rate of the replication of DNA in *E. coli* is 300 binucleotides per second at $37°C$ and $H^{\ddagger} = 5.5$ kcal mole^{-1}, calculate the increase in rate at a fever temperature of $103°F$.

18.   Using the rate of 300 nucleotides per second, calculate the length of time required to synthesize one molecule of *m*RNA for chymotrypsinogen.

19.   Calculate in years the time required to synthesize one molecule of *m*RNA for chymotrypsinogen by a random process based on the rate of 300 nucleotides per second for a DNA-guided process.

20.   If information in the brain is stored in molecules with RNA structure, neglecting any redundancy, calculate the weight of the molecules necessary to contain the same amount of information found in an encyclopedia consisting of 20 volumes of 1,000 pages each of typical English.

REFERENCES

Shannon, C. E., and W. Weaver: "The Mathematical Theory of Communication," The University of Illinois Press, Urbana, 1949.

Quastler, H. (ed.): "Essays on the Use of Information Theory in Biology," The University of Illinois Press, Urbana, 1953.

Yockey, H. P., R. L. Platzman, and H. Quastler (eds.): "Symposium on Information Theory in Biology," Pergamon Press, New York, 1958.

Diamond, S.: "Information and Error," Basic Books, Inc., Publishers, New York, 1959.

Kullback, S.: "Information Theory and Statistics," John Wiley & Sons, Inc., New York, 1959.

Kac, M.: "Probability and Related Topics in Physical Sciences," Interscience Publishers, Inc., New York, 1959.

Pierce, J. R.: "Symbols, Signals and Noise," Harper & Brothers, New York, 1961.

Fano, R. M.: "Transmission of Information," John Wiley & Sons, Inc., New York, 1961.

Brillouin, L.: "Science and Information Theory," Academic Press, Inc., New York, 1962.

Wolfowitz, J.: "Coding Theorems of Information Theory," 2d ed., Springer-Verlag New York Inc., New York, 1964.

Watson, J. D.: "Molecular Biology of the Gene," W. A. Benjamin, Inc., New York, 1965.

Flanagan, D. (ed.): "Information," W. H. Freeman and Company, San Francisco, 1966.

Herdan, G.: "The Advanced Theory of Language," Springer-Verlag New York Inc., New York, 1966.

Frisch, L. (ed.): *Symp. Quant. Biol.,* vol. 31, Cold Spring Harbor Laboratory of Quantitative Biology, 1966.

Asimov, I.: "The Genetic Code," New American Library, Inc., New York, 1966.

Beadle, G., and M. Beadle: "The Language of Life," Doubleday & Company, Inc., Garden City, N.Y., 1966.

Ramsey, D. M. (ed.): "Molecular Coding Problems," The New York Academy of Sciences, New York, 1967.

Howland, J. L.: "Introduction to Cell Physiology," The Macmillan Company, New York, 1968.

In several of the preceding chapters we have noted that, in general, life processes do not take place at equilibrium. Particularly in the chemical reactions associated with cell growth, the forward reaction is faster than the back reaction. In order to have synthesis proceed, the compound in question must be synthesized faster than it is broken up. For this reason, there is irreversibility and the consequent production of entropy.

26-1 FUNDAMENTAL IRREVERSIBILITY THEORY

Physical irreversibility

It may be helpful to recall the example of the isothermal expansion of a gas under equilibrium

Chapter Twenty-Six

bioirreversibility

and nonequilibrium conditions in order to see the relation between irreversibility and the production of entropy. Figure 26-1a shows the isothermal expansion of a gas under equilibrium conditions. As the gas expands from the initial volume V_1 to the final volume V_2 at constant temperature, the relation between pressure and volume is as shown in Fig. 26-2. Suppose that the system consists of 1 mole of gas at 0°C and 1 atm so that the initial volume V_1 is 22.4 liters. As the expansion takes place and the gas passes from state 1 to state 2 along the line shown in the diagram, the pressure falls as the volume increases. Recalling the ideal-gas law,

$$PV = RT \tag{26-1}$$

it is clear that when the temperature remains constant, the right side of the equation is a constant. Therefore, if the volume is doubled, the pressure must fall to one-half its initial value. If, during this change, the downward force on the gas is regulated so that at every instant it is effectively in balance with the upward thrust due to the pressure of the gas on the piston, the process takes place reversibly. An infinitesimal decrease in the downward force will permit the piston to rise, or an infinitesimal decrease in the pressure will cause the piston to fall.

Fig. 26-1 (*a*) Reversible and (*b*) irreversible isothermal expansion of an ideal gas.

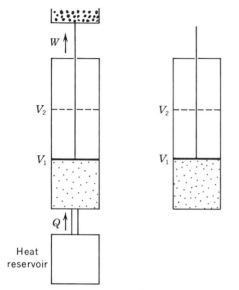

(*a*) Isothermal reversible expansion $S_2 - S_1 = Q/T$

(*b*) Isothermal irreversible expansion $S_2 - S_1 = I$

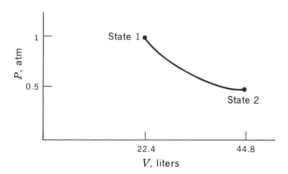

Fig. 26-2 State diagram for the isothermal expansion of an ideal gas.

As the piston rises, the gas does work in raising the weight, and energy in the form of work passes out at the top of the piston. This is given by

$$DW = -P\, dV \qquad (26\text{-}2)$$

where DW is energy out as work.

If the temperature is to remain constant, the internal energy must remain constant ($dE = 0$); and so according to the first law of thermodynamics, the loss of energy in the form of work must be exactly compensated by the gain of energy in the form of heat. For the total change from V_1 to V_2 the first law of thermodynamics takes the form

$$0 = dE = DQ + DW \qquad (26\text{-}3)$$
$$0 = E_2 - E_1 = Q_{1\to2} + W_{1\to2} \qquad (26\text{-}4)$$
$${+376\text{ cal}}\quad{-376\text{ cal}}$$

Thus, under the conditions specified (where the system consists of 1 mole of an ideal monatomic gas), 376 cal will leave the system in the form of work and 376 cal will enter the system in the form of heat.

According to the second law of thermodynamics,

$$dS = \frac{DQ_{\text{rev}}}{T} \qquad (26\text{-}5)$$

the increase in entropy dS is equal to the corresponding amount of heat DQ_{rev} flowing into the system in the reversible change divided by the absolute temperature T. For the complete change in going reversibly from state 1 to state 2, the increase in entropy is

$$S_2 - S_1 = \frac{Q_{\text{rev}}}{T} = \frac{376\text{ cal}}{273} = 1.38\text{ eu} \qquad (26\text{-}6)$$

Thus, 1.38 eu is the entropy increase of the system in passing from state 1 to state 2.

Suppose now that the change of the same system under the same conditions takes place from V_1 to V_2, not reversibly but against a vacuum with *no downward force* opposing the upward thrust due to the pressure of the gas on the piston. This could be brought about by removing weight from the piston and suddenly releasing the piston by pulling a pin, letting it shoot up under the pressure of the gas until it hits another pin which stops its motion when the gas has expanded to V_2. This is shown in Fig. 26-1*b*.

According to the second law of thermodynamics, entropy is a state function (true property) of the gas. Therefore, the value of the entropy change, $S_2 - S_1$, in passing from state 1 to state 2 must be the same irrespective of whether this change is brought about by the reversible path (*a*) or by the completely irreversible path (*b*). When the gas expands, there is no force resisting the expansion, and so no work is done. Therefore, if temperature remains constant, no heat enters the system and $Q_{\text{irrev}} = 0$.

In order to have a single formula to cover the reversible change, the irreversible change, and any intermediate partially irreversible changes, the second law can be written

$$dS = \frac{DQ}{T} + \frac{DI}{T} \qquad (26\text{-}7)$$

where DQ is any small amount of heat which enters the system and DI is the irreversibility, equivalent to the additional amount of heat which *would* have entered the system *if* the change had taken place with complete reversibility. For the finite change from state 1 to state 2, the second law of thermodynamics can be written

$$S_2 - S_1 = \frac{Q}{T} + \frac{I}{T} \qquad (26\text{-}8)$$

If the change takes place with complete reversibility, then $I = 0$ and Eq. (26-8) is in exactly the same form as Eq. (26-6). If, on the other hand, the change takes place with complete irreversibility, as shown in Fig. 26-1b, then $Q = 0$ and the equation becomes

$$S_2 - S_1 = \frac{I}{T} = \frac{376 \text{ cal}}{273} = 1.38 \text{ eu} \qquad (26\text{-}9)$$

where I is the heat which *would* have passed into the system *if* the change had taken place with complete reversibility (376 cal). Thus, the entropy change is $I/T = 376 \text{ cal}/273 = 1.38$ eu. To repeat, because entropy is a state function, the entropy change must be the same when the system goes from the same initial state to the same final state no matter by what process. Thus, if the system does not get the entropy by heat flowing in, the entropy is *generated* by the irreversibility of the process.

Chemical irreversibility

Exactly the same processes can be found when chemical changes take place. The free energy per mole of a substance is the chemical potential or *chemical pressure* which it exerts. If the sum of the free energies of the reactants is maintained equal to the sum of the free energies of the products, this change takes place at equilibrium. As a specific

illustration, let us consider the synthesis of ammonia from hydrogen and nitrogen, as shown in the expression

$$\underset{\text{reactants}}{\tfrac{3}{2}H_2 + \tfrac{1}{2}N_2} = \underset{\text{product}}{NH_3} \qquad (26\text{-}10)$$

The free-energy change ΔG° when 1 mole of ammonia is formed under standard conditions is related to the chemical potentials μ° of the ammonia and the reactants by the expression

$$\underset{\text{reactants}}{-(\tfrac{3}{2}\mu^\circ_{H2} + \tfrac{1}{2}\mu^\circ_{N2})} + \underset{\text{product}}{\mu_{NH3}} = \Delta G^\circ \qquad (26\text{-}11)$$

Let us consider, first, the free-energy profile of this reaction at 300°K as shown in Fig. 26-3. Under standard conditions at this temperature $\Delta G^\circ = -3.94$ kcal. The negative of ΔG° is frequently called the affinity ($+3.94$ kcal), as it is the thermodynamic driving force which tends to push the atoms from the form where they are linked as hydrogen and nitrogen molecules into the form where they are linked as ammonia molecules. Under the conditions as shown, the forward reaction tends to go faster than the backward reaction.

The rate for the forward reaction under general conditions J_f is given by the expression:

Rate forward: $\quad J_f = J_f^\circ \, e^{-G_f^{\ddagger}/RT} \qquad (26\text{-}12a)$

The rate for the equation backward J_b is given by the expression:

Rate backward: $\quad J_b = J_b^\circ \, e^{-G_b^{\ddagger}/RT} \qquad (26\text{-}12b)$

J_f° and J_b° are the rate when $G_f^{\ddagger} = 0$ and $G_b^{\ddagger} = 0$. The symbol J is used for rate in the current literature dealing with nonequilibrium thermodynamics.

Equations (26-12a) and (26-12b) for the forward and backward rates are related to the thermodynamic free energy by the equation

$$G_f^{\ddagger} - G_b^{\ddagger} = \Delta G \qquad (26\text{-}13)$$

Since the reaction does proceed irreversibly under standard conditions, ΔG° is a negative quantity and $G_b^{\ddagger\circ} > G_f^{\ddagger\circ}$, as shown in Fig. 26-3. However, under standard conditions, the free-energy barrier is so high that the reaction actually proceeds forward at a very slow rate.

Fig. 26-3 The free-energy barrier for the synthesis of ammonia under standard conditions (each reactant and the product at 1 atm).

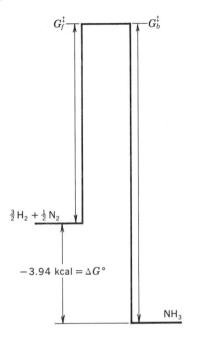

increasing the pressure of the NH_3, thereby decreasing the entropy as shown by

$$S_2 - S_1 = -R \ln \frac{P_2}{P_1} \qquad (26\text{-}14)$$

The decrease in the entropy of the ammonia increases its free energy:

$$G_2 - G_1 = (H_2 - H_1) - T(S_2 - S_1) \qquad (26\text{-}15)$$

Thus, as shown in Fig. 26-4, the line on the right side of the free-energy profile for ammonia is raised; and if the pressure is increased by just the right amount, the line on the right can be put at the same level as the line on the left. Under these conditions, the height of the free-energy barrier on the right $G_{b\,eq}^{\ddagger}$ becomes the same as the height on the left $G_{f\,eq}^{\ddagger}$. Therefore, the backward reaction has the same rate as the forward reaction, as shown by the expressions

$$J_f^e = J_f^1 \, e^{-G_f\ddagger e/RT} \qquad (26\text{-}16)$$

$$J_b^e = J_b^1 \, e^{-G_b\ddagger e/RT} \qquad (26\text{-}17)$$

While there are some chemical reactions in living cells where equilibrium conditions are maintained at least for part of the time during the life cycle, most reactions have energy profiles of the irreversible type shown in Fig. 26-3 rather than free-energy profiles of the equilibrium type shown in Fig. 26-4.

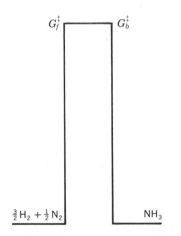

Fig. 26-4 The free-energy barrier for the synthesis of ammonia under equilibrium conditions.

Example 26-1

Suppose that $\frac{1}{2}$ mole of air per minute escapes from an astronaut's space suit into the space inside a lunar module. Assume air is an ideal gas. If the gas is originally at 1 atm and body temperature, and in the final state at 0.01 atm and the same temperature, what is the rate of entropy production per minute, assuming that the pressure in the space suit and the pressure in the module remain effectively constant during the first few minutes?

For $\frac{1}{2}$ mole of gas

$$\Delta S = 0.5 \ R \ln \frac{V_2}{V_1} = 2.29 \log \frac{P_1}{P_2}$$

$$= 2.29 \log 100 = 4.58 \text{ cal deg}^{-1} \text{ min}^{-1}$$

The synthesis of ammonia can also be carried out under different conditions of pressure which bring about chemical equilibrium ($\Delta G = 0$) at the temperature under consideration. This can be achieved by

Example 26-2

In the synthesis of ammonia, if the value of $G_f^{\ddagger}$ is lowered by 10,000 cal mole^{-1} at 300°C, what will be the ratio of the new value of the rate J_f^1 to the value of J_f before the value of $G^{\ddagger}$ was changed?

$$\frac{J_f^1}{J_f} = e^{-(G^{\ddagger 1} - G^{\ddagger})/RT}$$

$$\ln \frac{J_f^1}{J_f} = -\frac{G^{\ddagger 1} - G^{\ddagger}}{RT} = \frac{10,000}{RT}$$

$$\log \frac{J_f^1}{J_f} = \frac{10,000}{4.58 \times 573°} = 3.81$$

$$\frac{J_f^1}{J_f} = 6.46 \times 10^3$$

26-2 STEADY-STATE KINETICS

Steady-state reactions

Suppose that there is a tank of gas where the partial pressure of hydrogen, nitrogen, and ammonia are adjusted so that the equilibrium conditions are established for the reaction for the synthesis of ammonia, Eq. (26-10). Suppose that suddenly large additional quantities of hydrogen and nitrogen are injected into the tank so that the free-energy level of these substances is raised and the free-energy profile now looks like that in Fig. 26-3. There will be an increase in the forward rate with time as the system adjusts to the new conditions. Figure 26-5 is a plot of the time change of the rate of the reaction J_r, which is the difference between the forward rate J_f and the backward rate J_b. If, as the synthesis of ammonia proceeds, additional quantities of hydrogen and nitrogen are continually injected into the tank to maintain the difference in the levels of free energies of reactants and products constant at the value shown in Fig. 26-3, then when the ammonia is withdrawn, say by chemical adsorption, to maintain the partial pressure of ammonia constant, the rate will rise exponentially and level off to a constant value, as shown in Fig. 26-5.

The rate below the dotted line in this figure is called the transient rate; the rate at the dotted line is called

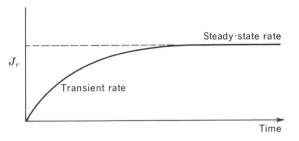

Fig. 26-5 The change of rate from transient values to a steady-state value with time.

the steady-state rate. This is the rate at which the reaction takes place under conditions of irreversibility with the generation of entropy; the driving force for the reaction (the affinity $= A$) is constant because the partial pressures of both reactants and product are constant. There is now a steady rate of "flow" of atoms from the state where the atoms are linked as hydrogen and nitrogen molecules to the state where they are linked as ammonia molecules.

Both transient rates and steady-state rates are found in the life process. When a man suddenly lifts his arm there are transient rates in the reactions as the various energy changes take place involved in this motion. However, if the man holds his arm steady in the raised position, these transient rates shift over rather quickly to the steady-state rates. In the interlinking of the great variety of biochemical reactions, the steady-state condition is far easier to define mathematically than the transient rate, though both play important roles.

Recently great advances have been made in our understanding of chemical reactions under steady-state conditions. It is therefore of considerable interest to examine some of the analogs of steady-state conditions found in *physical* systems in order to understand the way in which the mathematical equations have been developed for steady-state irreversible *chemical* reactions.

Hydrodynamic steady state

One of the simplest examples of a dynamic steady state is provided by the flow of a liquid like water through an orifice. Figure 26-6 shows an apparatus for the study of the hydraulic flow of water. Water

Fig. 26-6 Coupled flow reactions. (a) Physical flow; (b) variation of rates with time.

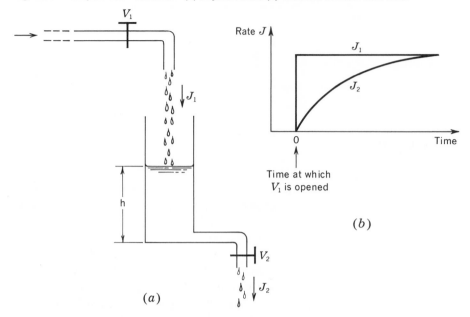

(b)

(a)

comes out of a tube at the top of the apparatus, where the rate of flow is controlled by a valve V_1. Suppose that this valve is set so that water flows at 1 cm³ sec⁻¹. This rate of flow is denoted by the symbol J_1. The water flows from this first tap into a cylindrical vessel from the bottom of which there protrudes a second tap equipped with a valve V_2. The rate at which the water comes out of this valve is J_2. Suppose that this latter vessel is empty at the beginning of the experiment. At the moment when we start counting time ($t = 0$), valve V_1 is opened to a fixed point and water starts flowing into the tank at the rate J_1. Suppose that valve V_2 is only partly open so that the orifice there is much smaller. Then some water will start flowing out through this valve; but if J_1 is larger than the rate of flow through V_2 *with no pressure head,* the height of the water will rise in the tank. As the height rises, the rate of flow through valve V_2 increases and finally, when the height is sufficiently high (head $= X_s$), J_2 will equal J_1. At this point, a steady state of flow has been attained. In the steady state, where L_w is the rate for unit head

$$J_1 = J_2 = J_s = L_w X_s \tag{26-18}$$

The values of J_1 and J_2 against time are plotted in Fig. 26-6b. J_2 rises through a transient condition and approaches the level of J_1 tangentially, finally effectively reaching this level and staying constant.

In the simplest type of flow, $a \, dX/dt = J_1 - L_w X = L_w(X_s - X)$, where X_s is the value of the head in the steady state and a is the cross-sectional area of the tube. During the transient period as the steady state is approached, the rate of change of X is given by

$$\frac{d(X_s - X)}{dt} = -L_w(X_s - X) \tag{26-19a}$$

This can be rearranged to give

$$\frac{d(X_s - X)}{X_s - X} = -L_w \, dT \tag{26-19b}$$

Upon integration this becomes

$$\ln\left(\frac{X_s - X}{X_s}\right) = -L_w t \tag{26-20a}$$

This equation can be put in the form

$$X = X_s(1 - e^{-L_w t}) \tag{26-20b}$$

which gives the variation of X with t as the steady state is approached. X_s can be determined from the relation

$$J_1 = L_w X_s \qquad (26\text{-}21)$$

under steady-state conditions.

Electrodynamic steady state

There are many examples of similar steady-state flow in other areas of physics and chemistry. For example, the rate of flow of electricity is given by the value of the current I in amperes which is the flow in coulombs per second (just like the flow of water in cubic centimeters per second). According to Ohm's law, I is equal to the conductance times the voltage drop across the part of the conductor through which the flow is taking place. This is the familiar Ohm's law:

$$I = c\mathcal{E} \qquad \text{or} \qquad \mathcal{E} = IR_e \qquad (26\text{-}22a)$$

where c is the conductance, which is the reciprocal of the electric resistance R_e. Let us put this equation in the same form as the expression we shall use for hydraulic flow and chemical flow:

$$J_e = L_e X_e \qquad (26\text{-}22b)$$

where J_e = electric flow
L_e = electric conductance
X_e = electric driving force

Thermal steady state

For the flow of heat through a conducting rod there is a similar expression

$$J_h = k_h \frac{dT}{dx} \qquad (26\text{-}23a)$$

Changing to the general symbols,

$$J_h = L_h X_h \qquad (26\text{-}23b)$$

where J_h = flow of heat, cal sec^{-1}
k_h = thermal conductance, sec^{-1} per unit temperature gradient = L_h
dT/dx = thermal gradient = X_h

This gradient could be the temperature drop per centimeter in a wire. In the terms we wish to use for considering a steady state under the most general conditions, L_h is the thermal conductance and X_h is the thermal driving force.

Chemical steady state

Consider the reaction

$$A + B \rightleftharpoons C + D$$

The rate of this reaction forward is given by

$$J_f = k_f[A][B] \qquad (26\text{-}24a)$$

and the rate backward by

$$J_b = k_b[C][D] \qquad (26\text{-}24b)$$

The reaction-rate constants are given by the Arrhenius expressions

$$k_f = k_1 e^{-H_f\ddagger/RT}\, e^{S_{0f}\ddagger/R} \qquad (26\text{-}25)$$
$$k_b = k_1 e^{-H_b\ddagger/RT}\, e^{S_{0b}\ddagger/R} \qquad (26\text{-}26)$$

The entropy difference between the system at concentrations [A][B] and unit concentration is

$$\Delta S_f = R \ln [A][B] \qquad (26\text{-}27)$$
$$\Delta S_b = R \ln [C][D] \qquad (26\text{-}28)$$

In exponential form

$$[A][B] = e^{\Delta S_f/R} \qquad (26\text{-}29)$$
$$[C][D] = e^{\Delta S_b/R} \qquad (26\text{-}30)$$

Thus

$$J_f = k_1 e^{-H_f\ddagger/RT}\, e^{S_{0f}\ddagger/R}\, e^{\Delta S_f/R}$$
$$= k_1 e^{-(H_f\ddagger - TS\ddagger)/RT} = k_1\, e^{-G_f\ddagger/RT} \qquad (26\text{-}31)$$

$$J_b = k_1 e^{-H_b\ddagger/RT}\, e^{S_{0b}\ddagger/R}\, e^{\Delta S_b/R}$$
$$= k_1 e^{-(H_b\ddagger - TS_b\ddagger)/RT}$$
$$= k_1 e^{-G_b\ddagger/RT} \qquad (26\text{-}32)$$

where $G_f^\ddagger$ and $G_b^\ddagger$ are the free-energy barriers for the forward and backward reactions at the concentrations [A][B] and [C][D].

Suppose that the reaction is proceeding forward under conditions of concentration which are only slightly removed from the concentrations corresponding to equilibrium conditions so that $G_b^\ddagger - G_f^\ddagger = \Delta G$ and ΔG is a very small quantity. Thus $G_b^\ddagger$ can be set equal to the term $G_f^\ddagger + \Delta G$. The rate at which the reaction is proceeding J_r can then be written

$$J_r = J_f - J_b = J_0 \, e^{-G_f{}^\ddagger/RT} - J_0 \, e^{-(G_f{}^\ddagger + \Delta G)/RT}$$

$$(26\text{-}33a)$$

Since $G_f{}^\ddagger$ is a constant, this equation can be rewritten

$$J_r = a - J_0 \, e^{-G_f{}^\ddagger/RT} \, e^{-\Delta G/RT}$$

$$= a - ae^{-\Delta G/RT}$$

$$= a(1 - e^{-\Delta G/RT}) \qquad (26\text{-}33b)$$

Since $e^x = 1 - x + x^2/2! + \cdots$, and since ΔG is small, this can be written

$$J_r = a\left(1 - 1 + \frac{\Delta G}{RT}\right) = \frac{a}{RT} \Delta G = L_c X_c \qquad (26\text{-}34)$$

where $L_c = J_0 \, e^{-G_f{}^\ddagger/RT}$ and $X_c = \Delta G$. Thus X_c is the driving force of the reaction, and L_c is the chemical conductance, the rate at which the reaction goes when X_c is equal to unity.

Entropy and steady state

In order to apply these relations in a way that will give us information about the production of entropy, several conditions must be fulfilled. First of all, the reaction must be proceeding slowly enough so that both the initial state and the final state are characterized by a Boltzmann distribution of the molecules over the various energy levels. This is necessary if the concept of temperature is to have meaning for this reaction. If this condition is fulfilled, the entropy in the initial state before the reaction takes place and in the final state after the reaction takes place is definable.

Under these conditions, the entropy produced by the irreversibility of the reaction is equal to the heat which *would* have been absorbed *if* the reaction had taken place reversibly divided by the absolute temperature at which the reaction takes place. Next, the deviation from the true equilibrium condition must be small enough so that the approximation expressed in Eqs. (26-33) and (26-34) is applicable. There is reason to believe that these conditions are fulfilled for a large majority of the biochemical reactions which together make up the process of life.

Example 26-3

If the flow in a state for the system in Fig. 26-6 is given by an expression with $L_w = 3.56$ cm³

sec⁻¹ per centimeter of head and the head X_w is 10.5 cm, what is the rate of flow J_w?

$$J_w = L_w X_w = 3.56 \text{ cm}^3 \text{ sec}^{-1} \text{ cm}^{-1}$$
$$\times \, 10.5 \text{ cm} = 37.4 \text{ cm}^3 \text{ sec}^{-1} \qquad Ans.$$

Example 26-4

In a certain chemical reaction the rate of reaction for $\Delta G = 1$ cal mole⁻¹ is 3.59×10^{-3} mole sec⁻¹ (cal mole⁻¹)⁻¹; calculate the rate of flow if $\Delta G = 6.53$ cal mole⁻¹.

$$J_c = L_c X_c$$
$$= 3.59 \times 10^{-3} \text{ mole sec}^{-1} \text{ (cal mole}^{-1}\text{)}^{-1}$$
$$\times \, 6.53 \text{ cal mole}^{-1}$$
$$= 2.34 \times 10^{-3} \text{ mole sec}^{-1} \qquad Ans.$$

26-3   COUPLED REACTIONS

One of the most characteristic features of the network of reactions that make up the life process is interdependence. For example, when adenosine triphosphate (ATP) is transformed to adenosine diphosphate (ADP), the free energy released in this reaction may be used to push some other reaction like the formation of creatine phosphate.

Phosphorylation

The reaction for the change of ATP into ADP plus P can be written

$$\text{ATP} \longrightarrow \text{ADP} + \text{P} \qquad \Delta G^\circ = -8.0 \text{ kcal} \qquad (26\text{-}35)$$

(Note that the usual convention is followed, using P to denote a phosphate group.) The reaction for the formation of creatine phosphate is

$$\text{Creatine}^+ + \text{HPO}_4{}^- \longrightarrow \text{creatine-PO}_4 + \text{H}_2\text{O}$$
$$\Delta G^\circ = +10.5 \text{ kcal} \qquad (26\text{-}36)$$

With the concentrations properly adjusted, the change of ATP into ADP provides enough free energy to make Eq. (26-36) go forward. This is one example of thousands of coupled reactions which occur in biochemistry.

Flow coupling

Coupled physical reactions have been of interest for many years. Lord Rayleigh carried out extensive investigations of couples in connection with his research on sound in the latter half of the nineteenth century. One of the simplest examples consists of two streams of gas flowing in directions at 90° to each other and interpenetrating. In Fig. 26-7, one of these gas streams is shown flowing between two plates separated by a small distance. The rate of flow is J_1 cm³ sec⁻¹; this is related to the "conductance" of the sheet orifice L_1 and the driving force X_1, which can be expressed in terms of torrs per centimeter, millimeters of mercury drop in pressure per centimeter along the path. The relation is

$$J_1 = L_1X_1 \qquad (26\text{-}37)$$

If a stream of gas of the same molecular species flows at right angles to the first stream, then for this stream

$$J_2 = L_2X_2 \qquad (26\text{-}38)$$

If the streams are molecular beams, occasionally molecules from one will collide with molecules from the other, but because of conservation of momentum, the rates of flow will not be changed, assuming elastic collisions. The two intersecting streams are shown in Fig. 26-8.

Now suppose that a vane at an angle of 45° is placed in the zone of intersection, as shown in Fig. 26-9. Some of the molecules in stream (1) will be deflected and come out in stream (2), while molecules in stream (2) will be deflected and come out in stream (1).

This action can be expressed by the relations

$$J_1 = L_{11}X_1 + L_{12}X_2$$
$$J_2 = L_{21}X_1 + L_{22}X_2 \qquad (26\text{-}39)$$

The amount of interaction depends proportionally on the length of the vane. Under these conditions the coefficient L_{12} denoting the action which increases stream (1) by adding molecules from stream (2) must be equal to the coefficient L_{21} which denotes the action of the vane in deflecting into stream (2) molecules from stream (1). L_{11} and L_{22} correspond to the simple conductances L_1 and L_2 in the previous equations.

26-4 THE ONSAGER RELATIONS

A crucial advance in interpreting coupled reactions was made by the American physical chemist L. On-

Fig. 26-7 The flow of a series of molecular beams between two infinite planes; the stream is denoted by the subscript (a). In (b) the vertical cross section of the two plates with gas flowing between is shown with the view at a right angle to the horizontal view of the plates shown in (a).

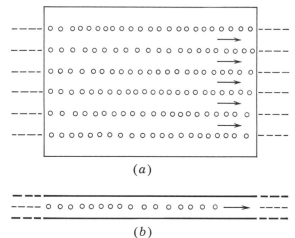

(a)

(b)

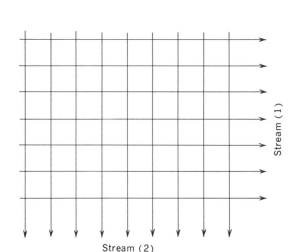

Fig. 26-8 The flow of two sets of molecular beams at right angles to each other.

sager, who pointed out that important relations could be deduced when the equations for flow are put in a form similar to that shown in Eqs. (26-39). In this special instance it turns out that the following relation between the coefficients is valid even for flows of different kinds like heat and electricity which influence each other

Fig. 26-9 The flow of two sets of molecular beams with interaction caused by a reflecting vane.

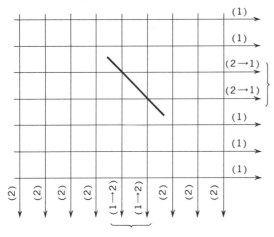

$$L_{21} = L_{12} \tag{26-40}$$

This expresses the fact that if there are components like vanes which bring about an interaction between a flow in one direction and a flow that normally would be independent in another direction, these interactions are interdependent.

Conjugate variables

Using basic theorems from statistical mechanics, Onsager proved that this situation holds where the two flows may be of a very different character. If these flows are expressed in variables which make the units in Eqs. (26-39) consistent, then Eq. (26-40) is valid. Two variables (X_1 and X_2) expressed in units which make Eqs. (26-39) and (26-40) valid are called *conjugate variables*. Such variables relate the emf of a thermocouple and the flow of heat along the thermocouple wire. Similarly, concentration gradients of ions can cause transport of matter, and the flow of matter can cause electric potentials.

Thermal anisotropy

One of the more interesting examples is heat flow through an anisotropic crystal in three dimensions. This is analogous to the flow of water across a plane surface, discussed above. If the crystal is anisotropic, a temperature gradient in the x direction affects the

flow of heat in the y direction at right angles to the x direction. If the lack of isotropy is three-dimensional, there are three equations similar to Eqs. (26-39).

The proof of this theorem also rests on an extension of the zeroth law of thermodynamics. Applied to the flow of heat governed by temperature, this law states that if the following pair of thermal equilibrium relations holds, then the third relation necessarily follows:

$$A \rightleftharpoons B \quad \text{and} \quad B \rightleftharpoons C \quad \text{implies} \quad A \rightleftharpoons C \tag{26-41}$$

Heat will not flow around in a circular pattern as shown in the expression

$$A \rightarrow B \atop \nwarrow_C\swarrow \tag{26-42}$$

Chemical anisotropy

The same considerations apply to chemical equilibria. If there is a chemical equilibrium between two species a and b and between two species b and c, there must be equilibrium between a and c. It is impossible to have a flow of atoms undergoing the reaction changing them from a to b, then from b to c, and then from c back to a. This is called the principle of microscopic reversibility. If a number of substances are in equilibrium in pairs, there cannot be any path by which there is a circular flow between three or more of these substances. Thus, for a pair of chemical reactions which are coupled like the reaction in Eq. (26-35) and the reaction in Eq. (26-36), the following expressions are found:

$$J_{\text{ATP}} = L_{11}X_{\text{ATP}} + L_{12}X_{\text{creatine}} \tag{26-43}$$
$$J_{\text{creatine}} = L_{21}X_{\text{ATP}} + L_{22}X_{\text{creatine}} \tag{26-44}$$

Example 26-5

If the values of the coefficients for Eq. (26-43) are $L_{11} = +0.68$ mole min^{-1} (kcal mole^{-1})$^{-1}$, $L_{12} = +0.31$ mole min^{-1} (kcal mole^{-1})$^{-1}$, and $L_{22} = +0.15$ mole min^{-1} (kcal mole^{-1})$^{-1}$, and if $X_{\text{ATP}} = 8.0$ kcal mole^{-1} and $X_{\text{creatine}} = -3.0$ kcal mole^{-1}, calculate the rates of the reaction when coupled in a steady state.

According to Onsager's law,

$$L_{21} = L_{12} = -0.31 \text{ mole min}^{-1} (\text{kcal mole}^{-1})^{-1}$$

Thus

$$J_{\text{ATP}} = 0.68 \times 8.0 + 0.31 \times (-3.0)$$
$$= 4.5 \text{ moles min}^{-1}$$

$$J_{\text{creatine}} = 0.31 \times 8.0 + 0.15 \times (-3.0)$$
$$= 2.03 \text{ moles min}^{-1} \qquad Ans.$$

Fig. 26-10   Irreversible flow of heat from a source to a sink, generating entropy.

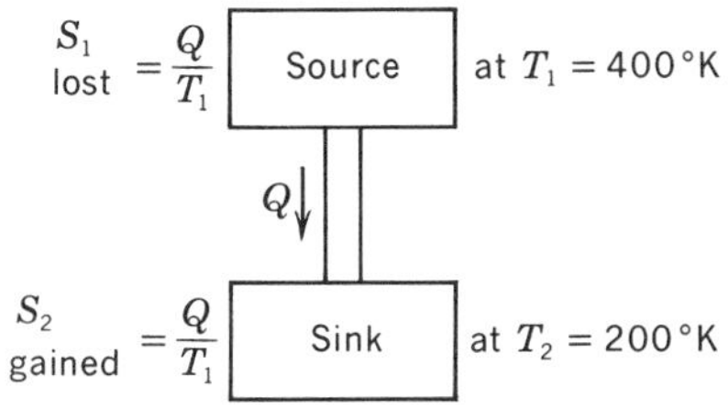

26-5   STEADY-STATE AND MINIMUM ENTROPY PRODUCTION

The Onsager reciprocal relation (26-40) has an important bearing on the analysis of how entropy is produced in coupled reactions.   In the simultaneous flow of heat and electricity, where the heat flow can take place in three dimensions and the electric flow can take place in three dimensions, the analysis shows that the entropy flow is equal to the product of the flux multiplied by the "force" conjugate to it.   Consider the simple case of heat flowing from one block of metal at temperature t_1 into another block of metal at temperature t_2 at a rate J' cal sec^{-1}.   The rate of production of entropy σ depends on the rate of heat flowing, the temperature of the source out of which the heat flows, and the temperature of the source of the sink into which the heat goes.   Consider the heat flow shown in Fig. 26-10.   In 1 sec if 1000 cal of heat flows out of the reservoir at the top, which is at 400°K, the entropy produced is 1000/400° = 2.5 eu.   When the same amount of heat flows into the sink at 200°K, the amount of entropy gained by the sink is 1000/200° = 5 eu.   These relations are summarized as follows:

Entropy lost: $S_1 = \dfrac{Q}{T_1} = \dfrac{1000 \text{ cal}}{400°} = 2.5 \text{ eu}$

$$(26\text{-}45)$$

Entropy gained: $S_2 = \dfrac{Q}{T_2} = \dfrac{1000 \text{ cal}}{200°} = 5.0 \text{ eu}$

If 1000 cal of heat flows, the entropy produced is

$$S_2 - S_1 = 2.5 \text{ eu} \qquad (26\text{-}46)$$

Thus, we can write

$$\sigma = \frac{S_2 - S_1}{\Delta t} = 2.5 \text{ eu sec}^{-1}$$

$$(26\text{-}47)$$

$$J' = \frac{Q}{\Delta t} = 1000 \text{ cal sec}^{-1}$$

where σ is the rate of production of entropy and J' is the rate of heat flow.

This can be put in the form

$$\sigma = J'\left(\frac{1}{T_2} - \frac{1}{T_1}\right) = 2.5 \text{ eu sec}^{-1} \qquad (26\text{-}48)$$

Now the expression in parentheses $1/T_2 - 1/T_1$ is the measure of the driving force which is causing the production of entropy in the system.   Using the symbol X' for this driving force, we have

$$\sigma = J'X' \qquad (26\text{-}49a)$$

For a system where $T_1 - T_2 = \Delta T$ is small, the driving force can be measured equally well by ΔT which can be denoted by the symbol X.   Since $1/T_2 - 1/T_1 = \Delta T/T^2$, the term T^2 can be incorporated into the thermal conductivity coefficient denoted by L, and Eq. (26-49a) can be put in the form

$$\sigma = JX = LX^2 \qquad (26\text{-}49b)$$

since $J = LX$ where the driving force $X = \Delta T$.   For two heat-flow processes which are coupled together

$$\sigma = J_1X_1 + J_2X_2 = L_{11}X_1{}^2 + L_{12}X_1X_2$$
$$+ L_{21}X_2X_1 + L_{22}X_2{}^2 \qquad (26\text{-}50)$$

For coupled chemical reactions, X_1 and X_2 represent the decrements of free energy which cause the reactions to proceed irreversibly.   Using the expression for J in terms of X and L, we can write for the entropy production in the coupled chemical reactions

$$\sigma = \sigma_{ATP} + \sigma_{creatine} = L_{11}X^2_{ATP} + L_{12}X_{ATP}X_{creatine}$$
$$+ L_{21}X_{creatine}X_{ATP} + L_{22}X^2_{creatine} \quad (25\text{-}51)$$

where the L's are entropy production rate coefficients. The general mathematical form for such an expression is

$$\sigma = \sum_i \sum_j L_{ij}X_iX_j \quad (25\text{-}52)$$

When a process like the flow of water or the flow of heat discussed above reaches a steady state, there is a kind of stability where small disturbances can produce a transient departure from the steady-state flow which tends to disappear, decaying exponentially with time. This is the kind of stability associated with the potential energy of a ball lying in the bottom of a concave dish. If the ball is pushed a little way up the side of the dish, it tends to roll back to the bottom. The steady state represents a minimum rate of production of entropy as compared with the production in any of the transient states which tend to change back into the steady state.

Homeostasis

The tendency of a system in a steady state when disturbed by an imposed force to return to the steady state is closely related to the theorem of Le Châtelier that when a system is disturbed, it yields in a way that tends to adsorb the disturbing force. For example, if the temperature of a system is suddenly raised, the equilibrium readjusts in the direction that adsorbs heat. In the same way, a steady state when disturbed moves toward a type of flow that represents a minimum production of entropy. This condition is called *homeostasis*.

In an earlier chapter we discussed the fact that biological systems from the thermodynamic point of view are open systems. The life process involves not only a flow of energy but also a flow of matter. If biological systems were regarded as closed, one would expect the various chemical and physical reactions ultimately to bring all the intensive properties to a common level. This would mean reducing all parts of the system to the same temperature, to the

same pressure, and to concentrations throughout the system corresponding to complete chemical equilibrium. This, of course, would mean stagnation; no process would be taking place; and to all intents and purposes the system would have no life. Actually, the living organism does not tend to an equilibrium state but to a steady state of minimum production of entropy.

The principal factor in the production of entropy is metabolism. In the growth process, some assimilated food is changed partly into protoplasm and partly into waste matter. There is a kind of coupling of reactions that decrease entropy to reactions that increase entropy. Thus, when amino acids are absorbed, presumably they are in a most disorganized state; entropy is lost when they are synthesized into the organized arrangement found in the proteins.

There is evidence that at the adult stage of growth the entropy itself and the production of entropy both attain their minimum values. When the organism has reached the limit of growth, its organization is at a maximum. The processes which require energy to bring about the growth have attained a minimum rate.

Schrödinger has suggested that the most informative point of view from which to regard the life process is the flow of entropy. In order to stay alive the organism has to have a certain amount of metabolism which produces entropy. However, the organism takes in from the surroundings certain kinds of food like proteins where molecules have been put together into a high degree of order. One can say that the organism is absorbing negentropy as it eats. The flow of this entropy is a crucial factor in the control of the network of reactions which constitute life.

Problems

1. If 35 moles of an ideal gas at 310°K expand isothermally from a volume of 100 liters to a volume of 1,000 liters with complete irreversibility in 15.89 sec, calculate the average time rate of production of entropy.

2. If 15.6 moles of an ideal gas at 25°C and a pressure of 10.2 atm expand isothermally during 11 sec under a pressure of 3 atm until internal pressure effectively balances external pressure, what is the average time rate of production of entropy due to irreversibility?

3. A chemical reaction proceeds at a steady state with the production of 0.568 mole min^{-1} of the product at 37°C. At 43°C the rate is 0.896 mole min^{-1}. Calculate the value of $H^{\ddagger}$, $S^{\ddagger}$, and $G^{\ddagger}$ for the particular concentrations used at 37°C. Assume that the back reaction is negligibly slow.

4. If the reaction described in Prob. 3 is operated at 37°C in the presence of a catalyst which reduces $H^{\ddagger}$ by 2.6 kcal mole^{-1}, what will be the rate?

5. If the reaction described in Prob. 3 takes place at 37°C in the presence of an enzyme which increases $S^{\ddagger}$ by 3.54 cal deg^{-1} mole^{-1}, what will the rate be?

6. If the back reaction complementary to the forward reaction in Prob. 3 has a value of $H_b^{\ddagger} = 23.7$ kcal mole^{-1} and a value of $S_b^{\ddagger}$ of -35.2 kcal mole^{-1} under the same conditions of concentration as in Prob. 3, by how much must $S_b^{\ddagger}$ be increased in order to establish equilibrium at 37°C?

7. If the flow rate through valve 1 in Fig. 26-6 is 37.4 cm^3 min^{-1} and the value of L_w for value 2 is 3.56 cm^3 sec^{-1} per centimeter of head, make a plot of J_2 as a function of time from the time of opening valve 1 to the establishing of steady state.

8. An aqueous solution of H_2SO_4 at 25°C is contained in a well-stirred vessel of 1,000-cm^3 capacity and is at a concentration of 1.56×10^{-2} mole liter^{-1}. A solution of $Ba(OH)_2$ flows into the flask at the rate of 1 cm^3 min^{-1} and has a concentration of 2.37×10^{-6} mole liter^{-1}. Assume that the rate of crystallization is proportional to $[Ba^{++}] = 6.92 \times 10^{-5}$ mole liter^{-1}. Make a plot of the rate of crystallization against time using the value of L of 0.237 mole min^{-1} (mole)$^{-1}$. Plot the rate of crystallization as a function of time up to the steady state.

9. The rate at which a certain reaction proceeds is 4.592 mole sec^{-1} when $\Delta G^{\ddagger} = 0$ at 37°C. Calculate the rate at 37°C when $\Delta G = 562$ cal mole^{-1} and the rate at this temperature if the concentration is reduced by 50 percent.

10. The reaction of A + B → C + D is coupled to the reaction E + F → G. The coefficients are $L_{11} = 2.96$, $L_{12} = 3.03$, and $L_{22} = 3.96$ mole hr^{-1} (kcal mole^{-1})$^{-1}$. If $X_1 = -3.2$ kcal mole^{-1} and $X_2 = 5.6$ kcal mole^{-1}, calculate the rates of the reactions under steady-state conditions.

11. Calculate the time rate of production of entropy when the reactions in Prob. 10 proceed under steady-state conditions if dS/dn is 15.2 eu mole^{-1} for reaction 1 and 18.3 eu mole^{-1} for reaction 2.

12. Heat flows along an insulated rod from a heat reservoir at 35°C to a reservoir at 15°C. If the rate of flow is 13.68 cal sec^{-1}, calculate the time rate of entropy production.

13. A Dewar flask of liquid oxygen absorbs heat from its surroundings at room temperature (25°C), and the absorbed heat makes the liquid evaporate at 90°K at the rate of 0.1567 cm^3 min^{-1} of liquid. The density of the liquid is 1.43 g cm^{-3}. The heat of vaporization is 50.9 cal g^{-1}. Calculate the time rate of entropy production.

14. If a molecule of DNA is replicated at the rate of 24.3 binucleotides per minute, calculate the time rate of decrease of entropy due to the ordering of the nucleotides, as contrasted with a random state (neglect redundancy).

15. At 37°C, what is the time rate of increase in free-energy entropy change produced by the process described in Prob. 14?

16. If the reaction in Eq. (26-35) is used to produce the free energy to drive the replication of DNA described in Prob. 14, at what rate will moles of ATP be converted into moles of ADP?

17. Assume the same values of L_{11}, L_{12}, and X_1 for the reaction in Eq. (26-35) that are given in Example 26-5. If the value of L_{22} for the DNA replication is 0.195 mole min^{-1} kcal^{-1} per mole of binucleotide and $X_2 = -2.6$ kcal per mole of binucleotide, calculate the rate at which the coupled reactions will proceed in the steady state.

18. Calculate the time rate of decrease of order entropy for the DNA synthesis described in Prob. 17.

19. If X for the ATP reaction is doubled, calculate the approximate amount by which the rate of replication of DNA will be increased, using the value of constants from Prob. 17.

20. According to Watson, a protein chain in the ribosome in *E. coli* is synthesized at the rate of 10 sec^{-1} for a molecule with a molecular weight of 40,000. Calculate the rate of entropy decrease per second on the basis that the amino acids are changed from a state of random order to a state of complete order.

REFERENCES

Denbigh, K. G.: "The Thermodynamics of the Steady State," John Wiley & Sons, Inc., New York, 1951.

DeGroot, S. R.: "Thermodynamics of Irreversible Processes," Interscience Publishers, Inc., New York, 1951.

Cox, R. T.: "Statistical Mechanics of Irreversible Change," The Johns Hopkins Press, Baltimore, 1955.

Prigogine, I.: "Introduction to Thermodynamics of Irreversible Processes," Charles C Thomas, Publisher, Springfield, Ill., 1955.

Prigogine, I.: "Transport Processes in Statistical Mechanics," Interscience Publishers, Inc., New York, 1958.

Prigogine I.: "Non-equilibrium Statistical Mechanics," Interscience Publishers, Inc., New York, 1962.

Fitts, D. D.: "Nonequilibrium Thermodynamics," McGraw-Hill Book Company, New York, 1962.

Rysselberghe, P. van: "Thermodynamics of Irreversible Processes," Blaisdell Publishing Company, New York, 1963.

Katchalsky, A., and P. F. Curran: "Non-equilibrium Thermodynamics in Biophysics," Harvard University Press, Cambridge, Mass., 1965.

Yourgrau, W., A. van der Merwe, and G. Rah: "Treatise on Irreversible and Statistical Thermophysics," The Macmillan Company, New York, 1966.

Mazo, R. M.: "Statistical Mechanical Theories of Transport Processes," Pergamon Press, New York, 1967.

One of the most important advances in science during the last 25 years has been the development of a body of mathematical and physical theory by which the concept of a simple thermodynamic system has been expanded into a broadly generalized system concept applicable to the interplay of energy and entropy as found all the way from mechanical and electrical systems for guiding space rockets to the electrophysicochemical complexes that constitute the human body and brain. As this theory has matured, there has been a gradual emergence of a common pattern in all the mechanisms that involve the complex channeling and interplay of energy and entropy factors.

Chapter Twenty-Seven

biocybernetics

Here in this final chapter we survey some of these systems, particularly those concerned with the life process, in order to show how so many of the principles considered in the earlier chapters of this book come to a common focus most helpful in understanding the nature of living matter. This relatively new area of science concerned with these common patterns of guidance and control is called *cybernetics*. The word comes from the original Greek root *kybernetes* meaning *steersman* or *controller* and is akin to the Latin *gubernates*, or governor.

27-1　SYSTEM CONCEPTS

Inclusive systems

In order to show the analogy between a thermodynamic system and a cybernetic system, the diagrams in Figs. 27-1 and 27-2 emphasize the common features of each. In the *narrow sense* of the term, the *thermodynamic system* is that *portion of matter* in the interior of the container; the system is the subject of study and changes its pattern by passing from one thermodynamic state to another. In many investigations this system consisted solely of a measured portion of an ideal gas; in others, a single molecular species is present in two or more phases like the gaseous phase, liquid phase, and

Fig. 27-1 A servotransducer unit.

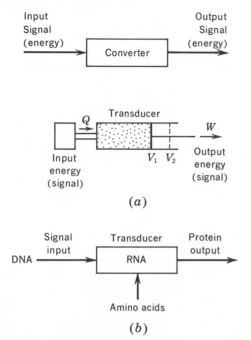

(a)

(b)

Fig. 27-2 A thermodynamic compression unit (*a*) and a ribosome (*b*) regarded as transducer units.

crystalline phase; in still other systems, several molecular species are present at once, as in the case of a solute and a solvent.

In the *broader sense* of the term, a system includes not only the material to be studied but the means for studying it. These *means* are the channels by which energy can be put in or taken out, like the work channel and the heat channel; also occasionally, there is an electric channel for introducing and taking out energy as carried by electrons; and there may be sometimes a matter channel, where known portions of molecular species can be put into or taken out of the container. The changes of state brought about by the passage of energy, or electrons, or atoms and molecules through these channels are studied by the observance of properties. In other words, the system is equipped with *information* channels from which come the signals that indicate the volume of the system, the pressure to which it is subjected, and its temperature. When the system consists of an

electrochemical cell, the signals also provide information with regard to the electric potentials. These information channels are also part of the *inclusive system.*

In the simple *cybernetic system,* the heart of the system is also a portion of matter in a container. Just as the thermodynamic container is sometimes referred to as the black box, so the heart of the cybernetic system also is frequently called the black box or *transducer*. This box is equipped with channels by which the introduction (or withdrawal) of energy brings about changes of state within the black box and causes an output of information. As in the thermodynamic system, this output of information can consist of pressure changes, temperature changes, or electrical changes.

Cybernetics and thermodynamics contrasted

The principal difference between the thermodynamic system and the cybernetic system lies in the greater degree of organization of the latter. In nearly all the thermodynamic systems we encounter, the portion of matter inside the black box consists of molecules in the form of a single phase or occasionally in two or three phases; the organization of these phases is extremely simple. True, there may be a few interfaces; and in a colloidal system there usually is a somewhat more complex degree of organization; but there is little if any complex channeling of energy and entropy inside the thermodynamic black box. By way of contrast, in the cybernetic system there is always a channeling of energy and entropy inside the box, frequently in a complicated network.

In spite of these differences, it is important to remember that in both the thermodynamic and the cybernetic system the common media for action are energy and entropy (frequently in the form of negentropy or information in the cybernetic system). Moreover, the cybernetic system often has thermodynamic aspects. This is especially true of biocybernetic systems such as the mechanism controlling the rate of breathing and the cardiovascular system, which relates the action of the heart to the functioning of the blood; thermodynamics also enters into the cybernetic action inside the biological cell as growth and replication take place.

In the study of cybernetic systems, there are several different possible attitudes. The engineer, confronted with the problem of designing a cybernetic system for antiaircraft fire or rocket guidance, starts with the requirements which must be met and puts into the black box the best materials and best organization suited for meeting these requirements. The biologist or physician studying a cell or a patient has before him the black box, sometimes operating and fulfilling its normal function and sometimes in a pathological condition and partially or completely failing to function. To a certain extent he can peer inside the black box and observe some of the grosser features of the matter and organization there; but, by the very nature of matter as embodied in the principle of indeterminacy, he is prevented from seeing the finer details, which consist of the interplay of individual molecules, atoms, or electrons. He must gather all the information he can by observing what goes into the box in the way of energy and negentropy and what comes out of the box in the form of controls, signals, and information; and then from this fragmentary knowledge he tries to construct a hypothetical model of the organization inside the box.

In nearly all cases, it is helpful to try to visualize the material and the organization of the cybernetic system in simple terms even though it may involve a gross oversimplification. For if the principal features of the *total* pattern are understood, this knowledge provides a better perspective in which to attempt to discern the finer details. Because so many of the larger features of pattern are shared alike by all cybernetic systems, e.g., mechanical, electrical, and physicochemical networks, the consideration of analogs is almost always helpful. We now turn to survey a few of the simpler types of systems as a preparation for studying some of the more complex biochemical cybernetic networks.

27-2   SYSTEM PATTERNS

Time factors

One of the primary differences between thermodynamic and cybernetic systems lies in the degree of involvement with time. A simple thermodynamic system, when left to itself, settles down to an equilibrium state in which the intensive quantities like temperature, pressure, and concentration take on constant values invariant with the passage of time. By way of contrast, a cybernetic system frequently is characterized by a repetitive variation of intensive properties. There is a cyclic repetition of pressure changes in breathing, a cyclic repetition of volume changes as the heart beats. Again, the length of each portion of the cycle can vary; e.g., the length of a single breath is shortened as the result of exercise. Also, the amplitude of the portions of the cycle can vary; deeper breaths are taken when the need for oxygen increases.

This cyclic pattern frequently originates in feedback coupling. When a thermostat is linked to a house furnace, there is a cyclic variation of temperature in the room where the thermostat is located. When the room cools off, the thermostat signals the furnace and the pilot light ignites the fuel. But it takes an appreciable interval of time before the heat, coming from the burning of the fuel, gets up into the room and raises the temperature of the thermostat to the point where the off signal is sent back to the furnace. After this signal is received by the furnace, there is an appreciable interval of time before the heat stops flowing to the room and the room starts cooling down again. This factor of time intervals is completely missing from pure thermodynamic systems.

The harmonic oscillator

As an introduction to the complexities of cybernetic systems, it is helpful first to examine a few simpler physical systems. One of the simplest systems to illustrate the transformation of energy into information is a harmonic oscillator, such as a hydrogen atom attached by a covalent bond to the surface of an enzyme (Fig. 27-3). This chemical system is closely analogous to a mechanical system where a mass (the hydrogen atom) is attached by a spiral spring (the covalent bond) to a larger mass (the atomic lattice of the enzyme). Imagine that the surface of the enzyme is directly in contact with the matrix fluid of the protoplasm. For the moment, let us forget about the zero-point energy which keeps the hydrogen atom in constant vibration and let us assume that the hydrogen atom is in the zero vibrational state,

Fig. 27-3 A hydrogen atom attached to a nitrogen atom on an enzyme surface acts as a transducer, under the influence of the force of attraction caused by hydrogen bonding.

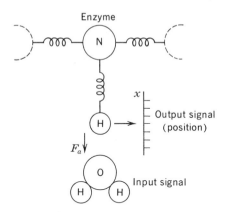

where, from the point of view of classical quantum theory, there will be no motion. Then, to be specific, let us further assume that the hydrogen atom under consideration is attached by a covalent bond to an atom of nitrogen which is part of the enzyme molecular lattice structure.

If a water molecule, properly oriented, approaches this hydrogen atom, it exerts a force that displaces the hydrogen atom from its normal position. We assume that the hydrogen atom will move only along the line through its nucleus and the nucleus of the nitrogen atom.

For small displacements, Hooke's law is obeyed

$$F_b = -k_h x \tag{27-1}$$

where F_b = force exerted on atom by springlike bond
k_h = elastic constant, or Hooke's law constant
x = the amount of displacement

Assume that the force F_a of the approaching water molecule pulls the hydrogen atom away from the nitrogen atom, producing a displacement x. We modify Eq. (27-1) and put it in the form

$$F_a \left(-\frac{1}{k_h} \right) = F_a c = x \tag{27-2}$$

and we denote $-1/k_h$ by the symbol c, because we think of this as a *conversion* factor; the input (signal) F_a is converted by the atom and spring into an out-

put signal x which is a kind of information about the effect that the force has produced. We write the block diagram for the cybernetic system as shown in Fig. 27-4. This shows how the input signal is converted inside the black box or *transducer* to produce the output signal.

The plot of input signal/output signal is frequently referred to as the gain diagram. Such a *gain diagram* is shown in Fig. 27-5, and the graph is called the *gain curve*. In this instance, the gain curve is actually a straight line, and the slope of this line is the conversion factor c.

The frequency ν of the vibration of a harmonic oscillator is related to the force constant k_h and the reduced mass μ by

$$\nu = \frac{1}{2\pi} \sqrt{\frac{k_h}{\mu}} \tag{27-3}$$

For a diatomic molecule made up of one atom with mass m_1 joined to another atom with mass m_2

$$\frac{1}{\mu} = \frac{1}{m_1} + \frac{1}{m_2} \tag{27-4}$$

When a single particle is attached by a springlike force to a fixed mass, as assumed in the enzyme example, $m_2 = \infty$, and $\mu = m_1 = m$, where m is the effective mass.

Damping

Suppose that the displacement of the hydrogen atom pushes away some other atoms in the vicinity so that there is a phenomenon like viscosity involved. In other words, there is *friction* when the hydrogen atom is displaced. Such a factor is called *damping*. Of course, in considering action at the molecular level, we are far removed from the scale of size where the ordinary equations of viscosity and friction apply; but in principle the force resulting from motion of the atom through a viscous fluid F_v is related to the velocity of motion dx/dt by the expression

$$F_v = -k_v \frac{dx}{dt} = b \frac{dx}{dt} \tag{27-5}$$

where $-k_v$ is denoted by the single symbol b. This force can be joined with the applied force to give the total force acting on the atom

$$b\frac{dx}{dt} + cx = F_a \qquad (27\text{-}6)$$

where $c = -k_h$. Denoting b/c by τ and d/dt by d, Eq. (27-6) can be written

$$F_a\left[\frac{1/c}{\tau d + 1}\right] = x \qquad (27\text{-}7)$$

showing that the quantity in brackets is a conversion factor. This new factor b alters the response curve. The diagrammatic representation of the system is shown in Fig. 27-6, the block diagram in Fig. 27-7, and the type of response curve in Fig. 27-8. The response curve now resembles the type of change when a system is displaced and returns to a steady state.

Inertial factors

If we are considering strictly the response, we must take into account also the mass of the hydrogen atom. In the Hooke's law example we considered a system moving with infinite slowness and did not allow for inertia involved in moving the hydrogen atom at finite speeds. If this term is included, the equation becomes

$$m\frac{d^2x}{dt^2} + R\frac{dx}{dt} + Kx = F_a \qquad (27\text{-}8)$$

Equation (27-3) for a simple harmonic oscillator related the frequency to the Hooke's law constant and the mass; the quantity $2\pi\nu$ can be denoted by ω

$$2\pi\nu = \omega = \left(\frac{k_h}{m}\right)^{1/2} \qquad (27\text{-}9)$$

where ω is the angular frequency. In analyzing systems of this sort, it also is helpful to define a quantity called the *damping ratio* ζ (zeta) defined by

$$\zeta = \frac{-k_v}{2(k_h m)^{1/2}} \qquad (27\text{-}10)$$

This puts the conversion expression in the form:

$$F_a\left[\frac{1/k_h}{\dfrac{1}{\omega^2}s^2 + \dfrac{2\zeta}{\omega}s + 1}\right] = x \qquad (27\text{-}11)$$

where the quantity in brackets is again just a conversion factor. The system shown in Fig. 27-9, the block diagram in Fig. 27-10, and some typical response curves in Fig. 27-11.

Fig. 27-4 The block diagram for the transducer action of the hydrogen atom on the enzyme in Fig. 27-3.

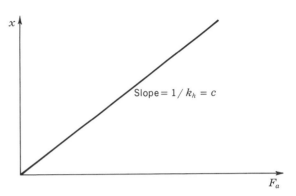

Fig. 27-5 The response curve for the enzyme hydrogen atom acting as a transducer (Fig. 27-3).

Feedback

The systems we have been discussing in the previous section are conversion systems where an input signal is changed into an output signal. The feature which makes a conversion system into a cybernetic system is a feedback circuit. The mathematics of feedback circuits has been developed into a useful form, applicable to biochemical problems, by the American physiologist F. S. Grodins (see References) and his generalized analytical methods have been used as the basis for the presentation of the specific applications discussed in this chapter.

From the output information a signal is fed back into the system to regulate the output signal in a manner which makes it approach a predetermined value. The system diagram shown in Fig. 27-12 suggests that the function of the black box is now to observe both the output signal x_0 and the predetermined desirable value of this signal x_1 and then to change the condition of the system so that the output signal x_0 becomes equal to x_1.

This situation can be represented by a block diagram of the sort shown in Fig. 27-13. The system

Fig. 27-6   Enzyme transducer action with damping component caused by viscosity.

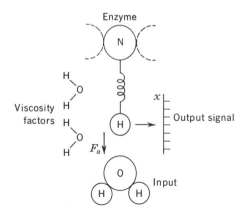

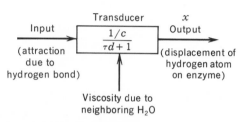

Fig. 27-7   Block diagram for enzyme transducer action with damping factor.

state system where entropy is being produced by means of chemical reactions there is a tendency to go to the minimum of rate of entropy production as the most stable steady state, a tendency which is, in essence, a cybernetic action.   If something disturbs the steady state, this is equivalent to the introduction of a disturbing force F_d.   When the steady state moves toward a higher production of entropy, the output x_0 no longer is equal to the state corresponding to the minimum rate of entropy production x_1.   The controlling system senses this error $(x_0 - x_1)$ and then produces an error signal x_e.   This in turn is con-

Fig. 27-8   Response curve for enzyme transducer action with damping factor.

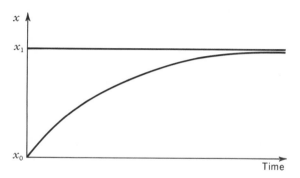

now consists of two principal parts, the block on the left, which receives the command x_1 to make the output on the far right x_0 equal to x_1.   When x_0 is not equal to x_1, the controlling system notes the difference between x_1 and x_0 by means of the error detector. This sends out an error signal which is converted by a controller into a signal to the controlling system shown in the block on the right.   This system, then, under the influence of the controlling signal, adjusts so that ultimately the output x_0 becomes equal to x_1. However, the whole purpose of this combination of units is to restore the value of x_0 to equality with x_1 if some disturbing influence is brought to bear.   This disturbing influence is shown as the force F_d.   The combination of these elements illustrates the simplest components which compose a cybernetic system.

   In the previous chapter we noted that in a steady-

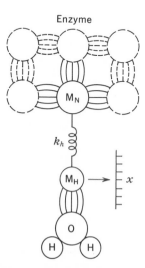

Fig. 27-9   Enzyme transducer action with inertial component caused by mass of the hydrogen atom.

trolling signal F_c, which, fed through the control system, restores x_0 to an equality with x_1 and places the steady state again in a condition of minimum rate of production of entropy.

The equations for the controlled and the controlling system can be combined. The first is

$$\tau\frac{dx_0}{dt} + x_0 = \frac{1}{c}(F_c + F_d) \qquad (27\text{-}12)$$

and the second is

$$F_c = c'(x_1 - x_0) \qquad (27\text{-}13)$$

In order to simplify the relations let us make the following definitions

$$\frac{F_c}{c} \equiv x_c \qquad (27\text{-}14)$$

$$\frac{F_d}{c'} \equiv x_d \qquad (27\text{-}15)$$

Fig. 27-10   The block diagram of an enzyme transducer with inertial and frictional components.

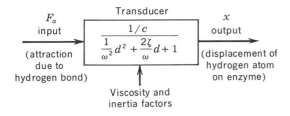

Fig. 27-11   Transient response curves for enzyme transducer with inertial and frictional components. (After F. S. Grodins, "Control Theory of Biological Systems," Columbia University Press, New York, 1963.)

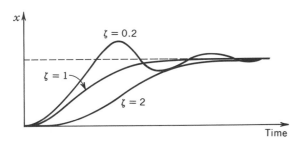

Fig. 27-12   Enzyme transducer with hydrogen bonding feedback.

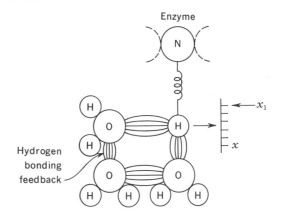

Then

$$\tau\frac{dx_0}{dt} + x_0 = x_c + x_d \qquad (27\text{-}16)$$

The ratio of the two constants can be written

$$\frac{c}{c'} = x_c \qquad (27\text{-}17)$$

Then

$$x_c = k(x_i - x_0) \qquad (27\text{-}18)$$

This gives the expression

$$\frac{\tau}{1+k}\frac{dx_0}{dt} + x_0 = \frac{k}{1+k}x_i + \frac{1}{1+k}x_d \qquad (27\text{-}19)$$

If this joint system undertakes to make x_0 follow x_i, the plot of these two quantities against time takes the form shown in Fig. 27-14. If we consider the error between x_0 and x_1 as a function of the disturbance which follows a time sequence similar to that of the command signal in Fig. 27-14, a plot against time has the form shown in Fig. 27-15.

In Fig. 27-16, the elements of control are shown in a block diagram. A single-loop feedback can be depicted as shown in Fig. 21-17. The joint servo and regulator operation is effectively that shown in Fig. 27-18. Finally, the regulator with a set point and reference value is shown in Fig. 27-19. These

Fig. 27-13 **Block diagram for feedback through hydrogen bonding.**

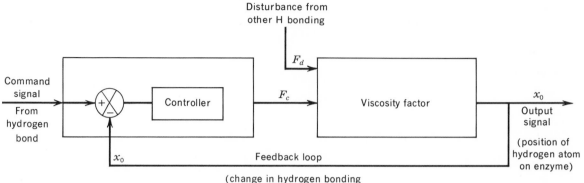

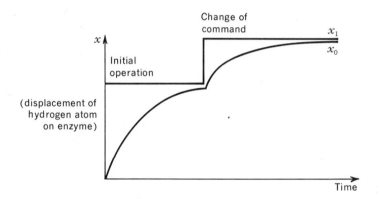

Fig. 27-14 **Transient response to change of command.**

diagrams are helpful in examining how chemical variations in the bloodstream control the rate of breathing and the relation between the heart action and the composition of the blood.

Transient response

When the transmission of signals and the response of components requires a significant interval of time, this may be the factor which determines the cyclic pattern of the steady state in a feedback system.

As shown in the previous chapter, one of the simplest types of response curve is based on an exponential function similar to that typical of first-order chemical reactions. When a radioactive isotope decays, the rate of decay dn/dt is given by a differential equation

$$\frac{dn}{dt} = -kn \tag{27-20}$$

where n = number of atoms
$\quad t$ = time
$\quad k$ = rate constant

The integrated form of this expression is

$$\frac{n}{n_0} = e^{-k_1 t} \tag{27-21}$$

or

$$\ln \frac{n}{n_0} = -k_1 t \tag{27-22}$$

where n_0 is the number of atoms present when $t = 0$. The half-life $t_{1/2}$, the time necessary for half of the atoms to decay, is equal to $(\ln 2)/k_1$.

When a chemical substance is decomposing, the same equation is applicable; in this case, n can be the number of moles per liter m or some other similar measure of concentration.

If part of a living cell receives a signal to act and the action involves a first-order chemical reaction, *then* the decrease in concentration may follow this same law. For example, consider a reaction activated by an enzyme

$$A \xrightarrow{\text{enzyme I}} B + C \qquad (27\text{-}23)$$

The relation between m and t is given by

$$m = m_0 e^{-k_1 t} \qquad (27\text{-}24)$$

where m_0 is the concentration at $t = 0$ and k_1 is the rate constant for the reaction. In logarithmic form, this is

$$\ln \frac{m}{m_0} = -k_1 t \qquad (27\text{-}25)$$

or

$$\log \frac{m}{m_0} = \frac{-k_1 t}{2.303} \qquad (27\text{-}26)$$

The plot of m as a function of t is shown in Fig. 27-20, and of $\log m$ as a function of t in Fig. 27-21, for $k_1 = 0.301$ min^{-1}. The half-life for this reaction is 2.3 min. Until the signal is received, the level of m is constant at the value m_1; immediately upon receiving the signal, the first-order reaction starts and the

decomposition begins. If the reaction then proceeds for an indefinite time, the concentration m falls tangentially to zero.

Suppose, however, that when the concentration falls to the value m_0, a feedback signal stops the enzyme action which is causing the decomposition and starts another process which regenerates A, such as

$$E \xrightarrow{\text{enzyme II}} A + F \qquad (27\text{-}27)$$

Then the concentration m may start to rise again in a pattern given by the equation

$$\frac{m}{m_2} = 1 - e^{-k_2 t} \qquad (27\text{-}28)$$

where k_2 is the reaction constant for the first-order reaction shown in Eq. (27-24) and m_2 is the concentration of A that would be present if this reaction

Fig 27-15 Regulator operation under the influence of disturbing force x_d at time t_d to maintain output at x_0.

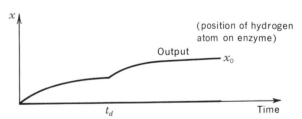

(secondary change in hydrogen-bonding)

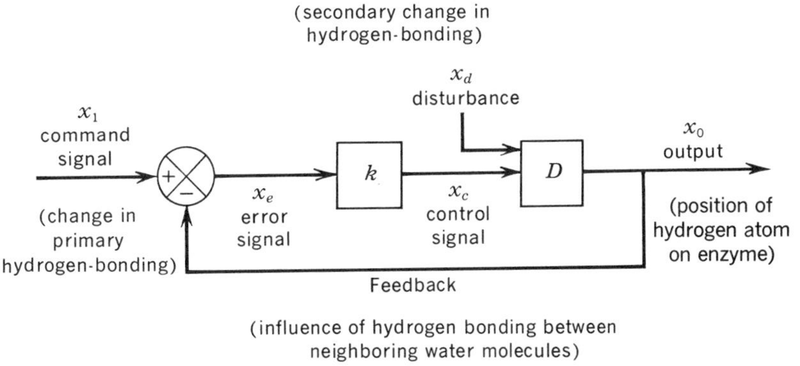

Fig. 27-16 Simplified signal diagram for regulator subject to disturbance. $D = (xd + 1)^{-1}$.

Fig. 27-17 Regulator with single feedback set to maintain output at x_0.

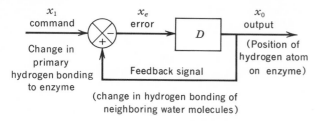

(change in hydrogen bonding of neighboring water molecules)

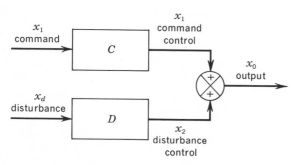

Fig. 27-18 Combined servo and regulator circuit.

proceeded to the point where E was completely decomposed. The graph of increasing concentration of A is shown in Fig. 27-22.

Now, suppose that the feedback signals are set to start the reaction in Eq. (27-23) when m_A attains a value of m_1 and stops the decomposition at m_0; concurrently the production of A by reaction (27-27) is started at m_0 and stopped at m_1. Under these conditions a typical cybernetic cycle takes place, which is a kind of alternating steady state of the kind illustrated in Fig. 27-23. This is the pattern which is characteristic of thermostat operations and of many other kinds of physical cybernetic systems for which the more complete equations were given in the preceding section.

Example 27-1

The elastic-force constant for the NH in many cases is 6.6×10^5 dynes cm^{-1}. If an approaching H_2O molecule attracts the H atom with a force of 1.53×10^{-4} dyne, what will be the displacement (the output signal) of the H atom

on the nitrogen atom, assuming that the latter is not moved?

According to Eq. (27-2), $x = F_a c$. Putting in the values for F_a and for c,

$$x = 1.53 \times 10^{-3} \text{ dyne} \frac{1}{6.6 \times 10^5 \text{ dynes cm}^{-1}}$$

$$= 0.23 \times 10^{-8} \text{ cm} = 0.23 \text{ Å} \quad Ans.$$

Example 27-2

In a cybernetic cycle for a pair of coupled reactions which synthesize and decompose a certain compound, the half-life of both reactions is 8.65 min. If the synthesis starts when the concentration of the compound is 0.354 mole liter^{-1} and stops when the concentration is 0.761 mole liter^{-1}, calculate the length of time of a single period of synthesis.

The value of k for the reaction can be calculated from the relation

$$k = \frac{0.693}{t_{1/2}} = \frac{0.693}{8.65} = 0.0801 \text{ min}^{-1}$$

The on and off values of m are related to the time by the expression

$$2.303 \log \frac{m_2}{m_1} = kt$$

so that

$$t = \frac{2.303}{k} \log \frac{m_2}{m_1} = \frac{2.303}{0.0801} \log \frac{0.761}{0.354}$$

$$= 28.8 \times 0.3324 = 9.57 \text{ min} \quad Ans.$$

27-3 BIOCYBERNETIC APPLICATIONS

In the human body there are many mechanisms which involve feedback and which therefore have a cybernetic pattern. Even in such a simple motion as raising the arm, there is a feedback between the nerves which sense the position of the arm and the nerves which control the muscles raising the arm. Other cybernetic patterns involve the relation between body temperature and metabolism, the action of the heart and the composition of the blood, the rate of breathing and blood composition. There are un-

Fig. 27-19   Regulator with setpoint and reference value.

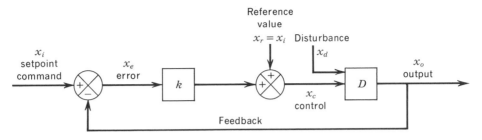

doubtedly cybernetic patterns in the action of the brain. And at the most basic level of all, there are cybernetic relationships within each cell governing the metabolism, growth, and replication.

Some cybernetic patterns are easier to analyze than others. If there are complicating factors, it is frequently possible to outline a simpler model to give the overall pattern of action. For example, there may be nonlinearity either in the control or in the response, but almost all nonlinear action can be approximated by linear action when the changes involved are sufficiently small. Another complicating factor is multiple feedback. When signals are coming in from several sources to the control mechanism, many of these cannot be measured by any simple algebraic function and the analysis of the result of their mixture is accordingly difficult. This is one of the aspects of brain cybernetics which has prevented any great advance in our understanding of the mechanism of thought.

Another aspect of biocybernetics is the difficulty of isolating any given cybernetic mechanism from its surroundings. The "wholeness" of the body is so pervasive that any physical attempt at isolating a cybernetic system within it is bound to alter significantly the pattern of behavior from normal.

In engineering cybernetics the focus of interest is on the response to transient disturbances. Many engineers regard the problem of the steady state as almost trivial. On the other hand, in biocybernetics, the first goal is the understanding of the steady-state mechanism. While there are many instances where transients play an important role in the total function, other transients may be regarded as pathological dis-

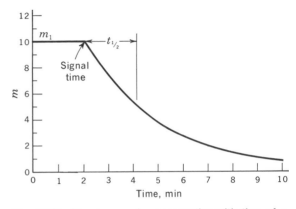

Fig. 27-20   The decrease in concentration with time of a compound which starts decomposing under enzyme action when signal is received at $t + 2$ min.

turbances which are of minor importance in trying to understand the total picture.

With the cybernetic mechanisms extending all the way from influences that involve the total body down to the cybernetic coupling of chemical reactions within the nucleus of the cell, the range of possible detail to include in any analysis is very wide. Actually, the ultimate microscopic level of feedback must lie somewhere in the interaction of the atomic orbitals. In trying to illustrate the nature of biocybernetics, a few examples are given which involve quantities like the concentration of carbon dioxide or oxygen in the blood without including finer details in the analysis.

The respiratory chemostat

There has been evidence available for many years indicating a close connection between the action of

Fig. 27-21   A semilogarithmic plot of the decrease in concentration of a compound decomposing according to a first-order reaction with $k_1 = 0.301$ min^{-1} ($t_{1/2} = 2.10$ min).

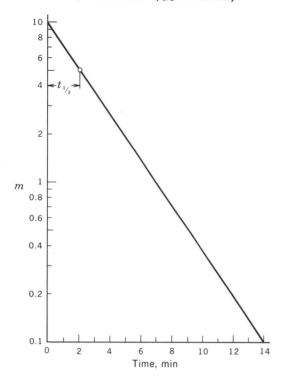

the lungs and the concentration of hydrogen ion, carbon dioxide, and oxygen in the bloodstream. Considerable progress has been made in analyzing the pattern of the interaction of these factors in terms of cybernetic analysis.

In the functioning of the body, there are normal levels for $[CO_2]$, $[H^+]$, and $[O_2]$ in the bloodstream. A cybernetic mechanism operates to restore the levels to normal when disturbances occur. For example, if the $[CO_2]$ in the air rises, then $[CO_2]$ and $[H^+]$ in the blood increase. This, in turn, increases the rate of breathing, which will tend to eliminate CO_2 from the blood and restore $[H^+]$ to normal. If a solution with a high $[H^+]$ is injected into the blood, this in itself will stimulate more rapid breathing. This, again, will eliminate CO_2 from the blood and lower $[H^+]$. If an animal breathes gas which is lower in the percentage of oxygen than normal air, the $[O_2]$ in the blood is lowered and this also stimulates a higher rate of breathing.

These relations can be summarized in a simple block diagram, as shown in Fig. 27-24. The input signals come from chemical effects that stimulate nerve action. The output signals change the values of $[CO_2]$, $[O_2]$, and $[H^+]$ in the blood, and these latter

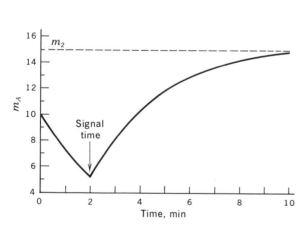

Fig. 27-22   The change of m_A with t for a first-order reaction based on the decomposition $E \rightarrow A + F$; $k_2 = 0.301$ sec^{-1}, $t_{1/2} = 2.3$ min.

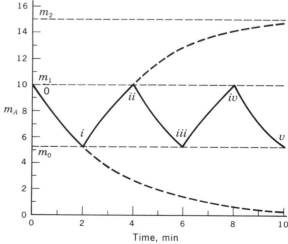

Fig. 27-23   The steady-state cyclic pattern of variation of m_A when controlled by enzymes which signal "decompose" at m_1 and "synthesize" at m_0.

signals are in turn relayed back to the controlling system.

One of the early attempts to understand the operation of this cybernetic system was made during the war. The American physiologist J. S. Gray made a study of the use of CO_2 to speed up breathing and counteract the lack of oxygen encountered by pilots at high altitudes. He proposed that the stimuli from the various types of signals were additive and that the relation could be expressed as

$$\dot{V}_a = 1.1[H^+] + 1.31[CO_2] - 90 + 10.6 \\ \times 10^{-8}(104 - [O_2])^{4.9} \quad (27\text{-}29)$$

where $\dot{V}_a$ = rate of breathing, liters min^{-1}

$[H^+]$ = arterial hydrogen-ion concentration, mμ-moles liter^{-1}

$[CO_2]$ = partial pressure of CO_2 in arterial blood, mm Hg

$[O_2]$ = partial pressure of O_2 in arterial blood, mm Hg

Expressions relating the partial vapor pressure in the bloodstream and the partial pressure in the air in the lungs are

$$[CO_2]_b = (CO_2)_t + \frac{KM_1}{\dot{V}_a} \quad (27\text{-}30)$$

$$[O_2]_b = (O_2)_t + \frac{KM_r}{\dot{V}_a} \quad (27\text{-}31)$$

where the subscripts b refer to the bloodstream, the subscripts t refer to the tracheal air, M_r is the metabolic gas exchange rate assumed equal for CO_2 and

O_2, and the constant K varies to include the effect of barometric pressure and makes the units consistent. The relation of the hydrogen-ion concentration in the blood $[H^+]_b$ is a linear function of the concentration of CO_2 in the bloodstream

$$[H^+]_b = a[CO_2]_b + b \quad (27\text{-}32)$$

In order to indicate that there are settings for normal levels and that the departures from these levels are detected, transmitted into nerve signals, and sent to the proper centers to stimulate appropriate breathing and then reduce stimulation as error is reduced, the block diagram for the respiratory chemostat can be drawn as shown in Fig. 27-25. To indicate a possible independent regulator functioning as a unit within the more complex regulator, the relations can be shown as in Fig. 27-26. Actual measurements of the responses are shown in Fig. 27-27.

The solution of the equations obtained in a reasonably complete analysis of such a cybernetic system is frequently difficult. Analog computers have been designed to help in such problems. The circuit of one of these is shown in Fig. 27-28. These studies have been helpful in furthering the understanding of the role of oxygen and carbon dioxide at this first stage where the oxygen enters the body to begin the enormously complicated series of reactions involved in metabolism. This approach should be even more useful in the future in solving many of the problems of space physiology, where human beings may be subjected to more and more abnormal situations with respect to the supply of oxygen for the body.

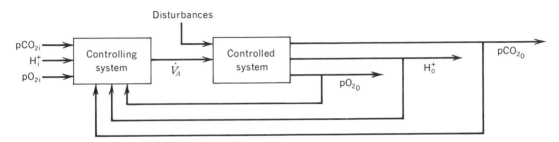

Disturbances

Fig. 27-24 The respiratory chemostat. (From F. S. Grodins, "Control Theory of Biological Systems," Columbia University Press, New York, 1963.)

Fig. 27-25   Single-loop respiratory chemostat.   (After F. S. Grodins, "Control Theory of Biological Systems," Columbia University Press, New York, 1963.)

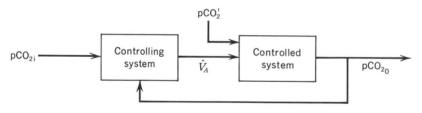

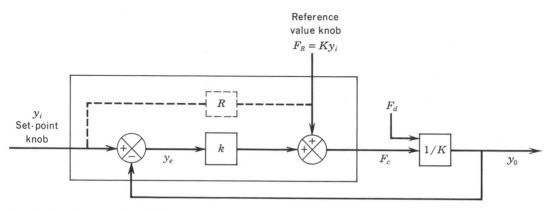

Fig. 27-26   Regulator with set-point and reference valve adjustments.   (After F. S. Grodins, "Control Theory of Biological Systems," Columbia University Press, New York, 1963.)

The cardiovascular regulator

Another equally important and equally basic cybernetic system in the body is made up of the relation between the heart and the pressure of the blood it circulates. This system is closely related to the respiratory chemostat because one of the functions of the circulation of the blood is to maintain the normal CO_2 and O_2 concentrations by variations in the rate of blood circulation.

From Fig. 27-29 one can see that the oxygen and carbon dioxide content in the blood must be taken into account both where the blood has entered and left the lungs and where it is circulating in the tissue. The blood in the tissue may be regarded as a tissue reservoir.

In the study of the hydrogen atom on the surface of an enzyme, discussed at the beginning of this chapter, a relation was established between the input signal (the force applied to the atom F_a), the conver-

sion factor (the reciprocal of the Hooke's law constant), and the output signal, which in this case was the shift in the position of the atom.   This gave the expression

$$\underset{\text{input signal}}{F_a} \times \underset{\substack{\text{conversion} \\ \text{factor}}}{\frac{1}{k_h}} = \underset{\substack{\text{output signal} \\ \text{(displacement)}}}{-x} \qquad (27\text{-}33)$$

The controller for both the respiratory chemostat and the circulatory chemostat have analogous equations, which may be put in the following form:

Controller:

$$\textit{Respiratory:} \quad \underset{\text{input signal}}{(CO_2)} \times \underset{\substack{\text{conversion} \\ \text{factor}}}{\phi} = \underset{\substack{\text{output signal} \\ \text{(breathing rate)}}}{\dot{V}_a} \qquad (27\text{-}34)$$

$$\textit{Circulatory:} \quad \underset{\text{input signal}}{(O_2)} \times \underset{\substack{\text{conversion} \\ \text{factor}}}{\phi'} = \underset{\substack{\text{output signal} \\ \text{(rate of blood} \\ \text{circulation)}}}{Q} \qquad (27\text{-}35)$$

For the respiratory chemostat, this output signal $\dot{V}_a$ goes to the controlled system, which in this case is the lung reservoir. For this and for the circulatory chemostat the basic equations are:

Controlled system:
  Respiratory:

$$\left[(O_2)_A - \frac{M_r}{\dot{V}_a}\right]\frac{1}{\dot{V}_a} = (O_2)_v$$

$$\left[(CO_2)_I + \frac{M_r}{\dot{V}_a}\right]\frac{1}{\dot{V}_a} = (CO_2)_v$$

(27-36)

Circulatory:

$$\left[(O_2)_A - \frac{M_r}{Q}\right]\frac{1}{Q} = (O_2)_v$$

$$\left[(CO_2)_A + \frac{M_r}{Q}\right]\frac{1}{Q} = (CO_2)_v$$

(27-37)

As might be expected, one can construct hydraulic analogs for which the equations can be analyzed and solved.

Example 27-3

If $[H^+]$ in the bloodstream is increased by 1.56×10^{-9} m, what is the rate of breathing increase?

The increase in breathing rate can be obtained from Eq. (27-29):

$$\dot{V} \text{ (liters min}^{-1}) = 1.1[H^+] + C$$
$$= 1.1[H_0^+ + \Delta H^+] + C$$
$$\Delta V = 1.1\Delta H^+ = 1.1 \times 1.56$$
$$= 1.7 \text{ liters min}^{-1} \qquad Ans.$$

Example 27-4

In the diagram given in Fig. 27-27, the response to an increase in alveolar pCO_2 is a kind of transient exponential curve. Assuming that the true form of the curve really is exponential, calculate the value of k_2 for the increased rate.

The equation for the increase in response r can be written

$$r = 1 - e^{-k_2(pCO_2)t}$$

The response increases with time in a way that indicates that the exponential term is cut to half its value in about 4 min. Therefore, the value of $t_{1/2} = 4$ min.

$$k_2 = \frac{0.693}{t_{1/2}} = 0.173 \text{ min}^{-1} \qquad Ans.$$

27-4 CYTOCYBERNETICS

One of the most remarkable processes in the complex scheme of biochemical reactions is the replication of the DNA molecule. In the language of genetics, a sequence of a few thousand cross-links on the DNA ladder constitute a single gene. Each complete DNA molecule comprises several thousand genes. Thus, at the very least, something between 10^7 and 10^8 cross-links must be reproduced in perfect sequence before a mother cell can form two daughter cells.

The American biophysicists A. Novick and L. Szilard carried out an experiment in which bacterial genes were duplicated in a steady state of growth. They estimated that duplication took place as many as 10^8 times before even one gene was altered. Considering the bombardment by cosmic rays and the many other disturbing influences that might upset the

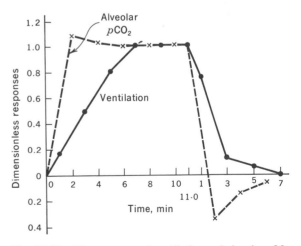

Fig. 27-27   The response of ventilation and alveolar pCO_2 to inhalation of 5.43% CO_2. [From F. S. Grodins, K. R. Schroeder, A. L. Norins, and R. W. Jones, *J. Appl. Physiol.*, 7:283 (1954).]

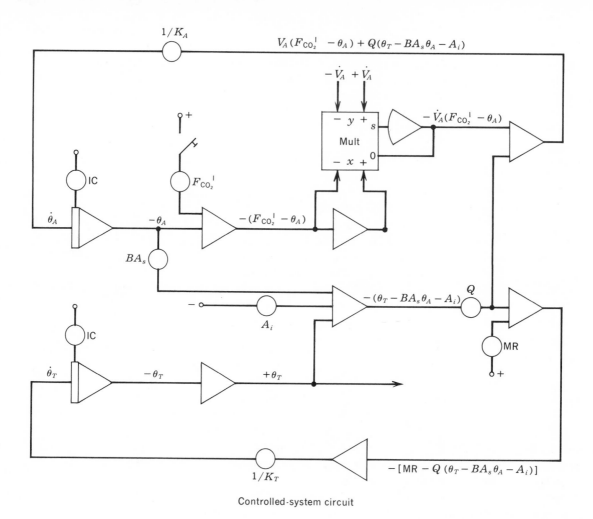

Controlled-system circuit

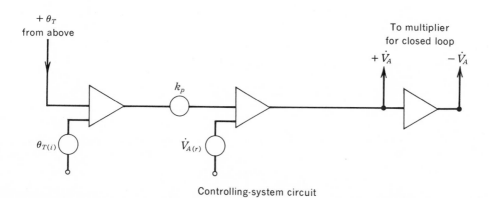

Controlling-system circuit

Fig. 27-28 An analog computer circuit for studying responses similar to those of the respiratory chemostat. (After F. S. Grodins, "Control Theory of Biological Systems," Columbia University Press, New York, 1963.)

replication process, this conclusion suggests that there is a cybernetic mechanism by which a molecule of DNA can repair itself and thus correct errors introduced by outside disturbing influences in much the same way that the respiratory chemostat and circulatory chemostat hold the breathing and the circulation of the blood on a steady course.

Cybernetic repairing

If one thinks of the DNA molecule with its two strands as being unzipped so that each strand is an independent entity, there is a redundancy in the storage of information since each independent ribbon contains the genetic message. In terms of this picture one can imagine a mechanism for repairing damage to one strand of the DNA by conveying the message from the other undamaged strand.

To a greater or lesser degree on the macroscopic scale we know that all living organisms have an ability to recover from injury. This recovery is in itself a cybernetic mechanism. If skin tissue is cut, the surrounding tissue grows back to repair the cut and somehow a signal is sent to the growing mechanism which stops the growth when the injury has been repaired.

Some years ago a discovery was made that ultraviolet radiation had a germicidal action. This action may be the formation of two spurious chemical bonds between pyrimidine bases that are adjacent to each other in DNA. If, in the genetic sequence, there are next to each other two links of the type thymine-adenine TA or two links of AT so that there are two thymine molecules side by side, ultraviolet light can cause a thymine dimer, where the two rings become linked together. Bacteria which contain abnormally large amounts of TA or AT links in their genetic message are especially susceptible to this type of damage. If even a few dimers are formed, the bacteria are unable to divide and form colonies. Dimers can also form between other similar pyrimidine rings, though not as readily as between two thymine rings. The effect of forming the thymine dimer may be compared to the fusion of two teeth on a zipper which would make the unzipping operation difficult if not impossible.

The damage done by ultraviolet light may be nulli-

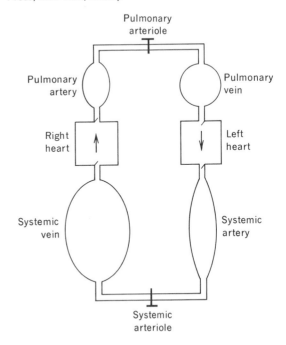

Fig. 27-29 The circulatory circuit. (After F. S. Grodins, "Control Theory of Biological Systems," Columbia University Press, New York, 1963.)

fied to a great extent by irradiation with visible light. The American biophysicist C. S. Rupert has shown that an enzyme in the protoplasm is selectively bound to DNA that has been irradiated by ultraviolet light. When activated by visible light, which serves as a source of energy, this enzyme splits the spurious bonds between the two thymine or cytosine rings and restores the DNA to its original form, in which the hydrogen bonds can unzip and replication can take place. Thus, the enzyme functions under the impetus of the error signal to produce a corrective force which restores the molecule to its normal state.

When one considers the whole pattern of activity within a biological cell, one realizes the necessity of hundreds or thousands of these chemical cybernetic linkages. In the process of cell duplication, there must be synthesized or brought in from the surrounding body fluid exactly the right amount of the thousands of different varieties of molecules necessary to produce each of the two daughter cells. Whenever

Fig. 27-30 **Suggested feedback pattern in the biosynthesis of amino acids of the aspartic family in** *Rhodopseudomonas capsulatus.* **Feedback inhibition of enzyme activity is indicated by hatched arrows and repression of enzyme synthesis (with glutamate as nitrogen source) by the dotted arrow.** **[From L. Burlant, P. Datta, and Howard Gest,** *Science,* **148:1351 (1965).** **Used by permission.]**

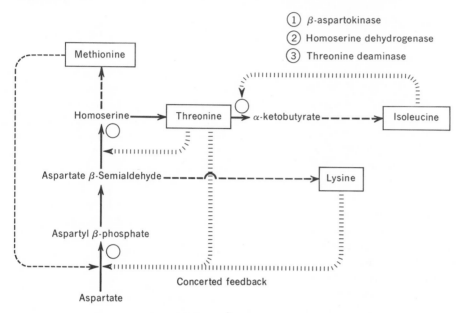

① β-aspartokinase
② Homoserine dehydrogenase
③ Threonine deaminase

there is a situation requiring not too little or not too much but just enough, there must be some cybernetic mechanism at work which senses when the requirement of synthesis has been met and prevents the production of an overabundance of the material under consideration. It is interesting to ask whether in a pathological condition like cancer, the cause of the pathology may be the failure of the cybernetic mechanism to stop the synthesis. Certainly, the pattern of synthesis deviates from the replicative path that would produce a normal, healthy cell.

The major part of the biochemical reactions inside a living cell are controlled by enzyme actions which are coupled in a cybernetic pattern. As a specific example of this type of coupling we consider a feedback system suggested by the American biophysicists L. Burlant, P. Datta, and H. Gest for the synthesis of the aspartic type of amino acids in bacteria. The reaction pattern is shown in Fig. 27-30. The $S^{\ddagger}$ profile is shown in Fig. 27-31. The enzymes are denoted

by the shaded boxes. A negative sign indicates that the signal originating in an excess concentration of the compound in the unshaded box slows down or inhibits the action of the enzyme. In other words, an excess of the compound in question cuts down the value of $S^{\ddagger}$ for the barrier and thus slows down the rate of the reaction. This reduction of $S^{\ddagger}$ is probably due to the occupation of active sites on the catalyst by the substrate compound acting as an inhibitor as shown in Fig. 27-32.

Biopatterns

In the latter half of this book, we have pursued the study of matter in its various forms from the elementary particles which make up the atom, through atomic molecular orbitals, through the more complex association of these forms into macromolecules, concluding with the observation of the functioning of macromolecules in patterns of the whole—in *holistic* behavior. Just as the electrons in an atom should not be regarded as independent particles but as fused into a pattern of collective behavior governed by a complex de Broglie wave, so the action of macromolecules at the cybernetic level must also be regarded in the perspective of collective action if the true nature of their functioning is to be understood. The cybernetic pattern of the synthesis of proteins is shown in Fig. 27-33 to suggest the feedback between DNA and RNA.

Ever since the scientific advances in the nineteenth century emphasized the *particulate* nature of matter, there has been a tendency to try to fit the behavior of matter all the way from electrons to biological cells into this particulate pattern. There is no doubt that

Fig. 27-32 Active sites on enzyme (open circle) blocked by inhibitor molecules (circle with cross).

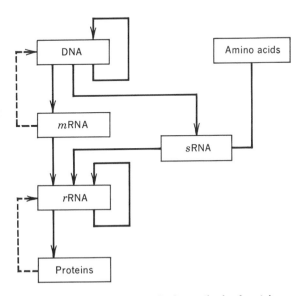

Fig. 27-33 Feedback pattern in the synthesis of proteins.

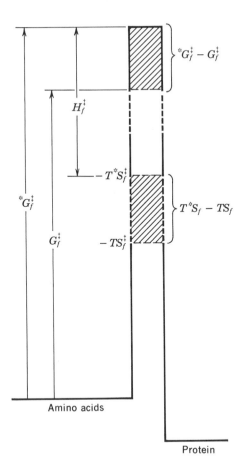

Fig. 27-31 Effect of inhibitor in raising height of entropy-factor barrier. Shaded portion is the result of inhibitor.

the representation of organic molecules in terms of *atoms* coupled together has been one of the most powerful tools in furthering our understanding of organic chemistry and the synthesis of new molecules; and the understanding of the replication of the cell in terms of the *atomic* structure of DNA has led to the exploration and understanding of whole new domains of genetics. However, it is also clear that in our enthusiasm for these successes we should not forget that the atoms of DNA, in essence, derive their significant patterns of behavior from the holistic patterns of the de Broglie waves, as Szent-Györgyi has repeatedly pointed out. Just as the most basic representation of these patterns emphasizes the significance of interaction, the cybernetic behavior of atoms when linked together by the thousands or millions also emphasizes the significance of the forces that derive from the total pattern. It is to be hoped that in a fusion of the particulate and the holistic perspective we may find a useful guide for furthering our understanding of matter both inanimate and animate.

Example 27-5

In Fig. 27-31, assume that for a formation of a single peptide bond the effect of the inhibitor is to decrease $S^{\ddagger}$ by 25 cal deg^{-1} mole^{-1}. By what ratio will the rate of protein synthesis be reduced with respect to this bond at 37°C?

$$\frac{^{*}r}{r} = e^{-T(S-^{*}S)/RT}$$

$$= e^{-310 \times 25}$$

$$\log \frac{r}{^{*}r} = \frac{310 \times 25}{4.5757 \times 310}$$

$$\log \frac{r}{^{*}r} = 5.46$$

$$^{*}r = 3 \times 10^{-5} r \qquad Ans.$$

Problems

1. The value of k_h for an OH bond is 8×10^5 dynes cm^{-1}. If a repulsive force of 6.02×10^{-3} dyne is brought to bear on a hydrogen atom attached to an oxygen atom by such a bond and the displacement is along the bond axis, what will be the transducer constant and the response signal of the H atom?

2. When a force 2.47×10^{-3} dyne is applied to a hydrogen atom, the atom moves along the bond axis by 3.8×10^{-9} cm. Calculate the transducer constant for the bond. Assuming that the atom on the other end of the bond does not move, identify this atom using the data in Fig. 20-34.

3. If after the displacement in Prob. 2, the atom vibrates freely, calculate the frequency of vibration, assuming that the mass of the atom to which the hydrogen atom is attached is so great that it does not move.

4. Calculate the value of m after 10.3 min if $m_0 = 3.59 \times 10^{-3}$ mole liter^{-1} and $t_{1/2} = 6.19$ min for the decomposition of a substance by an enzyme.

5. Calculate the value of k for a unimolecular decomposition reaction where 9.6 min is required for a change from a concentration of 0.057 to 0.032 mole liter^{-1}.

6. Using the value of k obtained for the decomposi-

tion in Prob. 5, calculate the increase in the value of s (entropy per mole) in going from the initial concentration to the final concentration.

7. Calculate the decrease in free energy per mole for the change stated in Prob. 6 at 37°C.

8. Calculate the time rate of increase in entropy per mole at the initial and final concentrations in Prob. 5.

9. Calculate the time rate of decrease of free energy per mole for the reaction in Prob. 5 at 37°C, both at the initial and at the final concentrations.

10. A substance is produced by a reaction

$$A \longrightarrow B + C$$

The reaction is first order in [A]. If the initial concentration of A is 1.3×10^{-2} mole liter^{-1} in a 1-liter flask and the initial concentration of B is zero, calculate the concentration of B after 6.95 min if the half-life of A is 3.21 min.

11. A substance HA changes slowly into HB, which then ionizes rapidly into H$^+$ and B$^-$. The change from HA to HB is controlled by an enzyme which ceases to function when [H$^+$] = 3.2×10^{-6} mole liter^{-1} and will not resume functioning until [H$^+$] drops to 1.9×10^{-6} mole liter^{-1}. If the half-life of HA is 15.2 min, calculate the time required for an initial concentration of HA of 1.3×10^{-5} mole liter^{-1} to stop changing into HB.

12. A cybernetic cycle is established between two reactions, one of which produces a compound AB by a first-order reaction and the other decomposes it, also by a first-order reaction. Both reactions have $t_{1/2} = 3.56$ min. The limits at which the reactions reverse are $m_0 = 1.46 \times 10^{-4}$ mole liter^{-1} and 4.35×10^{-4} liter^{-1}. Calculate the length of time required for one complete cycle.

13. If an inhibitor changes the value of the half-life in Prob. 12 to 4.2×10^5 min, calculate the increase in the length of cycle time.

14. If the value of k in Prob. 12 increases by a factor of 1.36 when the temperature is raised from 37 to 40°C, calculate the value of $H^{\ddagger}$ for these reactions.

15. Calculate the value of $S^{\ddagger}$ for the reactions in Prob. 12 at 37°K using the data from Prob. 14.

16. Calculate the change in the value of $S_f^{\ddagger}$ due to the inhibitor described in Prob. 13.

17. Calculate the change in the value of $G_f^{\ddagger}$ due to the inhibitor in Prob. 13.

18. Calculate the change in $S_f^{\ddagger}$ for each of the reactions described in Prob. 12 if the time of the cybernetic cycle is doubled.

19.   If the change in $S^{\ddagger}$ is due to a partial blocking of the active sites on the enzyme, what fractional increase is necessary to produce the change in $S^{\ddagger}$ calculated in Prob. 18?

20.   Draw the reaction-barrier profile for the production of compound AB described in Prob. 12 showing the changes due to the inhibitor.

REFERENCES

Weiner, N.: "Cybernetics," 2d ed., John Wiley & Sons, Inc., New York, 1961.

Grodins, F. S.: "Control Theory of Biological Systems," Columbia University Press, New York, 1963.

Barry, J. M.: "Molecular Biology: Genes and the Chemical Control of Living Cells," Prentice-Hall, Inc., Englewood Cliffs, N.J., 1964.

Gurland, J. (ed.): "Stochastic Models in Medicine and Biology," The University of Wisconsin Press, Madison, 1964.

Langley, L. L.: "Homeostasis," Reinhold Publishing Corporation, New York, 1965.

Walter, C.: "Steady-state Applications in Enzyme Kinetics," The Ronald Press Company, New York, 1965.

Chance, B., R. W. Estabrook, and J. R. Williamson (eds.): "Control of Energy Metabolism," Academic Press, Inc., New York, 1965.

Hanawalt, P. C., and R. H. Haynes: "The Repair of DNA," *Sci. Am.*, February, 1967.

Wynn, R. M., (ed.): "Fetal Homeostasis," The New Academy of Sciences, New York, 1967.

Prosser, C. L. (ed.): "Molecular Mechanisms of Temperature Adaption," American Association for the Advancement of Science, Washington, 1967.

Peachey, L. D. (ed.): "Conference on Cellular Dynamics," The New York Academy of Sciences, New York, 1967.

Weber, G. (ed.): "Advances in Enzyme Regulation," vol. 5, Pergamon Press, New York, 1967.

Mesarovic, M. D. (ed.): "Systems Theory and Biology," Springer-Verlag New York Inc., New York, 1968.

Cheek, D. B.: "Human Growth," Lea & Febiger, Philadelphia, 1968.

Tables

TABLE A1-1 Fundamental constants

Symbol	Name	Value
c	Speed of light[a]	2.997925×10^{10} cm sec^{-1}
h	Planck's constant[a]	6.6256×10^{-27} erg sec
k	Boltzmann's constant[a]	1.38054×10^{-16} erg deg^{-1} molecule^{-1}
e	Charge on the electron[a]	4.80298×10^{-10} esu
		1.60210×10^{-19} coulomb
m_e	Rest mass of electron[a]	9.1091×10^{-28} g
m_p	Rest mass of proton[a]	1.67252×10^{-24} g
		1.00727663 amu
m_n	Rest mass of neutron[a]	1.0086654 amu
N_A	Avogadro's number[a]	6.02252×10^{23} molecules mole^{-1}
R	Gas constant	82.055 cm^3 atm deg^{-1} mole^{-1}
		0.082054 liter atm deg^{-1} mole^{-1}
		8.3143 joules deg^{-1} mole^{-1}
		1.9872 defined cal deg^{-1} mole^{-1}
V_0	Standard gas volume[a]	2.24136×10^4 cm^3 mole^{-1}
	Ice point (0°C)	273.15°K
	Triple point of H_2O (Base of international temperature scale)	273.16°K
g	Standard gravity	980.665 cm sec^{-2}
$\mathcal{F}$	Faraday constant[a]	9.64875×10^4 abs coulomb equiv^{-1}
		2.3060×10^4 cal volt^{-1} equiv^{-1}
ev	Electron volt[a]	1.60210×10^{-12} erg molecule^{-1}
		2.3060×10^4 cal mole^{-1}

[a] Values of the physical constants are those recommended by the Committee on Fundamental Constants of the National Academy of Sciences–National Research Council and published in *Natl. Bur. Std. Tech. News Bull.*, **47**:175 (1963).

TABLE A1-2   Conversion factors and mathematical quantities

Length:

$1 \text{ cm} = 10^4\,\mu = 10^7\,\text{m}\mu = 10^8\,\text{Å}$
$= 0.3937 \text{ in.}$

$1 \text{ bohr radius} = 5.2922 \times 10^{-9} \text{ cm}$

Mass:

$1 \text{ g} = 2.2046 \times 10^{-3} \text{ lb} = 1.1023 \times 10^{-6} \text{ ton}$

Pressure:

$1 \text{ atm} = 1{,}013{,}250 \text{ dynes cm}^{-2} = 1.01325 \text{ bars}$
$= 1.0332 \text{ kg cm}^{-2} = 760 \text{ mm Hg}$
$= 29.921 \text{ in. Hg} = 14.696 \text{ lb-in.}^{-2}$

Energy:

$1 \text{ cal (defined)} = 4.1840 \text{ abs joules}$
$= 3.9657 \times 10^{-3} \text{ Btu}$
$= 4.12917 \times 10^{-2} \text{ liter atm}$

$1 \text{ cal mole}^{-1} = 6.9461 \times 10^{-17} \text{ erg molecule}^{-1}$
$= 4.3359 \times 10^{-5} \text{ ev molecule}^{-1}$
$= 0.34982 \text{ cm}^{-1} \text{ (wave number)}$

$1 \text{ ev} = 1.2398 \text{ (wavelength)}$
$= 8065.7 \text{ cm}^{-1} \text{ (wave number)}$

Mathematical quantities:

$\pi = 3.1415926536$
$\pi^2 = 9.869604$
$\pi^{-1} = 0.31831$
$\sqrt{\pi} = 1.77245$
$1/\sqrt{\pi} = 0.56419$

$$\frac{\ln x}{\log x} = 2.302585$$

$$\frac{R \ln x}{\log x} = 4.575697$$

$$e = 2.7183$$

$$\frac{1}{e} = 0.36787$$

TABLE A1-3   Thermodynamic data for compounds at 25°C[a]

Compound	ΔH_f°, kcal mole^{-1}	ΔG_f°, kcal mole^{-1}	Compound	ΔH_f°, kcal mole^{-1}	ΔG_f°, kcal mole^{-1}
$O_3(g)$	+34.0	+39.06	$(NH_4)_2SO_4(c)$	−281.86	−215.19
$H_2O(l)$	−68.32	−56.69	$PH_3(g)$	+2.21	+4.36
$H_2O_2(l)$	−44.84		$PCl_3(g)$	−73.22	−68.42
$HF(g)$	−64.2	−64.7	$PCl_5(g)$	−95.35	−77.59
$ClO_2(g)$	+24.7	+29.5	$POCl_3(g)$	−141.5	−130.3
$Cl_2O(g)$	+18.20	+22.40	$PBr_3(g)$	−35.9	−41.2
$HCl(g)$	−22.063	−22.769	$PN(g)$	−20.2	−25.3
$HBr(g)$	−8.66	−12.72	$CO(g)$	−26.4157	−32.8079
$HI(g)$	+6.20	+0.31	$CO_2(g)$	−94.0518	−94.2598
$SO_2(g)$	−70.96	−71.79	$CH_4(g)$	−17.889	−12.140
$SO_3(g)$	−94.45	−88.52	$CH_3OH(g)$	−48.08	−38.69
$H_2S(g)$	−4.815	−7.892	$CH_3OH(l)$	−57.02	−39.73
$SF_6(g)$	−262	−237	$CCl_4(g)$	−25.5	−15.3
$H_2Se(g)$	+20.5	+17.0	$CCl_4(l)$	−33.3	−16.4
$SeF_6(g)$	−246	−222	$CHCl_3(g)$	−24	−16
$H_2Te(g)$	+36.9	+33.1	$CHCl_3(l)$	−31.5	−17.1
$TeF_6(g)$	−315	−292	$CS_2(g)$	+27.55	+15.55
$NO(g)$	+21.6	+20.719	$CS_2(l)$	+21.0	+15.2
$NO_2(g)$	+8.091	+12.390	$HCN(g)$	+31.2	+28.7
$N_2O(g)$	+19.49	+24.76	$HCN(l)$	+25.2	+29.0
$N_2O_4(g)$	+2.309	+23.491	$C_2H_6(g)$	−20.236	−7.860
$NH_3(g)$	−11.04	−3.976	$C_2H_5OH(g)$	−56.24	−40.30
$NH_4Cl(c)$	−75.38	−48.73	$C_2H_5OH(l)$	−66.356	−41.77
$NOCl(g)$	+12.57	+15.86	$SiF_4(g)$	−370	−360
NOBR	+19.56	+19.70	$SiCl_4(g)$	−145.7	−136.9

[a] Data from *Natl. Bur. Std. Circ.* 500, 1952.

TABLE A1-4   Thermodynamic data for aqueous ions at 298.15°K[a]

Ion	$\bar{C}_p^\circ$, cal deg⁻¹ mole⁻¹	$\bar{S}^\circ$, cal deg⁻¹ mole⁻¹	$\Delta\bar{H}_f^\circ$, kcal mole⁻¹	$\Delta\bar{F}_f^\circ$, kcal mole⁻¹
H^+	0	0	0	0
OH^-	−32.0	−2.52	−54.96	−37.59
F^-	−29.5	−2.3	−78.66	−66.08
Cl^-	−30.0	13.2	−40.02	−31.35
ClO_2^-		24.1	−17.18	2.74
ClO_3^-	−18	39	−23.5	−0.6
ClO_4^-		43.2	−31.41	−2.47
Br^-	−31	19.29	−28.90	−24.57
I^-	−31	26.14	−13.37	−12.35
I_3^-		57.1	−12.4	−12.31
S^-		−4	7.8	20.6
HS^-		15.0	−4.10	3.00
SO_4^-	−66	4.1	−216.90	−177.34
HSO_4^-		30.52	−211.70	−179.94
SeO_3^-		3.9	−122.39	−89.33
SeO_4^-		5.7	−145.3	−105.42
$HSeO_4^-$		22.0	−143.1	−108.2
NH_4^+	16.9	26.97	−31.74	−19.00
$N_2H_6^{++}$		19	−4	22.5
$N_2H_5^+$		31	−1.7	21.0
$NH_2OH_2^+$		37	−30.7	−13.54
NO_2^-		29.9	−25.4	−8.25
NO_3^-	−18	35.0	−49.37	−26.43
PO_4^{3-}		−52	−306.9	−245.1
HPO_4^-		−8.6	−310.4	−261.5
$H_2PO_4^-$		21.3	−311.3	−271.3
$HAsO_4^-$		0.9	−214.8	−169.
$H_2AsO_4^-$		28	−216.2	−178.9
$HCOO^-$		21.9	−98.0	−80.0
HCO_3^-		22.7	−165.18	−140.31
CO_3^-		−12.7	−161.63	−126.22
CH_3COO^-		20.8	−116.84	−89.02
$C_2O_4^-$		10.6	−195.7	−159.4
CN^-		28.2	36.1	39.6
CNO^-		31.1	−33.5	−23.6
Sn^{++}		−5.9	−2.39	−6.27
Pb^{++}		5.1	0.39	−5.81
Tl^+		30.4	1.38	−7.75
Zn^{++}		−25.45	−36.43	−35.18
Cd^{++}		−14.6	−17.30	−18.58
Hg^{++}		−5.4	41.59	39.38
Cu^{++}		−23.6	15.39	15.53
Ag^+		17.67	25.31	18.43
$Ag(NH_3)_2^+$		57.8	−26.72	−4.16
$Ag(CN)_2^-$		49.0	64.5	72.05
$PtCl_4^-$		42	−123.4	−91.9
$PtCl_6^-$		52.6	−167.4	−123.1
Fe^{++}		−27.1	−21.0	−20.30
Fe^{3+}		−70.1	−11.4	−2.53

TABLE A1-4   Thermodynamic data for aqueous ions at 298.15°K[a] (Continued)

Ion	$\bar{C}_p^\circ$, cal deg^{-1} mole^{-1}	$\bar{S}^\circ$, cal deg^{-1} mole^{-1}	$\Delta\bar{H}_f^\circ$, kcal mole^{-1}	ΔF_f°, kcal mole^{-1}
$Fe(OH)^{++}$		-23.2	-67.4	-55.91
$FeNO^{++}$		-10.6	-9.7	1.5
Mn^{++}		-20	-53.3	-54.4
MnO_4^-		45.4	-129.7	-107.4
$H_2BO_3^-$		7.3	-251.8	-217.6
BF_4^-		40	-365	-343
Al^{3+}		-74.9	-125.4	-115
Gd^{3+}		$-43.$	-168.8	-165.8
Mg^{++}	4	-28.2	-110.41	-108.99
Ca^{++}	-9	-13.2	-129.77	-132.18
Sr^{++}		-9.4	-130.38	-133.2
Ba^{++}	-11	3	-128.67	-134.0
Li^+	14.2	3.4	-66.55	-70.22
Na^+	7.9	14.4	-57.28	-62.59
K^+	2.3	24.5	-60.04	-67.46
Rb^+	-8.7	28.7	-59.4	-67.65
Cs^+	-18.7	31.8	-62.6	-70.8
UO_2^{++}		-17	-250.4	-236.4

[a] From G. N. Lewis and M. Randall, "Thermodynamics," 2d ed., rev. by K. S. Pitzer and L. Brewer.   Copyright 1961.   McGraw-Hill Book Company. Used by permission.

TABLE A1-5   Vapor pressure of water at various temperatures

P, mm Hg	T, °C	P, mm Hg	T, °C
0.00001[a]	-100	55.3	40.0
0.0004[a]	-80	68.3	44.0
0.008[a]	-60	83.7	48.0
0.097[a]	-40	102.1	52.0
0.776[a]	-20	123.8	56.0
1.24[a]	-15	149.2	60.0
1.95[a]	-10	179.3	64.0
3.01[a]	-5	214.2	68.0
4.58[a]	0.0	254.5	72.0
6.1	4.0	301.4	76.0
8.0	8.0	355.0	80.0
10.5	12.0	416.7	84.0
13.6	16.0	487.0	88.0
17.5	20.0	567.0	92.0
22.4	24.0	657.5	96.0
28.3	28.0	760.0	100.0
35.7	32.0	1489.1	120.0
44.6	36.0		

[a] Vapor pressure of ice.

TABLE A1-6   Table of atomic weights; preliminary values (not official) recommended by the International Union of Pure and Applied Chemistry, 1970.[a]

Symbol	Element	Atomic number	Atomic weight	Symbol	Element	Atomic number	Atomic weight
Ac	Actinium	89	227	Hg	Mercury	80	200.59
Al	Aluminum	13	26.9815	Mo	Molybdenum	42	95.94
Am	Americium	95	243	Nd	Neodymium	60	144.24
Sb	Antimony	51	121.75	Ne	Neon	10	20.179
Ar	Argon	18	39.948	Np	Neptunium	93	237.0482
As	Arsenic	33	74.9216	Ni	Nickel	28	58.71
At	Astatine	85	210	Nb	Niobium	41	92.9064
Ba	Barium	56	137.34	N	Nitrogen	7	14.0067
Bk	Berkelium	97	247	No	Nobelium	102	(254)
Be	Beryllium	4	9.0122	Os	Osmium	76	190.2
Bi	Bismuth	83	208.9806	O	Oxygen	8	15.9994
B	Boron	5	10.81	Pd	Palladium	46	106.4
Br	Bromine	35	79.904	P	Phosphorus	15	30.9738
Cd	Cadmium	48	112.40	Pt	Platinum	78	195.09
Ca	Calcium	20	40.08	Pu	Plutonium	94	(242)
Cf	Californium	98	(251)	Po	Polonium	84	(210)
C	Carbon	6	12.011	K	Potassium	19	39.102
Ce	Cerium	58	140.12	Pr	Praseodymium	59	140.0977
Cs	Cesium	55	132.9055	Pm	Promethium	61	(147)
Cl	Chlorine	17	35.453	Pa	Protactinium	91	231.0359
Cr	Chromium	24	51.996	Ra	Radium	88	226.0254
Co	Cobalt	27	58.9332	Rn	Radon	86	(222)
Cu	Copper	29	63.546	Re	Rhenium	75	186.2
Cm	Curium	96	(247)	Rh	Rhodium	45	102.9055
Dy	Dysprosium	66	162.50	Rb	Rubidium	37	85.4678
Es	Einsteinium	99	(254)	Ru	Ruthenium	44	101.07
Er	Erbium	68	167.26	Sm	Samarium	62	150.4
Eu	Europium	63	151.96	Sc	Scandium	21	44.9559
Fm	Fermium	100	(253)	Se	Selenium	34	78.96
F	Fluorine	9	18.9984	Si	Silicon	14	28.086
Fr	Francium	87	(223)	Ag	Silver	47	107.868
Gd	Gadolinium	64	157.25	Na	Sodium	11	22.9898
Ga	Gallium	31	69.72	Sr	Strontium	38	87.62
Ge	Germanium	32	72.59	S	Sulfur	16	32.06
Au	Gold	79	196.9665	Ta	Tantalum	73	180.9479
Hf	Hafnium	72	178.49	Tc	Technetium	43	98.9062
He	Helium	2	4.00260	Te	Tellurium	52	127.60
Ho	Holmium	67	164.9303	Tb	Terbium	65	158.924
H	Hydrogen	1	1.0080	Tl	Thallium	81	204.37
In	Indium	49	114.82	Th	Thorium	90	232.0381
I	Iodine	53	126.9045	Tm	Thulium	69	168.9342
Ir	Iridium	77	192.22	Sn	Tin	50	118.69
Fe	Iron	26	55.847	Ti	Titanium	22	47.90
Kr	Krypton	36	83.80	W	Tungsten	74	183.85
La	Lanthanum	57	138.9055	U	Uranium	92	238.029
Lw	Lawrencium	103	(257)	V	Vanadium	23	50.9414
Pb	Lead	82	207.2	Xe	Xenon	54	131.30
Li	Lithium	3	6.941	Yb	Ytterbium	70	173.04
Lu	Lutetium	71	174.97	Y	Yttrium	39	88.9059
Mg	Magnesium	12	24.305	Zn	Zinc	30	65.37
Mn	Manganese	25	54.9380	Zr	Zirconium	40	91.22
Md	Mendelevium	101	(256)				

Note: Values in parentheses are from other sources.
[a] Courtesy of *Chem. and Eng. News,* p. 39, Jan. 26, 1970.

TABLE A1-7   Elastic-force constants
for chemical bonds[a]

Stretching constants

Bond	Molecule	k_h, dynes cm^{-1} × 10^5
H—F	HF	9.67
H—Cl	HCl	5.15
H—Br	HBr	4.11
H—I	HI	3.16
H—O	H_2O	7.8
H—S	H_2S	4.3
H—Se	H_2Se	3.3
H—N	NH_3	6.5
H—P	PH_3	3.1
H—As	AsH_3	2.6
H—C	CH_3X	~5.0
H—C	C_2H_4	5.1
H—C	C_6H_6	5.1
H—C	C_2H_2	5.9
H—Si	SiH_4	2.9
F—O	F_2O	5.6
F—C	CH_3F	5.6
Cl—C	CH_3Cl	3.4
Br—C	CH_3Br	2.8
I—C	CH_3I	2.3
F—B	BF_3	8.8
Cl—B	BCl_3	4.6
Br—B	BBr_3	3.7
P—P	P_4	2.1
Si—Si	Si_2H_6	1.7
S—S	S_2H_2	2.5
B—N	$B_3N_3H_6$	6.3
C⋯C	C_6H_6	7.62
N⋯O	N_2O	11.5
C—C		4.5–5.6
C=C		9.5–9.9
C≡C		15.6–17.0
N—N		3.5–5.5
N=N		13.0–13.5
N≡N		22.9
O—O		3.5–5.0
C—N		4.9–5.6
C=N		10–11
C≡N		16.2–18.2
C—O		5.0–5.8
C=O		11.8–13.4

TABLE A1-7   Elastic-force constants
for chemical bonds[a] (Continued)

Bending constants

Angle	Molecule	k_{hb}, dyne cm^{-1} × 10^5	
HOH	H_2O	0.69	between atoms
HSH	H_2S	0.43	spreading
HNH	NH_3	0.4–0.6	
HPH	PH_3	0.33	
HCH	CH_4	0.46	
HCH	C_2H_4	0.30	
FOF	F_2O	0.69	
ClOCl	Cl_2O	0.41	
FCF	CF_4	0.71	
ClCCl	CCl_4	0.33	
BrCBr	CBr_4	0.24	
FBF	BF_3	0.37	
ClBCl	BCl_3	0.16	
BrBBr	BBr_3	0.13	
HCF	CH_3F	0.57	
HCCl	CH_3Cl	0.36	
HCBr	CH_3Br	0.30	
HCI	CH_3I	0.23	
NNO	N_2O	0.49	
OCO	CO_2	0.57	
SCS	CS_2	0.23	
HCC	C_2H_2	0.12	
HCN	HCN	0.20	

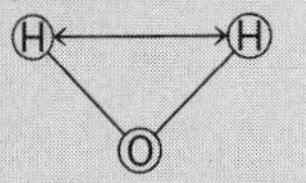

[a] From E. B. Wilson, Jr., J. C. Decius, and P. C. Cross,
"Molecular Vibrations." Copyright 1955. McGraw-
Hill Book Company. Used by permission.

TABLE A1-8 Thermodynamic data based partly on experimental measurements and partly on extrapolation with the help of spectroscopic data[a]

Substance	T, °K	C_p°, cal mole^{-1} deg^{-1}	S°, cal mole^{-1} deg^{-1}	$-\dfrac{F^\circ - H_{298}^\circ}{T}$, cal mole^{-1} deg^{-1}	$H^\circ - H_{298}^\circ$, kcal mole^{-1}	ΔH_f°, kcal mole^{-1}	ΔF_f°, kcal mole^{-1}	$\log K_p$
C (graphite)	298	2.037	1.363	1.363	0.000			
	500	3.496	2.788	1.650	0.569			
	1000	5.148	5.848	3.024	2.824			
	1500	5.677	8.051	4.352	5.549			
	2000	5.850	9.712	5.493	8.439			
$H_2(g)$	298	6.892	31.208	31.208	0.000			
	500	6.993	34.806	31.995	1.406			
	1000	7.219	39.702	34.758	4.944			
	1500	7.720	42.716	36.937	8.668			
	2000	8.195	45.004	38.678	12.651			
$N_2(g)$	298	6.961	45.760	45.760	0.000			
	500	7.070	49.376	46.551	1.413			
	1000	7.815	54.499	49.369	5.130			
	1500	8.327	57.775	51.656	9.179			
	2000	8.596	60.212	53.503	13.416			
$O_2(g)$	298	7.020	49.011	49.011	0.000			
	500	7.481	52.727	49.818	1.454			
	1000	8.336	58.197	52.771	5.427			
	1500	8.738	61.661	55.191	9.705			
	2000	9.029	64.216	57.141	14.148			
HCHO(g)	298	8.461	52.261	52.261	0.000	−27.400	−25.964	19.031
	500	10.460	57.077	53.281	1.898	−28.204	−24.764	10.824
	1000	14.817	65.806	57.485	8.332	−29.560	−20.718	4.528
	1500	17.013	72.284	61.386	16.346	−30.123	−16.153	2.353
	2000	18.095	77.343	64.767	25.153	−30.411	−11.450	1.251
$H_2O(g)$	298	8.025	45.105	45.105	0.000	−57.798	−54.635	40.047
	500	8.415	49.334	46.025	1.654	−58.277	−52.359	22.885
	1000	9.851	55.591	49.382	6.209	−59.246	−46.037	10.061
	1500	11.233	59.858	52.195	11.495	−59.824	−39.292	5.724
	2000	12.214	63.234	54.547	17.373	−60.150	−32.393	3.540
$NH_3(g)$	298	8.505	46.034	46.034	0.000	−11.040	−3.986	2.922
	500	9.920	50.763	47.047	1.858	−11.977	1.070	−0.468
	1000	13.400	58.739	51.026	7.713	−13.308	14.755	−3.225
	1500	15.800	64.668	54.625	15.065	−13.566	28.874	−4.207
	2000	17.190	69.424	57.750	23.347	−13.378	42.999	−4.699

[a] From D. R. Stull (ed.), "JANAF Thermochemical Data," Dow Chemical Company, Midland, Mich., 1963.

APPENDIX TWO

Wave-function Charts

A convenient set of formulas for atomic orbitals expresses the value of ψ in the units (electron charge bohr^{-3})$^{1/2}$. The *bohr* is the basic orbit radius in the hydrogen atom (0.52922×10^{-8} cm). The electron charge is 4.80298×10^{-10} esu. When these units are used, the value of ψ^2, expressing the density of charge, has the units electron charge per cubic bohr.

The wave function for the $1s$ orbital is frequently written

$$\psi_{1s} = \frac{1}{\sqrt{\pi}} \left(\frac{Z}{a_0}\right)^{3/2} e^{-\rho/2} \qquad \text{(A2-1)}$$

where $\rho = 2\alpha r$ and $\alpha = Z/a_0$.

For a hydrogen atom, the value of Z is unity. Equation (A2-1) reduces to

$$\psi_{1sH} = g e^{-r} \qquad \text{(A2-2)}$$

if r is expressed in bohrs. For $2s$ and $3s$ wave functions for H

$$\psi_n = g_n e^{-r/n}$$

where n is the quantum number and g is the preexponential factor. Figure A2-1 gives the value of the exponential factor for $n = 1$, 2, and 3; Fig. A2-2 gives the values of the exponential factors for $n = 3$ with still higher values of r. The preexponential factors for $2s$ and $2p$ functions are given in Fig. A2-3. These figures aid in the rapid estimation of the values of ψ and ψ^2 at various values of r.

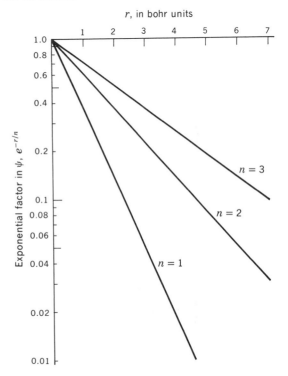

Fig. A2-1   Values for the exponential factor for the $1s$, $2s$, and $3s$ orbitals.

Fig. A2-2   Values of the exponential factor for the 3s orbital for higher values of r.

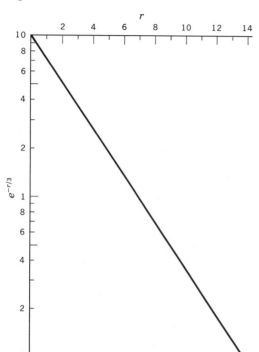

Fig. A2-3   The preexponential factors for 2s and 2p orbitals.

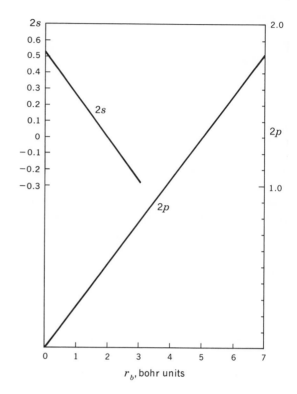

APPENDIX THREE

Symbols

Almost all symbols conform to the usage of G. N. Lewis and M. Randall, "Thermodynamics," 2d ed., rev. by K. S. Pitzer and L. Brewer, McGraw-Hill Book Company, New York, 1961. To aid the reader in identifying some of the symbols, references to equations where the symbols are used are sometimes included. Most rarely used symbols are omitted.

A ITALIC CAPITAL LETTERS

1. Extensive thermodynamic quantities

A	Helmholtz free energy
B, C, D	Virial coefficients for gases; number of equivalents in generalized chemical equations
C	Heat capacity
E	Energy content (2-15)
G	Gibbs free energy
H	Enthalpy (heat content)
J	Relative heat capacity
L	Relative enthalpy
Q	Quantity of heat transferred
S	Entropy
V	Volume

2. Other quantities

A	Area; Madelung constant (22-11); preexponential factor in the Boltzmann equation; equivalents
A_γ, A_H	Debye-Hückel coefficients
B	Spectroscopic rotational constant; ion interaction
C	Electric conductivity (13-12); ionic character

D	Dielectric constant; dissociation energy; inexact differential
F	Force (2-1).
I	Moment of inertia; constant of integration; ionic strength; irreversibility
J	Rotational quantum number; flux
K	Equilibrium constant; boiling- and freezing-point constant; information constant (25-5); cybernetic force
L	Rate constant; Onsager matrix coefficient; molecular quantum number; mean free path
M	Molecular weight; molarity; multiplicity
N	Number of molecules
N_A	Avogadro's number
P	Pressure
Q	Partition function; heat transferred
R	Gas constant; Rydberg constant (17-10)
T	Temperature, °K; tension (16-6); transition-moment integral (18-16)
U	Potential energy; mobility (13-7)
V	Potential energy in a molecule
$\dot{V}$	Rate of gas flow
W	Weight; statistical multiplicity
Y	Probability
Z	Partition function

B SMALL CAPITAL LETTERS

A	Molal free energy (Helmholtz)
B, C, D	Molal virial coefficients
C	Molal heat capacity
E	Molal energy content

G	Molal free energy (Gibbs)
H	Molal enthalpy (heat content)
J	Relative molal heat capacity
L	Relative molal enthalpy
S	Molal entropy
V	Molal volume
$\overline{G}, \overline{H}, \overline{S}$	Partial molal quantities

C SCRIPT CAPITAL LETTERS

$\mathscr{E}$	Galvanic cell electromotive force
$\mathscr{F}$	Faraday constant
$\mathscr{S}$	Solubility coefficient

D BOLDFACE CAPITAL LETTERS (VECTORS)

B	Electromagnetic induction
D	Electric displacement
E	Electric field
G	Reciprocal lattice vector
J	Current density
T	Translation vector

E LOWERCASE ITALIC LETTERS

a, b, c	Constants in Gay-Lussac law (1-3); van der Waals' constants (3-4); constants in heat-capacity equations
a	Activity
c	Concentration; velocity of light; conversion factor; number of components (11-97)
d	Density; degrees of freedom
e	Charge on the electron; thermodynamic efficiency
f	Fugacity; force; number of degrees of freedom
g	Acceleration of gravity; quantum degeneracy
h	Planck's constant; height
i	Van't Hoff factor
k	Boltzmann constant; Henry's law constant; rate constant; Hooke's law constant

l	Length; liquid; azimuthal quantum number
m	Mass; mass of single atom; magnetic quantum number
m_w	Molality
n	Number of moles; quantum number
p	Partial pressure; vapor pressure; number of phases
r	Rate; radius; electric resistance
s	Spin quantum number; sharp (spectra); solid; length (2-4)
t	Time; transference number; temperature, °C
u	Velocity (2-1)
v	Vibrational quantum number; velocity
x	Mole fraction; spectroscopic anharmonicity constant
z	Collisions per second; compressibility factor
z_+, z_-	Charge on a positive or negative ion
x, y, z	Linear coordinates

F BOLDFACE LOWERCASE VECTORS

k	Electromagnetic vector
r	Directional vector in crystal

G GREEK LETTERS

α	Coefficient of thermal expansion; degree of ionization; Bunsen absorption coefficient
β	Bohr magneton; coefficient of compressibility
γ	Activity coefficient; gyromagnetic ratio
Γ	Surface concentration
Δ	Difference
ϵ	Molecular energy level; energy per molecule
θ	Characteristic temperature in Einstein and Debye functions
κ	Compressibility; cell constant; specific conductance
λ	Wavelength; ionic conductance

Λ	Equivalent conductivity
μ	Chemical potential; Joule-Thomson coefficient; reduced mass
ν	Frequency; number of ions per molecule of electrolyte
$\bar{\nu}$	Wave number
ν^+, ν_+	Number of units of positive charge
ν^-, ν_-	Number of units of negative charge
π	Osmotic pressure; reduced pressure; 3.14159; molecular orbital
Π	Product of a series of terms; orbital
σ	Surface tension; symmetry number; wave number; rate of entropy production
Σ	Sum of terms
τ	Period; reduced temperature
ϕ	Wave function; electric potential (13-42); angle; reduced volume
χ	Magnetic susceptibility
ψ	Wave function; electric potential
ζ	Damping ratio
η	Viscosity
ω	Angular velocity; vibration frequency, cm^{-1}

H SUPERSCRIPTS

$\circ$	Standard state; degree
$*$	Specially selected volume; ideal-gas state
k	Kinetic; kinetic (energy)
e	Equilibrium
$\ddagger$	Activation
h	Henry's law; Hooke's law
M	Mixing
$'$	Standard reduction potential at pH = 7, as $\mathcal{E}_0'$
s	Statistical, as ${}^s s$ denoting statistical entropy
t	Thermodynamic, as ${}^t s$ denoting thermodynamic entropy

I SUBSCRIPTS

	(For Greek subscripts, see part G)
0	At $0°K$; equilibrium state
b	Back reaction; boiling point
c	Critical; crystal
e	Per equivalent
f	Forward reaction; fusion; formation from elements; freezing point
g	Gas; *gerade* (see Figs. 19-22 and 19-23)
h	Hooke's law
k	Kinetic
l	Liquid
m	Melting point
M	Transition-moment integral (18-16)
n	Normalized
p	Constant pressure
q	Quantum, (for example, ν_q is the frequency associated with the quantum or photon)
s	Surface quantity; sublimation point
σ	Symmetry term
T	Constant temperature
u	*Ungerade* (see Figs. 19-22 and 19-23)
v	Constant volume; vaporization
w	Water
$\pm$	Mean quantity for ions

J OTHER SYMBOLS

$!$	Factorial
$\bar{\nu}$	Wave number
∇	Double space differentiation
$(\ )$	Concentration in gas phase
$[\ ]$	Concentration or activity in liquid phase
∂	Partial differentiation
$\oint$	Circuit integral
pH	$-\log[H^+]$
pK	$-\log K$
$\equiv$	Defined as
$\simeq$	Approximately equal to
$\sim$	Proportional to
∞	Infinity
$|\xi|$	Absolute value
$\exp x$	e^x

Mathematics

A4-1 DERIVATION OF THE EINSTEIN HEAT-CAPACITY EQUATION

According to the Boltzmann distribution law, the ratio of the number of harmonic oscillators in the nth level to the number in the zero level is

$$\frac{n_i}{n_0} = e^{-\epsilon_i/kT} \tag{A4-1}$$

where $i = 0, 1, 2, 3, 4, \ldots$ and $\epsilon = h\nu$. Thus

$$\Sigma n_i = n_0(1 + e^{-\epsilon/kT} + e^{-2\epsilon/kT} + \ldots) \tag{A4-2}$$

and from the theory of infinite series

$$N = \Sigma n_i = \frac{n_0}{1 - e^{-\epsilon/kT}} \tag{A4-3}$$

where we denote Σn_i by N. Thus the total energy U of a collection of N oscillators is

$$U_N = 0 \times n_0 e^{-0/kT} + \epsilon \times n_0 e^{-\epsilon/kT}$$
$$+ 2\epsilon \times n_0 e^{-2\epsilon/kT} + \ldots \tag{A4-4}$$

Again this series can be transformed into a fraction

$$U = \frac{n_0 \epsilon e^{-\epsilon/kT}}{(1 - e^{-\epsilon/kT})^2} \tag{A4-5}$$

Using the relation in Eq. (A4-3), this expression becomes

$$U_N = \frac{N\epsilon}{e^{\epsilon/kT} - 1} = \frac{Nh\nu}{e^{h\nu/kT} - 1} \tag{A4-6}$$

The average energy per oscillator is

$$\overline{U} = \frac{h\nu}{e^{h\nu/kT} - 1} \tag{A4-7}$$

Denoting the ratio of Planck's constant to Boltzmann's constant by $\beta = 4.87 \times 10^{-11}$ cgs units, $\nu = (R/N)\beta\nu$. For a collection of $N_A = 6.02 \times 10^{23}$ oscillators

$$U = RT\frac{e^{\beta\nu/T}}{e^{\beta\nu/T} - 1} \tag{A4-8}$$

Thus

$$c_v = \frac{dU}{dT} = R\frac{e^{\beta\nu/T}(\beta\nu/T)^2}{(e^{\beta\nu/T} - 1)^2} \tag{A4-9}$$

This is Einstein's heat-capacity equation for a collection of N_A oscillators. The fraction following R is a function only of ν/T, which approaches zero as T approaches zero and approaches 1 as T goes to infinity. Thus denoting the fraction following R by $ein(\nu/T)$, a single plot of $ein(\nu/T)$ against $\log T$ serves as a graph by which the value of c_v can be determined at any temperature when a sliding scale of T is moved to the position corresponding to the appropriate value of ν. Since the quantity $\beta\nu$ has the dimension of temperature, it is frequently denoted by the symbol θ and is called the characteristic temperature; the Einstein function is then expressed as $ein(\theta/T)$.

A4-2 VECTORS

A *vector* is a quantity which has both *length* and *direction*. An example of a vector in two dimensions is shown in Fig. A4-1. The arrow has the length, and the direction in which it points is specified by the angle α between the arrow and the x coordinate axis. The vector is frequently denoted by a boldface letter such as **R**.

The projection of **R** on the x axis is called the x component of **R**; similarly the projection of **R** on the y axis is called the y component of **R**. The components are denoted by the symbols $\mathbf{R}_x$ and $\mathbf{R}_y$ (Fig. A4-2).

Fig. A4-1   A vector R in two dimensions.

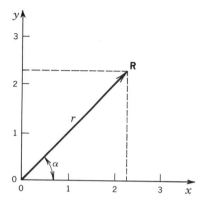

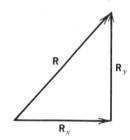

Fig. A4-2   The resolution of the vector R into two components R$_x$ and R$_y$.   R$_x$ + R$_y$ = R.

Two vectors are added by placing the tail of the second at the head (arrow end) of the first (Fig. A4-2).   The direction of the **R**$_y$ is not changed when it is moved to make its tail coincide with the head of **R**$_x$.   The vector drawn from the tail of the second to the head of the first then is the representation of the vector sum of the two vectors.

Examples of vectors are *velocity,* where the length is proportional to the numerical value of the velocity and the direction of the vector is in the direction of motion; *momentum,* where the length is proportional to the numerical value of the momentum and the direction of the vector coincides with the direction of motion; *angular momentum,* where the length of the vector is proportional to the numerical value of the angular momentum, the line of the vector is along the axis of spin, and the vector points in the direction of

the advance of a right-hand screw turning in the same direction as the moving body.

To distinguish vectors from quantities which have only numerical value and no direction like mass and volume, the latter are called *scalars.*

There are two kinds of vector multiplication.   The simpler kind is called the *inner* or *dot product* and is denoted by a boldface dot placed between the two vector symbols

$$\mathbf{R} \cdot \mathbf{S} = RS \cos \theta = Q \qquad \text{(A4-10)}$$

where R and S are the lengths of the vectors and θ is the angle between them.   This is equal to the product of the length of the component of **R** in the **S** direction multiplied by the length of **S**.   The dot product Q is a scalar.   If **R** and **S** point in the same direction, then $\mathbf{R} \cdot \mathbf{S} = R \times S$, the product of the lengths.   If **R** and **S** are at right angles to each other, then $\mathbf{R} \cdot \mathbf{S} = 0$.

In order to define the second kind of vector product, it is necessary to consider a vector in *three* dimensions.   Such a vector is shown in Fig. A4-3.   This vector **R** can be resolved into three components, each parallel to one of the coordinate axes, **R**$_x$, **R**$_y$, and **R**$_z$.   This second kind of vector product is called the *outer, cross,* or *skew* product and is denoted by a boldface cross ✕.   Thus

$$\mathbf{R} \times \mathbf{S} = \mathbf{p}RS \sin \theta = \mathbf{P} \qquad \text{(A4-11)}$$

where **p** is the vector of unit length pointing in a direction normal to the plane defined by **R** and **S** and

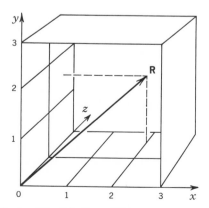

Fig. A4-3   A vector R in three dimensions.

θ is the angle between **R** and **S**. **p** points in the direction corresponding to the advance of a right-hand screw aligned parallel to **p** and turning in the direction that carries **R** into **S** through the smaller angle between **R** and **S**. Note that the numerical value of **P** is equal to the area of the parallelepiped with sides made of vectors corresponding to **R** and **S**. If **R** and **S** are in the same direction, the area vanishes and **P** = 0. If **R** and **S** are normal to each other ($\theta = 90°$), the length of **P** is just the length of **R** multiplied by the length of **S**.

A4-3   CRYSTAL-SOLUTION EQUILIBRIA

Equation (11-93) relates the logarithm of the mole fraction X of component 1 in an ideal solution to the temperature T at which the solution is in equilibrium with the crystalline phase of component 1. The heat of fusion ΔH_f of component 1 and its melting-point temperature T_f appear as constants. The derivation of this equation is as follows.

For component 1, the ratio of the vapor pressure of the crystal phase ^cP at T to the vapor pressure ^cP_f at T_f is related to the temperatures by the expression

$$R \ln \left(\frac{^cP}{^cP_f}\right) = -\Delta H_s\left(\frac{1}{T} - \frac{1}{T_f}\right) \tag{A4-12}$$

where ΔH_s is the heat of sublimation.

The ratio of the vapor pressure of the pure liquid ^lP at T to the vapor pressure ^lP_f at T_f is related to the temperatures by the expression

$$R \ln \left(\frac{^lP}{^lP_f}\right) = -\Delta H_v\left(\frac{1}{T} - \frac{1}{T_f}\right) \tag{A4-13}$$

where ΔH_v is the heat of vaporization.

From Raoult's law, the vapor pressure ^lP_s of component 1 in solution at mole fraction X is related to the vapor pressure of the pure liquid at the same temperature by the equation

$$^lP_s = {}^lPX \tag{A4-14}$$

At the temperature where the crystal phase is in equilibrium with this solution,

$$^cP = {}^lP_s \tag{A4-15}$$

The following equations also are applicable:

$$\Delta H_s = \Delta H_f + \Delta H_v \tag{A4-16}$$

$$^cP_f = {}^lP_f \tag{A4-17}$$

Substituting from these expressions in Eq. (A4-12) and subtracting Eq. (A4-13) from Eq. (A4-12), we get

$$R \ln \left(\frac{^lPX}{^lP_f}\right) - R \ln \left(\frac{^lP}{^lP_f}\right) = -\Delta H_f\left(\frac{1}{T} - \frac{1}{T_f}\right) \tag{A4-18}$$

and thus

$$R \ln \left(\frac{^lP}{^lP_f}\right) + R \ln X - R \ln \left(\frac{^lP}{^lP_f}\right) = -\Delta H_f\left(\frac{1}{T} - \frac{1}{T_f}\right) \tag{A4-19}$$

Canceling and changing to the common log,

$$\log X = \left(\frac{-\Delta H_f}{2.303R}\right)\left(\frac{1}{T} - \frac{1}{T_f}\right) \tag{A4-20}$$

which is Eq. (11-93).

name index

subject index